Ethnopharmacology of Wild Plants

Editors

Mahendra Rai
Department of Biotechnology
SGB Amravati University
Amravati, Maharashtra, India

Shandesh Bhattarai
Faculty of Science
Nepal Academy of Science and Technology
Khumaltar, Lalitpur, Nepal

Chistiane M. Feitosa
Department of Chemistry
Federal University of Piaui
Petronio Portela Campus, Teresina, Brazil

CRC Press
Taylor & Francis Group
Boca Raton London New York

CRC Press is an imprint of the
Taylor & Francis Group, an **informa** business
A SCIENCE PUBLISHERS BOOK

Cover credit: Fidel A. Torres-Guevara, Asociación para la Ciencia e Innovación Agraria de la Red Norte-AgroRed Norte, Mz. O Lote 20 Urb. Los Cocos del Chipe, Piura, Perú

First edition published 2021
by CRC Press
6000 Broken Sound Parkway NW, Suite 300, Boca Raton, FL 33487-2742

and by CRC Press
2 Park Square, Milton Park, Abingdon, Oxon, OX14 4RN

Library of Congress Cataloging-in-Publication Data

Names: Rai, Mahendra, editor. | Bhattarai, Shandesh, 1977- editor. | Feitosa, Chistiane Mendes, editor.
Title: Ethnopharmacology of wild plants / editors, Mahendra Rai, Department of Biotechnology, SGB Amravati University, Amravati, Maharashtra, India, Shandesh Bhattarai, Faculty of Science, Nepal Academy of Science and Technology, Khumaltar, Lalitpur, Nepal, Chistiane M. Feitosa, Department of Chemistry, Federal University of Piaui, Petronio Portela Campus, Brazil.
Description: First edition. | Boca Raton : CRC Press, 2021. | Includes bibliographical references and index.
Identifiers: LCCN 2020028365 | ISBN 9780367512026 (hardback)
Subjects: LCSH: Medicinal plants. | Ethnopharmacology. | Materia medica, Vegetable. | Wild plants, Edible--Therapeutic use.
Classification: LCC QK99.A1 E84 2021 | DDC 581.6/34--dc23
LC record available at https://lccn.loc.gov/2020028365

ISBN: 978-0-367-51202-6 (hbk)
ISBN: 978-1-003-05281-4 (ebk)

Typeset in Times New Roman
by Radiant Productions

Foreword

Animals, and subsequently humans, have used the local flora and fauna for healing purposes for eons; that desperate search for comfort and healing through trial and error continues. What healed? What didn't? What soothed pain? What caused pain? What was toxic? Which plant part? These questions were asked and answered, and the relevant information passed down, generation to generation. Take this, don't take this. Being added to, being refined, sometimes recorded locally, sometimes recorded by outsiders, sometimes never formally recorded, perhaps even until now. Or perhaps still unrecorded.

This history of plant usage for healing purposes is intensely valuable information, which, as Bhattarai, Feitosa, and Rai indicate in Chapter 1, has evolved into rich and extensive, science-based research programs, and in select instances, pharmaceutical and biological agents of immense contemporary value. Until 1899, when the first, semisynthetic drug, aspirin, based on a natural cure "for all pains", was introduced by Bayer, ALL medicinal agents were either from living sources, or from minerals. For many people around the world today, that is still the case. They RELY on nature to provide a medicament that works. Other people use medicinal agents more at their discretion, much less than as a dire, primary care need. So there is the duality of potential, a standardized traditional medicine or a fully developed drug.

Essentially every country in the world has evolved a system of ethnomedicine; some even have several systems, as the dynamic societal and religious beliefs impacted medicines and their preparation. In very few instances were these medicinal agents cultivated on a large scale. They were collected from the local region, "wild-crafted", and brought to market, or grown in a home medicinal plant garden, in the same way that herbs and vegetables might be. An integral aspect of family life. Grow your own medicines. Many still do of course, and the relationship between foods, especially spices, and medicine is also becoming more apparent as research progresses.

As populations expanded, and the urbanized demand for medicines in the ancient cities grew and were supplied by forest harvesting, so new locations for plants were identified, and the plants occasionally cropped for use. Most plants however, originated in the forest, the plains, the savannahs, the mountains, or the tropical jungles of the world. Over time, and in different cultures, more uses were found for a given plant, and the "value" of the plant was enhanced. Then along came chemistry and biology. Providing the ability to analyze the content and test samples for their biological properties in an increasingly advanced technological manner, in animals, in cells, and against enzymes and receptors, seeking the "arcanum", the powerful core, and a rationale, and explanation, for its use. Not surprisingly, the first plant to be studied was opium, the latex of *Papaver somniferum*.

From the early writings of the Sumerian, Egyptian, Chinese, Greek, Persian, Indian, Roman, and Andean cultures, we have extensive records of medicinal plant usage, and of the trade in those medicinal plants. A great tragedy is that this information, and all of the ethnopharmacologic information accrued in the past 1500 years or so, that vast treasure house of indigenous knowledge, is nowhere stored in a secure location in this world. What a gift to the future of humanity that would be! Which brings us to data and information, and its analysis and interpretation.

"Cyber" is critical, now, and for the foreseeable future. It is in your hand. Your future laboratory to take to the field for research work, is in your hand. Your decision-making capacity is in your hand. Literally, it is still called a phone...... although I'm not sure why anymore, actually. Everything you need to know is right there. All the ethnopharmacologic data you need. All the history and documentation, all the experiments, all the analysis, and therefore the ability to set priorities..... for systems, for research programs, and for designing experiments, and for getting to and finding your way in the "wild"!

There are two fundamental "myths" in ethnopharmacology which must be mentioned. The first is that if "a plant has been used for hundreds of years, that it is safe and effective". The second is that "the plant, and the knowledge associated with it, will always be there." Both statements are demonstrably false. When the extract of a plant is investigated for a biological activity, the goal is not to "validate" its use; that would invoke an experimental bias. It is the contemporary value of the historical record, and the potential for moving forward on either a drug discovery or a standardization mode, that is being assessed or evaluated.

From an "eco" perspective, the expanding use of medicinal plants, by a population close to 7.8 billion people, places extreme pressures on the continuing access to "wild" resources. The very core of ethnopharmacologic plant use and investigations is a sustainable supply. That process begins in research design. Root and bark samples are unlikely to be a sustainable resource, unless, in the former instance, the plant is fast-growing. There is much work to be done to seek alternative sources in that regard. Just because a plant has been used medicinally and not studied, does not mean it *should* be studied. Societal needs are critical. The chapters in this book relate to locally or globally assessed research needs, and consequently provide the important translation of the results back to society.

A plant is not a plant. Don't assume you know the activity of a particular plant sample. That means a plant collected from another location or habitat, or an alternative variety or strain will have a different chemical profile and therefore an altered property profile or potency. Is a plant "better" from Hunan or from Guangzhou Province? There is a long-standing urgency for standardization, based on chemistry and biology, having assured the botanical authenticity. Some of the science behind those determinations relate to metabolomics. What compounds are present? At what level? How does that relate to the observed biological effects? How is that being tested, and does the biological testing itself relate to the ethnomedical use? What is actually being tested? Is there an analysis? Are the data reproducible? There are four other "omics" of relevance. There is taxonomics, has the plant been unequivocally identified? And there is the genomics, the genetic profile for what the metabolite potential is for an organism. Finally, there are two other "omics" to be considered......agronomics for the sustainability and consistency of plant development, and economics..... the cost for the manufacturer and the affordable cost to the patient.

Creating systems for assessing medicinal plants and their sustainable use has been a goal of the World Health Organization for the past twenty years. The tools are available, and there is a well-defined need. Now, it is important to think in terms of the patient. What facets do we, as patients, want besides quality, safety, and efficacy? They were just indicated, consistency and accessibility......together they form QSECA, an important acronym. While there are food security programs in place through the FAO, there are no programs to assure "medicines security". Considering medicinal plants, as well as synthetic medicinal agents, as sustainable commodities, is a fundamental aspect of the future of this branch of the healing sciences.

The first part of this much-needed book is devoted to ethnopharmacology, its impact in the past, as well as contemporary studies, in particular areas of the world (Northern Peru, Latin America), with certain plant genera (*Pilocarpus*, *Eichhornia*), or for specific, globally demanding, disease states (diabetes, fungal infections, parasitic diseases, inflammation). The second part focuses more on the ethnobotanical and phytotherapeutic aspects of medicinal plants. It includes chapters on the ethnomedicine and phytochemistry of specific plants (*Bergenia*, *Chenopodium*, *Cenostigma*, *Rheum*, *Actinidia*, and those in the Apiaceae), as well as plants which produce signaling pathway effects in insulin therapy, and ritualistic and medicinal plants from Marajó Island in north-eastern Brazil.

These are compelling topics, and the editors and the authors are to be commended for dedicating their expertise to the assembly of this collection of valuable chapters. It is to be hoped that the presentations here will stimulate the development of new, sustainable research programs that will enhance the health care systems for patients in which ethnopharmacology and medicinal plants play an essential role, and promote the conservation of the resources on which the science relies. For after all, "We do not inherit the Earth from our ancestors. We borrow it from our children." (Attributed to Chief Seattle 1854). They, the young, will reap the benefits of what we do now and how we do it.

Geoffrey A. Cordell, Ph.D.

Professor Emeritus, University of Illinois,
Adjunct Professor, University of Florida, and
President, Natural Products Inc.

Preface

The research interest on wild plants has significantly increased due to the presence of beneficial secondary metabolites in them with a wide range of bioactivities. Traditional people have developed their knowledge regarding the use of wild plants as their major medicines. In the earliest period of human history, people used wild plants and animals as their basic food and medicine. These resources are culturally acceptable and easily accessible to local people throughout the globe, particularly in the remote areas and mountains. Several useful plant-derived drugs were discovered as a result of scientific follow up of well-known plants used in traditional medicine through the ethnopharmacological research. Many of today's major diseases like HIV and cancer are being treated with a group of potential products derived from plants. Thus, modern medicine has also benefited from wild plants that were originally used as home herbal remedies. Hence, wild plants should be investigated by modern scientific techniques to establish their safety and efficacy and to determine their potential as sources of new drugs.

The global diversity of wild plants is very high but little is known about their food/medicinal significance. Wild plants offer great potential for the discovery of unique molecules and new sources of active compounds. In this context, the present book provides a comprehensive overview of the wild plants, its usage, status, and threats, as well as the growing interest in ethnopharmacology research. The book comprises of important issues such as diversity of wild plants with emphasis on medicinal and food plants, threats to wild plants and traditional ethnobotanical knowledge, their uses in skin diseases, snake-bites, and in cosmeceuticals. Moreover, the ethnopharmacological relevance of wild plants in Latin America has been discussed.

This book will be useful for the researchers working in the areas of conservation biology, botany, ethnobiology, ethnopharmacology, national and international policymakers, etc. The principal aim of this book is to collect the dispersed global data about the wild plant resources and their usage. Some contributors from India, Nepal, Pakistan, Brazil, China, Bangladesh, Argentina, Chile, etc., have provided important information regarding the importance of their wild plant resources. Hence, the proposed publication will be considered as a main source of information and will be entirely different from other related publications.

We take the opportunity to thank all the contributors for their generous cooperation and efforts for offering up-to-date chapters. Further, we express our sincere thanks to the publisher and the authors of the chapters, whose research works have been cited in the present book. We are also thankful to the entire team of CRC Press for their cooperation, efforts, timely help and patience in the publication of this book. Finally, MKR and CMF thank CNPq (National Council for Scientific and Technological Development, Brazil) for financial support (Process number 403888/2018-2), and SB acknowledges NAST (Nepal Academy of Science and Technology), Kathmandu, Nepal for various support.

Mahendra Rai, India
Shandesh Bhattarai, Nepal
Chistiane M. Feitosa, Brazil

Contents

Section II: Ethnobotany and Phytotherapy

Section I
Ethnopharmacology

Chapter 1

An Introduction to the Ethnopharmacology of Wild Plants

Shandesh Bhattarai,[1,*] *Christiane Mendes Feitosa*[2] and *Mahendra Rai*[2,3,*]

Introduction

Natural sources including plants form the basis of modern medicine and contribute largely to the production of commercial drugs. Every tribal society has a traditional medicine system. The human use of plants as medicines may be traced back at least 60,000 years, reflecting their ability to produce secondary metabolites with a broad range of pharmacological properties (Fabricant and Farnsworth 2001, Shi et al. 2010, Buyel 2018). About 25% of drugs prescribed worldwide are derived from plants (Thillaivanan and Samraj 2014). With an annual global market worth 60 billion dollars, more than 1000 companies are involved in providing medicinal plant products (Newmaster et al. 2013), and an impressive number of modern drugs have been developed from terrestrial plants (Kaur et al. 2011).

The described and accepted number of plant species in the world is about 374,000 including 308,312 vascular plants. Of these, 295,383 are angiosperms (Christenhusz and Byng 2016). The largest vascular plant family is Orchidaceae followed by Asteraceae (Funk et al. 2009, Chase et al. 2015, Shrestha et al. 2018), but there may be several unexplored and unidentified plant species and thus the large parts of the world are still in need of additional botanical expeditions (Christenhusz and Byng 2016, Rai et al. 2021). In recent years, the research interest in wild plants is growing due to the emergence of several diseases like AIDS, cancer, Severe acute respiratory syndrome (SARS), and Coronavirus, etc. Currently, not only the local and traditional knowledge in remote regions but also those among immigrant communities are being studied (Heinrich and Casselman 2018, Yeung et al. 2019).

Around 80% of the people in the world depend on herbal medicinal products (Farnsworth et al. 1985, Ekor 2014), but the demand for herbal medicine is rising in both developed and developing countries (Ganie et al. 2015). About 52,885 plants are believed to be used in traditional medicine, but the exact number of medicinally useful bioactive metabolites in these plants is still unknown (Schippmann et al. 2002, Gewali 2008). Increasingly, medicinal species that exist in natural regions have received scientific and commercial attention, but we still know little about the treasure trove living among our wild places (Roberson 2008). Unfortunately, out of 52,885 plant species, about 15,000 are threatened due to habitat fragmentation, overexploitation, climate change and illegal trade in medicinal plants (Schippmann et al. 2006). According to the Royal Botanic Gardens at Kew, the UK, one in five species of plants in the world is threatened with the risk of extinction (Science Daily 2018, Shah and Bhat 2019). Experts estimate that the Earth is losing at least one potential major drug every two years (Roberson 2008).

[1] Faculty of Science, Nepal Academy of Science and Technology, Khumaltar, Lalitpur.
[2] Department of Chemistry, Federal University of Piaui, Petronio Portela Campus, Brazil.
[3] Sant Gadge Baba Amravati University, Amravati, Maharashtra, India.
* Corresponding authors: shandeshbhattarai@gmail.com; mahendrarai7@gmail.com

About two centuries ago, our medicinal practices were largely dominated by plant-based medicines (Cox and Ballick 1994). The consumption of plant-based medicines in the West has also increased in recent years and many developing nations continued to benefit from the rich knowledge of medicinal plants. For example, Siddha and Ayurveda medicines in India, Kampo Medicine in Japan, Traditional Chinese Medicine and Unani medicine in the Middle East and South Asia, still used by a majority of people, have been in practice for hundreds or even thousands of years (Mosihuzzaman and Choudhary 2008, Thillaivanan and Samraj 2014, Yuan et al. 2016). The herbal products form systems of knowledge and practice that have been transferred over centuries from one generation to another, but such indigenous knowledge is vanishing under the pressures of globalization (Heinrich 2010), including increasing popularity of Western medicine, cultural changes, etc. (Alves and Rosa 2007, Warude 2017).

Attempts are being made to isolate active constituents from natural sources that could be used to treat serious illness. Wild plants have abundant properties for the discovery of novel pharmacologically active molecules, mainly because of the environmental stress to which they are subjected (Eisner 1989, 1991, Mateo 2000, Svarstad 2000, Soejarto et al. 2005, Kaur et al. 2011). In this context, many useful drugs from plants were discovered as a result of scientific follow up of well-known plants used in traditional medicine (Cox and Ballick 1994).

Plants have a long history of use in the treatment of cancer and many medicinal plants have been found effective in experimental and clinical cases of cancers (Sivalokanathan and Ilayara 2005, Madhuri and Pandey 2008, Kaur et al. 2011). Still, cancer remains a major challenge to healthcare systems (Yabroff et al. 2011). Between 2003 and 2015, there was an increase from 6.2 to 8.8 million cancer-related deaths, which are approximately 13% of all deaths worldwide (Bray et al. 2012, Stewart and Wild 2014, WHO 2017). The cancer-related mortality rate is higher in developing countries than in industrialized countries (Sankaranarayanan 2014). Over 60% of currently used anticancer agents are derived from natural sources, including plants, marine organisms and micro-organisms (Kaur et al. 2011).

Plants continue to be important sources of drug discovery, especially with the rising popularity of the genomic approach (Atanasov et al. 2015, Harvey et al. 2015, Singh et al. 2018), and the ability of plants to produce medicines has been expended in the recent years (Buyel 2018). Herbs were the first medicines used by humans due to several pharmacologically active secondary metabolites produced by them (Buyel 2018). Thus, wild plants should be investigated by modern scientific techniques to establish their efficacy and to determine their potential as a source of new drugs.

This chapter introduces the ethnopharmacology of wild medicinal plants by some examples of the discovery of drugs from wild plants. Moreover, the challenges that have been faced by the researchers while conducting ethnopharmacological researches have also been discussed.

Guidelines to Conduct Ethnobotanical/Ethnopharmacological Research

Some selected guidelines to conduct ethnobotanical/ethnopharmacological study of wild plants following (Cox and Ballick 1994, Martin 1995, Taylor and Chaudhary 2003), with some modification are as follows:

(i) Selection of a society

Ethnobotanists choose a society that documents traditional knowledge of wild plants. Cox and Ballick (1994) provided guidelines for choosing a society. First, the society should be located in an area with rich floristic and cultural diversity. Secondly, the society should have remained in the region for many generations; and third, the culture must have a tradition in which healers or knowledgeable persons transmit their traditional knowledge from generation to generation (Taylor and Chaudhary 2003).

(ii) Prior informed consent

Researchers must obtain prior official permission before conducting any research with indigenous people. Fieldworkers should also obtain permission from community members before initiating research activities. Prior informed consent means that the interviewee must be made aware of the research that will be published for common people (Alexiades 1996).

(iii) Field visits, interview, and data collection

It is very important to establish a good relationship with several people in the community before starting the field works. Sometimes, the researchers forget to carry the necessary items that are required in the field. The researchers should take the proper equipment and materials needed for the ethnobotanical study. It is important to remember that the healer with whom the researcher wishes to speak may have a busy life and therefore the researcher should conduct interaction on the time convenient to the interviewee. For the conduction of interviews, it is essential to ask open-ended questions rather than questions that can be replied with one or two words (Martin 1995, Alexiades 1996, Chaudhary 1998, Jain and Mugdal 1999, Cunningham 2001). To approve the identity of the plant described by the interviewee, the interviewer must see the plant. This can be possible by visiting the forest or jungle alongwith the interviewee, where the plant can be identified in its natural environment. Many researchers like to take photographs of the research area if permission is granted. Audio and video recorders are essential for recording indigenous plant names and also a valuable aid during interviews (Alexaides 1996).

(iv) Collection of voucher specimens

The voucher specimen is a critical component of ethnobiological studies. Ethnobotanical information without suitable vouchered specimens has little scientific value (Alexaides 1996). It offers the documentation for the identity of plant material about which observations and data are documented (Lee et al. 1982). A good herbarium specimen comprises of a dried, pressed plant containing well-preserved vegetative and reproductive structures.

The plants should be collected ethically with conservation and local genetic diversity protection as well as with care. Plant samples should be collected from copious diverse places and not rooted as far as possible. Plant material needed for the bioassay test is best stored in cotton bags. Voucher specimens are collected for each species documented for proper identification and deposition.

(v) Bringing wild plants for laboratory research

It is necessary to study the wild plants in the laboratory. The analysis of medicinal plants for bioactive compounds is costly, time-consuming and requires expertise (Martin 1995). The preliminary laboratory study includes extraction of plants, identification of bioactive compounds and evaluation of bioactivities, etc. The promising wild ethnomedicinal plant can be analyzed and evaluated for their bioactivities.

(vi) Publishing the research findings

Finally, the bioactivity of the compound can be patented and the products can be developed. The innovative ethnopharmacological studies should be published in national or international journals for the benefit of other researchers. Fieldworkers should also confirm that the results of their research, including publications, are accessible to all. The effort should be made to spread the research findings through seminars and journals/books publication (Pearson 1985).

A Glimpse of Ethnopharmacology of Wild Plants

Ethnopharmacology was initiated in the 16th century in Latin America (Kuhn and Wang 2008). Ethnopharmacology of wild plants is connected to medicinal plant use and ethnobotany which is the scientific study of medicinal plants of any ethnic group and the use of plant compounds (Thomas and Carolyn 1996, Heinrich and Jäger 2015, www.wikipedia.org). The ethnopharmacological research observed a revival in the 1980s. Despite new awareness of the importance of intellectual property among the ethnic groups, numerous researchers started to engage in fieldwork throughout the 1990s and the beginning of the 21st century (Clapp and Crook 2002, Kim 2005, Gertsch 2009). Ethnobotany and ethnopharmacology are the starters of sustainable development and share the benefits derived from natural resources. Ethnobotanists perform research on uses and concepts of plants in cultures and societies and ethnopharmacologists research the medicinal uses of plants and their properties (Cox and Ballick 1994, Solis 2014).

Modern ethnopharmacology includes botanical, ethnobotanical, phytochemical, ethnopharmacological, pharmacological and toxicological fields (Solis 2014). The ethnomedicinal parts work to document the ethnomedical formula of individual medicinal plants into herbal medicines (Berlin and Berlin 2005), whereas the ethnopharmacological component conducts laboratory research and analyses on those medicinal plants which were selected by the ethnomedicinal study. The laboratory researches focus on discovering, isolating and pre-clinically evaluating the bioactive constituents. The steps in the ethnopharmacology research begin with the production of the compound through the identification and ethnotaxonomy of the natural material followed by the traditional preparation of the pharmaceutical forms. Afterward, bio-evaluation of the possible pharmacological action of such preparations should be carried out followed by their research on its potentiality for clinical effectiveness. At last, socio-medical aspects (medical anthropology) are applied for the use of these compounds (Thomas and Carolyn 1996, Heinrich and Jäger 2015, www.wikipedia.org).

Ethnopharmacological research greatly increases the chances of discovering new drugs (Soejarto et al. 2005). For example, one large-scale study provides some pointers to success rates with natural product screening. The National Cancer Institute's anti-cancer screening program publicized that screening of plants collected based on some ethnopharmacological information gave a hit rate of 19.9%, followed by random samples screening which gave a hit rate of 10.4% (Harvey 2002, Soejarto et al. 2005). Ethnopharmacologists typically use the working hypotheses resultant from anthropological fieldworks and therefore, ethnopharmacology study is trans-disciplinary (Etkin and Elisabetsky 2005, Gertsch 2009). Of thousands of claims of bioactive natural products with potential therapeutic applications, few claims are verified (Gertsch 2009). Natural product medicines have come from various sources including terrestrial plants, terrestrial microorganisms, marine organisms, and terrestrial vertebrates and invertebrates (Newman et al. 2000). Natural products are the most important source of drugs because ethnopharmacology is the source of all medicines.

Some novel examples of wild plants used for drug discovery

Plants continue to be a major source of raw material for drug discovery. Discoveries originate from traditional knowledge and have proven to be a useful tool in the search of new plant-based medicines. Neglected diseases are still treated with herbal medicines (Heinrich 2010), and thus ethnopharmacological approach is an important tool for the use of these herbal resources in medicines. In this context, there is a growing interest in the use of plants to manage or treat AIDS/HIV, chronic diseases like diabetes, chronic CNS disorders, and cancer or other emerging diseases (Heinrich 2010). Discoveries of plant-based novel molecules and products have revitalized concern in natural product research (Saslis-Lagoudakis et al. 2012, Solis 2014), which represent a unique and rich source of therapeutic agents and lead structures for new drug discovery (Newman and Cragg 2007, Bhargava 2015), but the natural products' research funds are small compared with other research areas (Laird 2000).

Plants with ethnopharmacological uses have been the primary sources of medicines for early drug discovery (Fabricant and Farnsworth 2001). Natural products and related drugs are used to treat 87% of all categorized human diseases including antibacterial, anticancer, anticoagulant, antiparasitic and immunosuppressant agents, among others (Newman et al. 2000, 2003, Newman and Cragg 2016), and there is still a renewed interest in developing new drugs from plants (Shah and Bhat 2019). About half of the anticancer drugs introduced since the 1940s are either natural products or their derivatives (Newman and Cragg 2007, 2016, Bhargava 2015). Out of the top 150 prescription drugs sold in the USA, at least 118 are based on natural sources (Daily et al. 1997, Roberson 2008). In the last 25 years, 70% of new drugs introduced in the USA were derived from natural products (Steenhuysen 2018). Plant-derived anti-cancer drugs save at least 30,000 lives per year in the United States alone (Roberson 2008).

Ethnopharmacology has provided some notable past successes (Gilani 2005). Early ethnopharmacologists made landmark discoveries, such as the elucidation of the pharmacological principles in Foxglove (*Digitalis* spp.), Poppy (*Papaver somniferum*), Curare (a mixture of plants including *Chondodendron tomentosum* and *Strychnos* spp.), the first antimalarial Quinine/quinidine for malaria and heart disease from *Cinchona* species (Tuxhill 1999, Online Medical Dictionary 2007), and the spectacular elucidation of the pharmacology of psychedelic plants (Cox and Ballick 1994, Gewali 2008). Otto Löwi discovered the pharmacological basis of nicotine action and for the first time described the nicotinic acetylcholine receptor (Karczmar 1996). Latin American ethnopharmacology has indirectly contributed to the discovery of the cholinergic system. Similarly, aspirin was developed with the bark of the willow tree (*Salix* spp.), which has been used to treat fever and inflammation in Europe, for pain relief, promotion of heart-health, and blood thinning (Tuxhill 1999). Another ethnopharmacology work on the Chinese antimalarial plant *Artemisia annua* resulted in the development of artemisinin into the new clinical semisynthetic antimalarial agent artemether. Other derivatives of artemisinin are in various stages of clinical development as antimalarial drugs in Europe (van Agtmael et al. 1999, Balunas and Kinghorn 2005, Kuhn and Wang 2008).

Terrestrial plants, Licorice (*Glycyrrhiza glabra*), Myrrh (*Commiphora* species), and Poppy capsule latex (*Papaver somniferum*), were recorded from Mesopotamia in 2600 BC, and are still in use today (Newman et al. 2000). A semisynthetic derivative of glycyrrhizic acid found in licorice (i.e., Hemisuccinate carbenoxolone sodium) is recommended for gastric and duodenal ulcers (Dewick 2002). Furthermore, morphine, codeine, noscapine (narcotine), and papaverine isolated from *P. somniferum* were developed as single chemical drugs and are still used clinically (Dewick 2002). Codeine is an alkaloid prepared from Opium or morphine by methylation. Opium is an isoquinolone alkaloid obtained from poppy plant [*Papaver somniferum*]. Codeine is used as a pain reliever, and for cough suppression (Online Medical Dictionary 2007, Bhandari et al. 2011).

Some of the approved drugs from terrestrial plants are Apomorphine hydrochloride, to treat Parkinson's disease, isolated from poppy (*Papaver somniferum*) (Deleu et al. 2004), followed by Tiotropium bromide for the treatment of bronchospasm associated with chronic obstructive pulmonary disease, from *Atropa belladonna* and related tropane alkaloids from other solanaceous plants (Koumis and Samuel 2005). Nitisinone, a new class of herbicides from the bottlebrush plant (*Callistemon citrinus*) (Hall et al. 2001), is used for the treatment of hereditary tyrosinaemia type 1 (HT-1) (Mitchell et al. 2001). Galantamine obtained from *Galanthus nivalis* has been used traditionally in Bulgaria and Turkey for neurological conditions (Howes et al. 2003, Heinrich and Teoh 2004), and for the treatment of Alzheimer's disease (Balunas and Kinghorn 2005).

Similarly, some other approved drugs are Aloe for treatment of burns and wounds isolated from [*Aloe* species] (Marinelli 2005), followed by Black cohoshv for the treatment of hormonal disorders, rheumatism, from Black cohosh [*Cimicifuga racemosa*] (Marinelli 2005), Bloodroot for the treatment of skin disorders and cancer from Bloodroot [*Sanguinaria canadensis*] (Robbins 1999). Likewise, Camphor for rheumatic pain relief from the wood of Camphor tree [*Cinnamomum camphore*] (Tuxhill 1999). Colchicine is used for cancer and gout treatment from Autumn crocus [*Colchicum autumnale*] (Tuxhill 1999), and Devil's club for infection cure, diabetes, and tuberculosis from *Oplopanax* species (Marinelli 2005, MPWG 2007). Species of the genus *Tabebuia* have a history of use in the treatment of several diseases including syphilis, fevers, malaria, cutaneous infections, and stomach disorders (Cragg et al. 2002).

In addition, some other approved drugs are Gamma-linolenic acid for the treatment of eczema, cancer and nerve-damage treatment isolated from Evening Primrose [*Oenothera biennis*] (Online Medical Dictionary 2007), followed by Hoodia for weight loss from Hoodia plant [*Hoodia gordonii*] (Online Medical Dictionary 2007); Madagascar periwinkle for the treatment of leukemia, Hodgkin's disease, and other cancers from Madagascar periwinkle [*Catharanthus roseus*] (Daily 1997); Pseudoephedrine (Pseudophed) for Bronchodilation, nasal decongestion, allergies from Mahuang shrub [*Ephedra sinica*] and other plants and fungi (Tuxhill 1999); Pilocarpine from Glaucoma treatment from Jaborandi plant [*Pilocarpus* species] (Online Medical Dictionary 2007); Scopolamine for sedation, motion sickness reduction from [*Datura/Solananceae* species]; Shikimic acid (e.g., Tamiflu) for the cure of viral diseases (influenza) from Star anise [*Illicium verum*] (Laurance 2005); Slippery elm bark lining for the relief of coughs, gastrointestinal ailments, skin irritations from Slippery elm [*Ulmusrubra*] (Roberson 2008); Tazopsine for malaria treatment from [*Strychnopsis thouarsii*] bark (Carraz et al. 2006); Tubocuranine for surgical muscle relaxation from Curare vine [*Chondrodendron tomentosum*] (Online Medical Dictionary 2007). Besides, Byetta for diabetes from Gila monster [*Heloderma suspectum*] (McClain 2005); Penicillin for bacterial infection cure from *Penicillium* fungus species (Abramovitz 1997).

Hartwell (1982) reported more than 3000 plant species from published and unpublished sources which have been reported to have alleged anticancer properties (Graham et al. 2000). Later, Graham et al. (2000) reviewed ethnomedicinal plants used against cancer using the NAPRALERT database and described over 350 species, distributed in over 280 genera and 100 families to be used against cancer. These species are not cited in the work of Hartwell (1982). The following families, namely: Actinidiaceae, Agavaceae, Balanaphoraceae, Buddlejaceae, Cephalotaxaceae, Chloranthaceae, Cycadaceae, Davalliaceae, Dichapetalaceae, Eriocaulaceae, Ganodermatacea, Globulariaceae, Gyrostemonacea, Hymenochaetace, Hypoxidaceae, Icacinaceae, Krameriaceae, Leeaceae, Lycoperdaceae, Olacaceae, Pandaceae, Pteridaceae, Sparganiaceae, Tremellaceae, and related plants reported in Graham et al. (2000), are not listed in Hartwell (1982). Similarly, following genera, namely: Acanthospermum, Achyrocline, Actinidia, Adenium, Aeonium, Anogeissus, Antrodia, Aphanamixis, Aspidosperma, Bolboschoenus, Bolbostemma, Cadaba, Calystegia, Camptotheca, Caragana, Carapa, Casearia, Cephalotaxus, Cerbera, Chloranthus, Cnidoscolus, Combretum, Coriolus, Crocosmia, Crotalaria, Cyanotis, Cyathea, Cyathostemma, Cycas, Davallia, Dichapetalum, Dillenia, Duchesnea, Dysosma, Eclipta, Entada, Epipremnum, Eriocaulon, Erythrophleum, Euclea, Fabiana, Fissistigma, Fouqueria, Ganoderma, Globularia, Glochidon, Glyptostrobus, Grifola, Gyrocarpus, Haplophyllum, Harrisonia, Heterostemma, Himatanthus, Holoptelea, Hura, Hydnocarpus, Icacina, Inonotus, Iodina, Isodon, Janakia, Jungia, Knema, Krameria, Kyllinga, Laggera, Laurentia, Leptadenia, Leptospermum, Leucas, Ligularia, Lycoperdon, Maackia, Mahonia, Melandrium, Melastoma, Merendera, Metopium, Minquartia, Monnina, Murraya, Nandina, Notholaena, Osyris, Parkinsonia, Pfaffia, Phymatosorus, Picria, Pleione, Plicosepalus, Pratia, Pulsatilla, Pygmaeopremna, Rabdosia, Raphidophora, Rauvolfia, Rhazya, Rogeria, Sarcandra, Sarcophyte, Sarcostemma, Selaginella, Semiaquilegia, Sempervivum, Sibthorpia, Sparganium, Stemona, Suregada, Thunbergia, Tripterygium, Tylophora, Vigna, Virola, Wedelia, Xeromphis, Ximenia, Zeyheria, Zhumeria, reported in Graham et al. (2000), are not listed in Hartwell (1982).

Graham et al. (2000) reported 350 species, distributed in over 280 genera and 100 families which were not mentioned in a compilation of Hartwell (1982). Among the described 100 families, Leguminosae (28 species) contained the highest number of species followed by Euphorbiaceae and Compositae (27 species each), Labiatae (15 species), Apocynaceae (14 species), Celastraceae (13 species), and Asclepiadaceae (10 species). The rest of the 93 families were represented by less than 10 species. Different parts of plants are used as medicine including root, stem, bark, entire plant, fruit, sap, resin, leaf, seed, latex, corm, aerial part, bulb, tuber, wood, flower, branch, shoot, inflorescence, protoplast, rhizome, thorn, essential oil, cotyledon, twig, pod, strobilus, and pericarp (Graham et al. 2000).

The first agents to advance into clinical use were the isolation of the vinca alkaloids, vinblastine, and vincristine from *Catharanthus roseus,* for the treatment of cancer followed by the discovery of paclitaxel from *Taxus brevifolia* (Kinghorn 1994, Kaur et al. 2011). Various parts of *the Taxus* have been used for the treatment of some noncancerous cases (Crag and Newmann 2005). *Taxus baccata* was also reported for the treatment of cancer. Paclitaxel is significantly active against ovarian cancer, advanced breast cancer, and lung cancer. Topotecan is used for ovarian and lung cancer from Xi shu tree [*Camptotheca acuminata*] (Steenhuysen 2007). Camptothecin, isolated from *Camptotheca acuminate,* is used for the treatment of ovarian and small cell lung cancers, and colorectal cancers (Kinghorn 1994, Creemer et al. 1996, Bertino 1997). Irinotecan is used for colon cancer treatment from Xishu tree [*Camptotheca acuminate*] (Online Medical Dictionary 2007). Epipodophyllotoxin was isolated as the active antitumor agent from the roots of *Podophyllum* species, *Podophyllum peltatum* and *Podophyllum emodi*, used in the treatment of lymphomas and bronchial and testicular cancers (Harvey 1999). The *Podophyllum peltatum* and *P. emodii* are used for the skin cancers.

The Combretastatins are a family of stilbenes that act as anti-angiogenic agents resulting in tumor necrosis (Cragg et al. 2002). Combretastatins isolated from the bark of *Combretum caffrum* showed to be active against colon, lung and leukemia cancers (Petit et al. 1987, Ohsumi et al. 1998). Homoharringtonine isolated from the *Cephalotaxus harringtonia* is at present in clinical use (Itokawa et al. 2005). Betulinic acid, primarily from *Betula* species (Cichewitz and Kouzi 2004), but also isolated from *Zizyphus mauritiana*, *Z. rugosa* and *Z. oenoplia* displayed selective cytotoxicity against human melanoma cell lines. Silvestrol from the fruits of *Aglaila sylvestre,* exhibited cytotoxicity against lung and breast cancer cell lines (Pisha et al. 1995). Other plant derived agents in clinical use are homoharringtonine isolated from *Cephalotaxus harringtonia* and elliptinium isolated from species of several genera of the Apocynaceae including *Bleekeria vitensis,* with reputed anti-cancer properties. Several *Terminalia* species have reportedly been used in the treatment of cancer.

Croton lechleri is an herb used in folk medicine as an anti-inflammatory, antimicrobial and anticancer properties (Hartwell 1969, Pieters et al. 1993, Lopes et al. 2004). Crude extracts from plants like *Colubrina macrocarpa, Hemiangium excelsum,* and *Acacia pennatula* possess a selective cytotoxic activity against human tumor cells (Popoca et al. 1998). In the Palestinian and Israeli territories, extracts of *Teucrium polium* and *Pistacia lentiscus* are known to treat liver disease, jaundice, diabetes, fertility problems and cancer (Howiriny and Sohaibani 2005). *Paris polyphylla* has been used to treat liver cancer and has been reported as a potent anticancer agent. *Salvia officinalis* are known for their antitumor effects (Liu et al. 2000), and *Solanum nigrum* to cure inflammation, edema, mastitis and hepatic cancer (Lee and Lim 2003). Until the development of penicillin in the early 1940s, most natural product-derived drugs were obtained from terrestrial plants. The success of penicillin in treating infection led to an expansion in the area of drug discovery from microorganisms. Unlike the long-standing historical medical uses of terrestrial plants, marine organisms have a shorter history of utilization in the treatment of human disease (Dewick 2002, Newman et al. 2003, Butler 2005, Sneader 2005).

The Challenges Ahead

There are numerous challenges aggravated by the multidisciplinary nature of ethnopharmacology but the main are scientific precision and accuracy. The scientific outcomes are scientifically reliable and reproducible (Heinrich and Verpoorte 2014). The botanical validity and details about how this was accomplished needs to be validated (Bennett and Balick 2014, Rivera et al. 2014). Ethnopharmacology is about linking traditional uses with biological and biomedical studies and hence it is essential to focus research on the uses and the importance of the plants (Heinrich et al. 2009). However, if it is not supported on a chemical and analytical base and the description of extracts, there will be particular challenges and as much information as possible on an extract and its composition is an essential element of all ethnopharmacological studies (Sheridan et al. 2012).

There are many challenges in the context of pharmacology, but again pharmacological research must use state-of-the-art approaches. The models and doses used must be of physiological and/or pharmacological significance, and the studies have to be directed based on the existing standards (Verspohl 2002, Cos et al. 2006). Ethnopharmacology has been influenced by studies on natural products (Malone 1983, Heinrich 2010). The primary challenge in pharmacology is the scientific study of the complex products resulting from indigenous knowledge and practices. Extracts obtained from plants, fungi, and animals poses some unique challenges. The extracts are multicomponent mixtures of substances including active, partially active and inactive substances and the activity is often not created on a single target (Heinrich 2010). Because of this difficulty, extracts often differ and thus the pharmacological outcomes cannot always be replicated. The phytochemical analysis presented a faster analysis and isolation of natural products and their identification/structure elucidation. But still, the complexity of mixtures found in extracts from natural sources continues to be a challenge in ethnopharmacology research (Heinrich 2010).

Thin Layer Chromatography (TLC) and later High-Performance Liquid Chromatography (HPLC) fingerprinting offered opportunities to characterize extracts and in the last two decades, hyphenated techniques directed scientists to identify quickly many of the metabolites in an extract (Heinrich 2010). It has been argued that metabolomics may offer unique opportunities to understand such complex mixtures (Verpoorte et al. 2005, Heinrich 2008, Schripsema 2010). Therefore, to explore the upcoming opportunities and challenges, all levels of research are necessary, i.e., from the critical study of traditional and local knowledge systems (Etkin 1988, Heinrich et al. 2009), the screening of such plants in primary assays, to the clinical evaluation of their effectiveness (Corson and Crews 2007, Graz et al. 2010) or studies on the public health implications of using (or not using) such medicines. In this context, some therapeutic areas have proven to be of particular relevance (Heinrich 2010). Molecular biology, pharmacogenomics, genomics, proteonomics, and related disciplines now offer such approaches (Efferth et al. 2008). When used in the context of the study of medicinal and food plants and their pharmacological activities, such approaches offer exciting new insights (Heinrich 2010, 2014).

In the 19th century, the original discoveries of natural products started hunting novel medicines (Heinrich 2010). But, numerous researches carried out during the 1970s and 1980s identified some novel compounds from plants or fungi without a linkage to ethnopharmacology. Frontiers in ethnopharmacology encourage truly trans-disciplinary approaches (Heinrich 2010, 2014). In areas like anti-cancer and anti-parasitic medications as well as in pain management, natural products form the core of human's treatment options for power and potential of turning traditional medicines into modern drugs (Corson and Crews 2007, Heinrich 2010). But, various researches are going on in the field of ethnopharmacology, which are poorly available (Heinrich 2010, 2014). A wide range of bioactive compounds with antimicrobial activity, monoamine oxidase inhibitors, angiotensin inhibitors, anti-inflammatory, and anti-allergic effects was explored (Van den Berghe et al. 1978, 1986, Ieven et al. 1979, Suzuki et al. 1981, Wagner 1989, Elbl and Wagner 1991) and thin-layer chromatography techniques with overlay techniques to detect antimicrobial activity were initiated (Houghton et al. 2007, Corson and Crews 2007, Graz et al. 2010).

Conclusion

The use of wild plants in fighting diseases is a common practice among the indigenous and local communities. The demand for herbal medicine is rising and also the research interest in wild plants is growing due to the emergence of several diseases. Natural sources form the basis of modern medicine and largely contribute to the production of commercial drugs. Ethnobotanists perform research on uses and concepts of plants in cultures and societies and ethnopharmacologists research the medicinal uses of plants and their properties. Plants with ethnopharmacological uses have been the primary sources of medicines for early drug discovery. Herbal medicines were the first medicines used by humans due to several pharmacologically active secondary metabolites produced by plants. Natural products discovered from ethnomedicinal plants have played an important role in the treatment of various diseases and thus medicinal plants maintain the health and life of people in different regions of the globe.

The analysis of medicinal plants for bioactive compounds is costly, tedious and time-consuming. The steps in the ethnopharmacology research begin with the production of the compound through the identification and ethnotaxonomy of the natural material followed by the traditional preparation of the pharmaceutical forms. Ethnopharmacological research greatly increases the chances of discovering new drugs. Of thousands of claims of bioactive natural products with potential therapeutic applications, few claims have been verified. Plants continue to be a major source of raw material for drug discovery. Neglected diseases are still treated with herbal medicines, and thus ethnopharmacological approach is an important tool for the use of these herbal resources in medicines. Early ethnopharmacologists made landmark discoveries, such as the elucidation of the pharmacological principles in foxglove, poppy, and curare.

There are numerous challenges aggravated by the multidisciplinary nature of ethnopharmacology but the main are scientific precision and accuracy. Ethnopharmacology is about linking traditional uses with biological and biomedical studies and hence it is essential to focus the research on the uses and the importance of the plants. Extracts obtained from the plants contain some unique challenges. The extracts are mixtures of substances including the active, partially active and inactive substances and the activity is often not produced on a single target. Because of this difficulty, extracts often differ and thus the pharmacological outcomes cannot always be replicated. The phytochemical analysis offers a faster analysis and isolation of natural products and their identification/structure elucidation. But still, the complexity of mixtures found in extracts from natural sources continues to be a challenge in ethnopharmacology research. The use of bioactive natural products in natural drug preparation dates back hundreds of years. However, their utilization as isolated and characterized compounds to modern drug development started only in the 19th century. But, numerous research carried out during the 1970s and 1980s identified some novel compounds from plants or fungi without a linkage to ethnopharmacology. Frontiers in ethnopharmacology encourage truly trans-disciplinary approaches.

Acknowledgments

MKR and CMF thank CNPq (National Council for Scientific and Technological Development, Brazil) for financial support (Process number 403888/2018-2). SB thanks NAST (Nepal Academy of Science and Technology, Kathmandu, Nepal) for various supports. Finally, we thank Professor Susana Zacchino, Rosario National University, Rosario, Argentina for critically going through the manuscript and offering helpful suggestions.

References

Abramovitz, J.N. 1997. Valuing nature's services. *In*: State of The World 1997: A World Watch Institute Report on Progress Towards Sustainable Society. New York: W.W. Norton and Co.

Alexiades, M.N. (ed.). 1996. Selected Guidelines for Ethnobotanical Research: A Field Manual. The New York Botanical Garden, Beonx, New York, USA.

Alves, R.R., Rosa, I.M. 2007. Biodiversity, traditional medicine and public health: where do they meet? J. Ethnobiol. Ethnomed., 3: 14.

Atanasov, A.G., Waltenberger, B., Pferschy-Wenzig, E.-M., Linder, T., Wawrosch, C., Uhrin, P., Temml, V., Wang, L., Schwaiger, S., Heiss, E.H. 2015. Discovery and resupply of pharmacologically active plant-derived natural products: A review. Biotechnol. Adv., 33(8): 1582–1614.

Balunas, M.J., Kinghorn, A.D. 2005. Drug discovery from medicinal plants. Life Sci., 78: 431–441.

Bennett, B.C., Balick, M.J. 2014. Does the name really matter? The importance of botanical nomenclature and plant taxonomy in biomedical research. J. Ethnopharmacol., 152: 387–392.

Berlin, E.A., Berlin, B. 2005. Some field methods in medical ethnobiology. Field Methods, 17(3): 235–268.

Bertino, J.R. 1997. Irinotecan for colorectal cancer. Semin Oncol., 24: S18–S23.

Bhandari, M., Bhandari, A., Bhandari, A. 2011. Recent updates on Codeine. Pharm. Methods, 2(1): 3–8.

Bhargava, K. 2015. Natural product drug discovery. J. Pharmacogn. Nat Prod., 1: e102.

Bray, F., Jemal, A., Grey, N., Ferlay, J., Forman, D. 2012. Global cancer transitions according to the human development index (2008–2030): a population-based study. The Lancet Oncol., 13(8): 790–801.

Butler, M.S. 2005. Natural products to drugs: natural products derived compounds in clinical trials. Nat. Prod. Rep., 22: 162–195.

Buyel, J.F. 2018. Plants as sources of natural and recombinant anti-cancer agents. Biotechno. Advances, 36: 506–520.

Carraz, M., Jossang, A., Franetich, J.-F., Siau, A., Ciceron, L., Hannoun, L., Sauerwein, R., Frappier, F., Rasoanaivo, P., Snounou, G., Mazier, D. 2006. A plant-derived morphinan as a novel lead compound active against malaria liver stages. PLoS Med., 3: e513.

Centre for Cancer Education. 2007. Online Medical Dictionary. University of Newcastle upon Tyne, UK.

Chase, M.W., Cameron, K.M., Freudenstein, J.V., Pridgeon, A.M., Salazar, G., Van den Berg, C., Schuiteman, A. 2015. An updated classification of Orchidaceae. Bot. J. Linn. Soc., 177: 151–174.

Chaudhary, R.P. 1998. Biodiversity in Nepal: Status and Conservation. S. Devi, Saharanpur, India and Tecpress Books, Bangkok, Thailand.

Christenhusz, M., Byng, J.W. 2016. The number of known plant species in the world and its annual increase. Phytot., 261(3): 201–217.

Cichewitz, R.H., Kouzi, S.A. 2004. Chemistry, biological activity, and chemotherapeutic potential of betulinic acid for the prevention and treatment of cancer and HIV infection. Med. Res. Rev., 24(1): 90–114.

Clapp, R.A., Crook, C. 2002. Drowning in the magic well: Shaman pharmaceuticals and the elusive value of traditional knowledge. J. Environ. Develop., 11: 79–102.

Corson, T., Crews, C. 2007. Molecular understanding and modern application of traditional medicines: triumphs and trials. Cell, 130(5): 769–74.

Cos, P., Vlietinck, A.J., Berghe, D.V., Maes, L. 2006. Anti-infective potential of natural products: How to develop a stronger *in vitro* 'proof-of-concept'. J. Ethnopharmacol., 106: 290–302.

Cox, P.A., Balick, M.J. 1994. The ethnobotanical approach to drug discovery. Sci. Am., 270(6): 82–7.

Cragg, G.M., Newmann, D.J., Holbeck, S. 2002. Natural products and derivatives as leads to cell cycle pathway targets in cancer chemotherapy. Curr. Cancer Drug Targets, 2(4): 279–308.

Cragg, G.M., Newman, D.J. 2005. Plants as a source of anti-cancer agents. J. Ethnopharmacol., 100(1–2): 72–9.

Creemers, G.J., Bolis, G., Gore, M., Scarfone, G., Lacave, A.J., Guastalla, J.P., Despax, R., Favalli, G., Kreinberg, R., VanBelle, S., Hudson, I., Verweij, J., Huinink, W.W.T. 1996. Topotecan, an active drug in the second-line treatment of epithelial ovarian cancer: results of a large European phase II study. J. Clin. Oncol., 14: 3056–61.

Cunningham, A.B. 2001. Applied Ethnobotany, People, Wild Plant and Conservation. Earths can Publications Ltd., London (UK) and Sterling VA (USA).

Daily, G. (ed.). 1997. Nature's Services: Societal Dependence on Natural Ecosystems. Covelo, CA: Island Press. 392 pp.

Daily, G.C., Alexander, S., Ehrlich, P.R., Goulder, L., Lubchenco, J., Matson, P.A., Mooney, H.A., Postel, S., Schneider, S.H., Tilman, D., Woodwell, G.M. 1997. Ecosystem services: benefits supplied to human societies by natural ecosystems. https://www.esa.org/esa/wp-content/uploads/2013/03/issue2.pdf.

Deleu, D., Hanssens, Y., Northway, M.G. 2004. Subcutaneous apomorphine: an evidence-based review of its use in Parkinson's disease. Drug Aging, 21: 687–709.

Dewick, P.M. 2002. Medicinal Natural Products: A Biosynthetic Approach. 2nd edition, Chichester, UK: John Wiley & Sons.

Efferth, T., Kahl, S., Paulus, K., Adam, M., Rauh, R., Boechzelt, H., Hao, X., Kaina, B., Baue, R. 2008. Phytochemistry and pharmacogenomics of natural products derived from traditional Chinese medicine and Chinese materia medica with activity against tumor cells. Mol. Cancer Ther., 7: 152–161.

Eisner, T. 1989. Prospecting for nature's chemical riches. Iss. Sci. Tech., 6(2): 31–34.

Eisner, T. 1991. Chemical prospecting: A proposal for action. pp. 196–202. *In*: Bormann, F., Kellert, R. (eds.). Ecology, Economics and Ethics: The Broken Circle. New Haven, CT: Yale University Press.

Ekor, M. 2014. The growing use of herbal medicines: issues relating to adverse reactions and challenges in monitoring safety. Front. Pharmacol., 4: 177.

Elbl, G., Wagner, H. 1991. A new method for the *in vitro* screening of inhibitors of angiotensin converting enzyme (ACE), using the chromophore- and fluorophore-labelled substrate, dansyltriglycine. Planta Med., 57: 137–141.

Etkin, N. 1988. Ethnopharmacology, biobehavioral approaches in the anthropological study of indigenous medicines. Ann. Rev. Anthropol., 17: 23–42.

Etkin, N.L., Elisabetsky, E. 2005. Seeking a transdisciplinary and culturally germane science: the future of ethnopharmacology. J. Ethnopharmacol., 100: 23–26.

Fabricant, D.S., Farnsworth, N.R. 2001. The value of plants used in traditional medicine for drug discovery. Environ. Health Perspect., 109: 69–75.

Farnsworth, N., Akerele, O., Bingel, A., Soejarto, D., Guo, Z. 1985. Medicinal plants in therapy. Bull. Wor. Health Organiz., 63(6): 965–981.

Funk, V.A., Susanna, A., Stuessy, T.F., Bayer, R.J. (eds.). 2009. Systematics, Evolution, and Biogeography of Compositae. International Association for Plant Taxonomy, Vienna.

Ganie, S.H., Upadhyaya, P., Das, S., Sharma, M.P. 2015. Authentication of medicinal plants by DNA markers. Plant Gene, 4: 83–99.

Gertsch, J. 2009. How scientific is the science in ethnopharmacology? Historical perspectives and epistemological problems. J. Ethnopharmacol., 122: 177–183.

Gewali, M.B. 2008. Aspects of Traditional Medicine in Nepal. Institute of Natural Medicine, University of Toyama, Toyama, Japan.

Gilani, A.H. 2005. Trends in ethnopharmacology. J. Ethnopharmacol., 100(1): 43–49.

Graham, J.G., Quinn, M.L., Fabricant, D.S., Farnsworth, N.R. 2000. Plants used against cancer—an extension of the work of Jonathan Hartwell. J. Ethnopharmacol., 73: 347–377.

Graz, B., Falquet, J., Elisabetsky, E. 2010. Ethnopharmacology, sustainable development and cooperation: the importance of gathering clinical data during field surveys. J. Ethnopharmacol. 130. doi: 10.1016/j.jep.2010.04.044.

Hall, M.G., Wilks, M.F., Provan, W.M., Eksborg, S., Lumholtz, B. 2001. Pharmacokinetics and pharmacodynamics of NTBC [2-(2-nitro-4-fl uoromethyl-benzoyl)-1,3-cyclohexanedione] and mesotrione, inhibitors of 4-hydroxyphenyl pyruvate dioxygenase (HPPD) following a single dose to healthy male volunteers. Br. J. Clin. Pharmacol., 52: 169–177.

Hartwell, J.L. 1969. Plants Used against Cancer: A survey. Lloyd., 32(2): 153–205.

Hartwell, J.L. 1982. Plants Used Against Cancer. 709 pp. Lawrence, Massachusetts.

Harvey, A.L. 1999. Medicines from nature: Are natural product still relevant to drug discovery. Trends Pharmacol. Sci., 20(5): 196–198.

Harvey, A.L. 2002. Natural products for high-throughput screening. pp. 39–44. *In*: Iwa and Wootton (eds.). Ethnomedicine and Drug Discovery. Elsevier Science B.A.

Harvey, A.L., Edrada-Ebel, R., Quinn, R.J. 2015. The re-emergence of natural products for drug discovery in the genomics era. Nat. Rev. Drug Discov., 14(2): 111–129.

Heinrich, M., Teoh, H.L. 2004. Galantamine from snowdrop—The development of a modern drug against Alzheimer's disease from local Caucasian knowledge. J. Ethnopharmacol., 92(2-3): 147–62.

Heinrich, M. 2008. Ethnopharmacy and natural product research—multidisciplinary opportunities for research in the metabolomic age. Phytochem. Lett., 1: 1–5.

Heinrich, M., Edwards, S., Moerman, D.E., Leonti, M. 2009. Ethnopharmacological field studies: a critical assessment of their conceptual basis and methods. J. Ethnopharmacol., 124: 1–17.

Heinrich, M. 2010. Ethnopharmacology in the 21st century—grand challenges. Front., Pharmacol., 1: 1–3. doi: 10.3389/fphar.2010.00008.

Heinrich, M. 2014. Ethnopharmacology: quo vadis? Challenges for the future? Rev. Bras. Farmacogn., 24: 99–102.

Heinrich, M., Verpoorte, R. 2014. Good practice in ethnopharmacology and other sciences relying on taxonomic nomenclature. J. Ethnopharmacol., 152: 385–386.

Heinrich, M., Jäger, A.K. (eds.). 2015. Ethnopharmacology. Wiley-Blackwell, pp. 462.

Heinrich, M., Casselman, I. 2018. Ethnopharmacology—From Mexican Hallucinogens to a 829 Global Transdisciplinary Science, Ethnopharmacologic Search for Psychoactive Drugs: 50th Anniversary Symposium. Synergetic Press, in Association with Heffter Research Institute.

Houghton, P.J., Howes, M.-J., Le, C.C., Steventon, G. 2007. Uses and abuses of *in vitro* tests in ethnopharmacology: visualizing an elephant. J. Ethnopharmacol., 110: 391–400.

Howes, M.-J.R., Perry, N.S.L., Houghton, P.J. 2003. Plants with traditional uses and activities, relevant to the management of Alzheimer's disease and other cognitive disorders. Phytother. Res., 17: 1–18.

Howiriny, T., Sohaibani, M. 2005. Effect of *Commiphora opobalsamum* (L.) Engl. (Balessan) on experimental gastric ulcers and secretion in rats. J. Ethnopharmacol., 98(3): 287–94.

Ieven, M., van den Berghe, D.A., Mertens, F., Vlietinck, A., Lammens, E. 1979. Screening of higher plants for biological activities. I. Antimicrobial activity. Planta Med., 36: 311–321.

Itokawa, H., Wang, X., Lee, K.H. 2005. Homoharringtonine and related compounds. pp. 47–70. *In*: Cragg, G.M., Kingston, D.G.I., Newman, D.J. (eds.). Anticancer Agents from Natural Products. Brunner-Routledge Psychology Press, Taylor and Francis Group. Boca Raton FL.

Jain, S.K., Mudgal, V. 1999. A Handbook of Ethnobotany. Bishen Singh Mahendra Pal Singh, Dehra Dun, India.

Karczmar, A.G. 1996. The Otto Loewi Lecture. Loewi's discovery and the XXI century. Prog Brain Res., 109: 1–27.

Kaur, R., Kapoor, K., Kaur, H. 2011. Plants as a source of anticancer agents. J. Nat. Prod. Plant Resour., 1(1): 119–124.

Kim, H.-S. 2005. Do not put too much value n conventional medicines. J. Ethnopharmacol., 100: 37–39.

Kinghorn, A.D. 1994. The discovery of drugs from higher plants. pp. 81–108. *In*: Gullo, V.P. (ed.). The Discovery of Natural Products with Therapeutic Potential. Boston, MA: Butterworth-Heinemann.

Koumis, T., Samuel, S. 2005. Tiotropium bromide: a new long-acting bronchodilator for the treatment of chronic obstructive pulmonary disease. Clin. Ther., 27: 377–392.

Kuhn, T., Wang, Y. 2008. Artemisinin—An innovative cornerstone for anti-malaria therapy. Nat. Compo. Drugs., 383–422.

Laird, S.A. 2000. Benefit-sharing 'best practice' in the pharmaceutical and botanical medicine industries. pp. 89–99. *In*: Svarstad, H., Dhillon, S.S. (eds.). Responding to Bioprospecting, from Biodiversity in the South to Medicines in the North, Spartacus Forlag AS, Oslo.

Laurance, J. 2005. Why an Exotic Fruit is the World's Only Weapon Against Bird Flu. The Independent (UK). Web resource: news.independent.co.uk/uk/health_medical/article319716.

Lee, S.J., Lim, K.T. 2003. Anti-oxidative effects of glycoprotein isolated from *Solanum nigrum* Linne on oxygen radicals and its cytotoxic effects on the MCF-7 cells. J. Food Sci., 68: 466–470.

Lee, W.L., Bell, B.M., Sutton, J.F. (eds.). 1982. Guidelines for acquisition and management of biological specimens. Association of Systematics Collections, Lawrence, Kansas.

Liu, J., Shen, H.M., Ong, C.N. 2000. *Salvia miltiorrhiza* inhibits cell growth and induces apoptosis in human hepatoma HepG2 cells. Cancer Lett., 153: 85–93.

Lopes, eLMI, Saffi, J., Echeverrigaray, S., Henriques, J.N.P., Salvador, M. 2004. Mutagenic and antioxidant activities of Croton lechlerisap in biological systems. J. Ethnopharmacol., 95(2-3): 437–445.

Madhuri, S., Pandey, G. 2008. Some dietary agricultural plants with anticancer properties. Plant Archives, 8: 13–16.

Malone, M.H. 1983. The pharmacological evaluation of natural products-general and specific approaches to screening ethnopharmaceuticals. J. Ethnopharmacol., 8: 127–147.

Marinelli J (ed). 2005. Plant: The Ultimate Visual Reference to Plants and Flowers of the World. New York: DK Publishing, Inc.

Martin, G.J. 1995. Ethnobotany: A Methods Manual. Chapman and Halls, London.

Mateo, N. 2000. Bioprospecting and conservation in Costa Rica. pp. 45–55. *In*: Svarstad, H., Dhillon, S.S. (eds.). Responding to Bioprospecting, from Biodiversity in the South to Medicines in the North. Spartacus Forlag AS, Oslo.

Mitchell, G., Bartlett, D.W., Fraser, T.E.M., Hawkes, T.R., Holt, D.C., Townson, J.K., Wichert, R.T. 2001. Mesotrione: a new selective herbicide for use in maize. Pest Manag. Sci., 57: 120–128.

Mosihuzzaman, M., Choudhary, M.I. 2008. Protocols on safety, efficacy, standardization, and documentation of herbal medicine. Pure Appl. Chem., 80(10): 2195–2230.

MPWG. 2007. Plant Conservation Alliance Medicinal Plant Working Group (MPWG): www.nps.gov/plants/medicinal/plants.htm.

Newman, D.J., Cragg, G.M., Snader, K.M. 2000. The influence of natural products upon drug discovery. Nat. Prod. Rep., 17: 215–234.

Newman, D.J., Cragg, G.M., Snader, K.M. 2003. Natural products as sources of new drugs over the period 1981–2002. J. Nat. Prod., 66: 1022–1037.

Newman, D.J., Cragg, G.M. 2007. Natural products as sources of new drugs over the last 25 years. J. Nat. Prod., 70(3): 461–77.

Newman, D.J., Cragg, G.M. 2016. Natural products as sources of new drugs from 1981 to 2014. J. Nat. Prod., 79(3): 629–61.

Newmaster, S.G., Grguric, M., Shanmughanandhan, D., Ramalingam, S., Ragupathy, S. 2013. DNA barcoding detects contamination and substitution in North American herbal products. BMC Med., 11: 222.

Ohsumi, K., Nakagawa, R., Fukuda, Y., Hatanaka, T., Morinaga, Y., Nihei, Y., Ohishi, K., Suga, Y., Akiyama, Y., Tsuji, T. 1998. New combretastatin analogues effective against murine solid tumors: design and structure-activity relationship. J. Med. Chem., 41: 705–706.

Pearson, D.L. 1985. United States biologist in foreign countries: The new ugly American? Bulletin of the Ecological Society of America, 66: 333–337.

Pettit, G.R., Singh, S.B., Niven, M.L., Hamel, E., Schmidt, J.M. 1987. Isolation, structure, and synthesis of combretastatin's A-1 and B-1, potent new inhibitors of microtubule assembly, derived from *Combretum caffrum*. J. Nat. Prod., 50: 119–20.

Pieters, L., Debruyne, T., Claeys, M. 1993. Isolation of a dihydrobenzofuran lignan from South American dragon's blood (*Croton* spp.) as an inhibitor of cell proliferation. J. Nat. Prod., 56: 899–906.

Pisha, E., Chai, H., Lee, I.S., Chagwedera, T.E., Farnsworth, N.R., Cordell, G.A., Beecher, C.W., Fong, H.H., Kinghorn, A.D., Brown, D.M., Wani, M.C., Wall, M.E., Hieken, T.J., Das Gupta, T.K., Pezzuto, J.M. 1995. Discovery of betulinic acid as a selective inhibitor of human melanoma that functions by induction of apoptosis. Nat. Med., 1: 1046–51.

Popoca, J., Aguilar, A., Alonso, D. 1998. Cytotoxic activity of selected plants used as antitumoral in Mexican traditional medicine. J. Ethnopharmacol., 59: 173–77.

Rai, M., Bhattarai, S., Christiane, M.F. (eds.). 2021. Wild Plants: The Treasure of Natural Healers. CRC Press/Taylor and Francis, USA.

Rivera, D., Allkin, R., Obón, C., Alcaraz, F., Verpoorte, R., Heinrich, M. 2014. What is in a name? The need for accurate scientific nomenclature for plants. J. Ethnopharmacol., 152: 393–402.

Robbins, C. 1999. Medicine from US Wildlands: An Assessment of Native Plant Species Harvested in the United States for Medicinal Use and Trade and Evaluation of the Conservation and Management Implications. TRAFFIC North America, the Nature Conservancy. Web resource: www.nps.gov/plants/medicinal/pubs/traffic.htm.

Roberson, E. 2008. Medicinal Plant at Risk, Nature's Pharmacy, Our Treasure Chest: Why We Must Conserve Our Natural Heritage, A Native Plant Conservation Campaign Report. Center for Biological Diversity. Tucson, AZ.

Sankaranarayanan, R. 2014. Screening for cancer in low- and middle-income countries. Ann. Glob. Health., 80(5): 412–417.

Saslis-Lagoudakis, C.H., Savolainen, V., Williamson, E.M., Forest, F., Wagstaff, S.J., Baral, S.R., Watson, M.F., Pendry, C.A., Hawkins, J.A. 2012. Phylogenies reveal the predictive power of traditional medicine in bioprospecting. PNAS, 109(39): 15835–15840.

Schippmann, U., Leaman, D.J., Cunningham, A.B. 2002. Impact of cultivation and gathering of medicinal plants on biodiversity: global trends and issues. Rome: Food and Agriculture Organizations.

Schippmann, U., Leaman, D., Cunningham, A.B. 2006. A comparison of cultivation and wild collection of medicinal and aromatic plants under sustainability aspects. pp. 75–95. *In*: Bogers, R.J., Craker, L.E., Lange, D. (eds.). Medicinal and Aromatic Plants. Netherlands: Springer.

Schripsema, J. 2010. Application of NMR in plant metabolomics: techniques, problems and prospects. Phytochem. Anal., 21: 14–21.

Science Daily. 2018. More than one-fifth of world's plants face threat of extinction, new analysis finds. https://www.sciencedaily.com/releases/2010/09/100928193452.htm.

Shah, S. and Bhat, J.A. 2019. Ethnomedicinal knowledge of indigenous communities and pharmaceutical potential of rainforest ecosystems in Fiji Islands. J. Integ. Med., 17: 244–249.

Sheridan, H., Krenn, L., Jiang, R., Sutherland, I., Ignatova, S., Marmann, A., Liang, X., Sendker, J. 2012. The potential of metabolic fingerprinting as a tool for the modernisation of TCM preparations. J. Ethnopharmacol., 140: 482–491.

Shi, Q.W., Li, L.G., Huo, C.H., Zhang, M.L., Wang, Y.F. 2010. Study on natural medicinal chemistry and new drug development. Chin. Tradit. Herb. Drugs., 41: 1583–1589.

Shrestha, K.K., Bhattarai, S., Bhandari, P. 2018. Handbook of Flowering Plants of Nepal (Volume 1. Gymnosperms and Angiosperms: Cycadaceae – Betulaceae). Scientific Publishers, Jodhpur, India, pp. 648.

Singh, S., Singh, D.B., Singh, S., Shukla, R., Ramteke, P.W., Misra, K. 2018. Exploring medicinal plant legacy for drug discovery in post-genomic era. pp. 1–11. Proceedings of the National Academy of Sciences, India Section B: Biological Sciences.

Sivalokanathan, S., Ilayara, M. 2005. Anticancer plants: A review. Ind. J. Expt. Biol., 43: 264–67.

Sneader, W. 2005. Drug Discovery: A History. Hoboken, NJ: John Wiley & Sons.

Soejarto, D.D., Fong, H.H., Tan, G.T., Zhang, H.J., Ma, C.Y., Franzblau, S.G., Gyllenhall, C., Riley, M., Kadushin, M.R., Pezzuto, J.M., Xuan, L.T., Hiep, N.T., Hung, N.V.,Vu, B.M., Loc, P.K., Dac, Le.X., Binh, le.t., Chien, N.Q., Nong, V.H., Bich, T.Q., Cuong, N.M., Southavong, B., Sydara, K., Bouamanivong, S., Ly, H.M., Thuy, T.V., Rose, W.C., Dietzman, G.R. 2005. Ethnobotany/ethnopharmacology and mass bioprospecting: issues on intellectual property and benefit-sharing. J. Ethnopharmacol., 100: 15–22.

Solis, I.M. 2014. Ethnobotany and ethnopharmacology, the first step towards bioprospecting. J. Biodivers. Biopros. Dev., 2(1): 1–2 (e107).

Steenhuysen, J. 2007. Mother Nature Still A Rich Source of New Drugs. Reuters Limited. March 19, 2007.

Steenhuysen, J. 2018. Mother Nature Still A Rich Source of New Drugs. https://www.reuters.com/article/us-drugs-nature/mothernature-still-a-rich-source-of-new-drugs-idUSN1624228920070319.

Stewart, B.W., Wild, C.P. 2014. World Cancer Report 2014. International Agency for Research on Cancer, Lyon.

Suzuki, O., Katsumata, Y., Oya, M., Chari, V.M., Vermes, B., Wagner, H., Hostettmann, K. 1981. Inhibition of type-A and type-B monamine oxidases by naturally occurring xanthones. Planta Med., 42: 17–21.

Svarstad, H. 2000. Reciprocity, biopiracy, heroes, villains and victims. pp. 19–35. *In*: Svarstad, H., Dhillion, S.S. (eds.). Responding to Bioprospecting, from Biodiversity in the South to Medicines in the North. Spartacus Forlag AS, Oslo.

Taylor, R.S.L., Chaudhary, R.P. 2003. Ethnobotanical field studies: considerations for researchers. Bot. Orient., pp. 31–34.

Thillaivanan, S., Samraj, K. 2014. Challenges, constraints and opportunities in herbal medicines—a review. Int. J Herb. Med., 2(1): 21–24.

Thomas, M.J., Carolyn, F.S. 1996. Ethnopharmacology: The Conjunction of Medical Ethnography and the Biology of Therapeutic Action. Medical Anthropology: Contemporary Theory and Method. Westport, Connecticut: Praeger Publishers; pp. 132–133.

Tuxhill, J. 1999. Nature's Conucopia: Our State in Plant Diversity. Worldwatch Paper #148. Washington, DC: Worldwatch Institute.

Van den Berghe, D.A., Leven, M., Mertens, F., Vlietinck, A.J. 1978. Screening of higher plants for biological activity, II. Antiviral activity. Lloydia, 41: 463–471.

VanAgtmael, M,A,, Eggelte, T.A., van Boxtel, C.J. 1999. Artemisinin drugs in the treatment of malaria: from medicinal herb to registered medication. Trends Pharmacol. Sci., 20: 199–205.

Verpoorte, R,, Choi, Y,H,, Kim, H.K. 2005. Ethnopharmacology and systems biology: a perfect holistic match. J. Ethnopharmacol., 100: 53–56.

Verspohl, E.J. 2002. Recommended testing in diabetes research. Planta Med., 68: 581–590.

Wagner, H. 1989. Search for new plant constituents with potential antiphlogistic and antiallergic activity. Planta Med., 55: 235–241.

Warude, D. 2017. Safety of traditional medicines. pp. 351–365. *In*: Innovative Approaches in Drug Discovery, Ethnopharmacology, Systems Biology and Holistic Targeting, Academic press.

WHO. 2017. Cancer. http://www.who.int/mediacentre/factsheets/fs297/en/ (accessed2018-01-23.2018).

WWW.wikepedia.org. Accessed on January 15, 2019.

Yabroff, K.R., Lund, J., Kepka, D., Mariotto, A. 2011. Economic burden of cancer in the United States: estimates, projections, and future research. Cancer Epidemiol. Biomarkers Prev., 20(10): 2006–2014.

Yeung, A.W.K., Heinrich, M., Kijjoa, A., Tzvetkov, N.T., Atanasov, A.G. 2019. The ethnopharmacological literature: An analysis of the scientific landscape. https://doi.org/10.1016/j.jep.2019.112414.

Yuan, H., Ma, Q., Ye, L., Piao, G. 2016. The traditional medicine and modern medicine from natural products. Molecules, 21(5): 559.

Chapter 2

Ethno-Pharmacological Relevance/ Features of Some Latin American Wild Medicinal Plants

Amner Muñoz-Acevedo,[1,*] *María C. González*,[1] *Alma Polo-Barrios*,[1] *Martha Cervantes-Díaz*,[2] *Ricardo D.D.G. de Alburquerque*,[3] *Ninoska Flores*,[4] *Alberto Giménez-Turba*,[4] *María E. Cazar*,[5] *Chabaco Armijos*,[6] *Omar Malagón-Avilés*,[6] *Feliza Ramón-Farias*,[7] *Leticia M. Cano-Asseleih*,[8] *Elsa Rengifo*[9] and *Gabriel Vargas-Arana* [9]

Introduction

The traditional herbal healers, along with wild medicinal plants and the knowledge they have about them, in certain regions of Latin America, are the prevalent alternatives for some people/communities to cover any primary health care requirements (e.g., treatment to prevent/improve/release pain/symptoms/ disorders/diseases). traditional (plant-based) medicine has gained worldwide acceptance because the plants used for this purpose have been apparently efficacious and safe for a long time (decades/centuries), these being some reasons why the WHO created the traditional, complementary and integrative unit. (WHO 2019).

Despite this, it is important to remember (i) what "ethnopharmacology" studies are about, which are, shortly, the uses, possible action modes, and bioproperties of plant-based preparations; and (ii) the contribution made by phytochemistry and pharmacology along with other related sciences (botany, pharmacy, microbiology, virology, parasitology, etc.) to find/check any biologically active drugs (molecules/fractions/extracts). In that sense, traditional medicine systems need more evidence-based studies that allow establishing the pharmacological potential together with the validation of therapeutic uses of medicinal plants and its chemical constituents [isolated or mixtures (fractions/extracts)] in such a way that safety and effectiveness are verified, as well as to reveal/understand their preventive and/or therapeutic mechanisms (Mukherjee et al. 2010, Schmidt and Klaser-Cheng 2017).

[1] Departamento de Química y Biología, Universidad del Norte, Barranquilla, Colombia.
[2] Facultad de Química Ambiental, Universidad Santo Tomás, Bucaramanga, Colombia.
[3] Laboratório de Tecnologia em Productos Naturais, Universidade Federal Fluminense, Niterói, Brasil.
[4] Instituto de Investigaciones Fármaco Bioquímicas, Universidad Mayor de San Andrés, La Paz, Bolivia.
[5] Grupo de Biotecnología y Biodiversidad, Departamento de Química Aplicada y Sistemas de Producción, Universidad de Cuenca, Ecuador.
[6] Departamento de Química y Ciencias Exactas, Universidad Técnica Particular de Loja, Ecuador.
[7] Instituto de Ecología, Xalapa, México.
[8] Centro de Investigaciones Tropicales, Universidad Veracruzana, Xalapa, México.
[9] Instituto de Investigaciones de la Amazonía Peruana, Iquitos, Perú.
* Corresponding author: amnerm@uninorte.edu.co

The 18 wild medicinal plants of interest were selected in agreement with the science literature revision and the know-how of the co-authors taking into account (i) the importance of these plants for the ethnic/afrodescendant/mestizo groups in Bolivia, Brazil, Colombia, Ecuador, Mexico and Peru; (ii) the therapeutic benefits (as treatment for different disorders/diseases); (iii) the pharmacological potential validated/determined (*in vitro*/*in vivo*); (iv) the active extracts/fractions and molecules obtained/isolated; and, (v) the useful parts of plants. Accordingly, the present chapter comprises the text mining of the subject, the general aspects of selected plants from Latin America, and pharmacological properties recognized of extracts/molecules from these plants.

Text Mining for the Topic "*Ethnopharmacology related to Latin American Wild Medicinal Plants*"

For the establishment of the research trend correlated with ethnopharmacological studies of some Latin American wild medicinal plants, a scientometric measurement was performed by using Scopus database (Elsevier 2019), whose data obtained were analyzed through the specialized Vantage Point software for text mining (Search Technology, http://www.thevantagepoint.com/). For this, the following search equation was structured: (*title-abs-key* (*ethnopharmacolog* *) *and title-abs-key* (*"medicinal plant *" and wild*)) *and doctype* (*ar*), with which 180 documents indexed in the database were obtained (Figure 2.1), and 2014 and 2015 being the years of greatest scientific activity (20 and 23 articles, respectively). Based on De Solla Price's Law (De Solla Price 1963), the annual growth rate of the articles published since 2000 related to this topic was calculated, which was ca. 23%/year (R^2 0.956); despite that, in the last four years (2016–2019) there was a decrease in the dynamic of publication.

In accordance with the mining analysis, the most important research/application areas were associated with toxicology and pharmacy, medicine, agriculture and biological sciences; these research areas are related to each other due to the interdisciplinary studies. Similarly, anti-inflammatory (25 articles), antioxidant (18), antiparasitic [malaria (10), protozoa (4), helminth (4)], and antimicrobial [bacteria (7), fungi (5)] were the most studied biological properties. On the other hand, Italy (24), Turkey (24), Pakistan (15), India (12) and USA (11) stood out for being the countries with the largest number of records, while in the top five of Latin America were Mexico (6), Brazil (4), Colombia and Peru (3), and Argentina (2).

When the plants of interest in the search equation were considered, *Guazuma ulmifolia* (198), *Calophyllum brasiliense* (130), *Amburana cearensi* (98), *Casimiroa edulis* (54), *Montanoa tomentosa* (46) had the highest number of records, while *Renealmia thyrsoidea* (3), *Huperzia compacta* (2),

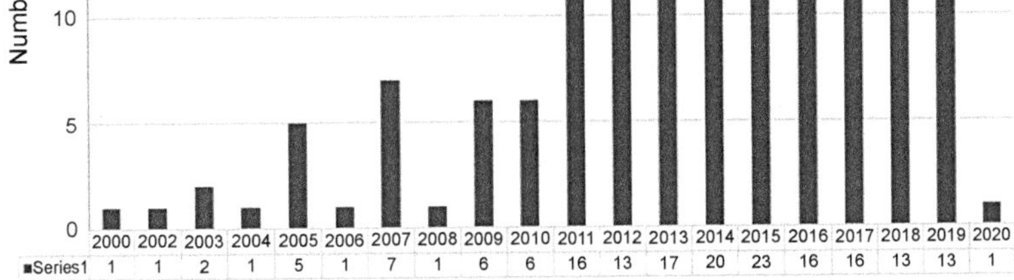

Figure 2.1. Distribution of publication by the number of registers per year (timeline 2000–2020), related to ethno-pharmacological studies of Latin American wild medicinal plants. Furthermore, it is included the trend of the growth rate by De Solla Price law. Source: Bibliometry Unit – CRAI-Library, Universidad Santo Tomas (Bucaramanga). Calculations base on Scopus information (Elsevir, B.V. 2019), processed with VantagePoint software (Search Technology).

H. espinosana (2), *H. tetragona* (1) and *Salacia impressifolia* (1) were the ones with the lowest number of records. Subsequently, Brazil and Mexico were the countries with the highest number of registers that involved the first three plants (*G. ulmifolia*, *Cal. brasiliense* and *Am. cearensis*), and the primary biological effects established for these plants were antioxidant, antibacterial and anti-inflammatory, each one with 36, 32 and 29 registers.

Selected Plants from Some Latin American Countries

Based on the review of the scientific literature and in the know-how of the co-authors, 18 plants (among herbs, shrubs and trees) from Bolivia, Brazil, Colombia, Ecuador, Mexico, and Peru were selected (three for each country), which have different/interesting ethnobotanical uses and some pharmacological results validated. These plants are *Adiantum cf. latifolium* Lam. (Pteridaceae), *Renealmia thyrsoidea* Poepp. (Zingiberaceae), *Salacia impressifolia* A.C. Sm (Hippocrateaceae), *Amburana cearensi* A.C. (Fabaceae), *Caesalpinia ferrea* (Mart. ex Tul.) L.P. Queiroz (Fabaceae), *Echinodorus macrophyllus* (Kunth) Micheli (Alismataceae), *Brosimum utile* Kunth. (Moraceae), *Guazuma ulmifolia* L. (Malvaceae), *Calophyllum brasiliense* Cambess. (Calophyllaceae), *Huperzia compacta* (Hook.) Trevis., *H. espinosana* B. Øllg., *H. tetragona* (Hook. & Grev.) Trevis. (Lycopodiaceae), *Agastache mexicana* (Kunth) Lint. & Epling. (Lamiaceae), *Casimiroa edulis* La Llave & Lex. (Rutaceae), *Montanoa tomentosa* Cerv. (Asteraceae), *Campsiandra angustifolia* Spruce ex Benth. (Fabaceae), *Copaifera paupera* (Herzog) Dwyer (Fabaceae) and *Sambucus peruviana* Kunth (Adoxaceae)—Figure 2.2. In other countries (from Latin America and/or Asia) different from those mentioned above, some of these plants have been found and certain indigenous/mestizo/afro-descendant communities have used them. In Table 2.1, each species has been described briefly in a general way.

Referring to Table 2.1, the plants under study are mainly (~ 56%) evergreen flowering trees (with fruits) with heights between 2–60 m, while the other species (~ 44%) are shrubs/fern up to 3 m. These plant species are mostly distributed in Latin American (tropical forest) although some of them (e.g., *Ad. latifolium*, *S. impressifolia*, *Cae. ferrea*, *B. utile*, *G. ulmifolia*) have been found in other regions of the world (e.g., India, Zimbabwe, China, Malaysia, Vietnam, Sri Lanka), where they were introduced. The different parts (flowers, leaves, fruits, seeds, barks, roots, oleoresins/exudates) of the trees/shrubs have some important ethnobotanical uses as medicine, food or source of ink by indigenous/mestizo/afro-descendant groups (e.g., Tacana, Bororos, Quilombolas, Witotos, Tikunas, Conibo, Saraguro, Otomi, Nahuatl, Totonaco, Quiché) from Amazon countries (Bolivia, Brazil, Colombia, Ecuador, Peru) and Mexico. The wide range of medicinal uses which these plants present should be noted, such as *antidote*, antiseptic, antipyretic, antifungal, antiherpetic, antiparasitic (leishmania and malaria), vermifuge, antiproliferative/cytotoxic/antitumor, hypotensor, diuretic, emollient, cleanser (against skin conditions), *purgative*, astringent, for treatment of respiratory (asthma, flu, expectorant), neuralgic and digestive (gastritis, diarrhea), circulatory, metabolic disorders, as well as for disorders of reproductive and muscle-bone (arthritis, rheumatism, osteoporosis) systems, against inflammation, diabetes, anemia, contusions, alopecia, *amoebiasis* and hemorrhoids, ulcers, for prevention of *tissue necrosis* (bite of the bushmaster snake), wound healing, among others.

Apart from this, some indigenous/afro-descendant healers (Tacana, Quilombolas, Saraguro) use certain plants for magical-religious healing rituals, for instance, *Ad. latifolium*, *E. macropyllus* and *Hurpezia* spp. ("wamingas/trencillas") have been applied to treat *susto* (folk illness in Bolivia), as protector (healer women in Brazil) and to cure "supernatural" ailments (*shuka/vaho* water/negative energy in Ecuador), respectively (Dixit et al. 1995, Bourbonnais-Spear et al. 2007, Maciel and Neto 2006, Armijos et al. 2012, 2014); other plants have been used by indigenous/peasant communities (Ingano, Coreguaje, Witotos, Huasteca (teenek), Bora-Miraña, Tikunas and Cocamas) as an important source of food, e.g., *B. utile* (exudate/resin) as a lactagogue (substitute milk), *G. ulmifolia* (fruit/seed/sap) to prepare flour/beverage/candy, and *S. impressifolia* and *Cas. edulis* (fruits) as edible fruit (Standley and Steyermark 1946, Bernal and Correa 1989, García-Barriga 1992, Argueta and Cano 1994, Castaño-Arboleda et al. 2007, Quattrocci 2012, Grandtner and Chevrette 2014, Texeira et al. 2019). For particular plants such as *B. utile*, *G. ulmifolia*, *Cal. brasiliense* and *Cop. paupera*, its exudates/oleoresins resulted to be the parts with the main ethnobotanical uses due to greater pharmacological effects.

A. Taken by IIFB 2007; Source: Bolivian co-authors
B. Taken by IIFB 2014; Source: Bolivian co-authors
C. Taken by IIFB 2014; Source: Bolivian co-authors
D. Source: https://www.ibflorestas.org.br/component/tags/tag/amburana-cearensis
E. Source: APNE/CNIP
F. Taken by B.R. Silva; Source: WGI 2009
G. Published by R. Aguilar; Source: Blog Arboles Peninsula de Osa, CR 2010
H. Taken by C. Delnatte; Source: INPN 2019
I. Taken by A. Muñoz-Acevedo 2018; Source: Colombian co-authors
J. Taken by C. Armijos 2016; Source: Ecuadorian co-authors
K. Taken by C. Armijos 2016; Source: Ecuadorian co-authors
L. Taken by C. Armijos 2016; Source: Ecuadorian co-authors
M. Taken by F. Ramón 2019; Source: Mexican co-authors
N. Taken by L. Cano-Alesseih 2011; Source: Mexican co-authors
O. Taken by L. Cano-Alesseih 2011; Source: Mexican co-authors
P. Taken by R.I. Barbosa; Source: Field Museum 2020
Q. Source: OSINFOR, Peru 2015
R. Source: Diffusion – Minagri/Ecoportal.net/unh.edu.pe

Figure 2.2. Images of the 18 interesting plants from Latin American countries. **A.** *Ad. latifolium*; **B.** *R. thyrsoidea*; **C.** *S. impressifolia*; **D.** *Am. cearenst*; **E.** *L. ferrea*; **F.** *E. macrophyllus*; **G.** *B. utile*; **H.** *G. ulmifolia*, **I.** *Cal. brasiliense*; **J.** *H. compacta*; **K.** *H. espinosana*, **L.** *H. tetragona*; **M.** *Ag. mexicana*; **N.** *Cas. edulis*; **O.** *M. tomentosa*; **P.** *Cam. angustifolia*; **Q.** *Cop. paupera*; **R.** *S. peruviana*.

Pharmacological Properties Recognized from Selected Plants from Latin America

The pharmacological properties (including *in vitro*/*in vivo* assessments) determined from extracts/isolated compounds obtained from the useful parts, according to the traditional uses reported for each plant and by indigenous/peasant/afro-descendant communities are included in Table 2.2.

Table 2.1. General information (including synonyms, common names, etc.) along with the ethnobotanical uses for some communities related to the 18 selected plants from Latin America.

Countries	Plants	Synonyms	Common names	Morphological characteristics	Communities	Ethnobotanical uses	References
Bolivia	*Adiantum latifolium*	*A. fovearum*, *A. humile*, *A. lucidum* var. *bipinnatum*	Atarisi huachidhi (Bolivian Tacana language), sorrapilla (Peru), shebebe (Shipibo/Conibo), shakushia (Amahuaca), montañero, cermanar (Colombia), avenca (Brazil)	Fern, 25–60 cm (height)	Ethnic groups from Bolivia, Brazil, Peru and Colombia	For skin infections, wounds, necrosis; as anxiolityc, analgesic, and antiinflammatory.	Dixit et al. 1995, Lopez et al. 2001, Bourbonnais Spear et al. 2007, Nonato et al. 2011, Arévalo-López et al. 2018
	Renealmia thyrsoidea	*R. cardenassi*, *Alpinia silvicola*, *Ethanium thyrsoideum*, *Amomum thyrsoideum*	Shiguango (Kichwa), nhaca nhaca (Bolivia), gönequemö, tëtemö (Huao, Ecuador), vira-cirindanga (Colombia), u´yume (Peru)	Flowering herb with fruits, 2–6 m	Ethnic groups from Bolivia, Ecuador	Against snakebites, leishmaniasis ulcers diarrhea; as healing and antipyretic	Davis et al. 1983, Muñoz et al. 2000, Odonne et al. 2013, Gómez-Betancur and Benjumea 2014, Cabanillas et al. 2014, Félix-Silva et al. 2017, Arevalo-López et al. 2018
	Salacia impressifolia	*Anthodon grandiflorum*, *S. grandiflora*, *Raddia impressifolia*	Panu/guapomó (Bolivia), miraruíra (Brazil), yuu pot (Colombia), chuchuhuasa (Perú), guoguo (Mexico), cachete viejo (Venezuela)	Flowering tree/liana with fruits, 5–10 m	Ethnic groups from Bolivia, Brazil and Peru	To treat diabetes, inflammation, general pain, rheumatism, cancer, dengue and renal affections	Bourdy et al. 2000, Grandtner and Chevrette 2014, Paz et al. 2018, Vásquez-Ocmín et al. 2018, Rodrigues et al. 2019
Brazil	*Amburana cearensis*	*Torresea cearensis*, *A. claudii*, *T. acreana*	Umburana-de-cheira, imburana (Brazil), cumaré (Paraguay), sorioco/tumi (Bolivia), caschoú (French Guiana), tsaik/tsantimatiqui (aguaruna/asháninca, Peru)	Perennial flowering tree, 4–10 m	Ethnic and afrodescendant groups from Brazil	As treatment for rheumatism, as well as for respiratory, neuralgic and digestive disorders	Lorenzi 1992, Cunha and Ferreira 2003, Almeida et al. 2005, Da Conceição et al. 2011, Quattrocci 2012, Silva et al. 2013a, Grandtner and Chevrette 2014, Braga 2015, Lima 2015
	Caesalpinia ferrea	*Libidibia ferrea*, *C. leiostachya*, *Apuleia ferrea*	Jucá, pau-ferro, ibirá-obi, imira-itá	Perennial flowering tree, 4–10 m	Ethnic groups from Brazil	Against respiratory and digestive disorders; to treat anemia, tuberculosis, diabetes and as healing and antiseptic, etc.	Di Stasi and Hiruma-Lima 2002, Maia et al. 2004, Queiroz 2009, Oliveira et al. 2010, Quattrocci 2012, Farias 2013, Grandtner and Chevrette 2014, Lewis 2015, De Souza 2015
	Echinodorus macrophyllus	*Alisma macrophyllum*, *E. scaber*, *Aquarius macrophyllus*	Chapéu-de-couro, chá-de-campanha, erva-do-pântano, erva-do-brejo, congonha-do-brejo	Flowering cryptophyte herb, with rhizome and fruit, 0.5–1.1 m	Ethnic and afro-descendant groups from Brazil	To release renal, respiratory, liver and skin disorders, as well as hypertension, diabetes, rheumatism	Stalcup 2000, Maciel and Neto 2006, De La Cruz 2008, Alonso 2016, Cajaiba et al. 2016, Nunes et al. 2016, Canalli and Bove 2017, Teixeira et al. 2017, Dos Santos et al. 2019, Pasa et al. 2019

Colombia	*Brosimum utile*	*B. allenii, B. foetidum, B. krukovii, B. humboldtii, Alicastrum utile, Galactodendron utile, Piratinera utilis*	Árbol de leche, sande (negro), yanchama colorada, Nw´wiri, noeme (Ticuna), avichuri, b(v)aco, amapa-murure, frutipán de monte, lechero, jihuigene/panguana, palo de hule	Perennial flowering tree (with milky exudate and fruits), up to 50 m	Ethnic and peasant groups from Colombia, Peru, Brazil	Against digestive, skin, respiratory and reproductive disorders; as food (latex is lactagogue)	Bernal and Correa 1989, García-Barriga 1992, Castaño-Arboleda et al. 2007, Quattrocchi 2012, Grandtner and Chevrette 2014
	Guazuma ulmifolia	*G. bubroma, Theobroma guazuma, Bubroma guazuma, Diuroglossum rufescens*	Aquiché, bolaina, cabeza de negrito, caulote, gua(á)c(s)imo(a), iumanasi, guasmo, jacocalalu, jenumara, tzuyui, ajya, cambaaca, chico-magro, pixoy, jumú, ya ana, yaga yana, mutamba	Perennial flowering and many-branched tree (with fruits), 5–15 m	Ethnic/peasant groups from Brazil, Panama, Peru, Colombia, Ecuador	As emollient, analgesic, diuretic; to treat malaria, skin conditions and gastrointestinal/respiratory disorders; also, as food	Bernal and Correa 1989, García-Barriga 1992, Roth and Lindorf 2002, Quattrocchi 2012, Grandtner and Chevrette 2014, UEIA 2014, Al Muqarrabun and Ahmat 2015, IUCN 2019, Pereira et al. 2020
	Calophyllum brasiliense	*C. antillanum, C. calaba, C. jacquinii, C. lucidum, C. piaroanum, C. chiapense*	Árbol de aceite, lagarto (Colombia), jacareuba, guanandi (Brazil), balsamaría (Bolivia), lagarto caspi (Peru), Santa María (Mexico), ocuje (Cuba), Barí, dame Marie (Haiti)	Perennial flowering tree (yellowish milky exudate and fruits), up to 50 m	Ethnic/peasant groups from Brazil, Peru, Colombia	As wound healing, vermifuge, diuretic, antifungal/antidysenteric; to treat skin conditions and relieve pain for lepra	Roth and Lindorf 2002, Quattrocchi 2012, García-Zebadúa et al. 2014, Grandtner and Chevrette 2014, Gómez-Verjan et al. 2015
Ecuador	*Huperzia compacta*	*Phlegmariurus compactus, Lycopodium compactum, L. jamesonii, Urostachys compactus*	Waminga verde	Perennial shrub, 10–20 cm	Ethnic groups from Ecuador	As purgative agents and for curing supernatural diseases (shuka, water vaho); to treat fractures	Lozano 2009, Bussman and Sharon 2015, Armijos et al. 2016a, Plant List 2019, Tropicos 2019
	Huperzia espinosana	*Phlegmariurus espinosanus*	Waminga oso hembra or musgo				
	Huperzia tetragona	*Lycopodium tetragonum, L. rubrum, Urostachys tetragonus, Phlegmariurus tetragona*	Trencilla roja				

Table 2.1 Contd. ...

...Table 2.1 Contd.

Countries	Plants	Synonyms	Common names	Morphological characteristics	Communities	Ethnobotanical uses	References
Mexico	*Agastache mexicana*	*Brittonastrum mexicanum*, *Cedronella mexicana*, *Dracocephalum mexicanum*, *Dekinia coccinea*, *Gordoquia betonicoides*	Toronjil morado, olotillo, abejera, cidronela, melisa, toronjil morado/rojo/colorado, hisopo	Flowering and fragrant erect herb, 50–60 cm	Ethnic/peasant groups from Mexico	As sedative, relaxant, antipyretic; to treat epilepsy, earache, high blood pressure, CNS disorder	Lint and Epling 1945, Santillán-Ramírez et al. 2008, Hernández-Abreu 2009, Ávila-Rosas 2013, Zielińska-Matkowski 2014, Tropicos 2019
	Casimiroa edulis	*C. pringlei*, *C. sapota*, *C. tetrameria*, *Fagara bombacifolia*, *Zanthoxylum araliaceum*	Matasano, tapaculo, zapote blanco, pera mejicana, ajachel, ahache, chapote, cacchique, cochitzapotl	Evergreen flowering tree (fruits), 2–18 m	Ethnic/peasant groups from Mexico	To relieve the symptoms related to metabolic, circulatory and central nervous system disorders, as well as food	Standley and Steyermark 1946, Argueta and Cano 1994, Grijalva 2006, Chízmar 2009, Quattrocchi 2012, Orellana 2014
	Montanoa tomentosa	*M. floribunda*, *M. pilosipalea*, *Eriocoma floribunda*, *E. fragrans*, *Montagnaea tomentosa*	Zoapatle, cihuapatli, náhuatl, nocuana-titete-xini-ni, cuana-xuana, ciguapactli, pirimo blanco, too, hierba de la perida	Flowering, branches-prominent shrub (fruits), up to 3 m	Ethnic groups from Mexico	To help childbirth and other disease of women (puerperium, uterotonic, antihemorrhage)	Gallegos 1983, Bejar et al. 2000, Tropicos 2019
Peru	*Campsiandra angustifolia*	*C. rosea*	Huacapurana, acapurana (Peru), acapu-do-igapo, caacapoc, pimakaru'ywa (Brazil), amanagüe, duíra, lavallaga (Colombia), chingo	Perennial flowering tree (fruits), 5–20 m	Ethnic groups from Colombia, Peru, Brazil	As treatment for rheumatism/joint pain, diarrhea, malarial, ulcers, cure sores, female reproductive disorders and as tonic	López-Camacho et al. 2006, Rengifo 2007, Quattrocchi 2012, Grandtner and Chevrette 2014, Lagos-Castillo 2015, Vásquez-Ocmín et al. 2018
	Copaifera paupera	*Copaiba paupera*, *C. langsdorffii* var. *peruviana*, *C. reticulata* var. *peruviana*	Copaiba, copaiba blanca, Bolivia copaifera, copal, copaiba balsam, copaiba resin, nanpiwara	Evergreen flowering tree-shrub (with fruits), up to 35 m	Ethnic groups from Peru, Bolivia, Brazil	As antimicrobial, blood purifier, expectorant, wound, healing, leishmanicidal; to treat bronchitis, dermatitis, herpes, malaria	Pérez 2002, Estevez et al. 2007, Duke 2008, Quattrocchi 2012, Odonne et al. 2013, Grandtner and Chevrette 2014, Vásquez-Ocmín et al. 2018
	Sambucus peruviana	*S. nigra* var. *peruviana*, *S. graveolens*, *S. laciniata*	Sauco (Colombia, Peru, Bolivia, Ecuador), kiola (Argentina), r'ayan (Quechua), sambuco, tilo, saucotillo, cinta de novia	Perennial flowering tree-shrub with fruits, 3–12 m	Ethnic groups from Peru	As febrifuge, diuretic, emetic, diaphoretic, disinfectant, purgative, analgesic; to treat rheumatism, malaria	Brack 1999, Quattrocchi 2012, Grandtner and Chevrette 2014, Bussman and Sharon 2015

Table 2.2. Pharmacological properties (including *in vitro/in vivo* assessments) reported for 18 interesting plants in the reviewed scientific literature.

Species	Plant useful parts/types of preparation based on ethnomedicine	Pharmacological effects - *in vitro/in vivo* evaluations			References
		Extracts/isolated compounds	**Tests**	**Results**	
Ad. latifolium	Whole plant, leaves (poultice/decoction/infusion/ compresses)	Fraction 3 (F3) and EtOH extract (EE) (whole plant)	Cytotoxic (LD_{50}, µg/mL)	HeLa line; F3: 36 ± 15; EE: 74 ± 2	Lopez et al. 2001, Nonato et al. 2011, Jhonson et al. 2017, Thomas 2017, Arévalo-López et al. 2018, Pradeep-Kumar et al. 2018, 2019
			Antiparasitic	*Leishmania* spp. (5 strains), IC_{50} (µg/ mL), EE: 25 ± 2–78 ± 3; F3: 15 ± 8 - 49 ± 15	
		Acetone (AE) and EtOH extracts (EE) (whole plant)	Antibacterial (ϕ inh. mm)	5 Strains; 1 mg/each extract, AE: 7.1 ± 0.5–9.1 ± 0.4, EE: 6.8 ± 0.3–14.4 ± 0.4	
		Acetone (AE) and MeOH extracts (ME) (whole plant)		3 Strains; 30 µg/each extract, AE: 8.7 ± 0.5–13.9 ± 0.4; ME: 13.4 ± 0.4–16.2 ± 0.5	
		MeOH extract (aerial part)	Antiviral	DNA HSV-1–MIC 12 µg/mL	
			Antiinflammatory	Dose 100–200 mg/kg/IP, 4 h	
			Antinociceptive	Writhing test, 1–100 mg/kg/IP, 200–400 mg/kg/OP	
		MeOH extract (ME) and ethyl acetate fraction (EAF) (aerial part)	Larvicidal	%Mortal.; LD_{50} (mg/kg) on *O. rhinoceros*, ME (1%): 93 ± 2, 5018; EAF (0.1%): 88 ± 2, 583	
		Adiantobischrysene (**1**) isolated of EAF		1–5 mg–~ 55–95% mort., LD_{50} 8.4 mg/kg	
		22-Hydroxyhopane (**2**) isolated of ME		50 mg/kg–93 ± 5% mort., LC_{50} 29 mg/kg	
R. thyrsoidea	Stem, leaves, rhizomes, roots, fruits (infusion/decoction/poultice)	EtOH extracts (rhizomes, leaves, roots)	Antiparasitic (µg/mL)	IC_{50}, *P. falciparum* chloroquine resistant: 7 ± 2; *L. amazonensis* (ax. amast.): 10.0 ± 0.8	Valadeau et al. 2009, Cabanillas et al. 2014, Noriega et al. 2016, 2017
		3–**9** isolated from leaves/rhizomes		*L. amazonensis*, EC_{50}, **3**: 20 ± 2, **4**: 23 ± 5, **5**: 36 ± 10, **6**–**8**: 119 ± 35–166 ± 47, **9**: 68	
		Essential oils (L-leaves/R-rhizomes)	Antioxidant IC_{50} (mg/mL)	L-$ABTS^{+}$: 1.4 ± 0.1; DPPH: 20.9 ± 0.4; R-DPPH: 17.1 ± 0.3; $ABTS^{+}$: 52.4 ± 0.4	
			Antimicrobial (MIC mg/mL)	4 Bacteria and 2 fungi; L: 0.4–1.4; R: 0.3–11	

Table 2.2 Contd. ...

...Table 2.2 Contd.

Species	Plant useful parts/types of preparation based on ethnomedicine	Pharmacological effects - *in vitro*/*in vivo* evaluations			References
		Extracts/isolated compounds	Tests	Results	
Sal. impressifolia	Bark (decoction/maceration/syrup-powder)	EtOH extract (peels)	Antifungal	*C. albicans* - MIC 10 µg/mL	Soares et al. 2018, Rodriguez et al. 2019
		EtOAc extract and fraction 3 from EAE	Anticancer/cytotoxic	Inh. tumor (HL-60) mice; 20 mg/kg, F3: 82%, EAE: 40%	
		EtOAc extract (stem bark)		8 Cell lines; IC_{50}: 8–30 µg/mL; 2 non-cancer cells; IC_{50}: 8–>50 µg/mL	
		Fraction 3 from EtOAc extract Fraction 1 from MeOH extract		8 Cell lines; IC_{50} (µg/mL) - F1: 0.5–1.0, F3: 0.1–0.8; 2 non-cancer cells - F1: 0–0.6, F3: 0.1–0.9	
		10-12 isolated from F3		8 Cell lines; IC_{50} (µM) - **10**: 0.8–2.2, **11**: 0.1–0.6, **12**: 2.9–13.6; 2 non-cancer cells - **10**: 0.8–8.6, **11**: 0.1–38.5, **12**: 0.6–3.5	
Am. cearensis	Bark, seeds, leaves (decoction/tea/infusion)	H_2O/EtOH extract (stem bark)	Antimicrobial	*S. aureus* - MIC: 512 µg/mL	Marinho et al. 2003, Silveira and Pessoa 2005, Oliveira et al. 2009, Figueredo et al. 2013, Lima et al. 2013, Gouveia et al. 2015, Dantas et al. 2016
			Antinociceptive	400 mg/kg, 41% red. contract (writhing test)	
			Immunomodulatory	400 mg/kg/v.o., 81% inh. (OVA-antibodies in asthma model)	
		H_2O/EtOH extract (leaves)	Ovarian follicle protection	Caprine ovarian tissue preantral follicles protection with 0.2 mg/mL	
		H_2O/EtOH:H_2O extracts (stem bark/seeds)	Anti-inflammatory	200 mg/kg, 42% inh. (carrageenan-induced cell migration)	
		H_2O extract (seeds)	Antiedematogenic	500 mg/kg, 50% inh. (carrageenan-induced paw edema)	
		Hexane extract (leaves)	Acaricidal	25 mg/mL on *R. microplus*, 67% efficacy	

Cae. ferrea	Stem, barks, leaves, fruits, seeds, flowers (decoction/macerate/powder/ointment)	H_2O extract (fruits)	Analgesic Anti-inflammatory	50 mg/kg, 67% inh. ear edema (ind. acetic acid); 74% inh. contortions number	Carvalho et al. 1996, Nakamura et al. 2002, Gonzalez 2005, Menezes et al. 2007, Nozaki et al. 2007, Cavalheiro et al. 2009, Sampaio et al. 2009, Oliveira et al. 2010, Vasconcelos et al. 2011, Lopes et al. 2013, Sawada et al. 2014, Nascimento et al. 2015, Biasi-Garbin et al. 2016
		EtOH extract (core stem/leaves)	Antiulcerogenic	10 mg/kg i.v., 15% oral, 37% red. (acute gastric lesion induced by EtOH/HCl test)	
		EtOH extract (core stem/fruits/leaves)	Antioxidant	0.8 µg/mL, 84–94% antioxidant effect (inh. autooxidation homogenate rat brain)	
		MeOH extract (fruits)	Antibacterial	MIC (µg/mL); *S. mutans*: 40, *S. salivarius*: 66, *S. oralis*: 100, *L. casei*: 66	
			Antifungal	*C. albicans*, MIC: 25 µg/mL	
		H_2O extract (bark)		*T. mentagrophytes* and *T. rubrum* MIC/MFC: 31–62 µg/mL	
			Hypotension	10 mg/kg/i.v., –9% $P_{arterial}$ reduction (measurement mean $P_{arterial}$. in rats)	
			Hypoglycaemic	300 mg/kg/day (streptozotocin-induced diabetics Wistar rats model)	
		Powder (bark)	Wound healing	Complete epithel. 1 mL ointment on goats	
		EtOAc extract (fruits)	Antimutagenic	100 µg/mL, 100% inh. Epstein-Barr antigen induced cells	
		H_2O extract (seeds)	Antinociceptive	1 mg/kg, 30% inh. contract (writhing test)	
			Larvicidal	16 % (v/v), *Aedes aegypti* 85% mortality	
			Enzimatic inhibition	10 % (w/v), 5% residual action on tripsine	
		SO_4^{-2} polysacch. of H_2O extract (seed)	Antiviral	IC_{50} (µg/mL), HSV: 405, Poliovirus: 2	

Table 2.2 Contd. ...

...*Table 2.2 Contd.*

Species	Plant useful parts/types of preparation based on ethnomedicine	Pharmacological effects - *in vitro*/*in vivo* evaluations			References
		Extracts/isolated compounds	**Tests**	**Results**	
E. macrophyllus	Leaves (infusion/macerate)	H_2O extract (leaves)	Immunosuppressive	0.5-5 mg/kg/b.w., 29-41% red. (hind paw edema mice model)	Pinto et al. 2007, Tanus-Rangel et al. 2010, Fernandes et al. 2012, Portella et al. 2012, Nascimento et al. 2014, Da Silva et al. 2016b
		Flavonoid-rich fraction from H_2O extract (leaves)/EtOH extract (leaves)	Antiinflammatory	F-RF: 2.5 mg/kg, red. 30% leukocytes and 90% leukotriene B4; EE: 75% leukotr. B4	
		Hexane extract (leaves)	Antiinflammatory Antinociceptive	25 mg/kg, 52% inh. # writh. (ind. acetic acid); 50 mg/kg, 60-90 min, antinociception	
		EtOH extract (leaves)	Diuretic	300 mg/kg, 30% urine decr. (rats)	
		NaCl 0.9 % solution (crude leaves)	Renoprotection	2 mg/kg, 65% red. urinary peroxides (CP-ind. acute renal lesions in rats)	
		Essential oil (leaves)	Antinociceptive	50 mg/kg, 65% inh. (writhing test)	
B. utile	Latex, fruits (decoction)	Latex suspension in $CHCl_3$/EtOAc	Cytotoxicity	*Artemia salina*, IC_{50}: 48-895 mg/L	Ferrari et al. 2005, Padilla et al. 2008, Rivera-Parada 2012, Chindoy 2013, Jiménez 2015, Acosta et al. 2017
		Alkaloid-rich fraction (**32**) from latex suspension in MeOH	Antimicrobial	4 Bacterial/1 fungal strains, MIC 1.6 mg/mL; 2 fungal strains, 2 mg/mL (ϕ inh. mm) - *T. rubrum*: 19.5±0.5, *E. floccosum*: 17.4±0.5	
		H_2O suspension from latex	Anticancer (IC_{50}, μg/mL)	K-562: 1.5	
		33 isolated from root bark		MCF-7: 26; PC-3: 28; HF: 130	
		$MeOH_{acid}/Me_2CO:H_2O$ (7:3) extract (bark)	Antioxidant	Red. power: 59.7±0.4 AsAE/g, AA: 61%, DPPH· - EC_{50}: 1.3; inh: 83%	

G. ulmifolia	Bark, mucilage, leaves, branches (decoction/infusion/poultice)	H_2O suspension from EtOH extract (leaves/flowers)	Gastroprotective	Wistar Rats - 125 mg/kg	Caballero-George et al. 2002, Navarro et al. 2003, Felipe et al. 2006, Lima et al. 2006, Berenguer et al. 2007, Funk et al. 2007, Alonso-Castro and Salazar-Olivo 2008, Hokche 2008, Magos et al. 2008, Kaneria et al. 2009, Jacobo-Salcedo et al. 2011, Boligon et al. 2013, Maldini et al. 2013, Lukman et al. 2014, Calixto Júnior et al. 2016, Da'i et al. 2016, Morais et al. 2017, Kumar and Gurunani 2019
		Proanthocyanidins isolated (BuOH fraction from bark extract)	Angiotensine II type 1 inhibition	CHO-K_1 cell line, IC_{50}: 13–286 μM	
		Procyanidin-rich fraction from acetone extract (bark)	Hypotensive Vasorelaxant	10 mg/kg, vasorelaxant on rats, IC_{50} (ng/mL) 35 ± 12, 101 ± 57	
		H_2O extract (bark)	Antidiabetic (T2-D)	Preadipocites, IC_{50} 1–70 μg/mL	
		MeOH extract (leaves)	Antioxidant	IC_{50} (μg/mL), DPPH·: 50.2 ± 0.2, OH·: 250 ± 11; $O_2^{\cdot-}$: 72.8 ± 0.4	
		EtOH extract (leaves)		DPPH· - EC_{50}: 120 ± 2 μg/mL	
			AChE inhibition	Inhibition zone, 1 cm	
		Me_2CO:H_2O extract EtOAc fraction (branch bark)	Antiviral	P-1 and BHV-1 viruses, 10–40 μg/mL	
		Three flavanocoumarins isolated (MeOH extract of bark)	Anti-STAT1 activity on THP-1 line	**43**,**44** K_D (M) 1–2 × 10^{-6}	
		Tiliroside (**46**) isolated (leaves)	Antiproliferative	IC_{50} (μg/mL) - T47-D: 68, MCF-7: 113	
		EtOH extract (leaves)	Antitumoral	HeLa line, IC_{50} 175 ± 1 μg/mL	
			Cytotoxicity	NCTC929 fibroblasts, LC_{50}: 413 μg/mL	
			Antiparasitic (500 μg/mL)	Promast. inh. *L. infantum*: 95.2 ± 0.4%, *L. brasiliensis*: 92.20 ± 0.04%; epimast. inh. *T. cruzi*: 61 ± 1%	
		EtOH extract (stem bark)	Antibacterial	*S. aureus* NORA and MSRA strains - susceptible	
		Essential oil (leaves)		4 Bacterial strains, MIC 62–750 μg/mL	
		H_2O lyophilized extract (bark)	Antifungal	*C. albicans*; 75 μg extr., ϕ inh.: 26 ± 2 mm	

Table 2.2 Contd. ...

...Table 2.2 Contd.

Species	Plant useful parts/types of preparation based on ethnomedicine	Pharmacological effects - *in vitro/in vivo* evaluations			References
		Extracts/isolated compounds	**Tests**	**Results**	
Cal. brasiliense	Bark (infusion)	Hexane/CH_2Cl_2 fraction/extract and mammea A/BB (**53**) from EtOH extract (leaves)	Antiparasitic	$IC_{50\text{-}90}$ (µg/mL); *L. amazonensis*, $DE_{promast.}$: 40–73, $DE_{amast.}$: 4–20, $HF_{promast.}$: 17–35, $HF_{amast.}$: 3–17, $\mathbf{53}_{promast.}$: 3–5, $\mathbf{53}_{amast.}$: 0.9–2	García-Barriga 1992, Da Silva et al. 2001, Mesía-Vela et al. 2001, Ito et al. 2002,2003, Cottiglia et al. 2004, Huertas-Reyes et al. 2004a,b, Isaias et al. 2004, Pretto et al. 2004, Reyes-Chilpa et al. 2004,2006,2008, Kimura et al. 2005, Yasunaka et al. 2005, Braga et al. 2007, Brenzan et al. 2007,2008a,b, Ruiz-Marcial et al. 2007, Suffredini et al. 2007,2014, Mesquita et al. 2009, Souza et al. 2009, Albernaz et al. 2010, Honda et al. 2010, García-Zebadúa et al. 2011, Lemos et al. 2012, Tiuman et al. 2012, Blanco-Ayala et al. 2013, Gonçalves et al. 2013, Kudo et al. 2013, Rea et al. 2013, Oliveira et al. 2014, Matsuda et al. 2015, Pires et al.
				LD_{50} (µg/mL); *L. braziliensis*, $DE_{promast.}$: 65, $\mathbf{53}_{promast.}$: 23	
		Mammeas mixture (**52–54**) and **52** from hexane extract (leaves)		3 Strains of *T. cruzi*; IC_{50} (µM), MM: 11–22, **52**: 18; growth inh. $MM_{epim.}$: 17–54, $\mathbf{52}_{\mathbf{epim.}}$: 10–18;	
		53 isolated from CH_2Cl_2 extract **58** isolated from hexane fraction		IC_{50} (µM); *L. amazonensis*, $\mathbf{53}_{promast.}$: 7.4 ± 0.3, $\mathbf{53}_{amast.}$: 14 ± 2, $\mathbf{58}_{promast.}$: 30 ± 4	
		Coumarin (**53**) from CH_2Cl_2 extract (leaves)		*L. amazonensis*, red. lesion mice, 18 mg/kg/d (i.m.): 60%; 0.2% (topical): 68%	
		Hexane fraction and CH_2Cl_2 extract from leaves		LD_{50} (µg/mL); promast. *L. amazonensis*, HF: 40, DE: 20	
		CH_2Cl_2 extracts from root wood (RW)/bark (RB)		IC_{50} (µg/mL); *P. falciparum* - RWE: 6.7 ± 0.9, RBE: 9.50 ± 0.02; *L. chagasi* - $RWE_{promat.}$: 27.6 ± 0.8; *T. cruzi* - $RBE_{epimast.}$: 97.1 ± 0.4	
		CH_2Cl_2 extract and **67** isolated from leaves/stems		IC_{50} (µg/mL); *P. falciparum* - DE: 8 ± 2, **67**: 55 ± 14	
		Different extracts from leaves/stems		IC_{50} (µg/mL); *P. falciparum* - EAE: 9 ± 1, DE/EAE: 8 ± 1	
		69 isolated of hexane/MeOH extract from stem bark		*T. cruzi*; IC_{50} (µM), $\mathbf{69}_{\mathbf{trypomast.}}$: 220, $\mathbf{69}_{\mathbf{amast.}}$: 210	

		73/76 isolated from MeOH extract (heartwood)	Antioxidant	**73/76** scavenged to ROS ($O_2^{\cdot-}$, $HO^{\cdot}$, $ONOO^-$), IC_{50}: 0.070 ± 0.03–336 ± 9 µM	2016, García-Niño et al. 2017, Klein-Júnior et al. 2017, Mesa-Vanegas et al. 2019a,b, Rodríguez-Hernández et al. 2019
		Soxhlet (hexane/CH_2Cl_2)/SFE-CO_2 extracts (leaves)		DPPH·, IC_{50} (µg/mL) - HE: 243, DE: 207, SFE: 132–149	
		EtOH extracts (leaves/stems)	Angiotensine converting enzyme inhibition	100 µg/mL, ACE inh., LE: 34 ± 2%, SE: 55 ± 16%	
		CH_2Cl_2 extracts from root wood/bark	Antifungal	MIC (µg/mL), *C. neoformans* RWE/RBE: 2; *C. albicans* RWE: 8, RBE: 16	
		PCA isolated from MeOH extract		2 Strains, MIC-PCA: 400–500 µg/mL	
		Different extracts and molecules from flowers/leaves/stems/fruits/roots	Antibacterial	9 Strains, $MIC_{Fl/L/S/Fr/RE}$: 100–900; PCA: 200–800, **70**: 200–800.	
		Soxhlet/maceration/SFE-CO_2 extracts (leaves) and isolated coumarin **61**		*M. tuberculosis*; MIC 62.5 µg/mL, for all extracts; MIC 125 µg/mL, for **61**	
		CH_2Cl_2:MeOH and H_2O extracts from leaves/stems/fruits		4 Strains, MIC/MBC (µg/mL), DME: 190–510/250–810, WLE: 280–600/300–840	
		Chromanones acids (**88**–**91**) isolated of hexane/EtOAc extract (bark)		MIC (µg/mL), *S. epidermis*: 16, *B. cereus*: 1–16	
		Hexane/$(CH_3)_2CO$/MeOH extracts and **52/72/73/76**/mammea A/AA isolated from leaves/heartwood		MIC (µg/mL), HE/AE/ME - *S. aureus*: 2–32, *E. coli*: 128–512, MSSA/MRSA: 8–32; **73/76** - *E. coli*: 128, **52/72/73/76**/mammea A/AA - MSSA: 1–256	
		CH_2Cl_2 fraction and hydroethanol extract from stem bark		*H. pylori*; 125–1000 µg/disk (ϕ inh. mm), WEE: 8–14, DF: 10–16	
		Hexane extract and chromanone acid-rich fraction from stem bark		*H. pylori*; 25–400 µg/disk (ϕ inh. mm) - HE: 7–14, CARF: 10–12; MIC 31 µg/mL, growth inh. CARF: 17–44%	

Table 2.2 Contd. ...

...*Table 2.2 Contd.*

Species	Plant useful parts/types of preparation based on ethnomedicine	Pharmacological effects - *in vitro/in vivo* evaluations			References
		Extracts/isolated compounds	Tests	Results	
			Antiulcer	Gastric ulcer red., 5–100 mg/kg HE: 63–97%; 20–100 mg/kg CARF: 67–95%	
		Xanthones and coumarins isolated of $(CH_3)_2CO$/MeOH extracts (heartwood)		H^+/K^+-ATPase inh., IC_{50} (μM), xanthones: 47 ± 8–173 ± 9, coumarins: 110 ± 8–638 ± 149	
		$CHCl_3$ fraction and **68/71** from MeOH extract (roots)	Antinociceptive	CF (mg/kg), ID_{50} 9–48 i.p. (132 p.o.), 41–94% inh. const. (≠ tests); **68:** ID_{50} 7 μmol/kg, **71:** ID_{50} 28 μmol/kg	
		Polar/non polar fractions, **50/70** and MeOH extracts, (flowers/fruits/roots)		ME (3–10 mg/kg i.p.), 70 ± 6–80 ± 4% inh. constr. (writhing test); PF - 81 ± 8%; NPF - 98 ± 6%; ID_{50} (μmol/kg) **50**: 12 (84 ± 4%), **70**: 30 (94 ± 2%)	
		62/63 isolated from stem bark	Cytotoxicity/antitumoral-anticancer	6 Human cell lines, IC_{50} (μM), $\mathbf{62_{BV173}}$: 3, $\mathbf{63_{BV173}}$: 9; **62** on 5 cell lines: 2–5	
		CH_2Cl_2 extract (root)		4 Human cell lines, IC_{50}: 3–18 μg/mL	
		CH_2Cl_2: MeOH extract (wood) H_2O extract (leaves)		100 μg/mL, inh. cell lines; KM-12 - DME: 100%, WE: 34%; RPMI8226 - WE: 84%	
		CH_2Cl_2 extract from root bark		NIH-3T3 line, IC_{50}; RBE: 137 μg/mL	
		Hexane extract from stem bark		CHO-k1 line, IC_{50} (μg/mL), HE: 120 ± 4 (24 h) - 8 ± 2 (72 h)	
		73 isolated from heartwood		Normal human PBMC, IC_{50}: 86 ± 12 μM	
		CH_2Cl_2: MeOH and H_2O extracts from leaves/stems		5 Cell lines, GI_{50} (μg/mL) - DME: 2-95, WLE: 0.6–49	
		52/53 mixture from CH_2Cl_2 extract (leaves)		BMK cells, 24 μg/mL - Inh.: 52 ± 5%	
		50,52,53,55–60 isolated from hexane-$(CH_3)_2CO$/MeOH extracts (leaves)		3 Cell lines, growth inh. (31 μM), $\mathbf{50_{PC3}}$:62%, remaining: 47–100%; IC_{50}: **52, 53/55–60**: 0.04 ± 0.02–11 ± 1.	

		Calanolides/mammea/brasixanthones-brasimarins isolated of $(CH_3)_2CO$ ext. (stem bark)	Cancer chemopreventive	TPA-ind. EBV-EA activ. Raji cells, IC_{50} (mole ratio), **56**, **62**,**64**, **66**, **71**, **72/77**–**87**: 120–525	
		Hexane/Me_2CO extract (leaves)	Antiviral	HIV-1 RT, IC_{50} (μg/mL), HE: 30, AE: 31	
				HIV-1 RT, inh. (50 μg/mL), HE: 77.9 ± 0.5% (IC_{50} 30 μg/mL), AE: 81 ± 2%	
		62 isolated from stem bark		Inh. HIV-1 replication on infected cells (PMA/TNF-α), IC_{50}: 8.4 μM	
				Inh. HIV-1 entry, 25–65% (10–100 μM)	
		Fractions from hexane/Me_2CO/MeOH extracts and **64**–**67** isolated from leaves		HIV-1 RT, inh. (50 μg/mL) fract. 18/19 (HE): 75 ± 4%, fract. 2 (AE): 60 ± 4%, EtOAc ins. fract. (ME): 74 ± 2%; IC_{50} (mM), **64**: 0.3, **65**: 0.5, **66**: 1, **67**: 0.7	
H. compacta	Whole plant (decoction/macerate)	Alkaloid-rich fraction from MeOH:H_2O extract (9:1)	Enzymatic inhibition	AChE inh., IC_{50}: 62 ± 17 μg/mL	Wu et al. 2009, Ham et al. 2012, Bishayee et al. 2013, Armijo et al. 2016b, Li et al. 2016
			Antitumoral	PC-3 cell line (dose: 148 μM at 12 h)	
		Tricin	Antioxidant	DPPH, EC_{50}: 90 μg/mL	
			Antiinflammatory	2 mg/kg induced paw edema (50%)	
H. espinosana		Selgin (**97**)	Enzymatic inhibition	α-Glucosidase inh.: 34–36 μM	
H. tetragona		Alkaloid-rich fraction from MeOH:H_2O extract (9:1)		50 μg/mL, inhibition AChE – 87%, 68%	
			Anti-inflammatory	45 mg/kg, on rats (24% inh.)	
Ag. mexicana	Stems and leaves (decoction)	CH_2Cl_2/MeOH extracts and tilianin (**98**) isolated from leaves	Vasorelaxant Antihypertensive Toxicity	Relax. effect (%) aorta rings (rat); DE: 76%, ME: 62%, **98**: 75-85%; IC_{50} (μg/mL) DE: 189, ME: 232, **98**: 240-276; 50 mg/kg, **98** dec. 38%/35% $P_{syst/diast}$, incr. heart rate 5%; mice, LD_{50}: 6624 mg/kg	Molina-Hernández et al. 2000, Hernández-Abreu 2009, 2013, Ibarra-Alvarado et al. 2010, Ávila-Rosas 2013, Verano et al. 2013, Estrada-Reyes et al. 2014,
		MeOH extract and **98** (whole plant)	Anxiolitic-like effect	Effect at 30 mg/kg p.i. for extract and **98**	

Table 2.2 Contd. ...

...*Table 2.2 Contd.*

Species	Plant useful parts/types of preparation based on ethnomedicine	Pharmacological effects - *in vitro*/*in vivo* evaluations			References
		Extracts/isolated compounds	Tests	Results	
		H_2O extract (aerial parts)	Anxiolitic-like effect Sedative	AEBT (< 10 mg/kg), BBT (10 mg/kg), HBT (> 10 mg/kg), FST (1–100 mg/kg) on mices; LD_{50} > 5 g/kg	González-Trujano et al. 2015, Flores-Flores et al. 2016, Esquivel-Gutiérrez et al. 2017, Navarrete et al. 2017, Ventura-Martínez et al. 2017, Cruz-Torres 2019
		98 isolated from whole plant (MeOH)	Antihypertensive	ED_{50}: 54 mg/kg, LD_{50}: 6.6 g/kg	
		CH_2Cl_2 extract and **99–101** isolated/mixture (aerial parts)	Vasorelaxant	Relax. effect aorta ring (rat), EC_{50} (µg/mL), DE: 174 ± 6, **99**: 211 ± 6, **100**: 40 ± 9, **99/101**: 33 ± 6, **99/100/101**: 34 ± 11	
		H_2O extract (whole plant)	Vasoactive	Rat aortic segment, EC_{50}: 234 ± 46 µg/mL, E_{max}: 25 ± 7% on ACh	
			Antioxidant	DPPH·, IC_{50}: 502 ± 1 µg/mL; TEAC: 927 µmol Trolox®/g	
		MeOH extract (aerial parts)		100 µg/mL, inhition DPPH·: 93%, ABTS·+: 99%, TBARS: 94%	
		MeOH extract from flowers	Spasmogenic	Guinea pig ileum contr. resp. 62 % (316 µg/mL)	
		Essential oil (aerial parts)	Vasorelaxant	Guinea pig trachea ring, 5–50 µg/mL relax. effect 100%; EC_{50}: 18 ± 1 µg/mL	
		Solid (precipitate) from whole plant		Relax. effect (%) aorta rings/smooth muscle (rat); solid: 100% (1 mg/mL), CE_{50}: 4.5 ± 0.7–157 ± 9 µg/mL	
			Hypoglycaemic	Rats, 100 mg/kg solid red. 50% sugar level	
			Antihypertensive	Rats, 100 mg/kg solid dec. 20%/40% $P_{syst/diast}$, incr. heart rate 5%	
		EtOAc extract and **100** isolated (aerial parts)	Antinociceptive	ED_{50} (mg/kg) i.p., EAE: 34, **100**: 2; FT: **100**, ED_{50}: 44; LD_{50}: 2828 mg/kg	
		H_2O extract (leaves)	Anxiogenic	Male rats (EPMT/OFT/FST), 3–12 mg/kg i.p., no change in OFT/FST; at 32 mg/kg incr. mobility in FST	

Cas. edulis	Leaves and seeds (Infusion/ drunk)	H_2O extract (leaves)	Hypotensive (angiotensin II)	Hypertensive rats decr. 17–33% SBP (100–600 mg/kg)	Magos and Vidrio 1991, Magos et al. 1995,1999, Navarro-Ruíz et al. 1995, Ito et al. 1998, Garzón-De-La-Mora et al. 1999, Molina-Hernández et al. 2004, Awaad et al. 2006, 2007, 2012, Vázquez-Cruz et al. 2009, Bertin et al. 2011,2014, Esposito et al. 2011, Froldi et al. 2011, Nagai et al. 2014
			Anxiolitic-like effect	Rats (EPMT/OFT), 25–35 mg/kg incr. exploration	
			Anticonvulsive (MES and METsc)	Rats (OD 100 mg/kg), dec. 50%/70% MES/METsc	
		H_2O and EtOH extracts (seed)		Rats, MES - WE: 70%, 10 mg/kg (2 h); EE: 70%, 100 mg/kg (4 h)	
		H_2O extract (kernel)	Relaxant/contractile effects	Inh. aorta vein constrict. on rats, 0.01–1 mg/mL increased 5HT > NA > $F_{2\alpha}$	
		Hexane and MeOH extracts (seed)	Vasorelaxation (20 µg/mL)	Red. constrict. arterial ring, HE: 86 ± 2%, ME: 95.4 ± 0.9%	
				Relax. arterial tissues of young/older rats, ME: 95 ± 2% (young), 54 ± 5% (older)	
		MeOH extract and molecules isolated (seed)	Vasodilation Antioxidant	Rat art. tissues, ME: 40 ± 2% dilat. (20 µg/mL); EC_{50} - **107**: 4 µM - **114** > **107**; DPPH, ME: 72 ± 3% inh. (600 µg/mL); IC_{50} - **107**: 2 mM - **107** > **114**	
			Hypotensive	Rats, extract induced hypotension	
		Imidazole/synephrine derivatives isolated from MeOH extract (kernel)		Rats, H, MMH and DMH showed hypotensive response lasting 3 min.	
		H_2O:EtOH extract (seed)	Antiviral Antiproliferative	HIV-1 RT (RDDP/RNase H), IC_{50} (µg/mL), RDDP: 0.3, RNase H: 2; K562, CC_{50}: 3×10^{-4} µg/mL	
		EO, EtOH extract/different fractions and coumarins isolated (leaves)	Toxicity Anticoagulant Antimicrobial	LD_{50}, EM > 5000 mg/kg; anticoagulant (s), EM/coumarins (60–400 mg/kg), PT: 22.6 ± 0.7–25.0 ± 0.5, APTT: 35.3 ± 0.8–40 ± 1; 8 bacterial and 3 fungi strains, 60 µg/ each sample (ϕ inh. mm), EO: 13–30; EE: 0–19, EIF: 0–10, EIIF: 0–15; MIC (µg/mL), EO: 12–135; EE: 0–321, EIF: 0–321, EIIF: 0–252	

Table 2.2 Contd. ...

...Table 2.2 Contd.

Species	Plant useful parts/types of preparation based on ethnomedicine	Pharmacological effects - *in vitro/in vivo* evaluations			References
		Extracts/isolated compounds	**Tests**	**Results**	
		EtOAc fraction and **103-106/110/111** isolated from MeOH extract (seeds)	Chemopreventive against mutagen DMBA	*S. typhimorium*, IC_{50} (µg/mL) **110**: 2, **103**: 4, **106**: 8, **111/104**: 10, **105**: 11; red. lesion mice mamma gland, 10 µg/mL EAF: 80%	
		Different extracts and flavonoids isolated (leaves)	Antioxidant Toxicity	$ABTS^{+}$ (µmol TE/g), ME: 842, EAF: 712, flavonoids: 452–772; LD_{50}: 6.0 ± 0.4 g/kg	
		Coumarins (**108/109**) isolated from MeOH extract (leaves)	Adipogenesis	3T3-L1 adipocyte, **108** and **109** (5 µM) incr. (1.5-fold) synthesis of palmitic acid	
		EtOH extract, alkaloid-rich fraction and alkaloids isolated (fruits)	Antihypertensive	Dog, dose (mg/kg), dec. hypert. 10–60 min. - EE: 500 (18%), ARF: 200 (30%), **118**: 50 (45%), **119**: 100 (36%), **120**: 200 (45%), **121**: 300 (33%)	
M. tomentosa	Leaves and flowers (infusión)	H_2O extract (leaves)	Aphrodisiac	Male rats incr. $\#_{eyac.}$ 100%, at 75 mg/kg	Senties and Amayo 1864, Landgren et al. 1979, Lozoya et al. 1983, Bejar et al. 1984, Campos-Bedolla et al. 1997, Carro-Juárez et al. 2004,2006, Sollozo-Dupont et al. 2015, Estrada-Camarena et al. 2019, Lagunes-Merino et al. 2019
			Oxytocinergic-like	Women ing. dose 2–15% causing vaginal dilation/uterine contr./ expulsion dead fetus	
			Abortifacient	Women ing. dose 1.1–1.3 g/kg p-o.a. causing cervix dilation, uterotonic effect	
			Anxiolytic/anxiogenic-like	Female rats, beneficial actions under low hormone conditions (PW challenge)	
			Anxiolytic-like	Male rats, 3 mg/kg ind. effect in EPMT, HBT, BBT,	
				Male rats, OFT, FST related OXT, 50 mg/kg incr. latency immobil., #Fos/ OXT	

		H_2O extract (whole plant)	Anxiolytic	Male rats, incr. oxytocin prod. at 50 mg/kg	
				Hypothal. OXT act. in male rats (HLT/OFT/FST), 50 mg/kg, decr. 32% non-mobility time.	
			Pro-ejaculatoy	Spinal system male rats, 25 mg/kg incr. $\#_{des}$: 13.3 ± 0.3, penile erection/ movement	
		H_2O extract and **136** isolated (leaves)	Uterine motility	Mot. uterine tubes (rats), 20 µL/mL (WE) incr. SAI: 40–290%; 12 µg/mL (**136**) incr. SAI: 87–357%	
		135/**136** isolated from leaves (hexane)	Uterine contraction	10–20 µg/mg, %inh. uter. contr. ind. ACh/OXT/ST, ACh: 53–80, OXT: 68–72, ST: 69–100	
		H_2O and hexane extracts, HE suspension and **136** isolated (leaves)		Uter. strips (rat/dog/guinea pig), WE (20 µL/mL), HE_{susp} (40 µg/mL), **136** (20 µg/mL) inhited contractions	
Cam. angustifolia	Bark (decoction)	MeOH extract (bark)	Enzymatic inhibition	PDE-5, 100–200 µg/mL: 81–89%	Kvist et al. 2006, Ravarocci-Quiroz and Carrasco-Huamán 2010, Ruiz et al. 2011, Vásquez-Ocmín et al. 2018, Schmeda-Hirschmann et al. 2019, Roumy et al. 2020
			Antimicrobial	15 Strains, IC_{50}: 0.3 ± 0.1–1.2 ± 0.4 mg/mL	
			Antioxidant	DPPH·, %inh. 94.5 ± 0.3 (2.5 mg/mL), IC_{50}: 0.069 ± 0.003 mg/mL	
		EtOH:H_2O extract (bark)	Antiparasitic (IC_{50} µg/mL)	IC_{50} (µg/mL), *P. falciparum*: 9, FBIT: 7	
		EtOH extract (bark)		*P. falciparum*, < 10	
				P. falciparum: >10, *L.. donovani*: > 20, *T. brucei gambiense*: > 10	
Cop. paupera	Oleoresin (direct use/topical application)			*L. amazonensis*, amast./prom.: > 100	Ticunsi et al. 2002, Estevez et al. 2007, Dos Santos et al. 2008, Santos et al. 2008, Izumi et al. 2012, Amorim et al.
		Oleoresin/kaurene-rich fraction/**137** and nanoemulsions containing to them		*L. amazonensis* - Norm: 17 ± 3–168 ± 7; Nanot.: 2 ± 1–35 ± 10; *L. infantum* - Norm: 11 ± 1–203 ± 18, Nanot.: 2.2 ± 0.8–55 ± 6	

Table 2.2 Contd. ...

...Table 2.2 Contd.

Species	Plant useful parts/types of preparation based on ethnomedicine	Pharmacological effects - *in vitro/in vivo* evaluations			References
		Extracts/isolated compounds	Tests	Results	
		135/140-142 isolated (oleoresin)		*T. cruzi*, epim.: 43 ± 6–168 ± 6 μM, amast.: 1.30 ± 0.06-28 ± 6 μM, $EC_{50\text{-}tryp.}$: 377 ± 88–965 ± 66 μM	2017, Furtado et al. 2018, Rodrigues et al. 2018, Vásquez-Ocmín et al. 2018, Liviac et al. 2019, Ribeiro et al. 2019, Roumy et al. 2020
		Oleoresin		*L. amazonensis*, 11.0 ± 0.4	
			Wound healing	Mice, 200 mg/kg, decr. wounds, NDM: 70%, DM: 43 ± 7%	
			Antimicrobial	5 Strains, IC_{50}: 0.15 ± 0.05–1.2 ± 0.4 mg/mL	
				5 Bacteria, MIC/MBC (μg/mL), MIC/MBC: 62–1000	
		135/140,141 isolated (oleoresin)		3 Bacteria, MIC (μg/mL), *B. subtilis*: 3.1–30, *S. aureus*: 6–50, *S. epidermidis*: 4–40	
		Volatile oil from oleoresin		6 Strains, MIC: 125–2000 μg/mL	
		EtOH extract (oleoresin)	Antioxidant Toxicity	FRAP: 2.43 ± 0.08 μmol/g, TBARS: 14.2 ± 0.4 mg/g; oral tox. rats, LD_{50} > 5 g/kg	
		Oleoresin/leaf EtOH:H_2O extract	Genotoxic/cytotoxic	Mice, OR/EWE: no genotoxic; V-79 line, IC_{50} (μg/mL), OR: 41 ± 8, EWE: 412 ± 66; OR/EWE: no cytotoxic *in vivo*	
		135/140-142 isolated (oleoresin)	Cytotoxic	$LLCMK_2$ and erythrocytes, CC_{50}: 39 ± 5–76 ± 5 μM, HC_{50}: 65.7 ± 0.2– > 1582 μM	
		142 isolated (oleoresin)		4 Cell lines, IC_{50}: 2.5–10 μg/mL	

Sam. peruviana	Leaves, flowers and fruits (infusion/decoction)	EtOH extract (leaves)	Antibacterial	7 Strains, ϕ inh. > 7 mm for all strains at 100 mg/mL	Hernández et al. 2000, Neto et al. 2002, Schmeda-Hirschmann et al. 2005, Bussman et al. 2011, Chirinos et al. 2013, Román-Farje et al. 2017
		Flavonoid-rich fraction from MeOH:H_2O extract (aerial parts)		3 Strains, CAEC (μg/mL): 38–61	
		MeOH:H_2O:HCl extract from dried fruits	Antioxidant	AOC (μmol TE/g), DPPH·: 155 ± 6, $ABTS^{+\cdot}$: 303 ± 7, ORAC: 361 ± 23.	
		MeOH extract from fruits		DPPH·, inh. 49 ± 4% at 100 μg/mL	
		H_2O and EtOH extracts (leaves)	Toxicity	*Artemia* sp., LC_{50} (μg/mL), WE: 168–10000, EE: 26–124	
		H_2O and EtOH extracts (leaf/bark/root)		*D. magna* (LC_{50}, g/L), *S. zeamais*/*C. koehleri* (mortality, %), LC_{50}: 0.4–23, %Mortal. (160 mg/g): 0–28 (120 h), %Mortal. (160 mg/g): 0–25 (48 h)	

In agreement to Table 2.2, a good antiparasitic effect on promastigotes of five strains of *Leishmania* spp. (*L. amazonensis*, *L. aethiopica*, *L. braziliensis* native, *L. lainsoni*, and *L. braziliensis*), with IC_{50} (µg/mL) values of 78 ± 3, 74 ± 22, 35 ± 5, 28 ± 9, and 25 ± 2, respectively, was exhibited for the total EtOH extract from *Ad. latifolium* (whole plant); even so, a remarkable pharmacological result was found for fraction three (F3) against the same strains whose IC_{50} (µg/mL) values were 37 ± 4, 49 ± 15, 32 ± 11, 19 ± 1, and 15 ± 8, severally. However, F3 turned out to be more cytotoxic on HeLa cells than the total extract (EE), according to the LD_{50} values (Arévalo-López et al. 2018). In consideration to the antibacterial potential for this plant, Thomas (2017) reported that the acetone and EtOH extracts (1 mg for each) from entire plant, tested on five bacterial strains (*Pseudomonas aeruginosa*, *Staphylococcus aureus*, *Klebsiella pneumoniae*, *Escherichia coli* and *Serratia marcescens*), presented acceptable diameters of inhibition zone (mm) as follows: *P. aeruginosa*: 9.1 ± 0.4–14.4 ± 0.4, *S. aureus*: 0–8.5 ± 0.6, *K. pneumoniae*: 0–6.8 ± 0.3, *E. coli*: 7.2 ± 0.7–10.1 ± 0.4, and *S. marcescens*: 7.1 ± 0.5–8.1 ± 0.3; *P. aeruginosa* being the most susceptible strain for both extracts. The values of minimum inhibitory and bactericidal concentrations (MIC/MBC) on *P. aeruginosa* were ~ 18 mg/mL and ~ 38 mg/mL, in that order, for EtOH extract. In contrast to above statement, Johnson et al. (2017) evaluated the acetone and MeOH extracts (30 µg for each) from whole plant against *E. coli*, *K. pneumoniae*, and *S. aureus*; these authors reported that the two extracts were highly effective on the most bacterial strains (*S. aureus* was the least susceptible strain to the acetone extract), with diameters of inhibition zone (mm) just like that: *E. coli*: 13.9 ± 0.4–16.2 ± 0.5, *K. pneumoniae*: 12.4 ± 0.3–13.4 ± 0.4, and *S. aureus*: 8.7 ± 0.5–13.6 ± 0.5. Furthermore, the MeOH extracts showed a notable antiviral (MIC 12 µg/mL), antinociceptive and anti-inflammatory (inhibition interleukin-1β production), larvicidal (0.1–1 mg extract/fraction, 88 ± 2–93 ± 2% mortality, LD_{50} 583–5018 mg/kg)/antibacterial (5–25 mg/mL extract/fraction, ϕ inh. 15–25 mm) effects using DNA herpes simplex virus type 1 (HSV-1), writhing/formalin/tail-flick tests (1–100 mg/kg/I.P., 200–400 mg/kg/O.P.), and carrageenan-induced/arachidonic acid-induced paw/ear edemas, and *Orycites rhinoceros* and eight bacterial strains, as reported by Lopez et al. (2001), Nonato et al. (2001), and Pradeep-Kumar et al. (2018). The last authors [Pradeep-Kumar et al. (2018, 2019)] isolated two triterpene: a type ecdysteroid [adiantobischrysene (**1**)] and other type hopanoid [22-hydroxihopane (**2**)] from ethyl acetate fraction or methanol extract which presented larvicidal (*O. rhinoceros*) and antibacterial activities.

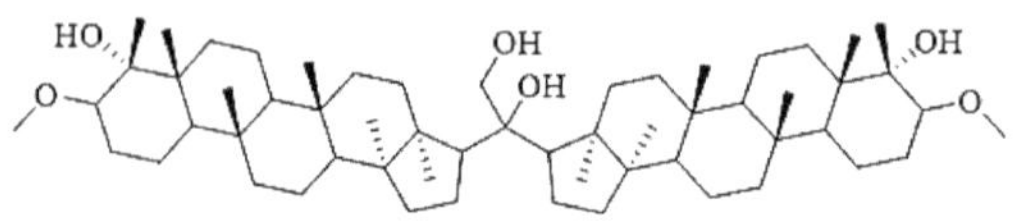

1

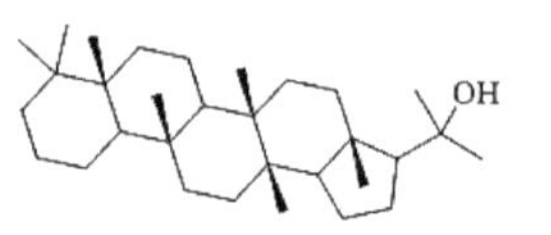

2

As to *R. thyrsoidea*, some indigenous communities (Yanesha/Chayahuita, Peru; Kichwa, Ecuador) have used this plant against malaria and leishmaniasis (Valadeau et al. 2009, Odonne et al. 2013, Noriega et al. 2017). Valadeau et al. (2009) validated the traditional use when they proved that the EtOH extract from rhizomes was active against *Plasmodium falciparum* chloroquine resistant and *L. amazonensis* amastigotes. Moreover, Cabanillas et al. (2014) isolated 11 and seven constituents (type sesquiterpenes, dihydrobenzofuranone, flavonoids, diarylheptanoids) from leaves and rhizomes, respectively, of which seven (**3–9**) were active (IC_{50}: 20 ± 2–166 ± 47 µg/mL) on *L. amazonensis* axenic amastigote form; the molecules **3–5** had the best IC_{50} values (20 ± 2–36 ± 10 µg/mL). Noriega et al. (2016, 2017) divulged the chemical composition of the essential oil (0.05% yield) from plant leaves/rhizomes and the radical-scavenging ($ABTS^{+\cdot}$ and $DPPH^{\cdot}$) and antimicrobial (*P. aeruginosa*, *E. coli*, *S. aureus* subsp. *aureus*, *Streptococcus mutans*, *Candida albicans*, *C. tropicalis*) capabilities; scilicet, terpinolene (~ 26%)/α-phellandrene (~ 17%), and β-pinene (~ 41%) were the main constituents for leaves and rhizomes, individually, while the IC_{50} values for radical-scavenging were among 1.4 ± 0.1–52.4 ± 0.4 mg/mL and MIC (mg/mL) values were: *P. aeruginosa*: 0–0.4, *E. coli*: 0–0.4, *S. aureus*: 1.4–10.8, *S. mutans*: 1.4–5.6, *C. albicans*: 0.4–0.7, and *C. tropicalis*: 0.3–1.4.

In another way, *Salacia* genus has been characterized by containing α-glucosidase inhibitors such as salacinol and kotalanol (1-deoxy-4-thioarabinofuranosyl cation derivatives) along with kotalagenin

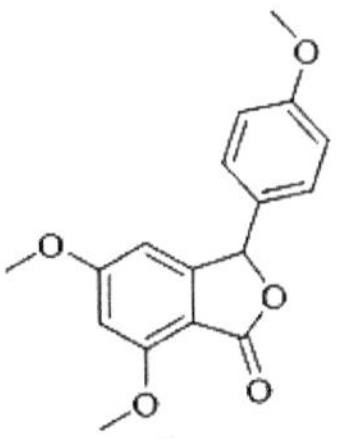
3

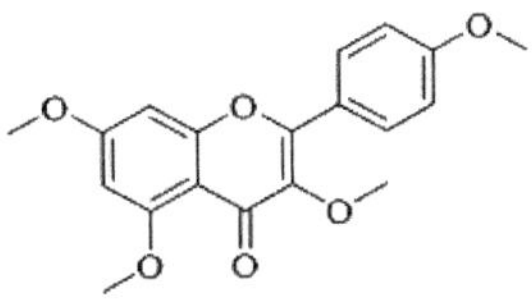
4

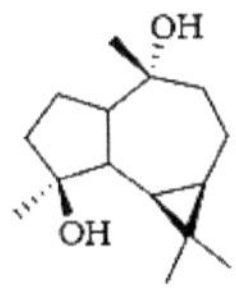
5

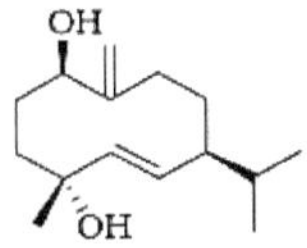
6

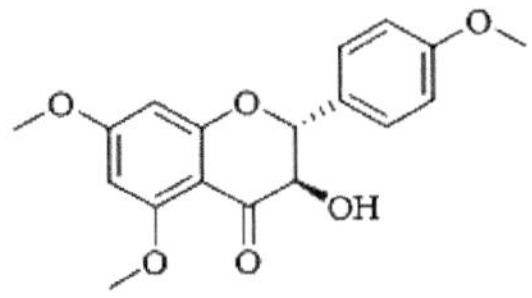
7

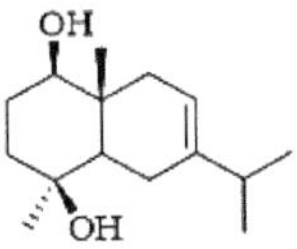
8

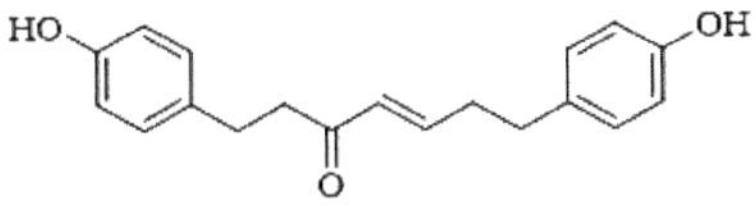
9

16-acetate (triterpenoid) as active principles (Deepak et al. 2014, Ripardo et al. 2015), as well as other triterpenoids belonging to series type ursanes, oleananes, friedelanes, lupanes, and quinonemethides (Ramakrishna et al. 2015, 2016, Da Silva et al. 2016a), all of them possibly responsible for the traditional use of *Salacia* spp. for the treatment of diabetes. Recently, *in vivo* and *in vitro* antileukemia and cytotoxic activities of extract (ethyl acetate, EAE)/fractions (F1 and F3)/isolated compounds (**10–12**, quinonemethide triterpenes) from stem bark were reported by Rodrigues et al. (2019). The IC_{50} values [μg/mL—extract/fractions, μM—isolated compounds] determined on cancer and non-cancer cells were: EAE—MCF-7: 8, HCT116: 13, HepG2: 27, SCC-4: 10, HSC-3: 10, HL-60: 29, K-562: 30, B16-F10: 18, and MRC-5: 8; F3 (from EAE)—MCF-7: 0.4, HCT116: 0.4, HepG2: 0.1, SCC-4: 0.2, HSC-3: 0.8, HL-60: 0.4, K-562: 0.6, B16-F10: 0.4, MRC-5: 0.1, and PBMC: 0.9; F1 (from ME)—HepG2: 0.5, HL-60: 0.6, K-562: 1.0, B16-F10: 0.5, and PBMC: 0.6; **10** (22-hydroxytingenone)—MCF-7: 2, HCT116: 2, HepG2: 1, SCC-4: 2, HSC-3: 0.8, HL-60: 1, K-562: 2, B16-F10: 1, MRC-5: 0.8, and PBMC: 9; **11** (tingenone)—MCF-7: 0.5, HCT116: 0.6, HepG2: 0.4, SCC-4: 0.6, HSC-3: 0.1, HL-60: 0.4, K-562: 0.5, B16-F10: 0.5, MRC-5: 0.1, and PBMC: 38; and **12** (pristimerin)—MCF-7: 8, HCT116: 9, HepG2: 8, SCC-4: 13, HSC-3: 3, HL-60: 9, K-562: 14, B16-F10: 6, MRC-5: 4, and PBMC: 0.6.

Based on the revised phytochemical studies (Bravo et al. 1999, Costa-Lotufo et al. 2003, Negri et al. 2004, Silveira and Pessoa 2005, Canuto et al. 2006, 2010) on *Am. cearensis*, it was found that alcohol/aqueous extracts from bark/leaves contained constituents type coumarins [e.g., coumarin (**13**), scopoletin, ayapin (**14**)], phenolic derivatives and flavonoids [e.g., 3,4-dihydroxybenzoic acid, protocatechuic acid, vanillic acid (**15**), quercetin, kaempferol, isokaempferide, afrormosine, catechol, guaiacol, 4-methoxy-physetin, amburosides A and B (**16/17**)], as well as anthraquinones (crysophanol), sterols (e.g., β-sitosterol, γ-sitosterol, stigmasterol, lupeol) and fatty acid derivatives (e.g., methyl/ethyl palmitate). Due to this diversity of secondary metabolites, the pharmacological effects related to the ethnomedicinal uses could be explained, among them: antibacterial (against *S. aureus*; MIC: 512 μg/mL), antinociceptive, antiinflammatory, analgesic, antispasmodic and bronchodilator activities of the hydroalcoholic extract from the stem bark, without any evidence of toxicity in usual doses (Silveira and Pessoa 2005, Oliveira et al. 2009, Figueredo et al. 2013), as well as immunomodulatory activity on OVA-antibodies in asthma model (400 mg/kg oral; 81% inhibition) (Marinho et al. 2004). An interesting

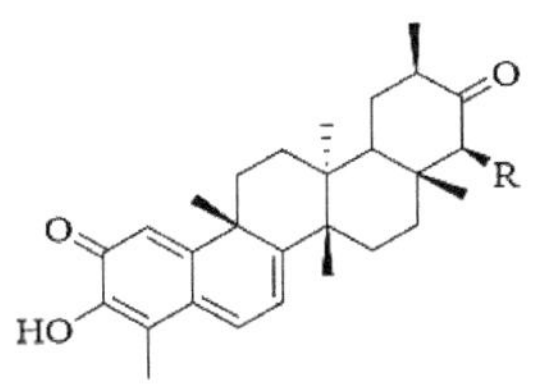
10 - R: OH
11 - R: H

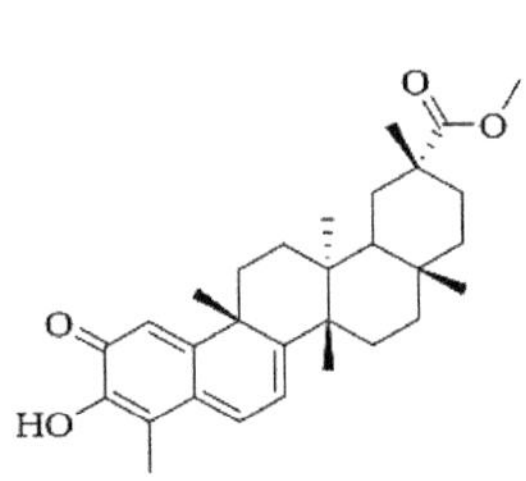
12

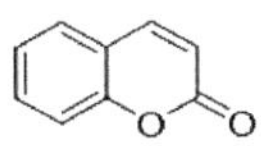
13

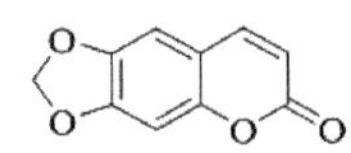
14

isolated compound from those mentioned above is the phenolic glycoside amburoside A (**16**), which would be responsible for the *in vitro* bioactivities such as anti-malarial (50 mg/kg; 24% inhibition on *P. berghei*), neuroprotective (0.1–100 μg/mL; 24–64% inhibition) and hepatoprotective activities (25–50 mg/kg) (Bravo et al. 1999, Leal et al. 2005).

Four other interesting manuscripts by Leal et al. (2003, 2011), Oliveira et al. (2017) and Nunes et al. (2018) reported, respectively, that: (i) the standardized ethanol extracts and vanillic acid (**15**) obtained from trunk barks from cultivated (4–9 month)/wild plants showed similar antiinflammatory activities on different models; (ii) 15 phenolic compounds were isolated including two novel molecules [dilmin (**18**) and lulin] of the 32 constituents (isoflavones, flavanonols, chalcones, isoflavonolignan and isoflavoquinone) characterized by HLPC/MS from resin; the principal compounds were evaluated against human DNA topoisomerase II-α which exhibited a good activity; and (iii) EtOH extract from *Am. cearenses* leaves showed a good solar protection factor (18) due to its content in the phenolic

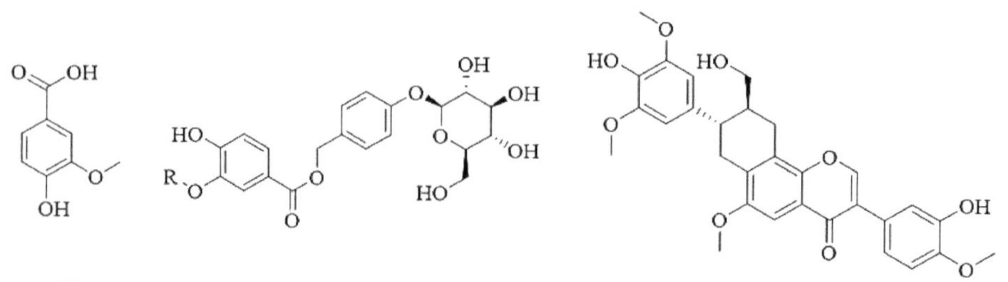

15 16 A - R: H 17 B - R: CH_3 18

compounds. In addition, the aqueous extract from seeds presented antiedematogenic activity (Lima et al. 2013); the leaf hexane partition (at 2.5%) also demonstrated acaricidal activity on *R. microplus* (Dantas et al. 2016) and protective activity on the ovarian follicle of goats (Gouveia et al. 2015). Likewise the EtOH extract and the hexane/CH_2Cl_2/ethyl acetate fractions from previous extract [containing coumarin, C_{16}–C_{18} fatty acids (saturated and unsaturated), sitosterol] were able to increase the viability for PC12 cells (despite the toxicity induced by glutamate), as a measurement of neuroprotective effects, without evidencing any toxicity (Pereira et al. 2017).

The chemical diversity, e.g., pauferrols A (**19**)/B (**20**)/C (stem), catechin/epicatechin (bark), gallic (stem/bark, fruits and leaves)/ellagic (**21**) acids (bark and fruits), lectin (fruits and leaves), lupeol, α-amyrin, quercetin/vitexin/orientin (**22**)/isoorientin (leaves) and galactomannans (seeds), of the *Cae. ferrea* organs would be related to its wide medicinal uses (Port's 2011, Lopes et al. 2013, Ohira et al. 2013, De Souza 2015). Several scientific studies have demonstrated the notable pharmacological potential of isolated compounds and plant extracts; for instance, Gonzalez (2005) reported the antiulcerogenic (37% reduction of lesion in treated rats with 1 mL 15% orally) and antioxidant (protector effect 84–94%) activities of the crude EtOH extract from core stem; like the stem, the leaf extract also showed antiulcerogenic property (81% reduction of affected area). In turn, Lima (2012, 2013) stated that the fruit

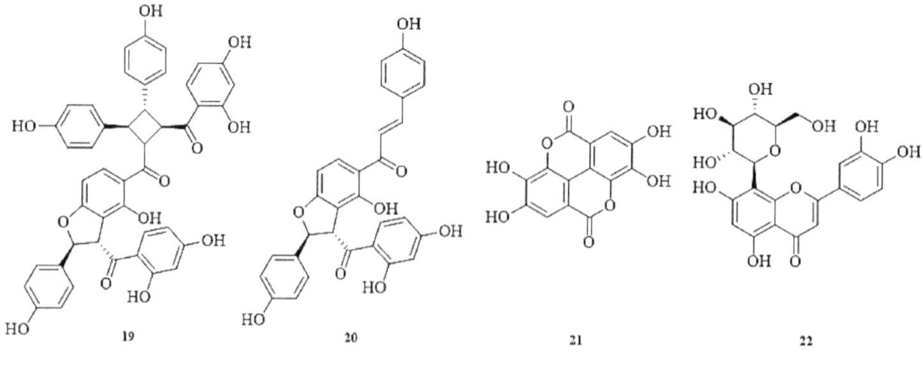

19 20 21 22

extract (ethanol, 50 mg/kg) showed anti-inflammatory/edematogenic (67% inhibition of ear edema by writhing test) and analgesic (74% reduction of the contortion number) activities. Nascimento et al. (2015) observed the antioxidant effects of ethanol extract by DPPH· (EC_{50}: 4.40 ± 0.05), $ABTS^{+\cdot}$ (EC_{50}: 2.50 ± 0.06) and β-carotene (47% inhibition) methods.

By comparison, Menezes et al. (2007) indicated that H_2O extract (bark) caused hypotension (in non-anesthetized rats) along with vasodilation in the mesenteric arteries; it also demonstrated hypoglycaemic activity in Wistar rats model (dose administered for seven days), reducing and improving their glucose levels in blood and the metabolic state. Likewise, Vasconcelos et al. (2011) found that protein kinase B in the liver/skeletal muscle of rats was increased, whereas acetyl-CoA carboxylase and AMP-activated protein kinase were reduced for both and only in the skeletal muscle, respectively. Still, the H_2O extract of the fruits showed anti-inflammatory activity (10–20 mg/kg, 88–92% reduction in number of writhing), i.e., the paw edema in rat induced by carrageenan was significantly inhibited (23–36% for 1–2 h) by oral administration of 300 mg/kg of the extract (Carvalho et al. 1996). In addition, the aqueous seed extract was active as an antinociceptive agent at 1 mg/kg, probably inhibiting the opioid and cholinergic receptors (Sawada et al. 2014); this extract also displayed larvicide activity against *Aedes aegypti* with 85% mortality (Cavalheiro et al. 2009).

According to Biasi-Garbin et al. (2016), the H_2O, EtOH and acetone extracts from barks presented antifungal activity on *Trichophyton rubrum* (MIC/MFC: 62 μg/mL) and *T. mentagrophytes* (MIC/MFC: 31 μg/mL); moreover, the MeOH extract exhibited antimicrobial activity (MIC: 25–100 μg/mL) against *C. albicans, S. mutans, S. salivarius, S. oralis* and *Lactobacillus casei* (Sampaio et al. 2009). Under other conditions, the ethyl acetate extract (fruits) had a prominent antitumor activity against Epstein-Barr virus early antigen (total inhibition of cell viability, at 100 μg/mL) (Nakamura et al. 2002). Among the isolated and pharmacologically active molecules could be mentioned, e.g., **19** (from acetone stem extract) which revealed a high inhibition (IC_{50}: 2.1 μM) on DNA topoisomerase II as described by Nozaki et al. (2007), and the sulphate polysaccharides (from H_2O seed extract) that disclosed the antiviral activity against herpes simplex virus (HSV, IC_{50}: 405 μg/mL) and poliovirus (IC_{50}: 2 μg/mL) (Lopes et al. 2013).

The Brazilian plant, *E. macrophyllus*, has been constituted by non-volatile [diterpenoids (chapecoderins A/B/C (**23–25**, labdanes), echinophyllins A (**26**)/B (**27**) (nitrogen-containing clerodanes), echinoic acid (**28**), echinodolides A/B (**29**, **30**) (cembranes) and flavonoids] (Kobayashi et al. 2000a, b, Shigemori et al. 2002) and volatile (essential oils—EO) secondary metabolites. All of them, both as individual components and in mixtures (extracts/essential oil), could be responsible for some pharmacological properties. For the case of the EO, they had different compositions according to the place of collection, i.e., from Nova Friburgo, the main constituents of EO were dillapiole (24%) and 2-tridecanone (15%) (Coelho et al. 2012), while from Belo Horizonte, the EO was constituted by methyl-palmitate (44%) and dihydroedulan (3%) (Silva et al. 2013b). With regard to the flavonoid content, they are also present in high concentrations (Da Silva et al. 2016b), and a new flavonoid [6-C-hexa-hydroxy-apigenin (**31**)] has been isolated for the species (Silva 2010). Moreover, other chemical groups are present in the plant, such as alkaloids, saponins, xanthones, triterpenoids and steroids (Silva 2010, Tanus-Rangel et al. 2010).

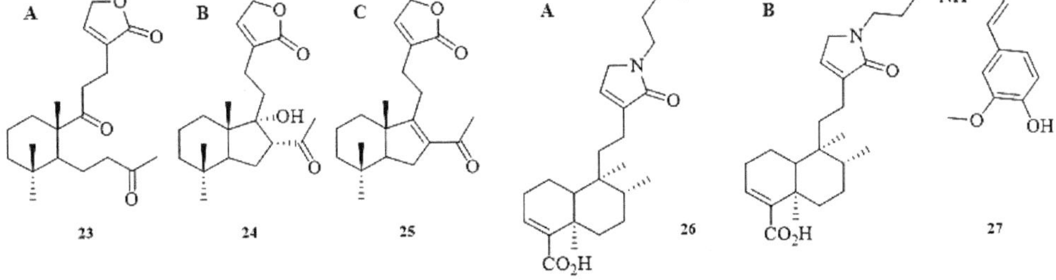

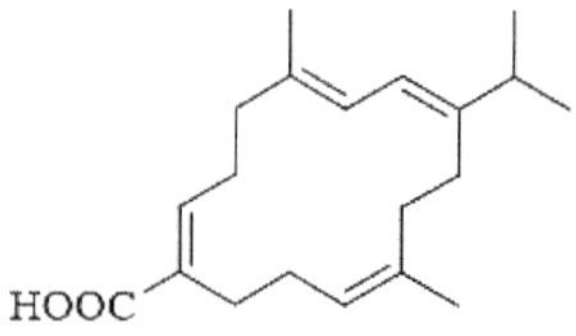

28

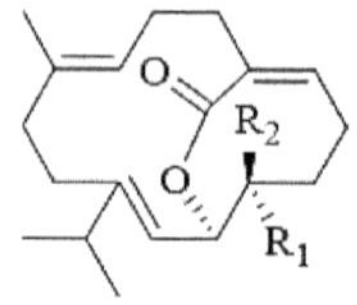

29 A - R_1:CH_3, R_2:OH
30 B - R_1:OH, R_2:CH_3

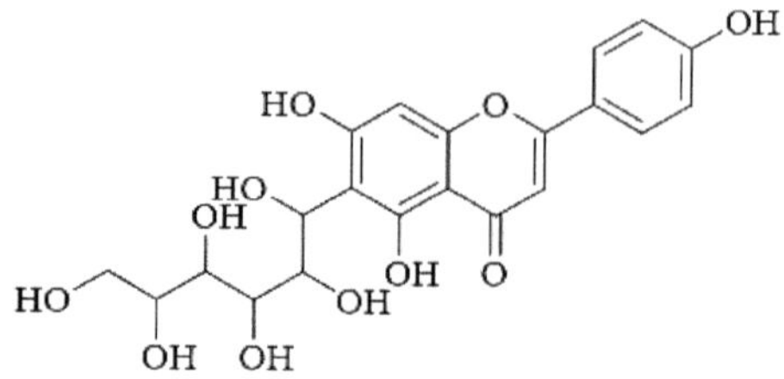

31

Concerning the pharmacological potential based on scientific studies, the species has presented a wide variety of effects; the aqueous extract showed immunosuppressive activity (on the antibody production by B cells, 0.5 mg/kg b.w.) on mice during seven days and delayed type hypersensitivity (on T cells, 0.5–5 mg/kg b.w.), by reducing the subcutaneous tissue infiltration. The extract also inhibited *in vitro* production of NO· (by stimulating J774 cells) (Pinto et al. 2007). Still, the flavonoid-rich fraction from H_2O extract exhibited antiinflammatory activity (*in vivo*/*in vitro*); at 2.5 mg/kg, it reduced total mice leucocytes (30%), showed inhibition of LTB4 (90%) and proteins (42%), as well as it caused *in vitro* inhibition of neutrophil migration and NO· production (Da Silva et al. 2016b). The EtOH extract (leaves) demonstrated significant and potent topical and systemic anti-inflammatory effect in rats by causing inhibition of paw edema and decreases in the leukocyte migration, as well as reduced the vascular permeability and ear edema in mice (Tanus-Rangel et al. 2010). Also, it presented diuretic action (10–300 mg/kg, reduction in urine elimination) and throwback of polyuria and glomerular filtration rate (reduction) (Portella et al. 2012). Interestingly, a renoprotective effect was also observed on rats treated with a leaf solution (2 mg/kg) (Nascimento et al. 2014). Finally, the plant EO from Nova Friburgo evidenced antinociceptive capability on mice treated orally (contortion inhibition of 65%, 50 mg/kg) (Coelho et al. 2012).

The reviewed scientific literature on *B. utile* showed that latex and bark were the most useful parts of plant from a pharmacological point of view due to its promising cytotoxic/antitumor properties [on *A. salina* (IC_{50} 48 μg/mL) and cancers of breast (IC_{50} 26 μg/mL), prostate (IC_{50} 28 μg/mL) and leukemia] (Ferrari et al. 2005, Chindoy 2013, Rivera-Parada 2013). Nonetheless, the antioxidant power (using different tests) of the bark extract from Venezuelan plant was determined by Padilla et al. (2008), who found that the extract was moderately active (reducing power: 59.7 ± 0.4 AsAE (ascorbic acid equivalent mg)/g, antioxidant activity: 61% (β-carotene bleaching), DPPH·—EC_{50} 1.3; %inh: 83); also, they measured the total phenolic [62 ± 4 GAE (galic acid equivalent mg)/g), flavonoid (8.9 ± 0.3 RTE (rutin equivalent mg)/g), and tannin (3.9 ± 0.4 CTE (catechin equivalent mg)/g] contents present in the extract. It is worth highlighting that an alkaloid-rich fraction [possibly containing bufotenin (**32**)] from latex suspension in MeOH of the Ecuadorian plant was obtained/reported by Jiménez (2015) and Acosta et al. (2017), which found it an antimicrobial potential based on the MIC values (1.6 mg/mL, for *P. aeruginosa*, *S. aureus*, *K. pneumoniae*, *E. coli*, and *C. albicans*) and the diameters of inhibition zone (2 mg/mL, *T. rubrum*: 19.5 ± 0.5 mm, *Epidermophyton floccosum*: 17.4 ± 0.5 mm). A particular chemistry has been associated to some isolated molecules (e.g., coumarins and isoflavones) from *B. utile*. Thus, Ferrari et al. (2005) purified seven isoflavones (**33**–**39**) from root bark; six of them were known molecules [lupalbigenin (**34**), alkenyl derivative of biochanin A (**35**), isowigtheone (**36**), ficuisoflavone (**37**), lupinisoflavone C (**38**), genistein (**39**)] and the last one was a new isoflavone [isowigtheone hydrate (**33**)], which was active on MCF7 and PC3 cell lines by *in vitro* cytotoxicity assays. In addition to isoflavones, the trunk bark was constituted by the coumarins xanthyletin (**40**) and luvangetin (**41**), according to the report by INPA (1972) on the chemical composition of Amazon plants.

The analysis of the consulted scientific literature on bioactive extracts from *G. ulmifolia* indicated that it is one of the wild medicinal plants with higher pharmacological evidences according to the *in vitro*/*in vivo* biological tries (Table 2.2). Among the remarkable bioactivities can be mentioned angiotensine II type 1 and acetylcholinesterase inhibitions, gastroprotective and vasorelaxing effects, antidiabetic, anti-

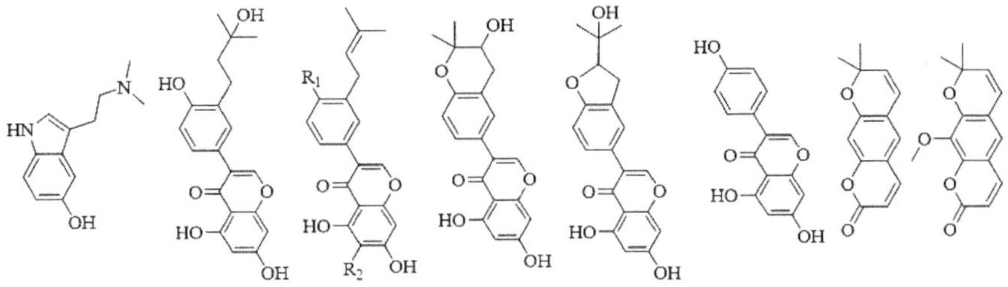

32 33 34 35 36 37 38 39 40 41

	R_1	R_2
34	OH	$CH_2CH{=}C(CH_3)CH_3$
35	OCH_3	H
36	H	H

STAT1, antiproliferative/antitumoral, antimicrobial, antiparasitic and antiviral. Nevertheless, promising bioproperties were vasorelaxing effect (IC_{50} 35 ng/mL), angiotensine II inhibition (IC_{50} 13-286 μM), antidiabetes (IC_{50} 1–70 μg/mL on T2-D), antiviral (IC_{50} 10 μg/mL on P-1 virus), anti-STAT1 (K_D 1–2 × 10^{-6} M), antiparasitic (500 μg/mL, strain inhibition epimastigotes of *L. infantum*: 95.2 ± 0.4%, *L. brasiliensis*: 92.20 ± 0.04%; epimastigotes *Trypanosoma cruzi*: 61 ± 1%) and antifungal (75 μg extract, ϕ inh.: 26 ± 2 mm for *C. albicans*). Regarding the chemical constituents from *G. ulmifolia* some important groups of secondary metabolites (proanthocyanidin, coumarins, terpenes, and flavonoids) have been identified (Pereira et al. 2019). Between them, proanthocyanidin oligomers and polymers [containing 1–15 units of epicatechin and catechin, e.g., **42** (three units)] from EtOH extract of bark were isolated by Hör et al. (1996) and these compounds inhibited the secretion of cholera toxin on rabbit distal colon set in an Using chamber. Additionally, specific flavanocoumarins (**43**–**45**) were isolated by Maldini et al. (2013) from plant bark and two of them (**43**, **44**) showed affinity for STAT1 (signal transducer and activator of inhibition of STAT1-DNA binding) which could be related to the anti-inflammatory properties attributed to *G. ulmifolia*.

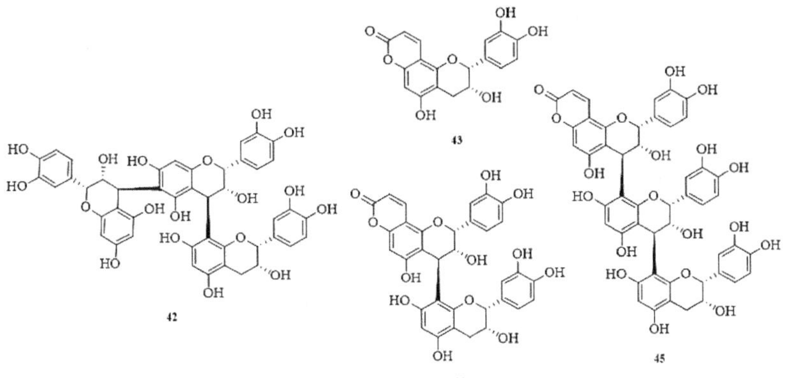

42 43 44 45

A glycoside flavonoid coumaroyl ester derivative, tiliroside (**46**), was also isolated from leaves as reported by Da'i et al. (2016); this ester derivative resulted to be a moderate antiproliferative agent against T47D and MCF lines via apoptosis mechanism by extrinsic pathways (caspases 8/9, p53, Bcl2 protein). Into the bargain, unusual molecules -2R-taxiphyllin (**47**) and dicoumarol (**48**)-insulated from leaves and roots were reported by Seigler et al. (2005) and Agarwal et al. (2010), respectively, which are a nitrile glucoside compound and a biscoumarin. Eventually, the leaf EO presented different main constituents: precocene I (**49**) (56%) and β-caryophyllene (14%) (Arriaga et al. 1997); citronellol (13%) and β-caryophyllene (12%) (Matulevich-Peláez and García-Rodríguez 2016); and thymol (21%), carvacrol (14%) and eugenol (10%) (Boligon et al. 2013). Particularly, the above mentioned EO presented a good radical-scavenging capacity (DPPH·, IC_{50} 7.61 ± 0.09 μg/mL) as well as favorable antibacterial effects against *P. aeruginosa* (MIC 62 μg/mL), *S. aureus* (MIC 125 μg/mL), *E. coli* (MIC 500 μg/mL) and *S. epidermidis* (MIC 750 μg/mL).

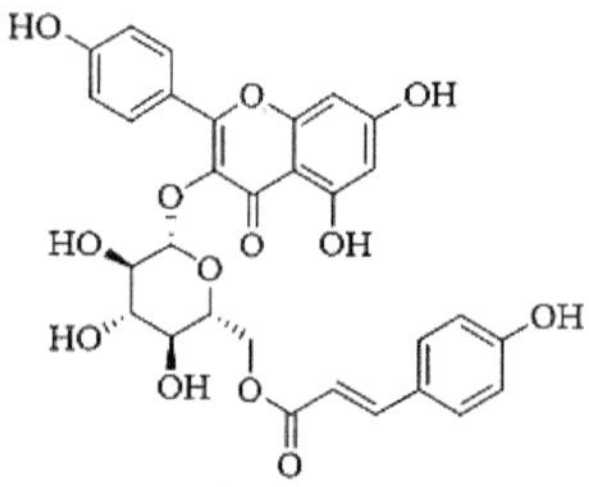

46

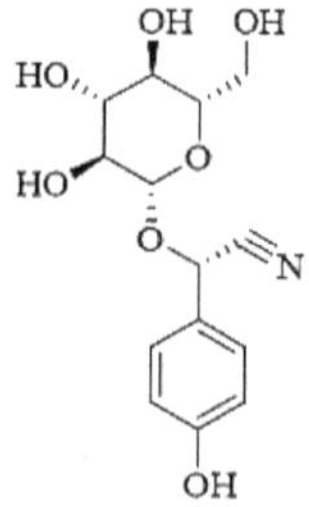

47

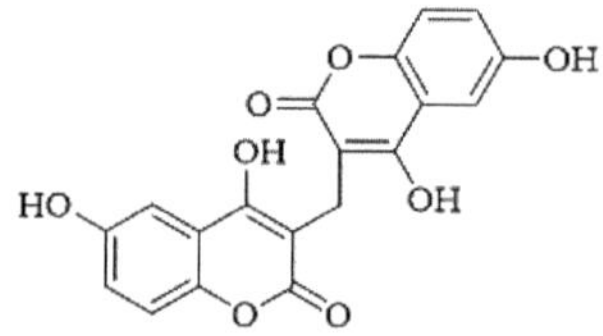

48

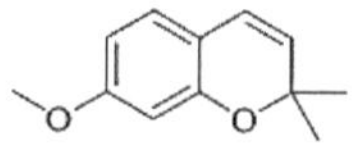

49

An astonishing Latin American plant is *Cal. brasiliense*: the extracts of the different parts (bark, leaves, resin) and the isolated/identified/tested chemical constituents (**50–91**, triterpenoids, coumarin, xanthones, chromanones, biflavonoids, etc.) were powerful biologically active substances as anti-HIV, anti-leishmaniasis/malarial/chagas, antinociceptive, anti-*Mycobacterium*, antibacterial, antifungal, cytotoxic, and antitumoral agents. Hence, hexane fraction and CH_2Cl_2 extract along with mammea A/BB (**53**) from EtOH extract (leaves) were active against *L. amazonensis* for promastigote (IC_{50} 3–40 μg/mL/IC_{90} 5–73 μg/mL) and axenic amastigote (IC_{50} 0.9–4 μg/mL/IC_{90} 2–20 μg/mL) forms (Brenzan et al. 2007). Furthermore, the selectivity index (SI) determined for **53** (on J774G8 macrophages) was highest for amastigote (SI: 29) form. In addition, **53** was able to reduce (60% and 68%) the lesion induced by infection subcutaneous with *L. amazonensis* on Balb/c mice (Tiuman et al. 2012); the CH_2Cl_2 extract and **53** were also active against promastigote form of *L. braziliensis* (Brenzan et al. 2008a), and coumarins **53**/**58** (mammea B/BB) were effective (IC_{50} 7.4 ± 0.3 μg/mL –30 ± 4 μg/mL) on *L. amazonensis* (intracellular amastigotes and promastigotes) (Brenzan et al. 2008b). Moreover, different extracts [ethyl acetate, CH_2Cl_2, CH_2Cl_2/MeOH (1:1), CH_2Cl_2/ethyl acetate (1:1)] and **67** isolated from leaves/stems restricted the growth of *P. falciparum* with IC_{50} values ranged amongst 8 ± 1 μg/mL and 55 ± 14 μg/mL (Mesa-Vanegas et al. 2019a, b). Other coumarins type mammea as **52** and **69** together with specific mixture (**52–54**) isolated of leaves and stem bark inhibited the growth of *T. cruzi* strains in some of its forms (epimastigotes, amastigotes and trypomastigotes) with IC_{50} values among 10–220 μM (Rea et al. 2013, Rodríguez-Hernández et al. 2019). Albernaz et al. (2010) determined the antimalarial (*P. falciparum*), antiChagas (*T. cruzi*) and antileishmanial (*L. chagasi*) power of CH_2Cl_2 extracts from root, wood and bark; the IC_{50} values were between 6.7 ± 0.9 μg/mL –97.1 ± 0.4 μg/mL.

An amazing pharmacological action determined for some fractions/extracts and isolated molecules from *Cal. brasiliense* has been the antiviral effectiveness on human immunodeficiency virus type 1 reverse transcriptase (HIV-1 RT). Accordingly, Huerta-Reyes et al. (2004a, b), García-Zebadúa et al. (2011), Kudo et al. (2013) and Matsuda et al. (2015) reported (i) the percentage inhibition (77.9 ± 0.5%–81 ± 2%) and IC_{50} value of hexane (30 μg/mL) and acetone (50 μg/mL) extracts on viral enzyme; (ii) the IC_{50} values (0.3–1 mM) for calanolides A-C (**64–66**) and soulattrolide (**67**), and the inhibition (60 ± 4%–75 ± 4%) for some fractions [18/19 (hexane extract), 2 (acetone extract) and ethyl acetate insoluble (MeOH

Mammeas

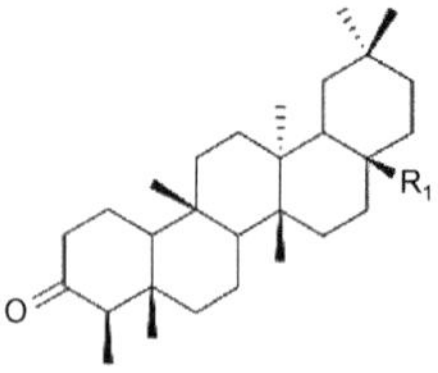

50 R_1: CH_3; **51** R_1: CH_2OH

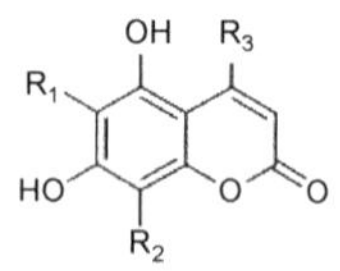

	R_1	R_2	R_3
52	$CH_2CH=C(CH_3)_2$	$(C=O)CH_2CH(CH_3)_2$	C_6H_5
53	$CH_2CH=C(CH_3)_2$	$(C=O)CH(CH_3)CH_2CH_3$	C_6H_5
54	$CH_2CH=C(CH_3)_2$	$(C=O)CH(CH_3)_2$	C_6H_5
55	$CH_2CH=C(CH_3)_2$	$(C=O)CH_2CH(CH_3)_2$	$(CH_2)_2CH_3$
56	$CH_2CH=C(CH_3)_2$	$(C=O)CH(CH_3)CH_2CH_3$	$(CH_2)_2CH_3$
57	H	$(C=O)CH_2CH(CH_3)_2$	$(CH_2)_4CH_3$
58	H	$(C=O)CH(CH_3)CH_2CH_3$	$(CH_2)_4CH_3$

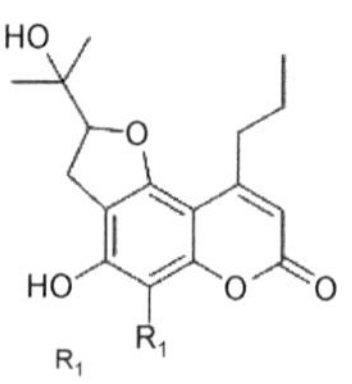

	R_1
59	$(C=O)CH_2CH(CH_3)_2$
60	$(C=O)CH(CH_3)CH_2CH_3$

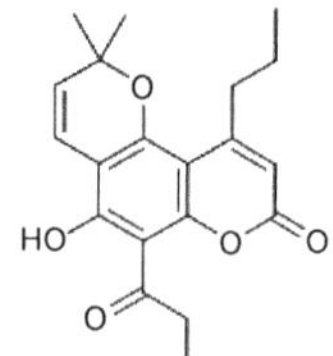

61

extract)] at 50 μg/mL on HIV-1 RT; (iii) the IC_{50} values for hexane (30–48 μg/mL) and acetone (31 μg/mL) extracts from leaves against the same enzyme; (iv) the inhibition (IC_{50} 8.4 μM) by **62** (isolated from stem bark) on HIV-1 replication of infected cells (U1 and Molt-4, in the presence of PMA and TNF-α); and, (v) the inhibition (25–65%) of HIV-1 entry (fusion cell) by **62** (10–100 μM) via decrease of cell membrane fluidity.

The validation of the use of "guanandi" as an analgesic in the ethnomedicine from Brazil, by means of *in vivo* antinociceptive evaluation, has been carried out by Da Silva et al. (2001), Isaias et al. (2004) and Klein-Júnior et al. (2017). These authors found that hexane/ethyl acetate/$CHCl_3$ fractions and MeOH extracts from leaves/flowers/fruits/roots, as well as some isolated molecules [friedelin (**50**), brasiliensic acid (**68**), 1,5-dihydroxanthone (**70**) and 1,2-dimethoxyxanthone (**71**)] were competent to inhibit the muscular constriction on mice measured by different methods [acetic acid-induced writhing, formalin-induced paw licking (phases I/II and edema), capsaicin-induced and glutamate-induced nociception] with the percentage inhibition values of 70 ± 6%–80 ± 4% at 3–10 mg/kg (i.p.) for MeOH extracts (flowers/fruits/roots), 41–94% at 60 mg/kg (i.p.) for $CHCl_3$ fraction (roots), 81 ± 8% for polar fractions (roots), 98 ± 6% for non polar fractions (roots), while the ID_{50} values were 9–48 mg/kg i.p. (132 p.o.) for $CHCl_3$ fraction (roots), 12 μmol/kg for **50**, 7 μmol/kg for **68**, 30 μmol/kg for **70** and 28 μmol/kg for **71**.

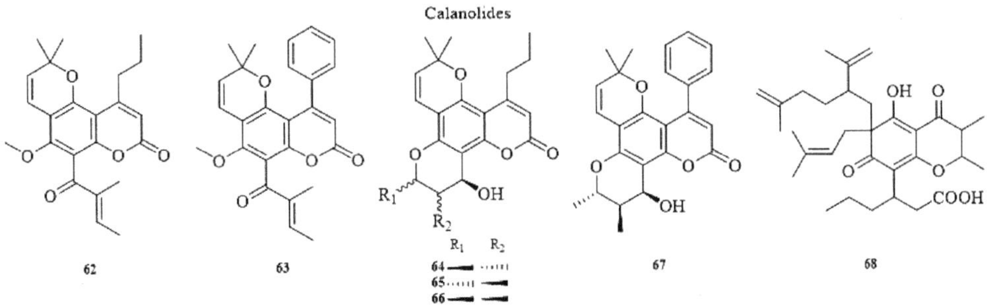

Some extracts/isolated molecules turned out to be moderate/good antimicrobials. For instance, certain extracts (Soxhlet, maceration and SFE-CO_2) and **61** (mammea B/BB cyclo D) isolated from leaves were useful against *Mycobacterium tuberculosis* with MIC values of 62.5 μg/mL and 125 μg/mL, respectively (Pires et al. 2016). Also, hexane, CH_2Cl_2, acetone, MeOH and H_2O extracts obtained from flowers/leaves/stems/fruits/roots and **52/70/72/73/76**, mammea A/AA, PCA (protocatechuic acid), chromanone acids (**80–83**) isolated from the leaves/bark/heartwood showed a good bacterial/fungi inhibition against 13 bacterial [*B. cereus* (MIC 1–900 μg/mL), *Enterobacter cloacae* (MIC 400 μg/mL), *E. faecalis* (MIC 300–510 μg/mL), *E. coli* (MIC 128–512 μg/mL), *Proteus mirabilis* (MIC 500 μg/mL), *P. aeruginosa* (MIC 190–800 μg/mL), *Salmonella typhimurium* (MIC 700 μg/mL), *S. aureus* (MIC 2–600 μg/mL), MSSA (MIC 1–256 μg/mL), MRSA (MIC 8–32 μg/mL), *S. saprophyticus* (MIC 400–900 μg/mL), *S. epidermis* (MIC 16 μg/mL), and *S. agalactiae* (MIC 100–700 μg/mL)] and three fungi [*C. albicans* (MIC 8–500 μg/mL), *C. tropicalis* (MIC 400 μg/mL), *Cryptococcus neoformans* (MIC 2 μg/mL)] strains (Cottiglia et al. 2004, Pretto et al. 2004, Albernaz et al. 2010, Suffredini et al. 2014). An important infectious bacterial agent causing gastric lesions is *Helicobacter pylori*, over which the CH_2Cl_2/chromanone acid-

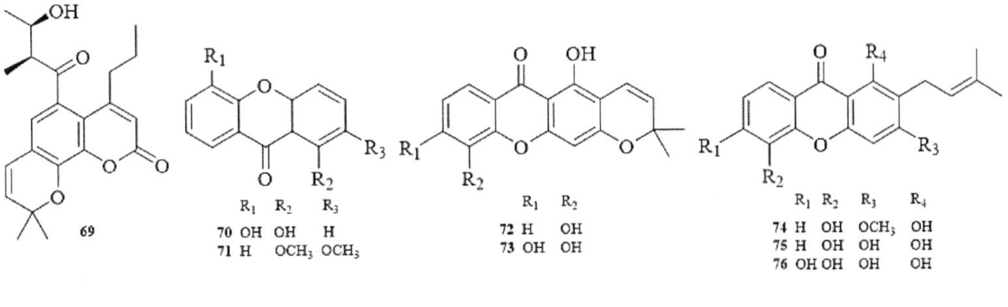

Brasixanthones

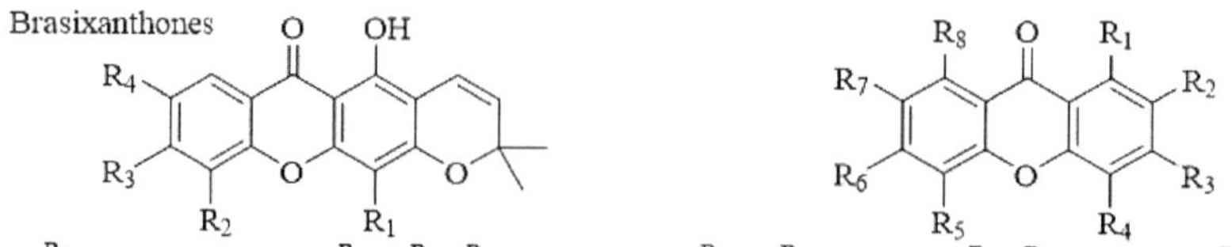

	R_1	R_2	R_3	R_4
77	H	OCH_3	OH	$CH_2CH{=}C(CH_3)_2$
78	$CH_2CH{=}C(CH_3)_2$	H	H	OH
79	$CH_2CH(OOH)C(CH_3){=}CH_2$	H	H	OH
80	$CH(OH)CH(epox.)C(CH_3)_2$	H	H	OH

	R_1	R_2	R_3	R_4	R_5	R_6	R_7	R_8
81	OCH_3	OCH_3	OH	H	H	H	H	OH
82	OH	$CH_2CH{=}C(CH_3)_2$	OH	$CH_2CH{=}C(CH_3)_2$	OH	H	H	H
83	OH	$CH_2CH{=}C(CH_3)_2$	OH	H	OCH_3	OH	$CH_2CH{=}C(CH_3)_2$	H
84	H	H	H	H	OH	H	H	H

rich fractions and hydroethanol/hexane extracts from stem bark have shown a potential; these extracts and fractions produced inhibition halos (ϕ inh.) between 8–16 mm (at 125–1000 μg/disk) or 7–14 mm (at 25–400 μg/disk, percentage inhibition 17–44%—MIC 31 μg/mL). Similarly, the extracts and fractions diminished gastric ulcer induced on rats with percentage inhibition of 63–97% (at 5–100 mg/kg) and 67–95% (at 20–100 mg/kg) (Souza et al. 2009, Lemos et al. 2012). Further, xanthones (**73**) and coumarins (**52**, **57**) isolated from acetone/MeOH extracts (heartwood) inhibited the H^+/K^+-ATPase enzyme as a measurement of its gastroprotective capacity; the IC_{50} values were (i) xanthones: 47 ± 8 μM–173 ± 9 μM, (ii) coumarins: 110 ± 8 μM–638 ± 149 μM (Reyes-Chilpa et al. 2006).

Besides, the antineoplastic/anticancer/antitumoral potential of extracts/molecules from "Bari" has been proved; thus, CH_2Cl_2 extract from root suppressed the growth of MDA-MB-435, HCT-8, HL-60 and SF-295 cell lines with IC_{50} values of 3 μg/mL, ~ 5 μg/mL, ~ 6 μg/mL, ~ 18 μg/mL, individually (Mesquita et al. 2009), whilst CH_2Cl_2:MeOH and H_2O extracts (100 μg/mL) inhibited the KM-12 and RPMI-8226 cancer cells with percentage values of 100% and 34–84%, one-to-one (Suffredini et al. 2007). The hexane extract obtained by Oliveira et al. (2014) from stem bark was effective against CHO-k1 line; similarly, the CH_2Cl_2:MeOH and H_2O extracts from leaves and stems were able to inhibit the growth of some cell lines (MCF7, PC3, NCIH460, KM12 and RPMI8226). Thus, the most active extract (aqueous of stem) inhited four lines (except RPMI8226) with GI_{50} values of 0.6 μg/mL, 49 μg/mL, 12 μg/mL and 1.2 μg/mL, respectively, followed by the H_2O extract from leaves, which inhibited three cell lines (except MCF7 and PC3) and with GI_{50} values of 16 μg/mL, 4 μg/mL, and 0.6 μg/mL, respectively (Suffredini et al. 2014). On the other hand, **50**, **52**, **53**, **55–60** isolated from hexane/acetone/MeOH extracts (leaves) inhibited three cancer lines (PC3, K562 and U251). Among all molecules tested on cell lines, coumarin 52 resulted the most active with IC_{50} values between 0.04 ± 0.02 μM–0.6 ± 0.2 μM, followed by a mixture **55**/**56** with IC_{50} values between 0.65 ± 0.09 μM–1.5 ± 0.1 μM (Reyes-Chilpa et al. 2004). Likewise, xanthones **62** and **63** from stem bark were evaluated on BV173, K562, NALM6, HL60, SEM and HCT116 cell lines; **62** (IC_{50} 2–5 μM) and **63** (IC_{50} 9 μM) inhibited the six lines and BV173 cell, in that order (Kimura et al. 2005). Also, calanolides (**62**, **64**, **66**), brasixanthones (**77–80**), xanthones (**81–84**), brasimarins (**85–87**), and mammeas (**56**) isolated from stem bark could be categorized as chemopreventive agents when assessed on TPA-EBV-EA (12-*O*-tetradecanoylphorbol-13-acetate induced Epstein-Barr virus early antigen activation) in Raji cells; the most active molecules were **78**, **56**, **79**, **80**, **62** and **64** with IC_{50} (mol ratio/32 pmol TPA) values of 120, 170, 200, 210, 268 and 290 (Ito et al. 2002, 2003). Moreover, a mixture of mammeas **52** and **53** (24 μg/mL) from leaves inhibited ca. 52% of BMK cell growth (Ruiz-Marcial et al. 2007), and **73** isolated from heartwood was cytotoxic,

Brasimarins

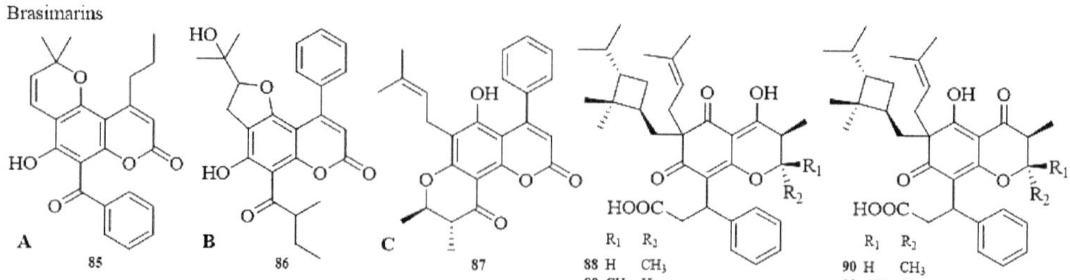

genotoxic and cytostatic by different tests (MTT, trypan blue cell cycle and proliferation, micronucleus, alkaline SCGE, aneuploidy, lethal dose 50) (García-Niño et al. 2017).

Lastly, Blanco-Ayala et al. (2013) determined the antioxidant potential (measuring the scavenging capability of $O_2^{-\cdot}$, HO· and $ONOO^-$) of two xanthones [jacareubin (**73**) and **76**] isolated from heartwood; the IC_{50} values were 336 ± 9 µM ($O_2^{-\cdot}$), 1.82 ± 0.06 µM (HO·) and 64 ± 12 µM ($ONOO^-$) for **73**, and 262 ± 8 µM ($O_2^{-\cdot}$), 0.070 ± 0.003 µM (HO·) and 30 ± 7 µM ($ONOO^-$) for **76**. In addition, Gonçalves et al. (2013) measured the antioxidant capacity of different extracts from leaves, which were effective against DPPH· radical with IC_{50} values of 243 µg/mL, 207 µg/mL, 132–149 µg/mL for hexane, CH_2Cl_2 and SFE extracts. According to the authors, the mammea A/BB (**53**) could be the responsible compound for this activity. Carvalho et al. (2016) reported that dry and granulated hydroethanol extract (500 mg/kg) orally administered to diabetic rats significantly reduced the glycemia levels as well as other clinical parameters (triacylglycerides, cholesterol, LDL, transaminases, urea and creatinine); Braga et al. (2007) found that EtOH extracts from leaves and stems (at 100 µg/mL) inhibited the angiotensine converting enzyme with values of 34 ± 2% and 55 ± 16%, respectively.

In consideration of medicinal species from Ecuador, Saraguros have used them in the health care systems (Armijos et al. 2012). Among the studies developed in 2016 by Armijos et al., they recorded the uses of *Huperzia* species into the magical-religious practices for the Saraguros; the healers use these plants as purgative agents, and to prepare psychotropic products, for ritual ceremonies. In the phytochemical analysis, a MeOH extract produced two fractions: (i) an alkaloid solution and (ii) a flavonoid/triterpene-rich precipitate. From the chemical analysis by GC-MS of the alkaloid-rich fractions from *Huperzia* spp., some constituents were presumably identified: lycodine (**92**), lycopodine (**93**), 6-hydroxy-lycopodine (**94**), des-N-methyl-α-obscurine (**95**) and flabelline (**96**) in *H. tetragona*; whilst **93** and **94** in *H. compacta*. Moreover, Gilardoni et al. (2014) and Malagon et al. (2016) identified/reported new hydroquinolinic alkaloids from *H. espinosana*. A non-common flavonoid [selgin (**97**)], previously isolated from *H. selago* (Voirin and Jay 1978), was positively identified in the non-alkaloid fraction from *H. espinosana*, as well as the flavonoid tricin in *H. tetragona* and *H. compacta*; this last compound was found in a high content in the extracts (Armijos et al. 2016b). The traditional use of species belonging to the genus *Huperzia* has been mainly to treat ailments possibly related to nervous disorders and this feature determined its potential as inhibitors of enzymes involved in central nervous system illness, e.g., Alzheimer's disease and the associated enzyme [acetylcholinesterase (AChE)]; evidently, the alkaloid-fractions from *Huperzia* spp. inhibited AChE with percentage values between 24–87%.

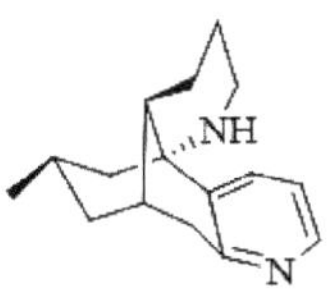

92

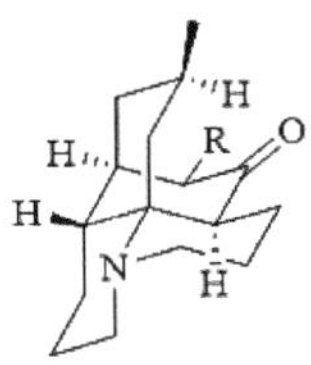

93 R: H; 94 R: OH

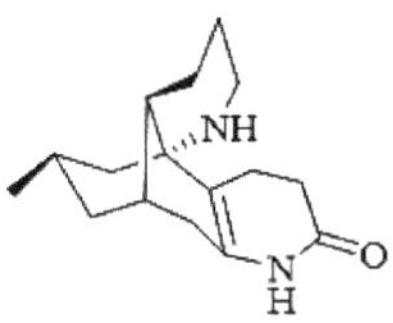

95

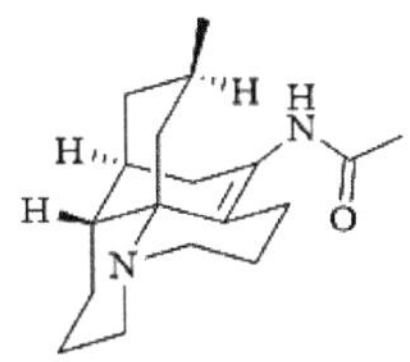

96

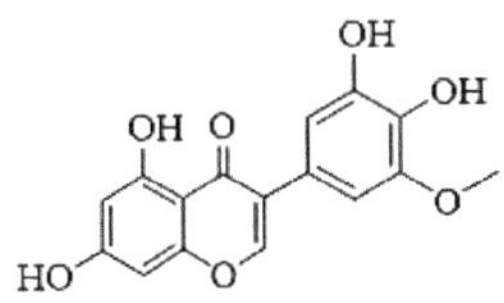

97

In Mexico, the ethnomedical use of *Ag. mexicana* is diverse. It is mainly used for "susto or nervios" and "espanto", a health condition known as cultural affiliation syndrome, dealing with fear and terror (Guzmán-Gutiérrez et al. 2014). It is not attended by modern allopathic doctors, but it is a well known illness among traditional medicine therapist, mainly in rural and indigenous communities. Different preparations (containing flavonoids and/or terpenoids) along with EO from this plant would be related to hypotensive, vasorelaxing, hypoglycaemic, depressant/anxiogenic, antioxidant, anti-inflammatory/antinociceptive, spasmogenic and relaxing effects based on the reports by Molina-Hernández et al. (2000), Hernández-Abreu et al. (2009, 2013), Ibarra-Alvarado et al. (2010), Ávila-Rosas (2013), Verano et al. (2013), Estrada-Reyes et al. (2014), González-Trujano et al. (2015), Flores-Flores et al. (2016), Esquivel-Gutiérrez et al. (2017), Navarrete et al. (2017), Ventura-Martínez et al. (2017) and Cruz-Torres (2019). In accordance with these authors, the CH_2Cl_2, MeOH and H_2O extracts resulted in antihypertensive, vasorelaxing/vasoactive, anxiogenic, anti-inflammatory/antinociceptive and antioxidant properties.

Thus, CH_2Cl_2 extracts showed a good relaxatory effect on contraction of aortic rings (with/without endothelium), and the H_2O extracts were from moderate to good anxiogenic, vasoactive and antioxidant agents. In the case of the antioxidant, MeOH extract (100 μg/mL, > 93% inh.) was partly (mainly) constituted by limonene, linalool, mentone, α-terpineol, pulegone and eugenol, whereas the EO (0.7% yield) with relaxing effect contained estragole (80%) and limonene (18%). The last chemical composition and yield differed in the amounts [estragole (87%)/limonene (11%) and yield (1.2%)] as described by Estrada-Reyes et al. (2004). An important flavonoid is tilianin (**98**) isolated from plant, which had an antihypertensive, vasorelaxant, depressant power, and acacetin (**99**) and ursolic acid (**100**) as vasorelaxant and antinociceptive, respectively.

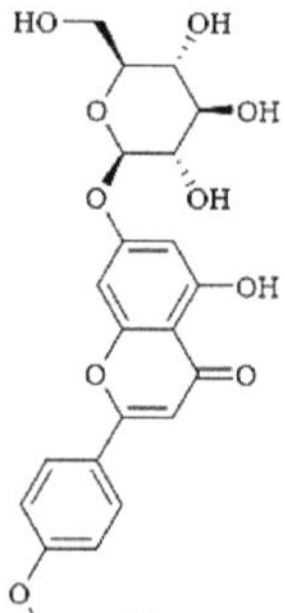

98

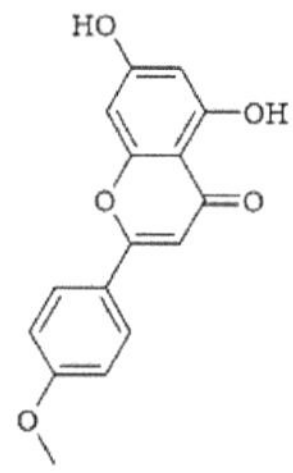

99

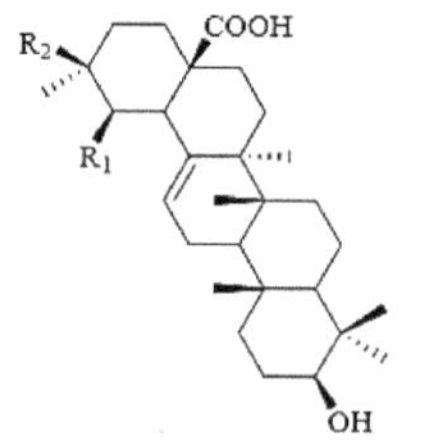

100 R_1: CH_3, R_2: H; 101 R_1: H, R_2: CH_3

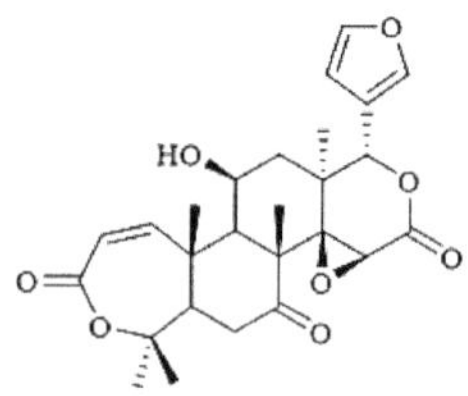

102

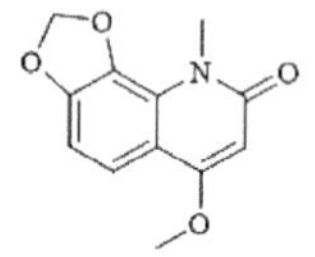

103

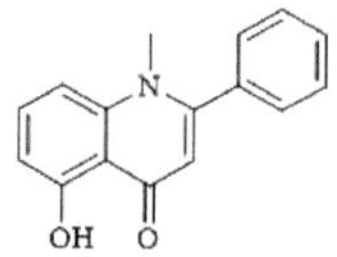

104

Another Mexican species of great potential is *Cas. edulis* due to its widely demonstrated pharmacological effects as hypotensive/vasorelaxant, anticonvulsive, antiviral, antioxidant, anticoagulant, antimicrobial, antiproliferative, chemopreventive (antimutagenic), non-toxic and relaxant, etc., from the extracts/fractions/isolated compounds of their leaves/seeds/kernel/fruits (Table 2.2). For instance, extracts (H_2O/hexane/MeOH/EtOH), fractions (ethyl acetate and alkaloid-rich) and isolated constituents [furocoumarin (**107**), flavonoid (**114**), alkaloids (quinolinones **118**–**121**) and imidazolic derivatives] from different parts (leaves/seeds/kernel/fruits) acted as effective hypotensors (Magos et al. 1995, 1999, Awaad et al. 2007, Froldi et al. 2011, Bertin et al. 2014). In a like manner, H_2O and EtOH extracts from leaves and seeds, respectively, showed satisfactory anticonvulsive action based on MES (maximal electroshock seizure) and METsc (subcutaneously injected metrazole) on rats (Navarro-Ruíz et al. 1995, Garzón-De La Mora et al. 1999). Four other interesting pharmacological properties [antiHIV-1 RT (on DNA polimerase and ribonuclease H activities), anticoagulant (based on prothrombin/activated parcial prothrombin times, PT/APTT), antimicrobial (against *B. subtilis*, *Micrococcus kristinae*, *S. aureus*, *Sarcina maxima*, *E. coli*, *S. typhi*, *Moraxella lacunata*, *Proteus vulgaris*, *C. albicans*, *Aspergillus niger* and *A. flavus*) and adipogenesis (on 3T3-L1 adipocyte)] revealed the EO, the hydroalcohol and EtOH extracts, and some fractions (EIF: ether:$CHCl_3$ and EIIF: ethyl acetate:butanol) and isolated coumarins (**107**–**109**) from seeds and leaves (Esposito et al. 2011, Awaad et al. 2012, Nagai et al. 2014). Considering the EO (leaves), it was constituted by *trans*-caryophyllene (28%) and benzyl-2-cyanoethyl ether (10%), which differed from the composition [germacrene D (16–22%) and (*E*)-caryophyllene (16–17%)] reported by Miller et al. (2009). Nonetheless, EO was effective against all microbial strains evaluated, while EtOH extract distinctively was not active against *A. flavus* and *S. typhi*. On the other hand, furocoumarins **108**/**109** were able to increase the glucose uptake, palmitic acid synthesis, and lipid accumulation in adypocites. By last, some extracts (MeOH), fractions (ethyl acetate) and isolated compounds [furocoumarins (**105**–**107**/**110**)/flavonoids (**111**/**114**/quercetin/rutin)/alkaloids (**103**/**104**)] were chemopreventive and antioxidant agents (Ito et al. 1998, Awaad et al. 2006), and the mucilage presented hypoglycemic effect (Ibrahim et al. 2019). It is worth noting that *Cas. edulis* has been a source of molecules with a widespread structural diversity, i.e., limonoid (**102**), furocoumarins (**105**–**110**), flavonoids (**111**–**115**) and alkaloids (**103**/**104**, **116**–**122**, γ-fagarine) such as zapoterin (**102**), casimiroin (**103**)/casimiroedine, isopimpinellin (**105**)/phellopterine (**106**)/imperatorin (**107**) and zapotinin/zapotin (**112**). This last molecule has been able to suppress *in*

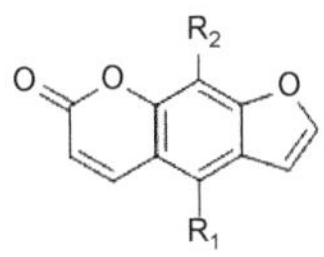

	R_1	R_2
105	OCH_3	OCH_3
106	OCH_3	$OCH_2CH{=}C(CH_3)_2$
107	H	$OCH_2CH{=}C(CH_3)_2$
108	H	$OCH_2CH{=}C(CH_3)CH_2O(C{=}O)CH_3$
109	OCH_3	$OCH_2CH{=}C(CH_3)CH_2O(C{=}O)CH_3$
110	OCH_3	$OCH_2CH{=}C(CH_3)(CH_2)_2CH(OH)C(OH)(CH_3)_2$

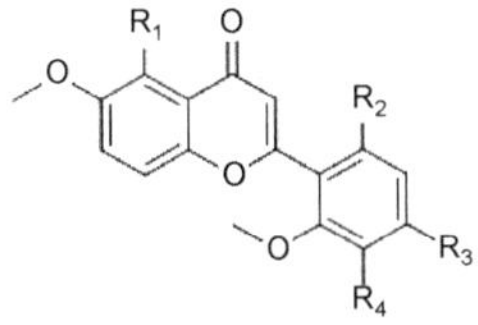

	R_1	R_2	R_3	R_4
111	OCH_3	H	H	H
112	OCH_3	OCH_3	H	H
113	OH	OCH_3	H	H
114	OCH_3	H	OCH_3	OCH_3
115	OCH_3	OCH_3	OCH_3	OCH_3

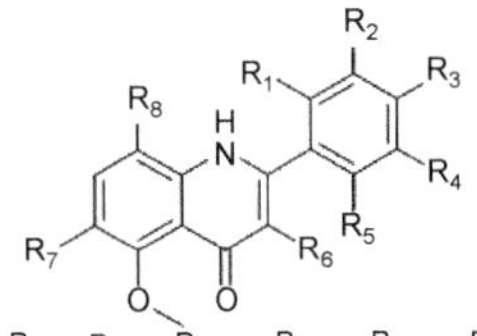

	R_1	R_2	R_3	R_4	R_5	R_6	R_7	R_8
116	H	OCH_3	H	H	H	H	OCH_3	H
117	H	OCH_3	OCH_3	H	H	H	OCH_3	H
118	OCH_3	H	H	OCH_3	OCH_3	H	OCH_3	H
119	OH	H	OCH_3	H	H	Propyl	H	OCH_3
120	H	OCH_3	H	H	H	Propyl	H	OCH_3
121	H	OCH_3	OCH_3	H	H	Propyl	H	OCH_3
122	H	OCH_3	H	OCH_3	OCH_3	H	OCH_3	H

vitro cell proliferation on the HT-29 (IC_{50} 192–212 ng/mL/2.7 × 10^{-1} μM), SW480 (IC_{50} 2.3 × 10^{-1} μM), and SW620 (IC_{50} 5.3 × 10^{-1} μM) lines, as well as autophagy and apoptosis on HeLaPKCεA/E sub-line (Sondheimer and Meisels 1960, Dreyer and Bertelli 1967, García-Argáez et al. 2000, Khaleel 2002, Murillo et al. 2007, Toton et al. 2016).

The last Mexican plant considered in this chapter is *M. tomentosa* (zoapatles or cihuapatli), which has been used in the traditional medicine since prehispanic times, mainly to help women to speed up labor (oxytocic property/uterotonic), and as an antifertility (anticonceptive) agent; in addition, it is also used as an "aphrodisiac" for men. The aforementioned properties have been validated along with anxiolityc-like/anxiogenic-like effects of the H_2O extracts from leaves/aerial parts (Table 2.2). Possibly, the molecules responsible for the oxytocic effect were the oxepane diterpenoids zoapatanol (**123**), montanol (**124**), and tomentol (**125**); other oxepane derivatives [tomexanthol (**126**) and tomexanthin (**127**)] were also isolated. Moreover, other purified molecules were sesquiterpene lactones (heliangolides/guaianolides) as tomentosin (**128**), zoapatanolides A-E (**129**–**133**)/pumilin (**134**) (Levine et al. 1979, Cotter 1981, Hahn et al. 1981, Smith et al. 1981, Kanojia et al. 1982, Quijano et al. 1982, 1984, 1985, Seaman et al. 1984, Oshima et al. 1986). Particular non-polar diterpenes as kaurenoic (**135**) and grandiflorenic (**136**) acids were also identified/isolated from plant, and the hexane or H_2O extracts along with diterpenes acted by increasing both uterine contraction/motility and anxiolytic/anxiogenic properties, which would be contradictory. Nonetheless, some studies (reports) have shown that the above mentioned properties depend on hormonal stage (endocrine conditions) and that nonpolar-factions/molecules would favor spontaneous contraction, whereas polar fractions/molecules would favor uterine stimulation. Lastly, in addition to finding diterpenic acids, some sesquiterpenoids [valencene (25–45%) and β-eudesmol (27–56%)] were identified in the glandular trichomes of plant leaves (Lozoya et al. 1983, Waller et al. 1987, Campos-Bedolla et al. 1997, Robles-Zepeda et al. 2009, Estrada-Camarena et al. 2019).

To conclude, the final group of plants dealt with in this chapter was of Peruvian origin. Based on the scientific review carried out, the first of them, *Cam. angustifolia*, has validated four pharmacological effects: antimicrobial, antioxidant, antiparasitic and enzymatic inhibition. Recently, Roumy et al. (2020) established the antimicrobial potential of the bark MeOH extract against 36 strains (some of them isolated from human infections); of these strains, 20 (e.g., *K. pneumoniae*, *Yersinia pseudotuberculosis*, *P. aeruginosa*, *Stenotrophomas maltophilia*, *Enterococcus* sp., *E. faecalis*, *S. aureus*, *S. epidermidis*, *S. lugdunensis*, *S. warneri*, *S. agalactiae*, *S. dysgalactiae*, *Corynebacterium striatum*, *M. smegmatis*, and *C. albicans*) showed various susceptibility to the extract, with MIC values between 0.3 ± 0.1 mg/mL and 1.2 ± 0.4 mg/mL. The majority of strains had MIC values of 0.6 ± 0.2 mg/mL and the most

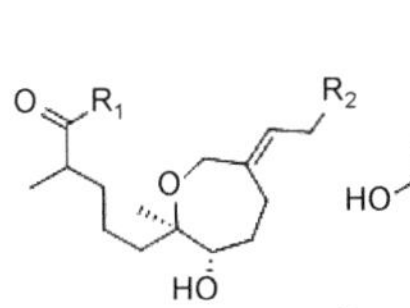

	R_1	R_2
123	$CH_2CH{=}C(CH_3)_2$	OH
124	$CH{=}C(CH_3)CH(CH_3)_2$	OH
125	$CH_2C({=}CH_2)CH(CH_3)_2$	OH
126	$CH{=}CHC(OH)(CH_3)_2$	OH
127	$CH{=}CHC(OH)(CH_3)_2$	$O(CO)CH_3$

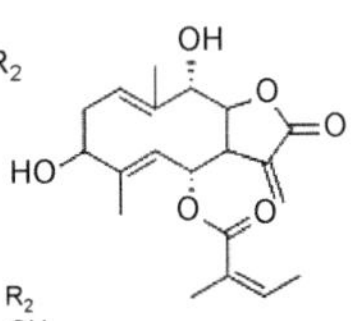

128

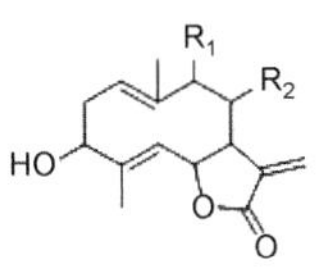

129 R_1: OH, R_2: $O(C{=}O)C(CH_3){=}CHCH_3$
130 R_1: $O(C{=}O)C(CH_3){=}CHCH_3$, R_2: OH

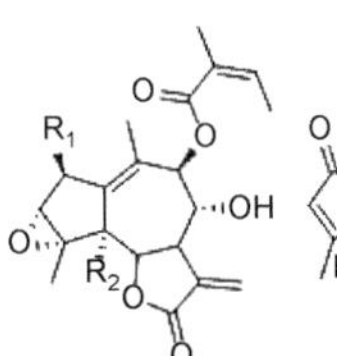

131 R_1: $O(C{=}O)CH_3$, R_2: OH
132 R_1: $O(C{=}O)CH_3$, R_2: H
133 $R_1{=}R_2$: OH

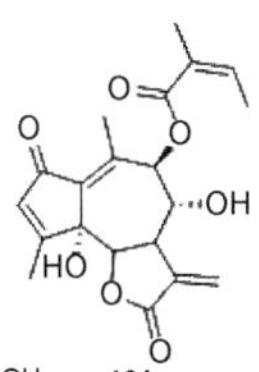

134

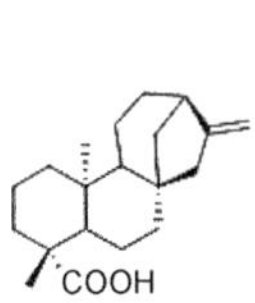

135

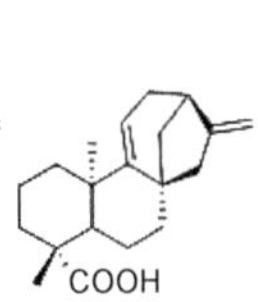

136

susceptible strain was *S. epidermidis* with a MIC value of 0.3 ± 0.1 mg/mL. In addition, Ravarocci-Quiroz and Carrasco-Huamán (2010), who studied the DPPH radical-scavenging capacity (IC_{50} 69 ± 3 µg/mL) of the same type of extract, found through the phytochemical screening that it contained saponins, flavonoids, and phenols/tannins. For its part, Schmeda-Hirschmann et al. (2019) described the probable chemical composition of a polar extract (MeOH) by HPLC-DAD-ESI-MS/MS and the inhibitory effect (81–89% at 100–200 µg/mL) on the phosphodiesterase-5 enzyme (PDE-5), which would represent one of the mechanisms involved in the adequate male erectile function (*Cam. angustifolia* is an ingredient of the most commercialized beverages as traditional male sexual enhancers of the Peruvian Amazon). The chemical analysis allowed the tentative identification of 31 constituents including mainly proanthocyanidins and gallotannins, flavonoids and caffeoylquinic acid. Vásquez-Ocmín et al. (2018), Ruiz et al. (2011) and Kvist et al. (2006) determined the effectivity of the EtOH and EtOH:H_2O extracts against *P. falciparum*, *L. donovani* axenic amastigotes and *T. brucei gambiense* trypomastigotes strains with IC_{50} values < 10 µg/mL and > 20 µg/mL. A last revised document (patent) on *Cam. angustifolia* mentioned the ability of an extract (ethyl acetate/MeOH) to inhibit COX-2 (30–60%) as a mechanism to treat inflammation or cancer (Koepke et al. 2009). To date, no bioactive compound has been isolated/structurally characterized for this plant.

Another Amazon plant with interesting and verified pharmacological properties is *Cop. paupera*, and particularly its oleoresin; some authors (Ticunsi et al. 2002, Estevez et al. 2007, Dos Santos et al. 2008, Santos et al. 2008, Izumi et al. 2012, Amorim et al. 2017, Furtado et al. 2018, Rodrigues et al. 2018, Vásquez-Ocmín et al. 2018, Liviac et al. 2019, Roumy et al. 2020) have examined the antiparasitic (*P. falciparum*, *L. amazonensis*, *L. donovani*, *L. infantum*, *T. cruzi* and *T. brucei gambiense*), antimicrobial (*S. aureus*, methicillin resistant *S. aureus*, *S. epidermidis*, *S. lugdunensis*, *B. subtilis*, *E. faecalis*, *S. agalactiae*, *S. dysgalactiae* and *C. striatum*), antioxidant (FRAP and TBARS), cytotoxic/genotoxic/toxic (P-388, A-549, HT-29, MEL-28, V79, $LLCMK_2$ cell lines and erythrocytes) and wound healing potential of pure oleoresin, certain oleoresin fraction (kaurene-rich), oleoresin/leaves/bark extracts, and isolated molecules (**135**, **137–142**). Thus, oleoresins (OR_1/OR_2), kaurene (**139**)-rich fraction (KRF) and isolated α-copaene (**137**) were able to inhibit the parasitic infection of the *L. amazonensis* and *L. infantum* promastigotes forms according to Rodrigues et al. (2018); these authors reported the IC_{50} values (µg/mL) which were OR_1/OR_2: 62 ± 8–104 ± 6, KRF: 168 ± 7 and **137**: 17 ± 3 against *L. amazonensis*; OR_1/OR_2: 66 ± 9–203 ± 18, KRF: 176 ± 4 and **137**: 11 ± 1 on *L. infantum*. Surprisingly, the nanoemulsions prepared by the same authors with the OR_1/OR_2/KRF/**137** increased the antiparasitic effectiveness between two-fold and seven-fold; viz., the IC_{50} values (µg/mL) were OR_1/OR_2: 17 ± 2–35 ± 10, KRF: 29 ± 4 and **137**: 2 ± 1 for *L. amazonensis*; OR_1/OR_2: 39.8 ± 0.3–47 ± 9, KRF: 55 ± 6 and **137**: 2.2 ± 0.8 for *L. infantum*. The oleoresins were constituted by OR_1: kaurene (33%), α-copaene (22%) and caryophyllene oxide (12%), and OR_2: α-copaene (39%) and β-caryophyllene (21%), whilst the fraction contained 65% of kaurene. Other oleoresin (Acre, Brazil), with promising antileishmaniasis effect (IC_{50} 11.0 ± 0.4 µg/mL), presented as main constituents to β-bisabolene (20%), α-zingiberene (19%) and kaurenoic acid (13%) (Santos et al. 2008). Besides, the OR with wound healing property was composed by α-copaene (23%), hardwickiic acid (8%) and δ-cadinene (7%) (Amorim et al. 2017). Otherwise, Dos Santos et al. (2008) and Roumy et al. (2020) also determined the antimicrobial efficacy on nine strains (*S. aureus*, methicillin resistant *S. aureus*, *S. epidermidis*, *S. lugdunensis*, *S. agalactiae*, *S. dysgalactiae*, *B. subtilis*, *E. faecalis* and *C. striatum*) by oleoresins, which had MIC values (in mg/mL) as follows: *S. epidermidis*: 0.6 ± 0.2–1.2 ± 0.4, *S. lugdunensis*: 0.6 ± 0.2, *S. agalactiae*: 0.3 ± 0.1–0.6 ± 0.2, *S. dysgalactiae*: 0.15 ± 0.05 and *C. striatum*: 0.3 ± 0.1 mg/mL, while MIC/MBC values (in µg/mL) were: *S. aureus*: 250/1000, methicillin resistant *S. aureus*: 250/500, *S. epidermidis*: 1000/1000, *B. subtilis*: 62/62 and *E. faecalis*: 62/500. While Ribeiro et al. (2019) measured the antimicrobial capacity of the volatile oil (43% yield) from oleoresin

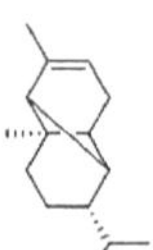

137

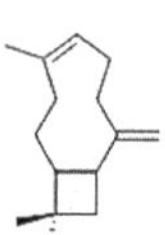

138

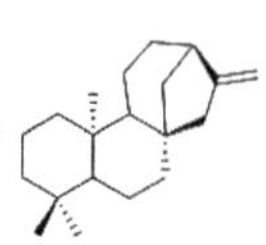

139

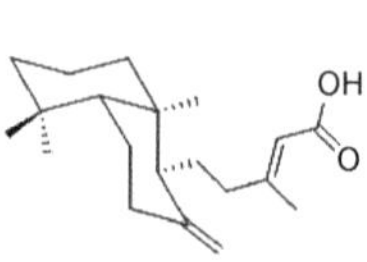

140

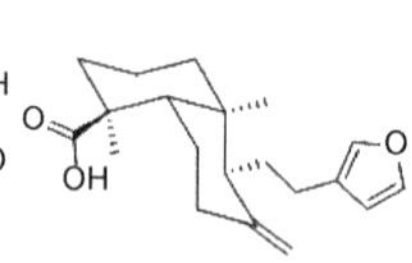

141

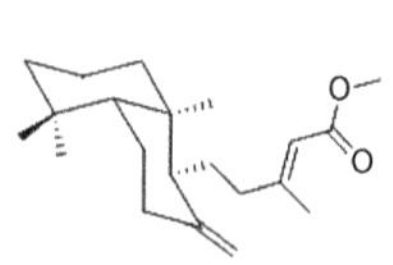

142

against six strains (*S. aureus*, *P. aeruginosa*, *Salmonella choleraesuis*, *C. albicans*, *C. tropicalis* and *C. krusei*), the most significant MIC values (μg/mL) were 125 and 500 on *S. choleraesuis* and *S. aureus*, in that order. The volatile oil was mainly constituted by α-copaene (35%), δ-cadinene (14%) and α-cubebene (10%). Last but not the least, an OR and hydroalcohol extract from leaves inhibited (*in vitro*) V79 cells with IC_{50} values individually of 41 ± 7 μg/mL and 412 ± 66 μg/mL, but they were not cytotoxic and genotoxic (*in vivo*) (Furtado et al. 2018).

Alternatively, some isolated sesquiterpenes (**137**)/kaurene type diterpenoids (**135**, **139–142**) inhibited the growth of microorganisms, replication of cancer cells and/or parasitic infection (*T. cruzi*) (Ticunsi et al. 2002, Isumi et al. 2012, Rodrigues et al. 2018). For example, IC_{50} values (96 h - μg/mL) of **135**, **140–142** on *T. cruzi* amastigote form were 16 ± 3, 1.30 ± 0.06, 28 ± 6 and 2.50 ± 0.06, respectively, while CC_{50}/HC_{50} values (96 h/3 h - μg/mL) against $LLCMK_2$ and erythrocyte cells were 76 ± 5/464 ± 56, 39 ± 5/65.7 ± 0.2, 60 ± 9/> 1582 and 69 ± 4/1597 ± 9, one-to-one. Other IC_{50} values for **142** as a measure of the anticancer potential were on P-388 (2.5 μg/mL), A-549 (5 μg/mL), HT-29 (5 μg/mL) and MEL-28 (10 μg/mL) lines. Meanwhile, the MIC values (μg/mL) calculated for **135**, **140** and **141** against *B. subtilis*, *S. aureus* and *S. epidermidis* were in turn: **135**: 5–2, 8–6 and 6–4; **140**: 6–3, 10–8 and 5–4; and **141**: 30–20, 50–40 and 40.

To sum up, the H_2O/MeOH/EtOH/hydroalcohol extracts or fractions from sauco (*Sam. peruviana*) were active as antimicrobial, antioxidant and toxic/insecticidal agents as described by Hernández et al. (2000), Neto et al. (2002), Schmeda-Hirschmann et al. (2005), Bussman et al. (2011), Chirinos et al. (2013) and Román-Farje et al. (2017). In this fashion, the EtOH extract or flavonoid-rich fraction (from H_2O:MeOH extract) were capable of inhibiting the growth of bacterial strains; that is, the EtOH extract (100 mg/mL) produced diameters of the inhibition zone > 7 mm on *Streptococcus bovis*, *S. pneumoniae*, *S. agalactiae*, *Clostridium histolyticum*, *C. diphtheria*, *Bacteroides fragilis* and *B. megaterium*, and the flavonoid-rich fraction (at 200 μg/mL) inhibited the growth of *S. epidermidis*, *B. subtilis* and *S. aureus* with equivalent concentrations of chloramphenicol (CAEC) of 38 μg/mL, 53 μg/mL and 61 μg/mL. Besides, the total extracts (MeOH or H_2O:MeOH:HCl) of fruits exhibited promising radical-scavenging/antioxidant capacities by $DPPH^{\cdot}$, $ABTS^{+\cdot}$ and ORAC methods: the total antioxidant capacity values (AOC), expressed as μmol TE/g, were between 155 ± 6 and 361 ± 23, and the percentage inhibition of 49 ± 4% on radical $DPPH^{\cdot}$ by MeOH extract, at 100 μg/mL. The preliminary phytochemical analysis of fruits from *Sambucus* spp. indicated the presence of anthocyanins such as cyanidin 3-*O*-sambubioside as principal constituent (Anton et al. 2013), but other secondary metabolites were also presumably identified in its leaves (e.g., flavonoids, lactones, triterpenes/steroids, saponins) (Ruiz-Reyes et al. 2013). Lastly, the extracts (H_2O and EtOH) from *Sam. peruviana* (leaf/bark/root) evidenced intermediate toxicity (2–28% of mortality) on *Sitophilus zeamais* and *Copidosoma koehleri*, whilst those extracts on crustaceans *Daphnia magna* and *A. salina* were moderate to highly lethal (LC_{50} 0.4–23 g/mL, 26–168 μg/mL).

Conclusion

The plants (tree/shrub/herb) selected/analyzed have been important as sources of medicine or food, and some of them with special purposes to cure "supernatural" diseases for the communities Tacana, Quilombolas, Saraguro, Ingano, Coreguaje, Witotos, Huasteca (teenek), Bora-Miraña, Tikunas and Cocamas, inter alia, from Bolivia, Brazil, Colombia, Ecuador, Mexico, and Peru. Among the relevant ethnobotanical uses, related to the therapeutic benefits of the 18 plants, could be mentioned: as treatment for disorders of the central nervous, circulatory, digestive, respiratory, reproductive and nervous systems, as well as treatment for diseases caused by microorganisms (bacteria/fungi/virus)/parasites (leishmania/Chagas/malaria)/tumors. Some benefits according to the traditional uses have been validated (others determined) through the establishment of the pharmacological effects (*in vitro*/*in vivo*), for instance, anti-leishmania/Chagas/malaria, anti-HIV/HSV/BHV, anti-tumoral/cancer/cytotoxic, anti-bacterial/fungal, anxiolytic/angiogenic/uterotonic/aphrodisiac and wound healing, from the EO/oleoresins and extracts/fractions (SFE-CO_2/hexane/ethyl acetate/$CHCl_3$/CH_2Cl_2/acetone/EtOH/MeOH/H_2O) obtained from the plant's useful parts (e.g., flowers/leaves/stems/barks/wood/heartwood/fruits/seeds/roots/rhizomes/latex/oleoresin). As a final point, specific/particular constituents (e.g., tingenones, amburosides, pauferrols, chapecoderins, echinophyllins, echinodolides, mammeas, calanolides, brasixanthones, brasimarins, furocoumarins, quinolinones, zoapatanolides, sesquiterpenes, diterpenes, among other), responsible for the pharmacological actions, were isolated and structurally characterized.

References

Acosta, K., Jiménez, A., Vinueza, D., Pilco, G., Abdo, S. 2017. *In vitro* evaluation of the antibacterial and anti-dermatological activities of the alkaloidal latex extract of *Brosimum utile* (Kunth.) Pittier. Perfiles, 18: 82–89.

Agarwal, O.P., Saxena, U., Jain, S.C., Jain, R. 2010. A new biscoumarin and other constituents from *Guazuma tomentosa* roots. Chem. Nat. Comp., 46: 713–715.

Al Muqarrabun, L.M.R., Ahmat, N. 2015. Medicinal uses, phytochemistry and pharmacology of family Sterculiaceae: a review. Eur. J. Med. Chem., 92: 514–530.

Albernaz, L.C., de Paula, J.E., Romero, G.A.S., Silva, M.R.R., Grellier, P., Mambu, L., Espindola, L.S. 2010. Investigation of plant extracts in traditional medicine of the Brazilian Cerrado against protozoans and yeasts. J. Ethnopharmacol., 131: 116–121.

Almeida, J.R.G.S., Moraes, A.C.A., Ribeiro, R.L., Góis, R.M.O., Quintansjúnior, L.J. 2005. Plantas medicinais comercializadas por raizeiros no Vale do São Francisco. In: Reunião regional da sociedade brasileira de plantas medicinais, Fortaleza. Resumo publicado em anais do evento. Fortaleza: Gráfica Universitária - UFC.

Alonso, J.R. 2016. Tratado de fitoterápicos e nutracêuticos. AC, São Paulo, Brazil. 1124 p.

Alonso-Castro, A.J., Salazar-Olivo, L.A. 2008. The anti-diabetic properties of *Guazuma ulmifolia* Lam. are mediated by the stimulation of glucose uptake in normal and diabetic adipocytes without inducing adipogenesis. J. Ethnopharmacol., 118: 252–256.

Amorim, J.L., Figueiredo, J.B., Amaral, A.C.F., Barros, E.G.O., Palmero, C., Palantinos, M.A.M., Ramos, A.S., Ferreira, J.L.P., Silva, J.R.A., Benjamim, C.F., Basso, S.L., Nasciutti, L.E., Fernandes, P.D. 2017. Wound healing properties of *Copaifera paupera* in diabetic mice. PloS One, 12: e0187380 (14 p.).

Anton, A.M., Pintea, A.M., Rugină, D.O., Sconţa, Z.M., Hanganu, D., Vlase, L., Benedec, D. 2013. Preliminary studies on the chemical characterization and antioxidant capacity of polyphenols from *Sambucus* sp. Dig. J. Nanomater. Bios., 8: 973–980.

Arévalo-Lopéz, D., Nina, N., Ticona, J.C., Limachi, I., Salamanca, E., Udaeta, E., Paredes, C., Espinoza, B., Serato, A., Garnica, D., Limachi, A., Coaquira, D., Salazar, S., Flores, N., Sterner O., Giménez, A. 2018. Leishmanicidal and citotoxic activity from plants used in Tacana traditional medicine (Bolivia). J. Ethnopharmacol., 216: 120–133.

Argueta, V.A., Cano, L.M. 1994. Atlas de las plantas de la medicina tradicional mexicana. Vol. II. Instituto Nacional Indigenista. Gobierno de México, pp. 1413–1414.

Armijos, C., Lozano, M., Bracco, F., Vidari, G. and Malagón, O. 2012. Plantas sagradas y psicoactivas usadas por los Saraguros en la región sur de Ecuador. 1ª Edición. Editorial de Universidad Técnica Particular de Loja, Ecuador.

Armijos, C., Cota, J., González, S. 2014. Traditional medicine applied by the Saraguro yachakkuna: a preliminary approach to the use of sacred and psychoactive plant species in the southern region of Ecuador. J. Ethnobiol. Ethnomed., 10: 26 (1–12).

Armijos, C., Gilardoni, G., Amay, L., Lozano. A., Bracco, F., Ramírez, J., Bec, N., Larroque, C., Vita Finzi, P., Vidari, G. 2016a. Phytochemical and ethnomedicinal study of *Huperzia* species used in the traditional medicine of Saraguros in Southern Ecuador; AChE and MAO inhibitory activity. J. Ethnopharmacol., 193: 546–554.

Armijos, C., Ponce, J., Ramírez, J., Gozzini, D., Vita Finzi, P. and Vidari, P. 2016b. An unprecedent high content of the bioactive flavone tricin in *Huperzia* medicinal species used by the Saraguro in Ecuador. Nat. Prod. Commun., 11: 273–274.

Arriaga, A.M.C., Machado, M.I.L., Craveiro, A.A., Pouliquen, Y.B.M., Mesquita, A.G. 1997. Volatile constituents from leaves of *Guazuma ulmifolia* Lam. J. Essent. Oil Res., 9: 705–706.

Ávila-Rosas, R.N. 2013. Determinación del mecanismo de acción relajante del aceite esencial de toronjil morado (*Agastache mexicana*) en la musculatura lisa bronquial de cobayo. Tesis de licenciatura. FES-Zaragoza. UNAM. México, D.F. 55 p.

Awaad, A.S., El-Sayed, N.H., Maitland, D.J., Mabry, T.J. 2006. Phenolic antioxidants from *Casimiroa edulis* leaves. Pharm. Biol., 44: 258–262.

Awaad, A.S., Maitland, D.J., Moneir, S.M. 2007. New alkaloids from *Casimiroa edulis* fruits and their pharmacological activity. Chem. Nat. Comp., 43: 576–580.

Awaad, A.S., Al-Jaber, N.A., Soliman, G.A., Al-Outhman, M.R., Zain, M.E., Moses, J.E., El-Meligy, R.E. 2012. New biological activities of *Casimiroa edulis* leaf extract and isolated compounds. Phytother. Res., 26: 452–457.

Bejar, E., Enriquez, R., Lozoya, X. 1984. The *in vitro* effect of grandiflorenic acid and zoapatle aqueous crude extract upon spontaneous contractility of the rat uterus during oestrus cycle. J. Ethnopharmacol., 11: 87–97.

Bejar, E., Reyes-Chilpa, R., Jiménez-Estrada, M. 2000. Chapter—Bioactive compounds from selected plants used in the XVI century Mexican traditional medicine. *In*: Attar-ur-Rahman (ed.). Studies of Natural Products Chemistry. Elsevier, London, 24: 799–844.

Berenguer, B., Trabadela, C., Sánchez-Fidalgo, S., Quílez, A., Miño, P., De La Puerta, R., Martín-Calero, M.J. 2007. The aerial parts of *Guazuma ulmifolia* Lam. protect against NSAID-induced gastric lesions. J. Ethnopharmacol., 114: 153–160.

Bernal, H.Y., Correa, J.M. 1989. Especies vegetales promisorias de los países del Convenio Andrés Bello. Tomo III. Programa de Recursos Vegetales del Convenio Andrés Bello. Editora Secretaria Ejecutiva del Convenio Andrés Bello, Bogotá.

Bertin, R., García-Argaéz, A., Martínez-Vázquez, M., Froldi, G. 2011. Age-dependent vasorelaxation of *Casimiroa edulis* and *Casimiroa pubescens* extracts in rat caudal artery *in vitro*. J. Ethnopharmacol., 137: 934–936.

Bertin, R., Chen, Z., Martínez-Vázquez, M., García-Argaéz, A., Froldi, G. 2014. Vasodilation and radical-scavenging activity of imperatorin and selected coumarinic and flavonoid compounds from genus *Casimiroa*. Phytomedicine, 21: 586–594.

Biasi-Garbin, R.P., Demitto, F.O., Amaral, R.C.R., Ferreira, M.R.A., Soares, L.A.L., Svidzinski, T.I.E., Baeza, L.C., Yamada-Ogatta, S.F. 2016. Antifungal potential of plant species from Brazilian caatinga against dermatophytes. Rev. Inst. Med. Trop., 58: 1–5.

Bishayee, K., Chakraborty, D., Gosh, S., Boujedani, N., Khuda-Bukhsh, A.R. 2013. *Lycopodine triggers* apoptosis by modulating 5-lipooxigenase, and depolarizing mitochondrial membrane potential in androgen sensitive and refractory prostate cancer cells without modulating p53 activity: signaling cascade and drug-DNA interaction. Eur. J. Pharmacol., 698: 110–121.

Blanco-Ayala, T., Lugo-Huitrón, R., Serrano-López, E.M., Reyes-Chilpa, R., Rangel-López, E., Pineda, B., Medina-Campos, O.N., Sánchez-Chapul, L., Pinzón, E., Cristina, T.S., Silva-Adaya, D., Pedraza-Chaverrí, J., Ríos, C., de la Cruz, V.P., Torres-Ramos, M. 2013. Antioxidant properties of xanthones from *Calophyllum brasiliense*: prevention of oxidative damage induced by $FeSO_4$. BMC Complem. Altern. M, 13: 262.

Boligon, A.A., Feltrin, A.C., Gindri, A.L., Athayde, M.L. 2013. Essential oil composition, antioxidant and antimicrobial activities of *Guazuma ulmifolia* from Brazil. Med. Aromat. Plants, 2: 126.

Bourbonnais-Spear, N., Awad, R., Merali, Z., Maquin, P., Cal, V., Arnason, J.T. 2007. Ethnopharmacological investigations of plants used to treat susto, a folk illness. J. Ethnopharmacol., 109: 380–387.

Bourdy, G., DeWalt, S.J., Chávez de Michel, L.R., Roca, A., Deharo, E., Muñoz, V., Balderrama, L., Quenevo, C., Giménez, A. 2000. Medicinal plants uses of the Tacana, an Amazonian Bolivian ethnic group. J. Ethnopharmacol., 70: 87–109.

Brack Egg, A. 1999. Diccionario enciclopédico de plantas utiles del Perú. Cusco. CBC, 222p.

Braga, R. 2015. Plantas do Nordeste, especialmente do Ceará. 3a ed. Mossoró: Série C, v. 1656.

Braga, F.C., Serra, C.P., Viana-Júnior, N.S., Oliveira, A.B., Côrtes, S.F., Lombardi, J.A. 2007. Angiotensin-converting enzyme inhibition by Brazilian plants. Fitoterapia, 78: 353–358.

Bravo, J.A.B., Sauvain, M., Gimenez, A., Muñoz, V., Callapa, J., Le Men-Olivier, L., Massiot, G., Lavaud, C. 1999. Bioactive phenolic glycosides from *Amburana cearensis*. Phytochemistry, 50: 71–74.

Brenzan, M.A., Nakamura, C.V., Dias Filho, B.P., Ueda-Nakamura, T., Young, M.C.M., Cortez, D.A.G. 2007. Antileishmanial activity of crude extract and coumarin from *Calophyllum brasiliense* leaves against *Leishmania amazonensis*. Parasitol. Res., 101: 715–722.

Brenzan, M.A., Ferreira, I.C.P., Lonardoni, M.V.C., Honda, P.A., Rodrigues Filho, E., Nakamura, C.V., Dias Filho, B.P., Ueda-Nakamura, T., Cortez, D.A.G. 2008a. Activity of extracts and coumarins from the leaves of *Calophyllum brasiliense* on *Leishmania braziliensis*. Pharm. Biol., 46: 380–386.

Brenzan, M.A., Nakamura, C.V., Dias-Filho, B.P., Ueda-Nakamura, T., Young, M.C.M., Côrrea, A.G., Alvim-Júnior, J., Santos, A.O., Cortez, D.A.G. 2008b. Structure-activity relationship of (-) mammea A/BB derivatives against *Leishmania amazonensis*. Biomed. Pharmacother., 62: 651–658.

Bussmann, R.W., Malca, G., Glenn, A., Sharon, D., Nilsen, B., Parris, B., Dubose, D., Ruiz, D., Saleda, J., Martinez, M., Carillo, L., Walker, K., Kuhlman, A., Townesmith, A. 2011. Toxicity of medicinal plants used in traditional medicine in Northern Peru. J. Ethnopharmacol., 137: 121–140.

Bussman, R.W., Sharon, D. 2015. Medicinal plants of the Andes and the Amazon. The magic and medicinal flora of Northern Peru, 292 p.

Caballero-George, C., Vanderheyden, P.M.L., Bruyne, T.D., Shahat, A.A., Heuvel, H.V., Solis, P.N., Vlietinck, A.J. 2002. *In vitro* inhibition of [3H]-angiotensin II binding on the human AT1 receptor by proanthocyanidins from *Guazuma ulmifolia* bark. Planta Med., 68: 1066–1071.

Cabanillas, B.J., Le lamer, A.C., Olagnier, D., Castillo, D., Arevalo, J., Valadeau, C., Coste, A., Pipy, B., Bourdy, G., Sauvain, M., Fabre, N. 2014. Leishmanicidal compounds and potent PPARγ activators from *Renealmia thyrsoidea* (Ruiz & Pav.) Poepp. & Endl. J. Ethnopharmacol., 157: 149–155.

Cajaiba, R.L., Gomes, A.F., Santos, M.C., Medeiros, R.R., Silva, W.B. 2016. Perfil dos comerciantes de plantas medicinais no município de uruará, Pará, Brasil. Enciclopédia Biosfera, Centro Científico Conhecer, Goiânia, GO, Brazil, 13: 1473–1477.

Calixto Júnior, J.T., Morais, S.M., Gomez, C.V., Molas, C.C., Rolon, M., Boligon, A.A., Athayde, M.L., Oliveira, C.D.M., Tintino, S.R., Coutinho, H.D.M. 2016. Phenolic composition and antiparasitic activity of plants from the Brazilian Northeast "Cerrado". Saudi J. Biol. Sci., 23: 434–440.

Campos-Bedolla, P., Campos, M.G., Valencia-Sánchez, A., Ponce-Monter, H., Uribe, C., Osuna, L., Calderón, J. 1997. Effect of kauranes from *Montanoa* spp. on rat uterus. Phytother. Res., 11: 11–16.

Canalli, Y.M., Bove, C.P. 2017. Flora do Rio de Janeiro: Alismataceae. Rodriguesia, 68: 17–28.

Canuto, K.M., Silveira, E.R. 2006. Constituintes químicos da casca do caule de *Amburana cearensis* A.C. Smith. Quim. Nova, 29: 1241–1243.

Canuto, K.M., Silveira, E.R., Bezerra, A.M.E. 2010. Estudo fitoquímico de espécimens cultivados de Cumaru (*Amburana cearensis* A.C. Smith). Quim. Nova, 33: 662–666.

Carro-Juárez, M., Cervantes, E., Cervantes-Méndez, M., Rodríguez-Manzo, G. 2004. Aphrodisiac properties of *Montanoa tomentosa* aqueous crude extract in male rats. Pharmacol. Biochem. Behav., 78: 129–134.

Carro-Juárez, M., Lobatón, I., Benitez, O., Espíritu, A. 2006. Pro-ejaculatory effect of the aqueous crude extract of cihuapatli (*Montanoa tomentosa*) in spinal male rats. J. Ethnopharmacol., 106: 111–116.

Carvalho, A.F.U. 2009. Atividades biológicas e enzimáticas do extrato aquoso de sementes de *Caesalpinia ferrea* Mart., Leguminosae. Rev. Bras. Farmacogn., 19: 586–591.

Carvalho, H.O., e Souza, B.S.F., dos Santos, I.V.F., Resque, R.L., Keita, H., Fernandes, C.P., Carvalho, J.C.T. 2016. Hypoglycemic effect of formulation containing hydroethanolic extract of *Calophyllum brasiliense* in diabetic rats induced by streptozotocin. Rev. Bras. Farmacogn., 26: 634–639.

Carvalho, J.C.T., Teixeira, J.R.M., Souza, P.J.C., Bastos, J.K., dos Santos, D., Sarti, S.J. 1996. Preliminary studies of analgesic and anti-inflammatory properties of *Caesalpinia ferrea* crude extract. J. Ethnopharmacol., 53: 175–178.

Castaño-Arboleda, N., Cárdenas-López, D., Otavo-Rodríguez, E. 2007. Ecología, aprovechamiento y manejo sostenible de nueve especies de plantas del departamento del Amazonas, generadoras de productos maderables y no maderables. Instituto Amazónico de Investigaciones Científicas -Sinchi-. Corporación para el Desarrollo Sostenible del Sur de la Amazonia, CORPOAMAZONIA, Bogotá, 266 p.

Cavalheiro, M.G., Farias, D.F., Fernandes, G.S., Nunes, E.P., Cavalcanti, F.S., Vasconcelos, I.M., Melo, V.M.M., Carvalho, A.F.U. 2009. Atividades biológicas e enzimáticas do extrato aquoso de sementes de *Caesalpinia ferrea* Mart., Leguminosae. Rev. Bras. Farmacogn., 19: 586–591.

Cavalheiro, M.G., Farias, D.F., Fernandes, G.S., Nunes, E.P., Cavalcanti, F.S., Vasconcelos, I.M., Melo, V.M.M., Coelho, M.G.P., Velozo, L.S.M., Fernandes, D.C., Alves, R.A., Siqueira, H.A.A., Silva, G.P., Santos, S.V.M., Gayer, C.R.M. 2012. Atividade antinociceptiva do óleo essencial de *Echinodorus macrophyllus* (Kunth.) Micheli (Alismataceae). Rev. Fitos., 7: 245–251.

Chindoy, L.Y. 2013. Evaluación de la actividad biológica de una fracción acuosa obtenida del látex de *Brosimum utile*, tradicionalmente utilizado en el tratamiento del cancer por la comunidad indígena Cämentsa. Trabajo de grado Bacteriología. Pontifica Universidad Javeriana, Bogotá (Colombia).

Chirinos, R., Pedreschi, R., Rogez, H., Larondelle, Y., Campos. D. 2013. Phenolic compound contents and antioxidant activity in plant with nutritional and/or medicinal properties from the Peruvian Andean region. Ind. Crop. Prod., 47: 145–152.

Chízmar, F.C. 2009. Plantas comestibles de Centroamérica. Instituto Nacional de Biodiversidad INBio. Costa Rica 360 p.

Coelho, M.G.P., Velozo, L.S.M., Fernandes, D.C., Alves, R.A., Siqueira, H.A.A., Silva, G.P., Santos, S.V.M., Gayer, C.R.M. 2012. Atividade antinociceptiva do óleo essencial de *Echinodorus macrophyllus* (Kunth.) Micheli (Alismataceae). Rev. Fitos., 7: 245–251.

Costa-Lotufo, L.V., Jimenez, P.C., Wilke, D.V., Leal, L.K.A.M., Cunha, G.M.A., Silveira, E.T., Canuto, K.M., Viana, G.S.B., Moraes, M.E.A., Moraes, M.O., Pessoa, C. 2003. Antiproliferative effects of several compounds isolated from *Amburana cearensis* A.C. Smith. Z. Naturforsch. C J. Biosci., 58: 675–680.

Cotter, M.L. 1981. ^{13}C NMR spectral studies of zoapatanol and montanol, novel diterpenes from *Montanoa tomentosa* (zoapatle), and their chemical derivatives. Org. Magn. Resonance, 17: 14–17.

Cottiglia, F., Dhanapal, B., Sticher, O., Heilmann, J. 2004. New chromanone acids with antibacterial activity from *Calophyllum brasiliense*. J. Nat. Prod., 67: 537–541.

Cruz-Torres, K.C. 2019. Caracterización farmacológica del extracto hidroalcohólico de *Agastache mexicana* en modelos murinos de diabetes, obesidad e hipertensión. Tesis maestría. UAEM. Morelos, México, 83 p.

Cunha, M.C.L., Ferreira, R.A. 2003. Aspectos morfológicos da semente e do desenvolvimento da planta jovem de *Amburana cearensis* (Arr. Cam.) A.C. Smith - Cumaru - Leguminosae Papilionoideae. Rev. Bras. Sementes, 2: 89–96.

Da Conceição, G.M., Ruggieri, A.C., Araújo, M.F.V., Conceição, T.T.M.M., Conceição, M.A.M.M. 2011. Plantas do cerrado: comercialização, uso e indicação terapêutica fornecida pelos raizeiros e vendedores, Teresina, Piauí. Scientia Plena, 7: 1–6.

Da Silva, K.L., dos Santos, A.R., Mattos, P.E., Yunes, R.A., Delle-Monache, F., Cechinel-Filho, V. 2001. Chemical composition and analgesic activity of *Calophyllum brasiliense* leaves. Therapie, 56: 431–434.

Da Silva, G.P., Fernandes, D.C., Vigliano, M.V., Fonseca, E.N., Santos, S.V.M., Marques, P.R., Justo, M.G., Sabino, K.C.C., Coelho, M.C.C. 2016b. Flavonoid-enriched fraction from *Echinodorus macrophyllus* aqueous extract exhibits high *in-vitro* and *in-vivo* anti-inflammatory activity. J. Pharm. Pharmacol., 68: 1584–1596.

Da Silva, F.M.A., Paz, W.H.P., Vasconcelos, L.S.F., da Silva, A.L.B., da Silva-Filho, F.A., de Almeida, R.A., de Souza, A.D.L., Pinheiro, M.L.B., Koolen, H.H.F. 2016a. Chemical constituents from *Salacia impressifolia* (Miers) A.C. Smith collected at the Amazon rainforest. Biochem. Syst. Ecol., 68: 77–80.

Da Trindade, R., da Silva, J.K., Setzer, W.N. 2018. Copaifera of the neotropics: a review of the phytochemistry and pharmacology. Int. J. Mol. Sci., 19: 1511 (33 p.).

Da'i, M., Wikantyasning, E., Wahyuni, A., Kusumawati, I., Saifudin, A., Suhendi, A. 2016. Antiproliferative properties of tiliroside from *Guazuma ulmifolia* lamk on T47D and MCF7 cancer cell lines. Natl. J. Physiol. Pharm. Pharmacol., 6: 627–633.

Dantas, A.C.S., Araujo, A.C., Pacheco, A.G.M., Branco, A., Sangioni, L.A., Almeida, J.R.G.S., Horta, M.C. 2016. Acaricidal activity of *Amburana cearensis* on the cattle tick *Rhipicephalus* (Boophilus) *microplus*. Cien. Rural., 46: 536–541.

Davis, E.W., Yost, J.A. 1983 The ethnomedicine of the Waorani of Amazonian Ecuador. J. Ethnopharmacol., 9: 273–297.

De Solla Price, D.J. 1963. Little Science, Big Science. Columbia University Press, New York.

De Souza, T.L. 2015. Levantamento etnoveterionário de plantas medicinais aplicadas à caprinocultura em assentamentos rurais de mossoró-rio grande do Norte. Master Thesis, Universidade Federal Rural do Semiárido, Mossoró, RN, Brazil, 142 p.

Deepak, K.G.K., Nageswara, R., Reddy, N., Surekha, C. 2014. Role of antidiabetic compounds on glucose metabolism—a special focus on medicinal plant: *Salacia* sps. Med. Chem., 4: 373–381.

Di Stasi, L.C., Hiruma-Lima, C.A. 2002. Plantas medicinais na Amazônia e na Mata Atlântica. São Paulo: Editora UNESP, 323–330.

Dixit, R.D., Ghosh, S.R., Ghosh, R.K. 1995. *Adiantum latifolium* Lam. an introduced fern in India. Bull. Bot. Surv. India, 37: 117–119.

Dos Santos, J.A., Silveira, A.P., Gomes, V.S. 2019. Knowledge and use of the flora in a Quilombola community of Northeastern Brazil. Floresta Ambient., 26: 1–12.

Dos Santos, A.O., Ueda-Nakamura, T., Dias Filho, B.P., Veiga Junior, V.F., Pinto, A.C., Nakamura, C.V. 2008. Antimicrobial activity of Brazilian copaiba oils obtained from different species of the *Copaifera* genus. Mem. Inst. Oswaldo Cruz, 103: 277–281.

Dreyer, D.L., Bertelli, D.J. 1967. The structure of zapotin. Tetrahedron, 23: 4607–4612.

Duke, J.A. 2008. Duke's Handbook of Medicinal Plants of Latin America. CRC Press, Boca Ratón, 962 p.

Esposito, F., Zinzula, L., Maxia, A. Tramontano, E., Sanna, C. 2011. Inhibition of HIV-1 reverse transcriptase associated activities by the hydroalcoholic extract of *Casimiroa edulis* seeds. Nat. Prod. Res., 25: 1067–1073.

Esquivel-Gutiérrez, E.R., Coria-Orozco, E., Torres-Martínez, R., Hernández-García, A., Ríos-Chávez, P., Manzo-Ávalos, S., Saavedra-Molina, A., Salgado-Garciglia, R. 2017. Antioxidant effects of *Agastache mexicana* extracts: an *in vitro* approach. FASEB J., 31 (abstract ib117).

Estevez, Y., Castillo, D., Tangoa Pisango, M., Arevalo, J., Rojas, R., Alban, J., Deharo, E., Bourdy, G., Sauvain, M. 2007. Evaluation of the leishmanicidal activity of plants used by Peruvian Chayahuita ethnic group. J. Ethnopharmacol., 114: 254–259.

Estrada-Camarena, E., Sollozo-Dupont, I., Islas-Preciado, D. González-Trujano, M.E., Carro-Juárez, M., López-Rubalcava, C. 2019. Anxiolytic- and anxiogenic-like effects of *Montanoa tomentosa* (Asteraceae): dependence on the endocrine condition. J. Ethnopharmacol., 241: 112006 (11 p.).

Estrada-Muñiz, E., García-Niño, W.R., Mireles-López, L., Reyes-Chilpa, R., Vega, L. 2016. Genotoxic potential of Jacareubein; a possible antineoplasic drug (abstracts - P20). Toxicol. Lett., 259: S192.

Estrada-Reyes, R., Aguirre-Hernández, E., García-Argáez, A., Soto-Hernández, M., Linares, E., Bye, R., Heinze, G., Martínez-Vázquez, M. 2004. Comparative chemical composition of *Agastache mexicana* subsp. *mexicana* and *A. mexicana* subsp. *xolocotziana*. Biochem. Syst. Ecol., 32: 685–694.

Estrada-Reyes, R., López-Rubalcava, C., Ferreyra-Cruz, O.A., Dorantes-Barrón, A.M., Heinze, G., Moreno-Aguilar, J., Martínez-Vázquez, M. 2014. Central nervous system effects and chemical composition of two subspecies of *Agastache mexicana*; an ethnomedicine of Mexico. J. Ethnopharmacol., 153: 98–110.

Farias, E.T.N. 2013. Atividade antimicrobiana *in vitro* do extrato etanólico de *Caesalpinia ferrea* Mart. (Leguminosae). Master Thesis, Universidade Federal Rural de Pernambuco, Recife, PE, Brazil, 36 p.

Felipe, A.M.M., Rincão, V.P., Benati, F.J., Linhares, R.E.C., Galina, K.J., de Toledo, C.E.M., Nozawa, C. 2006. Antiviral effect of *Guazuma ulmifolia* and *Stryphnodendron adstringens* on poliovirus and bovine herpesvirus. Biol. Pharm. Bull., 29: 1092–1095.

Félix-Silva, J., Silva-Junior, A.A., Zucolotto, S.M. and Fernandes-Pedrosa, M. F. 2017. Medicinal plants for the treatment of local tissue damage induced by snake venoms: an overview from traditional use to pharmacological evidence. Evid. Based Complement. Alternat. Med., 2017: 1–52. ID 5748256.

Fernandes, D.C., Martins, B.P., Medeiros, D.L.F., Santos, S.V.M., Gayer, C.R.M., Velozo, L.S.M., Coelho, M.G.P. 2012. Antinociceptive and anti-inflammatory activities of the hexanic extract of *Echinodorus macrophyllus* (Kunth) Micheli in mice. Braz. J. Health Biomed. Sci., 18: 25–32.

Ferrari, F., Delle Monache, F., Suárez, A.I., Arvelo, F., Compagnone, R.S. 2005. New cytotoxic isoflavone from the root bark of *Brosimum utile*. Nat. Prod. Res., 19: 331–335.

Figueredo, F.G., Ferreira, E.O., Lucena, B.F.F., Torres, C.M.G., Lucetti, D.L., Lucetti, E.C.P., Silva, J.M.F.L., Santos, F.A.V., Medeiros, C.R., Oliveira, G.M.M., Colares, A.V., Costa, J.G.M., Coutinho, H.D.M., Menezes, I.R.A., Silva, J.C.F., Kerntopf, M.R., Figueiredo, P.R.L., Matias, E.F.F. 2013. Modulation of the antibiotic activity by extracts from *Amburana cearensis* A.C. Smith and *Anadenanthera macrocarpa* (Benth.) Brenan. BioMed. Res. Int., 2013: 640682.

Flores-Flores, A., Hernández-Abreu, O., Rios, M.Y., León-Rivera, I., Aguilar-Guadarrama, B., Castillo-España, P., Perea-Arango, I., Estrada-Soto, S. 2016. Vasorelaxant mode of action of dichloromethane-soluble extract from *Agastache mexicana* and its main bioactive compounds. Pharm. Biol., 54: 2807–2813.

Froldi, G., Bertin, R., Secchi, E., Zagotto, G., Martínez-Vázquez, M., García-Argaéz, A. 2011. Vasorelaxation by extracts of *Casimiroa* spp. in rat resistance vessels and pharmacological study of cellular mechanisms. J. Ethnopharmacol., 134: 637–643.

Funk, V., Berry, P., Hollowell, T., Kelloff, C. 2007. Checklist of the plants of the Guiana Shield (Venezuela: Amazonas, Bolivar, Delta Amacuro; Guyana, Surinam, French Guiana), Dept of Botany National Museum of Natural History. Vol. 55, 584 p.

Furtado, R.A., de Oliveira, P.F., Senedese, J.M., Ozelin, S.D., de Souza, L.D.R., Leandro, L.F., de Oliveira, W.L., da Silva, J.J.M., Oliveira, L.C., Rogez, H., Ambrósio, S.R., Veneziani, R.C.S., Bastos, J.K., Tavares, D.C. 2018. Assessment of genotoxic activity of oleoresins and leaves extracts of six *Copaifera* species for prediction of potential human risks. J. Ethnopharmacol., 221: 119–125.

Gallegos, J.A. 1983. The zoapatle I: a traditional remedy from Mexico emerges to modern times. Contraception 27: 211–225.

García-Argáez, A.N., Toscano, R.A., Martínez-Vázquez, M. 2000. 11β-Hydroxyobacunone (zapoterin). J. Chem. Crystallogr., 30: 727–730.

García-Barriga, H.G. 1992. Flora medicinal de Colombia: Botánica médica (Tercer Mundo). Bogotá: Instituto de Ciencias Naturales, Universidad Nacional.

García-Niño, W.R., Estrada-Muñiz, E., Valverde, M., Reyes-Chilpa, R., Vega, L. 2017. Cytogenetic effects of Jacareubin from *Calophyllum brasiliense* on human peripheral blood mononucleated cells *in vitro* and on mouse polychromatic erythrocytes *in vivo*. Toxicol. Appl. Pharmacol., 335: 6–15.

García-Zebadúa, J.C., Margos-Guerrero, G.A., Mumbrú-Massip, M., Estrada-Muñiz, E., Contreras-Barrios, M.A., Huerta-Reyes, M., Campos-Lara, M.G., Jiménez-Estrada, M., Reyes-Chilpa, R. 2011. Inhibition of HIV-1 reverse transcriptase, toxicological and chemical profile of *Calophyllum brasiliense* extracts from Chiapas, Mexico. Fitoterapia, 82: 1027–1034.

García-Zebadúa, J.C., Reyes-Chilpa, R., Huerta-Reyes, M., Castillo-Arellano, J.I., Santillán-Hernández, S., Vázquez-Astudillo, B., Mendoza-Espinoza, J.A. 2014. The tropical tree *Calophyllum brasiliense*: a botanical, chemical and pharmacological review. Vitae, 21: 126–145.

Garzón-De La Mora, P., García-López, P.M., García-Estrada, J., Navarro-Ruíz, A., Villanueva-Michel, T., Villarreal-De Puga, L.M., Casillass-Ochoa, J. 1999. *Casimiroa edulis* seed extracts show anticonvulsive properties in rats. J. Ethnopharmacol., 68: 275–82.

Gilardoni, G., Armijos, C., Malagón, O., Lozano, M.A., Fronza, G., Vidari, G. 2014 Isolation of novel hydroquinolinic alkaloids from Ecuadorian plants of the genus *Huperzia*, employed in the traditional medicine of Saraguro people (OR7). Abstracts book of XXIII Italo-Latinoamerican Congress of Ethnomedicine. Pharmacol. online 2 (Supplementary): 49.

Gouveia, B.B., Barros, V.R.P., Gonçalves, R.J.S., Barberino, R.S., Menezes, V.G., Lins, T.L.B., Macedo, T.J.S., Santos, J.M.S., Rolim, L.A., Rolim-Neto, P.J., Almeida, J.R.G.S., Matos, M.H.T. 2015. Effect of ovarian tissue transportation in *Amburana cearensis* extract on the morphology and apoptosis of goat preantral follicles. Anim. Reprod., 12: 316–323.

Gómez-Betancur, I., Benjumea, D. 2014. Traditional use of the genus *Renealmia* and *Renealmia alpinia* (Rottb.) Maas (Zingiberaceae) a review in the treatment of snakebites. Asian Pac. J. Trop. Med., 7: S574–S582.

Gómez-Verjan, J., González-Sánchez, I., Estrella-Parra, E., Reyes-Chilpa, R. 2015. Trends in the chemical and pharmacological research on the tropical trees *Calophyllum brasiliense* and *Calophyllum inophyllum*, a global context. Scientometrics, 105: 1019–1030.

Gonçalves, R.M., Lemos, C.O., Leal, I.C., Nakamura, C.V., Cortez, D.A., da Silva, E.A., Cabral, V.F., Cardozo-Filho, L. 2013. Comparing conventional and supercritical extraction of (-)-mammea A/BB and the antioxidant activity of *Calophyllum brasiliense* extracts. Molecules, 18: 6215–6229.

Gonzalez, F.G. 2005. Estudo farmacognóstico e farmacológico de *Caesalpinia ferrea* Martius. Doctoral Thesis, Universidade de São Paulo, São Paulo, SP, Brazil, 155 p.

González-Trujano, M.E., Ponce-Muñoz, H., Hidalgo-Figueroa, S., Navarrete-Vázquez, G., Estrada-Soto, S. 2015. Depressant effects of *Agastache mexicana* methanol extract and one of major metabolites tilianin. Asian Pac. J. Trop. Med., 8: 185–190.

Grandtner, M.M., Chevrette, J. 2014. Dictionary of Trees, Volume 2: South America: Nomenclature, Taxonomy and Ecology (Elsevier's Dictionary of Trees), Academic Press.

Grijalva, P.A. 2006. Flora útil. Etnobotánica de Nicaragua. Gobierno de Nicaragua. Managua, Nicaragua. Marena, 352 p.

Guzmán-Gutiérrez, S.L., Reyes-Chilpa, R., Bonilla-Jaime, H. 2014. Medicinal plants for the treatment of "nervios", anxiety, and depression in Mexican traditional medicine. Rev. Bras. Farmacogn., 24: 591–608.

Hahn, D.W., Ericson, E.W., Lai, M.T., Probst, A. 1981. Antifertility activity of *Montanoa tomentosa* (Zoapatle). Contraception, 23: 133–140.

Ham, Y.M., Yoon, W.J., Park, S.Y., Jung, Y.H., Kim, D., Jeon, Y.J., Wijesingje, W.A.J.P., Kang, S.M., Kim, K.N. 2012. Investigation of the components of *Lycopodium serratum* extract that inhibits proliferation and mediates apoptosis of human HL-60 leukemia cells. Food Chem. Toxicol., 50: 2629–2634.

Hernández, N.E., Tereschuk, M.L., Abdala, L.R. 2000. Antimicrobial activity of flavonoids in medicinal plants from Tafí del Valle (Tucumán, Argentina). J. Ethnopharmacol., 73: 317–322.

Hernández-Abreu, O., Castillo-España, P., León-Rivera, I., Ibarra-Barajas, M., Villalobos-Molina, R., González-Christen, J., Vergara-Galicia, J., Estrada-Soto, S. 2009. Antihypertensive and vasorelaxant effects of tilianin isolated from *Agastache mexicana* are mediated by NO/cGMP pathway and potassium channel opening. Biochem. Pharmacol., 78: 54–61.

Hernández-Abreu, O., Torres-Piedra, M., García-Jiménez, S., Ibarra-Barajas, M., Villalobos-Molina, R., Montes, S., Rembao, D., Estrada-Soto, S. 2013. Dose-dependent antihypertensive determination and toxicological studies of tilianin isolated from *Agastache mexicana*. J. Ethnopharmacol., 146: 187–191.

Hokche, O., Fundación Instituto Botánico de Venezuela. 2008. Nuevo catálogo de la flora vascular de Venezuela. Caracas, Venezuela: Fundación Instituto Botánico de Venezuela Dr. Tobías Lasser.

Honda, P.A., Ferreira, I.C.P., Cortez, D.A.G., Amado, C.A.B., Silveira, T.G.V., Brenzan, M.A., Lonardoni, M.V.C. 2010. Efficacy of components from leaves of *Calophyllum brasiliense* against *Leishmania* (*Leishmania*) *amazonensis*. Phytomedicine, 17: 333–338.

Hör, M., Heinrich, M., Rimpler, H. 1996. Proanthocyanidin polymers with antisecretory activity and proanthocyanidin oligomers from *Guazuma ulmifolia* bark. Phytochemistry, 42: 109–119.

Huerta-Reyes, M., Basualdo, M.C., Abe, F., Jiménez-Estrada, M., Soler, C., Reyes-Chilpa, R. 2004a. HIV-1 inhibitory compounds from *Calophyllum brasiliense* leaves. Biol. Pharm. Bull., 27: 1471–1475.

Huerta-Reyes, M., Basualdo, M.C., Lozada, L., Jiménez-Estrada, M., Soler, C., Reyes-Chilpa, R. 2004b. HIV-1 inhibition by extracts of Clusiaceae species from Mexico. Biol. Pharm. Bull., 27: 916–920.

Ibarra-Alvarado, C. Rojas, A., Mendoza, S., Bah, M., Gutiérrez, D.M., Hernández-Sandoval, L., Martínez, M. 2010. Vasoactive and antioxidant activities of plants used in Mexican traditional medicine for the treatment of cardiovascular diseases. Pharm. Biol., 48: 732–739.

Ibrahim, N., El-Hawary, S., Mohammed, M., Ali, S., Kandil, Z., Refaat, E. 2019. Chemical compositions and hypoglycemic activities of the protein and mucilage of *Casimiroa edulis* (Llave & Lex) seeds and fruits. J. Appl. Pharm. Sci., 9: 84–91.

INPA - Instituto Nacional de Pesquisas da Amazônia. 1972. The chemical composition of Amazonian plants. A catalogue, edited by setor de fitoquímica, INPA, Manaus, Amazonas. Acta Amaz., 2: 91–92.

Isaias, D.E.B., Niero, R., Noldin, V.F., de Campos-Buzzi, F., Yunes, R.A., Delle-Monache, F., Cechinel-Filho, V. 2004. Pharmacological and phytochemical investigations of different parts of *Calophyllum brasiliense* (Clusiaceae). Pharmazie, 59: 879–881.

Ito, C., Itoigawa, M., Mishina, Y., Cechinel-Filho, V., Mukainaka, T., Tokuda, H., Nishino, H., Furukawa, H. 2002. Chemical constituents of *Calophyllum brasiliensis*: Structure elucidation of seven new xanthones and their cancer chemopreventive activity. J. Nat. Prod., 65: 267–272.

Ito, C., Itoigawa, M., Mishina, Y., Cechinel-Filho, V., Enjo, F., Tokuda, H., Nishino, H., Furukawa, H. 2003. Chemical constituents of *Calophyllum brasiliense*. 2. Structure of three new coumarins and cancer chemopreventive activity of 4-substituted coumarins. J. Nat. Prod., 66: 368–371.

Ito, A., Shamon, L.A., Yu, B., Mata-Greenwood, E., Lee, S.K., Van Breemen, R.B., Mehta, R., Farnsworth, N.R., Fong, H.H.S., Pezzuto, J.M., Kinghorn, A.D. 1998. Antimutagenic constituents of *Casimiroa edulis* with potential cancer chemopreventive activity. J. Agric. Food Chem., 46: 3509–3516.

IUCN. 2019. International Union for the Conservation of Nature. Species for restoration. https://www.iucn.org/es.

Izumi, E., Ueda-Nakamura, T., Veiga, V.F., Pinto, A.C., Nakamura, C.V. 2012. Terpenes from *Copaifera in vitro* antiparasitic synergic activity. J. Med. Chem., 55: 2994–3001.

Jacobo-Salcedo, M.R., Alonso-Castro, A.J., Salazar-Olivo, L.S., Carranza-Alvarez, C., González-Espíndola, L.A., Domínguez, F., Maciel-Torresa, S.P., García-Lujan, C., González-Martínez, M.R., Gómez-Sánchez, M., Estrada-Castillón, E., Zapata-Bustos, R., Medellin-Milán, P., García-Carrancá, A. 2011. Antimicrobial and cytotoxic effects of Mexican medicinal plants. Nat. Prod. Commun., 6: 1925–1928.

Jiménez, A.L. 2015. Estudio fitoquímico y evaluación de la actividad antidermatófica *in vitro* del látex de *Brosimum utile* (Kunth C.S.) (leche de Sandi). Trabajo de grado Bioquímica y Farmacia. Escuela Superior Politécnica de Chimborazo, Riobamba (Ecuador).

Johnson, M., Jeevas, Mahesh, M., Janakiranam, N. 2017. Antibacterial potentials of *Adiantum* species against the UTI pathogens. J. Microbiol. Exp., 4: 00104.

Kaneria, M., Baravalia, Y., Vaghasiya, Y., Chanda, S. 2009. Determination of antibacterial and antioxidant potential of some medicinal plants from Saurashtra Region, India. Indian J. Pharm. Sci., 71: 406–412.

Kanojia, R.M., Wachter, M.P., Levine, S.D., Adams, R.E., Chen, R., Chin, E., Cotter, M.L., Hirsch, A.F., Huettemann, R., Kane, V.V., Ostrowski, P., Shaw, C.J., Mateos, J.L., Noriega, L., Guzman, A., Mijarez, A., Tovar, L., Shefter, E. 1982. Isolation and structural elucidation of zoapatanol and montanol, novel oxepane diterpenoids from the Mexican plant zoapatle (*Montanoa tomentosa*). J. Org. Chem., 47: 1310–1319.

Khaleel, A.E.M. 2002. 2-Phenyl-4-quinolinone alkaloids from *Casimiroa edulis* Llave et Lex (Rutaceae). Monatsh. Chem., 133: 183–187.

Kimura, S., Ito, C., Jyoko, N., Segawa, H., Kuroda, J., Okada, M., Adachi, S., Nakahata, T., Yuasa, T., Cechinel-Filho, V., Furukawa, H., Maekawa, T. 2005. Inhibition of leukemic cell growth by a novel anti-cancer drug (GUT-70) from *Calophyllum brasiliense* that acts by induction of apoptosis. Int. J. Cancer, 113: 158–165.

Klein-Júnior, L.Z., Zambiasi, D., Salgado, G.R., Monache, F.D., Cechinel-Filho, V., Buzzi, F.C. 2017. The validation of *Calophyllum brasiliense* ("guanandi") uses in Brazilian traditional medicine as analgesic by *in vivo* antinociceptive evaluation and its chemical analysis. Naunyn-Schmiedeberg's Arch. Pharmacol., 390: 733–739.

Kobayashi, J., Sekiguchi, M., Shigemori, H., Ohsaki, A. 2000a. Chapecoderins A-C, new labdane-derived diterpenoids from *Echinodorus macrophyllus*. J. Nat. Prod., 63: 375–377.

Kobayashi, J., Sekiguchi, M., Shimamoto, S., Shigemori, H. and Ohsaki, A. 2000b. Echinophyllins C-F, new nitrogen-containing clerodane diterpenoids from *Echinodorus macrophyllus*. J. Nat. Prod., 63: 1576–1579.

Koepke, P., Subbiah, V., Burow, M.E. 2009. *Campsiandra angustifolia* extract and methods of extracting and using such extract (Patent 20090011060A1).

Kudo, E., Taura, M., Matsuda, K., Shimamoto, M., Kariya, R., Goto, H., Hattori, S., Kimura, S., Okada, S. 2013. Inhibition of HIV-1 replication by a tricyclic coumarin GUT-70 in acutely and chronically infected cells. Bioorg. Med. Chem. Lett., 23: 606–609.

Kumar, N.S., Gurunani, S.G. 2019. *Guazuma ulmifolia* Lam: A review for future view. J. Med. Plants, 7: 205–210.

Kvist, L.P., Christensen, S.B., Rasmussen, H.B., Mejía, K., Gonzalez, A. 2006. Identification and evaluation of Peruvian plants used to treat malaria and leishmaniasis. J. Ethnopharmacol., 106: 390–402.

Lagos-Castillo, C. 2015. Plantas medicinales utilizadas en el tratamiento de enfermedades ginecológicas en Leticia y Puerto Nariño (Amazonas, Colombia). Etnobiología, 13: 53–72.

Lagunes-Merino, O., Rodríguez-Landa, J.F., Caba, M., Carro-Juárez, M., García-Orduña, F., Saavedra-Vélez, M. Puga-Olguín, A., Rovirosa-Hernández, M.J. 2019. Acute effect of an infusion of *Montanoa tomentosa* on despair-like behavior and activation of oxytocin hypothalamic cells in Wistar rats. J. Tradit. Complement. Med., 2019: 1–7.

Landgren, B.M., Aedo, A.R., Hagenfeldt, K., Diczfalusy, E. 1979. Clinical effects of orally administered extracts of *Montanoa tomentosa* in early human pregnancy. Am. J. Obstet. Gynecol., 135: 480–484.

Leal, L.K.M.A., Nechio, M., Silveira, E.R., Canuto, K.M., Fontenele, J.B., Ribeiro, R.A., Viana, G.S.B. 2003. Anti-inflammatory and smooth muscle relaxant activities of the hydroalcoholic extract and chemical constituents from *Amburana cearensis* A.C. Smith. Phytother. Res., 17: 335–340.

Leal, L.K.A.M., Nobre Júnior, H.V., Cunha, G.M.A., Moraes, M.O., Pessoa, C., Oliveira, R.A., Silveira, E.R., Canuto, K.M., Viana, G.S.B. 2005. Amburoside A, a glucoside from *Amburana cearensis*, protects mesencephalic cells against 6-hydroxydopamine-induced neurotoxicity. Neurosci. Lett., 388: 86–90.

Leal, L.K.A.M., Pierdoná, T.M., Góes, J.G.S., Fonsêca, K.S., Canutoc, K.M., Silveira, E.R., Bezerra, A.M.E., Viana, G.S.B. 2011. A comparative chemical and pharmacological study of standardized extracts and vanillic acid from wild and cultivated *Amburana cearensis* A.C. Smith. Phytomedicine, 18: 230–233.

Lemos, L.M.S., Martins, T.B., Tanajura, G.H., Gazoni, V.F., Bonaldo, J., Strada, C.L., da Silva, M.G., Dall´Oglio, E.L., de Sousa-Júnior, P.T., Martins, D.T.O. 2012. Evaluation of antiulcer activity of chromanone fraction from *Calophyllum brasiliense* Camb. J. Ethnopharmacol., 141: 432–439.

Levine, S.D., Adams, R.E., Chen, R., Cotter, M.L., Hirsch, A.F., Kane, V.V., Kanojia, R.M., Shaw, C., Wachter, M.P., Chin, E., Huettemann, R., Ostrowski, P., Mateos, J.L., Noriega, L., Guzmin, A., Mijarez, A., Tovar, L. 1979. Zoapatanol and montanol, novel oxepane diterpenoids, from the Mexican plant zoapatle (*Montanoa tomentosa*). J. Am. Chem. Soc., 101: 3404–3405.

Lima, S.M.A., Araújo, L.C.C., Sitônio, M.M., Freitas, A.C.C., Moura, S.L., Correia, M.T.S., Malta, D.J.N., Gonçalves-Silva, T. 2012. Anti-inflammatory and analgesic potential of *Caesalpinia ferrea*. Rev. Bras. Farmacogn., 22: 169–175.

Lima, L.R., Cavalcante, R.R.L., Martins, M.C.C., Parente, D.M., Cavalcante, A.A.M.C. 2013. Avaliação da atividade antiedematogênica, antimicrobiana e mutagênica das sementes de *Amburana cearensis* (A. C. Smith) (Imburana-de-cheiro). Rev. Bras. Pl. Med., 15: 415–422.

Lima, M.R.F., Luna, J.S., dos Santos, A.F., Andrade, M.C.C., Sant'Ana, A.E.G., Genet, J.-P., Marquez, B., Neuville, L., Moreau, N. 2006. Anti-bacterial activity of some Brazilian medicinal plants. J. Ethnopharmacol., 105: 137–147.

Lint, H., Epling, C. 1945. A revision of *Agastache*. Am. Midl. Nat., 33: 207–208.

Liviac, D., Raunelli, P., Alvis, R., Puente, S., Best, I., Reategui, O. 2019. Phytochemical analysis, *in vitro* antioxidant capacity and toxicity assessment of *Copaifera paupera* oleoresin. Pharmacog. J., 11: 374–378.

Lopes, N., Faccin-Galhardi, L.C., Espada, S.F., Pacheco, A.C., Ricardo, N.M., Linhares, R.E., Nozawa, C. 2013. Sulfated polysaccharide of *Caesalpinia ferrea* inhibits herpes simplex virus and poliovirus. Int. J. Biol. Macromol., 60: 93–99.

Lopez, A., Hudson, J.B., Towers, G.H.N. 2001. Antiviral and antimicrobial activities of Colombian medicinal plants. J. Ethnopharmacol., 77: 189–196.

Lorenzi, H. 1992. Árvores brasileiras: manual de identificação e cultivo de plantas arbóreas nativas do Brasil. Nova Odessa: Ed. Plantarium, 352 p.

Lozano-Sarango, M.A. 2009. Estudio etnobotanico cuantitativo de especies alucinogenas de la familia Lycopodiaceae utilizadas por los Rikuyhampiyachakkuna (visionarios) de la etnia Saraguro en la provincia de Loja. Trabajo de Grado Ingeniería Industria Agropecuaria. Universidad Técnica Particular de Loja, Ecuador.

Lozoya, X., Enríquez, R.G., Bejar, E., Estrada, A.V., Girón, H., Ponce-Monter, H., Gallegos, A.J. 1983. The zoapatle V-the effect of kauradienoic acid upon uterine contractility. Contraception, 27: 267–279.

Lukman, M., Syahruni, R., Nisa, M., Fatmawati, A. 2014. Cytotoxic activity of three South Sulawesi medicinal plant extracts used in the treatment of HeLa cell line: Jati Putih (*Gmelina arborea* Roxb.), Jati Belanda (*Guazuma ulmifolia* Lamk.) and Lakka-lakka (*Curculigo orchioides* Gaerth). Int. J. Pharm. Sci., 5: 531–536.

Maciel, M.R.A., Neto, G.G. 2006. Um olhar sobre as benzedeiras de Juruena (Mato Grosso, Brasil) e as plantas usadas para benzer e curar. Ciênc. Hum., 1: 61–77.

Magos, G.A., Vidrio, H. 1991. Pharmacology of *Casimiroa edulis*; Part I. Blood pressure and heart rate effects in the anesthetized rat. Planta Med., 57: 20–24.

Magos, G.A., Vidrio, H., Enríquez, R. 1995. Pharmacology of *Casimiroa edulis*; III. Relaxant and contractile effects in rat aortic rings. J. Ethnopharmacol., 47: 1–8.

Magos, G.A., Vidrio, H., Reynolds, W.F., Enriquez, R.G. 1999. Pharmacology of *Casimiroa edulis* IV. Hypotensive effects of compounds isolated from methanolic extracts in rat and guinea pig. J. Ethnopharmacol., 64: 35–44.

Magos, G.A., Mateos, J.C., Páez, E., Fernández, G., Lobato, C., Márquez, C., Enríquez, R.G. 2008. Hypotensive and vasorelaxant effects of the procyanidin fraction from *Guazuma ulmifolia* bark in normotensive and hypertensive rats. J. Ethnopharmacol., 1: 58–68.

Maia, G.N. 2004. Caatinga: árvores e arbustos e suas utilidades. 1st Ed. São Paulo, D&Z Computação Gráfica e Editora. 540 p.

Malagón, O., Ramírez, J., Andrade, J., Morocho, V., Armijos, C., Gilardoni, G. 2016. Phytochemistry and ethnopharmacology of the Ecuadorian flora. A review. Nat. Prod. Commun., 11: 297–314.

Maldini, M., Di Micco, S., Montoro, P., Darra, E., Mariotto, S., Bifulco, G., Piacente, S. 2013. Flavanocoumarins from *Guazuma ulmifolia* bark and evaluation of their affinity for STAT1. Phytochemistry, 86: 64–71.

Marinho, M.G.V., Brito, A.G., Carvalho, K.A. 2004. *Amburana cearensis* e cumarina imunomodulam os níveis de anticorpos antígeno-específicos em camundongos Balb/c sensibilizados com ovalbumina. Acta Farm. Bonaerense., 23: 47–52.

Matsuda, K., Hattori, S., Kariya, R., Komizu, Y., Kudo, E., Goto, H., Taura, M., Ueoka, R., Kimura, S., Okada, S. 2015. Inhibition of HIV-1 entry by the tricyclic coumarin GUT-70 through the modification of membrane fluidity. Biochem. Biophys. Res. Commun., 457: 288–294.

Matulevich-Peláez, J., García-Rodríguez, J. 2016. Composición química del aceite esencial de hojas de *Guazuma ulmifolia* (Malvaceae). Scientia Tech., 21: 269–272.

Menezes, I.A.C., Moreira, Í.J.A., Carvalho, A.A., Antoniolli, A.R., Santos, M.R.V. 2007. Cardiovascular effects of the aqueous extract from *Caesalpinia ferrea*: involvement of ATP-sensitive potassium channels. Vascular Pharmacol., 47: 41–47.

Mesa-Vanegas, A.M., Blair-Trujillo, S., Peláez-Jaramillo, C. 2019a. Chemical standardization of extracts of *Calophyllum brasiliense* with antiplasmodial and cytotoxic activity. J. Med. Plants By-prod., 2: 171–179.

Mesa-Vanegas, A.M., Blair-Trujillo, S., Peláez-Jaramillo, C., Carda, M. 2019b. Antiplasmodials soulattrolide derivatives from *Calophyllum brasiliense* and its mechanism of activity. J. King Saud Univ. Sci., 31: 1208–1214.

Mesía-Vela, S., Sánchez, R.I., Estrada-Muñiz, E., Alavez-Solano, D., Torres-Sosa, C., Jiménez-Estrada, M., Kauffman, F.C. 2001. Natural products isolated from Mexican medicinal plants: novel inhibitors of sulfotransferases, SULT1A1 and SULT2A1. Phytomedicine, 8: 481–488.

Mesquita, M.L., de Paula, J.E., Pessoa, C., de Moraes, M.O., Costa-Lotufo, L.V., Grougnet, R., Michel, S., Tillequin, F., Espindola, L.S. 2009. Cytotoxic activity of Brazilian Cerrado plants used in traditional medicine against cancer cell lines. J. Ethnopharmacol., 123: 439–445.

Miller, S.L., Haber, W.A., Setzer, W.N. 2009. Chemical composition of the leaf essential oil of *Casimiroa edulis* La Llave & Lex. (Rutaceae) from Monteverde, Costa Rica. Nat. Prod. Commun., 4: 425–426.

Molina-Hernández, M., Téllez-Alcántara, P., Martínez, E. 2000. *Agastache mexicana* may produce anxiogenic-like action in the male rat. Phytomedicine, 7: 199–203.

Molina-Hernández, M., Téllez-Alcántara, N.P., Pérez-García, J., Olivera-López, J.I., Jaramillo, M.T. 2004. Anxiolytic-like actions of leaves of *Casimiroa edulis* (Rutaceae) in male Wistar rats. J. Ethnopharmacol., 93: 93–98.

Morais, S.M., Calixto-Júnior, J.T., Ribeiro, L.M., Sousa, H.A., Silva, A.A.S., Figueiredo, F.G., Matias, E.F.F., Boligon, A.A., Athayde, M.L., Morais-Braga, M.F.B., Coutinho, H.D.M. 2017. Phenolic composition and antioxidant, anticholinesterase and antibiotic-modulating antifungal activities of *Guazuma ulmifolia* Lam. (Malvaceae) ethanol extract. S. African J. Bot., 110: 251–257.

Mukherjee, P.K., Venkatesh, P., Ponnusankar, S. 2010. Ethnopharmacology and integrative medicine—let the history tell the future. J. Ayurveda Integr. Med., 1: 100–109.

Muñoz, V., Sauvain, M., Bourdy, G., Callapa, J., Rojas, I., Vargas, L., Tae, A., Deharo, E. 2000. The search for natural bioactive compounds through a multidiciplinary approach in Bolivia: antimalarial activity of some plants used by Mosetene Indians. J. Ethnopharmacol., 69: 139–155.

Murillo, G., Hirschelman, W.H., Ito, A., Moriarty, R.M., Kinghorn, A.D., Pezzuto, J.M., Mehta, R.G. 2007. Zapotin, a phytochemical present in a Mexican fruit, prevents colon carcinogenesis. Nutr. Cancer, 57: 28–37.

Nagai, H., Tanaka, T., Goto, T., Kusudo, T., Takahashi, N., Kawada, T. 2014. Phenolic compounds from leaves of *Casimiroa edulis* showed adipogenesis activity. Biosci. Biotechnol. Biochem., 78: 296–300.

Nakamura, E.S., Kurosaki, F., Arisawa, M., Mukainaka, T., Okuda, M., Tokuda, H., Nishino, H., Pastore, F. 2002. Cancer chemopreventive effects of *Caesalpinia ferrea* and related compounds. Cancer Lett., 177: 119–124.

Nascimento, P.L.A., Nascimento, T.C.E.S., Gomes, J.E.G., Silva, M.D.S., Souza, S.A., Silva, T.M.S., Falcão, R.A., Moreira, K.A. 2015. Antioxidant and antimicrobial properties of ethanolic extract of *Libidibia ferrea* pods. Rev. Fitos, 9: 207–216.

Nascimento, E.L., Watanabe, M., Fonseca, C.D., Schlottfeldt, F.S., Vattimo, M.F.F. 2014. Efeito renoprotetor do *Echinodorus macrophyllus* na lesão renal induzida. Acta Paul. Enferm., 27: 12–17.

Navarrete, A., Ávila-Rosas, N., Majín-León, M., Balderas-López, J.L., Alfaro-Romero, A., Tavares-Carvalho, J.C. 2017. Mechanism of action of relaxant effect of *Agastache mexicana* ssp. *mexicana* essential oil in guinea-pig trachea smooth muscle. Pharm. Biol., 55: 96–100.

Navarro, M.C., Montilla, M.P., Cabo, M.M., Galisteo, M., Cáceres, A., Morales, C., Berger, I. 2003. Antibacterial, antiprotozoal and antioxidant activity of five plants used in Izabal for infectious diseases. Phytother. Res., 17: 325–329.

Navarro-Ruíz, A., Bastidas-Ramírez, B.E., García-Estrada, J., García-López, P., Garzón, P. 1995. Anticonvulsant activity of *Casimiroa edulis* in comparison to phenytoin and phenobarbital. J. Ethnopharmacol., 45: 199–206.

Negri, G., Oliveira, A.F.M., Salatino, M.F.L., Salatino, A. 2004. Chemistry of the stem bark of *Amburana cearensis* (Allemão) (A.C.SM.). Rev. Bras. Plant. Med., 6. 1–4.

Neto, C.C., Owens, C.W., Langfield, R.D., Comeau, A.B., Onge, J., Vaisberg, A., Hammond, G.B. 2002. Antibacterial activity of some Peruvian medicinal plants from the Callejon de Huaylas. J. Ethnopharmacol., 79: 133–138.

Nonato, F.R., Oliveira Nogueira, T.M., de Almeida Barros, T.A., Lucchese, A.M., Cordeiro Oliveira, C.E., dos Santos, R.R., Pereira Soares, M.B., Villareal, C.F. 2011. Antinociceptive and antiinflammatory activities of *Adiantum latifolium* Lam.: evidence for a role of IL-1β inhibition. J. Ethnopharmacol., 136: 518–524.

Noriega, P.F., Paredes, E.A., Mosquera, T.D., Díaz, E.E., Lueckhoff, A., Basantes, J.E., Trujillo, A.L. 2016. Chemical composition antimicrobial and free radical scavenging activity of essential oil from leaves of *Renealmia thyrsoidea* (Ruiz &Pav.) Poepp. & Endl. J. Med. Plants Res., 10: 553–558.

Noriega, P.F., Paredes, E.A., Díaz-Gómez, E., Lueckhoff, A., Almeida, G.A., Suárez, S.E. 2017. Chemical composition and antimicrobial activity of the essential oil from rhizomes of *Renealmia thyrsoidea* (Ruiz & Pav) Poepp. & Eddl (shiwanku muyu). Rev. Cub. Plant. Med., 22: 1–22.

Nozaki, H., Hayashi, K.I., Kido, M., Kakumoto, K., Ikeda, S., Matsuura, N., Tani, H., Takaoka, D., Iinuma, M., Akao, Y. 2007. Pauferrol A, a novel chalcone trimer with a cyclobutane ring from *Caesalpinia ferrea* Mart exhibiting DNA topoisomerase II inhibition and apoptosis-inducing activity. Tetrahedron Lett., 48: 8290–8292.

Nunes, A.R., Rodrigues, A.L.M., Queiróz, D.B., Vieira, I.G.P., Neto, J.F.C., Calixto Júnior, J.T., Tintino, S.R., Morais, S.M., Coutinho, H.D.M. 2018. Photoprotective potential of medicinal plants from Cerrado biome (Brazil) in relation to phenolic content and antioxidant activity. J. Photoch. Photobiol. B, 189: 119–123.

Odonne, G., Valadeau, C., Alban-Castillo, J., Stien, D., Sauvain, M., Bourdy, G. 2013. Medical ethnobotany of the Chayahuita of the Paranapura basin (Peruvian Amazon). J. Ethnopharmacol., 146: 127–153.

Ohira, S., Takaya, K., Mitsui, T., Kido, M., Kakumoto, K., Hayashi, K.I., Kuboki, A., Tani, H., Ikeda, S., Iinuma, M., Akao, Y., Nozaki, H. 2013. New chalcone dimers from *Caesalpinia ferrea* Mart act as potent inhibitors of DNA topoisomerase II. Tetrahedron Lett., 54: 5052–5055.

Oliveira, A.F., Batista, J.S., Paiva, E.S., Silva, A.E., Farias, Y.J.M.D., Damasceno, C.A.R., Brito, P.D., Queiroz, S.A.C., Rodrigues, C.M.F., Freitas, C.I.A. 2010. Avaliação da atividade cicatrizante do jucá (*Caesalpinia ferrea* Mart. ex Tul. var. *ferrea*) em lesões cutâneas de caprinos. Rev. Bras. Plant. Med., 12: 302–310.

Oliveira, R.R.B., Góis, R.M.O., Siqueira, R.S., Almeida, J.R.G.S., Lima, J.T., Nunes, X.P., Oliveira, V.R., Siqueira, J.S., Quintans-Júnior, L.J. 2009. Antinociceptive effect of the ethanolic extract of *Amburana cearensis* (Allemão) A.C. Sm., Fabaceae, in rodents. Braz. J. Pharmacog., 19: 672–676.

Oliveira, M.C., Lemos, L.M.S., de Oliveira, R.G., Dall'Oglio, E.L., de Sousa-Júnior, P.T., Martins, D.T.O. 2014. Evaluation of toxicity of *Calophyllum brasiliense* stem bark extract by *in vivo* and *in vitro* assays. J. Ethnopharmacol., 155: 30–38.

Oliveira, G.P., Silva, T.M.G., Camara, C.A., Santana, A.L.B.D., Moreira, M.S.A., Silva, T.M.S. 2017. Isolation and structure elucidation of flavonoids from *Amburana cearensis* resin and identification of human DNA topoisomerase II-α inhibitors. Phytochem. Letters, 22: 61–70.

Orellana, P.A.D. 2014. Catálogo de frutales nativos de Guatemala. Instituto de Ciencia y Tecnología Agrícola. Gobierno de Guatemala. Ciudad de Guatemala, 95 p.

Oshima, Y., Cordell, G.A., Fong, H.H.S. 1986. Oxepane diterpenes from *Montanoa tomentosa*. Phytochemistry, 25: 2567–2568.

Padilla, F.J., Rincón, A.M., Bou Rached, L., Suárez, A. 2008. *Protium neglectum*, *Podocalyx loranthoides*, and *Brosimum utile* source of flavonoids and other phenolic compounds with antioxidant activity. Rev. Fac. Farm., 71: 8–14.

Pasa, M.C., Hanazaki, N., Silva, O.M.D., Agostinho, A.B., Zank, S., Esteves, M.I.P.N. 2019. Medicinal plants in cultures of Afro-descendant communities in Brazil, Europe and Africa. Acta Bot. Bras., 33: 340–349.

Paz, W.H.P., da Silva, F.M.A., da Silva-Filho, F.A., de Almeida, R.A., Pinheiro, M.L.B., Koolen, H.H.F. 2018. Triterpenes from the roots of *Salacia impressifolia*. Chem. Nat. Comp., 54: 200–201.

Pereira, G.A., Araujo, N.M.P., Arruda, H.S., Farias, D.P., Molina, G., Pastore, G.M. 2019. Phytochemicals and biological activities of mutamba (*Guazuma ulmifolia* Lam.): a review. Food Res. Int., 126: 108713.

Pereira, G.A., Arruda, H.S., de Morais, D.R., Araujo, N.M.P., Pastore, G.M. 2020. Mutamba (*Guazuma ulmifolia* Lam.) fruit as a novel source of dietary fibre and phenolic compounds. Food Chem., 310: 125857.

Pereira, E.P.L., Braga-de-Souza, S., Santos, C.C., Santos, L.O., Cerqueira, M.D., Ribeiro, P.R., Fernandez, L.G., Silva, V.D.A., Costa, S.L. 2017. *Amburana cearensis* seed extracts protect PC-12 cells against toxicity induced by glutamate. Rev. Brasil. Farmacogn., 27: 199–205.

Pérez, D. 2002. Etnobotánica medicinal y biocidas para malaria en la Región Ucayali. Folia Amaz., 13: 87–108.

Pinto, A.C., Rego, G.C.G., Siqueira, A.M., Cardoso, C.C., Reis, P.A., Marques, E.A., Coelho, M.G.P., Sabino, K.C.C. 2007. Immunosuppressive effects of *Echinodorus macrophyllus* aqueous extract. J. Ethnopharmacol., 111: 435–439.

Pinto, A.C., Rego, G.C.G., Siqueira, A.M., Cardoso, C.C., Reis, P.A., Marques, E.A., Coelho, M.G.P., Sabino, K.C.C., Portella, V.G., Cosenza, G.P., Diniz, L.R.L., Pacheco, L.F., Cassali, G.D., Caliari, M.V., Brandão, M.G.L., Vieira, M.A.R. 2012. Nephroprotective effect of *Echinodorus macrophyllus* Micheli on gentamicin-induced nephrotoxicity in rats. Nephron Extra., 2: 177–183.

Pires, C.T.A., Scodro, R.B.L., Brenzan, M.A., Cortez, D.A.G., Siqueira, V.L.D., Cardozo-Filho, L., Goncalves, R.M., Caleffi-Ferracioli, K.R., Cardoso, R.F. 2016. Anti-*Mycobacterium tuberculosis* activity of *Calophyllum brasiliense* extracts obtained by supercritical fluid extraction and conventional techniques. Curr. Pharm. Biotechnol., 17: 532–539.

Port's, P.S. 2011. Compostos fenólicos e potencial antioxidante de ervas consumidas na região amazônica brasileira. Master Thesis. Faculdade de Engenharia de Alimentos, Universidade Estadual de Campinas, Campinas, SP, Brazil, 82 p.

Portella, V.G., Cosenza, G.P., Diniz, L.R.L., Pacheco, L.F., Cassali, G.D., Caliari, M.V., Brandão, M.G.L., Vieira, M.A.R. 2012. Nephroprotective effect of *Echinodorus macrophyllus* Micheli on gentamicin-induced nephrotoxicity in rats. Nephron Extra., 2: 177–183.

Pradeep-Kumar, R., John, A., Kumar, P., Dinesh-Babu, K.V., Evans, D.A. 2018. Larvicidal efficacy of Adiantobischrysene from *Adiantum latifolium* against *Oryctes rhinoceros* through disrupting metamorphosis and impeding microbial mediated digestion. Pest Manag. Sci., 74: 1821–1828.

Pradeep-Kumar, R., Dinesh-Babu, K.V., Evans, D.A. 2019. Isolation, characterization and mode of action of a larvicidal compound, 22-hydroxyhopane from *Adiantum latifolium* Lam. against *Oryctes rhinoceros* Linn. Pestic. Biochem. Physiol., 153: 161–170.

Pretto, J.B., Cechinel-Filho, V., Noldin, V.F., Sartori, M.R.K., Isaias, D.E.B., Cruz, A.B. 2004. Antimicrobial activity of fractions and compounds from *Calophyllum brasiliense* (Clusiaceae/Guttiferae). Z. Naturforsch. C J. Biosci., 59: 657–662.

Quattrocchi, U. 2012. CRC World dictionary of medicinal and poisonous plants: common names, scientific names, eponyms, synonyms, and etymology (5 volume set). CRC Press, Boca Ratón, 4038 p.

Queiroz, L.P. 2009. Leguminosas da Caatinga. 1st Ed. Feira de Santana: Universidade Estadual de Feira de Santana, 443 p.

Quijano, L., Calderón, J.S., Gómez, F., Rios, T. 1982. Zoapatanolide A and B, two new heliangolides from *Montanoa tomentosa*. Phytochemistry, 21: 2041–2044.

Quijano, L., Gómez, F., Calderón, J.S., López, J., Rios, T. 1984. Zoapatanolides C and D, two guaianolides from *Montanoa tomentosa*. Phytochemistry, 23: 125–127.

Quijano, L., Calderón, J.S., Gómez, F., Rosario, V., Rios, T. 1985. Oxepane diterpenoids and sesquiterpene lactones from 'zoapatle' (*Montanoa tomentosa*), a Mexican plant with oxytocic activity. Phytochemistry, 24: 2337–2340.

Ramakrishna, D., Shashank, A.T., Shinomol Geopsrge, K., Kiran, S., Ravishankar, G.A. 2016. *Salacia* sps: A source of herbal drug for several human diseases and disorders. Int. J. Curr. Pharm. Rev. Res., 7: 122–133.

Ravarocci-Quiroz, C.A., Carrasco-Huamán, W. 2010. Tamizaje fitoquímico, perfil cromatográfico y evaluación de la actividad antioxidante *in vitro*, de las cortezas *de Erythrina fusca* L., *Campsiandra angustifolia* S.B. y *Swartzia polyphylla* DC. Informe Final Programa Farmacia y Bioquímica, Universidad Nacional de la Amazonía Peruana, Iquitos.

Rea, A., Tempone, A.G., Pinto, E.G., Mesquita, J.T., Rodrigues, E., Silva, L.G.M., Sartorelli, P., Lago, J.H.G. 2013. Soulamarin isolated from *Calophyllum brasiliense* (Clusiaceae) induces plasma membrane permeabilization of *Trypanosoma cruzi* and mytochondrial dysfunction. PLoS Negl. Trop. Dis., 7: e2556.

Rengifo, E. 2007. Las ramas floridas del bosque: experiencia del manejo en plantas medicinales amazónicas. Instituto de Investigaciones de la Amazonia Peruana.

Reyes-Chilpa, R., Estrada-Muñiz, E., Ramírez-Apan, T., Amekraz, B., Aumelas, A., Jankowski, C.K., Vázquez-Torres, M. 2004. Cytotoxic effects of mammea type coumarins from *Calophyllum brasilense*. Life Sci., 75: 1635–1647.

Reyes-Chilpa, R., Baggio, C.H., Alavez-Solano, D., Estrada-Muñiz, E., Kauffman, F.C., Sánchez, R.I., Mesía-Vela, S. 2006. Inhibition of gastric H^+,K^+-ATPase activity by flavonoids, coumarins and xanthones isolated from Mexican medicinal plants. J. Ethnopharmacol., 105: 167–172.

Reyes-Chilpa, R., Estrada-Muñiz, E., Vega-Avila, E., Abel, F., Kinjo, J., Hernández-Ortega, S. 2008. Trypanocidal constituents in plants. 7. Mammea-type coumarins. Mem. Inst. Oswaldo Cruz, 103: 431–436.

Ribeiro, V.P., Arruda, C., da Silva, J.J.M., Aldana Mejia, J.A., Furtado, N.A.J.C., Bastos, J.K. 2019. Use of spinning band distillation equipment for fractionation of volatile compounds of *Copaifera* oleoresins for developing a validated gas chromatographic method and evaluating antimicrobial activity. Biomed. Chromatogr., 33: e4412 (12 p.).

Ripardo, Filho, H.S., Costa, N.L.S., Pacheco, L.C., Andrade, E.S., Araújo, R.N.M., Mourão, R.H.V., Guilhon, G.M.S.P., Santos, L.S. 2015. Chemical constituents of the stem of *Salacia impressifolia* (Miers) A.C. Smith. Ew. J. Nat. Prod. Res., 1: 1–4.

Rivera-Parada, L.L. 2013. Caracterización fitoquímica, farmacéutica y alimenticia de Papa culebrera india (*Dracontium spruceanum* (Schott) G.H.Zhu, Araceae) y Sande (*Brosimum utile* (Kunth) Oken, Moraceae) del Jardín Botánico de Plantas Medicinales del CEA de CORPOAMAZONIA, Mocoa, Putumayo. Corporación para el Desarrollo Sostenible del Sur de la Amazonía CORPOAMAZONIA, Investigaciones, 11 p.

Robles-Zepeda, R.E., Lozoya-Gloria, E., López, M.G., Villarreal, M.L., Ramírez-Chávez, E., Molina-Torres, J. 2009. *Montanoa tomentosa* glandular trichomes containing kaurenoic acids chemical profile and distribution. Fitoterapia, 80: 12–17.

Rodrigues, A.C.B.D.C, de Oliveira, F.P., Dias, R.B., Sales, C.B.S., Rocha, C.A.G., Soares, M.B.P., Costa, E.V., da Silva, F.M.A., Rocha, W.C., Koolen, H.H.F., Bezerra, D.P. 2019. *In vitro* and *in vivo* anti-leukemia activity of the stem bark of *Salacia impressifolia* (Miers) A. C. Smith (Celastraceae). J. Ethnopharmacol., 23: 516–524.

Rodrigues, I.A., Ramos, A.S., Falcão, D.Q., Ferreira, J.L.P., Basso, S.L., Silva, J.R.A., Amaral, A.C.F. 2018. Development of nanoemulsions to enhance the antileishmanial activity of *Copaifera paupera* oleoresins. BioMed. Res. Int., 2018: 9781724 (9 p.).

Rodríguez-Hernández, K.D., Martínez, I., Agredano-Moreno, L.T., Jiménez-García, L.F., Reyes-Chilpa, R., Espinoza, B. 2019. Coumarins isolated from *Calophyllum brasiliense* produce ultrastructural alterations and affect *in vitro* infectivity of *Trypanosoma cruzi*. Phytomedicine, 61: ID152827.

Román-Farje, A., Iannacone, J., Alvariño, L. 2017. Efecto tóxico del saúco, *Sambucus peruviana* (Caprifoliaceae), en *Daphnia magna*, *Sitophilus zeamais* y *Copidosoma koehleri* en Perú. Chil. J. Agric. Anim. Sci., 33: 3–13.

Roth, I., Lindorf, H. 2002. South American medicinal plants. Botany, remedial properties and general uses. Springer-Verlag Berlin Heidelberg, 498 p.

Roumy, V., Ruiz-Macedo, J.C., Bonneau, N., Samaillie, J., Azaroual, N., Encinas, L.A., Rivière, C., Hennebelle, T., Sahpaz, S., Antherieu, S., Pinçon, C., Neut, C., Siah, A., Gutierrez-Choquevilca, A.L., Ruiz, L. 2020. Plant therapy in the Peruvian Amazon (Loreto) in case of infectious diseases and its antimicrobial evaluation. J. Ethnopharmacol., 249: 112411 (10 p.).

Ruiz, L., Ruiz, L., Maco, M., Cobos, M., Gutierrez-Choquevilca, A.L., Roumy, V. 2011. Plants used by native Amazonian groups from the Nanay River (Perú) for the treatment of malaria. J. Ethnopharmacol., 133: 917–921.

Ruiz-Marcial, C., Reyes-Chilpa, R., Estrada, E., Reyes-Esparza, J., Garrido-Fariña, G., Rodríguez-Fragoso, L. 2007. Antiproliferative, cytotoxic and antitumour activity of coumarins isolated from *Calophyllum brasiliense*. J. Pharm. Pharmacol., 59: 719–725.

Ruiz-Reyes, S.G., Venegas-Casanova, E., Ruidías-Romero, D., Horna-Acevedo, L., López-Cenizario, C.W. 2013. Capacidad antioxidante *in vitro* de los flavonoides totales obtenidos de las hojas de *Sambucus peruviana* H.B.K. (sauco) proveniente de la ciudad de Huamachuco. Pharmaciencia, 1: 57–64.

Sampaio, F.C., Perereira, M.S.V., Dias, C.S., Costa, V.C.O., Conde, N.C.O., Buzalaf, M.A.R. 2009. *In vitro* antimicrobial activity of *Caesalpinia ferrea* Martius fruits against oral pathogens. J. Ethnopharmacol., 124: 289–294.

Santillán-Ramírez M.A., López-Villafranco, M.E., Aguilar-Rodríguez, S., Aguilar-Contreras, A. 2008. Estudio etnobotánico, arquitectura foliar y anatomía vegetativa de *Agastache mexicana* ssp. *mexicana* y *Agastache mexicana* ssp. *xolocotziana*. Rev. Mex. Biodiv., 79: 513–524.

Santos, A.O., Ueda-Nakamura, T., Dias Filho, B.P., Veiga Junior, V.F., Pinto, A.C., Nakamura, C.V. 2008. Effect of Brazilian *Copaiba* oils on *Leishmania amazonensis*. J. Ethnopharmacol., 120: 204–208.

Sawada, L.A., Monteiro, V.S.C., Rabelo, G.R., Dias, G.B., Da Cunha, M., Nascimento, J.L.M., Bastos, G.N.T. 2014. *Libidibia ferrea* mature seeds promote antinociceptive effect by peripheral and central pathway: possible involvement of opioid and cholinergic receptors. BioMed. Res. Int., 2014: 1–10.

Schmeda-Hirschmann, G., Feresin, G., Tapia, A., Hilgert, N., Theoduloz, C. 2005. Proximate composition and free radical scavenging activity of edible fruits from the Argentinian Yungas. J. Sci. Food Agric., 85: 1357–1364.

Schmeda-Hirschmann, G., Burgos-Edwards, A., Theoduloz, C., Jiménez-Aspee, F., Vargas-Arana, G. 2019. Male sexual enhancers from the Peruvian Amazon. J. Ethnopharmacol., 229: 167–179.

Schmidt, B.M., Klaser-Cheng, D.M. 2017. Ethnobotany: A Phytochemical Perspective. John Wiley and Sons Ltd., Hoboken, NJ, USA.

Seaman, F.C., Malcolm, A.J., Fischer, N.H. 1984. Tomexanthin, an oxepane diterpene from *Montanoa tomentosa*. Phytochemistry, 23: 464–465.

Seigler, D.S., Pauli, G.F., Fröhlich, R., Wegelius, E., Nahrstedt, A., Glander, K.E., Ebinger, J.E. 2005. Cyanogenic glycosides and menisdaurin from *Guazuma ulmifolia*, *Ostrya virginiana*, *Tiquilia plicata*, and *Tiquilia canescens*. Phytochemistry, 66: 1567–1580.

Senties, L., Amayo, R. 1864. Symposium sobre avances en el estudio de la contractilidad uterina. IV. Efecto del cihuapatli sobre el útero humano grávido, Gac. Med. Méx. XOIV, 94: 343–350.

Shigemori, H., Shimamoto, S., Sekiguchi, M., Ohsaki, A., Kobayashi, J. 2002. Echinodolides A and B, new cembrane diterpenoids with an eight-membered lactone ring from the leaves of *Echinodorus macrophyllus*. J. Nat. Prod., 65: 82–84.

Silva, T. 2010. Avaliação da integridade de constituintes químicos de *Echinodorus macrophyllus* (chapéu-de-couro) tratada por irradiação gama. Master Thesis. Universidade Federal de Minas Gerais, Belo Horizonte, 121 p.

Silva, T.M., Miranda, R.R.S., Ferraz, V.P., Pereira, M.T., de Siqueira, E.P., Alcântara, A.F.C. 2013b. Changes in the essential oil composition of leaves of *Echinodorus macrophyllus* exposed to γ-radiation. Rev. Bras. Farmacog., 23: 600–607.

Silva, N.C.B., Regis, A.C.D., Almeida, M.Z. 2013a. Estudo etnobotânico em comunidades remanescentes de quilombo em Rio de Contas - Chapada Diamantina - Bahia. Rev. Fitos., 7: 99–109.

Silveira, E.R., Pessoa, O.D.L. 2005. Constituintes micromoleculares de plantas do Nordeste com potencial farmacológico. Fortaleza: Editora Expressão Gráfica, 216 p.

Smith, J.B., Smith, E.F., Lefer, A.M., Nicolaou, K.C. 1981. Spasmogenic effects of the anti-fertility agent, zoapatanol. Life Sci., 28: 2743–2746.

Soares, M.R.P.S., Caneschi, C.A., Chaves, M.G.A.M., Mota, M., Stroppa, P.H.F., Barbosa, W., Raposo, N.R.B. 2018. *In vitro* antifungal activity and cytotoxicity screening of dry crude extracts from Brazilian Amazonia plants. Afr. J. Tradit. Complement. Altern. Med., 15: 13–21.

Sollozo-Dupont, I., Estrada-Camarena, E., Carro-Juárez, M., López-Rubalcava, C. 2015. $GABA_A$/benzodiazepine receptor complex mediates the anxiolytic-like effect of *Montanoa tomentosa*. J. Ethnopharmacol., 162: 278–286.

Sondheimer, F., Meisels, A. 1960. Constituents of *Casimiroa edulis* Llave et Lex.-VI. 2′,5,6-Trimethoxyflavone, 2′,5,6,7-tetramethoxyflavone (zapotin) and 5-hydroxy-2′,6,7-trimethoxyflavone (zapotinin). Tetrahedron, 9: 139–144.

Souza, M.C., Beserra, A.M.S., Martins, D.C., Real, V.V., dos Santos, R.A.N., Rao, V.S., Martins, D.T. 2009. *In vitro* and *in vivo* anti-*Helicobacter pylori* activity of *Calophyllum brasiliense* Camb. J. Ethnopharmacol., 123: 452–458.

Stalcup, M.M. 2000. Plantas de uso medicinal ou ritual numa feira livre no Rio de Janeiro, Brasil. Master Thesis, Universidade Federal do Rio de Janeiro, Rio de Janeiro, Brazil, 212 p.

Standley, P.C., Steyermark, J.A. 1946. Malpighiaceae. *In*: Standley, P.C., Steyermark, J.A. (eds.). Flora of Guatemala, Part V. Fieldiana, Bot., 24: 468–500.

Suffredini, I.V., Paciencia, M.L.B., Varella, A.D., Younes, R.N. 2007. *In vitro* cytotoxic activity of Brazilian plant extracts against human lung, colon and CNS solid cancers and leukemia. Fitoterapia, 78: 223–226.

Suffredini, I.V., Varella, A.D., Younes, R.N. 2014. *Calophyllum brasiliensi* Cambess (Clusiaceae) extracts showed antimicrobial activity and cytotoxicity, *in vitro*. J. Health Sci. Inst., 32: 184–189.

Tanus-Rangel, E., Santos, S.R., Lima, J.C.S., Lopes, L., Noldin, V., Monache, F.D., Cechinel-Filho, V., Martins, D.T.O. 2010. Topical and systemic anti-inflammatory effects of *Echinodorus macrophyllus* (Kunth) Micheli (Alismataceae). J. Med. Food., 13: 1161–1166.

Teixeira, K., dos Santos, P., Citadini-Zanette, V., Dalbó, S., Amaral, P.A. 2017. Medicinal plants that can cause changes in blood pressure and interactions with antihypertensive agents. Am. J. Ethnomed., 4: 1–8.

Teixeira, N., Melo, J.C.S., Batista, L.F., Paula-Souza, J., Fronza, P., Brandão, M.G.L. 2019. Edible fruits from Brazilian biodiversity: A review on their sensorial characteristics *versus* bioactivity as tool to select research. Food Res. Int., 119: 325–348.

The plant list. 2019. Version 1.1. Published on the Internet; http://www.theplantlist.org/ (accessed august-november 2019).

Thomas, T. 2017. Antibacterial evaluation of gradient extracts of *Adiantum latifolium* Lam. towards pathogenic cutaneous bacteria. J. Glob Pharm. Tech., 06: 01–04.

Ticunsi, B.M., Jiménez, I.A., Bazzochi, I.L., Moujir, L.M., Mamani, Z.A., Barroso, J.P., Ravelo, A.G., Hernández, B.V. 2002. Antimicrobial terpenoids from the oleoresin of the Peruvian medicinal plant *Copaifera paupera*. Planta Med., 68: 808–212.

Tiuman, T.S., Brenzan, M.A., Ueda-Nakamura, T., Dias Filho, B.P., Cortez, D.A.G., Nakamura, C.V. 2012. Intramuscular and topical treatment of cutaneous leishmaniasis lesions in mice infected with *Leishmania amazonensis* using coumarin (-) mammea A/BB. Phytomedicine, 19: 1196–1199.

Toton, E., Romaniuk, A., Budzianowski, J., Hofmann, J., Rybczynska, M. 2016. Zapotin (5,6,2′,6′-tetramethoxyflavone) modulates the crosstalk between autophagy and apoptosis pathways in cancer cells with overexpressed constitutively active PKCε. Nutr. Cancer, 68: 290–304.

Tropicos org. 2019. Missouri Botanical Garden. August-november 2019. http://www.tropicos.org.

UEIA. 2014. Catálogo Virtual de flora del Valle de Aburrá. https://catalogofloravalleaburra.eia.edu.co. Consulted on-line: june-august 2019.

Valadeau, C., Pabon, A., Deharo, E., Albán-Castillo, J., Estevez, Y., Lores, F.A., Rojas, R., Gamboa, D., Sauvain, M., Castillo, D., Bourdy, G. 2009. Medicinal plants from the Yanesha (Peru): evaluation of the leishmanicidal and antimalarial activity of selected extracts. J. Ethnopharmacol., 123: 413–422.

Vasconcelos, C. 2011. Atividade anti-hiperglicemiante oral e segurança de uso do extrato aquoso da casca de *Caesalpinia ferrea* Martius Ex Tul (Leguminosae) em ratos Wistar. Doctoral Thesis, Universidade Federal de Pernambuco, Recife, PE, Brazil, 163 p.

Vásquez-Ocmín, P., Cojean, S., Rengifo, E., Suyyagh-Albouz, S., Amasifuen-Guerra, C.A., Pomel, S., Cabanillas B., Mejía, K., Loiseau, P.M., Figadère, B., Maciuk, A. 2018. Antiprotozoal activity of medicinal plants used by Iquitos-Nauta road communities in Loreto (Peru). J. Ethnopharmacol., 210: 372–385.

Vázquez-Cruz, B., Vázquez-Muñoz, M., Navarrete-Bastida, R., Trujillo-González, I.D., De Haro, R., Segura-Cobos, D., Muñoz-López, J.L. 2009. Antihypertensive and vasorelaxant effects of the aqueous extract of *Casimiroa edulis*. Pharmacologyonline, 3: 73–83.

Ventura-Martínez, R., Rodríguez, R., González-Trujano, M.E., Ángeles-López, G.E., Déciga-Campos, M., Gómez, C. 2017. Spasmogenic and spasmolytic activities of *Agastache mexicana* ssp. *mexicana* and *A. mexicana* ssp. *xolocotziana* methanolic extracts on the guinea pig ileum. J. Ethnopharmacol., 196: 58–65.

Verano, J., González-Trujano, M.E., Déciga-Campos, M., Ventura-Martínez, R., Pellicer, F. 2013. Ursolic acid from *Agastache mexicana* aerial parts produces antinociceptive activity involving TRPV1 receptors, cGMP and a serotonergic synergism. Pharmacol. Biochem. Behav., 110: 255–264.

Voirin, B., Jay, M. 1978. Apport de la biochimie flavonique á la systematique du genre *Lycopodium*. Biochem. System. Ecol., 6: 95–97.

Waller, D.P., Martin, A., Oshima, Y., Fong, H.H.S. 1987. Studies on zoapatle. V. Correlation between *in vitro* uterine and *in vivo* pregnancy interruption effects in guinea pigs. Contraception, 35: 147–1453.

World Heatlh Organization. 2019. WHO global report on traditional and complementary medicine. Geneva: World Health Organization, 228 p.

Wu, Q., Yang, X., Zou, L., Fu, D. 2009. Bioactivity guided isolation of alpha-glucosidase inhibitor from whole herbs of *Crossostephium chinense*. Zhongguo Zhong Yao Za Zhi, 34: 2206–2211.

Yasunaka, K., Abe, F., Nagayama, A., Okabe, H., Lozada-Pérez, L., López-Villafranco, E., Reyes-Chilpa, R. 2005. Antibacterial activity of crude extracts from Mexican medicinal plants and purified coumarins and xanthones. J. Ethnopharmacol., 97: 293–299.

Zielińska, S., Matkowski, A. 2014. Phytochemistry of aromatic and medicinal and bioactivity from the genus *Agastache* (Lamiaceae). Phytochem. Rev., 13: 391–416.

Chapter 3

Ethnopharmacology of Wild Plants Used in Diabetes

Patricia Landazuri,[1,*] *Nelsy Loango Chamorro*[2] and *Beatriz Restrepo Cortes*[1]

Introduction

Diabetes Mellitus is a chronic metabolic disease that can affect children and adults. The incidence and prevalence of diabetes has rapidly increased worldwide in recent decades, and this is currently emerging as a Pandemic. Globally, the number of people affected with DM increased from 108 million in 1980 to 500 million in 2014, and the number of people with the disease is predicted to rise beyond 430 million by 2030 and 592 million by the year 2035 (Ogurtsova et al. 2017). The WHO projects that diabetes will be the seventh leading cause of death in 2030 (WHO 2016).

Diabetes can be classified mainly in two major forms, diabetes type 1 (T1D) and diabetes type 2 (T2D). They have different origins, manifestations, and long-term outcomes (Rodriguez et al. 2002, van Duinkerken and Ryan 2020).

T1D is one of the most common chronic diseases of childhood. Globally, the incidence and prevalence of type 1 diabetes vary, so T1D makes up somewhat < 10% of all cases of diabetes, with an estimated prevalence in the United States of 1.3 million in 2017 (Bullard et al. 2018).

T2D, or non-insulin dependent DM, is the epidemic and major form of the disease; about 90–95% of DM cases are type 2 in adults between 20 and 69 years old. This is predicted to rise in 2030 (Wou et al. 2019, Furman et al. 2020), mainly due to the rise in obesity, sedentary lifestyle and other risk factors for T2D.

The primary treatment for both types of diabetes is glucose management to maintain it at proper levels, and glycated hemoglobin A1 (HbA1c) in normal values (less than 7%). Depending on the type of diabetes (type 1 or type 2), and the severity of the disease, strategies for managing blood glucose levels include dietary control, aerobic physical activity, oral hypoglycemic agents (OHA) and insulin injections. Oral hypoglycemic agents and insulin injections have been proved to be effective, but they have many side-effects, like hypoglycemia episodes, hypersensitivity, lactic acidosis reduced vitamin B12 absorption, hepatocellular and cholestatic hepatic injury and gastrointestinal side effects (Ghadge et al. 2014, Ghadge and Kuvalekar 2017). Plant extracts, or drugs from them, seem to be an alternative for the complementary management of diabetes, with few side effects.

Several ethnopharmacological studies describe numerous plant species used to reduce blood sugar levels (Neamsuvan et al. 2015, Chukwuma et al. 2019, Furman et al. 2020). Some of these studies indicate the hypoglycemic mechanisms of antidiabetic plants, such as the inhibition of the α-amylase

[1] Faculty of Health Sciences. Medicine and Biomedical Science Programs. Universidad Del Quindío. Armenia-Quindío, Colombia.
[2] Faculty of Basic Science and Technologies. Biology Program. Universidad Del Quindío. Armenia-Quindío, Colombia.
* Corresponding author: plandazu@uniquindio.edu.co

and α-glucosidase enzymes, or the restoration of pancreatic cells and insulin secretion and sensitivity restoration among others.

These mechanisms of action are mediated by secondary metabolites present in plants as polyphenols, sulfated polysaccharides, saponins, carotenoid pigments triterpenes, and elements such as metals Mg, Ca, K, Mn, Cu, Zn, Cr, and non-metals as P, S (Chinsembu 2019).

Thus, the objective of this chapter is to contribute to the review of some medicinal plants, whose extracts or active compounds have demonstrated efficacy on blood glucose reduction with their probable mechanisms behind the hypoglycemic activity.

Ethnopharmacology of Wild Plants and Diabetes

Ethnopharmacology is a multidisciplinary area of research for the study of the biological activities of any preparation used by humans; it includes the observation, perception, description, distinguishing proof, characterization and experimental investigation of native plants or other natural products (Heywood 2011, Mukherjee 2019).

According to another definition, ethnopharmacology is the scientific study of naturally occurring substances like plants, algae, and fungi that have been traditionally used by groups of people to treat various human diseases (Leonti et al. 2017, Zhao et al. 2018, Mukherjee 2019).

Ethnopharmacology relies on pharmacology, botany, chemistry, biochemistry, history anthropology, archaeology, and linguistics that contribute to and accumulate traditional knowledge for the discovery and use of natural products with biological activity (Mukherjee 2019).

Although humans have used medicinal plants for millennia, the history of ethnopharmacology is shorter (it begins around the 19th century), but its contribution to the knowledge of products and their medicinal use is large, as evidenced by the thousands of ethnopharmacological articles published each year; so ethnopharmacology has contributed to identify numerous plants reported anecdotally to treat diabetes and there are probably more than 800 such plants for which there is scientific evidence for their activity and action mechanisms (Furman et al. 2020).

Treatment with insulin alone, or with adjuvants, remains the cornerstone of treatment for type 1 diabetes, while for the treatment for type 2 diabetes, in addition to insulin, several oral hypoglycemic agents alone, or in combination with other drugs are available; in this OHA, currently, extensive research is being carried out. However, the natural products provide a vast source of secondary metabolites, the exploration of which may lead to new drugs (Furman et al. 2020), as was recognized in the report by the WHO Expert Committee on Diabetes Mellitus (1980), which recommended a methodical investigation of the natural components for the treatment of diabetes (WHO 1980).

Glucose Metabolism and Wild Medicinal Plants

Diabetes mellitus is a disorder of carbohydrates metabolisms, characterized by high blood glucose levels (hyperglycemia), that occurs when the pancreas does not produce enough insulin or there are defects in its action, or both (WHO 2006, Kavvoura and Owen 2019, Furman et al. 2020).

The WHO and diabetes associations define diabetes as having a fasting plasma glucose higher than 7 mmol/L (126 mg/dl) or any blood glucose of 11 mmol/L (200 mg/dL) or higher with symptoms of hyperglycemia, or an abnormal 2 h oral glucose-tolerance test (Rodriguez 2002, WHO 2006).

There are two major forms of diabetes (type 1 and type 2). Type 1 diabetes is caused by total insulin deficiency as a product of progressive, massive loss of insulin-secreting ß-cells. This is generally thought to be precipitated by an immune-associated, or immune-mediated, destruction of insulin-secreting ß-cells (Atkinson et al. 2014, van Duinkerken and Ryan 2020, Furman et al. 2020).

In type 2 diabetes, the ß-cells do not produce enough insulin to supply the high metabolic needs, demanded by peripheral insulin resistance, sight in obese T2D (Baeyens et al. 2018), leading to an imbalance of glucose homeostasis (GH); in both types of diabetes, GH is impaired. GH is the maintenance of plasma glucose concentrations within very narrow margins (between 3.0 and 9.0 mmol/L); this situation is mandatory for organism survival. GH is accomplished by the complex and coordinated interplay of metabolic pathways in different tissues such as the brain, liver, skeletal muscle, fat, and especially the pancreas (Pozo and Claret 2018), and an intricate system of hormones and receptors among others (Habtemariam 2019a, Gilon 2020).

At the level of enzymes, glucose metabolism is regulated at multiple points, three enzymes in glycolysis, hexokinase (HK), phosphofructokinase (PFK or PFK1) and pyruvate kinase (PK), four in gluconeogenesis (pyruvate carboxylase, phosphoenolpyruvate carboxykinase, fructose-1, 6-bisphosphatase, and glucose-6-phosphatase). In glycogenesis and glycogenolysis, the two key enzymes that are regulated include glycogen synthase and glycogen phosphorylase, respectively (Dallak et al. 2009, Mulukutla et al. 2010, Barghamdi et al. 2016, Habtemariam 2019a); all these enzymes, in turn, are regulated at the transcriptional level by signaling pathways (Mulukutla et al. 2010).

At the level of hormones, Insulin secreted by pancreatic β-cells, is a key hormone. In glucose metabolism, this hormone has a fundamental role, e.g., it regulates transporters (Glucose transporters (GLUT) family, enzymes such as glycogen synthase (active), glycogen phosphorylase (inhibits) (Habtemariam 2019a)). Figure 3.1 summarizes glucose homeostasis.

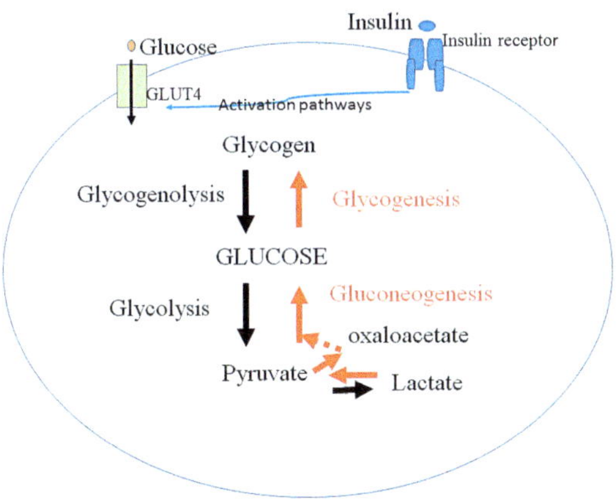

Figure 3.1. Summarizes some of the metabolic pathways, glucose transport, insulin and its receptor involved in glucose homeostasis.

Diabetes Treatment and Medicinal Plants

In addition to insulin, many OHA, such as Biguanides (Metformin), Thiazolidinediones (Troglitazone), Sulfonylureas (Chlorpropamide, Glimepiride), among others, are used separately or in combination to treat the DM (Ghadge et al. 2014, Ghadge and Kuvalekar 2017, Shao et al. 2020). Glucose metabolism enzymes, receptors, signaling pathways, or pancreas beta cells are their therapeutic targets; however, despite its effectiveness, these medications can cause serious side effects, as already mentioned.

These problems have motivated researchers to find new medicines from medicinal plants, to control diabetes and related diseases with fewer side effects. Recently, several ethnopharmacological and ethnobotanical studies showed that hundreds of plants are used worldwide for empirical diabetes treatment (Bhagour 2016, Chukwuma et al. 2019, Chinsembu 2019, Njume et al. 2019, Seetaloo et al. 2019, Furman et al. 2020); In these studies, in addition to describing the active components of some plants and their mechanisms of action and benefits, they also report animal studies and clinical trials.

In this part, we describe ten wild plants, from various regions of the world used in the treatment of diabetes. The information of plants was collected from different articles published in various journals (minimum three papers), and these papers should have described some mechanism of action or therapeutic target. These wild or native plants are *Artemisia herba alba, Brassica juncea, Catharanthus roseus, Citrullus colocynthis* L., *Eugenia jambolana* Lam, *Momordica charantia, Moringa oleífera, Nigella sativa Rosmarinus officinalis* and *Trigonella foenum-graecum.*

Artemisia herba alba (*A. herba-alba*; compositae); common name: wormwood

It is widely distributed in desert areas and 400 species or more have been cultivated or used by man (Al-waili 1986, Salido et al. 2004, Mohamed et al. 2010). Part used: aerial parts. Compounds: sesquiterpene lactones (Eudesmanolides, germacranolides), flavonoids (flavone and flavonol glycosides), Phenolic compounds (chlorogenic acid, 4,5-*O*-dicaffeoylquinic acid, isofraxidin, glucopyranosylcaffeic acid, rutin, schaftoside, isoschaftoside, and vicenin-2) and essential oils (1,8-cineole, chrysanthenone, chrysanthenol, davanone, 1,8-cineole), among others (Salido et al. 2004, Mohamed et al. 2010, Boudjelal et al. 2015)

Studies: In alloxan-induced diabetic rats, oral administration of *A. herba-alba* infusions (100, 200, and 300 mg/kg body weight), twice daily for 15 days, exhibited a significant reduction in blood glucose in plasma from 352.50 ± 1.87 to 100.50 ± 1.87 mg/dl with 300 mg/kg body weight (Boudjelal et al. 2015).

On the other hand, the results of Sekiou et al. (2020) in alloxan-induce diabetic rats, suggests that *A. herba-alba* aqueous extract (400 mg/kg/b.w.) by 30 days not only reduces blood glucose, but also the activities of antioxidant enzymes as Superoxide Dismutase (SOD), Catalase (CAT), Glutathione-S-transferase (GST), and Glutathione Peroxidase (GPx) in the kidney were significantly normalized in diabetic rats.

In another study, aqueous extract of aerial parts of *A. herba alba* (0.39 g/kg body weight) given to diabetic rats and rabbits treated daily for 4 weeks showed significant reduction in blood glucose concentration from 426.1 to 189.4 mg/dL ($p < 0.001$), in rats and 413.5 to 165.9 mg/dL in rabbits, ($p < 0.001$). The same treatment significantly reduced glycosylated hemoglobin in both groups of animals (32.02% in rats and 68.4% in rabbits) (Al-Shamaony et al. 1994).

In a human study, fifteen patients with diabetes mellitus were treated with *A. herba-alba* (±500 g of aerial parts in five liters of water). Each patient drank a half glass (about 50 ml) of this solution each twelve hours. The results showed that *A. herba-alba* had a potent therapeutic value in treating patients with diabetes because the extract of this plant reduced plasmatic glucose from patientes to near normal values (Al-waili 1986).

Brassica juncea (*B. juncea*; Cruciferae). Common names: brown mustard, chinese mustard, and oriental mustard

B. juncea has been consumed for centuries all over the world. Its leaves, roots, and shoots are known as vegetables, while its seed can be used as a condiment or a source of oil (Szollosi 2011). The parts which are used include leaves and seeds. These parts have glucosinolates, isothiocyanates, phenolics, Kaempferol, glycosides, flavonoid (Bhagour et al. 2016), sterols, and glyceride esters of fatty acids (Szollosi 2011, Sharma et al. 2018).

Thirumalai et al. (2011) showed that the aqueous seed extract (250, 350 and 450 mg/kg) given to STZ-induced diabetic rats achieved a significant reduction in blood glucose level and an increase in serum insulin in a dose-dependent manner.

Kumar et al. (2019) reported that methanolic extract of *B. juncea* (450 mg/kg), administered to STZ-rats by 21 days, achieved a significant difference in oxidative stress by increasing superoxide dismutase and glutathione and decreasing lipid peroxidation and nitric oxide in liver, kidney, pancreas, and muscle.

Catharanthus roseus (*C. roseus*; Apocynaceae)

Part used: leaf, flowers. Compounds: alkaloids as ajmalicine, serpentine, catharanthine, vindoline, vindolinine, vincristine, vinblastine, anhydrovinblastine and tetrahydroalstonine (Arumugam et al. 2013, Pan et al. 2016, Choudhury et al. 2018); saponins, tannins, flavonoids, anthraquinones triterpenoids and glycosides and reducing sugars (Ohadoma and Michael 2011); minerals like Zn, Fe, Mn, Mg, Na, K, Ca, Mg, Cr, Fe, Zn, Al, Cu, Ni, Pb, Cd and Mn (Chakraborty et al. 2016).

Studies: leaf juice of *C. roseus* in doses of 0.5, 0.75 and 1.0 ml/kg body weight was found to be hypoglycaemic as in normal and alloxan diabetic rabbits, the maximum reduction in blood glucose was 31.9% (20 h, $p < 0.01$) with 1.0 ml/ kg body weight (Nammi et al. 2003).

In alloxan-induced diabetic rats, leaf methanolic extract significantly reduces blood glucose levels of 260.49 ± 2.64 to 181.53 ± 1.26 and of 259.72 ± 1.99 to 171.75 ± 4.55, when it has been given in doses of 200 mg/kg and 400 mg/kg by 21 days, respectively, $p < 0.01$ compared with the corresponding value of diabetic control animals (Barkat and Mujeeb 2013).

The antioxidant enzyme activities (CAT and SOD) and glutathione (GSH) were also evaluated in this study in plasma and liver; increases of GSH levels and CAT and SOD activities were found in alloxan-induced diabetic rats, compared to those of the diabetic control group (Barkat and Mujeeb 2013).

Enzymic activities of glycogen synthase, glucose 6-phosphate-dehydrogenase, succinate dehydrogenase and malate dehydrogenase were significantly improved after treatment with extract at dose 500 mg/kg p.o. for 7 days in STZ-induced diabetic rat (Singh et al. 2001).

In summary, the mechanisms of action reported for *C. roseus* in diabetes are: enhancing intestinal glucose uptake through increasing glucose-6-phosphate dehydrogenase activity; inhibition of α-glucosidase in the gastrointestinal environment whereby it inhibits the carbohydrates hydrolyses; stimulating insulin release from the β cells; improvement of activity of glucokinase, and increasing the insulin sensitivity by elevating the phosphorylation of insulin receptor (Singh et al. 2001, Nammi et al. 2003, Ohadoma and Michael 2011, Arumugam et al. 2013, Barkat and Mujeeb 2013, Aziz et al. 2016, Chakraborty et al. 2016, Choudhury et al. 2018, de Almeida et al. 2019, Goboza et al. 2019).

***Citrullus colocynthis* L.** (*C. colocynthis*; Cucurbitaceae)

It is a desert plant, widely distributed in different parts of the world (Hussain et al. 2014, Heydari et al. 2019). Parts used: roots, fruits, seeds, rinds, and leaves; compounds: in fruit, several bioactive compounds have been recorded in the literature: glycosides, flavonoids, alkaloids, carbohydrates, fatty acids, and essential oils, cucurbitacins, and colocynthosides (Benariba et al. 2013, Hussain et al. 2014). Cucurbitacins have been reported as the main components of *C. colocynthis* fruits. In roots, the butanolic extract contains terpenoids, saponins flavonoids, and quinones (Chekroun et al. 2015); seeds extracts (aqueous, butanolic, hydromethanolic) contain: catechic tannins and flavonoids, terpenoids, Coumarins (Benariba et al. 2013).

The efficacy of *C. colocynthis* on diabetes mellitus has been evaluated in multiple studies (Dallak et al. 2009, Benariba et al. 2013, Hussain et al. 2014, Shafaei 2014, Shi et al. 2014, Heydari et al. 2019). These studies showed that *C. colocynthis* can significantly improve fasting blood glucose, glycated hemoglobin, and lipids of patients with T2D and animal models of diabetes.

Studies of *C. colocynthis* in patients with diabetes mellitus type 2 showed that, in daily doses from 125 to 1000 mg for 1–3 months, the extract decreases fasting blood glucose level, and HbA1c concentration, and too improvement the lipid profile in this patients, compared to placebo (Huseini 2009, Sebbagh et al. 2009, Hussain et al. 2014). For example, results reported by Huseini et al. (2009) showed a significant decrease in HbA1c ($10.4 \pm 1.9\%$ to $9.0 \pm 2.4\%$ ($p < 0.003$), and fasting blood glucose levels (189.2 ± 27 mg/dL to 174.0 ± 34 mg/dL ($p < 0.015$) at the beginning and at the end of the study, respectively, in *C. colocynthis* treated patients (300 mg/day in three doses of 100 mg whole dried fruit powder).

Similarly, Barghamdi et al. (2016) reported that in a group of patients treated with 125 of *C. colocynthis* once per day for 2 months led to a considerable decrease in the mean levels of HbA1c and fasting blood glucose without any side effects.

In animal studies, Sebbag and collaborators (2009) showed that in STZ-induced diabetic rats, after two months of feeding with 8% *C. colocynthis* oil-enriched diet, hyperglycaemia was significantly less pronounced in the group fed with a *C. colocynthis* oil-enriched diet compared to the control group (7.9 mM versus 12 mM). Pancreatic cell mass was also evaluated and it was found that pancreatic cell mass was similar in both groups of rats.

The possible mechanisms proposed for this plant are among others: insulinotropic effect, glucose intestinal absorption inhibition, decrease of glycogenolysis and gluconeogenesis (glucose 6-phosphatase and fructose-1,6-bisphosphatase inhibition); increase hexokinase activity and insulin level (Dallak et al. 2009, Shafaei et al. 2014, Shi et al. 2014, Barghamdi et al. 2016).

Eugenia jambolana **Lam.** (*E. jambolana*; Myrtaceae); (synonymous. *Syzigium cumini* (L.) SKEELS)

It is a plant native from India. Part used: seed and fruits (berry). Compounds: Berry contains flavonoids, terpenes, and anthocyanins (Baliga et al. 2011), and the seed has gallic acid, ellagic acid, corilagin, ellagitannins (Ravi et al. 2004). *Eugenia jambolana* seeds have been reported to be rich in flavonoids, a well-known antioxidant (Bhagour et al. 2016).

A study by Ravi et al. (2004), shows that the ethanolic extract of seed kernel of *E. jambolana* given to streptozotocin (STZ)-induced diabetic rats at the dose of 100 mg/kg for 30 days increased the activity of antioxidant enzymes (SOD, CAT, and GPx). On the other hand, a study revealed that a triterpenoid-enriched fruit extract (100 mg/kg by oral gavage for ten days) in STZ-induced type 1 diabetic mice, significantly attenuated STZ-induced hyperglycemia and glucose intolerance, suppressed the abnormal elevation of hepatic gluconeogenesis, and preserved the architecture and function of pancreatic islets (Xu et al. 2018).

Grover et al. (2000) reported a reduction of 73.51% in glucose levels in mice receiving 200 mg/kg per day of lyophilized powder of *E. jambolana* by 3 weeks of treatment. The results describe that treatment with *E. jambolana* increased glucose-6-phosphate, glucokinase, hexokinase, and phosphofructokinase content in the liver; thus, the mechanism of action of *E. jambolana* seems to have an overall effect in increasing glucose utilization.

Momordica charantia (*M. charantia*; Cucurbitaceae) common names: Bitter melon, karela, balsam pear, or bitter gourd

M. charantia grows in tropical or subtropical countries (Habtemariam 2019b). The fruits have been traditionally and extensively used as food and herbal medicine for type 2 diabetes mellitus in Asia, Brazil, east Africa, India, Nigeria and Bangladesh (Peter et al. 2019, Seetaloo et al. 2019). Compounds: Fruits: cucurbitane-type triterpenes, carotenoid, fatty acid, polysaccharides and protein (Poovitha and Parani 2017, Saini et al. 2017, Habtemariam 2019b).

The study by Cortez-Navarrete et al. (2018) found in patients with type 2 diabetes a significant reduction in HbA1c (7.8 ± 0.8 to 7.1 ± 1.3%), and a significant increase in areas under the curve (AUC) of total insulin secretion (0.29–0.18 vs. 0.41–0.29, P = .028), with 2000 mg daily of dried powder of the fruit pulp (two capsules orally of 500 mg twice daily), for three months.

In animal studies, Poovitha et al. (2017) found in STZ-diabetic rats treated with protein extracts (10 or 20 mg/kg body weight) from fruit pulp proteins, a significant reduction in blood glucose levels (72%, whit 20 mg/kg) when compared with diabetic controls. Also, improved oral glucose tolerance and reacquisition of pancreatic islet architecture; diabetic rats islets treated with *M. charantia* showed improvement in size and cellular integrity compared to no trated diabetic rats. Phosphorylated protein kinase B, glucose transporter4 (GLUT4), and phosphorylated AMP-activated protein kinase (pAMPK) expressions were significantly increased in skeletal muscle and adipose tissues.

In another animal study, Ma et al. (2017) showed that 400 mg/kg daily of ethanolic extract of *M. charantia* for 8 weeks produced a marked reduction in fasting blood glucose (7.41 versus 16.63 mmol/L), in high-fat STZ-diabetic rats.

Moringa oleifera (*M. oleifera*; Moringaceae)

It is a tree from India, but today it is present in tropical, subtropical and semi-arid regions of the world (Leone et al. 2015). Part used: leaves, fruits, roots and seeds, Compounds: these parts, in general, contain glucosinolates, isothiocyanates, flavonoids, anthocyanins, proanthocyanidins and cinnamates (Bennett et al. 2003, Amaglo et al. 2010); the pods contain tannins, steroids, triterpenoids, flavonoids, saponins, anthraquinones, alkaloids (Leone et al. 2015, Olayaki et al. 2015); leaves contain Carotenoids; tocopherols; flavonoids; phenolic acids; folate; polyunsaturated and fatty acids among others (Villarruel-López et al. 2018).

Villarruel-López et al. (2018) in their study with ethanol leaf power showed that giving it to Alloxan-induced rat model of diabetes, 50 mg/kg p.o. for 8 weeks, showed a hypoglycemic effect (<250 mg/dL) in the treated group. Meanwhile, Olayaki et al. (2015) found that oral administration of methanolic extracts of *M. oleifera* (300 or 600 mg/kg body weight by 6 weeks in Alloxan-induced rat), improved glucose tolerance, and increased serum insulin levels by 1.3–1.7-fold ($p < 0.01$) and glycogen synthase activities and glycogen contents were higher.

In humans studies, *M. oleifera* leaves have also been shown to reduce glycaemia; so Kumari et al. (2010), in a study with type 2 diabetes patients (between 30–60 years old), showed the hypoglycemic effect of 40-days administration of *M. oleifera* leaves powder (8 gm/day); fasting and post-prandial blood glucose concentrations were significantly reduced in the experimental group (28% to 26%, respectively, compared to the control group).

Some of these results showed that hypoglycemic effects of *M. oleifera* might be mediated through the stimulation of insulin release and sensitivity leading to enhanced glucose uptake and glycogen synthesis in the muscles and liver, decrease of gluconeogenesis enzymes (phosphoenolpyruvate carboxykinase and glucose-6-phosphatase), and inhibition of α-amylase and α-glucosidase activities (glucose oxidation enzymes inhibition of glucose uptake from the intestine) (Kumari 2010, Waterman et al. 2015, Ahmad et al. 2019, Samarakoon et al. 2019).

Nigella sativa (*N. sativa*; Ranunculaceae)

Nigella sativa is an annual flowering plant which grows up to 20–30 cm, mostly cultivated in the Middle Eastern countries and Southwest Asia.

Part used: powder, extract and oil from seed. Compounds: Thymoquinone (its main component), flavonoids, unsaturated fatty acids, nigellone, p-cymene and carvone (Heshmati and Namazi 2015, Choudhury et al. 2018); seed oil have polyphenols and tocopherols (Mohebbati and Abbasnezhad 2020).

Studies: Abbasnezhad et al. (2015) showed that *Nigella sativa* extract (200 mg/kg) for 45 days in STZ-induced diabetic rats significantly reduced serum glucose level compared to the diabetic control group.

In the study conducted by Kaleem et al. (2006) on STZ-induced diabetic rats, the oral administration of ethanolic extract of *N. sativa* seed (300 mg/kg) for 30 days significantly reduced the elevated levels of blood glucose, plasma insulin, and also improved altered levels in liver of antioxidant enzymes such as CAT (0.137 ± 0.019 to 0.210 ± 0.017 U/mg protein*10^3), SOD (14.26 ± 1.30 to 18.34 ± 3.06 U/mg protein), reduced glutathione (30.3 ± 2.34 to 52.7 ± 2.30 mg/100 g of tissue) and GPx (0.168 ± 0.038 to 0.170 ± 0.026 U/mg protein).

Human studies: Heshmati et al. (2015), in 36 patients with type 2 diabetes, showed that *Nigella sativa* oil (3 g/day (one three times a day) for 12 weeks), significantly reduced fasting blood sugar (–9.1%) and glycated hemoglobin (5.1%) in intervention group compared to the placebo group ($p < 0.05$).

Meanwhile, Bamosa et al. (2010) demonstrated that in patients with T2D intake of 2 g/day of *N. sativa* seed for 12 weeks decreased fasting blood sugar (189 ± to 162 ± 9.2 mg/dL), 2 hours postprandial glucose (286 ± 23.3 to 256 ± 28.1 mg/dL), glycated hemoglobin (1.52%), insulin resistance and improvement in the function of the beta cells of the pancreas without any renal or hepatic side effects.

In general, human and animal studies show that *N. Sativa's* action mechanism includes its effects on antioxidants, enzymes, insulin secretion, glucose absorption, gluconeogenesis, improvement in the function of the beta cells and gene expression (Heshmati and Namazi 2015, Farkhondeh et al. 2017, Daryabeygi-Khotbehsara et al. 2017, Hamdan et al. 2019, Mohebbati and Abbasnezhad 2020).

Trigonella foenum-graecum (Fenugreek; Fabaceae)

Though its origin is established in Asia Southwest, it is widely distributed throughout the world, be it wild or cultivated. Part used: leaves and seeds; compounds: seed phytochemical content includes carbohydrate, protein, polyphenol, phytic acid, saponin, and minerals (Zn, Fe, Mn, and Mg). Also contain trigonelline, tannic acid, diosgenin, gitogenin, and trigogenin (Singh et al. 2013, Riasat et al. 2017). Leaves' essential oils have principal compounds as (2E)-hexenal (26.61%), n-hexadecanoic acid (10.14%) and (E)-β-ionone (7.99%) among others (Riasat et al. 2017).

In human studies, Gupta et al. (2001) reported that 1000 mg daily administered for two months in patients newly-diagnosed with type 2 diabetes, no effect on fasting blood glucose was found but significantly improved oral glucose tolerance.

Tavakoly et al. (2018) showed that Fenugreek seed consumption resulted in a significant decrease in high-sensitivity C-reactive protein (p = 0.012) and a significant increase in SOD activity (p = 0.001) in patients with type 2 diabetes mellitus, compared to the control group.

In animal models, Joshi et al. (2015), in a study in rats, found that hydroethanolic extract of seeds of *Trigonella foenum-graecum*, 500–2000 mg/kg, p.o. for 31 days markedly reduced the fasting plasma glucose (7.6 vs 15.5 mmol/L) and HbA1c in alloxan-diabetic rats.

Hamden et al. (2011) also showed that administrated omega-3 fatty acid-rich fenugreek essential oil (5% in food) for 8 weeks in diabetic alloxan-rats was associated with a significant reduction a-amylase (46 and 52%) and maltase activities (37 and 35%), respectively, in pancreas and plasma and a reduction of blood glucose rate by 51% when these were compared to the control.

***Rosmarinus officinalis* L.** (*R. officinalis* Lamiaceae), common name: rosemary

R. officinalis is an aromatic evergreen herb shrub that is widely distributed in the Mediterranean region, but now this plant is cultivated and used worldwide. Parts used: leaves and stems, compounds: phenolic abietane diterpenes as carnosic acid; flavonoids, carnosol, caffeic acid, rosmarinic, alfa-pinene, camphor, eucalyptol, chlorogenic acid, monomeric acid, oleanolic acid, ursolic acid, rosmadial, rosmanol, rosmaquinones A and B, secohinokio, and derivatives of eugenol and luteolin (Cui et al. 2012, Ribeiro-Santos et al. 2015, Sedighi et al. 2015, Sánchez-Camargo and Herrero 2017, Oliveira et al. 2019).

Studies: Nazem et al. (2015) showed that exercise and 200 mg/kgbw of rosemary extract leaves supplied by gastric cannula at 8 AM daily for 8 weeks in STZ-induced-diabetic rats modified the antioxidant enzyme activities: superoxide dismutase (1442.1 ± 33.56 U/gHb), glutathione peroxidase (66.34 ± 3.36 U/gHb) and catalase (10.31 ± 1.02 U/gHb) until reaching the normal levels of those in the healthy group (1426 ± 22.57, 63.36 ± 2.55 and 11.04 ± 1.08 U/gHb, respectively); levels of SOD, GPx and CAT in diabetic and sedentary rats were 1189.1 ± 14.79; 39.83 ± 2.23 and 6.32 ± 0.97 U/gHb, respectively.

Bakırel et al. (2008), meanwhile, found that in alloxan-induced diabetic rabbits, 200 mg/kg of ethanolic extract of the leaves of *R. officinalis*, administered orally in a volume of 10 ml/kg, significantly lowered blood glucose level (383.57 ± 14.87 to 277.43 ± 16.01 mg/dL) after six hours, and increased serum insulin (6.21 ± 0.40 to 9.92 ± 0.47 μIU/ml) with $p < 0.01$ in both variables. Bakırel et al. (2008) also showed a gradual rise in serum CAT activity to reach a significant difference on 5th (32.4%) and 8th day (35.4%) as compared with diabetic control rabbits.

In general, the action *R. officinalis* in diabetes includes glucose reduction, increased activity of antioxidant enzymes and increased insulin or its effect (Sedighi et al. 2015, Selmi et al. 2017, Andrade et al. 2018, Borges et al. 2019, Karadağ et al. 2019). Figure 3.2 summarizes the mechanisms of action and therapeutic targets of the plants included in this study.

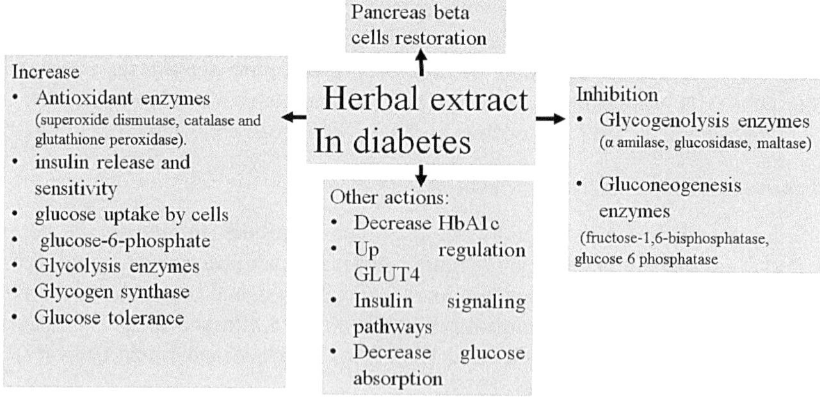

Figure 3.2. Summary of the mechanisms of action and therapeutic targets for *Artemisia herba alba, Brassica juncea, Citrullus colocynthis, Eugenia jambolana, Momordica charantia, Moringa oleífera, Nigella sativa, Rosmarinus officinalis* and *Trigonella foenum-graecum*.

Conclusion

Diabetes mellitus is a metabolic disease, characterized by high blood glucose levels. Maintaining glucose levels in normal ranges is the therapeutic objective in the treatment of DM. The traditional treatment of DM has been with insulin or oral hypoglycemic agents or both. Although it has been effective, it has some side effects. So the use of native plants as adjuvants is a proven alternative through many animal studies and human trials in hundreds of plants.

Many plants in folkloric or traditional medicine are used for diabetes management, some have clinical evidence, some have scientific evidence of their mechanism of action and others did not, but the use of medicinal plants to manage diabetes is a reality. This chapter provides knowledge of six medicinal plants used in diabetes treatment; it includes some active compounds from them and shows how these affect antioxidant enzymes, glycolysis, gluconeogenesis, glycogenolysis, glycogenesis, insulin release and restoration of beta pancreas cells. It is undeniable that wild plants are a source of drugs for diabetes and therefore need to be studied in depth.

References

Abbasnezhad, A., Hayatdavoudi, P., Niazmand, S., Mahmoudabady, M. 2015. The effects of hydroalcoholic extract of *Nigella sativa* seed on oxidative stress in hippocampus of STZ-induced diabetic rats. Avicenna J. Phytomed., 5(4): 333–340. https://doi.org/10.22038/ajp.2015.3921.

Ahmad, J., Khan, I., Blundell, R. 2019. *Moringa oleifera* and glycemic control: A review of current evidence and possible mechanisms. Phytother. Res., 33(11): 2841–2848. https://doi.org/10.1002/ptr.6473.

Al-Shamaony, L., Al-Khazraji, S.M., Twaij, H.A.A. 1994. Hypoglycaemic effect of *Artemisia herba alba*. II. Effect of a valuable extract on some blood parameters in diabetic animals. J. Ethnopharmacol., 43(3): 167–171. https://doi.org/10.1016/0378-8741(94)90038-8.

Al-waili, N.S.D. 1986. Treatment of diabetes mellitus by *Artemisia herba alba* extract: Preliminary study. Clin. Exp. Pharmacol. Physiol., 13(7): 569–73. doi:10.1111/j.1440-1681.1986.tb00940.x.

Amaglo, N.K., Bennett, R.N., Lo Curto, R.B., Rosa, E.A.S., Lo Turco, V., Giuffrida, A., Lo Curto, A., Crea, F., Timpo, G.M. 2010. Profiling selected phytochemicals and nutrients in different tissues of the multipurpose tree *Moringa oleifera* L., grown in Ghana. Food Chem., 122(4): 1047–1054. https://doi.org/10.1016/j.foodchem.2010.03.073.

Andrade, J.M., Faustino, C., Garcia, C., Ladeiras, D., Reis, C.P., Rijo, P. 2018. *Rosmarinus officinalis* L.: An update review of its phytochemistry and biological activity. Future Sci. OA, 4(4). https://doi.org/10.4155/fsoa-2017-0124.

Arumugam, G., Manjula, P., Paari, N. 2013. A review: Anti diabetic medicinal plants used for diabetes mellitus. J. Acute Dis., 2(3): 196–200. https://doi.org/10.1016/s2221-6189(13)60126-2.

Atkinson, M.A., Eisenbarth, G.S., Michels, A.W. 2014. Type 1 diabetes. Lancet, 383(9911): 69–82. https://doi.org/10.1016/S0140-6736(13)60591-7.

Aziz, S., Saha, K., Sultana, N., Nur, H.P., Ahsan, M.A., Ahmed, S., Hossain, M.K. 2016. Comparative studies of elemental composition in leaves and flowers of *Catharanthus roseus* growing in Bangladesh. Asian Pac. J. Trop. Biomed., 6(1): 50–54. https://doi.org/10.1016/j.apjtb.2015.10.003.

Baeyens, L., Lemper, M., Staels, W., De Groef, S., De Leu, N., Heremans, Y. German, M.S., Heimberg, H. 2018. (Re) generating human beta cells: Status, pitfalls, and perspectives. Physiol. Rev., 98(3): 1143–1167. https://doi.org/10.1152/physrev.00034.2016.

Bakirel, T., Bakirel, U., Keleş, O.Ü., Ülgen, S.G., Yardibi, H. 2008. *In vivo* assessment of antidiabetic and antioxidant activities of rosemary (*Rosmarinus officinalis*) in alloxan-diabetic rabbits. J. Ethnopharmacol., 116(1): 64–73. https://doi.org/10.1016/j.jep.2007.10.039.

Baliga, M.S., Bhat, H.P., Baliga, B.R.V., Wilson, R., Palatty, P.L. 2011. Phytochemistry, traditional uses and pharmacology of *Eugenia jambolana* Lam. (black plum): A review. Food Res. Int., 44(7): 1776–1789. https://doi.org/10.1016/j.foodres.2011.02.007.

Bamosa, A.O., Kaatabi, H., Lebda, F.M., Al Elq, A.M., Al-Sultan, A. 2010. Effect of *Nigella Sativa* seeds on the glycemic control of patients with type 2 diabetes mellitus. Indian J. Physiol. Pharmacol., 54(4): 344–354.

Barghamdi, B., Ghorat, F., Asadollahi, K., Sayehmiri, K., Peyghambari, R., Abangah, G. 2016. Therapeutic effects of *Citrullus colocynthis* fruit in patients with type II diabetes: A clinical trial study. J. Pharm. Bioallied Sci., 8(2): 130–134. https://doi.org/10.4103/0975-7406.171702.

Barkat, A., Mujeeb, M. 2013. Comparative study of *Catharanthus roseus* extract and extract loaded chitosan nanoparticles in alloxan induced diabetic rats. Int. J. Biomed. Res., 04(12): 670–678. https://doi.org/10.7439/ijbr.

Benariba, N., Djaziri, R., Bellakhdar, W., Belkacem, N., Kadiata, M., Malaisse, W.J., Sener, A. 2013. Phytochemical screening and free radical scavenging activity of *Citrullus colocynthis* seeds extracts. Asian Pac. J. Trop. Biomed., 3(1): 35–40. https://doi.org/10.1016/S2221-1691(13)60020-9.

Bennett, R.N., Mellon, F.A., Foidl, N., Pratt, J.H., Dupont, M.S., Perkins, L., Kroon, P.A. 2003. Profiling glucosinolates and phenolics in vegetative and reproductive tissues of the multi-purpose trees *Moringa oleifera* L. (Horseradish tree) and Moringa stenopetala L. J. Agr. Food Chem., 51(12): 3546–3553. https://doi.org/10.1021/jf0211480.

Bhagour, K., Arya, D., Gupta, R.S. 2016. A review: Antihyperglycemic plant medicines in management of diabetes. Acupunct. Relat. Ther., 4(4): 7–16. https://doi.org/10.1016/j.arthe.2016.11.001.

Borges, R.S., Ortiz, B.L.S., Pereira, A.C.M., Keita, H., Carvalho, J.C.T. 2019. *Rosmarinus officinalis* essential oil: A review of its phytochemistry, anti-inflammatory activity, and mechanisms of action involved. J. Ethnopharmacol., 229(September 2018): 29–45. https://doi.org/10.1016/j.jep.2018.09.038.

Boudjelal, A., Siracusa, L., Henchiri, C., Sarri, M., Abderrahim, B., Baali, F., Ruberto, G. 2015. Antidiabetic effects of aqueous infusions of Artemisia herba-alba and Ajuga iva in alloxan-induced diabetic rats. Planta Med., 81(9): 696–704. https://doi.org/10.1055/s-0035-1546006.

Bullard, K.M., Cowie, C.C., Lessem, S.E., Saydah, S.H., Menke, A., Geiss, L.S. Orchard, T.J., Rolka, D.B., Imperatore, G. 2016. Prevalence of diagnosed diabetes in adults by diabetes type-United States-2016. MWR Weekly, 2018; 67(12): 359–361.

Chakraborty, R., Roy, S., Mandal, V. 2016. Assessment of traditional knowledge of the antidiabetic plants of Darjeeling and Sikkim Himalayas in the context of recent phytochemical and pharmacological advances. J. Integr. Med., 14(5): 336–358. https://doi.org/10.1016/S2095-4964(16)60267-4.

Chekroun, E., Benariba, N., Adida, H., Bechiri, A., Azzi, R., Djaziri, R. 2015. Antioxidant activity and phytochemical screening of two Cucurbitaceae: *Citrullus colocynthis* fruits and Bryonia dioica roots. Asian Pac. J. Trop. Dis., 5(8): 632–637. https://doi.org/10.1016/S2222-1808(15)60903-3.

Chinsembu, K.C. 2019. Diabetes mellitus and nature's pharmacy of putative antidiabetic plants. J. Herb. Med., 15(August 2018): 100230. https://doi.org/10.1016/j.hermed.2018.09.001.

Choudhury, H., Pandey, M., Hua, C.K., Mun, C.S., Jing, J.K., Kong, L., Ern, L.Y., Ashraf, N.A., Kit, S.W., Yee, T.S., Pichika, M.R., Gorain, B., Kesharwani, P. 2018. An update on natural compounds in the remedy of diabetes mellitus: A systematic review. J. Tradit. Complement Med., 8(3): 361–376. https://doi.org/10.1016/j.jtcme.2017.08.012.

Chukwuma, C.I., Matsabisa, M.G., Ibrahim, M.A., Erukainure, O.L., Chabalala, M.H., Islam, M.S. 2019. Medicinal plants with concomitant anti-diabetic and anti-hypertensive effects as potential sources of dual acting therapies against diabetes and hypertension: A review. J. Ethnopharmacol., 235: 329–360. https://doi.org/10.1016/j.jep.2019.02.024.

Consultation Report of a WHO/IDF. 2006. Definition and diagnosis of diabetes mellitus and intermediate hyperglycemia. In World Health Organization.

Cortez-Navarrete, M., Martínez-Abundis, E., Pérez-Rubio, K.G., González-Ortiz, M., Méndez-Del Villar, M. 2018. *Momordica charantia* administration improves insulin secretion in type 2 diabetes mellitus. J. Med. Food, 21(7): 672–677. https://doi.org/10.1089/jmf.2017.0114.

Cui, L., Kim, M.O., Seo, J.H., Kim, I.S., Kim, N.Y., Lee, S.H. Park, J., Kim, J., Lee, HS. 2012. Abietane diterpenoids of Rosmarinus officinalis and their diacylglycerol acyltransferase-inhibitory activity. Food Chem., 132(4): 1775–1780. https://doi.org/10.1016/j.foodchem.2011.11.138.

Dallak, M., Bashir, N., Abbas, M., Elessa, R., Haidara, M., Khalil, M., Al-Khateeb, M.A. 2009. Concomitant down regulation of glycolytic enzymes, upregulation of gluconeogenic enzymes and potential hepato-nephro-protective effects following the chronic administration of the hypoglycemic, insulinotropic *Citrullus colocynthis* pulp extract. Am. J. Biochem. Biotechnol., 5(4): 153–161. https://doi.org/10.3844/ajbbsp.2009.153.161.

Daryabeygi-Khotbehsara, R., Golzarand, M., Ghaffari, M.P., Djafarian, K. 2017. *Nigella sativa* improves glucose homeostasis and serum lipids in type 2 diabetes: A systematic review and meta-analysis. Complement Ther. Med., 35(April): 6–13. https://doi.org/10.1016/j.ctim.2017.08.016.

de Almeida, V.L., Silva, C.G., Silva, A.F., Campana, P.R.V., Foubert, K., Lopes, J.C.D., Pieters, L. 2019. Aspidosperma species: A review of their chemistry and biological activities. J. Ethnopharmacol., 231(November 2018): 125–140. https://doi.org/10.1016/j.jep.2018.10.039.

Farkhondeh, T., Samarghandian, S., Borji, A. 2017. An overview on cardioprotective and anti-diabetic effects of thymoquinone. Asian Pac. J. Trop. Med., 10(9): 849–854. https://doi.org/10.1016/j.apjtm.2017.08.020.

Furman, B.L., Candasamy, M., Bhattamisra, S.K., Veettil, S.K. 2020. Reduction of blood glucose by plant extracts and their use in the treatment of diabetes mellitus; discrepancies in effectiveness between animal and human studies. J. Ethnopharmacol., 247(May 2019): 112264. https://doi.org/10.1016/j.jep.2019.112264.

Ghadge, A.A., Harke, S.M., Khadke, S.P., Diwan, A.G., Pankaj, M., Kulkarni, O.P., Ranjekar, P.K., Kuvalekar, A.A. 2014. Circulatory adipocytokines and lipid profile variations in type-2 diabetic subjects: Desirable side-effects of antidiabetic drugs. Diabetes Metabo. Syndr. Clin. Res. Rev., 8(4): 230–232. https://doi.org/10.1016/j.dsx.2014.09.010.

Ghadge, A.A., Kuvalekar, A.A. 2017. Controversy of oral hypoglycemic agents in type 2 diabetes mellitus: Novel move towards combination therapies. Diabetes Metab. Syndr. Clin. Res. Rev., 11: S5–S13. https://doi.org/10.1016/j.dsx.2016.08.009.

Gilon, P. 2020. The role of α-cells in islet function and glucose homeostasis in health and type 2 diabetes. J. Mol. Biol. https://doi.org/10.1016/J.JMB.2020.01.004.

Goboza, M., Aboua, Y.G., Chegou, N., Oguntibeju, O.O. 2019. Vindoline effectively ameliorated diabetes-induced hepatotoxicity by docking oxidative stress, inflammation and hypertriglyceridemia in type 2 diabetes-induced male Wistar rats. Biomed. Pharmacother., 112: 108638. https://doi.org/10.1016/j.biopha.2019.108638.

Grover, J.K., Vats, V., Rathi, S.S. 2000. Anti-hyperglycemic effect of *Eugenia jambolana* and *Tinospora cordifolia* in experimental diabetes and their effects on key metabolic enzymes involved in carbohydrate metabolism. J. Ethnopharmacol., 73(3): 461–470. https://doi.org/10.1016/S0378-8741(00)00319-6.

Gupta, A.K., Gupta, R. 2001. Effect of *Trigonella foenum-graecum* (fenugreek) seeds on glycaemic control and insulin resistance in type 2 diabetes mellitus: a double blind placebo controlled study. J. Assoc. Physicians India, 49: 1057–1061.

Habtemariam, S. 2019a. Glucose metabolism: Normal physiology, diabetic dysregulation, and therapeutic targets. Chapter 17 in Medicinal Foods as Potential Therapies for Type-2 Diabetes and Associated Diseases, 13–39. Academic Press. https://doi.org/10.1016/b978-0-08-102922-0.00002-x.

Habtemariam, S. 2019b. The chemical and pharmacological basis of bitter melon (*Momordica charantia* L.) as a potential therapy for type 2 diabetes and obesity. Chapter 2 in Medicinal Foods as Potential Therapies for Type-2 Diabetes and Associated Diseases. pg. 14–39. Academic Press. https://doi.org/10.1016/b978-0-08-102922-0.00008-0.

Hamdan, A., Idrus, R.H., Mokhtar, M.H. 2019. Effects of *Nigella sativa* on type-2 diabetes mellitus: A systematic review. Int. J. Environ. Res. Public Health, 16(24). https://doi.org/10.3390/ijerph16244911.

Hamden, K., Keskes, H., Belhaj, S., Mnafgui, K., Feki, A., Allouche, N. 2011. Inhibitory potential of omega-3 fatty and fenugreek essential oil on key enzymes of carbohydrate-digestion and hypertension in diabetes rats. Lipids Health Dis., 10(1): 226. https://doi.org/10.1186/1476-511X-10-226.

Heshmati, J., Namazi, N. 2015. Effects of black seed (*Nigella sativa*) on metabolic parameters in diabetes mellitus: A systematic review. Complement Ther. Med., 23(2): 275–282. https://doi.org/10.1016/j.ctim.2015.01.013.

Heshmati, J., Namazi, N., Memarzadeh, M.R., Taghizadeh, M., Kolahdooz, F. 2015. *Nigella sativa* oil affects glucose metabolism and lipid concentrations in patients with type 2 diabetes: A randomized, double-blind, placebo-controlled trial. Food Res. Int., 70: 87–93. https://doi.org/10.1016/j.foodres.2015.01.030.

Heydari, M., Hashempur, M.H., Ostovar, M., Shams, M. 2019. *Citrullus colocynthis* and Its Potential Role Against Diabetes and Its Complications. Bioactive Food as Dietary Interventions for Diabetes (2nd ed.). Elsevier Inc. https://doi.org/10.1016/b978-0-12-813822-9.00032-1.

Heywood, V.H. 2011. Ethnopharmacology, food production, nutrition and biodiversity conservation: Towards a sustainable future for indigenous peoples. J. Ethnopharmacol., 137(1): 1–15. https://doi.org/10.1016/j.jep.2011.05.027.

Huseini, H. Fallah, Darvishzadeh, F., Heshmat, R., Jafariazar, Z., Raza, Mohsin, Larijan, B. 2009. Clinical investigation of *Citrullus colocynthis* (L.) Schrad fruit in treatment of type II diabetic patients: A randomized, double blind, placebo-controlled clinical trial. Phytother. Res., 23: 1186–1189. Retrieved from http://www3.interscience.wiley.com/journal/117934759/abstract.

Hussain, A.I., Rathore, H.A., Sattar, M.Z.A., Chatha, S.A.S., Sarker, S.D., Gilani, A.H. 2014. *Citrullus colocynthis* (L.) Schrad (bitter apple fruit): A review of its phytochemistry, pharmacology, traditional uses and nutritional potential. J. Ethnopharmacol., 155(1): 54–66. https://doi.org/10.1016/j.jep.2014.06.011.

Joshi, D.V., Patil, R.R., Naik, S.R. 2015. Hydroalcohol extract of *Trigonella foenum-graecum* seed attenuates markers of inflammation and oxidative stress while improving exocrine function in diabetic rats. Pharm. Biol., 53(2): 201–211. https://doi.org/10.3109/13880209.2014.913296.

Kaleem, M., Kirmani, D., Asif, M., Ahmed, Q., Bano, B. 2006. Biochemical effects of *Nigella sativa* L. seeds in diabetic rats. Indian J. Exp. Biol., 44(9): 745–748.

Karadağ, A.E., Demirci, B., Çaşkurlu, A., Demirci, F., Okur, M.E., Orak, D., Sipahi, H., Başer, K.H.C. 2019. *In vitro* antibacterial, antioxidant, anti-inflammatory and analgesic evaluation of *Rosmarinus officinalis* L. flower extract fractions. S African J. Bot., 125: 214–220. https://doi.org/10.1016/j.sajb.2019.07.039.

Kavvoura, F.K., Owen, K.R. 2019. Monogenic diabetes. Medicine (United Kingdom), 47(1): 16–21. https://doi.org/10.1016/j.mpmed.2018.10.007.

Kumar, A., Rana, A.K., Singh, A., Singh, A. 2019. Bioactivity of methanolic extract of *Brassica juncea* in animal model of diabetes mellitus. Chin. Herb Med., 11(4): 434–437. https://doi.org/10.1016/j.chmed.2019.04.007.

Kumari, D.J. 2010. Hypoglycaemic effect of *Moringa oleifera* and Azadirachta indica in type 2 Diabetes mellitus. The Bioscan., 5(2): 211–214.

Leone, A., Spada, A., Battezzati, A., Schiraldi, A., Aristil, J., Bertoli, S. 2015. Cultivation, genetic, ethnopharmacology, phytochemistry and pharmacology of *Moringa oleifera* leaves: An overview. Int. J. Molecular Sciences, 16(6): 12791–12835. https://doi.org/10.3390/ijms160612791.

Leonti, M., Stafford, G.I., Cero, M.D., Cabras, S., Castellanos, M.E., Casu, L., Weckerle, C.S. 2017. Reverse ethnopharmacology and drug discovery. J. Ethnopharmacol., 198(December 2016): 417–431. https://doi.org/10.1016/j.jep.2016.12.044.

Ma, C., Yu, H., Xiao, Y., Wang, H. 2017. *Momordica charantia* extracts ameliorate insulin resistance by regulating the expression of socs-3 and jnk in type 2 diabetes mellitus rats. Pharm. Biol., 55(1): 2170–2177. https://doi.org/10.1080/13880209.2017.1396350.

Mohamed, A.E.H.H., El-Sayed, M.A., Hegazy, M.E., Helaly, S.E., Esmail, A.M., Mohamed, N.S. 2010. Chemical constituents and biological activities of *Artemisia herba-alba*. Rec. Nat. Prod., 4(1): 1–25.

Mohebbati, R., Abbasnezhad, A. 2020. Effects of *Nigella sativa* on endothelial dysfunction in diabetes mellitus: A review. J. Ethnopharmacol., 252: 112585. https://doi.org/10.1016/j.jep.2020.112585.

Mukherjee, P.K. 2019. Ethnopharmacology and ethnomedicine-inspired drug development. chapter 2, in Quality Control and Evaluation of Herbal Drugs, pg. 29–51. 1st Edition. Print Book & E-Book. Elsevier https://doi.org/10.1016/b978-0-12-813374-3.00002-8.

Mulukutla, B.C., Khan, S., Lange, A., Hu, W.S. 2010. Glucose metabolism in mammalian cell culture: New insights for tweaking vintage pathways. Trends Biotechnol., 28(9): 476–484. https://doi.org/10.1016/j.tibtech.2010.06.005.

Nammi, S., Boini, M.K., Lodagala, S.D., Behara, R.B.S. 2003. The juice of fresh leaves of *Catharanthus roseus* Linn. reduces blood glucose in normal and alloxan diabetic rabbits. BMC Complement Altern. Med., 3: 2–5. https://doi.org/10.1186/1472-6882-3-4.

Nazem, F., Farhangi, N., Neshat-Gharamaleki, M. 2015. Beneficial effects of endurance exercise with *Rosmarinus officinalis* Labiatae leaves extract on blood antioxidant enzyme activities and lipid peroxidation in streptozotocin-induced diabetic rats. Can J. Diabetes, 39(3): 229–234. https://doi.org/10.1016/j.jcjd.2014.11.003.

Neamsuvan, O., Madeebing, N., Mah, L., Lateh, W. 2015. A survey of medicinal plants for diabetes treating from Chana and Nathawee district, Songkhla province, Thailand. J. Ethnopharmacol., 174: 82–90. https://doi.org/10.1016/j.jep.2015.07.050.

Njume, C., Donkor, O., McAinch, A.J. 2019. Predisposing factors of type 2 diabetes mellitus and the potential protective role of native plants with functional properties. J. Funct. Foods, 53(December 2018): 115–124. https://doi.org/10.1016/j.jff.2018.12.001.

Ogurtsova, K., da Rocha Fernandes, J.D., Huang, Y., Linnenkamp, U., Guariguata, L., Cho, N.H., Cavan, D., Shaw, J.E., Makaroff, L.E. 2017. IDF diabetes Atlas: Global estimates for the prevalence of diabetes for 2015 and 2040. Diabetes Res. Clin. Pract., 128: 40–50. https://doi.org/10.1016/j.diabres.2017.03.024.

Ohadoma, S.C., Michael, H.U. 2011. Effects of co-administration of methanol leaf extract of *Catharanthus roseus* on the hypoglycemic activity of metformin and glibenclamide in rats. Asian Pac. J. Trop. Med., 4(6): 475–477. https://doi.org/10.1016/S1995-7645(11)60129-6.

Olayaki, L.A., Irekpita, J.E., Yakubu, M.T., Ojo, O.O. 2015. Methanolic extract of *Moringa oleifera* leaves improves glucose tolerance, glycogen synthesis and lipid metabolism in alloxan-induced diabetic rats. J. Basic Clin. Physiol. Pharmacol., 26(6): 585–593. https://doi.org/10.1515/jbcpp-2014-0129.

Oliveira, J.R. De, Esteves, S., Camargo, A. 2019. *Rosmarinus officinalis* L. (rosemary) as therapeutic and prophylactic agent. J. Biomed. Sci., 8: 1–22.

Pan, Q., Saiman, M.Z., Mustafa, N.R., Verpoorte, R., Tang, K. 2016. A simple and rapid HPLC-DAD method for simultaneously monitoring the accumulation of alkaloids and precursors in different parts and different developmental stages of *Catharanthus roseus* plants. J. Chromatogr. B: Analytl. Technol. Biomed. Life Sci., 1014: 10–16. https://doi.org/10.1016/j.jchromb.2016.01.034.

Peter, E.L., Kasali, F.M., Deyno, S., Mtewa, A., Nagendrappa, P.B., Tolo, C.U., Ogwang, P.E., Sesaazi, D. 2019. *Momordica charantia* L. lowers elevated glycaemia in type 2 diabetes mellitus patients: Systematic review and meta-analysis. J. Ethnopharmacol., 231(October 2018): 311–324. https://doi.org/10.1016/j.jep.2018.10.033.

Poovitha, S., Parani, M. 2017. Protein extract from the fruit pulp of *Momordica charantia* potentiate glucose uptake by up-regulating GLUT4 and AMPK. J. Funct. Foods, 37: 507–512. https://doi.org/10.1016/j.jff.2017.08.022.

Pozo, M., Claret, M. 2018. Hypothalamic control of systemic glucose homeostasis: the pancreas connection. Trends Endocrinol. Metab., 29(8): 581–594. https://doi.org/10.1016/j.tem.2018.05.001.

Ravi, K., Ramachandran, B., Subramanian, S. 2004. Effect of *Eugenia jambolana* seed kernel on antioxidant defense system in streptozotocin-induced diabetes in rats. Life Sci., 75(22): 2717–2731. https://doi.org/10.1016/j.lfs.2004.08.005.

Riasat, M., Jafari, A.A., Bahmanzadegan, A., Hatami, A., Zareiyan, F. 2017. The constituents of essential oil in leaves of Karaj accession of *Trigonella foenum graecum*. Natu. Prod. Res., 31(14): 1709–1712. https://doi.org/10.1080/14786419.2017.1286484.

Ribeiro-Santos, R., Carvalho-Costa, D., Cavaleiro, C., Costa, H.S., Albuquerque, T.G., Castilho, M.C., Ramos, F., Melo, N.R., Sanches-Silva, A. 2015. A novel insight on an ancient aromatic plant: The rosemary (*Rosmarinus officinalis* L.). Trends Food Sci. Technol., 45(2): 355–368. https://doi.org/10.1016/j.tifs.2015.07.015.

Rodriguez, B.L., Abbot, R.D., fujimoto, W., Waitzfelder, B., Chen, R., Masaki, K., Schatz, I., Petrovitch, H., Ross, W., Yano, K., Blanchette, P.L., Curb, J.D. 2002. The american diabetes association classifications for diabetes and World Health Organization. Diabetes Care, 25(6): 951–955.

Saini, R.K., Assefa, A.D., Keum, Y.S. 2017. Fatty acid and carotenoid composition of bitter melon (*Momordica charantia* L.) seed arils: a potentially valuable source of lycopene. J. Food Meas. Charact., 11(3): 1266–1273. https://doi.org/10.1007/s11694-017-9504-7.

Salido, S., Valenzuela, L.R., Altarejos, J., Nogueras, M., Sánchez, A., Cano, E. 2004. Composition and infraspecific variability of *Artemisia herba-alba* from southern Spain. Biochem. Syst. Ecol., 32(3): 265–277. https://doi.org/10.1016/j.bse.2003.09.002.

Samarakoon, D.N.A.W., Uluwaduge, D.I., Siriwardhene, M.A. 2019. Mechanisms of action of Sri Lankan herbal medicines used in the treatment of diabetes: A review. J. Integr. Med., 18(1): 14–20. https://doi.org/10.1016/j.joim.2019.11.003.

Sánchez-Camargo, A., del, P., Herrero, M. 2017. Rosemary (*Rosmarinus officinalis*) as a functional ingredient: recent scientific evidence. Curr. Opin. Food Sci., 14: 13–19. https://doi.org/10.1016/j.cofs.2016.12.003.

Sebbagh, N., Cruciani-Guglielmacci, C., Ouali, F., Berthault, M.F., Rouch, C., Sari, D.C., Magnan, C. 2009. Comparative effects of Citrullus colocynthis, sunflower and olive oil-enriched diet in streptozotocin-induced diabetes in rats. Diabetes Metab., 35(3): 178–184. https://doi.org/10.1016/j.diabet.2008.10.005.

Sedighi, R., Zhao, Y., Yerke, A., Sang, S. 2015. Preventive and protective properties of rosemary (*Rosmarinus officinalis* L.) in obesity and diabetes mellitus of metabolic disorders: A brief review. Curr. Opin. Food Sci., 2: 58–70. https://doi.org/10.1016/j.cofs.2015.02.002.

Seetaloo, A.D., Aumeeruddy, M.Z., Rengasamy Kannan, R.R., Mahomoodally, M.F. 2019. Potential of traditionally consumed medicinal herbs, spices, and food plants to inhibit key digestive enzymes geared towards diabetes mellitus management—A systematic review. S African J. Bot., 120: 3–24. https://doi.org/10.1016/j.sajb.2018.05.015.

Sekiou, O., Boumendjel, M., Taibi, F., Tichati, L., Boumendjel, A., Messarah, M. 2020. Nephroprotective effect of *Artemisia herba alba* aqueous extract in alloxan-induced diabetic rats. J. Tradit. Complement Med., https://doi.org/10.1016/j.jtcme.2020.01.001.

Selmi, S., Rtibi, K., Grami, D., Sebai, H., Marzouki, L. 2017. Rosemary (*Rosmarinus officinalis*) essential oil components exhibit anti-hyperglycemic, anti-hyperlipidemic and antioxidant effects in experimental diabetes. Pathophysiology, 24(4): 297–303. https://doi.org/10.1016/j.pathophys.2017.08.002.

Shafaei, H., Rad, J., Behjati, M., Delazar, A. 2014. The effect of pulp and seed extract of *Citrullus Colocynthis*, as an antidaibetic medicinal herb, on hepatocytes glycogen stores in diabetic rabbits. Adv. Biomed. Res., 3(1): 258. https://doi.org/10.4103/2277-9175.148230.

Shao, J.W., Jiang, J.L., Zou, J.J., Yang, M.Y., Chen, F.M., Zhang, Y.J., Jia, L. 2020. Therapeutic potential of ginsenosides on diabetes: From hypoglycemic mechanism to clinical trials. J. Funct. Foods, 64(July 2019): 103630. https://doi.org/10.1016/j.jff.2019.103630.

Sharma, A., Rai, P.K., Prasad, S. 2018. GC–MS detection and determination of major volatile compounds in *Brassica juncea* L. leaves and seeds. Microchem. J., 138: 488–493. https://doi.org/10.1016/j.microc.2018.01.015.

Shi, C., Karim, S., Wang, C., Zhao, M., Murtaza, G. 2014. A review on antidiabetic activity of *Citrullus colocynthis* schrad. Acta Pol. Pharm., 71(3): 363–367.

Singh, K.P., Nair, B., Kumar Jain, P., Kumar Naidu, A., Singh, K., Vigyan Kendra, K. 2013. Variability in the nutraceutical properties of fenugreek (*Trigonella foenum-graecum* L.) seeds. Rev. Colomb. Cienc. Hortic., 7(2): 228–239.

Singh, S.N., Vats, P., Suri, S., Shyam, R., Kumria, M.M.L., Ranganathan, S., Sridharan, K. 2001. Effect of an antidiabetic extract of *Catharanthus roseus* on enzymic activities in streptozotocin induced diabetic rats. J. Ethnopharmacol., 76(3): 269–277. https://doi.org/10.1016/S0378-8741(01)00254-9.

Szollosi, R. 2011. Indian Mustard (*Brassica juncea* L.) Seeds in Health. Nuts and Seeds in Health and Disease Prevention. Elsevier Inc. https://doi.org/10.1016/B978-0-12-375688-6.10078-7.

Tavakoly, R., Maracy, M.R., Karimifar, M., Entezari, M.H. 2018. Does fenugreek (*Trigonella foenum-graecum*) seed improve inflammation, and oxidative stress in patients with type 2 diabetes mellitus? A parallel group randomized clinical trial. Eur. J. Integr. Med., 18: 13–17. https://doi.org/10.1016/j.eujim.2018.01.005.

Thirumalai, T., Therasa, S.V., Elumalai, E.K., David, E. 2011. Hypoglycemic effect of *Brassica juncea* (seeds) on streptozotocin induced diabetic male albino rat. Asian Pac. J. Trop. Biomed., 1(4): 323–325. https://doi.org/10.1016/S2221-1691(11)60052-X.

van Duinkerken, E., Ryan, C.M. 2020. Diabetes mellitus in the young and the old: Effects on cognitive functioning across the life span. Neurobiol. Dis., 134(March 2019). https://doi.org/10.1016/j.nbd.2019.104608.

Villarruel-López, A., López-de la Mora, D.A., Vázquez-Paulino, O.D., Puebla-Mora, A.G., Torres-Vitela, M.R., Guerrero-Quiroz, L.A., Nuño, K. 2018. Effect of *Moringa oleifera* consumption on diabetic rats. BMC Complement Altern. Med., 18(1): 1–10. https://doi.org/10.1186/s12906-018-2180-2.

Waterman, Carrie, Rojas-Silva, Patricio, Tumer, Tugba Boyunegmez, Kuhn, Peter, Richard, Allison J., Wicks, Shawna, Stephens, Jacqueline M. Wang, Zhong, Mynatt, Randy, Cefalu, William and Raskin, I. 2015. Isothiocyanate-rich

Moringa oleifera extract reduces weight gain, insulin resistance and hepatic gluconeogenesis in mice. Mol. Nutr. Food Res., 59(6): 1013–1024. https://doi.org/10.1002/mnfr.201400679.
WHO Technical Report Series (TRS 646) Who expert committee on diabetes mellitus. Second report. 1980. https://apps.who.int/iris/bitstream/handle/10665/41399/WHO_TRS_646.pdf?sequence=1&isAllowed=y.
WHO Global Report. 2016. Global Report on Diabetes. Isbn, 978: 6–86. Retrieved from http://www.who.int/about/licensing/.
Wou, C., Unwin, N., Huang, Y., Roglic, G. 2019. Implications of the growing burden of diabetes for premature cardiovascular disease mortality and the attainment of the Sustainable Development Goal target 3.4. Cardiovasc. Diagn-Ther., 9(2): 140–149. https://doi.org/10.21037/cdt.2018.09.04.
Xu, J., Liu, T., Li, Y., Yuan, C., Ma, H., Seeram, N.P., Liu, F., Mu, Y., Huang, X. Li, L. 2018. Hypoglycemic and hypolipidemic effects of triterpenoid-enriched Jamun (*Eugenia jambolana* Lam.) fruit extract in streptozotocin-induced type 1 diabetic mice. Food Funct., 9(6): 3330–3337. https://doi.org/10.1039/c8fo00095f.
Zhao, C., Yang, C., Liu, B., Lin, L., Sarker, S.D., Nahar, L. Yu, H., Cao, H., Xiao, J. 2018. Bioactive compounds from marine macroalgae and their hypoglycemic benefits. Trends Food Sci. Technol., 72(December 2017): 1–12. https://doi.org/10.1016/j.tifs.2017.12.001.

Chapter 4

Ethnopharmacology of Wild Plants from the Tropical Mountains of Northern Peru

Fidel A. Torres-Guevara,[1,2] *Mayar L. Ganoza-Yupanqui*,*[,2,3]
Luz A. Suárez-Rebaza,[4] *Gonzalo R. Malca-García*[5] and *Rainer W. Bussmann*[6]

Introduction

In harnessing biodiversity as an opportunity for the Peruvian economy, there is a gap in inequity in the benefits that the industry derives from the results of knowledge-based and plant-based pharmacological research provided by rural societies, which in most cases do not get any return from the profits obtained by the pharmaceutical industry.

In a closer relationship between traditional and scientific knowledge needed to respond to the challenges of public health, science has created the concepts of ethnobotany, ethnomedicine, traditional medicine and ethnopharmacology. These can help to guide and delimit its practical actions and expected results, to achieve its usefulness to determine the boundaries of the meanings of research results, and properly focus the strategies of their implementation and the factors that are required in relation to the conditions of specific actions.

Ethnopharmacology has traditional medicine as a scenario of interaction that is not restricted to the use of plants as medicines. It also includes a set of traditional knowledge, practices, scenarios, routes, and meanings of wild plants that are the basis of a cultural health system. These, because they belong to a socially established territory, constitute an institution only understood considering all the beliefs and modes of conduct agreed and respected by the local population, and implicitly accepting conditions, relationships and interactions between the people who make up the population (Cardenas and Ojeda 2002, Kuramoto 2007, Ropero Beltrán 2016).

In the Andean-Amazon region of Northern Peru, research has been carried out addressing the different aspects that contribute to the culture of traditional medicine in regions where traditional medicine is especially commonly practiced and has a long tradition, such as Cajamarca (Sánchez and Sánchez 2012), as well as the north coast from La Libertad to Tumbes (Vásquez 2010) and the Northern Andean and Amazonian regions (Bussmann and Sharon 2007, 2014, 2016).

[1] The Mountain Institute INC, Vargas Machuca 408 Urb. San Antonio, Lima, Perú.
[2] Asociación para la Ciencia e Innovación Agraria de la Red Norte-AgroRed Norte, Mz. O Lote 20 Urb. Los Cocos del Chipe, Piura, Perú.
[3] Departamento de Farmacología, Facultad de Farmacia y Bioquímica, Universidad Nacional de Trujillo, Av. Juan Pablo II S/N, Ciudad Universitaria, Trujillo, Perú.
[4] Departamento de Farmacotecnia, Facultad de Farmacia y Bioquímica, Universidad Nacional de Trujillo, Av. Juan Pablo II S/N, Ciudad Universitaria, Trujillo, Perú.
[5] UIC/NIH Center for Botanical Dietary Supplements Research, Department of Medicinal Chemistry and Pharmacognosy, College of Pharmacy, University of Illinois at Chicago, 833 S. Wood St., Chicago, IL 60612, USA.
[6] Department of Ethnobotany, Institute of Botany, Ilia State University, Tbilisi, 0105, Georgia.
* Corresponding author: mganoza@unitru.edu.pe

In this chapter, we have focused on ethnopharmacology of the tropical mountains of Northern Peru, which geographically refers to the Andes of the Piura Region of Northern Peru, with an altitude between 2700 and 3500 masl, latitude between 4° 58' 7" and 5° 67' 25" S, and longitude 79° 29' 51" and 79° 30' 55" W (Figure 4.1). This region is part of the Huancabamba depression, an environmental mosaic recognized as a primary source of biodiversity for many plant groups (Sagástegui et al. 1999). This high biodiversity, and its corresponding use by humans, is one of the fundamental factors that determined its constitution as the Andean "health axis", which in the approach of territorial development can also be considered as a "health territory" located in the Ayabaca and Huancabamba Provinces, nationally recognized as a route of healing in popular thinking.

The research on the potential of traditional medicine based on wild species that has been identified scientifically, classified and pharmacologically analyzed is an important contribution to available knowledge. However, the next step—that the results of pharmacological research are really used in the industrial production process—is still to be taken in most cases. As a starting point, the traditional knowledge from the community to lead this research was required, and a bunch of knowledge from the community was collected that without their support it could not be possible; therefore, native communities must be included as co-authorship. Additionally, any new knowledge generated during this research must be shared with communities. Moreover, the aim of this study was also to establish a territorial ethnopharmacological strategy based on an alliance between scientific entities (National University of Trujillo) and communal organizations. On the other hand, it was also supported by the regional governments of Ayabaca and Huancabamba, to foster joint research on wild plants with medicinal potential and the immediate use of results in innovations implemented locally in the environment of the Paramos and cloud forests located in the Huancabamba depression.

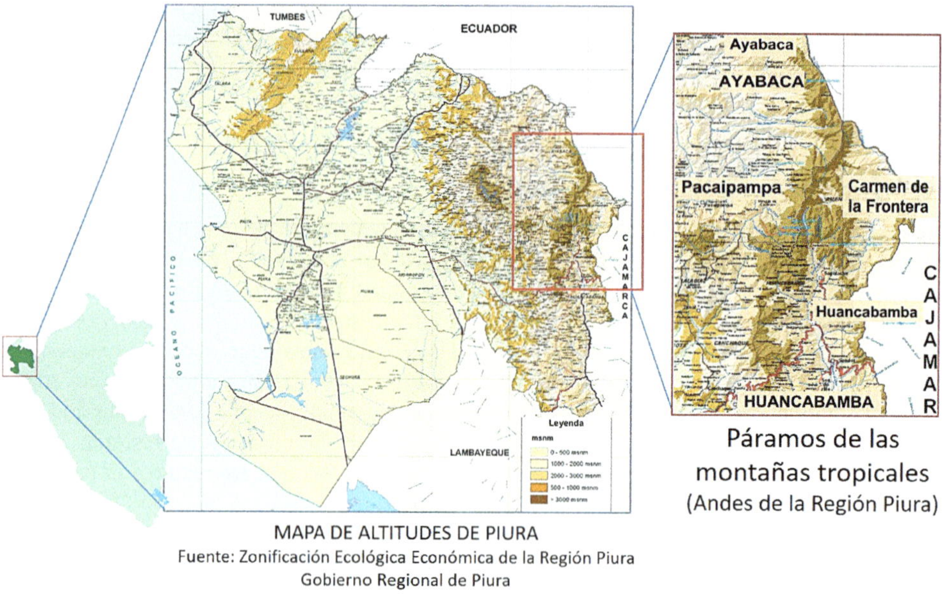

Figure 4.1. Tropical mountains of Northern Peru (Paramo and cloud forest), located in the provinces of Ayabaca and Huancabamba (Piura).

Ethnobotany, Ethnomedicine and Traditional Medicine

Ethnobotany is a broad term that refers to the study of the use of plants by any kind of society, including the use of plants as food, medicine, building materials and any other useful application, including all cultural expressions, domestication, management of varieties and dispersion of species (Fernández and Rodríguez 2007, Lagos-Witte et al. 2011). Ethnomedicine refers to the use of plants by societies as medicine (Ribeiro et al. 2017). Traditional medicine integrates a set of medical practices different from the concept of "Western medicine" and forms a system of knowledge, skills, and practices based on theories, beliefs and experiences of different cultures, whether or not explainable, used for the maintenance of health, as well as for the prevention, diagnosis, improvement or treatment of physical or mental diseases (León Barua 1996, Mukherjee 2019, World Health Organization [WHO] 2020). There are different traditional medicinal systems, some of them legitimized by a complex social agreement, and culturally supported by theories integrated into formal education. Non-Western medicinal systems can be codified with a documented written history, such as traditional Chinese medicine or Ayurvedic medicine from India. Such systems, very much like allopathic medicine, are periodically reviewed in light of their experiences, and update their guiding thinking, which gives them greater credibility, compared to other systems that do not codify their knowledge and are based on specialized personal knowledge that is transmitted from person to person. Despite Western science's skepticism even of codified non-allopathic and traditional medicine, it is not possible to ignore the contributions of these systems as they represent the primary health care of the majority of the world's population. It is, therefore, the task of ethnopharmacology to build systematic and scientific health development programs based on traditional medicine (Farnsworth 1994, Bussmann and Sharon 2014).

Traditional medicine in Northern Peru

A variety of interdisciplinary studies have contributed to the understanding of traditional medicine in Northern Peru, including taxonomy and ethnobotany (Sánchez 2011, Sánchez and Sánchez 2012), phytochemical profiles, and marketing system of 90 medicinal species commonly used in the markets from Jalca (3000 to 4300 masl) of Cajamarca, one of the centers of greatest biodiversity of medicinal species in a region characterized by its great culture and in the use of medicinal plants, which also supplies the northern regional markets. Regarding the traditional medicine of Cajamarca, Seminario (2014) pointed out the important issue of sustainability threats due to human activities, the marketing based on the extraction of species from Jalca (high altitude shrublands), the expansion of agricultural activities, and climate change (Torres and Lucio 2014). However, the main threat to local biodiversity is posed by large-scale open-scale metal mining that is promoted in this ecosystem. Other macroregional research in Northern Peru has mostly focused on the coastal area, including aspects of ethnobotany, taxonomy, ethnomedicine, and phytochemicals, e.g., Vásquez (2010) in La Libertad, Lambayeque, Piura, Tumbes, Cajamarca, Amazonas, and San Martín regions. The most far-reaching interdisciplinary studies (ethnobotanical, biochemical, toxicological, cultural, economic and political) has been carried out over two decades by Bussmann and Sharon (2006, 2011, 2014, 2016) in the Andean-Amazonian area of Northern Peru.

From different perspectives, nuances, and emphasis of the aforementioned authors, it is understood that traditional medicine represents not only the use of species for the cure or prevention of diseases but very importantly, systems of beliefs, practices, dynamics of access and communication, management and extraction, care and deterioration of ecosystems, education and culture, crises and potentials of these health axes or territories.

Curanderism and Traditional Medicine in the Andes of Piura

When focusing on the ethnopharmacology in the tropical mountains of Northern Peru, geographically, we refer to the Andes of the Piura Region of Northern Peru, which is part of the area called Huancabamba deflection, an environmental mosaic of high complexity and source of primary biodiversity of various groups of plants (Sagástegui et al. 1999).

Traditional medicine is not restricted to the use of plants as medicines but incorporates a set of knowledge, practices, scenarios, routes, and meanings of wild plants that collectively contribute to a cultural health system, and which also, belonging to a socially established territory, constitute an institution, which can be understood as all the beliefs and modes of conduct agreed and respected by the population, that in turn are established as explicitly and implicitly accepted rules that condition the interactions between the people who make up the collective (Cardenas and Ojeda 2002, Kuramoto 2007, Ropero Beltrán 2016).

Within the framework of this approach, Camino (1992) developed his interpretation of medicinal traditions of Northern Peru with specific reference to the Huancabamba mountains, which bases its livelihood on a health system in which master healers, bonesetters, midwives, communities, local protection committees, herbalists and patients interact, sharing ideas and meanings of health and disease, food system and types of medicinal plants, the function of the species, and the environments that contribute to their properties, e.g., the influence and power of mountains and lagoons. These complex beliefs as well as attitudes and intercalation forms (Polia 1996) that they share as members of a space molds a "health axis" or "health territory", represents a socially constructed space with its own history that originally comprised Quito, Zamora, and Loja in Ecuador, Las Huarinjas (Las Huaringas), Huancabamba, Sondor, Sondorillo, Huarmaca, Salas, the Olmos desert, Mochumí, Monsefú and possibly Puerto Etén in Peru. This space, despite the changes induced by modernity, is still visible in the travel routes of those seeking healing in Las Huaringas, stretching along the Northern Peruvian coast and southern Ecuador to Huancabamba. This territory is also delimited by wild medicinal species used in healing ceremonies, especially "San Pedro Cactus" (*Trichocereus pachanoi*) which demarcates a plant territorial boundary, in which "coca" and "ayahuasca" are not used, a territory where different wild medicinal species are exchanged from one ecological floor to another—"cold" plants (from the lower parts) and "hot" plants (from the upper parts) that comprise the ecosystems of cloud forests (1800 to 2900 masl), with a great abundance of medicinal species (Stepp and Moerman 2001) and Paramo (3000 to 3500 masl). Mescaline was determined as the major compound found in "San Pedro Cactus". On the other hand, there is evidence that this psychotropic compound has been determined in complex mixtures of "San Pedro Cactus", and related it with healing ceremonies in Northern Peru; therefore, "San Pedro Cactus" is deeply linked with the Peruvian traditional medicine, which has been even more important than allopathic medicine (Willsky et al. 2020). Health Territories, such as the Pacaipampa District (Ayabaca Province), Carmen de la Frontera and Huancabamba District (Huancabamba Province), where community organizations, such as the Asociación de Productores Conservadores de los Páramos y Bosques de Neblina de Pacaipampa (ACOBOSPA) and the Comunidad Campesina Segunda y Cajas de Huancabamba, demand a valorization from the research entities of the medicinal biodiversity that they possess in their Paramos and cloud forests, and the knowledge that they have of it, to use it as an environmental defense strategy against the threat of intensive open-pit extraction metal mining and as an opportunity for a productive diversification based on the local biodiversity in order to integrate it into a market economy that currently affects them negatively.

Beginnings of Territorial Participatory Ethnopharmacology in the Andes of Northern Peru

Ethnopharmacology represents a multidisciplinary effort, which has two axes that give it sustenance and meaning: traditional knowledge and western scientific knowledge, which in both cases represent cognitive, economic, productive, social and cultural systems (Patwardhan 2005). Ethnopharmacology has, as its starting point, the understanding of traditional medicine existing in a given territory, with specific environmental characteristics, and is generally held by rural agricultural societies that have accumulated, for generations, knowledge of the medicinal use of wild plants in their surrounding ecosystems. Traditional medicines are different, depending on the complexity of the society to which they belong.

The widespread idea of ethnopharmacology is that of a type of multidisciplinary scientific research on traditional therapeutics, aimed at the detection of biologically active substances used in the context of traditional culture (Bruhn and Holmstedt 1981), with the aim of providing the laboratory with the natural raw material from which new drugs might be developed. For this purpose, researchers in ethnopharmacology visit local populations in search of medicinal species and associated traditional knowledge. The main achievement of ethnopharmacology has been the discovery of new compounds derived from plants and animals used by traditional medicinal practitioners (Cunningham and Menezes

2011). The ethnopharmacological approach then leads to a research strategy combining the information acquired from the users of a medicinal flora (communities and "traditional specialists"), with chemical-pharmacological studies carried out in specialized laboratories (Sales et al. 2015). The frequent and sad occurrence is, however, that the owners of traditional knowledge that orients scientific research are mostly invisible in the process, in the product, in its application and in the benefits derived from their very knowledge.

For instance, the original research process in ethnopharmacology closely resembles mineral extraction that promises the return of a benefit that the owners of the territory from which the resource was extracted never manage to obtain as a positive and verifiable change in their well-being. An alternative ethnopharmacological approach is participatory research of traditional knowledge on medicinal plant species, in an alliance of community organizations with academic entities, in which the local counterparts act as co-authors, with equal rights to the research results. Such an investigative venture would be based on common interests between community organizations and academic entities, united to improve or add value to what they produce or offer. On the one hand, the communities provide knowledge and raw materials for the production of raw drugs (raw extracts), i.e., high quality herbal or natural products, and the western scientific counterparts provide the added value of knowledge of their chemical composition and pharmacological activity for further processing, and on the other hand, offer new knowledge about the biodiversity of strategic ecosystems.

Because traditional medicine of the Northern Peruvian Andes is based on the use of wild plants from Paramos and cloud forests, ethnopharmacology or integrative medicine, implementing participatory research by demand from the communities, allows the discovery of new compounds that may lead to the development of products that allow a profitable and sustainable use of their biodiversity, oriented towards the conservation of ecosystems to which peasant communities belongs (Gilani and Rahman 2005). Both approaches of non-governmental organizations (Asociación para la Ciencia e Innovacion Agraria de la Red Norte -AgroRed Norte- and The Mountain Institute INC) have been implemented since 2012 through projects co-financed by international cooperation (United Nations Development Program and International Union for Conservation of Nature), also supported by Peruvian government through the Programa Nacional de Innovación Agraria (PNIA) of the Ministry of Agriculture and Irrigation during period 2016–2019.

Paramos and Cloud Forests as Physical and Social Space for a Territorial Ethnopharmacology

Any ethnopharmacology needs to refer to the specific physical and social space in which a specific traditional medicine is developed. Each territory is defined by its environmental, economic and social characteristics (Berdegué and Favareto 2019). The Northern Peruvian Andes represents the most notable "health axes" (Camino 1992, Bussmann and Sharon 2016), or traditional medicine territory, recognized throughout Peru, which uses the most important plant species of Paramo and cloud forests. Ecosystems belonging to the Huancabamba depression represent an environmental mosaic with one of the most varied biodiversity combinations in the South American Andes (Sagástegui et al. 1999, Sagástegui et al. 2003, Emck et al. 2006, Bussmann and Sharon 2006). This territory also includes farmer communities with their communal organizations (where local healers and experts are located), which constitute the social management units (Figures 4.2 and 4.3).

Proposing to establish an ethnopharmacological strategy in this territory, based on the principle of participatory research and the economic use of its results by the community organizations providing traditional knowledge of wild medicinal plants, means that to implement this strategy, it must be built from the bottom up, i.e., from community organizations and local governments in alliance with academic entities that can operationally and logistically promote the effort. The original models, often the public governing entities of science and technology, designing imaginary scenarios, and focusing on urban or industrial-only societies and without consulting the priorities of the local population, are a trend that highlights the gap between the results of pharmacological research and its use by rural societies that provide the knowledge that guided such scientific research.

Figure 4.2. Community organizations of the Paramos deciding how to enrich their knowledge in order to value the biodiversity of the Paramos and cloud forests.

Figure 4.3. Healers information center—communal clinic "Las Huaringas-Huancabamba".

Importance of Participatory Research by the Communities in the use of New Knowledge Generated by Research

In order to develop an ethnopharmacological perspective that coincides with the vision of integrative medicine (Mukherjee 2019) in interdisciplinary research, it is necessary that the communities which know and use wild species and have an interest in valorizing them for use for economic and conservation purposes in a sustainable way are directly involved. Such investigations in the Northern Peruvian Andes (Piura) have already been done by the local communities. The knowledge that has to be valued is not hold by a single healer at most but it represents a collective knowledge of the whole peasant community, which leaders can organize research if it can receive co-financing from the Peruvian State, to invest in research whose results have collective, non-personal benefits, which in fact can then foster innovation on a commercial scale (Figures 4.4 and 4.5).

Traditional knowledge of biodiversity is part of rural family farming, mainly in the Andean and Amazonian regions, and the above approach provides new knowledge and products for public health, and directly benefits the sector that socially accounts for 95% of national agricultural activity, rather than pharmaceutical or natural products' companies in capital cities.

Figure 4.4. Ethnobotanical survey *in situ*—Ayabaca Province (Pacaipampa).

Figure 4.5. Wild plant collection by local experts (Pedro Ruiz, head of ACOBOSPA—Asociación de Productores Conservadores de los Páramos y Bosques de Neblina de Pacaipampa).

Ethnobotany and Taxonomy of Wild Plants on the Paramo

A noteworthy aspect of ethnobotanical and taxonomic studies is the answer to the need for communal organizations to better know the value of their traditional knowledge, the possibilities for management and cultivation of important medicinal species, an abundance of promising species and the possibility of *in situ* processing that allows them to use their biodiversity advantageously and sustainably. This consolidates local knowledge because it is based on the communities' priorities, not that of the scientist, who may have a priority agenda that does not necessarily coincide with local interests.

Ethnobotanical and taxonomic documentation should contribute to optimizing the communities capacities and competencies to harness the potential of local endemic wild species, giving them added value, in addition to avoiding the loss of knowledge or recovering it, and incorporating it at the educational level, and enabling them to offer specialized services either as a tour guide or in the exchange with research entities (Table 4.1).

Methods for Phytochemistry Analysis and Biological Activity from Wild Plants

The quality and appropriate technique of phytochemical analyses ensure the certainty of the presence of a wealth of bioactive molecules in species of interest. The methodological relevance based on logistical means and technical support allows estimating the economic feasibility for the establishment

Table 4.1. Ethnobotanical description of wild plants from the tropical mountains of Northern Peru (according to local herbalist experts).

Scientific name	Family	Common names	Province	Ecosystem	Traditional use	Medicinal properties
Berberis sp.	Berberidaceae	"espino amarillo"	Huancabamba	Cloud forest	malaria and internal infections	Antibiotic
Alnus acuminata Kunth	Betulaceae	"aliso"	Ayabaca	Cloud forest	improves digestion, to heal inflammation of the kidneys	Digestive, anti-inflammatory
Bejaria resinosa Mutis ex L. f.	Ericaceae	"payana"	Ayabaca	Paramo	inflammations, regulates vaginal drops	Anti-inflammatory, antibiotic
Cavendishia bracteata (Ruiz & Pav. ex J. St.-Hil.) Hoerold	Ericaceae	"guaguña"	Huancabamba	Cloud forest	edible fruit	Nutritious
Gaultheria erecta Vent.	Ericaceae	"ushpa"	Huancabamba	Paramo	fruits consumed by children	Nutritious
Vaccinium floribundum Kunth	Ericaceae	"ushpa"	Ayabaca	Paramo	fruit in children's food	Nutritious
Senna multiglandulosa (Jacq.) H.S. Irwin & Barneby	Fabaceae	"hierba de la disipela"	Huancabamba	Cloud forest	skin infections	Antiseptic
Gentianella crassicaulis J.S. Pringle	Gentianaceae	"shagapa"	Huancabamba	Paramo	internal infections and inflammations	Antibiotic, anti-inflammatory
Gentianella sp.	Gentianaceae	"shagapa morada"	Ayabaca	Paramo	improving liver function, cleanses the blood	Hepatic regulator
Gunnera pilosa Kunth	Gunneraceae	"shingor"	Ayabaca	Paramo	internal infections of the body	Antibiotic
Gunnera sp.	Gunneraceae	"shingor"	Huancabamba	Paramo	gives energy to walk, fever	Energizing, Antipyretic
Cuphea ciliata Ruiz & Pav.	Lythraceae	"hierba del toro"	Ayabaca	Cloud forest	against vaginal drops postpartum	Antibiotic
Cuphea strigulosa Kunth	Lythraceae	"hierba del toro"	Huancabamba	Cloud forest	intestinal infections	Antibiotic
Brachyotum angustifolium Wurdack	Melastomataceae	"zarcilleja"	Huancabamba	Paramo	cough with phlegm	Expectorant
Myrcianthes myrsinoides (Kunth) Grifo	Myrtaceae	"lanche"	Ayabaca	Cloud forest	improving digestion, maintains strength	Digestive, depurative
Myrcianthes rhopaloides (Kunth) McVaugh	Myrtaceae	"lanche colorado"	Ayabaca	Cloud forest	colds	Decongestant
Oenothera rosea L'Hér. ex Aiton	Onagraceae	"hierba de la rabia"	Ayabaca	Paramo	soothes people in a bad mood	Calmative
Stelis sp.	Orchidaceae	"congona del cerro"	Huancabamba	Cloud forest	heals bad pimples and other skin	Antiseptic

Castilleja virgata Dombey ex Wedd.	Orobanchaceae	"hierba de la rabia"	Huancabamba	Paramo	calm anger and headaches	Analgesic
Muehlenbeckia hastulata (Sm.) I.M. Johnst.	Polygonaceae	"chupicaure"	Ayabaca	Paramo	fever and chickenpox	Antibiotic
Muehlenbeckia tamnifolia (Kunth) Meisn.	Polygonaceae	"chupicaure"	Huancabamba	Paramo	control fevers and infections	Antibiotic
Acaena ovalifolia Ruiz & Pav.	Rosaceae	"pega-pega"	Ayabaca	Paramo	infections	Antibiotic
Symplocos fimbriata B. Ståhl	Symplocaceae	"chimicuna"	Ayabaca	Cloud forest	fruit in juice or soft drink from toasted fruit	Nutritious

of a reliable ethnopharmacological system in the Peruvian north that can be used both by the industry and the community organizations partnering in the research. The identification of secondary metabolites (Table 4.2), acute toxicity by the bioassay of *Artemia franciscana* expressed in mean lethal concentration (Table 4.3), the use of five extraction systems (decoction, infusion, ethanol 96%, 70%, and 45%) in order to be more efficient during the evaluation of phenolic compounds based on the Folin-Ciocalteu method, expressed in Galic acid equivalents (Table 4.4) and the antioxidant activity by ABTS (2,2'-azino-bis(3-ethylbenzothiazoline-6-sulfonic acid) method expressed in Trolox equivalents (Table 4.5), have allowed the verification and expansion of the properties assigned to the species by the local knowledge (Bussmann et al. 2011, Torres-Guevara and Ganoza-Yupanqui 2017).

Table 4.2. Preliminary test for the identification of secondary metabolites from wild plants selected corresponding to ethnobotanical study.

Scientific name	Common name	Province	Part	Al	An	PC	Co	St/Tr	Fl	Sa	Ta
Acaena ovalifolia	"pega-pega"	Ayabaca	leaves	(–)	(+)	(+)	(+)	(–)	(+)	(+)	(+)
Alnus acuminata	"aliso"	Ayabaca	leaves	(–)	(+)	(+)	(+)	(–)	(–)	(+)	(+)
Bejaria resinosa	"payana"	Ayabaca	leaves	(–)	(+)	(+)	(+)	(+)	(+)	(+)	(+)
Berberis sp.	"espino amarillo"	Huancabamba	leaves	(–)	(+)	(+)	(+)	(+)	(+)	(–)	(–)
Brachyotum angustifolium	"zarcilleja"	Huancabamba	leaves	(–)	(+)	(+)	(+)	(+)	(+)	(–)	(+)
Castilleja virgata	"hierba de la rabia"	Huancabamba	leaves	(–)	(–)	(+)	(+)	(–)	(+)	(+)	(+)
Cavendishia bracteata	"guaguña"	Huancabamba	leaves	(–)	(+)	(+)	(+)	(+)	(+)	(+)	(+)
Cuphea ciliata	"hierba del toro"	Ayabaca	leaves	(–)	(+)	(+)	(+)	(+)	(+)	(+)	(+)
Cuphea strigulosa	"hierba del toro"	Huancabamba	leaves	(–)	(+)	(+)	(+)	(–)	(+)	(+)	(–)
Gaultheria erecta	"ushpa"	Huancabamba	leaves	(–)	(+)	(+)	(+)	(–)	(+)	(+)	(+)
Gentianella crassicaulis	"shagapa"	Huancabamba	leaves	(–)	(–)	(+)	(+)	(+)	(+)	(+)	(+)
Gentianella sp.	"shagapa morada"	Ayabaca	leaves	(+)	(+)	(+)	(+)	(+)	(+)	(+)	(+)
Gunnera pilosa	"shingor"	Ayabaca	inflorescence	(–)	(+)	(+)	(+)	(+)	(+)	(–)	(+)
Gunnera sp.	"shingor"	Huancabamba	leaves	(–)	(+)	(+)	(+)	(–)	(–)	(+)	(+)
Muehlenbeckia hastulata	"chupicaure"	Ayabaca	leaves	(+)	(–)	(+)	(+)	(–)	(+)	(+)	(+)
Muehlenbeckia tamnifolia	"chupicaure"	Huancabamba	leaves	(–)	(+)	(+)	(–)	(+)	(+)	(–)	(–)
Myrcianthes myrsinoides	"lanche"	Ayabaca	leaves	(–)	(+)	(+)	(+)	(–)	(+)	(+)	(+)
Myrcianthes rhopaloides	"lanche colorado"	Ayabaca	leaves	(–)	(+)	(+)	(+)	(+)	(+)	(–)	(+)
Oenothera rosea	"hierba de la rabia"	Ayabaca	leaves	(–)	(–)	(+)	(+)	(–)	(+)	(–)	(+)
Senna multiglandulosa	"hierba de la disipela"	Huancabamba	leaves	(–)	(+)	(+)	(–)	(–)	(+)	(–)	(+)
Stelis sp.	"congona del cerro"	Huancabamba	leaves	(–)	(–)	(+)	(+)	(+)	(+)	(–)	(–)
Symplocos fimbriata	"chimicuna"	Ayabaca	fruits	(–)	(–)	(–)	(–)	(+)	(–)	(–)	(–)
Vaccinium floribundum	"ushpa"	Ayabaca	leaves	(+)	(+)	(+)	(+)	(+)	(+)	(–)	(+)

(–) absence, (+) presence, Alkaloids (Al), Anthraquinone (An), Phenolic compounds (PC), Coumarins (Co), Steroids/triterpenoids (St/Tr), Flavonoids (Fl), Saponins (Sa), Tannins (Ta).

Table 4.3. Acute toxicity by *Artemia franciscana* bioassay of wild plants collected from the tropical mountains of Northern Peru (Ayabaca Province).

Scientific name	**Common name**	**Part**	**Median lethal concentration (µg/mL)**				
			Decoctions	Infusions	Cañazo* (Ethanol 45%)	Ethanol 70%	Ethanol 96%
Acaena ovalifolia	"pega-pega"	leaves	625 ± 110	1440 ± 31	784 ± 31	714 ± 24	2769 ± 916
Alnus acuminata	"aliso"	leaves	5884 ± 1908	251 ± 15	3881 ± 471	8359 ± 783	1567 ± 361
Bejaria resinosa	"payana"	leaves	1100 ± 291	450 ± 56	365 ± 40	3951 ± 839	1895 ± 567
Calceolaria sp.	"hierba dulce"	leaves	2414 ± 384	11393 ± 1272	2977 ± 817	3737 ± 682	7195 ± 2354
Gentianella sp.	"shagapa morada"	leaves	617 ± 62	559 ± 50	3432 ± 903	1696 ± 921	4667 ± 1667
Gunnera pilosa	"shingor"	inflorescence	719 ± 28	1262 ± 518	799 ± 54	751 ± 77	466 ± 30
Myrcianthes myrsinoides	"lanche"	leaves	2901 ± 243	1825 ± 818	616 ± 28	1464 ± 157	799 ± 75
Myrcianthes rhopaloides	"lanche colorado"	leaves	715 ± 198	1009 ± 268	597 ± 72	457 ± 19	401 ± 63
Symplocos fimbriata	"chimicuna"	fruits	11866 ± 1321	10800 ± 2626	2136 ± 327	7893 ± 604	3692 ± 670
Vaccinium floribundum	"ushpa"	leaves	441 ± 69	756 ± 14	599 ± 43	8065 ± 3541	523 ± 47

* Cañazo: Sugar cane fermentated alcohol.

Table 4.4. Phenolic compounds (PC) of five extraction systems of wild plants collected from the tropical mountains of Northern Peru.

Scientific name	**Common name**	**Province**	**PC (mg Gallic acid equivalent/g of dry plant material)**				
			Decoctions	Infusions	Ethanol 45%	Ethanol 70%	Ethanol 96%
Bejaria resinosa	"payana"	Ayabaca	22,11 ± 0,84	21,83 ± 0,56	25,40 ± 1,36	34,88 ± 0,12	13,42 ± 0,51
Bejaria resinosa	"payana"	Huancabamba	69,73 ± 0,63	41,69 ± 1,13	57,57 ± 6,38	53,11 ± 3,5	14,55 ± 2,38
Brachyotum angustifolium	"zarcilleja"	Ayabaca	37,46 ± 0,98	23,67 ± 0,61	54,10 ± 5,82	53,36 ± 2,44	12,91 ± 0,05
Brachyotum angustifolium	"zarcilleja"	Huancabamba	86,13 ± 4,73	48,47 ± 1,02	86,76 ± 5,58	67,48 ± 4,18	15,20 ± 1,2
Cuphea ciliata	"hierba del toro"	Ayabaca	50,63 ± 1,85	47,03 ± 2,03	69,16 ± 3,40	53,89 ± 0,90	10,74 ± 0,56
Cuphea ciliata	"hierba del toro"	Huancabamba	50,41 ± 3,38	32,57 ± 1,31	79,3 ± 0,83	51,66 ± 2,04	16,17 ± 0,85
Muehlenbeckia hastulata	"chupicaure"	Ayabaca	53,18 ± 1,50	36,77 ± 2,72	96,94 ± 9,90	49,84 ± 3,91	17,46 ± 2,14
Muehlenbeckia hastulata	"chupicaure"	Huancabamba	49,07 ± 5,78	41,40 ± 0,69	106,72 ± 6,39	74,63 ± 4,52	25,25 ± 0,77
Vaccinium floribundum	"ushpa"	Ayabaca	77,62 ± 7,07	45,69 ± 2,26	87,88 ± 5,98	76,34 ± 4,03	19,66 ± 1,11
Vaccinium floribundum	"ushpa"	Huancabamba	97,63 ± 1,38	60,60 ± 3,4	86,23 ± 7,52	67,56 ± 3,51	43,98 ± 1,14

Table 4.5. Antioxidant activity of five extraction systems of wild plants collected from the tropical mountains of Northern Peru.

Scientific name	Common name	Province	ABTS (mg Trolox equivalent/g of dry plant material)				
			Decoctions	Infusions	Ethanol 45%	Ethanol 70%	Ethanol 96%
Bejaria resinosa	"payana"	Ayabaca	53,27 ± 2,20	55,17 ± 4,22	74,28 ± 3,86	94,17 ± 1,27	45,88 ± 0,89
Bejaria resinosa	"payana"	Huancabamba	188,62 ± 3,42	105,13 ± 5,51	182,25 ± 5,68	172,53 ± 7,05	68,02 ± 3,65
Brachyotum angustifolium	"zarcilleja"	Ayabaca	106,35 ± 1,28	60,08 ± 2,77	164,50 ± 9,51	172,79 ± 9,15	45,41 ± 1,74
Brachyotum angustifolium	"zarcilleja"	Huancabamba	281,60 ± 8,91	159,17 ± 1,99	276,62 ± 9,81	232,73 ± 7,99	57,79 ± 2,16
Cuphea ciliata	"hierba del toro"	Ayabaca	152,94 ± 8,55	135,54 ± 1,21	143,14 ± 17,45	170,83 ± 7,67	39,15 ± 1,04
Cuphea ciliata	"hierba del toro"	Huancabamba	168,52 ± 4,78	111,17 ± 1,57	273,82 ± 23,70	179,28 ± 12,37	60,60 ± 1,84
Muehlenbeckia hastulata	"chupicaure"	Ayabaca	147,17 ± 4,12	102,84 ± 3,50	247,54 ± 16,70	182,25 ± 2,38	68,57 ± 5,09
Muehlenbeckia hastulata	"chupicaure"	Huancabamba	152,66 ± 4,07	127,55 ± 0,82	316,51 ± 22,15	206,93 ± 5,50	95,41 ± 7,34
Vaccinium floribundum	"ushpa"	Ayabaca	170,21 ± 15,58	101,15 ± 4,12	206,76 ± 15,39	154,98 ± 17,30	77,31 ± 2,68
Vaccinium floribundum	"ushpa"	Huancabamba	188,55 ± 15,99	145,49 ± 2,74	171,92 ± 8,36	171,14 ± 7,30	114,60 ± 2,10

Importance of Chemical Fingerprinting for Wild Plant Identification

The establishment of an ethnopharmacological system of wild species requires the concordance between the local nomenclature with western taxonomic identity, i.e., botanical nomenclature. The scientific taxonomy of Paramo and cloud forest species still remains fragmentary, creating uncertainty in the results of phytochemical research. Chemical fingerprint determination is a complementary tool to ensure the identity of species from ethnobotanical studies.

Pharmacovigilance is essential to develop reliable information on the identity of medicinal herbs with certain properties assigned to them. Traditional medicine from different cultures often uses specific traceability systems which increase the complexity and difficulty to assure the quality of the products to be obtained. Botanical verification using its chemical fingerprint can be invaluable for the complete evaluation and verification of any species traceability report on the quality of a natural or pharmacological product. Systematic pharmacovigilance is essential for building reliable information on the safety of wild medicinal plants, and the development of appropriate guidelines for safe and effective use (Shaw et al. 2012).

From the determination of fingerprints by using high-performance liquid chromatography (HPLC) of the species "chupicaure" (*Muhelenbeckia hastulata*), "hierba del toro" (*Cuphea ciliata*), "payana" (*Bejaria resinosa*), "ushpa" (*Vaccinium floribundum*), "zarcilleja" (*Brachyotum angustifolium*), pega-pega (*Acaena ovalifolia*), "lanche" (*Myrcianthes myrsinoides* and *Myrcianthes rhopaloides*), it has been possible to verify the metabolome identity of one interesting species "ushpa" even when it is growing in two completely different habitats (Figure 4.6), as well as differences between two species of the same site (Pacaipampa-Ayabaca) with the same use as the "lanches", and the difference of two species coming from two different sites (Ayabaca and Huancabamba) that were taxonomically identified as the same species in the case of "zarcilleja". Such indicators are of great importance to build a reliable territorial ethnopharmacology system (Torres-Guevara et al. 2020).

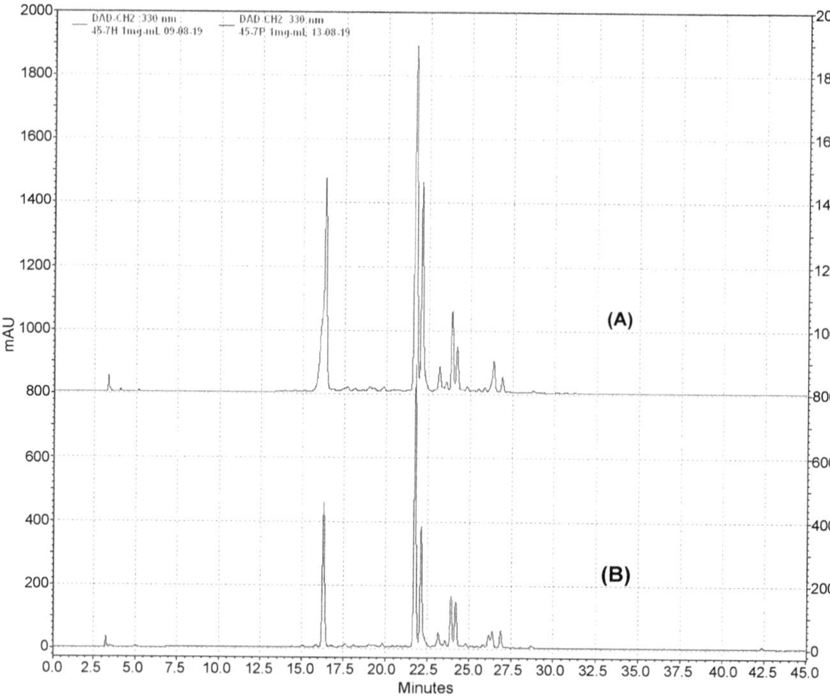

Figure 4.6. HPLC fingerprint comparing *Vaccinium floribundum* "ushpa" collected in Huancabamba provinces (A) and Ayabaca (B) with the characteristic peaks of their phenolic compounds purified with Amberlite XAD7HP from 45% ethanol extract at 330 nm.

Importance of the Determination of Molecular Structures for Wild Plants Innovation

The determination of the chemical structures of greater presence by liquid chromatography coupled to mass spectroscopy (LC-MS/MS) in wild species is of great importance to give added value to the raw drug, herbal or natural product that an organization considers relevant to offer. The concrete way for the farmers to receive the benefit of their participation in an ethnopharmacological effort is to offer a potentiated reward for the new properties discovered by scientific research. In each study of wild plants from the tropical mountains of Northern Peru, a very important point for discussion related a thorough phytochemical investigation of each constituent (Table 4.6) (Figures 4.7 and 4.8), this information acquired gives an extra value to the products made based on these species, such as a scientific support for their raw material (Torres Guevara 2019).

Table 4.6. Compounds identified in wild plants selected for ethnobotanical study from tropical mountains of Northern Peru.

Wild plants	RT (min)	[M+H] max (m/z)	Fragment Ions (m/z)	UV max absorption (nm)	Compounds
Bejaria resinosa	5.2	465	465, 303	258, 351	Quercetin-3-*O*-glucoside
Brachyotum angustifolium	3.9	355	355, 163	219, 327	Chlorogenic acid
Cuphea ciliata	5.7	479	479, 317	255, 351	Isorhamnetin-3-*O*-glucoside
Muehlenbeckia hastulata	5.0	611	611, 303	256, 352	Rutin
Vaccinium floribundum	5.2	465	465, 319	260, 346	Myricetin-3-*O*-Rhamnoside

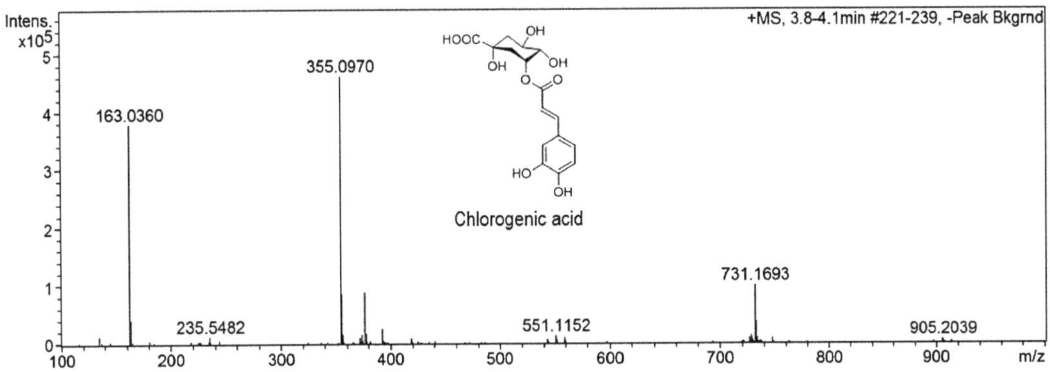

Figure 4.7. Determination of the chemical structure of Chlorogenic acid by LC-MS/MS from *Vaccinium floribundum* "ushpa", nutraceutical species from the tropical mountains of Northern Peru.

Bejaria resinosa "payana" contains Myricetin-3-*O*-rhamnoside, which shows strong antioxidant activity against DPPH, which can be of potential interest in the control of inflammatory conditions and may constitute an attractive molecule of interest for the development of new analgesic drugs.

Brachyotum angustifolium "zarcilleja" contains Quercetin-3-*O*-glucoside, whose Quercetin aglycone is a flavonoid that can be consumed in the diet. It has antioxidant, anti-inflammatory, antiproliferative and anticarcinogenic properties.

Cuphea ciliata "hierba del toro" contains Isorhamnetin-3-*O*-glucoside, which is widely found in many vegetables and rice, and after digestion, reaches its pharmacological effect (Du et al. 2014).

Muhelenbeckia hastulata "chupicaure" has a significant content of Rutin, which has a synergistic effect with vitamin C, allowing its regeneration, and thereby increases the period of antioxidant protection of food (Guo and Wei 2008).

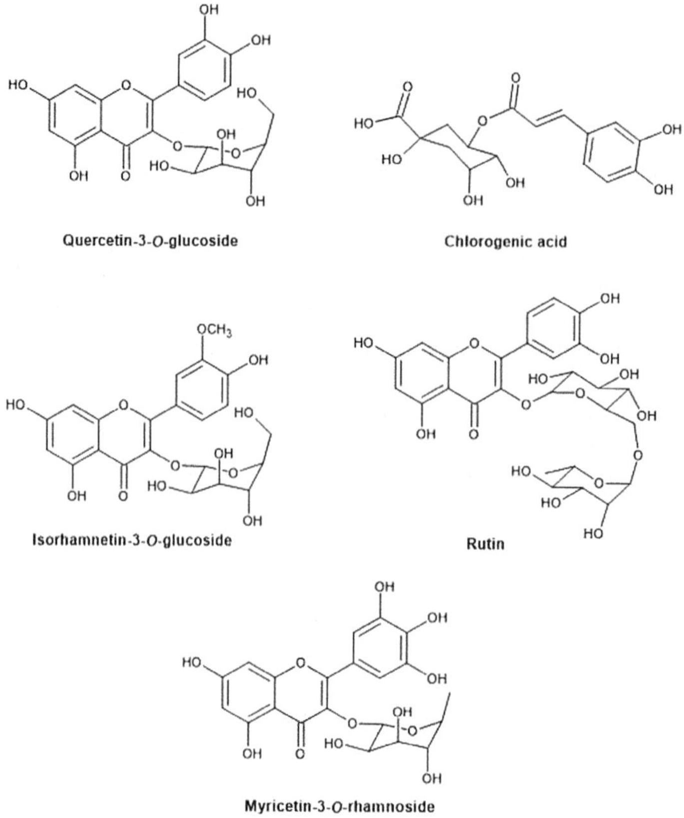

Figure 4.8. Chemical structures identified in wild plants used by community organizations from the tropical mountains of Northern Peru.

Vaccinium floribundum "ushpa" shows an important content of Chlorogenic Acid, which is one of the main polyphenols of coffee, and according to Martínez (2014), it is due to this polyphenol that regular consumption of coffee reduces the risk of diabetes up to 30%, by activating the action of insulin and improving glucose tolerance.

Scientific information gives an important added value to natural products that community organizations like the local Youth Association, secondary students from one of the Schools of the Comunidad Campesina Segunda y Cajas, and the Association ACOBOSPA of Pacaipampa-Ayabaca have started to manufacture (e.g., antimicrobial gels, skin creams, syrups, toothpaste, syrups, essential oils).

Rural Territorial Development (RTD) as a Focus on Building Ethnopharmacology or Integrative Medicine in the Territory of the Tropical Mountains of Northern Peru

The possibility of establishing a territorial ethnopharmacological strategy or integrative medicine system in specific regions (Paramo and cloud forests) uses rural territorial development as an operational implementation tool, based on which the actors interact, learn to know each other, and develop specific agendas of common convenience based on their particular interests.

The objective of the RTD is to create productive, institutional and social transformations in a particularly rural area (a "territory"), in order to incorporate rural people into the income and employment opportunities generated by decentralized growth (Mendieta and Agreda 2010).

Promoting RTD through the implementation of participatory and territorial ethnopharmacology involves simultaneously working with the social, institutional and productive aspects of the territories.

Socially, it means a social policy of establishing an integrative medicine system that strengthens competent local organizations, such as social capital formation. Institutionally, it involves promoting the empowerment and participation of local organizations and strengthening the capacities of local governments, to turn the strategy of territorial participatory ethnopharmacology into local and regional development policy. Productively, it means to articulate in a competitive and sustainable way the economy that can generate the strategy of dynamic marketplace for jobs and services.

Importance of Management for the Economic use of Research Results by the Member Communities of Participatory Research (Innovation Manager)

Because the main trend in ethnobotanical research in northern Peru has focused on documenting traditional knowledge in which both men and women have the role of informants, it is important to integrate a management approach, as this requires interacting with the real actors of traditional and integrative medicine systems in order to gain an understanding of the interests that rural communities may have in exchanging knowledge and participating as authors of research processes with scientific research entities, and sustainable biodiversity management aimed at the conservation of fragile ecosystems, with a horizon of innovation.

This requires an intercultural communication effort that seeks to value the knowledge of all partners. This is a fundamental condition for improving the management capacities of those with less access to media resources (community organizations), leading them to convert discourses and actions of culture, science, and innovation into sustainable local development policies through management, from local governments in their territories, as specific cases of efficient and effective governance exercise.

Research-based on an ethnopharmacological approach, aimed at sustainable innovation in the ecosystems of Paramos and cloud forests in Northern Peru, has no future if we do not work in parallel with social or institutional innovations so that projects become development policies. In the reality of northern Peru, converting the potential of research into its practical application, faces an operational gap that is the scarcity of articulation agents between the offers and demands of the two main actors; rural organizations and scientific-technological entities and other actors in the direct and indirect environment of traditional and conventional medicine systems.

Initial cases of this process are the agreements that the municipalities of the Ayabaca Province and Carmen de la Frontera District (Huancabamba Province) have signed with other partners, committing to their Local Development Plan, and investing in this strategy of ethnopharmacology as a policy of provincial and district development according to the level of each municipality.

Ethnopharmacology and Integrative Medicine as a New System of Innovation based on the Wild Plant Biodiversity of Sensitive Ecosystems (Paramo and Cloud Forests of Northern Peru)

Biodiversity, and particularly the wild plant diversity that represents a huge opportunity for the Peruvian economy, is manifested in a specific way according to the varied territories in which it is distributed in specific ecosystems (Galán de Mera et al. 2019). Such biodiversity can represent a comparative advantage, but this advantage is only feasible because of the traditional knowledge that local societies have. An expression of the comparative and competitive advantages of a territory can be based on its traditional medicine to restore and maintain its health.

Using biodiversity as a source of new economic opportunities is possible based on the specific socio-economic units in a territorial development approach.

In the face of the growing demand for health alternatives, an inescapable fact is that to turn biodiversity and its knowledge into assets for innovation (sustainable businesses), it requires the integration of knowledge of traditional medicine with allopathic medicine scientists (Villar López et al. 2016), based on mutual interest, positioning both actors as leading agents of economic development. Both represent specific agents or economic and cultural groups (communal organizations, healers, healers' associations, local governments, commercial articulators, universities, private and public research institutions, companies, regional governments, regulators, developers, ministries) with their specific interests allied for common purposes.

The construction of new ethnopharmacology or integrative medicine involves the progressive construction of an alternative health system, in which rural agrarian societies possessing traditional medicines and associated knowledge, through new organizational arrangements, acquire a new role offering their traditional knowledge, raw drugs, high-quality herbal and natural products for scientific research and innovation chains of phytopharmaceuticals based on wild plants. They will only be able to achieve this using the added value provided by scientific research that in turn is positioned as a reliable activity, generating collective well-being. This process is, therefore, a matter of public demand for collective development (Figures 4.9, 4.10 and 4.11).

Understanding this approach of ethnopharmacology or integrative medicine as a chained system of agents or interest groups directly and indirectly involved, collective action necessarily induces the creation of convenient and profitable rules of play for its members who hold them together to participate in the income that this approach generates, which could not be achieved individually. These collectively constructed rules of the game allow participating players to know the interests behind their agreements and disagreements to remain united for the good of the collective, and this is what the institutionalization of the system means. Institutionalization reduces the uncertainty and transaction costs required by the integration of traditional medicines with conventional medicine (Glave and Jaramillo 2007).

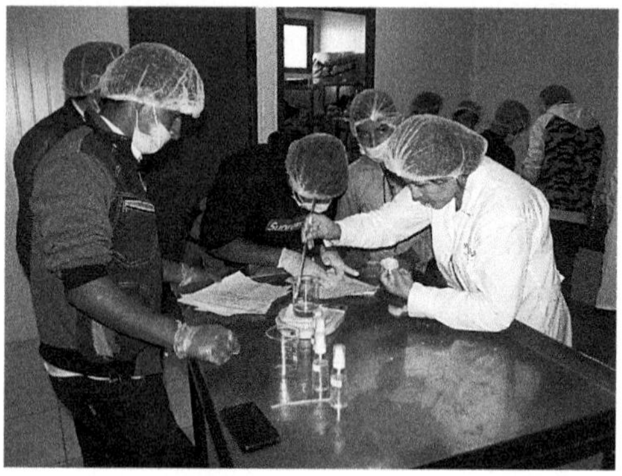

Figure 4.9. Training in phyto-preparations based on the targeted species.

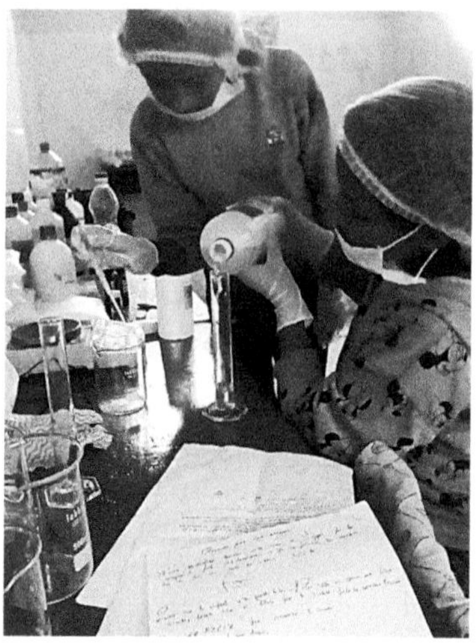

Figure 4.10. Students from local schools performing phyto-preparations of the targeted species for future commercialization.

Figure 4.11. *In situ* specialist advisors carry out pilot tests of innovation using results of the research.

Conclusions

Ethnopharmacology focused on the use of wild plants refers to specific ecosystems such as Paramos and cloud forests, in which, in addition to the community organizations possessing traditional knowledge, other agents act, directly and indirectly, forming a "health axes" as a territorial unit of interaction and planning of research and innovation based on biodiversity.

New ethnopharmacology has sustainability if implemented by the demand of societies that want to optimize the traditional medicine of their territory.

Participatory ethnopharmacology, in order to generate new knowledge and innovations based on wild plants from Paramos and cloud forests, commits community organizations to share the risks, responsibilities as well as the benefits of any investigation.

The phytochemical and pharmacological results obtained from wild plants used and provided by community organizations, which show presence and concentration of important biomolecules in antioxidant capacity and activity, with anti-inflammatory and antibacterial activities, confirm the importance of the participatory nature of this type of research.

Phytochemical and pharmacological processes and results shared in an intercultural way encourage the social appropriation of science by the communities that use them to add value to the phyto-preparations developed, based on wild plants used by local communities. This induces the communities to become permanent claimants of new knowledge, and to access new opportunities for innovation.

Management is needed to turn positive research achievements into sustainable processes such as development policies, underpinned by territorial biodiversity and the competitive capacities of rural community organizations and institutions National Universities.

References

Berdegué, J.A., Favareto, A. 2019. Desarrollo territorial rural en América Latina y el Caribe. 2030-Alimentación, agricultura y desarrollo rural en América Latina y el Caribe, No. 32. 18 p.

Bruhn, J.G., Holmstedt, B. 1981. Ethnopharmacology: objectives, principles and perspectives. pp. 405–430. *In*: Beal, J.L., Reinhard, E. (eds.). Natural Products as Medicinal Agents. Hippokrates, Stuttgart Germany.

Bussmann, R.W., Sharon, D. 2006. Traditional medicinal plant use in Northern Peru: tracking two thousand years of healing culture. J. Ethnobiol. Ethnomed., 2: 47.

Bussmann, R.W., Malca, G., Glenn, A., Sharon, D., Nilsen, B., Parris, B., Dubose, D., Ruiz, D., Saleda, J., Martinez, M., Carillo, L., Walker, K., Kuhlman, A., Townesmith, A. 2011. Toxicity of medicinal plants used in traditional medicine in Northern Peru. J. Ethnopharmacol., 137(1): 121–40.

Bussmann, R.W., Sharon, D. 2014. Medicina tradicional y medicina moderna en México y el Perú: valorización y explotación. pp. 425–458. *In*: Limón, S., Millones, L. (eds.). Por la mano del hombre: prácticas y creencias sobre chamanismo y curandería en México y el Perú. Fondo Editorial de la Asamblea Nacional de Rectores, Lima Perú.

Bussmann, R.W., Sharon, D. 2016. Plantas medicinales de los Andes y la Amazonía - La flora mágica y medicinal del Norte del Perú. Ethnobot. Res. Appl., 15(1): 1–293.

Camino, L. 1992. Cerros, plantas y lagunas poderosas: La medicina al norte del Perú. CIPCA, Lima Perú; 296 p.

Cárdenas, E., Ojeda, J. 2002. La Nueva institucionalidad y la teoría de la implementación. Revista de Economía Institucional. 4(6): 153–169.

Cunningham, F., Menezes, F.S. 2011. Ethnopharmacology in Dublin: surveys on the medicinal plants use profile. Rev. Bras Farmacogn., 21(5): 814–817.

Du, L.-Y., Zhao, M., Xu, J., Qian, D.-W., Jiang, S., Shang, E.-X., Guo, J.-M., Duan, J.-A. 2014. Analysis of the metabolites of isorhamnetin 3-*O*-glucoside produced by human intestinal flora *in vitro* by applying ultraperformance liquid chromatography/quadrupole time-of-flight mass spectrometry. J. Agric. Food Chem., 62(12): 2489–2495.

Emck, P., Moreira-Muñoz, A., Richter, M. 2006. El clima y sus efectos en la vegetación. pp. 11–36. *In*: Moraes, M., Øllgaard, R.B., Kvist, L.P., Borchsenius, F., Balslev, H. (eds.). Botánica Económica de los Andes Centrales Universidad Mayor de San Andrés, La Paz Bolivia.

Farnsworth, N.R. 1994. Ethnopharmacology and drug development. pp. 42–59. *In*: Chadwick, D.J., Marsh, J. (eds.). Ethnobotany and the Search for New Drugs. Ciba Foundation Symposium 185, Chichester England.

Fernández, A., Rodríguez, E. 2007. Etnobotánica del Perú Pre-Hispano. Ediciones Herbarium Truxillense (HUT), Universidad Nacional de Trujillo, Trujillo Perú, 253 p.

Galán de Mera, A., Linares-Perea, E., Martos, F., Montoya-Quino, J. Rodríguez-Zegarra, C., Torres-Marquina, I. 2019. Distribución bioclimática de plantas medicinales y sus principios activos en el Departamento de Cajamarca (Perú). BLACPMA, 18(2): 130–143.

Gilani, A.H., Rahman, A.U. 2005. Trends in ethnopharmacology. J. Ethnopharmacol., 100(1-2): 43–49.

Glave, M., Jaramillo, M. 2007. Perú: Instituciones y Desarrollo. Avances y agenda de Investigación. pp. 301–349. *In*: Teillier, C. (ed.). Investigación, políticas y desarrollo en el Perú. GRADE, Lima Perú.

Guo, R., Wei, P. 2008. Studies on the antioxidant effect of rutin in the microenvironment of cationic micelles. Microchimica Acta, 161: 233–239.

Kuramoto, J. 2007. Sistemas de Innovación Tecnológica. pp. 103–133. *In*: Teillier, C. (ed.). Investigación, políticas y desarrollo en el Perú. GRADE, Lima Perú.

Lagos-Witte, S., Sanabria, O., Chacón, P., García, R. (eds.). 2011. Manual de Herramientas Etnobotánicas relativas a la Conservación y el Uso Sostenible de los Recursos Vegetales Una contribución de la Red Latinoamericana de Botánica a la Implementación de la Estrategia Global para la Conservación de las Especies Vegetales hacia el logro de las Metas 13 y 15. Red Latinoamericana de Botánica (RLB), Santiago Chile, 134 p.

León Barua, R., Berendson Seminario, R. 1996. Medicina teórica. Definición de la medicina y su relación con la biología. Rev. Med. Hered., 7: 1–3.

Martínez, S. 2014. Cacao y café, alimentos ricos en fitoquímicos con propiedades beneficiosas en salud. Estudios de biodisponibilidad de metilxantinas y efectos en salud en voluntarios sanos y con riesgo cardiovascular (Doctor's thesis). Universidad Complutense de Madrid, Facultad de Farmacia, Madrid, España, 375 p.

Mendieta, C., Agreda, V. 2010. Programa de Compensaciones para la Competitividad: lecciones aprendidas en programas afines. Ministerio de Agricultura del Perú.

Mukherjee, P.K. 2019. Ethnopharmacology and ethnomedicine-inspired drug development. pp. 29–51. *In*: Mukherjee, P.K. (ed.). Quality Control and Evaluation of Herbal Drugs. Elsevier, Massachusetts USA.

Patwardhan, B. 2005. Ethnopharmacology and drug discovery. J. Ethnopharmacol., 100(1-2): 50–52.

Polia, M. 1996. Despierta, remedio, cuenta...: adivinos y médicos del Ande. PUCP-Fondo Editorial, Lima Perú, 893 p.

Ribeiro, R.V., Bieski, I.G.C., Balogun, S.O., Martins, D.T.O. 2017. Ethnobotanical study of medicinal plants used by Ribeirinhos in the North Araguaia microregion, Mato Grosso, Brazil. J. Ethnopharmacol., 205: 69–102.

Ropero Beltrán, S. 2016. La institucionalidad rural en Colombia: Reflexiones para su análisis y fortalecimiento. Mundo Agrario., 17(35): e022.

Sagástegui, A., Dillon, M., Sánchez, I., Leiva, S., Lezama, P. 1999. Diversidad Florística del Norte de Perú (tomo I). Fondo Editorial Universidad Privada Antenor Orrego, Trujillo Perú, 228 p.

Sagástegui, A., Sánchez, I., Zapata, M., Dillon, M. 2003. Diversidad Florística del Norte del Perú (tomo II). Fondo Editorial de la Universidad Privada Antenor Orrego; Trujillo, Perú, 305 p.

Sales, M.D.C., Sartor, E.B., Gentilli, R.M.L. 2015. Ethnobotany and ethnopharmacology: traditional medicine and the bioprospection of phytotherapics. Salus J. Health Sci., 1(1): 17–25.

Sánchez, I. 2011. Especies medicinales de Cajamarca I: contribución etnobotánica, morfológica y taxonómica. Universidad Privada Antonio Guillermo Urrelo, Lumina Cooper Fondo Editorial, Cajamarca Perú, 227 p.

Sánchez, I., Sánchez, A. 2012. La diversidad biológica en Cajamarca: visión étnico-cultural y potencialidades. Visual 47 Ediciones, Gobierno Regional de Cajamarca, Cajamarca Perú, 205 p.

Seminario, A. 2008. Diagnóstico situacional y factores de riesgo de la biodiversidad de especies vegetales medicinales en el centro poblado de Combayo, Cajamarca (Master's tesis). Universidad Nacional de Cajamarca, Cajamarca, Perú, 194 p.

Shaw, D., Graeme, L., Pierre, D., Elizabeth, W., Kelvin, C. 2012. Pharmacovigilance of herbal medicine. J. Ethnopharmacol., 140(3): 513–518.

Stepp, J.R., Moerman, D.E. 2001. The importance of weeds in ethnopharmacology. J. Ethnopharmacol., 75(1): 19–23.

Torres Guevara, F.A. 2019. Etnobotánica y tamizaje fitoquímico de especies vegetales con potencial económico de los páramos de Ayabaca y Huancabamba, Piura-Perú (Doctor's thesis). Universidad Nacional de Trujillo, Trujillo, Perú, 87 p.

Torres, F., Lucio, L. 2014. Conocimientos tradicionales y potencialidades de la Jalca: alternativa de desarrollo al modelo actual y estrategias de conservación. Associació Catalana d'Engyneria Sense Fronteres, Cajamarca Perú, 68 p.

Torres-Guevara, F.A., Ganoza-Yupanqui, M.L. 2017. Etnobotánica y sistemas de extracción para compuestos fenólicos, actividad antioxidante y toxicidad de plantas de páramos y bosques nublados del norte peruano. Revista Peruana de Medicina Integrativa, 2(2): 101–109.

Torres-Guevara, F.A., Ganoza-Yupanqui, M.L., Suárez-Rebaza, L.A., Malca-García, G.R., Bussmann, R.W. 2020. Wild plants of Northern Peru: Traditions, scientific knowledge and innovation. pp. 35–63. *In*: Rai, M., Bhattarai, S., Feitosa, C.M. (eds.). Wild Plants: The Treasure of Natural Healers. CRC Press, USA.

Vásquez, L, Escurra, J., Aguirre, R., Vásquez, G., Vásquez, L. 2010. Plantas medicinales del norte del Perú. FINCyT, Lambayeque Perú, 382 p.

Villar López, M., Ballinas Sueldo, Y., Soto Franco, J.N., Medina Tejada, N. 2016. Conocimiento, aceptación y uso de la medicina tradicional, alternativa y/o complementaria por médicos del seguro social de salud. Revista Peruana de Medicina Integrativa, 1(1): 13–18.

Willsky, G.R., Bussmann, R.W., Ganoza-Yupanqui, M.L., Malca-Garcia, G., Castro, I., Sharon, D. 2020. Integrating traditional and modern medicine: perspectives from ethnobotany, medical anthropology, microbiology, and pharmacy. pp. 301–314. *In*: Smith, K.H., Ram, P.K. (eds.). Transforming Global Health. Springer, Cham Switzerland.

World Health Organization. 2020. Traditional, Complementary and Integrative Medicine. Retrieved from https://www.who.int/health-topics/traditional-complementary-and-integrative-medicine#tab=tab_1 (visited website on 03/14/2020).

Chapter 5
Ethnopharmacology of the Genus *Pilocarpus*

Ronaldo dos Santos Sousa Junior,[1,*] *Mahendra Rai*,[2] *Chistiane Mendes Feitosa*,[1] *Leiz Maria Costa Veras*[3] and *Pedro Vitor Oliveira S. Furtado*[1]

Introduction

The use of plants for medicinal purposes to treat, cure and prevent diseases is one of the oldest forms of medicinal practice for mankind. People prepared their medicines with plants obtained from the forests, especially healers, who together with myths and rituals, formed an important part of local cultures (Lima Saraiva et al. 2015). Thus, during the history of mankind people began to use natural resources for their benefit, such as animals and plants, not only as a source of food but also for their survival and to treat their illnesses. This practice has spread among many people and is currently common worldwide.

Brazil is considered one of the most diverse countries in terms of biodiversity. Ranked as the fifth largest country in the world, it is composed of nine different ecosystems, with an average of 46 thousand species of plants (Carvalho et al. 2017). In the country, knowledge about medicinal plants is a rich mixture of information from indigenous, European and Africanized people, especially regarding tropical and exotic species acclimatized since the colonization process that survives to the present day (Paixão et al. 2013).

Medicinal plants are also associated with different stages of urbanization, from rural communities to large urban centers where medicinal plants can be found in street markets, popular and in residential backyards. Generally, backyards in urban areas serve as a source of food and medicine production, which is common among the poorest sections of the population (Macedo et al. 2011).

Brazilian biodiversity, in addition to attracting companies targeting the drug market, inspires the use of medicinal plants and their derivatives, strengthening the implementation of programs developed in the field of phytotherapy, which emerged in order to parameterize guidelines and encourage research in the sector of medicinal plants and herbal medicines (Lima and Gomes 2014). This rich biodiversity in Brazil also gives the country a privileged space to research natural products, since an enormous amount of bioactive compounds present in plant and animal species can be promising in the development of new medicines.

Many studies have shown that phytotherapy is not always the choice between the popular and the conventional therapy, but because it is the only source of medication used in several villages. People who live in rural areas have little access to scientifically tested drugs (Niehues et al. 2011).

As a strategy in the investigation of medicinal plants, ethnopharmacology has the objective of rescuing and evaluating the effectiveness of "traditional" techniques on the use of plants for therapeutic purposes, combining information acquired from communities that use medicinal flora with ethnodirected

[1] Post-graduate Programs in Chemistry, Federal University of Piaui, Center of Nature Sciences, Teresina-PI, Brazil.
[2] Department of Biotechnology, SGB Amravati University, Amravati, 444 602, Maharashtra State, India.
[3] Biodiversity and Biotechnology Research Center, BIOTEC, Federal University of Piaui, Parnaíba-PI, Brazil.
* Corresponding author: ronaldojunior2000@hotmail.com

studies, that is, with ethnobotanical and ethnopharmacological studies that aim to contribute to the discovery of new drugs (Borcard et al. 2015).

Ethnopharmacology is the science that studies human-plant interaction and seeks, among others, the rescue of knowledge transmitted from one generation to another. As the access to industrialized medicines and health services is not always possible for economically disadvantaged populations, their movement in search of medicinal plants is natural, often easy to obtain and with a tradition of being beneficial and safe (Cavalcante et al. 2017).

Distribution and Morphology

Popularly known in Brazil as "jaborandi", the genus *Pilocarpus*, of the family Rutaceae, is considered one of the most important plants of the Brazilian flora, presenting native species in the north and northeast of the country. *Pilocarpus* is a neotropical genus of tree plants, comprising 16 species, ranging from southern Mexico and Central America to southern South America. 13 species occur in Brazil, 11 of which are endemic. Most species are found in the north and northeast of Brazil (Skopura 1996).

Although many species are named 'jaborandi', those belonging to the genus *Pilocarpus* are recognized as the 'true' jaborandi. The official species named in most pharmacopeias, however, *Pilocarpus* jaborandi Holmes and *Pilocarpus microphyllus* Stapf ex Wardleworth (Rutaceae) are important (Sá and Elisabetsky 2012).

There are several known species of the genus *Pilocarpus*, among them *Pilocarpus microphyllus* (Figure 5.1), is the most studied species, *Pilocarpus alatus*, *Pilocarpus spicatus*, *Pilocarpus pennatifolius, Pilocarpus racemosus, Pilocarpus alvaradoi, Pilocarpus grandifloras*, *Pilocarpus trachylophus* among others (Santos and Moreno 2004).

Pilocarpus species are characterized by being bushes or small trees with long racemes that reach 2 to 3 meters in height (Skorupa 1996). The name is derived from the Tupi-Guarani language, which means a plant that causes saliva, the flowers are tiny, pentamerous with a strong and unpleasant odor, with reduced green sepals, the petals and stamens are red and the anthers are yellow. Flowering occurs from February to July, but it is also possible to find some species with buds in other months of the year (Souza et al. 2003). The leaves are composed of leathery leaflets about 8–15 cm long (Duarte et al. 2009).

Figure 5.1. *Pilocarpus microphyllus* (jaborandi). Source: Veras (2012).

Secondary Metabolites and Bioactivities

Pilocarpus spp. has been studied widely, due to its composition of secondary metabolites and great economic importance for Brazil, mainly because it contains in its leaves a high content of alkaloids, such as pilocarpine, an imidazolic alkaloid used worldwide in the control of primary glaucoma (Andrade-Neto 1997). The *Pilocarpus* genus is the only known species that contain this compound, and many of the species of this genus have been threatened with extinction due to over-harvesting of wild populations (Pinheiro 2002).

As mentioned before, the species of the genus *Pilocarpus* have different types of bioactive compounds; in addition to being rich in alkaloids, they also have flavonoids, coumarins, terpenoids among others (Conceição 2004, Davies et al. 2009, Ferreira et al. 2010, Madeiro et al. 2017). Figure 5.2 shows some chemical structures of substances found in Jaborandi species according to the secondary metabolite group to which it belongs.

Pilocarpine, a natural plant alkaloid derived from the leaves of *Pilocarpus jaborandi*, is a muscarinic agonist. It binds to M3 muscarinic receptors in salivary glands and increases the salivary secretion for an adequate length of time; in addition, pilocarpine can be administered by mouth and has tolerable side effects (Fox et al. 1991). Therefore, this drug has been widely used clinically for the treatment of xerostomia caused by various diseases and medical conditions including Sjogren's syndrome and radiation-induced xerostomia after radiotherapy for head and neck cancer (Derek and Vivino 2004, Mercadante et al. 2017).

The extraction of jaborandi is responsible for a large mass of resources in the North and Northeast regions of Brazil. In 2014, it handled around 252 tonnes in these regions, amounting to approximately R$ 1 million out of a total of R$ 20.8 billion referring to plant extraction carried out in Brazil that year (IBGE 2014).

The therapeutic use of Jaborandi has long been known by indigenous tribes in northern Brazil as an antidote to poisons and toxins. Currently, this medicinal plant is used intensively in the pharmaceutical

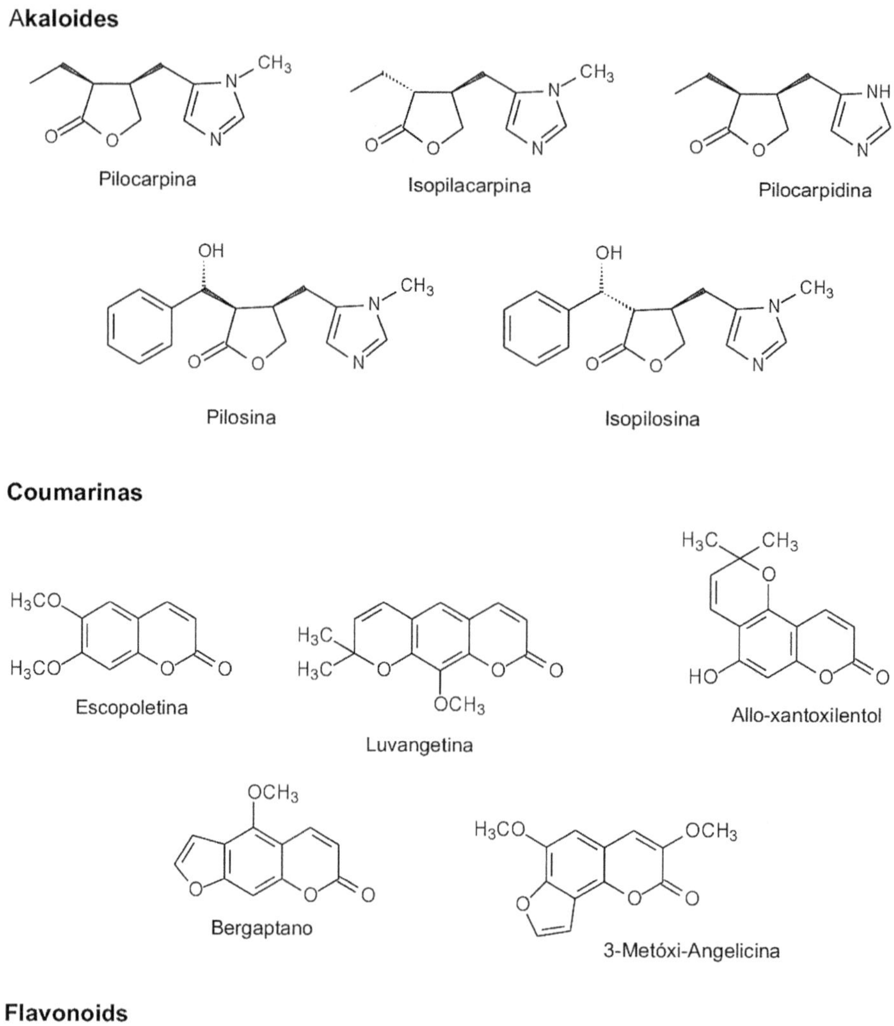

Figure 5.2. Primary metabolites present in the *Pilocarpus* species.

industry (Vital and Acco 2006). Jaborandi was used in the past to stimulate the scent of hunting dogs. It is also indicated in the treatment of diseases of the respiratory system and used in the formulation of shampoos used in hair loss (Pinheiro 1997). It can be found in backyards, orchards, fairs, municipal markets, in houses that sell natural products, on sites that sell seedlings and also sold in the form of cosmetics.

According to Corrêa (1969) and Conceição (2004), the species of *Pilocarpus* began to be studied for their pharmacological properties only in 1873, when it was proven that jaborandi was a potent stimulant of the secretory system, presenting activities of sialagogue, diuretic and diaphoretic.

The whole plant or different parts, as well as isolated compounds obtained from the genus *Pilocarpus*, are traditionally used in medicine for a variety of human ophthalmic and geriatric diseases (Bertrand et al. 2001, Andrade-Neto et al. 2002, Auffarth et al. 2002, Avancini et al. 2003, Taveira et al. 2003, Agra et al. 2007). *Pilocarpus* leaves are used in infusions to treat fever, stomatitis, bronchitis and psoriasis (Holmsted et al. 1979).

Pilocarpus microphyllus Stapf ex Wardlew is the most studied. It is native to the North and Northeast of Brazil occurring in the east of the State of Pará, west, and north of Maranhão and north of Piauí, moving the extraction of tons every year in these states (Lima et al. 2015).

P. microphyllus is one of the most prominent Brazilian medicinal species in pharmacopeias worldwide, due to the use of its active ingredient pilocarpine (Lewis and Elvin-Lewis 1977), and is still of great relevance for the pharmaceutical industries. From the leaves, pilocarpine hydrochloride salts and pilocarpine nitrate are processed, in addition to their free form (base pilocarpine) used in the formulation of eye drops for the treatment of glaucoma (Nordstrom et al. 2005, Lima 2008).

Its leaves are also used in Brazil as a diuretic, sudorific, digestive, sialogogue, hair tonic and eye inflammation (Taveira et al. 2003). In traditional medicine, this species is used for earache, colic in children, sinusitis and even "evil eye" and "broken shield", epilepsy, cardiac depression, asthma, inflammation, diarrhea, wound-healing, alopecia, Bright's disease, deafness, flu, gastritis, gonorrhea, hepatitis and others (Duke 2002). The infused and macerated leaves are used in the form of juice.

In ethnobotanical research carried out by Monteles and Pinheiro (2007) in a traditional community in the state of Maranhão, it was found that the juice of *P. microphyllus* leaves was used to treat skin diseases. Maia et al. (2011) reported bioactivities of this species such as induction of sweating, constriction of the pupil (miosis), acceleration of peristalsis, induction of salivation and in the treatment of bronchitis, dry skin, fever, acne, hair loss, and seborrhea.

The importance of *P. microphyllus* lies in the research and development of the pharmacological and biotechnological application of other imidazolic alkaloids present in this species, such as the recent application of the alkali epiisopiloturina and epiisopilosina in combating neglected diseases (Veras et al. 2012, Rocha 2013), presenting significant anti-hyperglycemic activity in the induction of *Streptozotocin* (Devi 2010) and effective activity against *Schistosoma mansoni* (Veras et al. 2012).

The antinociceptive and anti-inflammatory activities of epiisopiloturine, an imidazole alkaloid found in the leaves of *Pilocarpus microphyllus,* were investigated by Silva and collaborators (2013). The anti-inflammatory activity was evaluated using several agents that induce paw edema and peritonitis in Swiss mice. The antinociceptive activity was evaluated by acetic acid-induced writhing, the hot plate test, and pain induction using formalin. The tested alkaloid showed satisfactory results in terms of anti-inflammatory and antinociceptive activity.

Morais et al. (2018) evaluated the antileishmania activity of the essential oil of *P. microphyllus*. Leishmaniasis is a group of infectious diseases caused by different species of the *Leishmania* genus. The authors evaluated the antileishmania activity against promastigote forms of *Leishmania infantum*. In the results, it was reported that the essential oil of this species proved to be very promising against promastigote forms of *L. infantum*.

In another study, the activity of aqueous, ethanolic extracts and pilocarpine hydrochloride, extracted and isolated from *Pilocarpus microphyllus*, on *Rhipicephalus (Boophilus) microplus* was evaluated. Pilocarpine hydrochloride showed the highest tick activity on larvae. These results suggest the possibility of using these extracts as acaricides (Castro et al. 2016).

Almeida Rocha et al. (2017) reported the antibacterial and anthelmintic activity of five alkaloids found in leaves of *P. microphyllus*, namely, pilosina, epiisopilosina, isopilosina, epiisopiloturina e macaubina. Regarding anthelmintic activity, alkaloids were studied using *in vitro* assays to assess survival time and damaged integuments for adult *Schistosoma mansoni* worms. The potent anthelmintic activity of epiisopilosina indicates the potential of this natural compound as an antiparasitic agent.

The synergistic effects of *in vitro* combinations of piplartine, epiisopiloturina an alkaloid of *P. microphyllus* and praziquantel against *Schistosoma mansoni* were studied by Campelo et al. (2017), in which superficial changes were observed in the integument of adult schistosomes after the microscopic treatments. Synergistic combinations between the substances were cited to create the possibility of reduced doses to be used against *Schistosoma mansoni*.

The protective effect of epiisopiloturine hydrochloride, an imidazole alkaloid of *P. microphyllus*, has been studied in gastrointestinal damage induced in rats. Pretreatment with the alkaloid prevented induced macro and microscopic gastric damage with a maximum effect of 10 mg/kg. Histological analysis revealed that epiisopiloturine decreased the damage score by showing its protection against induced gastric and intestinal damage, reducing pro-inflammatory cytokines and oxidative stress (Nicolau et al. 2017).

The literature reports the nematicidal activity of essential oils from the leaves of native *P. microphyllus* and five cultivated varieties of this species collected from the state of Maranhão in Brazil. The bioassays were carried out with juveniles of *Meloidogyne incognita* in the second stage. The oils of *P. microphyllus* native and of two varieties showed high nematicidal activity while the other oils showed no activity. The oils that showed activity had high concentrations of 2-tridecanone and 1-hydroxy-2-tridecanone, suggesting that these two ketones are responsible for the nematicidal activity of essential oils (Romero et al. 2006).

Another species that stands out in this genus is *Pilocarpus spicatus* Saint-Hilaire, a species found only in Brazil, in the states of São Paulo, Rio de Janeiro, Espírito Santo, and Bahia. The phytochemical study of this species demonstrated the presence of several terpenes and the chalepine coumarin, which showed trypanomicidal activity (Pavão et al. 2002, Santos and Moreno 2004).

Bioassays against *Rhodnius prolixus* nymphs were conducted by Mello et al. (2007) using *Pilocarpus spicatus* essential oil extracted by hydrodistillation. The main results show high levels of toxicity and paralysis associated with slight inhibition of seedling induced by topical application of essential oil on insect, partial phage inhibition, high levels of seedling inhibition, prolonged intermittent period and high number of paralyzed insects, but absence of toxicity was observed after oral treatment of *P. spicatus* essential oil per mL of ingested blood.

Silva et al. (2014) evaluated the antiparasitic activity of *P. spicatus* extracts against the epimastigote forms of *Trypanosoma cruzi*. The hexanic and methanolic extracts of leaves and roots were tested *in vitro* against *Trypanosoma cruzi* cepa Y and showed relevant trypanomicidal activity.

The antibacterial and immunomodulatory activities of *P. spicatus* leaf extract were studied, as well as the activity of crude stem extracts and their fractions, against the trypomastigote form of *T. cruzi* with results above 80% of parasite inhibition were reported (Mafezoli et al. 2000, Costa et al. 2010).

Oliveira et al. (2010) examined the essential oil extracted by steam dragging the aerial parts of *P. spicatus* and examined the GC-MS identifying 17 components that correspond to 96.06% of the chemical composition of the oil. In this study, the essential oil of *P. spicatus* had an inhibitory effect on bacterial growth (*Escherichia coli* and *Staphylococcus aureus*) and exhibited anticholinesterase activity in a test. In addition, the volatile oil has proven to be toxic to *Artemia salina* larvae.

Santos et al. (1997) evaluated the essential oil of the genus of *P. spicatus* for antibacterial activity *in vitro* using five different bacterial strains. Gram-negative organisms, *S. typhi* and *E. coli* were, however, insensitive to this oil. *P. spicatus* also showed high activity against *S. positive*, Gram-positive.

The aerial parts of *P. spicatus* were subjected to processes of extraction, partitioning and isolation of its chemical constituents. The antibacterial activity was evaluated by using the Minimum Inhibitory Concentration (MIC) and modulating drug resistance in *Staphylococcus aureus* from the crude ethanolic extract of *P. spicatus,* and isolated coumarins (bergaptene, xanthotoxin, isopimpinelin, imperatorine) was recorded. The extract and coumarins did not show relevant antibacterial activity. However, in the drug resistance modulation assay, isopimpineline reduced erythromycin MIC up to 4 times, and imperatorin showed the best results with a reduction in tetracycline MIC (up to 2 times), erythromycin (up to 4 times) and norfloxacin (up to 4 times) (Madeiro et al. 2017).

Regarding the species *Pilocarpus pennatifolius* Lemaire, it is native to some countries in South America, such as Brazil, Argentina, Uruguay and Paraguay. In Rio Grande do Sul (Brazil), it is popularly known as "jaborandi" or "agouti-branca" (Santos et al. 2004). Do Carmo et al. (2018) reported that leaf infusions, dry leaves, and extracts are used in folk medicine for the treatment of asthma, diabetes, liver disease and in the form of cosmetics for hair treatment. However, this plant is still commercialized to treat other diseases, since its dry leaves have gastrostimulating, peristaltic and secretagogic activities, suitable for the treatment of urinary, gastrointestinal disorders, and lung disorders, etc. (Ferreres et al. 2015).

Several other studies have already been carried out on *P. pennatifolius* in which the presence of coumarins was determined with antibacterial and antifungal properties (Santos and Moreno 2004). Extracts and pure compounds of *P. pennatifolius* were tested against several strains of bacteria and fungi, showing promising antimicrobial activities. The activities of the crude methanol extract from the bark, the basic acetate fraction of the leaf and a compound against Gram-negative bacteria *Shigella flexneri*, *Enterococcus fecalis* Gram-positive and two Gram-negative bacteria *Salmonella enteritidis* and *Pseudomonas aeruginosa* were highlighted (Do Carmo et al. 2018).

When assessing the ability to inhibit enzymes related to neurodegenerative diseases, namely α-glucosidase, AChE and BuChE, Ferreres et al. (2015) concluded that *P. pennatifolius* herbal tea can be used as a nutraceutical ingredient by the food and pharmaceutical industries.

Another species, *Pilocarpus racemosus,* is widely distributed in the Antilles, South America. There are only collections from Venezuela and Mexico, according to the location of the world's dry forests (Bezaury 2010). *P. racemosus* is a species with a lot of variation in its characters, one of the most visible being the types of leaves. The character that distinguishes the specimen identified as *Racemosus* is the presence of translucent hair on the terminal shoots and eyelashes (Kaastra 1982).

Few studies report the pharmacological activities of this species; however, the presence of pilocarpine among its constituents attributes *P. racemosus* as a promising source of drugs (Schrittwieser 2013). Sawaya et al. (2011) observed that the pilocarpine content may be higher in species of *P. racemosus* than in *P. microphyllus* allowing greater economic exploitation compared to other species. In another study, Castillo-Bautista et al. (2018) reported the use of *P. racemosus* in the treatment of headache and nerve disorders.

The antimycobacterial activity of pilocarpine extracted from *P. racemosus* was evaluated by Rastogi et al. (1998). The authors tested pilocarpine against *Mycobacterium tuberculosis, M. avium* and *M. kansasii* using the agar medium. Pilocarpine did not show significant antimycobacterial activity.

Pilocarpus alatus has few reports of study in the available literature. There was a phytochemical study of leaves with the study of volatile composition and isolation of flavonoids and one of taxonomic importance with the study of leaf greases (Skorupa et al. 1998, Bezerra 2004).

In a study by Bezerra (2009), the antioxidant activity of compounds isolated from *P. alatus* was evaluated based on the scavenging effect of free radicals on DPPH. Only phenolic compounds of this species showed antioxidant activity, with two compounds being more active.

Antimycobacterial activity and cytotoxicity of alkaloids of *Pilocarpus alatus* were analyzed in *Mycobacterium tuberculosis*. The results for the alkaloids are indicative of the promising antimycobacterial activity for the mentioned species and there was no evident cytotoxicity (De Souza et al. 2007).

Regarding the *Pilocarpus alvaradoi* species, several metabolites were isolated from the ethanolic extract of the leaves, including lactones such as benzoyl-ɣ-valerolactone, pentacyclic triterpenes known as lupeol and epibetulin, and furanocoumarins such as baptapene and psoralen (Cuca and Avila 2007).

Marín et al. (2017) evaluated the antioxidant and leishmanicidal activity of this species. The results of this study showed that the ethyl acetate extract of the leaves and the ethanol extract of the bark demonstrated a potential leishmanicidal activity compared to the control amphotericin B. The authors reported that the *in vitro* activity of these extracts against *L. parasite. (V) panamensis* provides a basis for ongoing studies with the aim of having candidate molecules that can become an alternative for the treatment of leishmaniasis.

Another species is *Pilocarpus grandiflorus*—it has alkaloids such as dictamine with a structural relationship with coumarins and, probably because of this similarity, this substance exhibited a photosensitizing activity (Ashwood et al. 1982). Studies carried out by Zancul (2001) about the effect of different parts of *P. grandiflorus* on the development of the symbiotic fungus, showed that both the stem and branch dichloromethane extract, at a concentration of 1000 mg/mL, showed 100% inhibition.

Souza et al. (2005a) reported the isolation of 18 compounds. The structural elucidation of a new alkaloid imidazole was evaluated for its activities as growth inhibitors of the symbiotic fungus *Leucoagaricus gongylophorus* and leaf-cutting ants, *Atta sexdens rubropilosa*. The dichloromethane extract of the stems, as well as fractions of *P. grandiflorus*, showed strong inhibitory activity on the growth of *L. gongylophorus* and the results of the search confirmed the antifungal activity of uroquinoline alkaloids against *L. gongylophorus*.

Another species, *Pilocarpus trachylophus* Holmes, is a shrub occurring in Caatinga Forests. Due to the pressure for extraction, the decline in area and the quality of habitat, this species is threatened with extinction. In traditional medicine, it is used as a sudorific and sialagogue for the treatment of bronchitis. It is also well known for the production of pilocarpine (Bertrand et al. 2001).

Pilocarpus goudotianus is a shrub found predominantly in Neotropical countries, such as Colombia and Venezuela. According to Amaro-Luis et al. (1990) and Macías et al. (1993), the leaves extract of this species is rich in three exclusive coumarins: Allo-Xanthoxilletol, (+)-Elisin, and DonatinIn. In addition, it was reported by the aforementioned authors that the crude leaf extract was significantly active in the germination of *Lactuca sativa* seeds. Given the number of coumarins present in the species, it showed anticoagulant, cytostatic, antibacterial and fungicidal properties. Also, for the *Pilocarpus goudotianus* species, the effects of food deterrence of secondary metabolites of this plant were observed in insects of the genus *Spodoptera* (Ayil-Gutiérrez et al. 2018).

Pilocarpus riedelianus is also a sub-shrub species, native to tropical America. According to Guerreiro et al. (2005), the species was studied in order to obtain compounds effective in fighting fungi and bacteria. Among the fractions that were analyzed, the hexane fraction of the dichloromethane extract showed activity against several fungi and bacteria. In addition, through CG-MS analysis, it was possible to identify the presence of 6 sesquiterpenes (α-Calarone, β-Calacorene, γ-Calacorene, Cadalene, Sesquichamaenol and 1-hydroxy-1,3,5-bisabolatrien-10-one) in the hexane fraction. In this context, another sub-species *Pilocarpus pauciflorus* presents the allopatric geographic distribution and is found mainly in Santa Catarina, Paraná, São Paulo, Rio de Janeiro and Espírito Santo, in ombrophilous coastal forests and in semideciduous seasonal forests towards the interior. Additionally, it can be identified by its leaf blades, which are generally obovate to narrowly obovate with obtuse and rounded tips (Skorupa 2003).

The essential oil of *P. pauciflorus* has a profile similar to the oil of *P. pennatifolius*, as demonstrated by Conceição (2004). In these species, the major constituents are substances derived from the mechanism via acetate, and terpenes are produced in a lower proportion. It should also be noted that such oils have high bactericidal efficiency against *Bacilus subtilis, Salmonella Typhimirum, Escherichia coli* and *Pseudomonas aeruginosa*.

In *Pilocarpus giganteus* Engler, also known as *Pilocarpus macrocarpus* Engler, the shrub shape is also predominant, presenting a yellowish-brown or grayish-brown foliage, being found in primary and secondary forest, in the regions of Minas Gerais, São Paulo and Rio de Janeiro.

P. macrocarpus is an inflorescent carat plant, with flowers and fruits. Among its properties used in popular medicine, diuretics stand out. Allevato et al. (2019) reported that the species has a low amount of alkaloids in contrast to a significant amount of coumarins, and indicates the absence of pilocarpine, evidenced by the absence of orange color in the foliage.

Thus, the research has shown that the different species of the genus *Pilocarpus* have been very promising for the development of new phytomedicines, due to a wide range of pharmacological activities. No other studies were found in the literature that report the pharmacological activities of other species of the *Pilocarpus* genus, paving the way for further research with this medicinal plant.

Conclusion and Future Perspectives

It is understood that the knowledge about medicinal plants is of great importance to society. Often these plants represent the only alternative for many communities in the treatment of diseases. The importance of different species of *Pilocarpus* in traditional medicine has been emphasized in the chapter. This plant has historically been widely used in the treatment of a variety of diseases in various regions.

Recent literature reports that the different species of jaborandi contains a variety of chemical constituents and has demonstrated several pharmacological activities. It is hoped that the studies presented will motivate researchers to investigate promising new sources of medicines from different species of the genus *Pilocarpus*.

Acknowledgment

The authors would like to thank CAPES and CNPq for scholarships and financial supports.

References

Agra, M.F., França, P.F., Barbosa-Filho, J.M. 2007. Synopsis of the plants known as medicinal and poisonous in Northeast of Brazil. Rev. Bras. Farmacogn., 17: 114–140.

Allevato, D.M., Groppo, M., Kiyota, E., Mazzafera, P., Nixon, K.C. 2019. Evolution of phytochemical diversity in *Pilocarpus* (Rutaceae). Phytochemistry, [s.l.] 163: 132–146.

Almeida Rocha, J. 2017. Anthelmintic, antibacterial and cytotoxicity activity of imidazole alkaloids from *Pilocarpus microphyllus* leaves. Phytother. Res., 31: 624–630.

Amaro-Luis, J., Massanet, G., Pando, E., Rodriguez-Luis, F., Zubia, E. 1990. New coumarins from *Pilocarpus* goudotianus. Plant Med., 56(3): 304–306.

Andrade-Neto, M. 1997. Contribuição ao conhecimento químico de *Pilocarpus* spp. Tese (Doutorado em Química Orgânica) – Universidade Federal do Ceará, Fortaleza.

Andrade-Neto, M., Cunha, U.D., Mafezoli, J., Silveira, E.R. 2002. Volatile constituents of different populations of *Pilocarpus spicatus* Saint Hill. (Rutaceae) from the Northeast of Brazil. J. Essent. Oil Res., 14: 319–324.

Ashwood-Smith, M.J., Torres, G.H., Abramowski, Z., Poulton, G.A., Liu, M. 1982. Photobiological studies with dictamnine, a furoquinoline alkaloid. Mutat. Res., 102(4): 401–412.

Auffarth, G.U., Martin, H., Fuchs, H.A., Rabsilber, T.M., Becker, K.A., Schmack, E.U. 2002. Validity of anterior chamber depth measurements for the evaluation of accommodation after implantation of an accommodative humanoptics 1CU intraocular lens. Ophthalmologe., 99: 815–819.

Avancini, G., Abreu, I.N., Saldana, M.D.A., Mohamed, R.S., Mazzafera, P. 2003. Induction of pilocarpine formation in jaborandi leaves by sialicylic acid and methyljasmonate. Phytochemistry, 63: 171–175.

Avila, M., Cuca, L.E. 2007. Metabolitos fenólicos aislados de *Pilocarpus alvaradoi* (Rutaceae). Scientia et Technica., 13: 67–69.

Ayil-Gutiérrez, B.A., Sánchez-Teyer, L.F., Vazquez-Flota, F., Monforte-González, M., Tamayo-Ordóñez, Y., Tamayo-Ordóñez, M.C., Rivera, G. 2018. Biological effects of natural products against Spodoptera spp. Crop Protection, 114: 195–207.

Baptista Neto, C., Sugaya, N.N. 2004. Treatment of xerostomia in patients irradiated in the head and neck. Rev. Biociên., 10(3): 147–151.

Bertrand, C., Fare, N., Moulis, C. 2001. Constituents of *Pilocarpus trachyllophus*. Fitoterapia., 72: 844–847.

Bezaury, C.J. 2010. The dry jungles of the Mexican Pacific in the global context. pp. 21–40. *In*: Diversity, Threats and Priority Areas for the Conservation of the Dry Forests of the Pacific of Mexico.

Bezerra, F.S. 2004. Contribuição ao Conhecimento Químico de Plantas do Nordeste do Brasil: Pilocarpus alatus Joseph ex Skorupa. Dissertação (Mestrado em Química Orgânica) – Universidade Federal do Ceará, Fortaleza.

Bezerra, F.S., Sousa de, A.H., Andrade Neto, M., Almeida, M.M.B., Fonseca, A.M., Oliveira de, M.C.F., Lemos, T.L.G. 2006. Isolamento e Atividade Antioxidante de Flavonóides de *Pilocarpus alatus* (Rutaceae). 29a Reunião Anual da Sociedade Brasileira de Química. Águas de Lindóia – SP.

Bezerra, F.S. 2009. Estudo fitoquímico de *Pilocarpus alatus* Joseph ex Skorupa. Tese de Doutorado. Universidade Federal do Ceará. Fortaleza-CE.

Borcard, G.G., Conde, B.E., Alves, M.J.M., Chedier, L.M., Pimenta, D.S. 2015. Ethnopharmacological study around urban forest as a subsidy for the implementation of Phytotherapy in the Unified Health System. Rev. Bras. Plantas Med., 17(4): 928–936.

Campelo, Y.D.M., Mafud, A.C., Véras, L.M.C., Guimarães, M.A., Yamaguchi, L.F., Lima, D.F., Arcanjo, D.D.R., Kato, M. J., Mendonça, R.Z., Pinto, P.L.S. 2017. Synergistic effects of *in vitro* combinations of piplartine, epiisopiloturine and praziquantel against *Schistosoma mansoni*. Biomed. Pharmacother., 88: 488–499.

Carvalho, A.C.B., Lana, T.N., Perfeito, J.P.S., Silveira, D. 2017. The Brazilian market of herbal medicinal products and the impacts of the new legislation on traditional medicines. J. Ethnopharmacol., 212: 29–35.

Castro, K.N.C., Lima, D.F., Wolschick, D., Andrade, I.M., Santos, R.C., Santos, F.J.S., Veras, L.M.C., Costa-Júnior, L.M. 2016. *In vitro* effects of *Pilocarpus microphyllus* extracts and pilocarpine hydrochloride on *Rhipicephalus (Boophilus) microplus*. Rev. Bras. Parasitol. Vet., 25(2): 248–253.

Cavalcante, J.W., Cavalcante, V.M.G., Bieski, I.G.C. 2017. Traditional and ethnopharmacological knowledge of the medicinal plant copaiba (*Copaifera langsdorffii* Desf.). Biodiversidade., 16(2): 123.

Calcagno, M.P., Coll, J., Lloria, J., Faini, F., Alonso-Amelot, M.E. 2002. Evaluation of synergism in the food deterrence of some furanocoumarins in *Spodoptera littoralis*. J. Chem. Ecol., 28(1): 175–191.

Castillo-Bautista, C.M. 2018. Efecto neuroprotector de especies vegetales utilizadas en la medicina tradicional Maya. Centro de Investigación Científica de Yucatán, A.C. Posgrado en Ciencias Biológicas. Mérida, Yucatán, México.

Conceição, A.P.S. 2004. Estabelecimento de culturas *in vitro* de *Pilocarpus pennatifolius* Lemmaire e estudos iniciais sobre a biossíntese do alcaloide pilocarpina. Tese de Doutorado. Instituto de Química. São Paulo.

Costa, J.F.O., Juiz, P., São Pedro, A., David, J.P.L., David, J.M., Giulietti, A.M., França, F., Santos, R.R., Soares, M.B.P. 2010. Immunomodulatory and antibacterial activities of extracts from Rutaceae species. Rev. Bras. Farmacogn., 20(4): 502–505.

De Souza, A.O., Galetti, F.C.S., Silva, C.L., Bicalho, B., Parma, M.M., Fonseca, S., F., Marsaioli, A.J., Trindade, A.C.L.B., Gil, R.P.Freitas., Bezerra, F.S. 2007. Antimycobacterial and cytotoxicity activity of synthetic and natural compounds. Química Nova, [s.l.]., 30(7): 1563–1566.

Derk, C.T., Vivino, F.B. 2004. A primary care approach to Sjögren's syndrome. Helping patients cope with sicca symptoms, extraglandular manifestations. Postgrad. Med., 116: 49–54, 59, 65.

Devi, R. 2010. Antidiabetic activity of *Pilocarpus microphyllus* extract on Streptozotocin-induced diabetic mice. Int. J. Pharm. Sci. Rev. Res., 5: 87–92.

Do Carmo, G., Fernandes, T.S., Pedroso, M., Ferraz, A., Neto, A.T., Silva, U.F., Mostardeiro, M.A., Back, D.F., Dalcol, I.I., Morel, A.F. 2018. Phytochemical and antimicrobial study of *Pilocarpus pennatifolius* Lemaire. Fitoterapia., 131: 1–8.

Duke, J.A. 2002. Handbook of Medicinal Herbs. 2nd edition. CRC Press, New York, 415.

Duarte, M.R., Golambiuk, G., Costa, R.E. 2009. Illustration of microscopic characters of plant drugs for pharmacognostic quality control. IV. Jaborandi. (*Pilocarpus pennatifolius* Lem.; Rutaceae). Visão Acadêmica, 10(2): 7–14.

Ferreira, M.E., de Arias, A.R., Yaluff, L., de Bilbao, N.V., Nakayama, H., Torres, S., Schinini, U.M., Guy, I., Heinzen, H., Fournet, U.M.A. 2010. Antileishmanial activity of furanoquinolines and coumarins from *Helietta apiculata*. Phytomedicine, 17: 375–378.

Ferreres, F., Grosso, C., Gil-Izquierdo, A., Fernandes, A., Valentão, P., Andrade, P.B. 2015. Comparing the phenolic profile of *Pilocarpus pennatifolius* Lem. by HPLC–DAD–ESI/MSn with respect to authentication and enzyme inhibition potential. Ind. Crop Prod., 77(23): 391–401.

Fox, P.C., Atkinson, J.C., Macynski, A.A., Wolff, A., Kung, D.S., Valdez, I.H., Jackson, W., Delapenha, R.A., Shiroky, J., Baum, B.J. 1991. Pilocarpine treatment of salivary gland hypofunction and dry mouth (xerostomia). Arch. Intern. Med., 51: 1149–1152.

Guerreiro, G., Lima, J.E.S. de, Fernandes, J.B., Silva, M., Fátima, G.F., Viera, P.C., Magalhães, C.M.I., Pagnocca, F.C., Bueno, O.C., Hebling, M.J.A., Pirani, J.R. 2005. Sesquiterpenos do caule de *Pilocarpus riedelianus* e atividades sobre microorganismos. Quím Nova., 8(6): 986–990.

Holmsted, B., Wassén, H.S., Shultes, R.E. 1979. Jaborandi: An interdisciplinary appraisal. J. Ethnopharm., 1: 3–21.

IBGE. Instituto Brasileiro de Geografia e Estatistica. 2014. Sistema IBGE de recuperação automática: produção da extração vegetal e da silvicultura/PEVS.

Kaastra, R. 1982. Flora neotropica - monograph 33. Pilocarpinae. New York Botanical Garden, New York.

Lewis, W.H., Elvin-Lewis, M.P.F. 1977. Medical Botany—Plants Affecting Man's Health. New York: John Wiley & Sons. 515 p.

Lima, D.F. 2008. Alcalóides de interesse industrial: aspectos químicos e biológicos. 56 f. Monografia (Especialização em Plantas Medicinais) – Pós-Graduação em Plantas Medicinais, Universidade Federal de Lavras, Lavras.

Lima, D.F., Silva, R., Marques, L.G.A., Veras, L., Simões, E., Leite, J.R.S.A., Santos, M.R.M.C., Pessoa, C. 2015. Technological prospecting of jaborandi (*Pilocarpus microphyllus*): economically important species in the north and northeast of Brazil. Revista GEINTEC, 5(1): 1626–1638.

Lima, L.O., Gomes, E.C. 2014. Food or medicine? Vegetable species according to Brazilian legislation. Rev. Bras. Pl. Med., 16(3): 771–782.

Lima Saraiva, S.R.G., Saraiva, H.C.C., Oliveira-Júnior, R.G., Silva, J.C., Damasceno, C.M.D., Almeida, J.R.G., Amorim, E.L.C. 2015. The implantation of the program of medicinal plants and herbal medicines in the public health system in Brazil: a literature review. Rev. Interdisc. Pesq. Inov. Sergipe., 1(1): 1–11.

Macías, Francisco A., Galindo, J.C.G., Massanet, G.M., Rodriguez-Luis, F., Zubia, Eva. 1993. Allelochemicals from *Pilocarpus goudotianus* leaves. J. Chem. Ecol., 19(7): 1371–1379.

Madeiro, S.A.A., Borges, N.H., Souto, A.L., de Figueiredo, P.T., Siqueira-Junior, J.P., Tavares, J.F. 2017. Modulation of the antibiotic activity against multidrug resistant strains of coumarins isolated from Rutaceae species. Microb. Pathog., 104: 151–154.

Macías, F.A. 1993. Allelochemicals of leaves *Pilocarpus goudotianus*. J. Chem. Ecol., 19(7): 1371–1379.

Macedo, M., Pereira, M.L.S., Silva, F.H.B. 2011. Plants with probable antifungal action used by residents of the Cidade Verde neighborhood, Cuiabá, Mato Grosso. FLOVET, 1: 3.

Mafezoli, J., Vieira, P.C., Fernandes, J.B., Silva, M.F.G.F., Albuquerque, S. 2000. *In vitro* activity of Rutaceae species against the trypomastigote form of *Trypanosoma cruzi*. J. Ethnopharm., 73: 335–340.

Maia, E.A., Francisco, J., Pires, T., Manfredi-Coimbra, S. 2011. The use of plant species for medicinal purposes by two communities in the Serra Catarinense, Santa Catarina, Brasil. Rev. Biol. Ciênc. Terra., 11(1): 54–74.

Marin, F.J., Torres, O.L., Robledo, S.M., Doria, E.D. 2018. Estudio Fitoquímico y Evaluación de la Actividad Antioxidante y Leishmanicida de la Especie *Pilocarpus alvaradoi* (Rutaceae). Inf. tecnol., La Serena, 29(3): 177–186.

Mello, C.B., Uzeda, C.D., Bernardino, M.V., Mendoça-Lopes, D., Kelecom, A., Fevereiro, P.C.A., Guerra, M.S., Oliveira, A.P., Rocha, L.M., Gonzalez, M.S. 2007. Effects of the essential oil obtained from *Pilocarpus spicatus* Saint-Hilaire (Rutaceae) on the development of *Rhodnius prolixus* nymphae. Revista Brasileira de Farmacognosia, [s.l.] 17(4): 514-520.

Mercadante, V., Al Hamad, A., Lodi, G., Porter, S., Fedele, S. 2017. Interventions for the management of radiotherapy-induced xerostomia and hyposalivation: A systematic review and meta-analysis. Oral Oncol., 66: 64–74.

Miura, L.M.C.V. 2009. Isolamento do alcalóide epiisopiloturina a partir da biomassa do jaborandi (pilocarpus microphyllus) e avaliação de suas atividades antibacteriana, antileishmania, antiviral, antiesquistosoma e sialagoga. Dissertação (Mestrado em Ciência Animal) – Pós-Graduação em Ciência Animal, Universidade Federal do Piauí, Teresina.

Monteles, R., Pinheiro, C.U.B. 2007. Medicinal plants in a quilombo in Maranhão: an ethnobotanical perspective. Rev. Biol. Ciênc. Terra., João Pessoa, 7: 2.

Moraes Alves, M.M., Cruz, L.P.L., Freitas, R.I.C., Costa, A.M.S., Lima, J.E.O., Reis R.L.R., Véras, L.M.C., Carvalho, F.A.A. 2018. Essential oil of *Pilocarpus microphyllus* Stapf. against promastigotes forms of Leishmania infantum. J. Soil Plant Bio., 1: 40–43.

Nicolau, L.A.D., Carvalho, N.S., Pacífico, D.M., Lucetti, L.T., Aragão, K.S., Véras, L.M.C., Souza, M.H.L.P., Leite, J.R.S.A., Medeiros, J.V.R. 2017. Epiisopiloturine hydrochloride, an imidazole alkaloid isolated from *Pilocarpus microphyllus* leaves, protects against naproxen induced gastrointestinal damage in rats. Biomed. Pharmacother., 87: 188–195.

Niehues, J., Bonetti, P., Souza, M.R., Maia, A.L., Piovezan, A.P., Peters, R.R. 2011. Ethnopharmacological survey and botanical identification of medicinal plants in communities assisted by a health service. Arquivos Catarinenses de Medicina, 40: 1.

Nordstrom, B.L., Friedman, D.S., Mozaffari, E., Quigley, H.A., Walker, A.M. 2005. Persistence and adherence with topical glaucoma therapy. Am. J. Ophthalmol., 140(4): 598–606.

Oliveira, A.P., Cruz, R.A.S., Botas, G.S., Gonzalez, M.S., Santos, M.G., Teixeira, L.A., Rocha, L.M. 2010. Chemical and biological investigations of *Pilocarpus spicatus* essential oils. Bol. Latinoam. Caribe Plantas Med. Aromát., 9(3): 206–211.

Paixão, J.L.F., Humberto, D., Oliveira, J.E.Z.E. 2013. Organic garden of medicinal herbs: social inclusion in the Barra community in Muriaé/MG - Brasil. Rev. Agrogeoamb., 5(2): 19–30.

Pavão F., Castilho, M.S., Pupo, M.T., Dias, R.L., Correa, A.G., Fernandes, J.B., Silva, M.F., Mafezoli, J., Vieira, P.C., Oliva, L. 2002. Structure of *Trypanosoma cruzi* glycossomal glyceraldehyde-3-phosphate dehydrogenase complexed with chalepin, a natural product inhibitor, at 1.95 Angstron resolution. Febs Letters, 520: 13–7.

Pinheiro, C.U.B. 1997. Jaborandi (*Pilocarpus* sp., Rutaceae): a wild species and its rapid transformation into a crop. Econ Bot. 51(1): 49–58.

Pinheiro, C.U.B. 2002. Extractivism, cultivation and privatization of jaborandi (*Pilocarpus microphyllus* Stapf ex Holm.; Rutaceae) in Maranhão. Acta Bot. Bras., 16: 141–150.

Rastogi, N., Abaul, J., Goh, K.S., Devallois, A., Philogène, E., Bourgeois, P. 1998. Atividade antimicobacteriana de substâncias naturais quimicamente definidas da flora do Caribe em Guadalupe. FEMS Immunol. Med. Microbiol., 20(4): 267–273.

Rocha, J.A. 2013. Variabilidade genética e avaliação anti-schistosoma dos alcalóides pilosina, epiisopilosina, isopilosina e macaubina de jaborandi (Pilocarpus microphyllus Stapf ex Wardlew.). 145f. Dissertação (Mestrado em Biotecnologia) – Programa de Pós-Graduação em Biotecnologia – Universidade Federal do Piauí, Parnaíba.

Rocha, T.M. Efeito anti-inflamatório e antinociceptivo de alcaloides Imidazólicos de pilocarpus microphyllus: estud*o in vitro, in vivo e in Silico.* 2017. 145f. Dissertação (Mestrado em em Farmacologia) – Programa de Pós-graduação em Farmacologia–Universidade Federal do Ceará, Fortaleza, 2017.

Romero, N.R., Neto, M.A., Lopes, E.L., Sousa, A.H., Nascimento, R.R.G., Bezerra, J.N.S., Oliveira, M.C.F. 2006. Atividade Nematicida de Pilocarpus microphyllus Stapf (Rutaceae). *29a Reunião Anual da Sociedade Brasileira de Química.* Águas de Lindóia – SP.

Sá, I.M., Elisabetsky, E. 2012. Medical knowledge exchanges between Brazil and Portugal: An ethnopharmacological perspective. J. Ethnopharm., 142: 762–768.

Santos, F.A., Cunha, G.M.A., Viane, G.S.B., Rao, V.S.N., Manoel, A.N., Silveira, E.R. 1997. Antibacterial activity of essential oils of plant species *Psidium* and *Pilocarpus*. Phytother. Res., 11: 67–69.

Santos, A.P., Moreno, P.H.R. 2004. *Pilocarpus* spp.: A survey of its chemical constituents and biological activities. Braz. J. Pharm. Sci., 40: 2.

Santos, A.P., Lopes, M.C., Limberger, R.P., Apel, M.A., Henriques, A.T., Moreno, P.R.H. 2004. Analysis of the volatile oil from Pilocarpus pennatifolius Lemmaire (Rutaceae) leaves by GC–MS. Flavour. Frag. J., [s.l.]., 19(4): 325–326.

Sawaya, A.C.H.F., Vaz, B.G., Eberlin, M.N., Mazzafera, M. 2011. Screening species of *Pilocarpus* (Rutaceae) as sources of pilocarpina and other imidazole alkaloids. Genet. Resour. Crop Evol., 58: 471–480.

Silva, C.V., Bomfim, F.C.C., Santos, M.A.V. Dos, Velozo, E.S. 2014. Estudo fitoquímico e avaliação in vitro da atividade anti-Trypanosoma cruzi cepa Y de *Pilocarpus spicatus* St. Hil. (Rutaceae). Rev. Bras. Plantas Med., 16(4): 812–818.

Silva, V.G., Silva, R.O., Damasceno, S.R.B., Carvalho, N.S., Prudêncio, R.S., Aragão, K.S., Guimarães, M.A., Campos, S.A., Véras, L.M.C., Godejohann, M. 2013. Anti-inflammatory and antinociceptiva activity of epiisopiloturina, an imidazole alkaloid isolated from *Pilocarpus microphyllus*. J. Nat. Prod., 76(6): 1071–7.

Skopura, L.A. 1996. Taxonomic Review of *Pilocarpus* Vahl (Rutaceae) PhD Dissertation. Institute of Biosciences, University of São Paulo. São Paulo.

Skopura, L.A., Salatini, M.L.F., Salatino, A. 1998. Hydrocarbons of leaf epicuticular waxes of *Pilocarpus* (Ruteaceae): Taxonimic meaning. Biochem. Syst. Ecol., 26: 655–662.

Skorupa, L.A. 2003. Novos táxons infraespecíficos e combinações em *Pilocarpus* Vahl (Rutaceae). Rev. Bras. Botân., 26(2): 263–270.

Souza, R.C., Fernandes, J.B., Vieira, P.C., Da Silva, M.F.G., Godoy, M.F.P., Pagnocca, F.C., Bueno, O.C., Hebling, M.J.A., Pirani, J.R. 2005a. A new imidazole alkaloid and other constituents from *Pilocarpus grandiflorus* and their antifungal activity. Z. Naturforsch, 60: 787–791.

Souza, S.M., Monache, F.D., Junior, A.S. 2005b. Antibacterial activity of coumarins. Z. Naturforsch, 60: 693–700.

Souza, L.A., Mourão, K.S.M., Moscheta, I.S., Rosa, S.M. 2003. Flower morphology and anatomy of *Pilocarpus pennatifolius* Lem. (Rutaceae). Rev. Bras Bot., 26: 175–184.

Taveira. F.S.N., Andrade, E.H.A., Lima, W.N., Maia, J.G.S. 2003. Seasonal variation in the essential oil of *Pilocarpus microphyllus* Stapf. An. Acad. Bras Cienc., 75: 27–31.

Veras, L.M., Guimarães, M.A., Campelo, Y.D., Vieira, M.M., Nascimento, C., Lima, D.F., Vasconcelos, L., Nakano, E., Kuckelhaus, S.S., Batista, M.C., Leite, J.R., Moraes, J. 2012. Activity of epiisopiloturine against *Schistosoma mansoni*. Curr. Med. Chem., 19: 2051–58.

Vital, M.A.B.F., Acco, A. 2006. Cholinergic agonists and antagonists. *In*: Spinosa, H.S., Górniak, S.L., Bernardi, M.M. (eds.). Farmacologia aplicada à medicina veterinária. Rio de Janeiro: Guanabara Koogan. 7: 84–85.

Zancul, M.S. 2001. Toxicidade de extratos de Pilocarpus grandiflorus Engl. Para o fungo simbionte de Atta sexdens. Monografia (Trabalho de Conclusão de Curso Licenciatura e Bacharelado em Ciências Biológicas) Instituto de Biociências, Universidade Estadual Paulista, Rio Claro.

Chapter 6

Antifungal Activity Validation of Wild Plants used in Argentine Ethnomedicine

A Revision of *Polygonum*

Norma Hortensia Álvarez,[1,2] *Laura Noemí Fernandez,*[1,3] *Gisela Marisol Seimandi,*[1] *María Inés Stegmayer,*[1] *Verónica Eugenia Ruiz*[1,4] and *Marcos Gabriel Derita*[1,5,*]

Introduction

Fungal diseases are life-threatening to human health and cause an important burden around the world being associated with high numbers of morbidity and mortality. The current antifungal treatment comprises drugs such as azoles, echinocandins and polyenes but most of the time the cure is not guaranteed because it has severe side effects and can last for a long period of time.

Aspergillus is a broad fungal genus comprising more than 300 different species distributed ubiquitously worldwide. The genus is currently organized in different subgenera and sections such as *Fumigati, Flavi, Terrei* and *Nigri* are the most clinically relevant due to their great impact as human opportunistic pathogens. *Aspergillus* infections show a wide range of clinical manifestations, including infections that drive hypersensitivity reactions such as allergic *Aspergillus* sinusitis, infections via the skin that result in cutaneous aspergillosis and chronic conditions such as chronic pulmonary aspergillosis. In general, approximately 200,000 cases of infections are estimated to occur every year worldwide and the recommended treatment for aspergillosis is voriconazole. It is believed that azoles may disrupt the role of ergosterol in stimulating growth and proliferation but, nevertheless, the incidence of azole-resistant *Aspergillus* isolates has increased alarmingly in recent years, which is thought to result from the use of azoles both in clinical and agricultural settings, directly contributing to therapeutic failures (Pérez-Cantero et al. 2020).

Among all the threatening species that belongs to *Candida* genera, the emergence of *C. auris* that has been associated with invasive infections (e.g., bloodstream infections), has raised a significant impact

[1] ICiAgro Litoral, Universidad Nacional del Litoral, CONICET, Facultad de Ciencias Agrarias, Kreder 2805, Esperanza, 3080HOF, Argentina.
[2] Cultivos Intensivos, Facultad de Ciencias Agrarias, Universidad Nacional del Litoral, Kreder 2805, Esperanza, 3080HOF, Argentina.
[3] Fitopatología, Facultad de Ciencias Agrarias, Universidad Nacional del Litoral, Kreder 2805, Esperanza, 3080HOF, Argentina.
[4] Fisiología Vegetal, Facultad de Ciencias Agrarias, Universidad Nacional del Litoral, Kreder 2805, Esperanza, 3080HOF, Argentina.
[5] Farmacognosia, Facultad de Ciencias Bioquímicas y Farmacéuticas, Universidad Nacional de Rosario, Suipacha 531, S2002LRK, Rosario, Argentina.
* Corresponding author: mgderita@hotmail.com

on healthcare. Furthermore, up to 90% of strains are resistant to fluconazole, and *C. auris* is able to adapt and become resistant to other antifungal agents. Additionally, since its 'discovery' in 2009, *C. auris* has been reported from every continent in the world (Kenters et al. 2019).

Despite some yeasts, like *Saccharomyces cerevisiae*, which have always been considered as innocuous to human beings and are commonly used for food production, alcoholic beverages and different biotechnological process, it has recently been detected that they could act as human opportunistic pathogens causing a variety of infections on immunocompromised individuals. The acquisition of infections by *S. cerevisiae* has been recently linked to the use of probiotics or dietary supplements, both of which could represent a source of this agent. According to the current guidelines, treatment with amphotericin B or 5-fluorocytosine is recommended but sooner or later, resistant strains will become resistant strains (Pérez-Cantero et al. 2019).

Thus, setting new routes for the discovery of effective and safe antifungal drugs should be a priority within the health care system (Castelli et al. 2016). The discovery of new alternative, efficient and antifungal drugs with fewer side effects is very slow and remains a challenge. Plants produce metabolites that can inhibit pathogen growth and these compounds have been intensely explored. Natural products can be a source of antifungals and, used in combinatorial therapy, could be an important advance (Butassi et al. 2015). Several proteins, enzymes and metabolic pathways could be targets for the discovery of efficient inhibitor compounds and recently heat shock proteins calcineurin, salinomycin, trehalose biosynthetic pathway and the glyoxylate cycle have been investigated in several fungal species. Transcriptomic and proteomic approaches have advanced antifungal drug discovery and pointed to new important specific-pathogen targets (Butassi et al. 2019). Thus, the development of new antifungal strategies might reduce the therapeutic time and increase the quality of life of patients, the reports of plants used in the folk medicine of different cultures being an important resource to be considered (Svetaz et al. 2010).

The present chapter attempts to follow this line of thought and provide information about the reported use of some Argentinian wild plant species acting as antifungals as well as its validation based on *in vitro* studies performed in our laboratory. In addition, a revision of *Polygonum* genus, which represents a valuable source of antifungal compounds, was also performed.

Problems Caused by Fungi in Human Health

There are approximately 250,000 known fungal species and at least 100 of which cause diseases in human beings. In the last three decades, an increasing number of fungal infections have been observed due to an increase in the population of severely immunocompromised patients (Pérez-Cantero et al. 2020, Kenters et al. 2019), such as patients with Human Immunodeficiency Virus, hematological disorders (for example, leukemia), who suffer organ transplants and extreme age groups (elders and neonates). Many clinical procedures and treatments such as surgery, catheter use, parenteral nutrition, radiation and chemotherapy, among others, are risk factors for fungal infections (Selitrennikoff 1992). The causative agents of these infections can be true pathogenic fungi or opportunistic pathogens. The latter has very low virulence but in immunocompromised patients, they show their pathogenicity, often causing lethal infections (Armstrong 1989).

Fungal infections can be classified, according to their location, into three large groups: superficial, subcutaneous and systemic. Superficial infections occur in the outermost layers of the skin, nails, hair, and mucous membranes and most are produced by dermatophytes (especially *Trichophyton*, *Microsporum* and *Epidermophyton* genera), although they can also be caused by *Candida* species (Figure 6.1). The clinical manifestations are mild and do not compromise the patient's life but affect their quality of life, being able to become invasive or expand to other individuals. Subcutaneous infections are restricted to tropical and subtropical regions and are mainly caused by *Sporothryx schenckii* and *Fonsecae pedrosoi* (Murphy et al. 1997). Systemic infections are mainly caused by species of the genus *Candida* or *Aspergillus* (Figure 6.1) and by *Cryptococcus neoformans* that is the causative agent of cryptococcal meningitis, one of the infections that cause high mortality in patients with AIDS (Armstrong 1989).

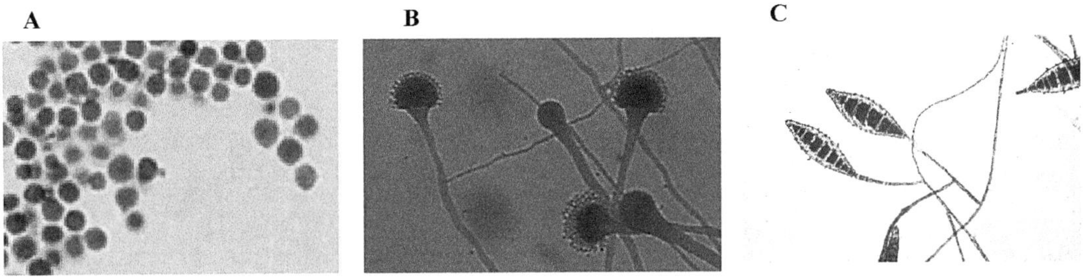

Figure 6.1. Fungi observed with an optical microscope (40x) of (A) *Candida albicans* (B) *Aspergillus niger* and (C) *Trichophyton mentagrophytes*.

Current Antifungal Therapies

Although there appear to be many drugs for the treatment of fungal infections, there are actually a limited number of effective antifungal drugs (Denning 2004). They belong to only five classes of compounds: pyrimidines (e.g., 5-Fluocytosine); polyenes (e.g., Amphotericin B and Nystatin); azoles (e.g., Ketoconazole and Miconazole belonging to imidazole group and Fluconazole and Itraconazole belonging to triazole group); allylamines (e.g. Naftifine and Terbinafine) and the newly introduced lipopeptides Caspofungin (Cancidas®, Merck), Micafungin (Mycamine®, Funguard®, Fujisawa) and Anidulafungin (Vicuron®) (Castelli et al. 2016). Table 6.1 shows the most common antifungal agents clinically used and classified by their target of action.

Among the antifungal drugs in clinical use, many of them are not broad-spectrum of action or have undesirable side effects. This can be explained by the great similarity between fungal and mammalian cells. All of them are eukaryotic and share a large amount of enzymatic and biochemical machinery, making it difficult to develop antifungal agents that have specificity for fungus structures (Kontoyiannis et al. 2003).

Another drawback of the antifungals currently used is that some of them are fungistatic but not fungicidal causing recurrence (azoles). In addition, since it is difficult and usually takes time to diagnose a mycosis, especially if it is invasive, antifungals are often used preventively for long times. As a result, resistance to antifungal drugs has been growing alarmingly, constituting a serious problem for the very near future (Marr and Borden 1998).

From all the explanation above, it is remarkable that despite there being considerable progress in antifungal therapy, many unsolved problems remain. So, there is a general consensus that new antifungal

Table 6.1. Different antifungal drugs in clinical use grouped according to their target of action.

1. Cell nucleus
Deficient synthesis of RNA/DNA (5-Fluocytosine)
2. Cellular membrane
2.1. Ergosterol binding (Polyenes: Amphotericin B and Nystatin)
2.2. Ergosterol biosynthesis inhibition:
2.2.1. Squalene epoxidase inhibition (Allylamines: Naphthifine and Terbinafine)
2.2.2. Lanosterol 14-α-demethylase inhibition:
a) Imidazoles (Ketoconazole, Miconazole)
b) Triazoles (Fluconazole, Itraconazole)
3. Cellular wall
Biosynthesis inhibition of (1,3)-β-D-glucan (Caspofungin, Micafungin, Anidulafungin)

compounds that may act through a selective mechanism of action are necessary, and that there should be alternatives for the development of new, more effective and possibly safer antifungal drugs (Kontoyiannis et al. 2003).

Bioassays to Detect Antifungal Activity

For the search of plant extracts or compounds with antifungal activity, a test scheme is established starting with one that has a great capacity, so as to perform a screening of all extracts and compounds in a short time (first-order tests). In this way, extracts or non-active compounds are discarded and those that are active continue to be evaluated. These are subjected to second-order trials in which the activity observed in the first trial is confirmed. Then, some samples are selected to be subjected to studies of the mechanism of action. Thus, from a large number of extracts or compounds subjected to first-order tests, a few candidates arise to continue their study (Hostettmann et al. 2001). Among the first order tests (without specific target), we should mention the following: diffusion, bioautographic and dilution tests (Fakuda et al. 1991, Li et al. 1995, Vasques da Silva et al. 2002).

Plants as Sources of Bioactive Compounds

The plant kingdom represents a huge source of novel molecules. It is estimated that there are between 500,000 and 700,000 species of plants, although only about 250,000 are known. Among these, only a low percentage has been studied from a phytochemical point of view and an even smaller percentage has been subjected to biological tests (Hostettmann 1991).

Among the different natural sources used as medicines since ancient times, plants have occupied the first place as natural sources of medicines since ancient times. There are reports from primitive civilizations regarding to their use for the cure of different illness. Even today, at least 80% of the population in developing countries uses plant materials for primary health care (Cordell and Colvard 2005).

The study of medical treatments of different cultures is known as Traditional Medicine or Ethnopharmacology and represents very valuable information that in some cases has led to the discovery of drugs currently in clinical use (Raza 2006). Many other traditional uses of plants caused great curiosity among scientists, who submitted them to pharmacological studies in the laboratory and then to the isolation of their active ingredients. The isolation of bioactive molecules from plants is a laborious and long process, which fortunately has been greatly improved in recent decades thanks to the development of refined separation and isolation methods that make it possible to isolate, from a complex mixture, not only majority but also minority compounds which, many times, possess biological properties (Hostettmann 1991).

The Argentine Republic has a great plant diversity of which only a small percentage has been investigated. Many of the plants that inhabit our region are native and a large percentage of them are endemic. Particularly in the Argentine territory, there are 1,914 endemic species, out of a total of 9,689 species, which represents approximately 20% of the total species present in the country. This high degree of endemism represents a huge genetic reservoir, still little known in terms of its usefulness. It is important to bear in mind that there is a risk of extinction of some species due to the anthropic impact, losing an incalculable biodiversity potential (Del Vitto et al. 1997a, Scarpa 2004).

There are different criteria for selecting plant species to be subjected to biological studies. Among them, the most used are: (1) Selection based on ethnopharmacological data: it has been shown that 77% of the total drugs derived from plants that are used currently for different ailments were discovered from the information provided by folk medicine (Torssell 1983, Gottlieb and Kaplan 1993, Soejarto and Gillenhaal 1997). (2) Random selection: in this case the search for active plant species is at random, evaluating all the species to which researchers have access, including those that have been collected for other purposes (Svetaz et al. 2010).

Comparison of Antifungal Activity Results Between a Group of Selected Plants According to their Ethnopharmacological use and Another Group of Randomly Selected Plants

To carry out this study, the results generated in our laboratory in recent years were used, dividing the plants that had been evaluated by antifungal activity into two groups:

- Group I: 43 Argentinian plants with ethnopharmacological use related to fungal infections.
- Group II: 43 Argentinian plants selected at random.

With respect to the plants that are used in Traditional Medicine, different groups of researchers have made important efforts to gather the ethnopharmacological information of Argentina. Regional ethnobotanical studies have been mainly carried out and these were taken into account to classify the first group of plants that correspond to those that present some use related to fungal infections (Ruiz Leal 1972, Bustos et al. 1996, Del Vitto et al. 1997a, b, Roig 2001, Scarpa 2004, Goleniowski et al. 2006). The following uses were considered: vulnerary, antiseptic, antidandruff, antialopecic, antiseborrheic, antipruritic, disinfectant, to heal wounds, ulcers, psoriasis, leucorrhea and venereal diseases. These are all ailments that may be caused by some type of fungus (Petenatti et al. 2008).

Methanolic extracts were made by macerating the aerial parts of all the selected species and they were evaluated for antifungal activity against a panel of true and opportunistic human pathogenic fungi. The Minimum Inhibitory Concentrations (MICs) for each extract were determined using the broth microdilution method following the CLSI guidelines. Those extracts with MICs ≤ 1000 μg/mL were considered active (Petenatti et al. 2008).

Table 6.2 shows the antifungal activity of the 43 plants with traditional use related to antifungals (group I) against yeasts, species of the genus *Aspergillus* and dermatophytes. The traditional use of them is also indicated.

Table 6.2. Antifungal activity (MICs expressed in μg/mL) of plants with reported ethnopharmacological uses to treat fungal infections or related conditions. I: inactive (MIC > 1000 μg/mL).

	Family	Plant species	Ethnopharmacological use (References)	Antifungal activity (MIC in μg/mL)		
				Yeasts	*Aspergillus* spp.	Dermatophytes
1	Anacardiaceae	*Schinus fasciculata* (Griseb.) I.M. Johnst	Vulnerary (Del Vitto et al. 1997)	I	I	1000
2	Anacardiaceae	*Schinus molle* L.	Vulnerary (Del Vitto et al. 1997)	I	I	500–1000
3	Aristolochiaceae	*Aristolochia argentina* Griseb.	Antifungal (Del Vitto et al. 1997)	250–1000	250–1000	125–250
4	Bignoniaceae	*Jacaranda mimosifolia* D.Don.	Healing, venereal diseases (Roig 2001)	I	I	I
5	Cannabaceae	*Celtis pallida* Torr.	Desinfectant (Arenas 1981)	I	I	I
6	Cannabaceae	*Celtis tala* Gillies ex Planch.	Desinfectant (Arenas 1981)	I	I	I
7	Combretaceae	*Terminalia australis* Cambess.	Healing wounds (Arenas 1981)	100–1000	250–1000	10–1000
8	Combretaceae	*Terminalia triflora* (Griseb.) Lillo	Desinfectant (Novara 1984)	50–1000	250–1000	25–1000
9	Compositae	*Conyza bonariensis* (L.) Cronquist	Vulnerary, healing (Alonso et al. 2006)	I	I	I

Table 6.2 Contd. ...

...Table 6.2 Contd.

	Family	Plant species	Ethnopharmacological use (References)	Antifungal activity (MIC in μg/mL)		
				Yeasts	*Aspergillus* spp.	Dermatophytes
10	Compositae	*Flaveria bidentis* (L.) Kuntze.	Antifungal, antiseptic (Marzocca 1997)	I	I	I
11	Compositae	*Gaillardia megapotamica* (Spreng.) Baker	Antialopecic, antidandruff, antiseborrhoeic (Del Vitto et al. 1997)	I	I	250–1000
12	Compositae	*Gaillardia pulchella* var. *drummondii* (Hook.) B.L. Turner	Antialopecic, antidandruff, antiseborrhoeic (Del Vitto et al. 1997)	I	I	250–1000
13	Compositae	*Gaillardia megapotamica* var. *scabiosoides* (Arn. ex DC.) Baker.	Antialopecic, antidandruff, antiseborrhoeic (Del Vitto et al. 1997)	I	I	250–1000
14	Compositae	*Gochnatia glutinosa* (D.Don) D.Don ex Hook. & Arn.	Antifungal (Fenner et al. 2006)	500–1000	I	50–62.5
15	Compositae	*Mikania periplocifolia* Hook. & Arn.	Vulnerary, healing wounds, venereal diseases (Alonso et al. 2006)	I	I	50–1000
16	Compositae	*Pterocaulon virgatum* (L) DC.	Antifungal (Fenner et al. 2006)	I	I	1000
17	Compositae	*Schkuhria pinnata* (Lam.) Kuntze ex Thell.	Antifungal (Alonso et al. 2006)	I	I	I
18	Compositae	*Solidago chilensis* Meyen	Healing infected wounds (Fenner et al. 2006)	I	I	I
19	Compositae	*Tagetes filifolia* Lag.	Antipruritic, healing infected wounds (Alonso et al. 2006)	I	I	I
20	Compositae	*Xanthium spinosum* L.	Vulnerary, skin diseases (Fenner et al. 2006)	I	I	80–1000
21	Convolvulaceae	*Ipomoea cairica* (L.) Sweet.	Vulnerary (Novara 1984)	I	I	I
22	Euphorbiaceae	*Acalypha communis* Müll. Arg.	Skin diseases (Novara 1984)	I	I	250–500
23	Lamiaceae	*Marrubium vulgare* L.	Antiseptic (Fenner et al. 2006)	I	I	I
24	Leguminosae	*Acacia caven* (Molina) Molina.	Antifungal, oral antiseptic (Alonso et al. 2006)	I	I	I
25	Leguminosae	*Inga urugüensis* Hook & Arn.	Healing infected wounds (Zacchino et al. 1998)	I	I	I
26	Leguminosae	*Senna alata* (L.) Roxb.	Antiseptic, healing wounds (Hieronymus 1982)	I	I	I

Table 6.2 Contd. ...

...Table 6.2 Contd.

	Family	**Plant species**	**Ethnopharmacological use (References)**	**Antifungal activity (MIC in μg/mL)**		
				Yeasts	*Aspergillus* spp.	Dermatophytes
27	Leguminosae	*Zuccagnia punctata* Cav.	Antifungal (Del Vitto et al. 1997)	500–1000	500–1000	250–500
28	Lythraceae	*Heimia salicifolia* (Kunth) Link.	Vulnerary (Hieronymus 1982)	I	I	250–1000
29	Malvaceae	*Luehea divaricata* Mart.	Healing wounds (Zacchino et al. 1998)	I	I	I
30	Myrtaceae	*Blepharocalyx salicifolius* (Kunth) O.Berg	Leucorrhea (Fenner et al. 2006)	I	I	I
31	Papaveraceae	*Argemone subfusiformis* G.B. Ownbey.	Oral antiseptic, healing, antialopecic (Bourdy 2002)	I	I	250–500
32	Ranunculaceae	*Clematis campestris* A.St.-Hil	Vulnerary (Del Vitto et al. 1997)	I	I	125–250
33	Rutaceae	*Fagara coco* (Gillies) Engl.	Antifungal (Del Vitto et al. 1997)	250–1000	500–1000	250–500
34	Santalaceae	*Acanthosyris falcate* Griseb.	Vulnerary (Toursarkissian 1980)	I	I	I
35	Scrophulareacea	*Buddleja cordobensis* Griseb.	Vulnerary (Del Vitto et al. 1997)	I	I	I
36	Solanaceae	*Cestrum parqui* (Lam.) L'Hér.	Vulnerary, skin diseases, antiseptic (Hieronymus 1982)	I	I	125–250
37	Solanaceae	*Nicotiana glauca* Graham.	Antifungal, vulnerary, infected wounds (Bourdy 2002)	I	I	I
38	Solanaceae	*Physalis viscosa* L.	Vulnerary (Toursarkissian 1980)	I	I	I
39	Urticaceae	*Parietaria debilis* G. Forst.	Vulnerary (Del Vitto et al. 1997)	I	I	I
40	Winteraceae	*Drimys brasiliensis* Miers.	Leucorrhea (Fenner et al. 2006)	I	I	I
41	Winteraceae	*Drimys winteri* J.R. Forst. & G. Forst.	Vulnerary, psoriasis and wounds (Fenner et al. 2006)	250	I	125–500
42	Zygophyllaceae	*Larrea cuneifolia* Cav.	Healing wounds (Alonso et al. 2006)	500–1000	I	25–100
43	Zygophyllaceae	*Porlieria microphylla* (Baill.) Descole, O'Donell & Lourteig.	Vulnerary, venereal diseases (Del Vitto et al. 1997)	I	I	500–1000

Among the 43 plants with ethnopharmacological uses related to antifungal activity evaluated, 22 were active against at least one type of fungus. This amount represents 51% of the total plants tested with ethnopharmacological use related to antifungals.

Table 6.3 shows the results of the antifungal activities evaluation of the 43 species collected at random (group II).

Table 6.3. Antifungal activity (MICs expressed in μg/mL) of randomly selected plants. I: inactive (MIC > 1000 μg/mL).

			Antifungal activity (MIC in μg/mL)		
	Family	**Species**	**Yeasts**	***Aspergillus* spp.**	**Dermatophytes**
1	Acanthaceae	*Justicia gilliesii* (Nees) Benth.	I	I	I
2	Acanthaceae	*Justicia squarrosa* Griseb.	I	I	I
3	Acanthaceae	*Justicia tweediana* (Nees) Griseb.	I	I	I
4	Achatocarpaceae	*Achatocarpus praecox* Griseb.	I	I	500–1000
5	Amaranthaceae	*Iresine diffusa* Humb. & Bonpl. ex Willd.	I	I	I
6	Apocynaceae	*Tabernaemontana catharinensis* A. DC.	I	I	500–1000
7	Berberidaceae	*Berberis microphylla* G. Forst	I	I	I
8	Boraginaceae	*Ehretia corymbosa* Bojer	I	I	I
9	Bromeliaceae	*Aechmea disticantha* Lem.	I	I	I
10	Calyceraceae	*Boopis anthemoides* Juss.	I	I	I
11	Capparaceae	*Cynophalla retusa* (Griseb.) Cornejo & LLtis	I	I	I
12	Combretaceae	*Combretum laxum* Jacq.	I	I	I
13	Compositae	*Pseudognaphalium gaudichaudianum* DC. Anderb.	I	I	I
14	Compositae	*Grindelia pulchella* var. *pulchella* Hook. & Arn.	I	I	I
15	Compositae	*Mutisia retusa* J. Rémy	I	I	I
16	Compositae	*Plazia daphnoides* Wedd.	I	I	I
17	Compositae	*Thymophylla pentachaeta* (DC) Small	I	I	I
18	Cucurbitaceae	*Abobra tenuifolia* (Gillies ex Hook. & Arn.) Cogn.	I	I	I
19	Euphorbiaceae	*Croton sarcopetalus* Müll. Arg.	I	I	I
20	Lamiaceae	*Clinopodium gillesii* (Benth.) Kuntze	I	I	I
21	Lamiaceae	*Teucrium grisebachii* Hieron. ex. Epling	I	I	I
22	Lamiaceae	*Vitex megapotamica* (Spreng.) Moldenke.	I	I	I
23	Leguminosae	*Adesmia macrostachya* Benth.	I	I	1000
24	Leguminosae	*Astragalus pehuenches* Niederl.	I	I	I
25	Leguminosae	*Coursetia hassleri* Chodat.	I	I	I
26	Leguminosae	*Inga vera* Willd.	I	I	I
27	Leguminosae	*Collaea argentina* Griseb.	I	I	I
28	Malvaceae	*Pseudabutilon virgatum* (Cav.) Fryxell	I	I	I
29	Malvaceae	*Ayenia lingulata* Griseb.	I	I	I
30	Nictaginaceae	*Boungainvillea stipitata* Griseb.	I	I	I
31	Orobanchaceae	*Agalinis communis* (Cham. & Schltdl.) D´Arcy.	I	I	I
32	Phytolaccaceae	*Trichostigma octandrum* (L.) H. Walter.	I	I	I
33	Phytolaccaceae	*Seguieria americana* L.	I	I	I
34	Phytolaccaceae	*Phytolacca tetramera* Hauman	250–1000	500	250–500

Table 6.3 Contd. ...

...Table 6.3 Contd.

			Antifungal activity (MIC in μg/mL)		
	Family	**Species**	**Yeasts**	***Aspergillus* spp.**	**Dermatophytes**
35	Plantaginaceae	*Maurandya antirrhiniflora* Humb & Bonpl. ex Willd.	I	I	I
36	Polygalaceae	*Polygala cyparissias* A. St.-Hill.	I	I	I
37	Rhamnaceae	*Scutia buxifolia* Reissek.	I	I	I
38	Rubiaceae	*Galium latoramosum* Clos.	I	I	250–1000
39	Salviniaceae	*Salvinia adnata* Desv.	I	I	I
40	Simaroubaceae	*Ailanthus altissima* (Mill.) Swingle.	I	I	I
41	Solanaceae	*Lycium cuneatum* Dammer.	I	I	I
42	Verbenaceae	*Lippia integrifolia* (Griseb.) Hieron.	I	I	250–1000
43	Verbenaceae	*Xeroaloysia ovatifolia* (Moldenke) Tronc.	I	I	I

Among the 43 plant species randomly selected, only 6 were active against at least one type of fungus. This amount represents 14% of the total plants tested in this group. Figure 6.2 clarifies the results obtained in Tables 6.2 and 6.3.

The results showed that the percentage of active extracts in the group of plants with reported use related to fungal infections in Traditional Medicine is significantly higher than the percentage of active extracts obtained in the group of plants randomly selected. Based on these results, we could conclude that selecting plants for the study of antifungal activity based on ethnopharmacological use gives better probabilities to obtain positive results than when the plants are chosen at random.

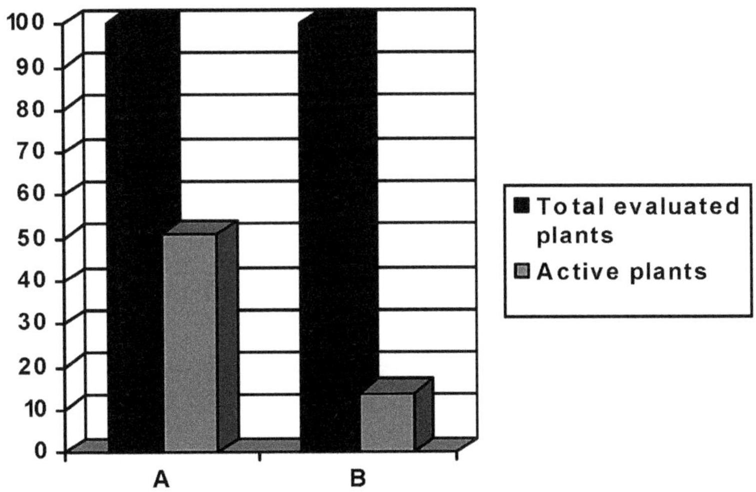

Figure 6.2. Percentage of active methanolic extracts obtained from a screening of: (A) a group of species traditionally used to treat diseases related to fungal infections that showed 51% of active extracts (B) a group of randomly selected species that showed 14% of active extracts.

Polygonum Genus: One of the Main Sources of Antifungal Compounds in Argentina

Taking into account the previous analysis, one of the plant genera most used by Argentinian Traditional Medicine to treat fungal conditions is the genus *Polygonum*. This genus comprises about 250 species that have a wide geographical distribution, from Polar Regions to the tropics of all continents. Formerly it was divided into five sections: *Echinocaulon*, *Amblygonum*, *Persicaria*, *Tiniaria* and *Polygonum* but nowadays, due to its botanical and phytochemical complexity, species are classified into two true genera: *Polygonum* and *Persicaria*. According to the plant list (www.plantlist.org), most of them have been regrouped with accepted names but their synonyms are still used. Many works have been published about these plants and here we will develop a summary of them.

In Argentina, the genus is represented by 20 species which are detailed in Table 6.4, which also shows the regions of Argentina in which each species grows.

As can be seen in Table 6.4, there are five species (*P. convolvulus*, *P. lapathifolia*, *P. maculosa*, *P. brasiliense* and *P. aviculare*) that grow throughout the country; three species (*P. amphybia*, *P. maritimun* and *P. arenarium*) that grow only in Patagonia; one species (*P. minor*) grows in the province of Misiones; one species (*P. orientalis*) in the Paraná River delta and the remaining 10 are distributed between the north and the center of the country (Cialdella 1989, Gattuso 2000).

Table 6.4. Species of *Polygonum* and *Persicaria* genera that grow in Argentina. The region in which each species grows is also indicated.

	Argentinian regions				
Plant species	**Northwest and central-west**	**Northeast and central-east**	**Misiones province**	**Delta of Paraná River**	**Patagonia**
Polygonum stelligerum Cham.		X			
Persicaria meisneriana (Cham. & Schltdl.) M. Gómez		X			
Persicaria orientalis L. Spach				X	
Polygonum convolvulus L.	X	X	X	X	X
Persicaria hispidum (Kunth) M.Gómez	X				
Persicaria lapathifolia (L.) Delarbre	X	X	X	X	X
Persicaria ferruginea (Wedd.) Soják		X			
Persicaria glabra (Willd.) M.Gómez		X			
Persicaria paraguayensis (Wedd.) S.T. Kim & Donoghue		X			
Persicaria amphibia (L.) Delarbre					X
Persicaria acuminata (Kunth) M. Gómez	X	X		X	
Persicaria punctata (Elliott) Small	X			X	
Persicaria minor (Huds.) Opiz			X		
Persicaria maculosa Gray	X	X	X	X	X
Persicaria hydropiperoides (Michx.) Small	X			X	
Polygonum maritimun L.					X
Polygonum stypticum Cham. et Schldl.	X				
Polygonum arenarium Waldst. et Kit.					X
Polygonum brasiliense K. Koch	X	X	X	X	X
Polygonum aviculare L.	X	X	X	X	X

Ethnopharmacological uses of *Polygonum* and *Persicaria* Species that Grow in Argentina

Among the 20 species present in Argentina, 11 (55%) have been reported for their ethnopharmacological uses in the NaPrAlert® (Natural Products Alert) database. They are *P. acuminata*, *P. amphibia*, *P. aviculare*, *P. convolvulus*, *P. glabra*, *P. hydropiperoides*, *P. lapathifolia*, *P. minor*, *P. orientalis*, *P. maculosa* and *P. punctata*. Figure 6.3 shows the number of *Polygonum* and *Persicaria* species that grow in Argentina reported for each ethnopharmacological use mentioned. Thus, three species (*P. glabra*, *P. lapathifolia* and *P. punctata*) are used to treat venereal diseases; three (*P. aviculare*, *P. glabra* and *P. orientalis*) for liver disorders; three (*P. aviculare*, *P. lapathifolia* and *P. punctata*) for varicose veins; four (*P. amphibia*, *P. aviculare*, *P. lapathifolia* and *P. punctata*) for gout or rheumatism; nine (*P. acuminata*, *P. aviculare*, *P. convolvulus*, *P. glabra*, *P. hydropiperoides*, *P. minor*, *P. orientalis*, *P. maculosa* and *P. punctata*) for menstrual pains; one (*P. amphibia*) is used as a sedative; five (*P. amphibia*, *P. aviculare*, *P. lapathifolia*, *P. maculosa* and *P. punctata*) are used to heal wounds; four (*P. amphibia*, *P. aviculare*, *P. maculosa* and *P. punctata*), as diuretics; three (*P. acuminata*, *P. hydropiperoides* and *P. punctata*) as antiseptic and three (*P. acuminata*, *P. aviculare* and *P. punctata*) as vermifuges. Some of these uses (venereal diseases, wound healing, antiseptic, vermifuge) may be related to the antifungal activity (Rios et al. 1987, Schmeda-Hirschmann and Rojas de Arias 1992, Del Vitto et al. 2003).

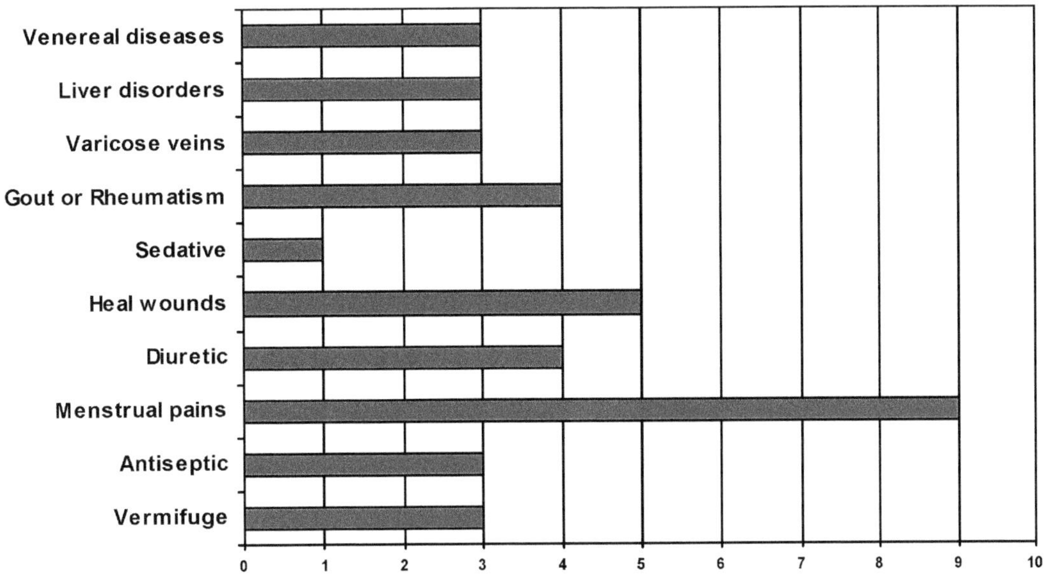

Figure 6.3. Number of *Polygonum* and *Persicaria* species present in Argentina reported in folk medicine to treat different ailments.

Presence of Compounds in Species of *Polygonum* and *Persicaria* that Grow in Argentina

Among the 20 *Polygonum* and *Persicaria* species present in Argentina, there are 10 (50%) that have been studied phytochemically according to the NaPrAlert® database. They are *P. amphibia*, *P. aviculare*, *P. convolvulus*, *P. glabra*, *P. lapathifolia*, *P. minor*, *P. orientalis*, *P. ferruginea*, *P. punctata* and *P. stelligerum* (Yaacob 1990, Sartor et al. 1999, Takasaki et al. 2001, López et al. 2006). Figure 6.4 shows the number of *Polygonum* and *Persicaria* species of Argentina in which different types of compounds were found: alcohols and long chain aldehydes (*P. minor*), tannins (*P. punctata*), triterpenes (*P. orientalis*); monoterpenes (*P. minor*); sesquiterpenes (*P. glabra*, *P. minor* and *P. punctata*); compounds derived

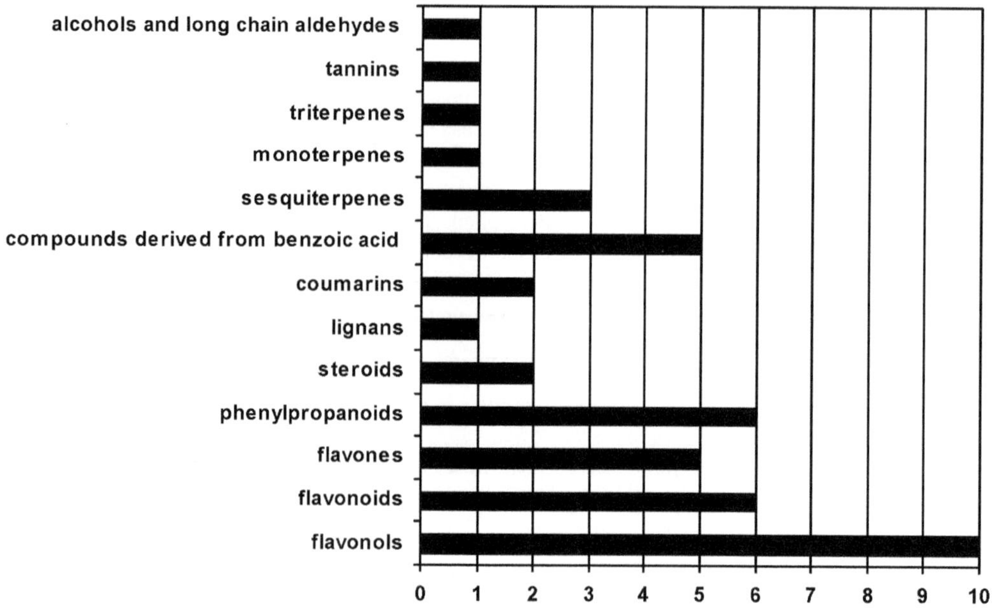

Figure 6.4. Number of *Polygonum* and *Persicaria* species present in Argentina that have previous phytochemical studies which are indicated on the y-axis.

from benzoic acid (*P. amphibia*, *P. aviculare*, *P. convolvulus*, *P. glabra* and *P. punctata*); coumarins (*P. convolvulus* and *P. orientalis*); lignans (*P. aviculare*); steroids (*P. amphibia* and *P. aviculare*); phenylpropanoids (*P. amphibia*, *P. aviculare*, *P. convolvulus*, *P. glabra*, *P. lapathifolia* and *P. punctata*); flavones (*P. amphibia*, *P. aviculare*, *P. lapathifolia*, *P. minor* and *P. orientalis*); flavonoids (*P. amphibia*, *P. aviculare*, *P. ferruginea*, *P. glabra*, *P. lapathifolia* and *P. orientalis*) and flavonols (*P. amphibia*, *P. aviculare*, *P. convolvulus*, *P. glabra*, *P. lapathifolia*, *P. minor*, *P. orientalis*, *P. ferruginea*, *P. punctata* and *P. stelligerum*).

Biological Activities Found in Species of *Polygonum* and *Persicaria* Genus that Grow in Argentina

Among the 20 species of both genera present in Argentina, there are 13 (65%) that have been studied for their biological activities according to the NaPrAlert® database. They are *P. acuminata*, *P. aviculare*, *P. convolvulus*, *P. ferruginea*, *P. glabra*, *P. hydropiperoides*, *P. lapathifolia*, *P. meinsneriana*, *P. minor*, *P. orientalis*, *P. maculosa*, *P. punctata*, and *P. stelligerum* (Gottshall et al. 1949, Oliveira Simoes et al. 1989, Lee et al. 1994). Figure 6.5 shows the number of *Polygonum* and *Persicaria* species which have been studied for their biological properties: Central Nervous System depressing activity (*P. convolvulus*, *P. hydropiperoides* and *P. lapathifolia*); molluscicide (*P. glabra*, *P. meisneriana* and *P. punctata*); analgesic or anti-inflammatory (*P. glabra* and *P. punctata*); hypotensive (*P. glabra*, *P. minor* and *P. punctata*); anthelmintic (*P. glabra*); anti-viral (*P. glabra*, *P. lapathifolia*, *P. maculosa* and *P. punctata*); anti-oxidant (*P. aviculare* and *P. lapathifolia*); anti-hepatotoxic (*P. aviculare*, *P. convolvulus*, *P. orientalis* and *P. punctata*); anti-bacterial (*P. acuminata*); insecticide (*P. acuminata* and *P. aviculare*) and antifungal (*P. amphibia*, *P. aviculare*, *P. ferruginea* and *P. punctata*).

Revision of the Species *Persicaria acuminata* syn. *Polygonum acuminatum*

This species, which formerly belonged to *Persicaria* section of *Polygonum* genus and whose now accepted named is *Persicaria acuminata*, did not present previous phytochemical studies and there were few reports regarding its biological activities, among which its insecticidal activity against *Lutzomyia*

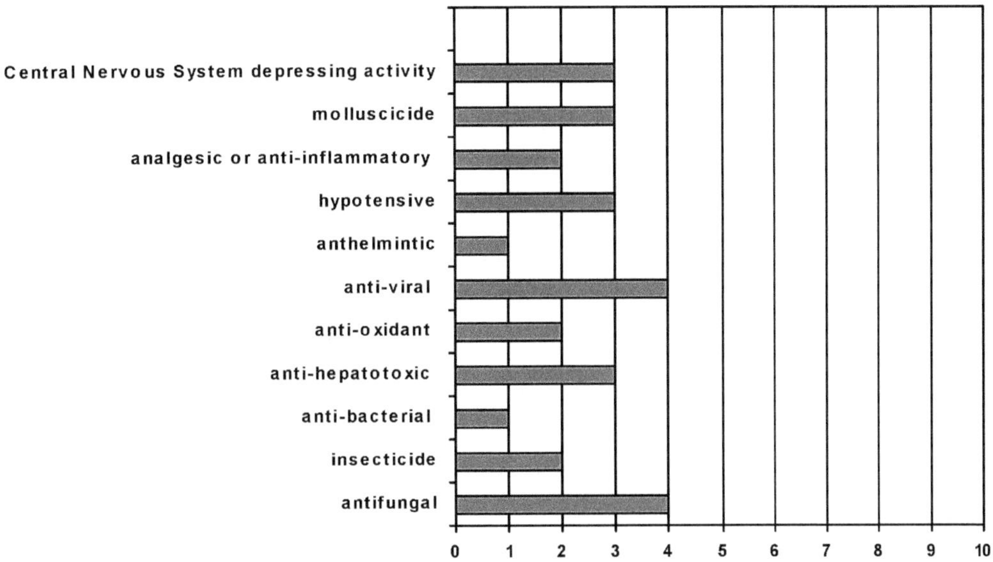

Figure 6.5. Number of *Polygonum* and *Persicaria* species present in Argentina which have been studied for their biological activities which are detailed on the y-axis.

longipalis was highlighted (Chiappeta et al. 1983, Rojas de Arias et al. 1992), hence our greatest interest in it.

P. acuminata (Kunth) M. Gómez is a native herb that grows in Argentina, whose vernacular name is "big catay". It is used in Traditional Medicine as an antiseptic and to treat menstrual disorders (Schmeda-Hirschmann and Rojas de Arias 1992). This herb grows in the central-north region of Argentina and mainly in the provinces of the Argentinian Litoral (Santa Fe, Corrientes, Entre Ríos and Misiones provinces), usually on the banks of rivers and in ditches or swamps (Figure 6.6).

From the bio-guided fractionation of its aerial parts, five sesquiterpenes with a drimane skeleton were isolated: drimenol (**1**), isopolygodial (**2**), confertifoline (**3**), polygodial (**4**) and 1-β-(p-methoxycinnamoyl) polygodial (**5**) (Figure 6.7). Among them, the most active one was **4**, inhibiting the growth of *C. albicans*, *C. neoformans* and the dermatophytes *M. gypseum*, *T. rubrum* and *T. mentagrophytes* displaying MICs between 3.9 and 62.5 μg/mL and MFCs between 7.8 and 125 μg/mL, which showed a fungicidal activity in addition to a strong fungistatic one (Derita et al. 2013).

The finding of polygodial (**4**) in *P. acuminata* (which had also been previously isolated from *P. punctata* and *P. hydropiper*) led us to carry out a study related to the chemosystematic of these genera. Hagendoorn et al. (1994) had demonstrated that compound **4** was produced and stored in specialized epidermal cavities, called irritant glands in *P. hydropiper*. Gattuso (2000) proposed that these glands could be a distinctive sign for the botanical delimitation of both genera. From these previous works and after our finding of polygodial in *P. acuminata*, we decided to study five more species that grow

Figure 6.6. (A) Natural habitat where *P. acuminata* herb grows (B) Photo of the species.

1 2 3 4 5

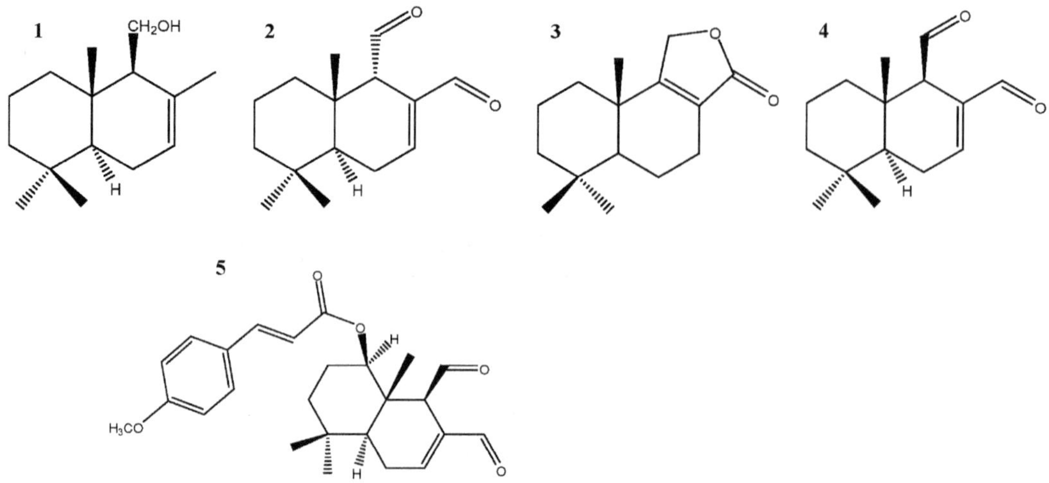

Figure 6.7. Chemical structures of drimenol (**1**), isopolygodial (**2**), confertifoline (**3**), polygodial (**4**) and 1-β-(p-methoxycinnamoyl) polygodial (**5**).

in Argentina (*P. hydropiperoides*, *P. punctata*, *P. maculosa*, *P. ferruginea* and *P. lapathifolia*) with the objective of determining whether or not they contained polygodial and whether or not they contained irritant glands. Since in all cases where polygodial was found, the glands were also present, we suggest that the presence of **4** (which was produced in the glands and easily detected by GC-MS) could be of diagnostic value for the chemistry delimitation of these genera, together with the glands that provided a botanical delimitation (Derita et al. 2008).

In order to determine which part of the plant, which extract type and which season of collection will be the best conditions for improving the antifungal activity, the evaluation of different extract types from leaves, stems and fruits collected at different seasons of the year of *P. acuminata* was performed. The results showed that dichloromethane extracts of *P. acuminata* leaves collected in autumn had the highest polygodial content and the lowest MICs against the fungi tested, thus being the most suitable for treating conditions related to fungal infections. Polygodial was not found in stems and its presence in fruits was almost null (Derita et al. 2009).

Revision of the Species *Persicaria maculosa* syn. *Polygonum persicaria*

This species, which formerly belonged to *Persicaria* section of *Polygonum* genus and whose now accepted named is *Persicaria maculosa*, is a Europe native herb but is currently widely distributed throughout the world. In Argentina country, it is mainly found at the borders of streams and lagoons of the Northern provinces and in the ditches of Mendoza and Neuquén ones. It is used in Traditional Medicine to heal wounds, cuts or hits (Rios et al. 1987). A picture of the species in its natural habitat is shown in Figure 6.8.

This species had few phytochemical studies and among them, we could mention some publications that reported the presence of flavonols (Isobe and Noda 1987), phenylpropanoids (Smolarz 2000) and stilbenes (Smolarz and Potrzebowski 2002) in the methanolic extracts of its aerial parts. In previous studies of biological activities, it was proved that these extracts were active against *Staphylococcus aureus* (Rios et al. 1987) but showed no activity against *C. albicans* and *Mycobacterium tuberculosis* (Gottshall et al. 1949).

The bio-guided fractionation of *P. maculosa* also led to the isolation of the sesquiterpenes **2** and **4** in addition to the flavanone pinostrobin (**6**) and the chalcones flavokawin B (**7**) and cardamonin (**8**) (Figure 6.9). The latter compounds were active against *C. albicans*, *C. neoformans* and the dermatophytes *M. gypseum*, *T. rubrum* and *T. mentagrophytes* displaying MICs between 15.6 and 500 μg/mL. Unlike *P. acuminata*, the antifungal activity of *P. maculosa* was due to not only polygonal and isopolygodial but also to flavonoids, which added interesting chemotaxonomical data (Derita and Zacchino 2011a).

A B

Figure 6.8. (A) Natural habitat where *P. maculosa* herb grows (B) Photo of the species.

6

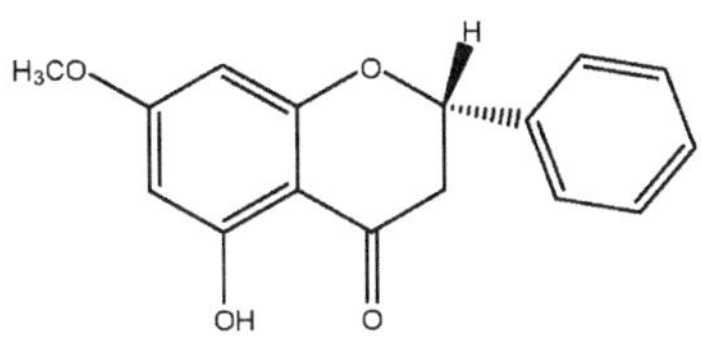

7

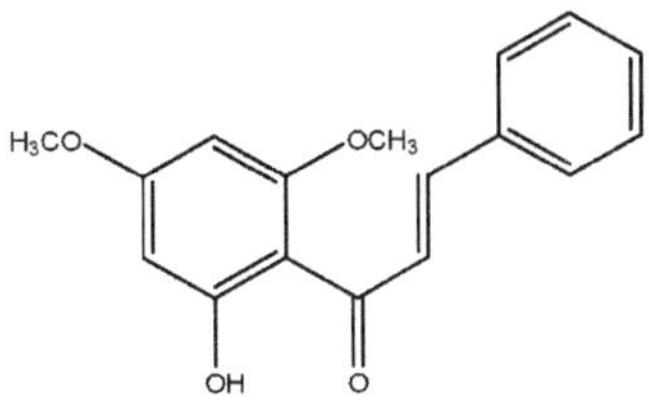

8

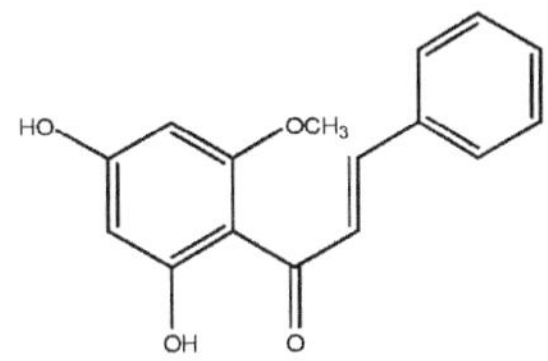

Figure 6.9. Chemical structures of pinostrobin (**6**), flavokawin B (**7**) and cardamon (**8**).

Flavonoids have played an important role in the systematics of *Polygonum* species, being proposed as chemotaxonomic markers of the genus. However, as stated before, the finding of polygodial within the *Persicaria* section (together with the presence of irritant glands) caused a sesquiterpene (not a flavonoid) to be proposed as a chemotaxonomic marker for the *Persicaria* section. Consequently, with the aim of finding compounds that could be suggested as new chemical markers (in addition to **4**) for the delimitation of *Persicaria* section of the genus, a comparative study about the presence and quantification of the sesquiterpenes **1**, **2**, **3**, **4** and the flavonoids **6**, **7** and **8** in six species belonging to the section was carried out. It was found that *P. hydropiperoides*, *P. lapathifolia* and *P. ferruginea*, which were the three species that did not contain **4**, showed the presence of the flavonoids **6**, **7** and **8**. In turn, *P. punctata* and *P. acuminata* contained the sesquiterpenes **1–4** but not flavonoids **6–8**. Surprisingly, *P. persicaria* was the only species belonging to this section that presented both types of compounds (Derita and Zacchino 2011b). All these findings contributed to the regrouping of these species into the current *Polygonum* and *Persicaria* genera.

Regarding the relationship between antifungal activity and the content of sesquiterpenes and/or flavonoids, it was concluded that both types of compounds contribute to the antifungal activity, but among all the compounds present in the different species, polygodial proved to be the most active one since it was not only fungistatic but also fungicidal (Derita and Zacchino 2011a, b).

Conclusions

Given the imperative need to find new antifungal drugs against the severe pathologies that cause fungi in humans, considering the plant diversity of different regions of the planet as a source of active therapeutic principles could be very useful. In this sense, it was possible to demonstrate, once more in this chapter, that the ethnopharmacological information provided by the different communities or cultural groups that inhabit Argentina is of great value to find positive results during *in vitro* antifungal assays. The results of the antifungal activity of two groups of plants consisting in 43 samples each were collected, and the percentage of each group of extracts that yielded positive inhibition results against some type of fungus (yeasts, *Aspergillus* spp. or dermatophytes) was analyzed. Furthermore, the revision of ethnomedicinal uses and biological activities reported for 20 species of *Polygonum* and *Persicaria* genus that grow in Argentina indicated the potential of these plants as a source of antifungal compounds. Particularly, some phytochemical and morphological results obtained in previous studies in our laboratory were shared, emphasizing the importance of considering sesquiterpenes and flavonoids for the chemotaxonomical delimitation of both genera. *P. acuminata* and *P. maculosa* were the most active against the fungi panel evaluated and the bio-guided fractionation of extracts showed that both types of compounds (sesquiterpenes and flavonoids) were responsible for the antifungal activities, although some species only contained sesquiterpenes and others only flavonoids.

Acknowledgments

Authors gratefully acknowledge Consejo Nacional de Investigaciones Científicas y Técnicas (CONICET) and Agencia Nacional de Promoción Científica y Tecnológica (ANPCyT) for financial support (PIP N° 2015-0524, PICT N° 2015-2259). NHA, LNF, GMS and MIS are also thankful to CONICET for their fellowships.

References

Alonso, J., Desmarchelier, C. 2006. Plantas medicinales autóctonas de la Argentina. Bases científicas para su aplicación en atención primaria de la salud. Fitociencia, Buenos Aires, 663 pp.

Arenas, P. 1981. Etnobotánica Lengua Maskoy. Fundación para la educación, la ciencia y la cultura, Buenos Aires, Argentina, 351 pp.

Armstrong, D. 1989. Problems in management of opportunistic fungal diseases. Rev. Infect. Dis., 2: 1591–1599.

Bourdy, G. 2002. Plantas del Chaco II. Usos Tradicionales Isoceño-Guaraní. Impresiones Sirena Color, Santa Cruz, 120 pp.

Bustos, D., Tapia, G., Feresin G., Ariza Espinar, L. 1996. Ethnopharmacobotanical survey of Bauchazeta district, San Juan province, Argentina. Fitoterapia, 67(5): 411–415.

Butassi, E., Svetaz, L., Ivancovich, J., Feresin, G., Tapia, Alejandro A., Zacchino, S. 2015. Synergistic mutual potentiation of antifungal activity of *Zuccagnia punctata* Cav. and *Larrea nitida* Cav. extracts in clinical isolates of *Candida albicans* and *Candida glabrata*. Phytomedicine, 22(6): 666–678.

Butassi, E., Svetaz, L., Sortino, M., Quiroga, A., Carvalho, V., Cortes, J.C., Ribas, J.C., Zacchino, S. 2019. Approaches to the mechanism of antifungal activity of *Zuccagnia punctata-Larrea nitida* bi-herbal combination. Phytomedicine, 54: 291–301.

Castelli, M.V., Derita, M., López, S. 2016. Novel antifungal agents: a patent review (2013–present). Expert Opin. Ther. Pat., 27(4): 415–426.

Chiappeta, A., Francisco de Melo, J., Maciel, G. 1983. Higher plants with biological activity plants of Pernambuco I. Universidad Federal Pernambuco, Recife, Rev. Inst. Antibiot., 21: 43–50.

Cialdella, A. 1989. Revisión de las especies Argentina de *Polygonum* (Polygonaceae). Darwiniana, 29: 179–246.

Clinical and Laboratory Standards Institute (CLSI). 2008. Reference method for broth dilution antifungal susceptibility testing for filamentous fungi (M38 A2), 2nd ed., Vol. 28. USA: Wayne, pp. 1–35.

Cordell, G., Colvard, M. 2005. Some thoughts on the future of ethnopharmacology. J. Ethnopharmacol., 100(1): 5–14.

Del Vitto, L., Petenatti, E., Petenatti, M. 1997a. Recursos herbolarios de San Luis (Argentina). Primera parte: plantas nativas. Multequina 6:49-66.

Del Vitto, L., Petenatti, E., Petenatti, M. 1997b. Plantas medicinales nativas de Mendoza y San Luis. Servicio Técnico Herbario UNSL 5: 135.

Del Vitto, L.A., Petenatti E., Petenatti, M. 2003. Materia Médica vegetal. Plantas medicinales nativas y exóticas empleadas en fitomedicinas, homeopáticos, galénicos. Servicio Técnico Herbario UNSL, 14: 1–63.

Denning, D. 2004. Antifungals: where are we headed? Drug Plus Int., 3: 11–13.

Derita, M., Gattuso, S., Zacchino, S. 2008. Occurrence of polygodial in species of *Polygonum* genus belonging to *Persicaria* section. Biochem. Syst. Ecol., 36: 55–58.

Derita, M., Leiva, M., Zacchino, S. 2009. Influence of plant part, season of collection and content of the main active constituent, on the antifungal properties of *Polygonum acuminatum* Kunth. J. Ethnopharmacol., 124: 377–383.

Derita, M., Zacchino, S. 2011a. Validation of the ethnopharmacological use of *Polygonum persicaria* for its antifungal properties. Nat. Prod. Comm., 6(7): 931–933.

Derita, M., Zacchino, S. 2011b. Chemotaxonomic importance of sesquiterpenes and flavonoids in Argentinian species of *Polygonum* genus. J. Ess. Oil Res., 23: 11–14.

Derita, M., Montenegro, I., Garibotto, F., Enriz, R., Cuellar Fritis, M, Zacchino, S. 2013. Structural requirements for the antifungal activities of natural drimane sesquiterpenes and analogues, supported by conformational and electronic studies. Molecules, 18: 2029–2051.

Fenner, R., Betti, A., Mentz, L., Rates, S. 2006. Plantas utilizadas na medicina popular brasileira com potencial atividade antifúngica. Braz. J. Pharmac. Sc., 42: 369–395.

Fukuda, D., Nakatsukasa, W., Yao, R., Hunt, A., Gordee, R., Zeckner, D., Mynderse, J. 1991. A 45507, a complex of fungal cell wall inhibitors produced by a mold. 31st, ICAAC Meeting, Chicago, Poster No. 211.

Gattuso, S. 2000. Caracteres generales morfo-anatómicos del vástago de las especies del género *Polygonum* (Polygonaceae) presentes en la Argentina. Bol. Soc. Argent. Bot., 35: 91–105.

Goleniowski, M., Bongiovanni, G., Palacio, L., Núñez, C., Cantero, J. 2006. Medicinal plants from the "Sierra de Comechingones" Argentina. J. Ethnopharmacol., 107: 324–341.

Gottlieb, O., Kaplan, M. 1993. Das plantas medicinais aos fármacos naturais. Ciência Hoje, 15(89): 51–54.

Gottshall, R., Lucas, E., Lickfeldt, A., Roberts, J. 1949. The occurrence of antibacterial substances active against *Mycobacterium tuberculosis* in seed plants. J. Clin. Investig. 28: 920–923.

Hagendoorn, M., Geelen, T., van Beek, T., Jamar, D., Tetteroo, F., van der Plas, L. 1994. Occurrence of polygodial in plant organs and tissue culture of *Polygonum hydropiper*. Physiol. Plant. 92: 595–600.

Hieronymus, J. 1882. Plantae diaphoricae florae argentinae. Boletín Academia Nacional de Ciencias (Buenos Aires), 4: 199–598.

Hostettmann, K. 1991. Methods in Plant Biochemistry, vol. 6. Dey, P., Harborne, J. (eds.). Academic Press, London, 350 pp.

Hostettmann, K., Wolfender, J., Terreaux, C. 2001. Modern screening techniques for plants extracts. Pharm. Biol., 39: 18–32.

Isobe, T., Noda, Y. 1987. A chemotaxonomic study of flavonoids from Japanese *Polygonum* species. Yakugaku Zasshi, 107: 1001–1004.

Kentners, N., Kiernan, M., Chowdhary, A., Denning, D., Pemán, J., Saris, K., Schelenz, S., Tartari, E., Widmer, A., Meis, J., Voss, A. 2019. Control of *Candida auris* in healthcare institutions: Outcome of an International Society for Antimicrobial Chemotherapy expert meeting. Int. J. Antimicrob. Agents, 54: 400–406.

Kontoyiannis, D., Mantadakis, E., Samonis, G. 2003. Systemics mycoses in the immunocompromised host: an unpdate in antifungal therapy. J. Hosp. Infect., 53: 243–258.

Lee, C., Li, N., Hong, N., Kwon, C.H. 1994. Anti-lipid peroxidation and liver protective effects of *Polygonum aviculare*. Korean J. Pharmacogn., 25(1): 59–69.

Li, E., Clark, A., Hufford, C. 1995. Antifungal evaluation of pseudolaric acid B, a major constituent of *Pseudolaryx kaempferi*. J. Nat. Prod., 58(1): 57–67.

López, S., González Sierra, M., Gattuso, S., Furlán, R., Zacchino, S. 2006. An unusual homoisoflavanone and a structurally-related dihydrochalcone from *Polygonum ferrugineum*. Phytochemistry, 67: 2152–2157.

Marr, K., Borden, R. 1998. Clinical, cellular and molecular factors that contribute to antifungal drug resistance. Clin. Microbiol. Rev., 11(2): 382–402.

Marzocca, A. 1997. Vademecum de Malezas Medicinales (Nativas y Exóticas). Buenos Aires. Orientación Gráfica Editora, Argentina, 363 pp.

Murphy, M., Bernard, E., Ishimaru, T., Armstrong, D. 1997. Activity of voriconazole (UKA-109,496) against clinical isolates of *Aspergillus* spp. and its effectiveness in an experimental model of invasive pulmonary aspergillosis. Antimicrob. Agents Chemother., 41(3): 696–698.

Natural Products Alert NaPralert. Program for Collaborative Research in the Pharmaceutical Sciences. College of Pharmacy, University of Illinois at Chicago. Chicago, Ill, USA.

Novara, L. 1984. Las Utilidades de los Géneros de Antófitas del Nordeste del Valle de Lerma (Salta, República Argentina). Facultad de Ciencias Naturales, Universidad Nacional de Salta, 130 pp.

Oliveira Simoes, C., Ribeiro do Vale, R., Poli, A., Nicolau, M., Zanin, M. 1989. Pharmacological investigation on *Polygonum punctatum* Elliot extracts. J. Pharm. Belg., 4: 275–284.

Pérez-Cantero, A., Thomson, P., Paredes, K., Guarro, J., Capilla, J. 2019. Antifungal susceptibility of *Saccharomyces cerevisiae* and therapy in a murine model of disseminated infection. Rev. Iberoam. Micol., 36(1): 37–40.

Pérez-Cantero, A., López-Fernández, L., Guarro, J., Capilla, J. 2020. Azole resistance mechanisms in *Aspergillus*: update and recent advances. Int. J. Antimicrob. Agents, 55: 105807.

Petenatti, E., Gette, M., Derita, M., Petenatti, M., Solís, C., Zuljan, F., Zacchino, S. 2008. Importance of the ethnomedical information for the detection of antifungal properties in plant extracts from the Argentine flora. pp. 15–38. *In*: Martino, V., Muschietti, L. (eds.). South American Medicinal Plants as Potential Source of Bioactive Compounds. Transworld Research Network, Kerala.

Raza, M. 2006. A role for physicians in ethnopharmacology and drug discovery. J. Ethnopharmacol., 104(3): 297–301.

Rios, J., Recio, M., Villar, A. 1987. Antimicrobial activity of selected plants employed in the spanish mediterranean area. J. Ethnopharmacol., 21(2): 139–152.

Roig, F. 2001. Flora medicinal mendocina. Ediunc, Mendoza, 305 pp.

Rojas de Arias, A., Schmeda-Hirschmann, G., Falcao, A. 1992. Feeding deterrency and insecticidal effects of plant extracts on *Lutzomyia longipalis*. Phytother. Res., 62(2): 64–67.

Ruiz Leal, P. 1972. Flora popular mendocina. Deserta, 3: 3–296.

Sartor, C., Da Silva, C., De Souza, M. 1999. Flavonoid glycosides of *Polygonum stelligerum* Cham. Biochem. Syst. Ecol., 27: 303–304.

Scarpa, G. 2004. Medicinal plants used by the criollos of northwestern Argentine Chaco. J. Ethnopharmacol., 91(1): 115–135.

Schmeda-Hirschmann, G., Rojas de Arias, A. 1992. A screening method for natural products on triatomine bugs. Phytother. Res., 62: 68–73.

Selitrennikoff, C.P. 1992. Screening for antifungal drugs. pp. 189–217. *In*: Biotechnology of Filamentous Fungi. Technology and Products. Butterworth Heinemann, Boston.

Smolarz, H. 2000. Chromatographical analysis of phenolic acids in some species of *Polygonum* L. genus. Acta Soc. Bot. Pol., 69(1): 21–23.

Smolarz, H., Potrzebowski, M. 2002. Persilben, a new carboxystilbene from *Polygonum persicaria*. J. Mol. Struct., 605(2-2): 151–156.

Soejarto, D., Gillenhaal, C. 1997. Información etnomédica en el descubrimiento y desarrollo farmacéutico. VI Congreso Italo-Latinoamericano de Etnomedicina, Guatemala, Conferencia Plenaria.

Svetaz, L., Zuljan, F., Derita, M., Petenatti, E., Tamayo, G., Cáceres, A., Cechinel Filho, V., Giménez, A., Pinzón, R., Zacchino, S., Gupta, M. 2010. Value of the ethnomedical information for the discovery of plants with antifungal properties. A survey among seven Latin American countries. J. Ethnopharmacol., 127: 137–158.

Takasaki, M., Konoshima, T., Kuroki, S., Tokuda, H., Nishino, H. 2001. Cancer chemopreventive activity of phenylpropanoid esters of sucrose, vanicoside B and lapathoside A, from *Polygonum lapathifolium*. Cancer Lett., 173(2): 133–138.

Torssell, K. 1983. Natural Product Chemistry. John Wiley & Sons, London, 387 pp.

Toursarkissian, M. 1980. Plantas Medicinales de la Argentina. Sus Nombres Botánicos, Vulgares, Usos y Distribución Geográfica. Editorial Hemisferio Sur, Buenos Aires, 300 pp.

Vasques da Silva, R., Debonsi Navickiene, H.M., Kato, M.J., Bolzani, V., Méda, C.I., Young, M.C., Furlan, M. 2002. Antifungal amides from *Piper arboreum* and *Piper tuberculatum*. Phytochemistry, 59: 521–527.

Yaacob, K. 1990. Essential oil of *Polygonum minus* Huds. J. Ess. Oil Res., 2: 167–172.

Zacchino, S., Santecchia, C., López, S., Gattuso, S., Muñoz, J. de D., Cruañes, J., Vivot, E., Cruañes, M. del C., Salinas, A., Ruiz, R.E., de, Ruiz, S. 1998. *In vitro* antifungal evaluation and studies on mode of action of eight selected species from the Argentine flora. Phytomedicine, 5: 389–395.

Chapter 7
Antiprotozoal Effects of Wild Plants

Muhammad Subbayyal Akram,[1,*] *Rao Zahid Abbas*[1] and *José L. Martinez*[2]

Introduction

For centuries, many traditional medicinal wild plants and their extracts have been used by the quacks to treat diseases, which they are facing locally, especially in developing countries. This knowledge is transmitted from one generation to another verbally (Atindehou et al. 2004). Phytochemical evaluation proves the presence of secondary metabolites in the various extracts of wild plants which exhibit properties like antiprotozoal, anthelmintic, insecticidal, larvicidal, antibacterial, antiviral, antifungal, anticarcinogenic, sedative and many more.

Protozoa is one of the subkingdoms of the kingdom Protista. Protozoa can survive in all habitats of the world: most species are free-living and some species need an intermediate host for its survival. Fifty thousand species of it have been discovered, all of them are unicellular and seen only by microscope. However, they have complex internal structure and metabolic activities as compared to other unicellular eukaryotes. The metabolic pathway of protozoa is like large animals and requires both organic as well as inorganic compounds. According to modern classification, subkingdom protozoa is classified into six phyla, and members of phylum *Sacromastigophora* and phylum *Apicomplexa* are responsible for most of the protozoal diseases in human. In humans, parasitic protozoa multiply (both sexually and asexually), and actively feed on its host; this stage is typically known as trophozoites or many other terms have been used for this stage in other species (Neva et al. 1994). That stage can be extracellular and intracellular in the host and cause host cell damage during feeding and multiplication, which leads the host to life-threatening stage or even death due to organ dysfunction (Yaeger 1996). That's why parasitic protozoans are one of the largest public health problems in the world and infect people from tropical to subtropical and from developed countries to developing countries. Morbidity and mortality of Plasmodium, Leishmania, Trypanosoma and Trichomonas protozoans are very high. *Plasmodium* spp. produced 219 million cases of malaria and 435,000 deaths only in 2017 (WHO 2019b). Similarly, it causes one million deaths of children yearly only in Africa alone (WHO 2011). *Leishmania* spp. and *Trypanosoma* spp. are hemoflagellates of *Trypanosomatidae* family and cause the three most dreadful human diseases, which are Chagas disease (American trypanosomiasis), sleeping sickness (African trypanosomiasis) and leishmaniasis (cutaneous, mucocutaneous and visceral leishmaniasis) (Teixeira et al. 2012). According to the World Health Organization, only *Trichomonas vaginalis* alone infects 250 million people yearly. These statistics show the influence of these diseases worldwide.

Many synthetic drugs have been used by the physicians to treat these diseases, but many drug-resistant strains of protozoans are developed which show resistance to the drugs used against them. These synthetic drugs used to treat these protozoal diseases which have miserable side effects. In this scenario,

[1] Department of Parasitology, University of Agriculture, Faisalabad.
[2] Vicerrectoria de Investigación, Desarrollo e Innovación, Universidad de Santiago de Chile, Santiago, Chile.
* Corresponding author: 2013ag2739@uaf.edu.pk

plant extract which has secondary metabolites has shown promising results against protozoans and can lead to the development of new synthetic drugs (Polonio and Efferth 2008).

Potential of plant extract is due to secondary metabolites which are Alkaloids, Anthraquinone, Cardiac glycosides, Flavonoid, Phenol, Phlobatannin, Pyrrolizidine alkaloids, Reducing sugar, Saponin, Steroids, Tannin, Terpenoid, Volatile oil and many more (Sasidharan et al. 2011). For the extraction from the plants, various methods are used including Soxhlet extraction, Sonification, and maceration. Methanol, ethanol and sometimes a mixture of alcohol and water are used as solvents in all these methods (Zygmunt and Namiesnik 2003, Phrompittayarat et al. 2007). After extraction of plants material, different methods such as thin-layer chromatography (TLC), column chromatography, flash chromatography, immunoassay, phytochemical screening assay, Fourier-transform infrared spectroscopy (FTIR) and many more (Sasidharan et al. 2011) are used to evaluate how many compounds or secondary metabolites are present in the extract.

Plant extracts have structures that have high diversity in chemicals, biochemical specificity and many other properties, which make them optimum to be used for the treatment or lead to the development and discovery of new drugs (Clardy and Walsh 2004). The diversity of structures in plant extract can be understood as 400 alkaloids have been isolated and many have been proven for their several biological activities (Unver 2007). Similarly, urgency in the discovery of new drugs can be understood as the development of drug resistance in *Plasmodium* spp., is one of the major problems in controlling it and treatment of malaria chiefly relies on quinine and artemisinin derivatives. Both compounds, first isolated from extracts of medicinal plants, still show results even against drug-resistant strains of *Plasmodium falciparum*. That's why plant extracts are still a major source of development of drugs, not only for parasites but also for the treatment of other infectious diseases including bacterial, fungal, lipid disorders, immunomodulation and even cancer (Altmann 2001).

Effect of Wild Medicinal Plants against *Plasmodium* spp.

Plasmodium spp.

For centuries, many species of *Plasmodium* has been causing diseases in humans and animals worldwide, with huge mortality and socioeconomic impacts. According to (WHO 2019b), 219 million cases were reported globally only in 2017 with 435,000 deaths due to it, where livestock and wild animal losses went unregistered. These numbers represent an 18% decrease in malaria cases and a 28% decrease in deaths since 2010. These encouraging facts result from disease awareness, prophylactic treatment, and management.

According to World Health Organization (WHO), Algeria, Armenia, Morocco, Turkmenistan, Uzbekistan, Argentina, Maldives, Sri Lanka, and the United Arab Emirates have become certified malaria-free countries between 2010 and 2019, but more than half of the world is still at risk (WHO 2019b). About five species of *Plasmodium* (*P. falciparum, P. malariae P. ovale, P. vivax* and *P. knowlesi*) cause malaria in humans. Among these five species, *P. falciparum* is much more virulent by causing tentatively 200 million cases each year, while *P. vivax* produces 13.8 million cases each year (Ngoubangoye et al. 2016). Different species of mosquitos transmit this intracellular protozoan, as its life cycle requires two hosts, for sexual stages in mosquitoes and asexual stages in humans (Molina-Cruz et al. 2015).

The life cycle starts with the bite of an infected female mosquito, transmitting sporozoites into the host during ingesting blood meal. Sporozoites then move to the liver and tentatively 30,000–40,000 merozoites escape in blood after sporogony in the liver. Merozoites in blood then transform to trophozoites and these trophozoites invade in erythrocytes. These trophozoites are differentiated into microgametocytes as male and macrogametocytes as female, which can be moved to already infected or non-infected female mosquitoes while ingesting blood meal from an infected host. After ingestion, sexual stages occur in mosquitoes, starting with the development of zygote from microgametocytes and macrogametocytes, then zygote converts into ookinete and oocyst which again releases sporozoites into the host (Negi et al. 2014).

For the treatment of *P. falciparum,* WHO recommends the use of Artemisinin-based Combination Therapy (ACT) in endemic regions. With the passage of time, artemisinin-resistant *P. falciparum* developed in Africa and Southeast Asia due to mutation in the (*kelch13*) gene at 20 different sites (Ashley et al. 2014). Mosquitoes modified themselves genetically to develop chloroquine-resistant gene (*pfcrt*) and multidrug-resistant gene (*pfmdr1*). That's why control and treatment of malaria is still a challenge that can be overcome by the development and use of new antimalarial agents with different mechanisms of action, so the world can combat emerging species that have developed multi drug-resistant genes (Long and Zavala 2016).

Pertinent Studies about the Effectiveness of Wild Plants against *Plasmodium* spp.

For centuries, in many countries quacks have been using hundreds of species of wild plants and their extracts to treat malaria, from which several are proven, pharmacologically and experimentally, to be effective against *Plasmodium* spp.

Derris trifoliata

Derris trifoliata belongs to family *Fabaceae* and is commonly found in the coastal swamps of Konkan, India, and Kenya. Many rotenoloids, spirohomo-carotenoid and unique isoflavonoid derivatives have been isolated from the roots (Yenesew et al. 2005) and seeds (Yenesew et al. 2006) of *Derris trifoliata* which show anti-parasitic and larvicidal activities (Figure 7.1).

For the fraction dichloromethane, methanol extract was used and extracts were tested against two strains of *Plasmodium falciparum* D6 and W2. Seed pods extracts showed weak efficacy, as compared to extracts of other parts of plants, with IC_{50} values of 12.2 ± 2.4 µg/ml and 13.4 ± 2.6 µg/ml against D6 and W2, respectively. Similarly, chloroquine extract from plants gives IC_{50} values of 0.008 ± 0.004 µg/ml and 0.051 ± 0.010 µg/ml against D6 and W2 strains of *P. falciparum*, respectively. Mefloquine is also one of the extracts of *D. trifoliata* which is potentially active against D6 and W2 with IC_{50} values of 0.042 ± 0.008 µg/ml and 0.015 ± 0.002 µg/ml, respectively. Flavanone (Lupinifolin), which is extracted from plant, shows moderate activity against D6 and W2 strains with IC_{50} values of 2.6 ± 0.8 µg/ml and 3.7 ± 1.4 µg/ml, respectively. Various parts' extracts also show larvicidal activity against *Aedes aegypti* at second instar larval. The seed extracts are more active than others with IC_{50} value of 0.7 ± 0.2 µg/ml at 24 hours (h). Seed extracts also show larvicidal activity against third and fourth instar larval stages of *Aedes aegypti* with IC_{50} values of 3.7 ± 1.5 µg/ml and 8.5 ± 2.3 µg/ml, respectively (Peter et al. 2009).

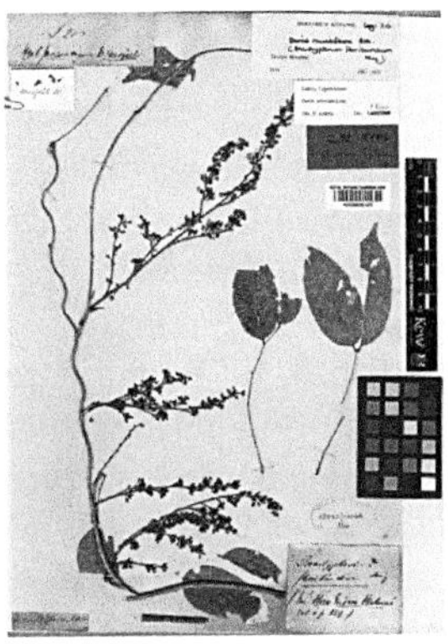

Figure 7.1. Scientific name: *Derris trifoliata*, Specimen: K000898325. Image credit to: "© copyright of the Board of Trustees of the Royal Botanic Gardens, Kew".

Hintonia latiflora

Hintonia belongs to family *Rubiaceace*, mostly found in the forests of Mexico. *Hintonia latiflora* (Figure 7.2) is proven to be effective against *Plasmodium berghei. Ethyl acetate extraction of H. latifolra* stem bark and High-performance liquid chromatography reveals the presence of two compounds which are

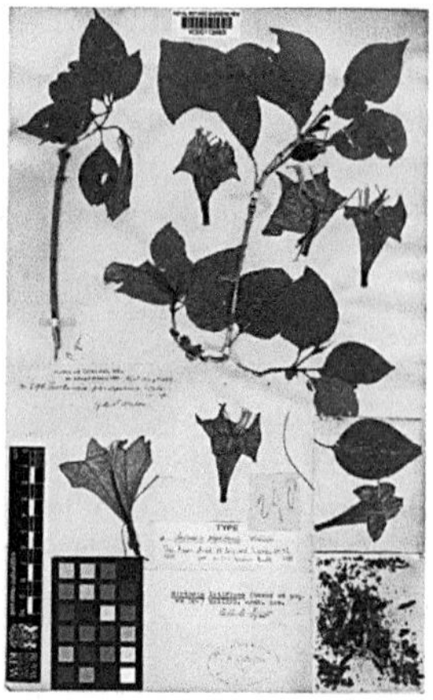

Figure 7.2. Scientific name: *Hintonia latiflora*, Specimen: K000173483. Image credit to: "© copyright of the Board of Trustees of the Royal Botanic Gardens, Kew".

5-O-β-D-glucopyranosyl-7,40-dimethoxy-30-hydroxy-4-phenylcoumarin and 5-O-β-D-glucopyranosyl-7-methoxy-30,40-dihydroxy-4-phenylcoumarin. The effectiveness of these molecules is tested *in vitro* that results in 70.8% suppression in the development of schizonts when its extract was used by the dosage of 40 mg/kg in *in vitro* trials. Both molecules are proven to be effective at IC50, which are 24.7 and 25.9 μm against *P. berghei,* especially in schizonts development in *in vitro* trials (Argotte-Ramos et al. 2006).

Another compound, HIMeOHe, was recently found in *H. latiflora* by methanolic extract which shows it's an action against Plasmodium. According to toxicity assay, the median lethal dose (LD 50) was 2783.71 mg/kg. The researcher Rivera et al. (2014) conducted an experiment on mice infected by *Plasmodium voelii* and treated it by HIMeOHe by the dose of 300 and 600 mg/kg but they died after 6 and 7 days with the parasitemia level around 45%, but 70% in untreated mice. The most interesting part is that using a dose of 1200 mg/kg, mice died after 23 days with a parasitemia level of 23.6%. HIMeOHe is mutagenic in nature as the number of micronuclei remarkably increased from untreated mice which are 0.9 and 4.8 micronuclei in untreated and treated mice groups, respectively and found HIMeOHe as a chemical which should help to reduce its genotoxic potential.

Bidens pilosa

Bidens L. is known for the beautiful flowering; it belongs to the family of *Asteraceae,* which is distributed in tropical, subtropical and mostly in the warm temperature regions. Traditionally in many countries, it was used for the treatment of malaria fever, anti-inflammatory agent, antibiotic and also for the treatment of nematodes (Wat et al. 1980).

Its roots are taken, dried and turned into powder, for the extraction of active agent which is 1-phenyl-3,5,7-heptatriyn, α-L-rhamnopyranosyl-(1-6)-β-D-glucopyranose and β-D-glucopyranose. For the extraction and separation of a required compound, series of agents are used just like formic acid, acetic acid, ethyl acetate and water system and at the end reagent, NP/PEG, is used (Wagner and Bladt 1996). This compound shows its activity against *P. falciparum, P. berghei* both in *in vitro* and *in vivo* (Dounga et al. 1983). It reduces the parasitemia of *Plasmodium berghei* up to 60% if used with the dose rate of 1000 mg/kg; similarly, its (IC50) dose is 485 mg/kg, which statistically helps in the reduction of parasitemia (Andrade-Neto et al. 2004).

Exostema caribaeum and *Exostema mexicanum*

Genus *Exostema* is of flowering plants, belonging to family *Rubiaceae,* which is mostly found in the forests of the West Indies (Mabberley 2008). *E. caribaeum* and *E. mexicanum* are the most prominent species of Exostema, which have been proven to be effective against many drug-resistant species of

Plasmodium spp. *E. carbaeum* has been traditionally used by Mexican people for the treatment of malarial fever.

Extracts of its stem bark show the potential of antimalarial *in vitro*. It's due to the phenylcoumarium derivative which was isolated from its stem bark by hydrolyzed ethyl acetate extracts that give results against *Plasmodium falciparum*, which was sensitive to pyrimethamine and chloroquine (Noster and Kraus 1990). After that, *E. mexicanum* stem bark fractionation by hydrophilic and lipophilic extracts revealed two new 4-phenylcoumariuns in them, which were named as (exomexin A) 4',8-dihydroxy-5,7-dimethoxy-4-phenylcoumarin and (exomexin B) 4',8-dihydroxy-5,7-dimethoxy-4-phenylcoumarin. *In vitro* trials confirmed the (exomexin A) strong activity against *Plasmodium falciparum* strains (poW), which are sensitive to chloroquine and also effective on chloroquine-resistant strains (Dd2), with inhibitory concentration (IC50) values of 3.6 and 1.6 μg/ml, respectively (Köhler et al. 2001).

Artemisia annua

Artemisia annua belongs to the diverse and large family *Asteraceae*, commonly known as the sunflower family (Figure 7.3). In Mexico, old inhabitants used Artemisia leaves for the treatment of diarrhea; later on, modern science revealed its anti-parasitic properties. Not only *A. annua* leaves are beneficial for the treatment of malaria, but also effective against many parasitic diseases as *Giardia lamblia*, *Entamoeba histolytica* and many more (Fernandez et al. 2005).

Ethanolic extract reveals the presence of Artemisinin, which is majorly found in *A. annua* about 2% by dry weight (Trigg 1989). Artemisinin was very effective against major *Plasmodium* species and its synthetic derivatives are developed and sold in the market, which is defiantly cheaper as compared to a natural extract of artemisinin from the plant. Artemisinins are toxic to Plasmodium at nanomolar concentration, whereas mammalian cells are at risk if it's administered at a micromolar concentration (Lai et al. 1995). The key to its very good result is hidden in the uptake of artemisinin by the parasite or the cell which host parasite. The concentration of it is 100 folds more in erythrocytes which are infected by Plasmodium as compared to non-infected erythrocytes (Kamchonwongpaisan et al. 1994). The uptake of active agents is reversible and dependent partially on metabolic energy (Gu et al. 1986).

Artemisinin tackles *Plasmodium* spp. by oxygen free radical as it is lacking an endoperoxide bridge. The results of the *in vivo* trial of artemisinin give 98.6% inhibition of *Plasmodium yoelii* (Klayman 1985, Brossi et al. 1988).

Figure 7.3. Scientific name: *Artemisia annua*, Specimen: K000891905. Image credit to: "© copyright of the Board of Trustees of the Royal Botanic Gardens, Kew".

Effect of Wild Medicinal Plants against *Trypanosoma* spp.

Trypanosoma spp.

In genus *Trypanosoma*, species are flagellated, unicellular and parasitic in nature. Only *Trypanosoma cruzi* is the third cause of deaths after *Plasmodium* spp. and schistosomiasis in Latin America. *Trypanosoma cruzi* is transmitted by a bug *Triatoma infestans* causing Chagas disease or American trypanosomiasis (Braz et al. 2016). Approximately 16 to 18 million people are annually infected by *T. cruzi* around the globe and millions are still at risk of having an infection (WHO 2016).

An infection transmitted to humans by the bite of infected *Triatoma infestans,* commonly known as Kissing bug. Infected bug releases feces near the bite wound while ingesting meal and that feces contains trypomastigotes. At the bite wound site, trypomastigotes penetrate various cells and transform into amastigotes. Amastigotes multiply in the infected cell and convert into trypomastigotes, then cell bursts and trypomastigotes are released into the bloodstream. Trypomastigotes in bloodstream of infected person again ingested by the bite of bug while ingesting blood meal from host and its life cycle will repeat in this fashion again and again (CDC 2019b).

Trypanosoma cruzi infection has an acute and chronic phase, which lasts for a few weeks to months. In acute phase, disease signs are seen immediately after the host is infected by the infection, which includes fever, and swelling at the bite site and rarely, the acute phase ends with inflammation of the brain, heart muscles and behavioral abnormalities (Vilar-Pereira et al. 2015). During this phase, *Trypanosoma cruzi* is found in the bloodstream and is usually mild or maybe asymptomatic in nature. Predominantly, the acute phase enters in a prolonged asymptomatic phase or chronic phase during which hardly any or no parasites are present in the bloodstream. As in chronic phase, host never develops symptoms related to Chagas disease which results in life-threatening health problems with the passage of time (Hotez et al. 2013).

Pertinent Studies about the Effectiveness of Wild Plants against *Trypanosoma cruzi*

Nifurtimox and benznidazole are drug of choice against *Trypanosoma cruzi* but both of them only effective in acute phase of infection and induce grievous side effects on patient (Porcal et al. 2008). That's why it is urgent to have an alternative treatment for it and extract of many wild plants proves to be sufficiently effective against *T. cruzi* with zero side effects on humans and animals.

Aristolochia taliscana

Aristolochia taliscana (Figure 7.4) belongs to one of the large genus *Aristolochia,* which produces two to three meters long stems that scramble over plants or on the ground (Standley 1926). After the methanolic extraction of roots of *Aristolochia taliscana,* phytochemical study reveals six secondary metabolites that show activities against *T. cruzi*. Four out of six secondary metabolites are neolignans (eupomatenoid-7, licarin-A, eupomatenoid-1, and licarin B) and two lignans (austrobailignan-7 and fragransin E_1) (Schmidt et al. 2012). Austrobailignan-7, licarin-A and epimastigotes show remarkable results as compared to other neolignans in *A. taliscana*. Eupomatenoids can show results at a very low concentration of 0.5 mg/ml on epimastigotes of *Trypanosoma cruzi*. These lignans cause loss of hydroxyl group which halt trypanocidal activity determined by using Structure-Activity Relationship analysis (SAR). These compounds also affect

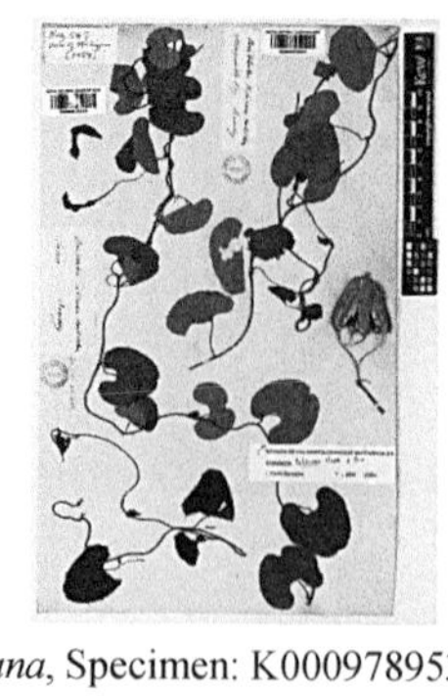

Figure 7.4. Scientific name: *Aristolochia taliscana*, Specimen: K000978953. Image credit to: "© copyright of the Board of Trustees of the Royal Botanic Gardens, Kew".

the biological behavior of it. Minimal Inhibitory Concentration or MIC_{100} values of austrobailignan-7 and licarin-A are 25 and 40 μg/ml, respectively (Ramirez-Moreno et al. 2017). According to another research, eupomatenoid-7, licarin A, austrobailignan-7 and fragransin E_1 show IC_{50} values of 25–75 μg/ml against *T. cruzi* after incubation of 48 h. However, eupomatenoid-1 and licarin B were inactive (Abe et al. 2002).

Piper regnellii

Many *Piper* species are traditionally used for the treatment of parasitic disease. They are mostly found in the forests of Africa, Latin America, and Asia. These are some plants from the genus *Piper,* which are proven to be effective against *Trypanasoma cruzi. Pepper hooked* (Gutiérrez et al. 2016), leaves of *Piper cubeba* (Esperandim et al. 2013), *Piper jericoense* (Hamedt et al. 2014) and *Piper regnellii* (Figure 7.5) are one of them. These species of Piper commonly have a compound named eupomatenoid-5 which shows it's an action against *T. cruzi.* The fractionation of eupomatenoid-5 is performed from ethyl-acetate crude extract by using dry leaves of the plant (Luize et al. 2006). After fractionation of eupomatenoid-5, its purification and identification can be done by absorption-chromatographic methods and UV spectral analyses, respectively (Chauret et al. 1996). Eupomatenoid-7 causes alteration of ultra-structures, dysfunction, depolarization and oxidative damage of *T. cruzi* mitochondria and cell membrane, and lipoperoxidation leads to the death of parasite by using a dose of EC_{50} 40.5 mM (Pelizzaro-Rocha et al. 2011).

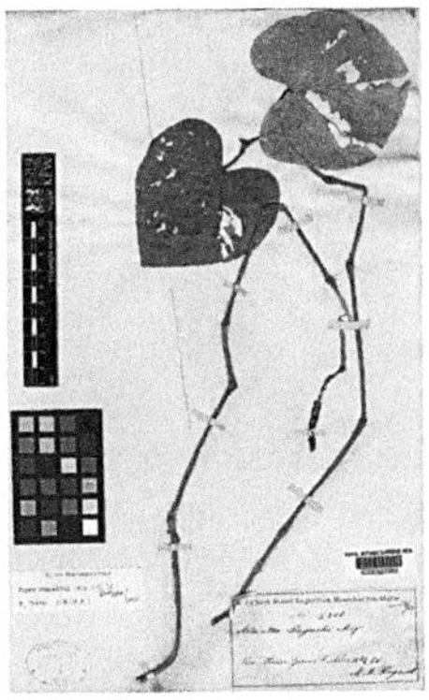

Figure 7.5. Scientific name: *Piper regnellii*, Specimen: K000323965. Image credit to: "© copyright of the Board of Trustees of the Royal Botanic Gardens, Kew".

Senna villosa

Senna villosa is one of the leguminous plants mostly found in the forests of the tropical regions of the world. The seed of *Senna* spp. is found in the fossils of Rancho clay in the timeline of the middle Eocene epoch (Herendeen 1992). Traditionally, *S. villosa* has been used as antifungal, antimicrobial and for the treatment of stomach disorders. Its phytochemical analysis reveals the presence of alkaloids, anthraquinones, sterols, and flavonoids. Similarly, crude chloroformic extract of leaves of *S. villosa* (Jimenez-Coello et al. 2010) showed the presence of (8-hydroxymethylen)-trieicosanyl acetate in it and it acts on epimastigotes of *T. cruzi* with the concentration of 1.6 mg/ml. The reduction of amastigotes in the cardiac tissue is seen at a concentration of 3.3 mg/g as compared to the untreated group of mice (Jimenez-Coello et al. 2011). (8-hydroxymethylen)-trieicosanyl acetate is 3.4 times more active than allopurinol against blood trypomastigotes of *T. cruzi*, when both are used with the same concentration of 8.5 μg/g (Jimenez-Coello et al. 2010).

Haematoxylum brasiletto

Traditionally, 10 wild plants have been used in Mexico for the treatment of parasitic infection. Molina-Garza et al. (2014) evaluated these 10 plant extracts for the treatment of *T. cruzi,* which are *Schinus molle, Lippia graveolens, Artemisia mexicana, Ruta chalepensis, Castela texana, Persea americana, Cymbopogon citratus, Marrubium vulgare, Eryngium heterophyllum,* and *Haematoxylum brasiletto.*

Metholic extract 150 mg/ml of *H. brasiletto* (Figure 7.6) produces inhibition (88–100%) of *T. cruzi.* However, other plants, *A. mexicana* and *C. citratus,* halt growth up to 83%. *L. graveolens* and *C. texana* show the least effect of 33% only. *H. brasiletto* shows the highest trypanocidal activity among all other wild plants which are traditionally used in Mexico. Phytochemical analysis of *H. brasiletto* reveals the presence of 5-methoxypsoralen, 4-hydroxycinnamic acid, hematoxylin, phloroglucinol, brazilin, methyl gallate, caffeic acid and gallic acid (Argotte-ramos et al. 2006). These compounds show that it's action on epimastigotes when used with a concentration of 7.92 mg/ml against *T. cruzi*. The extract of this plant also has unsaturated compounds, such as triterpenes, carboxyl groups, carbonyl groups, a sesquiterpene lactone, tannins, flavonoids and quinones (Pérez-Treviño et al. 2016). The mode of action and identification of active compounds from the extract is still unknown.

Figure 7.6. Scientific name: *Haematoxylum brasiletto*, Specimen: K000264530. Image credit to: "© copyright of the Board of Trustees of the Royal Botanic Gardens, Kew".

Trichilia emetica

Trichilia emetica belongs to family *Meliaceae* and is natively found in the forests of Africa. It is traditionally used for the treatment of abdominal pain, jaundice, skin problems, chest pain and many more (Komane et al. 2015). Atindehou et al. (2004) evaluated 88 plants of Côte d'Ivoire and their 101 crude ethanol extracts to verify antitrypanosomal activity and he founded *T. emetica* to be the most promising one. The root bark extract shows IC_{50} value of 0.04 μg/ml against *Trypanosoma brucei rhodensiense.* Its leaf extracts also exhibited satisfactory results against *Trypanosoma brucei brucei* and *Trypanosoma brucei rhodesiense* with IC_{50} value of 14.9 μg/ml and 8.6 μg/ml, respectively (Hoet et al. 2004). Phytochemical analyses reveal the presence of limonoids such as trichilins and seco-liminoids which show antitrypanosomal activity by damaging DNA (Atindehou et al. 2004).

Effect of Wild Medicinal Plants against *Lesihmania* spp.

Lesihmania spp.

Visceral, cutaneous and mucocutaneus Leishmaniasis are predominantly caused by 20 species of leishmania, some of them are *Leishmania chagasi*, *Leishmania infantum,* and *Leishmania donovani.* All species of Leishmania are almost identical in morphology, but they are different in their biochemical and molecular structures by which their identification is possible through isoenzyme analysis, monoclonal antibodies and by molecular methods (Murray et al. 2005). Worldwide, 88 countries, i.e., 67 countries from the old world and 21 countries from the new world, are affected by the infection of this set of parasites. About 90% visceral leishmaniasis cases were reported only from five countries, which are India, Brazil, Sudan, Bangladesh, and Nepal, and cases of cutaneous leishmaniasis were mainly reported in seven countries—Syria, Saudi Arabia, Afghanistan, Algeria, Iran, Peru and Brazil ((Dey and Singh 2006). Tentatively, 700,000 to 1 million new cases and deaths of 26,000 to 65,000 people are reported annually (WHO 2019a).

Female *Phlebotomine* sand flies transmit leishmaniasis by biting and injecting promastigotes into their host while ingesting blood through the proboscis. These promastigotes are then phagocytized from the wounded area by macrophages and other mononuclear phagocytic cells. In macrophages, promastigotes multiply themselves and transform into amastigotes and then cell bursts. Amastigotes then do multiplications in various tissues of the body and spread the infection to other cells. The cycle of leishmaniasis continues when the noninfected sand fly takes a blood meal from an infected host and ingests macrophages which are infected by amastigotes. Then, amastigotes transform into promastigotes in the gut of sand fly and promastigotes migrate to proboscis so it can again transmit it to another host (CDC 2019a).

Mucocutaneous leishmaniasis is characterized by severe invasive infection of nasopharyngeal mucosa, while cutaneous leishmaniasis is associated with skin ulcers. In visceral leishmaniasis, host show symptoms of fever, enlargement of the liver, weight loss, swelling of the spleen and decreased blood count. The treatment of leishmaniasis depends on immune status, clinical signs and the geographical location of a host. Usually, visceral leishmaniasis and severe cases of mucocutaneous and cutaneous leishmaniasis are treated with liposomal amphotericin B and in developed countries, drugs like miltefosine, paromomycin and pentavalent antimony are used for its treatment. These drugs are unaffordable due to their very high cost and show severe side effects, including renal insufficiency. So, many patients stop their doses during treatment which results in the development of drug-resistant strains (WHO 2016).

Pertinent Studies about the Effectiveness of Wild Plants against *Leishmania* spp.

Pentalinon andrieuxii

Pentalinon andrieuxii (Figure 7.7) belongs to family *Apocynaceae* and is natively found in the forests of Mexico. In Mayan Folk medication, it's used for the treatment of lesions related to cutaneous leishmaniasis (Pan et al. 2012). The aqueous and organic extraction methods are used to get extracts from its leaves and roots. Its extract contains numerous secondary metabolites, including triterpenoids, cardenolides, flavonoids, trinorsesquiterpenoids and pregnane sterols (Lezama-Davila et al. 2007). Phytochemical analysis reveals the presence of one triterpenoid, three coumarins, and 16 sterol derivatives. Five sterols proved to be more active and have higher inhibit effects than the Sodium stibogluconate (it's a drug which is referred for leishmania treatment) after 48-hours of exposure with IC_{50} of 1.4 to 3.5 μM and 2.7 μM, respectively. Among these derivatives, 6,7-dihydroneridienone and cholest-4-en-3-one are the

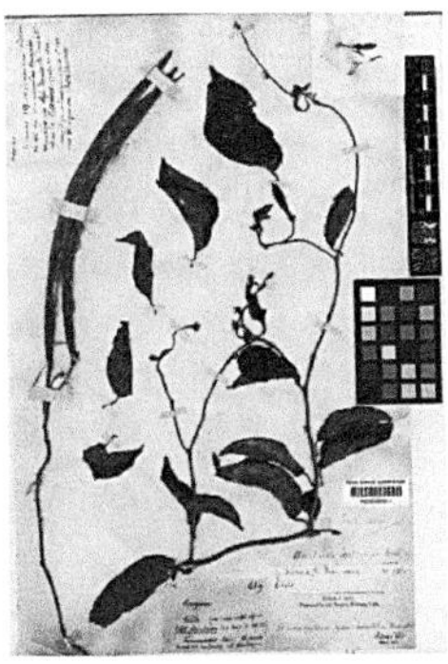

Figure 7.7. Scientific name: *Pentalinon andrieuxii*, Specimen: K000582911. Image credit to: "© copyright of the Board of Trustees of the Royal Botanic Gardens, Kew".

most active metabolites with IC_{50} values of 9.2 μM and 0.03 μM. These compounds were also proven to be non-cytotoxic to bone marrow-derived macrophages. These compounds halt biosynthesis of sterol, which causes an alteration in the membrane of Leishmania spp. and results in destruction and abnormalities of amastigotes morphology (Pan et al. 2012).

Vitex negundo

Vitex negundo is traditionally used in folk medicines, mostly in south and southeast Asia. It's a large shrub that has an aroma with tomentose, quadrangular and whitish branchlets (Figure 7.8). *V. negundo* is now scientifically proven to have anti-leishmanial potentials and compounds as flavonoids (Luteolin and Quercetin) were identified and separated from the extract of *V. negundo,* which shows anti-leishmanial activities. Luteolin and quercetin are flavonoids that show IC_{50} value of 12.5 μM against promastigotes and intracellular amastigotes of *Leishmania donovani* (Polonio and Efferth 2008). Both compounds have the capability to induce topoisomerase 2 in kinetoplastid stage of leishmania at DNA minicircle cleavage in intracellular amastigotes and promastigotes of *L. donovani.* Luteolin and quercetin, ultimately halting the cell cycle of promastigotes in G0/G1 phase, which then leads to cell death by apoptosis. *In vivo* results have shown that luteolin and quercetin cause 80% and 90% reduction in the splenic load of the parasite, when used by 3.5 mg/kg and 14 mg/kg body weight, respectively (Mittra et al. 2000). In another study, it has been testified that quercetin and luteolin are also halted topoisomerase 1 (Das et al. 2006).

Figure 7.8. Scientific name: *Vitex negundo*, Specimen: K001114060. Image credit to: "© copyright of the Board of Trustees of the Royal Botanic Gardens, Kew".

Maesa balansae

Maesa balansae (Figure 7.9) is a shrub that belongs to *Primulaceae* family and grows naturally in the forests of the northern areas of Vietnam. Traditionally, it has been used for the treatments of headache, allergies, worm infections, sprains and skin ulcers in Vietnamese medicines. Polar extracts and bioassay-guided fractionation of *M. balansae* leaves help in isolation and characterization of its compounds, which divulges the presence of six oleane triterpenoid saponins mesabalides I–VI. These six triterpenoid saponins were separated by liquid chromatography and each was separately evaluated against leishmaniasis (Leonard et al. 2003). Each mesabalides has the capability to halt specifically leishmaniasis but most potent are Maesabalide III and IV with IC_{50} values of 7 and 14 ng/ml against intracellular amastigotes. *In vivo* studies show more than 90% reduction in the burden of amastigotes just after one week of subcutaneous single dose at 0.2–0.4 mg/kg. No cytotoxicity was observed in the human fibroblast (MRC-5) cell line even at $CC_{50} > 32$ μg/ml (Germonprez et al. 2005).

Haplophyllum bucharicum

Haplophyllum bucharicum is one of the endemic plants of Uzbekistan. Many species of genus *Haplophyllum* were studied for their antiviral and anticancer properties; among them, *H. bucharicum*

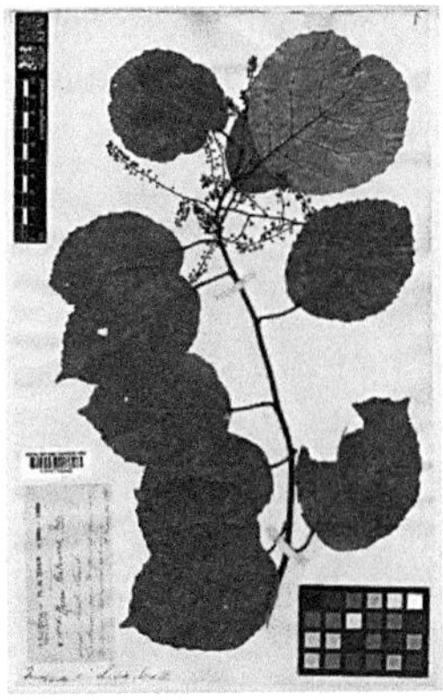

Figure 7.9. Scientific name: *Maesa balansae*, Specimen: K000756245. Image credit to: "© copyright of the Board of Trustees of the Royal Botanic Gardens, Kew".

has been proven for its anti-leishmanial activity. Diphyllin is a lignan which is extracted from this plant by methanolic extract from its aerial parts. Diphyllin shows strong potential against amastigote activity by just IC_{50} value of 0.2 μM and halts growth of promastigotes at IC_{50} value of 14.4 μM. However, diphyllin has moderate antiproliferative potential against human monocytes, which are infected by promastigotes with IC_{50} value of 35.2 μM (Giorgio et al. 2005). Diphyllin mainly halts promastigote growth by hindering protein synthesis and cell cycle in S-phase of promastigotes by inducing arrest in DNA synthesis (Pradheepkumar et al. 1999). Diphyllin interferes with the specific mechanisms of promastigotes by which they recognize and penetrate within host macrophages. It's due to the modulation of molecules (glycoproteins) on the surface, which is responsible for the attachment of promastigotes to macrophages (Sibley and Norma 2000). It results in a strong reduction of promastigote burden in macrophages.

Swertia chirata

It belongs to family *Gentianaceae* and is naturally found in the mountains of the Himalayas at 1200 m to 3000 m altitudes from Kashmir to Bhutan (Figure 7.10). Its common name in Pakistan is Chirata talkh and is traditionally used as anti-pyretic, anti-inflammatory, anti-parasitic, stomachic, ulcer, and diabetes treatment (Iqbal et al. 2006). Phytochemical analysis reveals the presence of amarogentin in niosomal and liposomal forms in this plant (Medda et al. 1999). Amarogentin shows its action by inhibiting DNA topoisomerase 1 by exerting its inhibitory potential on enzymes and halts the binary complex formation of DNA and topoisomerase 1. This action of amarogentin results in the death of *L. donovani* (Ray et al.

Figure 7.10. Scientific name: *Swertia chirata*, Specimen: K000854477. Image credit to: "© copyright of the Board of Trustees of the Royal Botanic Gardens, Kew".

1996). The potential of amarogentin may lead to the development of synthetic similar drug development against leishmaniasis. Amarogentin in niosomal form reduces the 90% parasitic load only after six doses at the rate of 2.5 mg/kg. Spleen histological tissue staining, blood pathology examination and liver enzymes evaluation of animals in the hamster model show no toxic effect of amarogentin on them (Medda et al. 1999).

Effect of Wild Medicinal Plants against *Trichomonas vaginalis*

Trichomonas vaginalis

In the genus *Trichomonas,* only *Trichomonas vaginalis* and *Trichomonas tenax* are causative species in humans and some other species infect animals. *Trichomonas vaginalis* is a protozoan, pear shaped trophozoites, 10 to 20 μm in length, anaerobic flagellated, which lives in the urogenital tract of humans, commonly vulva, vagina, and urethra of female and epididymis, prostate and urethra of male and replicates in these parts by binary fission. *T. vaginalis* is the most widespread non-viral sexually transmitted disease in the world, which tentatively infects 250 million people per annum (WHO 2016). Even extragenital infections of *T. vaginalis* are also reported from respiratory samples of neonates and from the patients suffering from human immunodeficiency virus (HIV) infection (Press et al. 2001). Females are more prone to infection than males with a ratio of 36:1. Females aged 25–44 years were infected in large numbers, with tentatively 60,000 cases of females in 2011 (Salud et al. 2011).

Almost 50% of individuals who are infected by *T. vaginalis* remained asymptomatic, so they were neither detected nor treated; that's why this protozoal infection has been spreading all over the world silently (Carlton et al. 2007). A female suffering from its infection shows signs of abdominal pain, vaginitis, dysuria, cervicitis, unpleasant smell and seropurulent discharge from the vagina and ultimately infertility. Preterm birth, respiratory problems and low weight of newborns can also be the outcome of its infection. Moreover, its infection increases the chance of sexually transmitted diseases just like HIV, herpes simplex virus 2 (HSV-2) and papillomavirus (Sorvillo and Kerndt 1998, Boselli et al. 2005). *T. vaginalis* does not survive well in the external environment; it is usually transmitted among humans by sexual intercourse and humans are its only host. Its incubation period is between 5 days to 28 days. In men urethritis, prostatitis and epididymitis usually occur after its infection (CDC 2017).

Until recently, metronidazole, and its derivatives secnidazole and tinidazole, was the only drug of choice for the treatment of trichomoniasis (Wendel and Workowski 2007). However, it's not safe to use metronidazole in pregnancy as it may result in fetal malformation, and increase the risk of preterm birth, specifically when used in the first trimester (Klebanoff et al. 2001). Persistent and overuse of metronidazole and its derivatives result in the development of drug-resistant strains of *T. vaginalis;* chiefly, parasite develops resistance against 5-nitomidazole derivatives, which was observed in up to 20% of patients (Krashin et al. 2010). The resistance of metronidazole and tinidazole in *T. vaginalis* were 9.6% and 0.56%, respectively (Schwebke and Barrientes 2006). In this scenario, development and research of new drugs is a necessity and wild plants' extracts show promising results against resistant strains of *T. vaginalis.*

Pertinent Studies about the Effectiveness of Wild Plants against *Trichomonas vaginalis*

Polygala decumbens

Polygala decumbens belongs to family *Polygalaceae* commonly found in the forest of Caatinga, Brazil. In folk medicines, it's used for the treatment of sexually transmitted diseases such as blennorrheas and as an expectorant, diuretic and emetic agent (Agra et al. 2008). In 2012, Amanda Piccoli Frasson and team gave the first report on anti-*Trichomonas vaginalis* potential of *P. decumbens*. They evaluated 23 Caatinga plants and their 44 aqueous extracts for the treatment of *T. vaginalis* which are used in traditional medicines. Only *P. decumbens* root extract was potentially effective even against metronidazole resistant *T. vaginalis.* Aqueous extract of *Polygala decumbens* have significant potential as compare to other plants used in this trail. Minimum inhibitory concentration (MIC) of *P. decumbens* against metronidazole-resistant *Trichomonas vaginalis* was at the dose of 1.56 mg/ml. The growth and morphology of its trophozoites were evaluated by kinetic growth experiments. Its extracts completely halt the growth after the first hour of incubation. 35% viable trophozoites were found after 30 mins and after 2 hours no parasites were observed. In a nutshell, *P. decumbens* root aqueous extracts are highly cytotoxic to *T. vaginalis*, freshly collected samples from the clinic and also for the metronidazole resistant *T. vaginalis* (Frasson et al. 2012).

Dalbergia cochinchinensis

Dalbergia cochinchinensis (Figure 7.11) belongs to family *Fabaceae,* mostly found in the forests of Vietnam, Thailand, Laos, Cambodia, and China. Traditionally, its wood is used in the furniture. Arthan et al. (2008) evaluated the activity of *D. cochinchinensis* against *T. vaginalis* and he concluded that β-glaycosides and many aglycones in its extract show selective inhibitory activity against *T. vaginalis* without affecting mammalian cells. Svasti et al. (1999) isolated and structurally studied the extracts of *D. cochinchinensis* and confirmed the presence of Isoflavonoids glucoside, dalcochinin-8'-β-D-glucoside and its aglycone. β-glycosides show its biological activity against *T. vaginalis* due to the presence of the hydroxyl group which is linked with β anomeric center of glucose. These extracts from *D. cochinchinensis* show minimal inhibitory concentration (MIC) of 12.5 μM after a period of 24 hours. However, further studies are required to understand the mode of action of its extracts and mechanism of selective inhibition of *T. vaginalis* without harming mammalian cells (Arthan et al. 2008). β-glycosides are less effective as compared to MIC value of metronidazole which is 0.05 μM (Chavalitshewinkoon-Petmitr et al. 2003). Metronidazole users are at potential risk of mutagenicity, carcinogenicity, and development of drug resistance in parasite, but β-glycosides are not cytotoxic to mammalian cell lines, even at various concentrations and these extracts of *D. cochinchinensis* lead to the development of compounds similar

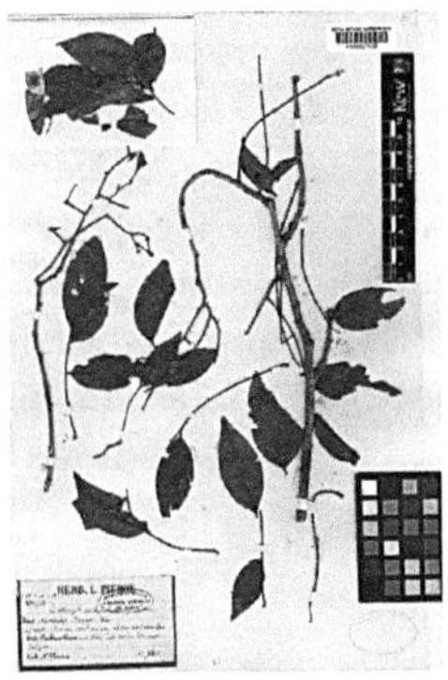

Figure 7.11. Scientific name: *Dalbergia cochinchinensis*, Specimen: K000827939. Image credit to: "© copyright of the Board of Trustees of the Royal Botanic Gardens, Kew".

to β-glycosides, which will show more MIC against *T. vaginalis* as compared to β-glycosides (Arthan et al. 2008).

Lavandula angustifolia

Lavandula angustifolia belongs to family *Lamiaceae* and it is native to countries around the Mediterranean Sea (Spain, France, Italy, Turkey, Syria, Egypt, etc.). For centuries, extracts of genus *Lavandula* were used by the natives as antibacterial, tranquilizer, antifungal, antidepressive and were also effective for insect bites and burn wounds (Cavanagh and Wilkinson 2002). Moon et al. (2006) reported the significant potential of *Lavandula angustifolia* and Lavandula × intermedia oils extract on three protozoal parasites including *Trichomonas vaginalis*, *Giardia duodenalis* and *Hexamita inflata*. Oil extracted from these plants is used in three various concentrations of 1%, 0.5% and 0.1% against *T. vaginalis* and both show almost same results. *L. angustifolia* oil at 1% concentration caused 100% death of *T. vaginalis* after 20 min and at 0.5% concentration after 30 to 35 min, respectively. However, 0.1% gave less satisfactory results as compared to other concentrations and require 65 to 80 min of exposure to kill *T. vaginalis*. Microscopic examination of *T. vaginalis* shows the lysis of cells, but further studies are required to understand the exact mode of action of these oils and for identification and separation of the active molecule for its action on *T. vaginalis* (Moon et al. 2006).

Hippeastrum santacatarina

Lycorine was isolated from *Hippeastrum santacatarina* and belongs to family *Amaryllidaceae* which are the native plant of Brazil (Giordani et al. 2008). Lycorine is one of the most common Amaryllidaceae alkaloids found in *H. santacatarina*. The *T. vaginalis* in the peak growth culture was exposed to lycorine at 250 1M, which halted growth of parasite as compared to untreated growth culture. Analysis indicates the arrest in cell cycle of *T. vaginalis* at the G1 phase after 6 hours of treatment and lycorine also has significant potential to halt the cell cycle of *T. vaginalis* in the G2/M phase after 24 hours as the cells in the G2/M phase increased in the culture, from 47.37% to 62.31%, which are treated by it (Giordani et al. 2011). This cell cycle arrest confirms the DNA damage of cells and it tries to repair the damage (Weirnet and Hartwell 1990). Lycorine induces alteration in the original shape of trophozoites, depression can be observed on its surface, the endoplasmic reticulum of parasitic cell become undisrupted, and cytoplasmic degeneration and abnormal shape of nucleus were accrued after 24 hours. Lycorine causes paraptosis of *T. vaginalis* when used with a concentration of 2.5 to 1000 1M (Giordani et al. 2011).

Verbascum thapsus

Verbascum thapsus belongs to family *Scorphulariaceae* and it's native to America, Australia, Africa and in many Asian and European countries (Figure 7.12). It has anti-bacterial (Prakash et al. 2016), anti-inflammatory (Dar et al. 2019) and skin healing properties (Jafarirad et al. 2019), which have already been studied and in traditional medicines, its extract is used for the treatment of diarrhea and for genitals as well as urinary organs infection by quacks. *V. thapsus* extracts' various concentrations are tested on the *T. vaginalis* trophozoites. Ethanolic extracts of this plant induce apoptosis in *T. vaginalis,* which is determined by Flow cytometry. Its ethanolic extracts give inhibitory concentration 50% (IC_{50}) of 39.17 μg/ml after 24 hours, while IC_{50} value of metronidazole is 0.0326 μg/ml. However, metronidazole induces carcinogenic effects, drug resistance in parasites and teratogenic effects on the fetus. Ethanolic extracts of *V. thapsus* induce programmed cell death (apoptosis) in *T. vaginalis* (Zohreh et al. 2015).

Figure 7.12. Scientific name: *Verbascum thapsus*, Specimen: K000806423. Image credit to: "© copyright of the Board of Trustees of the Royal Botanic Gardens, Kew".

Conclusion

In a nutshell, loss of efficacy of the currently available drugs, against most of the dreadful protozoal diseases, is due to the development of drug-resistant strains. Such as development of metronidazole resistance in *Trichomonas vaginalios*, chloroquine resistantce in *Plasmodium falciparum* lead researches to search and develop new drugs which can be used against these drug resistant protozoans. Similarly, present drugs show miserable side effects, as the use of metronidazole results in fetal malformation, increase chances of preterm birth and users are at potential risk of mutagenicity, and carcinogenicity. Side effects of other drugs which are used to treat these protozoan diseases are also grievous. In this scenario, extracts of the wild plants have proven their significant effects on even drug-resistant strains. Relevant studies show wild plant extract to have the potential to fill that gap. Plants extracts and their metabolites used to make drugs, just like quinine and artemisinin are more effective against multi-drug resistant *Plasmodium falciparum* then any other drugs. Pertinent studies of medicinal wild plants show no side effects in mammalian cells and even show more effective results than recommended drugs, as *Pentalinon* andricuxii extract gave less half-maximal inhibitory concentration as compared to referred drug against *Leishmania* spp. This shows the potential of wild plant extracts, but identification of active agents in some medicinal wild plants and mode of action of that active agents still has to be done. It contains several secondary metabolites which can lead to the development of new drugs.

Abbreviations

TLC	Thin-layer Chromatography
FTR	Fourier-transform infrared spectroscopy
ACT	Artemisinin-based combination therapy
EC_{50}	Effective dose 50
IC_{50}	Half maximal inhibitory concentration
SAR	Structure-activity relationship
UV	Ultraviolet
CC_{50}	50% Cytotoxic Concentration
LD_{50}	Median lethal dose
HSV-2	Herpes Simplex Virus 2
MIC	Minimum Inhibitory Concentration
WHO	World Health Organization
CDC	Centers for Disease Control and Prevention

References

Abe, F., Nagafuji, S., Yamauchi, T., Okabe, H., Maki, J., Higo, H., Akahane, H., Aguilar, A., Jiménez-Estrada, M., Reyes-Chilipa, R. 2002. Trypanocidal constituents in plants 1. Evaluation of some Mexican plants for their Trypanocidal activity and active constituents in Guaco, roots of *Aristolochia taliscana*. Biol. Pharm. Bull., 25: 1188–1191.

Agra, M.F., Silva, K.N., Basílio, I.J.L.D., Freitas, P.F., Barbosa-Filho, J.M. 2008. Survey of medicinal plants used in the region Northeast of Brazil. Rev. Bras. Farmacogn., 18: 472–508.

Altmann, K.H. 2001. Microtubule-stabilizing agents: a growing class of important anticancer drugs. Curr. Opin. Chem. Bio., 15: 424–431.

Andrade-Neto, V.F., Brandão, M.G.L., Oliveira, F.Q., Casali, V.W.D., Njaine, B., Zalis, M.G., Oliveira, L.A., Krettli, A.U. 2004. Antimalarial activity of *Bidens pilosa* L. (Asteraceae) ethanol extracts from wild plants collected in various localities or plants cultivated in humus soil. Wiley InterScience. Res., 18: 634–639.

Argotte-Ramos, R., Ramírez-Avila, G., Rodríguez-Gutiérrez, M.C., Ovilla-Muñoz, M., Lanz-Mendoza, H., Rodríguez, M.H., Gonzalez-Cortazar, M. 2006. Antimalarial 4-phenylcoumarins from the stem bark of *Hintonia latiflora*. J. Nat. Prod., 69: 1442–1444.

Arthan, D., Sithiprom, S., Thima, K., Limmatvatirat, C., ChavalitshewinkoonPetmitr, P., Svasti, J. 2008. Inhibitory effects of Thai plants betaglycosides on *Trichomonas vaginalis*. Parasitol. Res., 103: 443–448.

Ashley, E.A., Dhorda, M., Fairhurst, R.M., Amaratunga, C., Lim, P., Suon, S., Sreng, S. 2014. Tracking Resistance to Artemisinin Collaboration (TRAC). Spread of artemisinin resistance in Plasmodium falciparum malaria. N. Engl. J. Med., 371: 411–423.

Atindehou, K.K., Schmid, C., Brun, R., Kon'e, M.W., Traore, D. 2004. Antitrypanosomal and antiplasmodial activity of medicinal plants from Cote d'Ivoire. J. Ethnopharmacol., 90: 221–227.

Boselli, F., Chiossi, G., Bortolamasi, M., Gallinelli, A. 2005. Prevalence and determinants of genital shedding of herpes simplex virus among women attending Italian colposcopy clinics. Eur. J. Obstet. Gynecol. Reprod. Biol., 118: 86–90.

Braz, M.C., Piccoli, F.A. and Tiana, T. 2016. "Trichomoniasis—are we giving the deserved attention to the most common non-viral sexually transmitted disease worldwide?". Microb. Cell. Fact., 3: 404–418.

Brossi, A., Venugopalan, B., Gerpe, L.D., Yeh, H.J.C., Flippen-Anderson, J.L., Buchs, P., Luo, X.D., Milhous, W. and Peters, W. 1988. Arteether, A new antimalarial drug: Synthesis and antimalarial properties. J. Med. Chem., 31: 645–650.

Carlton, J.M., Hirt, R.P., Silva, J.C., Delcher, A.L., Schatz, M., Zhao, Q., Wortman, J.R., Bidwell, S.L., Alsmark, U.C.M., Besteiro, Sébastien. 2007. Draft genome sequence of the sexually transmitted pathogen *Trichomonas vaginalis*. Science, 315: 207–212.

Cavanagh, H.M.A., Wilkinson, J.M. 2002. Biological activity of lavender essential oil. J. Pharmacognosy Phytother., 16: 301–308.

CDC. 2017. DPDx-Laboratory identification of parasites of public health concern [internet]. Available from: https://www.cdc.gov/dpdx/trichomoniasis/index.html [Reviewed: 2017-12-31].

CDC. 2019a. Parasites-Leishmaniasis [internet]. Available from: https://www.cdc.gov/parasites/leishmaniasis/biology.html [Accessed: 2019-7-17].

CDC. 2019b. Parasites-American trypanosomiasis (also known as Chagas Disaese) [internet]. Available from: https://www.cdc.gov/parasites/chagas/biology.html [Accessed: 2019-03-6].

Chauret, D.C., Bernad, C.B., Arnason, J.T., Durst, T. 1996. Insecticidal neo-lignans from Piper decurrens. J. Nat. Prod., 59: 152–155.

Chavalitshewinkoon-Petmitr, P., Ramdja, M., Kajorndechakiat, S., Ralph, R.K., Denny, W.A., Wilairat, P. 2003. *In vitro* susceptibility of *Trichomonas vaginalis* to AT-specific minor groove binding drugs. J. Antimicrob. Chemoth., 52: 287–289.

Clardy, J., Walsh, C. 2004. Lessons from natural molecules. Nature, 432: 829–837.

Dar, M.A., Bhat, M.F., Hassan, R., Masoodi, M.H., Mubashir, H., Mir, S.R., Mohiuddin, R. 2019. Extensive phytochemistry, comprehensive traditional uses, and critical pharmacological profile of the great mullein: *Verbascum thapsus* L. J. Nat. Prod., 9: 158–171.

Das, B.B., Sen, N., Roy, A., Dasgupta, S.B., Ganguly, A., Mohanta, B.C., Dinda, B., Majumder, H.K. 2006. Differential induction of *Leishmania donovani* bi-subunit topoisomerase I-DNA cleavage complex by selected flavones and camptothecin: activity of flavones against camptothecin-resistant topoisomerase I. Nucleic Acids Res., 34: 1121–1132.

Dey, A., Singh, S. 2006. Transfusion transmitted leishmaniasis: a case report and review of literature. Indian J. Med. Microbiol., 24: 165–170.

Dounga, M., Balansard, G., Babadjamian, A., David, Gasquet, P.T.M. 1983. Contribution a l'e´tude deBidens pilosaL. Identification and activite´ antiparasitaire de la phe´nyl-1-heptatriyn-1,3,5. Plantes Medicinales et Phytothe´rapie, 2: 64–75.

Esperandim, V.R., da Silva Ferreira, D., Sousa Rezende, K.C., Magalhães, L.G., Medeiros Souza, J., Pauletti, P.M., Januario, A.H., Laurentz, R., da Silva, Bastos, J.K., Símaro, G.V., Cunha, W.R., Silva, M.L.A. 2013. *In vitro* antiparasitic activity and chemical composition of the essential oil obtained from the fruits of *Piper cubeba*. Plant Med., 79: 1653–1655.

Fernandez, S.S., Guerra, M.C.R., Mata Cardenas, B.D., Villarreal, J.V., Trevino, L.V. 2005. *In vitro* antiprotozoal activity of the leaves of Artemisia ludoviciana. Phytother Res., 76: 466–468.

Frasson, A.P., Santos, O.D., Duarte, M., Trentin, D.D.S., Giordani, R.B., Silva, A.G.D., Silva, M.V.D., Tasca, T., Macedo, A.J. 2012. First report of anti-*Trichomonas vaginalis* activity of the medicinal plant Polygala decumbens from the Brazilian semi-arid region, Caatinga. Parasitol., 110: 2581–2587.

Germonprez, N., Maes, L., Puyvelde, L.V., Tri, M.V., Tuan, D.A., Kimpe, N.D. 2005. *In vitro* and *in vivo* anti-leishmanial activity of Triterpenoid saponins Isolated from *Maesa balansae* and some chemical derivative. J. Med. Chem., 48: 32–37.

Giordani, R.B., Pagliosa, L.B., Henriques ,A.T., Dutilh, J.E., Zuanazzi, J.A.S. 2008. Investigação do potencial antioxidante e anticolinesterásico de Hippeastrum (Amaryllidaceae). Quim. Nova, 31: 2042–2046.

Giordani, R.B., Vieira Pde, B., Weizenmann, M., Rosemberg, D.B., Souza, A.P., Bonorino, C., De Carli, G.A., Bogo, M.R., Zuanazzi, J.A., Tasca, T. 2011. Lycorine induces cell death in the amitochondriate parasite, *Trichomonas vaginalis*, via an alternative non-apoptotic death pathway. Phytochemistry, 72: 645–650.

Giorgio, C.D., Delmas, F., Akhmedjanova, V., Ollivier, E., Bessonova, I., Riad, E., Timon-David, P. 2005. *In vitro* antileishmanial activity of diphyllin isolated from *Haplophyllum bucharicum*. Planta. Med., 71: 366–369.

Gu, H.M., Warhurst, D.C., Peters, W. 1986. Hemolysis induced by artemisinin and its derivatives *in vitro*. Acta. Pharmacol. Sin., 7: 269–272.

Gutiérrez, Y., Montes, R., Scull, R., Sánchez, A., Cos, P., Monzote, L., Setzer, W.N. 2016. Chemodiversity associated with cytotoxicity and antimicrobial activity of *Piper aduncum* var. *ossanum*. Chem. Biodivers., 13: 1715–1719.

Hamedt, A.L., Ortiz, I.C., García-Huertas, P.A., Sáenz, J., de Araujo, A.C., De Mattos, J.C.P., Rodríguez-Gazquez, M.A., Triana-Chávez, O. 2014. Cytotoxic, mutagenic and genotoxic evaluation of crude extracts and fractions from *Piper jericoense* with trypanocidal action. Acta. Trop., 131: 92–97. DOI: 10.1016/j.actatropica.2013.12.008.

Herendeen, P.S. 1992. The fossil history of Leguminosae from the Eocene of Southeastern North America. 303–316. *In*: Herendeen, P.S., Dilcher (eds.). Advances in Legume Systematics: Part 4, The Fossil Record. The Royal Botanic Gardens, Kew, ISBN 0 947643 40 0.

Hoet, S., Opperdoes, F., Brun, R., Adjakidjé, V., Quetin-Leclercq, J. 2004. *In vitro* antitrypanosomal activity of ethno pharmacologically selected Beninese plants. J. Ethnopharmacol., 91: 37–42.

Hotez, P.J., Dumonteil, E., Cravioto, M.B., Bottazzi, M.E., Tapia-Conyer, R., Meymandi, S., Karunakara, U., Ribeiro, I., Cohen, E.M., Pecoul, B. 2013. An unfolding tragedy of chagas disease in North America. PLoS Neglect. Trop. D., 7(10): e2300. DOI:10.1371/journal.pntd.000 2300.

Iqbal, Z., Lateef, M., Khan, M.N., Jabbar, A., Akhtar, M.S. 2006. Anthelmintic activity of Swertia chirata against gastrointestinal nematodes of sheep. Elsevier, Fitoterapia., 77: 463–465.

Jafarirad, S., Rasoulpour, I. 2019. Pharmaceutical ethnobotany in the Mahabad (West Azerbaijan) biosphere reserve: ethno-pharmaceutical formulations, nutraceutical uses and quantitative aspects. Braz. J. Pharm. Sci., 55: 2175–9790.

Jimenez-Coello, M., Acosta-Viana, K.Y., Guzman-Marin, E., Perez, G.C., Perez, G.M.S. 2010. Anti-trypanosomal activity of (8-hydroxymethylen)-trieicosanyl acetate against infective forms of *Trypanosoma cruzi*. Pharm. Biol., 48: 666–671.

Jimenez-Coello, M., Guzman-Marin, E., Perez-Gutierrez, S., Polanco-Hernandez, G.M., Acosta-Viana, K.Y. 2011. Antitrypanosomal activity of Senna villosa in infected BALB/c mice with *Trypanosoma cruzi* during the subacute phase of infection. AJTCAM, 8: 164–169.

Kamchonwongpaisan, S., Chandra-ngam, G., Avery, M.A., Yuthavong, Y. 1994. Resistance to artemisinin of malaria parasites (*Plasmodium falciparum*) infecting a-thalassemic erythrocytes *in vitro*. Competition in drug accumulation with uninfected erythrocytes. J. Clin. Invest., 93: 467–473.

Klayman, D.L. 1985. Qinghaosu (artemisinin): an antimalarial drug from China. Science, 228: 1049–1055.

Klebanoff, M., Carey, J., Hauth, J., Hillier, S.L., Nugent, R., Thom, E., Ernest, J., Heine, R., Wapner, R., Trout, W., Moawad, A., Leveno, K., Miodovnik, M., Sibai, B., Van Dorsten, J., Dombrowski, M., O'Sullivan, M., Varner, M., Langer, O., McNellis, D., Roberts, J. 2001. Failure of metronidazole to prevent preterm delivery among pregnant women with asymptomatic *Trichomonas vaginalis* infection. N. Engl. J. Med., 345: 487–493.

Köhler, I., Jenett-Siems, K., Mockenhaupt, F.P., Siems, K., Jakupovic, J., González, J.C., Hernández, M.A., Ibarra, R.A., Berendsohn, W.G., Bienzle, U., Eich, E. 2001. *In vitro* antiplasmodial activity of 4-phenylcoumarins from Exostema mexica-num. Planta Medica, 67: 89–91.

Komane, B.M., Olivier, E.I., Viljoen, A.M. 2015. *Trichilia emetica* (Meliaceae)—A review of traditional uses, biological activities and phytochemistry. Phytochem., 4: 1–9.

Krashin, J.W., Koumans, E.H., Bradshaw-Sydnor, A.C., Braxton, J.R., Evan Secor, W., Sawyer, M.K., Markowitz, L.E. 2010. *Trichomonas vaginalis* prevalence, incidence, risk factors and antibiotic- resistance in an adolescent population. J. Sex. Transm. Dis., 37: 440–444.

Lai, H., Singh, N.P. 1995. Selective cancer cell cytotoxicity from exposure to dihydroartemisinin and holotransferrin. Cancer Lett., 91: 41–46.

Leonard, S., Capote, R., Germonprez, N., Puyvelde, L.V., Kimpe, N.D., Vermeersch, H., Rosier, J., Maes, L., Roets, E., Hoogmartens, J. 2003. Liquid chromatographic methods for analysis of saponins in *Maesa balansae* extract active against leishmaniasis. J. Chromatogr. A, 1012: 39–46.

Lezama-Davila, C.M., Isaac-Márquez, A.P., Zamora-Crescencio, P., Uc-Encalada Mdel, R., Justiniano-Apolinar, S.Y., del Angel-Robles, L., Satoskar, A., Hernandez-Rivero, L. 2007. Leishmanicidal activity of Pentalinon andrieuxii. Fitoterapia, 78: 255–257.

Long, C.A., Zavala, F., 2016. Malaria vaccines and human immune responses. Can. J. Microbiol., 32: 96–102.

Luize, P.S., Ueda-Nakamura, T., Dias Filho, B.P., Cortez, D.A.G., Nakamura, C.V. 2006. Activity of neolignans isolated from *Piper regnellii* (MIQ.) C.DC. var. pallescens (C.DC.) YUNCK against *Trypanosoma cruzi*, Biol. Pharm. Bull., 10: 2126–2130.

Mabberley, D.J. 2008. Mabberley's Plant Book (3 ed.). Cambridge University Press. ISBN 978-0-521-82071-4.

Medda, S., Mukhopadhyay, S., Basu, M.K. 1999. Evaluation of the *in vivo* activity and toxicity of amarogentin, an antileishmanial agent, in both liposomal and niosomal forms. J. Antimicrob. Chemother., 44: 791–794.

Mittra, B., Saha, A., Chowdhury, A.R., Pal, C., Mandal, S., Mukhopadhyay, S., Bandyopadhyay, S., Majumder, H.K. 2000. Luteolin, an abundant dietary component is a potent anti-leishmanial agent that acts by inducing topoisomerase II-mediated kinetoplast DNA cleavage leading to apoptosis. Mol. Med., 6: 527–541.

Molina-Cruz, A., Canepa, G.E., Kamath, N., Pavlovic, N.V., Mu, J., Ramphul, U.N., Ramirez, J.L., Barillas-Mury, C. 2015. Plasmodium evasion of mosquito immunity and global malaria transmission: the lock-and-key theory. PNAS, 112: 15178–15183.

Molina-Garza, Z.J., Bazaldúa-Rodríguez, A.F., Quintanilla-Licea, R., Galaviz-Silva, L. 2014. Anti-*Trypanosoma cruzi* activity of 10 medicinal plants used in northeast Mexico. Acta. Tropica., 136: 14–18.

Moon, T., Wilkinson, J.M., Cavanagh, H.M.A. 2006. Antiparasitic activity of two Lavandula essential oils against *Giardia duodenalis*, *Trichomonas vaginalis* and *Hexamita inflate*. Parasitol. Res., 99: 722–728.

Murray, H.W., Berman, J.D., Davies, C.R., Saravia, N.G. 2005. Advances in leishmaniasis. Lancet., 366: 1561–1577.

Negi, A.S., Gupta, A., Hamid, A.A. 2014. Combating malaria with plant molecules: a brief update. Curr. Med. Chem., 21: 458–500.

Neva, F.A., Brown, H. 1994. Basic Clinical Parasitology, 6th edition, Appleton & Lange, Norwalk, CT.

Ngoubangoye, B., Boundenga, L., Arnathau, C., Mombo, I.M., Durand, P., Tsoumbou, T.A., Otoro, B.V., Sana, R., Okouga, A.P., Moukodoum, N., Willaume, E., Herbert, A., Fouchet, D., Rougeron, V., Tidiane, Ba C., Ollomo, B., Paupy, C., Leroy, E.M., Renaud, F., Pontier, D., Prugnolle, F. 2016. The host specificity of ape malaria parasites can be broken in confined environments. Int. J. Parasitol., 46: 737–744.

Noster, S., Kraus, L.J. 1990. *In vitro* antimalarial activity of *Coutarea latiflora* and *Exostema caribaeum* extracts on *Plasmodium falciparum*. Planta Medica, 56: 63–65.

Pan, L., Lezama-Davila, C.M., Isaac-Marquez, A.P., Calomeni, E.P., Fuchs, J.R., Satoskar, A.R., Kinghorn, A.D. 2012. Sterols with antileishmanial activity isolated from the roots of *Pentalinon andrieuxii*. Phytochem., 82: 128–135.

Pelizzaro-Rocha, K.J., Veiga-Santos, P., Lazarin-Bidoia, D., Ueda-Nakamura, T., Filho, B.P.D., Ximenes, V.F., Silva, S.O., Nakamura, C.V. 2011. Trypanocidal action of eupomatenoid-5 is related to mitochondrion dysfunction and oxidative damage in *Trypanosoma cruzi*. Microbes Infect., (12-13): 1018–24. doi: 10.1016/j.micinf.2011.05.011.

Pérez-Treviño, K.C., Molina-Garza, Z.J., Galaviz-Silva, L. 2016. Evaluation of antitrypanosomal activity of methanol extract from medicinal plants. Entomología Mexicana, 3: 656–659.

Peter, M.G., Yenesew, A., Twinomuhwezi, H., Kabaru, J.M., Akala, H.M., Kiremire, B.T., Heydenreich, M., Eyase, F., Waters, N.C., Walsh, D.S. 2009. Antiplasmodial and larvicidal flavonoids from *Derris trifoliata*. B. Chem. Soc. Ethiopia, 23: 409–14.

Phrompittayarat, W., Putalun, W., Tanaka, H., Jetiyanon, K., Wittaya-areekul, S., Ingkaninan, K. 2007. Comparison of various extraction methods of *Bacopa monnier.* Naresuan Univ. J., 15: 29–34.

Polonio, T., Efferth, T. 2008. Leishmaniasis: Drug resistance and natural products (review). Int. J. Mol. Med., 22: 277–286.

Porcal, W., Hernández, P., Boiani, L., Boiani, M., Ferreira, A., Chidichimo, A., Cazzulo, J.J., Olea-Azar, C., Gonzalez, M., Cerecetto, H. 2008. New trypanocidal hybrid compounds from the association of hydrazone moieties and benzofuroxan heterocycle. Bioorg. Med. Chem., 16: 6995–7004.

Pradheepkumar, C.P., Shanmugam, G. 1999. Anticancer potential of cleistanthin A isolated from the tropical plant *Cleistanthus collinus*. Oncol. Res., 11: 225–32.

Prakash, V., Rana, S., Sagar, A. 2016. Studies on antibacterial activity of *Verbascum thapsus*. J. Med. Plants, 4: 101–10.

Press, N., Chavez, V.M., Ticona, E., Calderon, M., Apolinario, I.S., Culotta, A., Arevalo, J., Gilman, R.H. 2001. Screening for sexually transmitted diseases in human immunodeficiency virus-positive patients in Peru reveals an absence of Chlamydia trachomatis and identifies *Trichomonas vaginalis* in pharyngeal specimens. Clin. Infect. Dis., 32: 808–81.

Ramirez-Moreno, E., Soto-Sanchez, J., Rivera, G., Marchat, L.A. 2017. Mexican medicinal plants as an alternative for the development of new compounds against protozoan parasites. Intechopen http://dx.doi.org/10.5772/67259.

Ray, S., Majumder, H.K., Chakravarty, A.K., Mukhopadhyay, S., Gil, R.R., Cordell, G.A. 1996. Amarogentin, a naturally occurring secoiridoid glycoside and a newly recognized inhibitor of topoisomerase I from *Leishmania donovani*. J. Nat. Prod., 59: 27–29.

Ridley, R.G. 2002. Medical need, scientific opportunity and the drive for antimalarial drugs. Nature, 415: 686–693.

Rivera, N., López, P.Y., Rojas, M., Fortoul, T.I., Reynada, D.Y., Reyes, A.J., Rivera, E., Beltran, H.I., Malagon, F. 2014. Anti-malarial efficacy, cytotoxicity, and genotoxicity of methanolic stem bark extract from *Hintonia latiflora* in a *Plasmodium yoelii yoelii* lethal murine malaria model. Parasitol. Res., 113: 1529–1536.

Salud, S.D. SUIVE/DGE/SALUD/Información Epidemiológica de Morbilidad, Anuario 2011 (SUIVE/DGE/SALUD/ Epidemiológic Information on Morbility, Yearbook 2011). Versión Ejecutiva [Internet]. 2012. Available from: http://www.epidemiologia.salud.gob.mx/doctos/infoepid/publicaciones/2012/ver_ejecutiva_2011.pdf [Accessed: 2016-09].

Sasidharan, S., Chen, Y., Saravanan, D., Sundram, K.M., Latha, L.Y. 2011. Extraction, isolation and characterization of bioactive compounds from plants' extracts. Afr. J. Tradit. Complement Altern. Med., 8: 1–26.

Schmidt, T.J., Khalid, S.A., Romanha, A.J., Alves, T.M.A., Biavatti, M.W., Brun, R., Da Costa, F.B., de Castro, S.L., Ferreira, V.F., de Lacerda, M.V.G. 2012. The potential of secondary metabolites from plants as drugs or leads against protozoan neglected diseases—part I. Curr. Med., 19: 2128–2175.

Schwebke, J.R., Barrientes, F.J. 2006. Prevalence of *Trichomonas vaginalis* isolates with resistance to metronidazole and tinidazole. Antimicrob. Agents Chemother., 50: 4209–4210.

Sibley, D.L., Norma, A. 2000. Cell invasion by un-palatable parasites. Traffic, 1: 100–6.

Sorvillo, F., Kerndt, P. 1998. *Trichomonas vaginalis* and amplification of HIV-1 transmission. Lancet., 351: 213–214.

Standley, P.C. 1926. Trees and Shrubs of Mexico. Contributions from the United States National Herbarium Vol. 23.

Svasti, J., Srisomsap, C., Techasakul, S., Surarit, R. 1999. Dalcochinin-8′O-β-D-glucoside and its β-glucosidase enzyme from *Dalbergia cochinchinensis*. Phytochemistry, 50: 739–743.

Teixeira, S.M., de Paiva, R.M., Kangussu-Marcolino, M.M., Darocha, W.D. 2012. Trypanosomatid comparative genomics: contributions to the study of parasite biology and different parasitic diseases. Genet. Mol. Biol., 35: 1–17.

Trigg, P.I. 1989. Qinghaosu (artemisinin) as an antimalarial drug. Econ. Med. Plant Res., 3: 21–55

Unver, N. 2007. New skeletons and new concepts in Amaryllidaceae alkaloids. Phytochem. Res., 6: 125–135.

Vilar-Pereira, G., Ruivo, L.A. de, S., Lannes-Vieira, J. 2015. Behavioural alterations are independent of sickness behaviour in chronic experimental Chagas disease. Memórias do Instituto Oswaldo Cruz, 110: 1042–1050.
Wagner, H., Bladt, S. 1996. Plant drug analysis. *In*: A Thin Layer Chromatography Atlas. 2nd edn. Springer Verlag: Berlin, 384.
Wat, C.K., Johns, T., Towers, G.H.N. 1980. Phototoxic and antibiotic activities of plants of Asteraceae used in folk medicine. J. Ethnopharmacol., 2: 279–290.
Weirnet, T.A., Hartwell, L.H. 1990. Characterization of RAD9 of *Saccharomyces cerevisiae* and evidence that its function acts post translationary in cell cycle arrest after DNA damage. Mol. Cell. Biol., 10: 6554–6564.
Wendel, K.A., Workowski, K.A. 2007. Trichomoniasis: challenges to appropriate management. Clin. Infect. Dis., 3: S123–129.
WHO. 2011. World Malaria Report 2011 [Internet]. Available from: http://www.who.int/ iris/handle/10665/44792 [Accessed: 2016-09-02].
WHO. 2012. Global incidence and prevalence of selected curable sexually transmitted infections—2008 [Internet]. Available from: http://www.who.int/reproductivehealth/ publications/rtis/stisestimates/en/[Accessed: 2016-10-10].
WHO. 2016. Leishmaniasis [Internet]. Available from: http://www.who.int/leishmania- sis/en/[Accessed: 2016-09-11].
WHO. 2019a. Leishmaniasis [Internet]. Available from: https://www.who.int/news-room/fact-sheets/detail/leishmaniasis [Accessed: 2019-03-14].
WHO. 1955–2019. Certification of malaria elimination [Internet]. Available from: http://www.int/malaria/areas/elimination/ malaria-free-countries/en/ [Accessed: 2019-05-22].
WHO. 2019b. Malaria [Internet]. Available from: https://www.who.int/news-room/fact-sheets/detail/malaria [Accessed: 27-3-2019].
Yaeger, R.G. 1996. Protozoa: Structure, Classification, Growth and Development. Medical Microbiology, 4th edition. ISBN-10: 0-9631172-1-1.
Yenesew, A., Mushibe, E.K., Induli, M., Derese, S., Midiwo, J.O., Kabaru, J.M., Heydenreich, M., Koch, A., Peter, M.G. 2005. 7a-O-methyldeguelol, a modified rotenoid with an open ring-C, from the roots of *Derris trifoliata*. Phytochemistry, 66: 653–657.
Yenesew, A., Kiplagat, J.T., Derese, S., Midiwo, J.O., Kabaru, J.M., Heydenreich, M., Peter, M.G. 2006. Rotenoid derivatives form Kenyan Millettia and Derris species as larvicidal agents. Phytochemistry, 67: 161–168.
Zohreh, F.K., Mohsen, A., Mahdi, D., Hossein, H., Mohsen, T., Zeynab, J. 2015. Effect of *Verbascum thapsus* ethanol extract on induction of apoptosis in *Trichomonas vaginalis in vitro*. J. Infect., 15: 125–130.
Zygmunt, J.B., Namiesnik, J. 2003. Preparation of samples of plant material for chromatographic analysis. J. Chromatogr. Sci., 41: 109–116.

Chapter 8

Anti-Inflammatory Activity of Wild Medicinal Plants of Piauí State-Brazil

Valdiléia Teixeira Uchôa,[1,*] *Mahendra Rai,*[2] *Gilmânia Francisca Sousa Carvalho,*[1] *Herbert Gonzaga Sousa,*[1] *Patrícia e Silva Alves,*[1] *Renata da Silva Carneiro,*[1] *Ariane Maria da Silva Santos Nascimento,*[3] *Felipe Pereira da Silva Santos*[3] and *Gabriel e Silva Sales*[3]

Introduction

The use of medicinal plants as an alternative or complementary treatment to cure diseases is common in the Brazilian population, especially in rural communities (Matos 2007, Roque et al. 2010). Its use is based on the rich knowledge of the population, passed from generation to generation and allied to synthetic drugs substitution, which can be justified, for example, by the side effects that the use of these drugs entails (Jeengar et al. 2014, Adebayo et al. 2015). The pharmacological activities of plants are related to the presence of bioactive compounds produced in the secondary metabolism of these plant species (Matos 2009, Souza et al. 2014). In this context, the search for safe natural agents has become the key to the development of drugs derived or synthesized from plant materials with effective action in disease treatment (Oliveira et al. 2010).

The pathophysiology of various diseases involves the inflammation process as the body's direct response to damage and disturbance caused by external agents. This process includes the production and recruitment of chemical mediators such as cells, cytokines, histamines, leukotrienes, and prostaglandins that act as an inflammatory response to fight infection (Lawal et al. 2019). Inflammation, although initially necessary, when not controlled, induces prolongation of the patient's diseased condition, impairing their life quality (Dobrian et al. 2011). Currently, the reference drugs used are represented by non-steroidal anti-inflammatory drugs (NSAIDs), which by inhibiting pro-inflammatory enzymes, end up blocking the biosynthesis of gastric cytoprotection, triggering adverse effects (Rang et al. 2007).

In literature, several medicinal plants have shown considerable anti-inflammatory and immunomodulatory action through lymphocyte activation and apoptosis spread, acting similarly to synthetic drugs (Tasneen et al. 2018). Its mechanisms of action involve the modulation of cytokines inherent to the inflammatory process, inhibition of enzymes such as cyclooxygenase and lipoxygenase in their various isoforms and decreased nitric oxide production (Coutinho et al. 2009). The anti-inflammatory capacity of these plants is mainly associated with the presence of phenolic compounds, such as flavonoids, which through antioxidant activity, control the levels of reactive species and facilitate the healing process (Kurahashi and Fujii 2015, Bernatoniene and Kopustinskiene 2018).

[1] PPGQ-GERATEC, Department of Chemistry-State University of Piauí, Teresina, Piauí, Brazil.
[2] Nanobiotechnology Lab., Department of Biotechnology, SGB Amravati University, Amravati-444602, Maharashtra.
[3] PPGQ, Department of Chemistry-Federal University of Piauí, Teresina, Piauí, Brazil.
* Corresponding author: vtuquimica@yahoo.com.br

Brazil is considered as a megadiverse country and has one of the richest flora on the planet, accounting for approximately 20% of the world's biodiversity (Garcez et al. 2016, Macêdo et al. 2018). Brazilian vegetation is distributed in the most different biomes. Piauí, the third-largest state in northeastern Brazil with 251,529 km^2 of territorial area, has its vegetation belonging to Caatinga and Cerrado (Santos-Filho et al. 2015). In this state, some plant species belonging to the Fabaceae (Santos et al. 2013), Combretaceae (Horinouchi et al. 2013), Sapotaceae (Leite et al. 2015), Xanthorrhoeaceae (Fehrmann-Cartes et al. 2019), Solanaceae (Rios et al. 2017a) and Oleaceae families (Carvalho et al. 2020) have demonstrated prominent anti-inflammatory activity. In this chapter, we have discussed the use of 22 wild medicinal plants found in abundance in the state of Piauí-Brazil used in place of usual remedies or the concomitant use of medications, taking into consideration popular reports and scientific investigations, as shown in Table 8.1.

Table 8.1. Plants with anti-inflammatory action found in abundance in the Piauí state-Brazil.

Family	Scientific name	Local name	Parts used	Use in folk medicine	References
Adoxaceae	*Sambucus nigra* Linn.	Elderberry	Flower, bark, leaf and fruit	Inflammations, skin disorders, diuretics, colds, fever and other respiratory disorders	Blumenthal et al. 2000, Weiss and Fintelmann 2000, EMA 2008, Basas-Jaumandreu and De Las Heras 2019
Anacardiaceae	*Anacardium occidentale* L.	Cajueiro	Fruit and bark	Astringent, antiseptic, vermifuge, tonic in weaknesses; used for gargling, throat irritation and thrush; it enjoys hypoglycemic properties	Silva et al. 2009, Grandi 2015, Alencar et al. 2018, Tufail et al. 2019
Anacardiaceae	*Schinopsis brasiliensis* Engl	Braúna-do-sertão	Leaf, bark and fruits	Inflammation, pain, flu, diarrhea, impotence, animal worms, natural antiseptic, treatment of wounds and superficial mycoses	Júnior et al. 2011, Saraiva et al. 2013, Santos et al. 2014, Donati et al. 2015
Anacardiaceae	*Spondias tuberosa* A.	Umbu	Bark and fruits	Diabetes mellitus, infections, digestive disorders, diarrhea, menstrual disorders and placental delivery disorders	Albuquerque et al. 2007, Zielinsk et al. 2014, Siqueira et al. 2016b, Martins et al. 2019
Xanthorrhoeaceae	*Aloe vera* L.	Babosa	Leaf	Laxative, anti-inflammatory effect, skin related problems and for wound healing purposes; used externally in sprains, bruises and rheumatic pain	Yagi 1997, Hussain 1998, Abílio 2011, Grandi 2015, Majumder et al. 2019

Table 8.1 Contd. ...

...Table 8.1 Contd.

Family	Scientific name	Local name	Parts used	Use in folk medicine	References
Bignoniaceae	*Handroanthus impetiginosus* (Mart. Ex DC) Mattos	Ipê-roxo, pau-d'arco-roxo, Pau d'arco and ipê-roxo-da-mata	Bark and leaf	Diabetes, inflammation, ulcer, malaria, cancer, kidney and hepatic disorders, syphilis, stomach infections and has anti-inflammatory properties and antimicrobial	Moraes et al. 2007, Jausoro et al. 2010, Lima et al. 2014, Fonseca Filho et al. 2017
Urticaceae	*Cecropia pachystachya* Trécul.	Embaúba	Leaf	Used for respiratory diseases, hyperglycemia and bronchitis	Soraru and Bandoni 1978, Lorenzi and Matos 2002b, Aragão et al. 2010
Combretaceae	*Combretum leprosum* Mart.	Mofumbo or mufumbo	Leaf, flower, roots and fruits	Pertussis, uterine bleeding, flu, hemorrhages, inflammations, prevention of rashes, wound cleaning, antidiarreic, expectorant and anti-teússico	Loiola et al. 2009, Soares Neto et al. 2014, Viau et al. 2014, Cavalcanti et al. 2019
Leguminoseae	*Caesalpinia ferrea* Mart.	Ironwood or Juca	Bark, fruits and roots	Diabetes; Fruits, pods and roots have been used as infusions to treat various inflammatory disorders such as bruising, rheumatism, scarring, pulmonary hemoptysis, fever, enterocolitis	Braga 1976, Pio Correa 1984, Balbach 1988, Ueda et al. 2001, Vasconcelos et al. 2011
Leguminoseae	*Caesalpinia pyramidalis* Tul.	Catingueira	Bark and leaf	Anti-inflammatory, diarrhea, stomach pain and dysentery	Bahia et al. 2005, Agra et al. 2007, 2008, Santos et al. 2011
Leguminoseae	*Erythrina velutina* Willd	Mulungu, mulungu-da-flor-vermelha ou bucaré	Bark, leaf and flowers	Treatment as sedative and suppressor of cough, asthma, bronchitis treatment, muscle aches, anticoagulant, nervous crisis, for anxiety, insomnia, seizures, menopause, gingivitis, hepatitis, respiratory diseases, general anti-inflammatory, hemorrhoids	Galvão and Medeiros 2002, Dantas et al. 2004, Carvalho et al. 2008, Ximenes et al. 2018

Table 8.1 Contd. ...

...*Table 8.1 Contd.*

Family	Scientific name	Local name	Parts used	Use in folk medicine	References
Leguminoseae	*Tamarindus indica* L.	Tamarind	Bark and fruit	Inflammation, stomachache, throat and rheumatism, diarrhea, dysentery, fever, malaria, respiratory disease, constipation, gonorrhea	Dighe et al. 2009, Kuru 2014, Rai et al. 2018, Komakech et al. 2019
Leguminosae	*Bauhinia forficata* Link.	Cowpaw, cow's hoof, ox's paw and mororó	Leaf and bark	Pain, inflammation or microbial infections and antidiabetic	Silva et al. 2002, 2012, Marques et al. 2013, Farag et al. 2015
Leguminoseae	*Anadenanthera colubrina* (Vell.) Brenam	Angico manso	Bark, fruit and leaf	Uterine hemorrhages; respiratory conditions such as cough, bronchitis, asthma and pharyngitis; antioxidant capacity; diseases associated with inflammation and pain	Corrêa 1984, Monteiro et al. 2012, Rodrigues et al. 2014, Xavier 2018
Leguminoseae	*Mimosa tenuiflora* Willd.	Black jurema	Bark	Antimicrobial, antifungal, antiviral, antidiarrheal, antiseptic and effective in combating gastrointestinal diseases such as gastritis, ulcers, poor digestion and reflux	Albuquerque 1999, Bezerra 2011, Maia-Silva et al. 2012
Anacardiaceae	*Myracrodruom urundeuva* Allemão	Aroeira	Fruits and bark	Fevers, urinary tract problems, cystitis, urethritis, cough, bronchitis, menstrual problems with excessive bleeding, inflammation in general, antiseptic for open fractures and wounds	Lorenzi 2002, Monteiro et al. 2006, Brandão 2014
Myrtaceae	*Eucalyptus glubulus* Labill	Eucalyptus, blue bud, fever tree, white mahogany	Leaf	Respiratory diseases such as asthma, bronchitis or even a simple flu or cold	Andrade 1961, Mora and Garcia 2000
Olacaceae	*Ximenia americana* L.	Bush plum, ababa, plum, Bahia plum, umbu-bravo	Fruit, bark and leaf	Anti-inflammatory, antimicrobial, antifungal, antineoplastic, antitrypsal, antirheumatic, antioxidant, analgesic, molluscicidal, pesticidal activities	Uchôa et al. 2016, Shettar et al. 2017, De Menezes et al. 2019, Carvalho et al. 2020

Table 8.1 Contd. ...

...Table 8.1 Contd.

Family	Scientific name	Local name	Parts used	Use in folk medicine	References
Rhamnaceae	*Ziziphus joazeiro* Mart.	Juazeiro, juá, juá-thorn, juá-fruit, cowboy orange, juá-mirim	Bark and leaf	Gingivitis and pain caused by tooth extraction, gastric problems and flu and cough syrups	Matos 2000, Lorenzi and Matos 2002a, Romão et al. 2010
Rubiaceae	*Morinda citrofolia* Linn.	Noni, like iada, ninth, canary	Fruit, bark, leaf, flower, roots	Antiviral, antibacterial, antifungal, antitumor, analgesic, hypotensive and anti-inflammatory	Carniello et al. 2010, Correia et al. 2011, Barbosa et al. 2017
Sapotaceae	*Sideroxylon obtusifolium* (Roem. & Schult.)	Quixabeira, quixaba or rompegibão	Bark and leaf	Treatment of gastritis, duodenal ulcer, genital damage, chronic inflammation, colic, kidney problems, cough treatment, digestive and heart disease, ovarian inflammation, diabetes, gastritis	Figueiredo et al. 2015, Leite et al. 2015, Aquino et al. 2016, Silva et al. 2017
Solanaceae	*Solanum paniculatum* L.	Jurubeba, juribebe, juripeba	Roots, bark, fruit and leaf	Carminative, diuretic, herbal, antipyretic, analgesic, anti-inflammatory, stomach and liver dysfunction	Vieira et al. 2010, Clementino-Neto et al. 2016, Rios et al. 2017a, Valerino-Díaz et al. 2018, De Souza et al. 2019

Angico Manso (*Anadenanthera colubrina* (Vell.) Brenam.) (Figure 8.1)

Angico is a tree present in the Brazilian phytogeographic domains of the Atlantic Forest, Caatinga, Cerrado and Amazon. It is a straight or tortuous tree from 13 to 20 m high, and belongs to the subfamily Mimosoideae of the family Leguminoseae. It is also known as yellow-angico, cedar-angico, tannery, angico-dos-montas, dirty-angico, and bathed-angico. In addition, it is also found in Argentina, Bolivia, Paraguay, and Peru (Pacheco et al. 2017, Oliveira et al. 2019).

In the Northeast, it occurs in the sandy, calcareous and alluvial soils (Oliveira et al. 2019). This species has high natural durability and medium mechanical strength. Its wood is used in construction, commonly used as a source of energy and in folk medicine (Vieira et al. 2019).

The bark of angico is bitter, purifying, hemostatic and astringent tonic, and has tannins, resinous and mucilaginous materials. Its bark is indicated for rickets, in uterine hemorrhages, being used in dysmenorrhea and bleeding in general. The fruit contains saponins, organic acids, anthocyanins, flavones, and alkaloids. A gummy exudate that is used in respiratory problems mainly consists of galactose and glucuronic acid. It is also used in respiratory problems such as cough, bronchitis, asthma and pharyngitis, as well as providing a gum replacement for Arabica, which is a polymer widely used in pharmaceutical industries (Corrêa 1984, Rodrigues et al. 2014).

The phytochemical analysis carried out with the leaves of this plant showed antioxidant activity through the iron reduction method and these secondary metabolites are very important because they are responsible for several biological activities. Regarding the antioxidant activity shown by angico extracts,

Figure 8.1. *Anadenanthera colubrina* (Vell.) Brenam (Angico manso) tree (A), leaves (B), flowers (C) and fruits (D). Adapted from http://qplantaeessa.blogspot.com/2015/04/farinha-seca-camisa-fina-ou-angico.html#.XkWfTGhKhPY and https://www.arvores.brasil.nom.br/new/angicobranco/index.htm.

it can be suggested that it may play an important role in the anti-inflammatory action. Ethanolic extracts obtained from angico leaves, branches and bark and their fractions were used to evaluate antimicrobial activity (Rodrigues et al. 2014).

Pharmacological studies of the plant *Anadenanthera colubrina* Brenam has shown that the ethanolic extract of its bark has an important anti-inflammatory effect, demonstrated by its ability to inhibit neutrophil migration in a dose-dependent manner and decreased vascular permeability. These results provide experimental evidence for the efficacy of traditional use of *A. macrocarpa* in treating various diseases associated with inflammation and pain (Rodrigues et al. 2014, Xavier et al. 2018).

Regarding the use of this plant by ancient people, it is believed that when the leaves wither on the plant, they become toxic to cattle due to the formation of hydrocyanic acid, but when deliberately dried, they constitute good fodder. The wood is of high durability and high rot resistance. The plant is suitable for rural, naval and civil construction. It is used for the manufacture of sleepers, furniture, tables, rafters, slats, frames, floor joists, ceilings, jambs, posts, beams, crosspieces, supports, floors, fences, stakes, poles, wagons, furniture, and wheels, etc. It produces good quality firewood and charcoal. The peel is used in home medicine, infusion, syrup, maceration, and tincture (Monteiro et al. 2012).

Aroeira (*Myracrodruom urundeuva* Allemão) (Figure 8.2)

Shrub 60 cm to 2–3 m high or tree 7–8 m, with young branches almost always pubescent and old glabrous in most cases. Leaves composed of 2–7 pairs of leaflets, membranous, oblong-elliptical or oval. Red globular drupe fruit with the reniform seed (Grandi 2015).

Naturally sourced products have long been used worldwide by alternative medicine. Data from the World Health Organization (WHO) show that around 80% of the world's population has used weed to relieve some painful or unpleasant symptoms and it is estimated that local communities use about 10% of the herbs. Brazil is endowed with immense biodiversity and traditional knowledge which brings plant utilization closer to available therapeutic resources. This practice is recommended by the government (Abílio 2011).

The active principle is understood as the compound or mixture of compounds synthesized by plants through their nutrients, water and light received, which can cause positive or negative reactions in living organisms (Abílio 2011). Therefore, the aroeira consists of essential oil: rich in mono and sesquiterpenes, with a content of 1% for leaves and 5% for fruits. It is also composed of tannins, resins, alkaloids, flavonoids, steroidal saponins, steroids and triterpenes. For seeds, the fixed oil content of around 14% is quoted (Lorenzi 2002).

Fruit ingestion can cause vomiting and diarrhea in children. Many people use the mastic against fevers, urinary tract problems, cystitis, urethritis, cough, bronchitis, menstrual problems with excessive

Figure 8.2. *Myracrodruom urundeuva* Allemão (Aroeira) tree (A), tree and fruits (B), flowers (C) and fruits (D). Adapted from https://www.greenme.com.br/ and http://www.abelhasjatai.com.br/plantas-meliferas/a-aroeira-pimenteira-schinus-terebinthifolius-e-seu-potencial-como-planta-melifera/.

bleeding, flu and inflammation in general. Its resin is indicated for the treatment of rheumatism, besides serving as purgative and to fight respiratory diseases. Its resin oil is used externally for healing and toothache. The light yellow resin (which hardens in the air turning bluish and then brown), coming from the lesions of the barks, is a drug of wide application among the sertanejos, as a tonic, when they use barks. In other times, the aroeira was used by the Jesuits, who with their resin, prepared the "Balm of the Missions", famous in Brazil and abroad (Lorenzi 2002, Brandão 2014).

The entire plant is used externally as an antiseptic for open fractures and wounds. Essential oil is primarily responsible for various activities of this plant, especially antimicrobial action against various types of bacteria and fungi and plant viruses, as well as repellent activity against house fly. This oil, rich in monoterpenes, is also indicated for respiratory disorders and is effective for mycoses, candidiasis (local use) and some cancers (carcinoma, sarcoma, etc.). It has tissue regenerating action and is useful in burns and skin problems (Lorenzi 2002).

The aroeira is highly exploited as a source of wood, fuel and medicine in the semiarid region of northeastern Brazil. Folk medicine indicates the use of mastic as an external treatment in "sitting" baths against uterine inflammation and is used in the mouthwash to prevent mouth and throat infections (Monteiro et al. 2006, 2012).

Babosa (*Aloe vera* L.) (Figure 8.3)

The popular name of this plant is *Aloe vera*, and is also known as Medicinal Aloe, Aloe, Cape Aloe, or Azerbaijani. *Aloe vera* is a juicy and perennial xerophyte belonging to the genus *Aloe*. Xerophyte plants have the versatility and adaptability to survive in hot dry areas, and particularly, the succulence property of these plants is one of the important xerophytic adaptations. In hot-dry areas, they are forced to collect and store water for their long-term survival. This type of adaptation makes it a rich source of various phytochemicals (Majumder et al. 2019).

Aloe vera L. is a native plant from sub-Saharan Africa, the Saudi Arabian Peninsula and various islands of the Indian Ocean. The term "aloe" originates from the Arabic word alloeh, which implies "bitter and brilliant substance" and "vera" means truth in Latin (Majumder et al. 2019). It grows on dry, sandy soil and does not require much water. It grows best in dry weather, but sometimes also grows in moist. In summer, it should be watered because the soil is almost dry and leaves may dry (Abílio 2011).

It has a long history of medicinal use and provides health benefits. Ancient cultures such as Indian and Chinese registered it as a medicinal plant. Greeks, Romans, and Babylonians use *A. vera* leaves as a skin ointment. Traditionally, Aloe latex is used as a laxative to treat constipation and *Aloe vera* L.

Figure 8.3. *Aloe vera* L. (Babosa) plant and flowers (A) and leaves (B and C). Adapted from https://www.greenme.com.br/.

gel is used as anti-inflammatory, for skin related problems and wound healing purposes (Grandi 2015, Majumder et al. 2019).

In modern times, aloe vera is used in pharmaceuticals, functional foods and cosmetics. The anti-cancer properties of *Aloe vera* L. are claimed to be mainly due to two separate mechanisms: antiproliferative and immune-stimulating activities. Antiproliferative activity is exhibited due to the presence of aloe latex, a bitter flavor component, while the immune-stimulating effect is due to the polysaccharides (Grandi 2015, Chacó et al. 2019).

Aloe latex is produced in the thin middle layer of *A. vera* leaf, consisting of chlorenchyma and vascular bundles. It contains 75 bioactive potential phytochemicals. The main phytoconstituents are anthraquinones, naphthalenones, polysaccharides, proteins, enzymes and organic acids (Abílio 2011).

Active ingredients include Anthraquinonic glycosides (especially alloin), and mucilage consisting of a complex polysaccharide, aloeferon. It has vitamins B, C and E. Its internal use is not recommended because it has a series of anthraquinone compounds that have high toxicity when ingested in high doses. Thus, teas, syrups and other prepared medicines can cause severe acute nephritis crises when taken in higher doses than recommended, causing, especially in children, intense water retention in the body that can be fatal (Nascimento and Vieira 2014, Grandi 2015).

In relation to popular use, it is used externally in sprains, bruises and rheumatic pain. The parts used are leaves, pulp and sap. It can be prepared as juice and tinctures. Due to its toxicity, women should not use it orally during menstruation or pregnancy. It should also be avoided in hemorrhoidal states. It should not be given orally to children (Yagi 1997, Hussain 1998, Abílio 2011).

Black Jurema (*Mimosa tenuiflora* (Willd). Poir) (Figure 8.4)

The black jurema, popular name of *Mimosa tenuiflora* (Willd). Poir is a native species of the semiarid region and belongs to the Leguminoseae of the typical order of the caatinga fabales. It occurs in almost all of northeastern Brazil, and can also be found in other countries such as Mexico, Panama, El Salvador, Honduras, Colombia and Venezuela. This plant has some important features, such as its forked shrub size, and low branches reaching a height of around 4.5 meters (Maia 2004).

As for its physical aspect, it has a tortuous trunk that does not reach a large diameter, and being a deciduous species, loses its leaves in cold or rainy periods. The family to which the black jurema belongs has a characteristic that is the presence of acuminus, similar to thorns, in the branches and stem. Another important observation is that its fruits are small and pod-shaped. Black jurema is of great ecological importance, as its small, fragrant white flowers provide floral resources such as pollen and nectar for many species of bees, flies, wasps, and other insects, so it is considered an important plant for maintenance of biodiversity (Maia-Silva et al. 2012).

Figure 8.4. *Mimosa tenuiflora* (Willd). Poir (Black Jurema) tree (A), leaves (B), flowers (C) and fruits (D). Adapted from http://ainfo.cnptia.embrapa.br/digital/bitstream/item/146566/1/cartilha-plantas-caatinga-11-8.pdf.

In indigenous tribes and in Afro-Brazilian cults, jurema wines are prepared. Indians who do not use jurema for beverage production, as is the case with some tribes, only worship the plant in their religious rituals. The use of this plant in rituals is quite different and in cults practiced in rural and urban areas, such as xango and candomblés, where an element of the ritual characteristic of several Afro-Brazilian cults "juremeiros" is a practice of inserting a seed of *M. tenuiflora* (jurema) under the skin of novices. One of the catimbó rituals, known as Juremação, the implantation of the seed consists of planting, magically or really, a seed of the jurema tree under the skin of the disciple's body. According to reports, the drink prepared from the bark of the plant is considered energetic, because those who consume it during the rituals mentioned are more willing to dance. Studies indicate that the plant has narcotic and hallucinogenic effects attributed to the presence of the substance N, N-dimethyltryptamine (Albuquerque 1997).

Because it is a fast-growing plant and has a high regeneration power after cutting and fire, it is a great source of good quality firewood and coal for power generation, especially in forges and foundries. Its wood has high strength and is also used in small buildings and in the manufacture of rustic looking furniture. The stem bark of this plant is also used in folk medicine to treat diseases and infections (Maia-Silva et al. 2012).

The black jurema stem bark is used in folk medicine by infusing the bark to wash wounds, gargle and rinse in case of mouth injuries or inflammations, and can even be used to eliminate parasites from the human body. The bark is also crushed in some cases and placed on wounds, burns and leg ulcers, and its use has shown effective results in folk medicine (Albuquerque 1999).

According to some studies, it is found that black jurema has about 18% of condensed tannins that can act as free radical scavengers. The plant also shows antimicrobial, antifungal, antiviral, antidiarrheal, and antiseptic properties. In addition to all these properties, black jurema also has antioxidants capable of reducing the number of free radicals present in humans, preventing the oxidation of cells. Even with all these applications that benefit human beings, black jurema is still frequently used only in religious rituals in indigenous tribes (Bezerra et al. 2011).

Baraúna-Do-Sertão (*Schinopsis brasiliensis* Engl.) (Figure 8.5)

The genus *Schinopsis* (Anacardiaceae) is composed of fourteen species. *Schinopsis brasiliensis* Engl. is a large tree, reaching up to 12 m in height, and 20 to 60 cm in diameter, with branches provided with thorns. The leaves are dark green at the top and pale at the bottom (Dantas et al. 2008). It is popularly known as baraúna (Cardoso et al. 2015), native to the Caatinga and Atlantic Forest and found in semi-arid regions of Brazil (De Souza Santos et al. 2018).

Different parts of *S. brasiliensis* such as leaves, bark, stems and fruits are used in folk medicine as anti-inflammatory and antioxidant agents for various diseases such as fever, cough, diarrhea and as a

Figure 8.5. *Schinopsis brasiliensis* Engl (Baraúna-do-Sertão) tree (A), leaves (B), flowers (C) and fruits (D). Adapted from https://www.naturalista.mx/taxa/745716-Schinopsis-brasiliensis/browse_photos and https://faunaefloradorn.blogspot.com/2019/06/barauna-schinopsis-brasiliensis-engl.html.

natural antiseptic to treat wounds (Saraiva et al. 2013, Santos et al. 2014, Donati et al. 2015). Its bark is widely used by the northeastern population for the treatment of various diseases, such as inflammation, pain, flu, diarrhea, impotence and animal worms, a natural antiseptic and the treatment of wounds and superficial mycoses (Júnior et al. 2011).

S. brasiliensis leaves contain tannins and polyphenols, such as methyl gallate, gallic acid, and ellagic acid. It contains alkylphenol methyl 6-eicosanoyl-2-hydroxy-4-methoxybenzoate isolates and 5α, 8α-epidioxy ergosta-6,22-dien-3β-ol. In the essential oil, a good amount of mircene and low amounts of other compounds such as β-caryophyllene, eucalyptol and guaiol are present (Fernandes et al. 2015). The major substances include hydrolysable tannins and flavonoids (Dos Reis Luz et al. 2018).

Bush Plum (*Ximenia americana* L.) (Figure 8.6)

The genus *Ximenia* belongs to the Olacaceae and comprises about eight species (De Menezes et al. 2019). *Ximenia americana* L. was named after Francisco Ximenes (Aragão et al. 2018). It is commonly found in Africa, India, New Zealand, Central America, and South America. It is a wild-occurring cosmopolitan tropical plant that is widely distributed in northeastern Brazil, where it has numerous popular synonyms, including bush plum, ababa, plum, Bahia plum, umbu-bravo (Brasileiro et al. 2008).

In folk medicine, it is mainly used for the treatment of stomach pain, syphilis, rheumatism, cancer, and mouth infections and inflammation (De Menezes et al. 2019). Investigations over the past 10 years

Figure 8.6. *Ximenia americana* L. (Bush plum) tree (A), leaves (B), flowers (C) and fruit (D). Adapted from http://sweetgum.nybg.org/science/world-flora/taxon-details/?irn=152296 and http://tropical.theferns.info/image.php?id=Ximenia+americana.

have shown that its constituents have anti-inflammatory, antimicrobial, antifungal, antineoplastic, antitrypsin, antirheumatic, antioxidant, analgesic, molluscicidal, pesticidal activities, as well as hepatic and haematological effects (Silva et al. 2017).

All parts of this species have reports of use in folk medicine. The fruits in the green ripening stage have a higher content of yellow flavonoids, total anthocyanins, polyphenols, starch, antioxidant enzyme activity and increased antioxidant activity. The antioxidant activity attributed to the fruit was related to the content of polyphenols, flavonoids, anthocyanins and antioxidant enzymes (Almeida et al. 2016). The leaves have antibacterial activity, used to treat fever, tuberculosis, caries and wounds. The roots are used in the treatment of leprosy, syphilis and dysentery. The stem bark shows antioxidant and anti-inflammatory activity (Shettar et al. 2017), and flavonoids (-) *epi*-catechin and quercetin have already been identified in ethanol extracts of stem bark and leaves of *X. americana*, with high antioxidant activity, and high total phenol content (Uchôa et al. 2016).

The results of *in silico* studies concerning the interaction between the flavonoid (-) *epi*-catechin, present in the aqueous extract of *X. americana* stem barks and the inflammatory enzyme 5-lipoxygenase, demonstrate efficacy in leukotriene release, suggesting decrease in inflammatory process (Carvalho et al. 2020).

Cajueiro (*Anacardium occidentale* L.) (Figure 8.7)

Cashews belong to the Anacardiaceae family, including 60 genera and 400 species of trees and shrubs. The common synonyms are beach cashew and cashew (Grandi 2015). The cashew tree (*Anacardium occidentale*) is a tropical evergreen tree widely cultivated in Brazil. The stem, also called the cashew apple, represents over 90% of the total edible parts of cashew but is considered an industrial waste (Mog et al. 2019). Cashew is a fruit of social and economic importance for several countries, such as India, some African countries and Brazil, whose cultivation is concentrated in the Northeast region (Abílio 2011).

The chestnut (true fruit) is still the main product in the commercial exploitation of cashew. Although cashew apples can also be consumed as juice, ice cream and other foods, only 10% of their production is used by industry (Grandi 2015). Since medicinal plants are classified as natural products, the law allows them to be freely traded and can be cultivated by those with the necessary minimum conditions. This facilitates targeted self-medication in cases considered the simplest and most common, which reduces demand for health professionals, facilitating and reducing further the cost of public health service (Abílio 2011).

Thus, the trunk bark is used in leather tanning, as it is astringent, with high tannin content, and as a dye in the dyeing of clothes, nets and fishing lines. Medicinal resin is used as expectorants. The wood of this plant is resistant to seawater and is used for firewood and coal. Pseudo fruit is rich in vitamin C used

Figure 8.7. *Anacardium occidentale* L. (Cajueiro) tree (A), leaves and flowers (B), flowers (C) and fruits (D). Adapted from https://candidegardening.com/.

in food for humans and animals and can be consumed as beverages (liquor, wine, and brandy), jams, and raisins (Grandi 2015).

The true fruit, the nut, after being processed is used as a snack in the preparation of cakes, sweets, chocolates, chocolates and ice cream. The cashew nut liquid (LCC) is also extracted from the cashew nut, and used in the chemical and automotive industries. It is also used as an antiseptic and vermifuge, destined largely to the foreign market, and is mainly responsible for the commercial support of the product. It is also made up of other active ingredients such as acetylphenone, amirine, anacardol and limonene (Silva et al. 2009).

There are many fatty acids present in seeds. They are used as antidiabetic. Dosage forms include infusion, tincture, fluid extract, elixir, syrup or wine (Silva et al. 2009, Grandi 2015, Tamiello-Rosa et al. 2019). Due to the presence of harmful oil, cashews should be cooked before eating because the essential oil contains sensitizing substances that cause dermatitides such as pentadecylresorcinol and pentadecylferol. Also, oil vapor is irritating if inhaled. The substantial importance is demonstrated by the cashew tree due to the presence of various components of great economic value (Grandi 2015, Alencar et al. 2018, Tufail et al. 2019).

It is astringent, tonic and stimulator in weakness. Externally, it is used for gargling, throat irritation and thrush. It contains hypoglycemic, antimicrobial, antitumor and cytotoxic properties (Grandi 2015).

Catingueira (*Caesalpinia pyramidalis* Tul.) (Figure 8.8)

C. pyramidalis Tul. (Leguminoseae), popularly known as "catingueira", is a native tree from northeastern Brazil, found mainly in the caatinga region (Albuquerque et al. 2007). Some parts of this plant such as bark and leaves are generally used due to their anti-inflammatory, antidiuretic and digestive activity (Bahia et al. 2005, Agra et al. 2007).

The use of *C. pyramidalis* in folk medicine is accomplished by macerating a handful of its bark and mixing it with one liter of wine or sugar cane brandy (cachaça), being consumed twice a day before meals, such as diarrhea, stomach pain, dysentery and other gastrointestinal disorders (Agra et al. 2008, Santos et al. 2011).

One study used ethanol extract from the *C. pyramidalis* bark for oral treatment in mice to decrease acetic acid-induced abdominal contractions and formalin-induced nociception, also in mice. The extract showed antinociceptive and anti-inflammatory activity in carrageenan-induced edema and peritonitis in mice (Santos et al. 2011).

In another study, to evaluate the effect of *C. pyramidalis* bark ethanol extract on acute pancreatitis (AP), an inflammatory disease with symptoms of abdominal pain was induced to AP in male rats, and the results showed a significant reduction of inflammation in rats treated with the extract (Santana et al. 2012). In another research or ethanolic extract from the catingueira bark, oral doses were applied to rats, which showed a gastroprotective response by reducing inflammation (Diniz et al. 2015).

Figure 8.8. *Caesalpinia pyramidalis* Tul. (Catingueira) tree (A), leaves (B), flowers (C) and fruits (D). Adapted from http://www.cnip.org.br/banco_img/Catinga%20de%20Porco/caesalpiniapyramidalistul3.html.

Studies on *C. pyramidali* extracts reinforce its use in folk medicine for anti-inflammatory and gastrointestinal diseases, thus favoring the use of this plant for various gastrointestinal disorders (Santana et al. 2012, Ribeiro et al. 2013).

Cowpaw (*Bauhinia forficata* Link.) (Figure 8.9)

The genus *Bauhinia* comprises over 300 species (Pinafo et al. 2019). The Brazilian orchid, *Bauhinia forficata* Link, popularly known as cow paw, due to the shape of the leaves, belongs to the Leguminosae family (Palsikowski et al. 2019). The tree grows naturally in the rainforests and tropical parts of Peru and Brazil, as well as the tropical areas of Asia, eastern Paraguay and northeastern Argentina (Lim et al. 2006).

Several species belonging to this genus are recognized for providing health benefits. Most often, bark of leaves and stems are used as remedies for pain, inflammation or microbial infections (Farag et al. 2015). *B. forficata* Link is widely used in folk medicine as antidiabetic because it contains inhibitors that decrease inflammation and prolong blood clotting (Silva et al. 2012). A simple leafy tea helps to balance glucose levels (Silva et al. 2002). This property is attributed to quercetin and caemperol flavonol glycosides, especially 3,7-di-O-α-L-ramnopyranosylkanferol (caempferitrin), reported in the literature as a chemical marker for this herbal drug (Marques et al. 2013).

In qualitative and quantitative analyses, dozens of phenolic secondary metabolites including alkaloids, flavonoids, mucilage, essential oils and tannins were isolated (Volpato et al. 2008). Moreover, campferol and quercetin O-glycosides were isolated from the leaves of *B. forficata* (Da Cunha et al. 2010).

Figure 8.9. *Bauhinia forficata* Link (Cowpaw) tree (A), leaves (B), flower (C) and fruit (D). Adapted from https://www.arvores.brasil.nom.br/new/patavaca/index.htm, http://www.mudasnativaslof.com.br/especies/detalhes/pata-de-vaca; https://link.springer.com/chapter/10.1007/978-94-024-1552-0_11 and https://herbariohbei.wixsite.com/hbei/bauhinia-forficata?lightbox=image_1duk.

Elderberry (*Sambucus nigra* Linn.) (Figure 8.10)

The *S. nigra* Linn., is a tall, deciduous, evergreen shrub, or small tree belonging to the Adoxaceae Family, originally from Asia, North Africa, and Europe. This plant has narrow, dark-green leaves arranged opposite, 10 to 30 cm long (Basas-Jaumandreu and De Las Heras 2019). The common names of this plant are elderberry, black elder, European elder, European elderberry and European black elderberry (ITIS 2016).

All parts of this plant (flower, bark, leaf and fruit) are rich sources of dietary phytochemicals such as carbohydrates, lipids, terpenoids, flavonoids, phenolic acids, alkaloids, etc. (Agalar 2019). The flowers

Figure 8.10. *Sambucus nigra* Linn. (Elderberry) tree (A), leaves (B), flowers (C) and fruits (D). Adapted from Cech (2000) and http://7flechasejurema.blogspot.com/2018/02/sabugueiro.html.

are used in traditional medicine to treat inflammation, skin and urinary disorders, colds, fever and other respiratory disorders (Blumenthal et al. 2000, Weiss and Fintelmann 2000, EMA 2008).

Scientific evidence suggests that elderberry flowers can be used for both prevention and therapy of various diseases due to their anti-inflammatory (Olejnik et al. 2015), antioxidant (Dawidowicz et al. 2006) and antimicrobial activities (Arjoon et al. 2012). In addition, elderberry extracts are rich in polyphenols with therapeutic and medicinal advantages (Ribeiro et al. 2019).

Its leaves are mainly composed of sugars, organic acids, anthocyanins and quercetin (Veberic et al. 2009). These last two types of compounds (anthocyanins and quercetin) are part of the flavonoid groups that provide this plant with antioxidant properties (Ribeiro et al. 2019). Its medicinal properties are commonly associated with the presence of bioactive anthocyanins (mainly cyanidine3-mono- and disaccharides), quercetin, kaempferol, myricetin and their derivatives (Sidor and Gramza-Michałowska 2015, Porter and Bode 2017).

Studies have shown the benefits of elderberries as antiviral drugs (Zakay-Rones et al. 1995, Roschek et al. 2009), and efficient supplements in the treatment and prevention of diabetes, cardiovascular diseases and cancer (Jing et al. 2008, Ciocoiu et al. 2009). *S. nigra* berry nutraceutical products (syrups, drops, tablets, emulsions, suspensions, extracts) are commercially available (Młynarczyk et al. 2018), for example, Samubucol®, is effective against different strains of influenza virus (Barak et al. 2019).

Embaúba (*Cecropia pachystachya* Trécul) (Figure 8.11)

Species of the genus *Cecropia* (Urticaceae) are native to Central and South America, being fast-growing and short-lived trees, typical of forest margins (Almeida et al. 2006). They are used in folk medicine to treat asthma and hypertension (Delarcina et al. 2007, Tanae et al. 2007, Cruz et al. 2013), and also as a potent bronchodilator and vasodilator (Bergeron et al. 2007, Demougeot et al. 2007).

The species *Cecropia pachystachya* Trécul (*C. pachystachya*), belonging to the genus *Cecropia* and popularly known as embaúba or ambaíba or sloth tree, is a native South American plant that is popularly used for respiratory diseases, hyperglycemia and bronchitis (Soraru and Bandoni 1978, Lorenzi and Matos 2002b). In folk medicine, its leaves are boiled in water, and consumed during the day (Aragão et al. 2010).

C. pachystachya leaves are rich in flavonoids and polyphenols (Lacaille-Dubois et al. 2001). In a study by Cruz and collaborators (2013), nine flavonoids were identified (chlorogenic acid, orientin, isovitexin, apigenin, (+)-catechin, (-)-epicatechin, isoquercitrin, lutelin and quercetin) in ethanol extract from *C. pachystachya* leaves, in which the characterization of the extract was done by liquid chromatography coupled with mass spectrometry. Flavonoids have an anti-leishmanial activity and are

Figure 8.11. *Cecropia pachystachya* Trécul (Embaúba) tree (A), leaves and flowers (B) and fruits (C). Adapted from http://chaves.rcpol.org.br/profile/species/eco/eco:pt-BR:Cecropia%20pachystachya.

inhibitors of arginase, in this case by inhibiting arginase that can block the initial synthesis of polyamines essential for maintenance and multiplication of the antioxidant mechanism.

Ethnobotanical studies show that *C. pachystachya* is commonly used to treat kidney disease. Studies carried out for methanol extract from *C. pachystachya* leaves demonstrated anti-inflammatory (Aragão et al. 2013) and hypoglycemic activities (Aragão et al. 2010). The aqueous extract of *C. pachystachya* leaves showed cardiotonic, sedative, hypotensive (Consolini and Migliori 2005) and genotoxic effects on brain tissue (Mendonça et al. 2016). In the hexane extract of *C. pachystachya*, leaves exhibited sitosterol which showed significant anti-inflammatory effect (Hikawczuk et al. 1998).

In a study, anti-inflammatory activity of methanol extract of *C. pachystachya* leaves was investigated in mice with acute inflammation in the ear, induced by croton oil, capsaicin, phenol and others. The results were compared to some reference drugs and it was found that the extract had a significant anti-inflammatory effect (Pacheco et al. 2014). Another study evaluated the anti-inflammatory activity, also using *C. pachystachya* leaf methanol extract, inducing inflammation in the ear of mice by croton, which showed similar results to indomethacin, an anti-inflammatory drug used as a reference (Aragão et al. 2013).

Eucalyptus (*Eucalyptus globulus* Labill) (Figure 8.12)

Eucalyptus, scientifically named *Eucalyptus globulus* Labill and popular names such as eucalyptus, blue bud, fever tree, white mahogany, among others, belong to the Myrtaceae Family, one of the most important Angiosperm families in Brazil, concentrated in a single tribe, Myrteae and three subtribes Myrciinae, Eugeniinae and Myrtinae. This family is one of the most representatives concerning the floristic survey, especially in areas of Atlantic Forest and Cerrado, since there are approximately 1000 species and 23 genera occurring in Brazil. They are trees or shrubs represented mainly by fruit plants (Landrum and Kawasaki 1997).

The genus *Eucalyptus* originated from Australia, with over 700 botanically recognized species. However, no more than twenty of these species are currently used for commercial purposes in the world. These species have very diverse physical and chemical properties, allowing their use for various purposes. An exact date of introduction of eucalyptus in Brazil is not known, but the first plantation took place around 1868 in the Rio Grande do Sul (Lorenzi and Matos 2002a).

As has been said, the chemical and physical properties of eucalyptus make it versatile and can be used for various purposes, such as for energy, sawmill, pulp and paper, and in healthcare, such as the production of medicines, cosmetics and food. Important factors such as the rapid growth of eucalyptus species, their ability to adapt to different ecological regions and the low cost of productivity confer great

Figure 8.12. *Eucalyptus globulus* Labil (Eucalyptus) tree (A), leaves (B), flowers (C) and fruits (D). Adapted from https://ainfo.cnptia.embrapa.br.

fascination to eucalyptus cultivation. In addition to the economic benefits resulting from cultivation, it is possible to highlight the improvement in air quality, pleasant thermal sensation in the planting regions, reduction of soil erosion and noise pollution level, among many other environmental benefits (Lorenzi and Matos 2002a).

The use of eucalyptus as a source of energy through charcoal has been growing due to the replacement of native wood by planted forests to generate power for iron production. Larger diameter eucalyptus wood is widely used in rustic buildings such as fences and poles, and in some cases it is used in the manufacture of rustic furniture. However, the smaller diameter wood is used in construction as easels and props. Eucalyptus is one of the most widely used species in the world for the production of pulp, paper, paper packaging and cardboard. For this purpose, young trees with a diameter of about 18 cm are used. The use of eucalyptus in the production of essential oils that are widely used for the production of cosmetics (focused on hygiene and cleaning), drugs and food is noteworthy. It is also important to note that apiculture products, such as honey and propolis from flowering eucalyptus are very popular (Teske and Trentini 1997).

Before discussing the medicinal use of eucalyptus, specifically leaves and oil, it is important to be aware of its chemical composition. The essential oil contains compounds such as 1,8-cineole, p-cymene, α-pinene, limonene, geraniol and camphene. In the leaves, it is possible to find substances present in the essential oil, as well as some others such as pinene, mirtenol, pinocarvone, terpenes, aliphatic aldehydes, tannins, flavonoids (eucaliptrin, hyperoside, quercetin and rutin), polyphenols, resin and waxes. These active ingredients present in eucalyptus provide it a characteristic aroma. Its essential oil, being volatile, is used in the treatment of respiratory diseases such as asthma, bronchitis or even the simple flu or cold. These bioactive substances can be absorbed by the human body through inhaling the steam generated when the leaves are cooked. With this, the substances act directly on the lungs (Mora and Garcia 2000).

In addition to its use in the treatment of respiratory problems presented, eucalyptus is also considered as a good antiseptic and can be used to wash wounds and ulcers, which has a disinfectant and healing effect. The essential oil also has astringent, feverish and tonic properties. However, it is contraindicated for children as it can cause poisoning even with small amounts of oil and the sign of this is observed by lowering of blood pressure, choking and circulatory problems. For individuals using diabetes medications, eucalyptus can potentiate its effects as it is able to lower blood glucose levels (Andrade 1961).

Ironwood (*Caesalpinia ferrea* Mart.) (Figure 8.13)

Caesalpinia ferrea (*C. ferrea*) Martius, popularly known as ironwood or juca, is a large tree belonging to the Leguminosae family. It can be found mainly in northern and northeastern Brazil (Vasconcelos et al.

Figure 8.13. *Caesalpinia ferrea* Mart. (Ironwood) tree (A), leaves (B), flowers (C) and fruits (D). Adapted from https://appverde.wordpress.com/2015/09/18/juca-libidibia-ferrea/.

2011). It has great medicinal and ornamental potential, and its wood is used in construction and carpentry (Lima et al. 2006).

C. ferrea is a subspecies of Caesalpinia *sensu lato* which is part of a pantropical group of about 150 species of lianas, shrubs and trees, many growing in arid soils and especially in the seasonally dry rainforests of southern and central America (Gagnon et al. 2013). The Caesalpinia Group species are karyotypically characterized by numerical stability (2n = 24) and small chromosomes (~ 2 μm) with metasystem/subcentric and acrocentric morphologies (Van-Lume et al. 2017, Rodrigues et al. 2018).

In folk medicine, *C. ferrea* stem bark tea is used to treat diabetes (Vasconcelos et al. 2011). Already fruits, pods and roots have been used for years, as infusions, to treat various inflammatory disorders, such as bruising, rheumatism, scarring, pulmonary hemoptysis (Braga 1976), fever (Pio Correa 1984), enterocolitis (Balbach 1988) and diabetes (Ueda et al. 2001). Its wide use in traditional medicine is due to its diversity of bioactive compounds such as terpenoids and phenolics present in different parts of the plant, such as bark, fruits, seeds, leaves and roots (Cavalheiro et al. 2009).

Previous studies have shown that species of the genus *Caeslpinia* exhibited a wide range of pharmacological activities such as anticancer, anti-inflammatory, antipyretic, antimicrobial, antimalarial, antirheumatic, anti-ulcer and anti-diabetic activities. In addition, chemical investigation of this plant revealed the presence of triterpenoids, diterpenes, flavonoids, steroids and phenolic compounds. Isolated diterpenes of this genus are a group of highly diverse composite structures and exhibited a variety of pharmacological activities (Wei et al. 2015).

Souza et al. (2006) also performed studies with crude extract of *C. ferrea* and found that the species contains anthraquinones, alkaloids, depsides, depsidones, flavonoids, lactones, saponins, sugars, tannins, sesquiterpenes and triterpenes; however, the tannins were considered the main component.

Juazeiro (*Ziziphus joazeiro* Mart.) (Figure 8.14)

Caatinga is known for its true Brazilian biome, taking into account the wide variety of animals and plants. Juazeiro belongs to the Rhamnaceae family. It is an endemic Caatinga species, very important economically and biologically because of its medicinal properties. Juazeiro is also known by popular names such as juá, juá-thorn, juá-fruit, cowboy orange, juá-mirim, among many others. The term Juá, the name given to the fruit, derives from the tupi "a-ju-á", meaning "fruit obtained from thorns" (Silva et al. 2003).

Juazeiro is a perennial, heliophytic and autotrophic plant with pivoting roots that favor water absorption from the subsoil and this keeps the plant always green, even in a period of long drought. There are no natural juazeiro forests, as it occurs in isolation in arid-adapted forests. The plant is also

Figure 8.14. *Ziziphus joazeiro* Mar (Juazeiro) tree (A), leaves (B), flowers (C), and fruits (D). Adapted from https://www.portalsaofrancisco.com.br/biologia.

characterized by its size ranging from medium to high, trunk aligned and curved, due to the slow growth and presence of resistant thorns with tortuous branches (Lucena et al. 2009).

Juazeiro provides wood of great strength and durability, which attracts the extractive activity. The main uses of this wood are the construction of fences, corrals and firewood and coal for the supply of energy. In addition, after burning their ashes are used in soap making because it contains a high potassium content. This plant, in addition to bringing benefits to humans, is also used for animal feed, as goat fodder mainly in periods of drought (Rocha 2012).

Due to its drought resistance, juazeiro is very important economically for the northeast region of Brazil, and practically all parts of the plant are used by the population. In addition to what has already been said about its use, we can add the fact that juazeiro is also used by the population to clean teeth, gingivitis and pain caused by tooth extraction, since its bark is rich in saponin and contains an anti-caries principle. It is also used against hair loss, asthma, flu and all types of fevers and expectorant. The fruits of juazeiro are edible, being much consumed by the northeastern sertanejo, as is the case of mature juá that is highly esteemed by children and adults, because in times of drought it relieves hunger and thirst. Due to the popular use of juazeiro, several studies were conducted with the perspective of substantiating its properties and actions (Romão et al. 2010).

Juazeiro is chemically composed of saponin, carbohydrates, cellulose and vitamin C, and the latter is present in high concentration in its fruits. In addition to the consumption of fruits, the peels and leaves are used in popular medicine in the northeast in the form of water extract, to relieve gastric problems and for cosmetic and hygienic use in cleaning hair, teeth and lightening the skin of the face (Matos 2000).

Some studies show that the use of the juazeiro plant for medicinal purposes stands out from other uses, as it is one of the only sources to obtain raw material for the production of home remedies. A good example is the use of juazeiro zest for the production of flu and cough syrups. Juá, as it is also called juazeiro, is used to combat seborrhea, justified by the presence of saponins which is a substance used in the production of anti-dandruff shampoo and hair tonic. The wide use of this plant in home practices of medicine, cosmetics and dentistry applied by the people is of great interest for future studies, aiming at the best use of this plant (Lorenzi and Matos 2002a).

Jurubeba (*Solanum paniculatum* L.) (Figure 8.15)

The Solanaceae family comprises more than 3000 species distributed in 106 genera, and their species have a cosmopolitan distribution. Among the species of the genus *Solanum*, *Solanum paniculatum* L. is the most widely used for popular use (Tenório et al. 2016). Known by the name of Jurubeba, it occurs in Brazil, Paraguay, Bolivia and Argentina, and is considered by the Ministry of Health of Brazil as a

Figure 8.15. *Solanum paniculatum* L. (Jurubeba) tree (A), leaves (B) and fruit (C). Adapted from https://www.flickriver.com/photos/tags/solanumpaniculatum/interesting/ and http://www.aplantadavez.com.br/2018/03/jurubeba-solanum-scuticum-m-nee.html.

medicinal herb (Rios et al. 2017a). In the Northeast of Brazil, it is represented by 80 species, including 32 that are endemic to Brazil and 20 that are endemic to the Northeast region, widely used in folk medicine (Silva et al. 2007).

The plant is rustic, drought-resistant, has no fertility requirement adapting to different soil types (Campos et al. 2015) and is a component of various pharmaceutical formulations including syrups, infusions or decoctions, ethanolic extracts, dyes and elixirs (Mesia-Vela et al. 2002). In the floral and reproductive biology of species of the genus, pollination by bees is a common character (Buchmann et al. 1977). Preparations of *S. paniculatum* roots, stems, fruits and leaves are used as tonic, carminative, diuretic, herbal, antipyretic, analgesic and anti-inflammatory, and to relieve the symptoms of stomach and liver dysfunction (Vieira et al. 2010, Clementino-Neto et al. 2016, Rios et al. 2017b, Valerino-Díaz et al. 2018, De Souza et al. 2019).

In this species, bioactive compounds and secondary metabolites, such as steroids, saponins, alkaloids and glycosides of pharmacological importance are considered a good source of polyphenol oxidase (Antunes et al. 2018). Phytochemical screening performed of *S. paniculatum* L. extract revealed the presence of alkaloids, lactones, as well as a high content of phenols (flavonoids and tannins). Qualitative analysis of phenols by TLC showed a suggestive point of isovitexin (flavonoid) and the presence of tannic acid (tannin) in the extract (Garcia et al. 2008).

The presence of proven anti-inflammatory (Rios et al. 2017c) and antioxidant (Kaur et al. 2014) metabolites provide evidence on the popular use of this plant.

Mofumbo (*Combretum leprosum* Mart.) (Figure 8.16)

The *Combretum leprosum* Mart. is a shrub belonging to the Combretaceae family, well distributed in the North and Northeast of Brazil. It is popularly known as mold, mufumbo or monkey comb (Cavalcanti et al. 2019). The Combretaceae family consists of 18 genera, and covers approximately 600 species of trees and shrubs. The largest representative of the family is genus *Combretum* with 370 species (Viau et al. 2016).

These plants are shrubs, trees, sloping shrub or liana that varies from 2.5 to 3.0 m in height, and has a trunk covered with whitish or yellowish lepidote induction. The inflorescences are in raceme panicles, with nectariferous disc, villous margin; rounded to oblate petals; densifloras, congestions, terminals and axillaries with essential oil flowers. The leaves are of the opposite and petiolate type; cardiac to subcordate, elliptical, broad-elliptical to rounded, acute to rounded base, acute to acuminate apex, slightly discolor to concolor; opaque on both sides; prominent ribs biaxially (Soares Neto et al. 2014). The fruits are dried, betuloid, 4-winged, largely elliptical, sparsely lepidoto in the wings, more compact in the central region,

Figure 8.16. *Combretum leprosum* Mart. (Mofumbo) tree (A), leaves (B), flowers (C) and fruits (D). Adapted from https://appverde.wordpress.com/2015/09/18/mofumbo-combretum-leprosum/.

coriaceous, greenish-white in immature fruit, dark-brown in mature fruits, leaf cotyledons, transverse and closely striated; fruits indehiscent, with winged and monospermic wings (Loiola et al. 2009).

In folk medicine, different parts of the plant such as leaves, shells, flowers, roots and fruits are employed by infusions, decoctions and syrups, such as sedatives and in the treatment of pertussis, uterine bleeding, flu, hemorrhages, inflammations, prevention of rashes, wound cleaning, antidiarréic, expectorant and antitússico (Longhi-Balbinot et al. 2012, Moraes et al. 2007). According to phytochemical studies, *C. leprosum* is rich in cycloaneans, triterpenes [arjunolic and malic acid, phenolic acid and 3β, 6β, 16β-trihydroxy-lup-20(29)-eno (TTHL) and flavonoids (3-0-methylquercetin, 5.3′-di-hydroxy-3,7,4′-trimethoxyflavone, 5.3′, 4′-tri-hydroxy-3,7-dimetoxyflavone, epicatechin and quercetin] (Lopes et al. 2010). In a previous study, they proved that extracts and compounds isolated from different parts of *C. leprosum* demonstrated biological activities such as antinociceptive, antimicrobial, antioxidants, anti-inflammatory, antiproliferative, gastroprotective, antidiarreic, anticholinesteratica, immunomodulatory, healing, cytotoxic and mutagenic, antienzymatic activity of Na^+, K^+-ATPase, leishmanicidal activity, neuropathic (Parkinson's), antiophydic activity and vasoactive activity with the relaxation of arterial rings isolated (Pietrovski et al. 2006, Lopes et al. 2010, Viau et al. 2016).

Mulungú (*Erythrina velutina* Willd) (Figure 8.17)

Erythrina velutina Willd, a species belonging to the Leguminoseae family, is a tree of great resistance to drought, endemic to Caatinga found in the coastal region of all northern and northeastern states, and is popularly known as mulungu, mulungu-da-flor-vermelha ou bucaré. This tree belongs to the order Fabales, subfamily Faboideae and genus *Erythrina* consisting of about 120 species, mainly trees and shrubs. It is commonly grown as an ornamental plant due to its attractive flowering and use in folk medicine (Menezes et al. 2016b).

This plant is characterized as an aculous or thorny arboreal tree with a deciduous leaf-changing behavior. Larger trees reach dimensions close to 15 m in height and 80 cm of DBH (diameter at breast height, measured at 1.30 m from the ground) in adulthood, with straight to the slightly tortuous trunk, slightly accumulated branches. Its branch is dichotomous, with its wide, open and rounded crown. The outer bark or rhytidome is smooth to slightly rough. The leaves are trifoliate, supported by petiole from 6 cm to 14 cm in length; the leaflets are orbicular, oval-rhomboid or triangular, of cardiac consistency, with the ventral face only powdery and dorsal, lighter green in color with dense feline hair, measuring from 6 cm to 12 cm long by 5 cm to 14 cm width. Its inflorescences occur in axillary fascicles, measuring from 12 cm to 20 cm in length and with three flowers. The flowers have orange or red, with almost orbicular lamina and spatial calyx. Curved fruits with apex and sharp bases, internally not septate, with

Figure 8.17. *Erythrina velutina* Willd. (Mulungú) tree (A), leaves (B), flowers (C), seeds and fruits (D). Adapted from Palumbo et al. (2016).

1 to 3 seeds. Bicolor seeds, called mimetics, are dark-red and orange-red. They are also sub-squared or oblong, with a short midline hilum (Galvão and Medeiros 2002, Carvalho et al. 2008).

In folk medicine, stem bark, leaves and flowers are used in the form of powder, infusion, decoction or fluid extract in treatment as sedative and suppressor of cough, asthma, bronchitis treatment, muscle aches, anticoagulant, nervous crisis, for anxiety, insomnia, seizures, menopause, gingivitis, hepatitis, respiratory diseases, general anti-inflammatory, hemorrhoids, parasitic intestinal infections, antibacterial action against *Staphylococcus aureus* and *Streptococcus pyogenes*, anti-hepatotoxic and fights oral infections (Dantas et al. 2004).

The phytochemical studies reveal the presence of alkaloids, terpenes and flavonoids. The flavonoids analyzed were two flavanones, 5,7-dihydroxy-4, 5-dimethoxy-flavanone (homoesperidin) and 5, 3-trihydroxy-4-O-methyl-5-prenylflavanone (4-O-methylsigmoidin B); two isoflavonoids, 3,9-dihydroxy-10prenylpterocarp(phaseolidine)and2-4-dihydroxy-6-prenyl-7-methoxy-isoisoflavanone (erivelutinone), 7-O-[α-rhamnopyranosyl-β-glucopyranoside]-genysteine and two triterpene: 3β-olean-12-ene-3,28-diol and 3β-1,2,6,7-tetradehydro-3,16-dimethoxy-eritrin-15-ol (erysovina) (Ximenes et al. 2018). The extracts and compounds isolated from different parts of *E. velutina* have biological activities such as neuroprotective, anxiolytic, sedative, anticonvulsant, antinociceptive, antioxidant and anti-inflammatory effects of neuromuscular blockade (Carvalho et al. 2009, Amorim et al. 2019).

Noni (*Morinda citrifolia* Linn.) (Figure 8.18)

The botanical name of noni is *Morinda* citrifolia Linn and popular names are iada, ninth, canary, tooth; it belongs to the Rubiaceae family and genus *Morinda.* It originates from trees native to Polynesia, Indonesia and Asia. Its adaptation to the Brazilian climate was so satisfactory that this fruit is produced throughout the year. This easy adaptation makes Noni grow in just about any environment, such as sandy, rocky and rainforest terrain. In addition, it resists soils with high salinity and dry regions, making it easy to find it in many regions of Brazil and the world (Pawlus and Kinghorn 2007).

The tree in its adult form can be 4 to 10 meters high. There are two varieties of *Morinda citrifolia* which include *M. citrifolia* var. *bracteata* and *M. citrifolia* var. *citrifolia*. However, in the literature, it is often not possible to see this distinction, since the botanical characteristics that differentiate the species are barely noticeable, especially when the observer is a layperson on the subject. As the use of Noni is recent in Brazil, mainly therapeutic use, there is not much evidence to prove its effectiveness. Recent studies have shown results obtained with *in vivo* and *in vitro* models, where all animals survived and showed no signs of toxicity or behavioral changes during acute toxicity studies, but the question that still causes concern among researchers is that it is not known under what conditions Noni is ideal and safe for human consumption (Rao and Subramanian 2009).

Figure 8.18. *Morinda citrofolia* Lin (Noni) tree (A), leaves (B), flowers (C) and fruits (D). Adapted from https://ainfo.cnptia.embrapa.br/.

In places where Noni is a native plant, people from different social classes use the fruit to make juices in order to alleviate pain. In addition to the fruit, bark, leaves, flowers, roots and stem are combined to make herbal remedies. There are over 40 records of these remedies in the literature. In Southeast Asia, Noni is also used non-medicinally and is applied to the dyeing of clothing and fabrics. Since the introduction and use of Noni in Brazil is relatively recent, only fruit juice mixed with some tastier juices, such as grape juice, is generally used to hide the taste in the treatment of headaches, arthritis, sexual impotence, weight loss and hypertension (Lima et al. 2007).

In many countries, Noni is used for different purposes, such as the treatment of cholesterol related diseases and diabetes in the United States. Like eucalyptus, heated Noni leaves placed on the chest are used in Malaysia and China to treat respiratory problems. In other countries, such as Asia and Australia, only fruits are used as daily food, being consumed only raw with salt or cooked with some spices. Despite the various applications in the treatment of some diseases, there are contraindications to use, as it can aggravate kidney disease due to the high potassium content of Noni fruit juice, making it difficult for the body to accept it at the time of transplantation due to the strengthening of the immune system and even aggravate heart problems (Carniello et al. 2010).

Noni has several classes of compounds belonging to triterpenoids, flavonoids, lignins, anthraquinones and iridoids, as well as benzophenone derivatives. Also noteworthy in Noni is the substance 6-methoxy-7-hydroxycoumarin, called scopoletin, which occurs in several plants. This substance is relevant because some studies already demonstrated interesting biological activities, such as antiangiogenic, antioxidant, hepatoprotective and antiproliferative in human prostate cancer. In addition to this substance, another important anthraquinone found in Noni is damnacanthal, which has antitumor potential, as it has the inhibitory activity of different tyrosine kinases (Correia et al. 2011).

The fruit juice of Noni is most commonly used as antiviral, antibacterial, antifungal, antitumor, analgesic, hypotensive and anti-inflammatory, although the leaves and roots are used to a lesser extent. Brazilians cultivate Noni a lot because of its easy adaptation, and due to the spread of reports of the benefits of Noni among the population, its consumption is only increasing in the regions of the country. Those who use them, usually older people, get benefits of being hypotensive, anti-inflammatory, healing, anti-allergic, pro rejuvenation, sexual stimulant, weight loss, among many others. Many people also consume Noni fruit juice by mixing it with alcohol to ensure better results. This situation is worrying because there are not enough scientific studies to prove all these beneficial properties reported by users (Barbosa et al. 2017).

Pau D'arco (*Handroanthus impetiginosus* Mart.) (Figure 8.19)

The *Handroanthus impetiginosus* (Mart. Ex DC) Mattos, belongs to the Bignoniaceae family, and is popularly known as ipê-roxo, pau-d'arco-roxo and ipê-roxo-da-mata. The ipe comprises a group of plants belonging to the family Bignoniaceae, with the most representative genera being *Handroanthus* and

Figure 8.19. *Handroanthus impetiginosus* Mart. (Pau D'arco) tree (A), leaves (B), flowers (C) and fruits (D). Adapted from https://appverde.wordpress.com/2015/09/30/ipe-roxo-handroanthus-impetiginosus-2/.

Tabebuia (Fernandes et al. 2018). The genus *Handroanthus* belongs to the Bignoniaceae family, which was segregated from the genus *Tabebuia* Gomes ex DC as a result of molecular phylogenetic studies (Felix et al. 2018). In Brazil, approximately 14 species of Pau D'arco are commonly found in native vegetation areas of the northeast, mainly in the Caatinga biome, where saline, sodic and water-deficient soils occur naturally in arid and semi-arid regions and the Amazon region (Lima et al. 2014, Santos et al. 2018).

Tree species native to Brazil that can reach 20 to 35 m tall, relatively smooth, hard, thick, grayish bark; dark and dense brown wood, twigs subterete, glabrescent, the apices mealy pubescent. Leaves are composed of 5 or 7 leaflets, often anisophilic, leaflets ovate to ellipticals, acuminate, coined to rounded or nearly undercut at base, leaflet 5–15 cm long, 1.5–8 cm wide, progressively smaller lateral views, attenuated to acuminate or rounded apex, membranous to cartilian, somewhat lepidotes above and below, pubescent with simple or bifurcated trichomes, at least in the axils of the lateral nerves below, sometimes pubescent along or over the median vein. The surface below of leaves are formed of terminal petiole 1 to 4.2 cm in length, progressively smaller sides, petiole 4 to 13 cm in long, lepidote and puberty (Lorenzi and Matos 2002a).

Inflorescence a terminal panicle, usually more or less congested, the flowers in groups of three, the whitish or tan branches of the fleshy pubescence of the stellate trichomes. Flowers with truncated or slightly 5-lobed cupular calyx, 4–6 (–9) mm long, 3–6 mm wide, mealy pubescent with thick stellate trichomes; corolla magenta, yellow throat in anthesis, fading to purplish, campanular tubular, 4–7.5 cm long, 1.2–5 cm wide at the mouth of the tube, the tube 2.5 to 5 cm long, the lobes 0.9 to 2 cm long, puberty outside, inside with some simple trichomes scattered in the tube, glandular-pubescent at the level of the stamen insertion; dynamic stamens, the teak divaricate, 2.5–3.5 mm in length; ovary linear, ribbed, lepidote 3–4 mm long, 1 mm wide, glabrous. 4-series in each locule; cupuliform disc, 1–1.5 mm long, 2 mm wide. The fruits are elongated-cylindrical capsule attenuates at both ends, 12–56 cm long, 1.3–2.6 cm wide, glabrous; thin, 1–1.6 cm long, 3.4–8 cm wide, with hyaline membranous wings, clearly visible on the seed body (Gentry 1992).

In folk medicine, the bark is used as tea or decoction for the treatment of diabetes, inflammation, ulcer, malaria, cancer, kidney and hepatic disorders, syphilis, stomach infections and has anti-inflammatory and antimicrobial properties. The leaves and roots possess anti-inflammatory, antiseptic, analgesic, antibiotic and antineoplastic action (Morais et al. 2007, Lima et al. 2014). In a previous study, the authors found that the extracts and compounds isolated from different parts of *H. impetiginous* have biological activities such as antibacterial, antifungal, anticoagulant, contraceptive, anti-inflammatory, antineoplastic, anti-parasitic, anti-thermal, anti-genetic, antiviral, healing, chondroprotective and diuretic (Jausoro et al. 2010, Fonseca Filho et al. 2017).

Quixabeira (*Sideroxylon obtusifolium* Roem. & Schult) (Figure 8.20)

The *S. obtusifolium* (Roem. & Schult.) is a tree belonging to the *Sapotaceae* family, native to the Caatinga, and is popularly known as "quixabeira", "quixaba" or "rompegibão", but also by many other different common names (Figueiredo and Lima 2015). It is found in the North, Northeast and South of Brazil, with great frequency from Ceará to Rio Grande do Sul.

It is a deciduous or semi-deciduous tree, with dense crown and height of 7 to 18 m, hanging and thorny branches and its trunk bark rough and superficially fissured, with gray or brownish color, after cutting presents sparse latex. The leaves are simple and leathery, spiral or opposite, oblanceolate, measuring between 1.5–6.5 cm long by 0.5–3.5 cm wide and 3–9 mm long petiole. Species bloom between October and November; its flowers are odoriferous and clustered, with 2–20 flowers per fascicle, bisexual, yellowish-green or cream, axillary, tiny and aromatic, puberulent pedicel 2–5 mm long. Five sepals, 2–3 mm in length. Glorious corolla, 4–4.5 mm in length. Stamens (four) five, glabrous (Barbosa et al. 2019).

The fruit is of the drupe type, smooth and shiny, ranging from ellipsoid, obovoid to globose, rounded or truncated apex, rounded base, glabrous. Pericarp of blackish color, fleshy and juicy flesh, with average length, diameter, thickness and weight of 12; 10; 0.5–2 mm and 1 g. The seeds may vary from globose to ellipsoid, with average length, diameter and weight of 8 and 5 mm and 0.15 g, respectively. Pericarp-free forehead, smooth, rigid and shiny, light-brown and waterproof. Basal hilum, vertical embryo, with flat-convex cotyledons and a thin endosperm sheath around them (Da Silva et al. 2017, Barbosa et al. 2019).

According to Figueiredo and Lima (2015), this species is used in folk medicine in the form of vials, infusions and teas, using water or rum as a solvent, to treat gastritis, duodenal ulcer, genital damage, chronic inflammation, colic, kidney problems, cough treatment, digestive and heart disease, ovarian inflammation, diabetes, flu, gastritis, kidney and stroke pain, diabetes, wounds and pain in general. It is also used as a sitz bath for genitourinary problems, eyewash, injuries and general inflammation (Leite et al. 2015).

The phytochemical studies revealed the presence of triterpenes, steroids, tannins, alkaloids, carotenoids, flavonoids (flavonosides, flavones, flavononols and xantonas), saponins, cyanogenic compounds, carbohydrates and fatty acids are organic compounds (Aquino et al. 2016, Salgado et al. 2017). The isolated compounds include the triterpene glycoside 3-O-(β-D-glucopyranosyl)-protobassic acid 28-O-β-D-apiofuranosyl-(1→3)-O-[O-β-D-apiofuranosyl-(1→3)-β-D-xylopyranosyl-(1→4)]-O-α-L-rhamnopyranosyl-(1→2)-α-L-arabinopyranosyl ester. The flavonoids analyzed were, 5.7-dihydroxy-4, 5–dimethoxy–flavanone (homoesperidine), 5, 3-trihydroxia-4–Methyl-5-prenylflavanone (4-O-methylsigmoidin B), quercetin-3-O-(O-α-L-rhamnopyranosyl-(1→2)-O-[β-D-glucopyranosyl-(1→3)]-β-D-galactopyranoside) and kaempferol-3-O-(O-α-L-rhamnopyranosyl-(1→2)-O-[β-Dglucopyranosyl-(1→3)]-β-D-galactopyranoside), quercetin, catechin, among others (Oliveira et al. 2012).

Studies show that quixabeira bark has anti-inflammatory, analgesic, antinociceptive and antioxidant action in the treatment of pain and inflammation (Leite et al. 2015). The leaves also demonstrated anti-

Figure 8.20. *Sideroxylon obtusifolium* Roem. & Schult. (Quixabeira) tree (A), leaves and fruits (B), flowers (C) and seeds (D). Adapted from Silva and Dantas (2017).

inflammatory and antinociceptive action, in addition to antifungal action (Aquino et al. 2016, Silva and Dantas 2017).

Tamarindo (*Tamarindus indica* L.) (Figure 8.21)

The *Tamarindus indica* L., commonly known as tamarind, is a tropical tree belonging to the family Leguminosae, subfamily Caesalpiniaceae (Rai et al. 2018). It is very decorative, but slowly growing tree, reaching up to 30 m in height, canopy diameter of 12 m, circumference of 17, trunk of 7 m and, under favorable conditions, reaching up to 200 years of life (Santos-Serejo et al. 2009). Its availability is seasonal; in Brazil, it occurs mainly during the dry season of the year (July to October), but can be on the market all year round in the form of frozen pulp (García et al. 2012).

T. indica is one of the highly marketed medicinal plants and is known for its potent anti-inflammatory activities (Street and Prinsloo 2012, Menezes et al. 2016a). This tropical tree has been used in traditional medicine to treat inflammation, stomachache, sore throat and rheumatism (Komakech et al. 2019). It has also been used to control a myriad of other diseases including wound healing, diarrhea, dysentery, parasites, infestation, fever, malaria, respiratory disease, conditions, helminth infections, constipation, cellular cytotoxicity, gonorrhea, eye disease and as aphrodisiac (Dighe et al. 2009, Kuru 2014), as well as a food supplement (Ranaivoson et al. 2015).

Its bark has astringent and tonic characteristics, serving as a therapeutic digestive in folk medicine. When embedded in lotions or poultices, the bark can soothe wounds, ulcers, boils and rashes and contribute to healing (Kuru 2014). The aqueous extract of the checked shell showed 25% inhibitory effect on *Salmonella typhi*. The bioactive compounds stimulate the immune system and act on parasitemia (amount of parasites present in the bloodstream) (Nguta and Mbaria 2013).

Medicinal activities and the use of *T. indica* are attributed to the presence of phytochemicals in different parts of the plant, including flavonoids, alkaloids, tannins, phenols, triterpenoids, fatty acids, saponins and steroids (Escalona-Arranz et al. 2010, Gomathinayagam et al. 2017). Tamarind also contains vitamins A, B and C and organic acids such as citric acid, tartaric acid and malic acid (Rai et al. 2018).

Studies report that tamarind pulp extracts showed antioxidant and cholesterol-fighting properties, effects that can be attributed to the presence of phytochemicals such as phenolic compounds (Lim et al. 2013). Antioxidants help to balance the production and elimination of reactive oxygen species and other related compounds by inhibiting and reducing free radical damage to cells by preventing diseases such as cancer (Vieira et al. 2011).

Figure 8.21. *Tamarindus indica* L. (Tamarindo) tree (A), leaves (B), flowers (C) and fruits (D). Adapted from Komakech et al. (2019).

Umbu (*Spondias tuberosa* Arr.) (Figure 8.22)

The umbuzeiro (*Spondias tuberosa* Arr.) belongs to the genus *Spondias* and family Anacardiaceae, which is formed by about 20 species, distributed in Neotropical, Asiatic and Madagascar. One of the most common species in Brazil is the Umbuzeiro. About its botany, almost all Spondias have fibrous endocarp and leaflets with intra-marginal veins (Martins et al. 2019).

The Anacardiaceae family is composed of 81 genera and 800 species, which are found in various types of environments, climates, altitudes, rainfall, but mainly in tropical, subtropical and temperate regions (Pell et al. 2011). It has important phytochemical groups, which include the presence of tannins, saponins, resins, sterols and triterpenes, flavonoids and alkaloids, of relevance to the pharmaceutical industry, because they have groups of substances with analgesic, hypoglycemic, anti-inflammatory effects in humans and animals (Ojewole 2005, Siqueira et al. 2016a, Sameh et al. 2018), antimicrobial effect (Melo et al. 2014) and mostly have no toxicological effects (Barbosa et al. 2016).

The species *S. tuberosa* can be found in dry forests known as Caatinga (Albuquerque et al. 2007). The stem bark and fruits are used in medicine to treat a wide variety of diseases including diabetes mellitus, infections, digestive disorders, diarrhea, menstrual disorders and placental delivery (Albuquerque et al. 2007, Siqueira et al. 2016a).

Several studies address the versatility of *S. tuberosa* in the literature, especially scientific research that includes anti-inflammatory activity of umbuzeiro leaf extract in living organisms (Dias et al. 2015), the antioxidant potential of leaves (Uchôa et al. 2015, Siqueira et al. 2016b) and the extraction of oils from trunk bark and umbuzeiro leaves (Guimarães et al. 2018), besides enzyme extraction for use as adjuvants in different types of industries (Pinto et al. 2015).

Studies on the fruit of this plant have also reported several biological activities, such as antioxidant, antiviral, anti-inflammatory, antimicrobial activities, acetylcholinesterase inhibition and cancer chemopreventive activities (Zielinsk et al. 2014, Zeraik et al. 2016). In addition, umbu seeds are important sources of byproducts. Generally, these wastes contain large amounts of various bioactive compounds that can be extracted to provide various products such as nutraceuticals, food additives and pharmaceuticals (Ayala-Zavala et al. 2011). Potentially extractable active substances from residues include sugars, polysaccharides, proteins, fibers, natural flavoring compounds and various phytochemicals such as free fatty acids and polyphenols (Baiano 2014).

Figure 8.22. *Spondias tuberosa* Arr. (Umbu) tree (A), leaves (B), flowers (C) and fruits (D). Adapted from De Lima et al. (2018) and http://chaves.rcpol.org.br/profile/species/eco/eco:pt-BR:Spondias%20tuberosa.

Future Perspectives

The medicinal plants described above, as well as the various other plant species found in state of Piauí, northeast region and throughout the Brazilian territory, are an important tool in the treatment of diseases, both in terms of use by the population in the form of preparations, herbal medicines or *in natura*, regarding its use in scientific laboratories in the isolation of active principles. In this context, phytotherapy has become one of the main approaches chosen by people affected with the most varied pathologies, this being an area that is also gaining prominence in the official health systems (Piriz et al. 2014). In Brazil, the Unified Health System (SUS) adopted an official document that lists medicinal plants with proven actions and safe use in human health (Brazil 2006, Brazil 2010). This measure, in addition to assisting health professionals in recommending complementary treatments, also contributes to the enhancement of therapeutic plant species available in the country, which directly reflects in the increase in planting, preservation and incentive to scientific research with these species.

The anti-inflammatory action of the listed plants brings out the importance of cataloging the therapeutic vegetation in the region and dissemination to the population and scientific community, aiming to expand the relevance of associated traditional knowledge and awaken the eyes of public authorities and educational institutions for development research with these species, which show promise in the development of anti-inflammatory products.

Conclusion

The available literature provides evidence that a large composition of species with plant diversity can be observed in the northeast region, having mainly members of Adoxaceae, Anacardiaceae, Bignoniaceae, Combretaceae, Leguminosae, Myrtaceae, Olacaceae, Rhamnaceae, Rubiaceae, Sapotaceae, Solanaceae, Urticaceae and Xanthorrhoeaceae. The various species have great economic value, both in the area of ornamentation and in folk medicine. Traditional knowledge based on several biological and pharmacological activities has been presented with emphasis on the anti-inflammatory potential of some important plants occurring in the northeast region.

References

Abílio, G.M.F. 2011. Plantas Medicinais. Editora Universitária da UFPB - Bananeiras - PB, 23(1).

Adebayo, S.A., Dzoyem, J.P., Shai, L.J., Eloff, J.N. 2015. The anti-inflammatory and antioxidante activity of 25 plant species used traditionally to treat pain in southern African. BMC Complement. Altern. Med., 15: 159. DOI: 10.1186/s12906-015-0669-5.

Agalar, H. 2019. Elderberry (*Sambucus nigra* L.). Nonvitamin and Nonmineral Nutritional Supplements. Acad. Press, 211–215. DOI: 10.1016/C2016-0-03546-5.

Agra, M.D.F., Freitas, P.F.D., Barbosa-Filho, J.M. 2007. Synopsis of the plants known as medicinal and poisonous in Northeast of Brazil. Rev. Bras. Farmacogn., 17(1): 114–140. DOI: 10.1590/S0102-695X2007000100021.

Agra, M.D.F., Silva, K.N., Basílio, I.J.L.D., Freitas, P.F.D., Barbosa-Filho, J.M. 2008. Survey of medicinal plants used in the region Northeast of Brazil. Rev. Bras. Farmacogn., 18(3): 472–508. DOI: 10.1590/S0102-695X2008000300023.

Alencar, P.G., Espindola, G.M., Carneiro, E.L.N. 2018. Dwarf cashew crop expansion in the Brazilian semiarid region: Assessing policy alternatives in Pio IX, Piauí. Land Use Policy. 79: 1–9. DOI: 10.1016/j.landusepol.2018.07.042.

Albuquerque, U.P. 1997. Etnobotânica de uma bebida cerimonial no nordeste do Brasil. Rev. Bras. Farm., 78(4): 86–89.

Albuquerque, U.P. 1999. Referências para o estudo da etnobotânica dos descendentes culturais do africano no Brasil. Acta Farm. Bom., 18(4): 299–306.

Albuquerque, U.P., Medeiros, P.M., Almeida, A.L.S., Monteiro, J.M., Lins Neto, E.M.F., Melo, J.G., Santos, J.P. 2007. Medicinal plants of the caatinga (semi-arido) vegetation of NE Brazil: a quantitative approach. J. Ethnopharmacol., 3: 325–354. DOI: 10.1016/j.jep.2007.08.017.

Albuquerque, U.P., Medeiros, P.M., Almeida, A.L.S., Monteiro, J.M., Lins-Neto, E.M.F., Melo, J.G., Santos, J.P. 2009. Plantas medicinais da Caatinga e Floresta Atlântica nordestina: aspectos químicos, ecológicos e culturais.

Almeida, M.L.B., de Souza Freitas, W.E., de Morais, P.L.D., Sarmento, J.D.A., Alves, R.E. 2016. Bioactive compounds and antioxidant potential fruit of *Ximenia americana* L. Food Chem., 192: 1078–1082. DOI: 10.1016/j.foodchem.2015.07.129.

Almeida, R.R., Raimundo, J.M., Oliveira, R.R., Kaplan, M.A.C., Gattas, C.R., Sudo, R.T., Zapata-Sudo, G. 2006. Activity of *Cecropia lyratiloba* extract on contractility of cardiac and smooth muscles in Wistar rats. Clin. Exp Pharmacol Physiol., 33(1): 109–113. DOI:10.1111/j.1440-1681.2006.04332.x.

Amorim, J., de Carvalho Borges, M., Fabro, A.T., Contini, S.H.T., Valdevite, M., Pereira, A.M.S., Carmona, F. 2019. The ethanolic extract from *Erythrina mulungu* Benth. flowers attenuates allergic airway inflammation and hyperresponsiveness in a murine model of asthma. J. Ethnopharmacol., 242: 111467. DOI: 10.1016/j.jep.2018.08.009.

Andrade, E.N. O eucalipto. 1961. Companhia Paulista de Estradas de Ferro, Jundiaí, SP, 667 p.

Antunes, R., Garcia, L., Somerset, V., Gil, E., Lopes, F. 2018. The use of a polyphenoloxidase biosensor obtained from the fruit of jurubeba (*Solanum paniculatum* L.) in the determination of paracetamol and other phenolic drugs. Biosensors, 8(2): 36. DOI: 10.3390/bios8020036.

Aquino, P., Gomes Figueredo, F., Pereira, N., Nascimento, E., Martin, A., Veras, H., Menezes, I. 2016. Evaluation of the topical anti-inflammatory activity and antibacterial activity of methanol extract in the *Sideroxylon obtusifolium* leaves. Acta Biol. Colombiana, 21(1): 131–140. DOI: 10.15446/abc.v21n1.48170.

Aragão, D.M., Guarize, L., Lanini, J., Costa, J.C., Garcia, R.M., Scio, E. 2010. Hypoglycemic effects of *Cecropia pachystachya* in normal and alloxan-induced diabetic rats. J. Ethnopharmacol., 128(3): 629–633. DOI:10.1016/j.jep.2010.01.008.

Aragão, D.M.O., Lima, I.V.A., da Silva, J.M., Bellozi, P.M.Q., da Costa, J.C. 2013. Efeitos anti-inflamatórios, antinociceptivos e citotóxicos do extrato de metanol de *Cecropia Pachystachya* Trécul. Pesq. Fitot., 27(6): 926–930. DOI:10.1002/ptr.4811.

Aragão, T., Prazeres, L., Brito, S., Neto, P., Rolim, L., Almeida, J., Wanderley, A. 2018. Contribution of secondary metabolites to the gastroprotective effect of aqueous extract of *Ximenia americana* L. (Olacaceae) stem bark in rats. Molecules, 23(1): 112. DOI: 10.3390/molecules23010112.

Arjoon, A., Saylor, C., May, M. 2012. *In vitro* efficacy of antimicrobial extracts against the atypical ruminant pathogen, Mycoplasma mycoides subsp. capri. BMC Complement. Altern. Med., 12: 169. DOI: 10.1186/1472-6882-12-169.

Ayala-Zavala, J.F., Vega-Vega, V., Rosas-Domínguez, C., Palafox-Carlos, H., Villa-Rodriguez, J.A., Siddiqui, M.W., Dávila-Aviña, J.E., GonzálezAguilar, G.A. 2011. Agro-industrial potential of exotic fruit byproducts as a source of food additives. Food Res. Int., 44: 1866–1874. DOI: 10.1016/j.foodres.2011.02.021.

Baiano, A. 2014. Recovery of biomolecules from food wastes—A review. Molecules, 19: 14821–14842. DOI: 10.3390/molecules190914821.

Bahia, M.V., Santos, J.B.D., David, J.P., David, J.M. 2005. Biflavonoids and other phenolics from Caesalpinia pyramidalis (Fabaceae). J Braz Chem Soc., 16(6): 1402–1405. DOI:10.1590/S0103-50532005000800017.

Balbach, A. 1988. As Plantas Que Curam. Três, São Paulo – SP, Brazil.

Barbosa, A.F., Costa, I.C.M., Zucolotto, S.M., Giordani, R.B. 2017. *Morinda citrifolia*: fatos e riscos sobre o uso do noni. Rev. Fitos, 11(2): 119–249. DOI: 10.5935/2446-4775.20170027.

Barbosa, D.A., Lucena, R.F., Cruz, D.D. 2019. Traditional knowledge as basis for phytochemical prospecting of *Sideroxylon obtusifolium* (Roem. & Schult.) TD Penn. aiming at conservation in the Brazilian semi-arid zone. Ethnobot. Res. Appl., 18: 1–10. DOI: 10.32859/era.18.3.1-10.

Barbosa, H.M., Nascimento, J.N., Araújo, T.A., Duarte, F.S., Albuquerque, U.P., Vieira, J.R., Lira, E.C. 2016. Acute toxicity and cytotoxicity effect of ethanolic extract of *Spondias tuberosa* Arruda Bark: Hematological, biochemical and histopathological evaluation. Na. Acad. Bras. Ciênc., 88: 993–2004.

Barak, V., Halperin, T., Kalickman, I. 2019. The effect of Sambucol, a black elderberry based, natural product, on the production of human cytokines: I. Inflammatory cytokines. Eur. Cytokine Netw., 12: 290–296.

Basas-Jaumandreu, J., De Las Heras, F.X.C. 2019. Allelochemicals and esters from leaves and inflorescences of *Sambucus nigra* L. Phytochem., 30: 107–115. DOI: 10.1016/j.phytol.2019.01.030.

Bergeron, C., Boulet, L.P., Laviolette, M., Zimmermann, N., Rothenberg, M.E., Hamid, Q. 2007. Influence of cigarette smoke on asthmatic airways: increased expression of arginase I. J. Allergy Clin. Immunol., 119(2): 391–397. DOI:10.1016/j.jaci.2006.10.030.

Bernatoniene, J., Kopustinskiene, D.M. 2018. The role of catechins in cellular responses to oxidative stress. Molecules, 23: 965–975. DOI:10.3390/molecules23040965.

Bezerra, D.A.C., Rodrigues, F.F.G., Costa, J.G.M., Pereira, A.V., Sousa, E.O., Rodrigues, O.G. 2011. Abordagem fitoquímica, composição bromatológica e atividade antibacteriana de Mimosa tenuiflora (Wild) Poiret e Piptadenia stipulacea (Benth) Ducke. Acta Scientiarum, 33: 99–106. DOI: 10.14295/CS.v8i1.1493.

Blumenthal, M., Goldberg, A., Brinckmann, J. 2000. Herbal Medicine: Expanded Commission E Monographs. ABC, 519.

Braga, R. 1976. Plantas do Nordeste, Especialmente do Ceará. DNOCS, 540.

Brandão, M.G.L. 2014. Plantas Medicinais E Fitoterápicos. Banco de Dados e Amostras de Plantas Aromáticas, Medicinais e Tóxicas Museu de História Natural e Jardim Botânico, (1).

Brasil. 2006. Ministério da Saúde. Política Nacional de Plantas Medicinais e Fitoterápicos. Brasília: Ministério da Saúde, 60 p.

Brasil. 2010. Ministério da saúde. Agência Nacional de Vigilância Sanitária. Resolução --RDC Nº 10 de 09 de março de 2010. Dispõe sobre a notificação de drogas vegetais. Brasília: Ministério da Saúde.

Brasileiro, M.T., Egito, A.A., Lima, J.R., Randau, K.P., Pereira, G.C., Neto, P.J.R. 2008. *Ximenia americana* L.: botânica, química e farmacologia no interesse da tecnologia farmacêutica. Rev. Bras. Farm, 89(2): 164–7.

Buchmann, S.L., Jones, C.E., Colin, L.J. 1977. Vibratile pollination of *Solanum douglasii* and *Solanum xanti* (Solanaceae) in southern California. Wasmann J. Biol., 35: 1–25.

Campos, L.F.C., Peixoto, J.V.M., De Oliveira, R.M., Seleguini, A., Dos Reis Nascimento, A. 2015. Propriedades físico-químicas de frutos de jurubeba de três regiões do Cerrado. J. Neotrop. Agricul., 2(4): 48–54. DOI: 10.32404/rean.v2i4.740.

Cardoso, M.P., Lima, L.S., David, J.P., Moreira, B.O., Santos, E.O., David, J.M., Alves, C.Q. 2015. A new biflavonoid from *Schinopsis brasiliensis* (Anacardiaceae). J. Braz. Chem. Soc., 26(7): 1527–1531. DOI: 10.5935/0103-5053.20150101.

Carniello, M.A., Silva, R.S., Cruz, M.A.B., Guarim-Neto, G. 2010. Quintais urbanos de Mirassol D'Oeste-MT Brasil: uma abordagem etnobotânica. Acta Amaz., 40(3): 451–470. DOI:10.1590/S0044-59672010000300005.

Carvalho, A.C.C., Almeida, D.S., Melo, M.G., Cavalcanti, S.C., Marçal, R.M. 2009. Evidence of the mechanism of action of *Erythrina velutina* Willd (Fabaceae) leaves aqueous extract. J. Ethnopharmacol., 122(2): 374–378. DOI: 10.1016/j.jep.2008.12.019.

Carvalho, G.F.S., Marques, L.K., Souza, H.G., Silva, L.R., Ferreira, D.C.L., Amaral, F.P.M.A.L., Filho, M.M., Figueredo-Silva, J., Alves, W.S., Oliveira, M.D.A., Júnior, J.S.C. Júnior, F.L.C.C., Ramos, R.M., Rai, M., Uchôa, V.T., 2020. Phytochemical study, molecular docking, genotoxicity and therapeutic efficacy of the aqueous extract of the stem bark of *Ximenia americana* L. in the treatment of experimental COPD in rats. J. Ethnopharmacol., 247: 1–12. DOI:10.1016/j.jep.2019.112259.

Carvalho, P.E.R. 2008. Mulungu (*Erythrina velutina*). Embrapa Florestas-Circular Técnica (INFOTECA-E).

Cavalcanti, P., Martins, M., Do Carmo, C., Nunes, P.H., Alves Filho, F.C., Silva, J.D., Cavalcanti, S. M. 2019. Antidiarrheal effect of extract from the bark of *Combretum leprosum* in mice. An. Acad. Bras. Cienc., 91(1). 1–13. DOI:10.1590/0001-3765201820170932.

Cavalheiro, M.G., Farias, D.F., Fernandes, G.S., Nunes, E.P., Cavalcanti, F.S., Vasconcelos, I.M., Melo, V.M.M., Carvalho, A.F.U. 2009. Atividades biológicas e enzimáticas do extrato aquoso de sementes de *Caesalpinia ferrea* Mart., Leguminosa. Rev. Bras. Farmacogn., 19: 586–591. DOI: 10.1590/S0102-695X2009000400014.

CECH, R. 2000. Making plant medicine. Horizon Herbs. ISBN 0970031203.

Chacón, O., Forno, N., Laoierre, L., Munõz, R., Fresno, M., Martín, B.S. 2019. Effect of *Aloe barbadensis* Miller (Aloe vera) associated with beta-lactam antibiotics on the occurrence of resistance in strains of *Staphylococcus aureus* and *Streptococcus uberis*. Eur. J. Integr. Med., 32: 109–196. DOI:10.1016/j.eujim.2019.100996.

Ciocoiu, M., Mirón, A., Mares, L., Tutunaru, D., Pohaci, C., Groza, M., Badescu, M. 2009. The effects of *Sambucus nigra* polyphenols on oxidative stress and metabolic disorders in experimental diabetes mellitus. J. Physiol. Biochem., 65: 297–304. DOI: 10.1007/BF03180582.

Clementino-Neto, J., Pereira, J.C., Vasconcelos, L.H., de Souza, I.L., Silva, A.D., Silva, T.M., De Cavalcante, F.A. 2016. Toxicological, antidiarrheal and spasmolytic activities of *Solanum paniculatum*. Planta Med., 82(01/02): 58–64. DOI: 10.1055 / s-0035-1557831.

Corrêa, M.P. 1984. Dicionário das Plantas Úteis do Brasil. Brasília: IBDF.

Correia, A.A.S., Gonzaga, M.L.C., Aquino, A.C., Souza, P.H.M., Figueiredo, R.W., Maia, G.A. 2011. Caracterização química e fisioquímica da polpa do noni (*Morinda citrifolia*) cultivado no estado do Ceará. Alimentos e Nutrição, Fortaleza, 22(4): 609–615.

Consolini, A.E., Migliori, G.N. 2005. Cardiovascular effects of the South American medicinal plant *Cecropia pachystachya* (Ambay) on rats. J. Etnofarmacol., 96(3): 417–422. DOI: 10.1016/j.jep.2004.09.030.

Cruz, E.M., Silva, E.R., Maquiaveli, C.C., Alves, E.S.S., Lucon Jr, J.F., Reis, M.B.G., Toledo, C.E.M., Cruz, F.G., Vannier-Santos, M.A. 2013. Leishmanicidal activity of *Cecropia pachystachya* flavonoids: Arginase inhibition and altered mitochondrial DNA arrangement. Phytochemistry, 89: 71–77. DOI:10.1016/j.phytochem.2013.01.014.

Coutinho, M.A.S., Muzitano, M.F., Costa, S.S. 2009. Flavonoides: Potenciais Agentes Terapêuticos para o Processo Inflamatório. Rev. Virtual Quim., 3: 241–256. DOI:10.5935/1984-6835.20090024.

Dantas, M.D., De Oliveira, F.S., Bandeira, S.M., Batista, J.S., Silva Jr, C.D., Alves, P.B., Marchioro, M. 2004. Central nervous system effects of the crude extract of *Erythrina velutina* on rodents. J. Ethnopharmacol., 94(1): 129–133. DOI: 10.1016/j.jep.2004.05.007.

Dantas, B.F., Soares, F.D.J., Lúcio, A.A., Aragão, C.A. 2008. Alterações bioquímicas durante a embebição de sementes de baraúna (*Schinopsis brasiliensis* Engl.). Embrapa Semiárido-Artigo em periódico indexado (ALICE).

Da Cunha, A.M., Menon, S., Menon, R., Couto, A.G., Bürger, C., Biavatti, M.W. 2010. Hypoglycemic activity of dried extracts of *Bauhinia forficata* Link. Phytomedicine, 17(1): 37–41. DOI: 10.1016/j.phymed.2009.06.007.

Dawidowicz, A.L., Wianowska, D., Baraniak, B. 2006. The antioxidant properties of alcoholic extracts from *Sambucus nigra* L. (antioxidant properties of extracts). LWT Food Sci. Technol., 39: 308–315. DOI: 10.1016/j.lwt.2005.01.005.

De Menezes, I.R.A., Da Costa, R.H.S., Boligon, A.A., Rolón, M., Coronel, C., Vega, C., De Albuquerque, T.R. 2019. *Ximenia americana* L. enhances the antibiotic activity and inhibit the development of kinetoplastid parasites. Comp. Immunol. Microbiol. Infect. Dis., 64: 40–46. DOI: 10.1016/j.cimid.2019.02.007.

De Souza, G.R., De Oliveira, A.C.A., Soares, V., Chagas, L.F., Barbi, N.S., Paumgartten, F.J.R., Da Silva, A.J.R. 2019. Chemical profile, liver protective effects and analgesic properties of a *Solanum paniculatum* leaf extract. Biomed. Pharmacother., 110: 129–138. DOI: 10.1016/j.biopha.2018.11.036.

De Souza Santos, C.C., Guilhon, C.C., Moreno, D.S.A., Alviano, C.S., Dos Santos Estevam, C., Blank, A.F., Fernandes, P.D. 2018. Anti-inflammatory, antinociceptive and antioxidant properties of *Schinopsis brasiliensis* bark. J. Ethnopharmacol., 213: 176–182. DOI: 10.1016/j.jep.2017.11.012.

Delarcina, S., Lima-Landman, M.T.R., Souccar, C., Cysneiros, R.M., Tanae, M.M., Lapa, A.J. 2007. Inhibition of histamine-induced bronchospasm in guinea pigs treated with *Cecropia glaziovi* Sneth and correlation with the in vitro activity in tracheal muscles. Phytomedicine, 14(5): 328–332. DOI:10.1016/j.phymed.2006.12.022.

De Lima, M.A.C., Silva, S.D.M., De Oliveira, V.R. 2018. Umbu – *Spondias tuberosa*. 427–433. *In*: Exotic Fruits. DOI: 10.1016/B978-0-12-803138-4.00057-5.

Demougeot, C., Prigent-Tessier, A., Bagnost, T., André, C., Guillaume, Y., Bouhaddi, M., Marie, C., Berthelot, A. 2007. Time course of vascular arginase expression and activity in spontaneously hypertensive rats. Life Sci., 80(12): 1128–1134. DOI:10.1016/j.lfs.2006.12.003.

Dias, J., Falcão, R., Albuquerque, H., Dantas, R., Medeiros, A., Simões, M., Santos, V. 2015. Potencial atividade anti-inflamatória de *Spondia tuberosa* Arr. Cam. (Umbú). BIOFARM, 10: 01–03.

Dighe, N.S., Pattan, S.R., Nirmal, S.A., Kalkotwar, R.S., Gaware, V.M., Hole, M.B. 2009. Analgesic activity of *Tamarindus indica*. J. Pharmacogn. Phytochem., 1: 69–71.

Diniz, P.B., Ribeiro, A.R.S., Estevam, C.S., Bani, C.C., Thomazzi, S.M. 2015. Possible mechanisms of action of *Caesalpinia pyramidalis* against ethanol-induced gastric damage. J. Ethnopharmacol., 168: 79–86. doi.org/10.1016/j.jep.2015.03.054.

Dobrian, A.D., Lieb, D.C., Cole, B.K., Taylor-fishwick, D.A., Chakrabarti, S.K., Nadler, J.L. 2011. Functional and pathological roles of the 12- and 15-lipoxygenases. Prog. Lipid Res., 50: 115–131. DOI: 10.1016/j.plipres.2010.10.005.

Donati, M., Mondin, A., Chen, Z., Miranda, F.M., do Nascimento Jr, B.B., Schirato, G., Froldi, G. 2015. Radical scavenging and antimicrobial activities of *Croton zehntneri*, *Pterodon emarginatus* and *Schinopsis brasiliensis* essential oils and their major constituents: estragole, trans-anethole, β-caryophyllene and myrcene. Nat. Prod. Res., 29(10): 939–946. DOI: 10.1080/14786419.2014.964709.

Dos Reis Luz, L., Porto, D.D., Castro, C.B., Silva, M.F.S., Alves Filho, E.G., Canuto, K.M., Zocolo, G.J. 2018. Metabolomic profile of *Schinopsis brasiliensis* via UPLC-QTOF-MS for identification of biomarkers and evaluation of its cytotoxic potential. J. Chromatogr. B, 1099: 97–109. DOI: 10.1016/j.jchromb.2018.09.019.

EMA. 2008. Assessment Report for the Development of Community Monographs and for Inclusion of Herbal Substance(s). Preparation(s) or combinations thereof in the list: *Sambucus nigra* L. London flows. EMEA/HMPC/283170/2007.

Escalona-Arranz, J.C., Pérez-Rosés, R., Jiménez, I.L., Rodríguez-Amado, J., Argota-Coello, H., Cañizares-Lay, J., Morris-Quevedo, H.J., Sierra-González, G. 2010. Chemical constituents of *Tamarindus indica* L. leaves. Rev. Cub. Quím., 22: 65–71.

Farag, M.A., Sakna, S.T., El-fiky, N.M., Shabana, M.M., Wessjohann, L.A. 2015. Phytochemical, antioxidant and antidiabetic evaluation of eight *Bauhinia* L. species from Egypt using UHPLC–PDA–qTOF-MS and chemometrics. Phytochem., 119: 41–50. DOI: 10.1016/j.phytochem.2015.09.004.

Fernandes, F.H., Batista, R.S.D.A., Medeiros, F.D.D., Santos, F.S., Medeiros, A.C. 2015. Development of a rapid and simple HPLC-UV method for determination of gallic acid in *Schinopsis brasiliensis*. Rev. Bras. Farmacogn., 25(3): 208–211. DOI: 10.1016/j.bjp.2015.05.006.

Fernandes, F.M., Graichen, F.A.S., Silva, A.R.D.D., Rêgo, N.H. 2018. Occurrence of rusts on *Handroanthus impetiginosus* and *Handroanthus vellosoi* in the state of Mato Grosso do Sul. Rodriguésia, 69(4): 2245–2252. DOI:10.1590/2175-7860201869449.

Figueiredo, F.J., Lima, V.L.A.G. 2015. Antioxidant activity of anthocyanins from quixabeira (*Sideroxylon obtusifolium*) fruits. Rev. Bras. Pl. Med., 17(3): 473–479. DOI:10.1590/1983-084X/14_005.

Felix, F.C., de Medeiros, J.A., Pacheco, M.V. 2018. Morphology of seeds and seedlings of *Handroanthus impetiginosus* (Mart. ex DC.) Mattos. Rev. Ciênc. Agrár., 41(4): 161–170. DOI:10.19084/RCA18126.

Fehrmann-Cartes, K., Coronado, M., Hernandéz, A.J., Allende, M.L., Feijoo, C.G. 2019. Anti-inflammatory effects of *Aloe vera* on soy meal-induced intestinal inflammation in zebrafish. Fish Shellfish Immun., 75: 564–576. DOI:10.1016/j.fsi.2019.10.075.

Fonseca Filho, I.C., Bomfim, B.L.S., Farias, J.C., Fábio José Vieira, F.J., Barros, R.F.M. 2017. Pau-d'arco-roxo (*Handroanthus impetiginosus* (Mart. ex DC.) Mattos): knowledge and use of wood in rural communities in northeast Brazil. Gaia Sci., 11(2): 57–70. DOI: 10.22478/ufpb.1981-1268.2017v11n2.34878.

Gagnon, E., Lewis, G.P., Sotuyo, J.S., Hughes, C.E., Bruneau, A. 2013. A molecular phylogeny of *Caesalpinia sensu* lato: Increased sampling reveals new insights and more genera than expected. S. Afr. J. Bot., 89: 111–127. DOI: 10.1016/j.sajb.2013.07.027.

Galvão, A.P.M., Medeiros, A.D.S. 2002. Restoration of the Atlantic Forest in areas of its original natural occurrence. Embrapa Florestas-Livro Científico (ALICE).

Garcez, F.R., Garcez, W.S., Yoshida, N.C., Figueiredo, P.O. 2016. A diversidade dos constituintes químicos da flora de Mato Grosso do Sul e sua relevância como fonte de substâncias bioativas. Rev. Virtual Quim., 8: 97–129. DOI: 10.5935/1984-6835.20160008.

García, J.C.V., Rodríguez, K.A.F., López, F.G., Pérez, E.G., Rosado, O.L., Rosas, F.H. 2012. Systems management and marketing of tamarind (*Tamarindus indica* L.) in three municipalities of Veracruz. Rev. Mex. Cienc. Agríc., 3: 1217–1230. DOI: 10.29312/remexca.v3i6.1373.

Garcia, J., Jacobson, T.K.B., Farias, J.G., De Fátima Boaventura, R. 2008. Effectiveness of methods to increase the germination rate of jurubeba (*Solanum paniculatum* L.) seeds. Pesqui. Agropecu. Trop., 38(3): 223–226.

Gomathinayagam, S., Tewari, B., Rekha, G. 2017. Heavy metal and phytochemical studies of crude leaf extract of tamarind (*Tamarindus indica*). Adv. Life Sci., 7: 1–4.

Gentry, A.H. 1992. Bignoniaceae: Part II (Tribe Tecomeae). Flora Neotropica, 1–370.

Grandi, T.S.M. 2015. Tratado da Plantas Medicinais Mineiras, Nativas e Cultivadas. Belo Horizonte: Adequatio Estudio. (1): 1204.

Guimarães, A.L., Oliveira, A.P., Silva, G.S.F.S., Bezerra, G.S., Sousa, I., Almeida, J.R.G.S., Rodrigues, J.D.S., Alencar Filho, J.M.T., Araújo, E.C.C. 2018. Gas chromatography mass spectrometry (GC-MS) analysis of the constituents of the fixed oils obtained from the barks, leaves and stems of *Spondias tuberosa* Arruda (Anacardiaceae). J. Med. Plant. Res., 12: 89–95.

Hikawczuk, V.J., Saad, J.R., Guardia, T., Juarez, A.O., Giordano, O.S. 1998. Anti-inflammatory activity of compounds isolated from *Cecropia pachystachya*. An Asoc Quím Argent., 86(3): 167–170.

Horinouchi, C.D.S., Mendes, D.A.G.B., Soley, B.S., Pietrovski, E.S., Facundo, V.A., Santos, A.R.S., Cabrini, D.A., Otuki, M.F. 2013. *Combretum leprosum* Mart. (Combretaceae): Potential as an antiproliferative and anti-inflammatory agente. J. Ethnopharmacol., 145: 311–319. DOI:10.1016/j.jep.2012.10.064.

Hussain, S.A. 1998. Effect of the local specie of Aloe on the healing process. Pakistan J. Pharmacol., 15(1): 31. DOI:10.1371/journal.pone.0164799.

ITIS. 2016. *Sambucus* L. Taxonomic Serial no.: 35315. Integrated Taxonomic Information System Taxonomy and Nomenclature. Accessed Google Scholar. http://www.itis.gov/servlet/SingleRpt/SingleRpt?search_topic=TSN&search_value=35315.

Jausoro, V., Lorente, B.E., Apóstolo, N.M. 2010. Structural differences between hyperhydric and normal *in vitro* shoots of *Handroanthus impetiginosus* (Mart. ex DC). J. Plant. Biotechnol., 101(2): 186–191.

Jeengar, M.K., Shrivastava, S., Nair, K., Singareddy, S.R., Putcha, U.K., Talluri, M.V.N.K., Naidu, V.G.M., Sistla, R. 2014. Improvement of bioavailability and antiinflammatory potential of curcumin in combination with Emu oil. Inflammat., 37: 2139–2155. DOI: 10.1007/s10753-014-9948-4.

Jing, P., Bomser, J.A., Schwartz, S.J., He, J., Magnuson, B.A., Giusti, M.M. 2008. Structure− function relationships of anthocyanins from various anthocyanin-rich extracts on the inhibition of colon cancer cell growth. J. Agr. Food. Chem., 56: 9391–9398. DOI: 10.1021/jf8005917.

Júnior, W.S.F., Ladio, A.H., De Albuquerque, U.P. 2011. Resilience and adaptation in the use of medicinal plants with suspected anti-inflammatory activity in the Brazilian Northeast. J. Ethnopharmacol., 138(1): 238–252. DOI: 10.1016/j.jep.2011.09.018.

Kaur, C., Nagal, S., Nishad, J., Kumar, R. 2014. Evaluating eggplant (*Solanum melongena* L.) genotypes for bioactive properties: A chemometric approach. Food Res. Inter., 60: 205–211. DOI: 10.1016/j.foodres.2013.09.049.

Komakech, R., Kim, Y.G., Matsabisa, G.M., Kang, Y. 2019. Anti-inflammatory and analgesic potential of *Tamarindus indica* Linn. (Fabaceae): a narrative review. Integr. Med. Res., 8: 181. DOI: 10.1016/j.imr.2019.07.002.

Kurahashi, T., Fujii, J. 2015. Roles of antioxidative enzymes in wound healing. J. Dev. Biol., 3: 57–70. DOI:10.3390/jdb3020057.

Kuru, P. 2014. *Tamarindus indica* and its health related effects. Asian Pac. J. Trop. Biomed., 4: 676–681. DOI: 10.12980/APJTB.4.2014APJTB-2014-0173.

Lacaille-Dubois, M.A., Franck, U., Wagner, H. 2001. Search for potential angiotensin converting enzyme (ACE)-inhibitors from plants. Phytomedicine, 8(1): 47–52. DOI:10.1078/0944-7113-00003.

Landrum, L.R., kawasaki, M.L. 1997. The genera of Myrtaceae in Brazil: an illustrate synoptic treatment and identification keys. Brittonia, 49(4): 508–536. DOI:10.2307/2807742.

Lawal, F., Bapela, M.J., Adebayo, S.A., Nkadimeng, S.M., Yusuf, A.A., Malterud, K.E., McGaw, L.J., Tshikalange, T.E. 2019. Anti-inflammatory potential of South African medicinal plants used for the treatment of sexually transmitted infections. S. Afr. J. Bot., 125: 62–71. DOI:10.1016/j.sajb.2019.06.023.

Leite, N.S., Lima, A.P., Araújo-Neto, V., Estevam, C.S., Pantaleão, S.M., Camargo, E. A., Thomazzi, S.M. 2015. Evaluation of the cicatrizing, topical anti-inflammatory and antioxidant activities of the ethanol extract of *Sideroxylon obtusifolium*. Rev. Bras. Pl. Med., 17(1): 164–170. DOI:10.1590/1983-084X/09_189.

Lima, J.D., Almeida, C.C., Dantas, V.A.V., Silva, B.M.S., Moraes, W., da S. 2006. Efeito da temperatura e do substrato na germinação de sementes de *Caesalpinia ferrea* Mart. ex Tul. (Leguminosae, Caesalpinoideae). Rev. Árvore., 30: 513–518. DOI: 10.1590/S0100-67622006000400003.

Lima, C.B.L., Bellettini, N.M.T., Silva, A.S., Cheirubim, A.P., Janani, J.K., Vieira, M.A.V., Amador, T.S. 2007. O uso de plantas medicinais pela população da zona urbana de Bandeirantes, PR. Rev. Bras. Biociências, 5(1): 600–602.

Lima, P.R., Horbach, M.A., Dranski, J.A.L., Ecco, M., Malavasi, M.D.M., Malavasi, U.C. 2014. Morphophysiological evaluation in seedlings of *Handroanthus impetiginosus* (Mart. ex DC.) Mattos during hardening. Floresta e Ambiente, 21(3): 316–326. DOI: 10.1590/2179-8087.058813.

Lim, C.Y., Junit, S.M., Abdulla, M.A., Aziz, A.A. 2013. *In vivo* biochemical and gene expression. Analyses of the antioxidante activities and hypocholesrolaemic properties of *Tamarindus indica* fruit pulp extract. Plos One, California, 8: e70058. DOI: 10.1371/journal.pone.0070058.

Lim, H., Kim, M.K., Lim, Y., Cho, Y.H., Lee, C.H. 2006. Inhibition of cell-cycle progression in HeLa cells by HY52, a novel cyclin-dependent kinase inhibitor isolated from *Bauhinia forficata*. Cancer let., 233(1): 89–97. DOI: 10.1016/j.canlet.2005.03.002.

Loiola, M.I.B., Rocha, E.A., Baracho, G.S., Agra, M.D.F. 2009. Flora da Paraíba, Brasil: Combretaceae. Act. Bot. Brasilica, 23(2): 330–342. DOI:10.1590/S0102-33062009000200005.

Longhi-Balbinot, D.T., Lanznaster, D., Baggio, C.H., Silva, M.D., Cabrera, C.H., Facundo, V.A., Santos, A.R. 2012. Anti-inflammatory effect of triterpene 3β, 6β, 16β-trihydroxylup-20(29)-ene obtained from *Combretum leprosum* Mart & Eich in mice. J. Ethnopharmacol., 142(1): 59–64. DOI: 10.1016/j.jep.2012.04.013.

Lopes, L.S., Marques, R.B., Pereira, S.S., Ayres, M.C.C., Chaves, M.H., Cavalheiro, A.J., Almeida, F.R.C. 2010. Antinociceptive effect on mice of the hydroalcoholic fraction and (-) epicatechin obtained from *Combretum leprosum* Mart & Eich. Braz. J. Med. Biol. Res., 43(12): 1184–1192. DOI: 10.1590/S0100-879X2010007500121.

Lorenzi, H. 2002. Árvores Brasileiras: Manual de identificação e cultivo de plantas arbóreas do Brasil. Instituto Plantarum de Estudos da Flora, (2): 368. ISBN: 85-8671407-0.

Lorenzi, H., Matos, F.J. 2002a. Plantas medicinais no Brasil: nativas e exóticas. Instituto Plantarum de Estudos da Flora, (1): 512. ISBN: 8586714186.

Lorenzi, H., Matos, F.J.A. 2002b. *Cecropia pachystachya* Trécul. Plantas Medicinais no Brasil: nativas e exóticas cultivadas. Instituto Plantarum de Estudos da Flora LTDA. Nova Odessa, 118–119.

Lucena, R.F.P. 2009. Avaliando a eficiência de diferentes técnicas de coleta e análise de dados para a conservação da biodiversidade a partir do conhecimento local do, Nordeste do Brasil. (Tese de Doutorado) Universidade Federal Rural de Pernambuco, Recife-PE.

Macêdo, M.J.F., Ribeiro, D.A., Santos, M.O., Macêdo, D.G., Macedo, J.G.F., Almeida, B.V., Saraiva, M.E., Lacerda, M.N.S., Souza, M.M.A. 2018. Fabaceae medicinal flora with therapeutic potential in Savanna áreas in the Chapada do Araripe, Northeastern Brazil. Rev. Bras. Farmacogn., 28: 738–750. DOI:10.1016/j.bjp.2018.06.010.

Maia, G.N. 2004. Caatinga: árvores e arbustos e suas utilidades. São Paulo: D&Z Computação Gráfica e Editora, 413p.

Maia-Silva, C., da Silva, C.I., Hrncir, M. 2012. Guia de plantas visitadas por abelhas na Caatinga, 1. ed. Fortaleza, CE: Editora Fundação Brasil Cidadão.

Majumder, R., Das, C.K., Mandal, M. 2019. Lead bioactive compounds of *Aloe vera* as potential anticancer agent. Pharmacol. Res., 148: 104–116. DOI:10.1016/j.phrs.2019.104416.

Marques, G.S., Leão, W.F., Lyra, M.A., Peixoto, M.S., Monteiro, R.P., Rolim, L.A., Soares, L.A. 2013. Comparative evaluation of UV/VIS and HPLC analytical methodologies applied for quantification of flavonoids from leaves of *Bauhinia forficata*. Rev. Bras. Farmacog., 23(1): 51–57. DOI: 10.1590/S0102-695X2012005000143.

Martins, G., Balbino, E., Marques, A., Almeida, C. 2019. Complete mitochondrial genomes of the *Spondias tuberosa* Arr. Cam and *Spondias mombin* L. reveal highly repetitive DNA sequences. Gene, 720: 144026. DOI: 10.1016/j.gene.2019.144026.

Matos, F.J.A. 2000. Plantas medicinais. Guia de seleção e emprego de plantas usadas em fitoterapia no Nordeste do Brasil. 2. ed. Ceará: UFC.

Matos, F.J.A. 2007. Plantas medicinais: guia de seleção e emprego das plantas usadas em fitoterapia no Nordeste do Brasil, 3a. ed., Imprensa Universitária: Fortaleza.

Matos, F.J.A. 2009. Introdução à fitoquímica experimental, 3a. ed., Editora UFC: Fortaleza.

Melo, M.C., Gadelha, D.N., Oliveira, T.K., Brandt, C.T. 2014. Alcohol extract of *Schinu sterebinthifolius* Raddi (Anacardiaceae) as a local antimicrobial agent in severe autogenously fecal peritonitis in rats. Acta Cir. Bras., 29: 52–56. DOI: 10.1590/S0102-86502014001300010.

Mendonça, E.D., da Silva, J., dos Santos, M.S., Carvalho, P., Papke, D.K.M., Ortmann, C.F., Picada, J.N., Reginatto, F.H., Ferraz, A.D.B.F. 2016. Genotoxic, mutagenic and antigenotoxic effects of *Cecropia pachystachya* Trécul aqueous extract using *in vivo* and *in vitro* assays. J. Ethnopharmacol., 193: 214–220. DOI:10.1016/j.jep.2016.07.046.

Menezes, A.P.P., Trevisan, S.C.C., Barbalho, S.M., Guiguer, E.L. 2016a. *Tamarindus indica* L.. A plant with multiple medicinal purposes. J. Pharm. Phytochem., 5: 50.

Menezes, K.A.S., Nunes, G.D.O., Sampaio, A.A., da Silva, A.F., de Souza, L.S.B., Gava, C.A.T., Fernandes Junior, P.I. 2016b. Diversity of new root nodule bacteria from *Erythrina velutina* Willd., a native legume from the Caatinga dry forest (Northeastern Brazil). Embrapa Semiárido-Artigo em periódico indexado (ALICE). DOI:10.19084/RCA15050.

Mesia-Vela, S., Santos, M.T., Souccar, C., Lima-Landman, M.T.R., Lapa, A.J. 2002. *Solanum paniculatum* L. (Jurubeba): potent inhibitor of gastric acid secretion in mice. Phytomedicine, 9(6): 508–514. DOI: 10.1078/09447110260573137.

Młynarczyk, K., Walkowiak-Tomczak, D., Łysiak, G.P. 2018. Bioactive properties of *Sambucus nigra* L. as a functional ingredient for food and pharmaceutical industry. J. Funct. Foods, 40: 377–390.

Mog, B., Janan, P., Nayak, M.G., Adiga, J.D., Meena, R. 2019. Manipulation of vegetative growth and improvement of yield potential of cashew (*Anacardium occidentale* L.) by Paclobutrazol. Sci. Hortic., 257: 108–148. DOI:10.1016/j.scienta.2019.108748.

Monteiro, J.M., Albuquerque, U.P., Neto, E.M.F. 2006. Use patterns and knowledge of medicinal species among two rural communities in Brazil's semi-arid northeastern region. J. Ethnopharmacol., 105: 173–186. DOI:10.1016/j.jep.2005.10.016.

Monteiro, J.M., Araújo, E.L., Amorim, E.L.C. 2012. Valuation of the Aroeira (*Myracrodruon urundeuva* Allemão): perspectives on conservation. Acta Bot. Brasilica., 26(1): 125–132. DOI:10.1590/S0102-33062012000100014.

Mora, A.L., Garcia, C.H. 2000. A cultura do eucalipto no Brasil. São Paulo: Sociedade Brasileira de Silvicultura (SBS), 112 p.

Moraes, L.S., Rohor, B.Z., Areal, L.B., Pereira, E.V., Santos, A.M., Facundo, V.A. Morais, S.K.R.D., Silva, S.G., Portela, C.N., Nunomura, S.M., Quignard, E.L.J., Pohlit, A.M. 2007. Bioactive dihydroxyfuranonaphthoquinones from the bark of *Tabebuia incana* AH Gentry (Bignoniaceae) and HPLC analysis of commercial pau d'arco and certified *T. incana* bark infusions. Acta Amaz., 37(1): 99–102. DOI:10.1590/S0044-59672007000100011.

Nascimento, I.G., Vieira, M.R.S. 2014. Manual De Plantas Medicinais. Farmácia Verde Católica Unisantos, 50 p.

Nguta, J.M., Mbaria, J.M. 2013. Brine shrimp toxicity and antimalarial activity of some plants traditionally used in treatment of malaria in Msambweni district of Kenya. J. Ethnopharmacol. Pretoria, 148: 988–992. DOI: 10.1016/j.jep.2013.05.053.

Ojewole, J.A. 2005. Anti-inflammatory, analgesic and hypoglycemic effects of *Magnifera indica* L. (Anacardiaceae) stem-bark aqueous extract. Methods Find Exp. Clin. Pharmacol., 27: 547–554. DOI: 10.1358/mf.2005.27.8.928308.

Olejnik, A., Kowalska, K., Olkowicz, M., Rychlik, J., Juzwa, W., Myszka, K., Dembczyński, R., Białas, W. 2015. Anti-inflammatory effects of gastrointestinal digested *Sambucus nigra* L. fruit extract analyzed in co-cultured intestinal epithelial cells and lipopolysaccharide-stimulated macrophages. J. Funct. Foods, 19: 649–660. DOI: 10.1016/j.jff.2015.09.064.

Oliveira, A.P., Raith, M., Kuster, R.M., Rocha, L.M., Hamburger, M., Potterat, O. 2012. Metabolite profiling of the leaves of the Brazilian folk medicine *Sideroxylon obtusifolium*. Planta Med., 78(07): 703–710. DOI: 10.1055/s-0031-1298269.

Oliveira, G.L., Oliveira, A.F.M., Andrade, L.H.C. 2010. Plantas medicinais utilizadas na comunidade urbana de Muribeca, Nordeste do Brasil. Acta Bot. Bras., 24: 571–577. DOI:10.1590/S0102-33062010000200026.

Oliveira, R.LC., Brito, S.O., Veridiana, L.F.P.A. 2019. Uso e extrativismo do angico numa comunidade indígena na savana de Roraima, norte da Amazônia brasileira. São Paulo: Ambiente: Gestão e Desenvolvimento, 4127(1): 47–52.

Pacheco, N.R., Pinto, N.C.C., da Silva, J.M., Mendes, R.F., da Costa, J.C., Aragão, D.M.O., Castañon, M.C.M.N., Scio, E. 2014. *Cecropia pachystachya*: uma espécie com expressivo *in vivo* anti-inflamatório tópico e *in vitro* Efeitos Antioxidantes. BioMed. Res. Internat., 2014: 1–10. DOI:10.1155/2014/301294.

Pacheco, G., Nogueira, C.R., Meneguim, A.B. 2017. Development and characterization of bacterial cellulose produced by cashew tree residues as alternative carbon source. Ind Crop Prod., 107: 13–19. DOI:10.1016/j.indcrop.2017.05.026.

Palumbo, C.F.G., Gardin, N.E., Nakamura, M.U. 2016. *Erythrina mulungu* Mart. ex Benth and *Erythrina velutina* Willd.—Pharmacological aspects and anthroposophic view of these Brazilian plants. Arte Méd. Ampl., 36(4): 152–61.

Palsikowski, P.A., Besen, L.M., Santos, K.A., da Silva, C., Da Silva, E.A. 2019. Supercritical CO_2 oil extraction from *Bauhinia forficata* link subsp. pruinosa leaves: Composition, antioxidant activity and mathematical modeling. J. Supercrit. Fluid., 153. DOI:10.1016/j.supflu.2019.104588.

Pawlus, A.D., Kinghorn, D.A. 2007. Review of the ethnobotany, chemistry, biological activity and safety of the botanical dietary supplement *Morinda citrifolia* (noni). J. Pharm. Pharmacol., 59: 1587–1609. DOI:10.1211/jpp.59.12.0001.

Pell, S.K., Mitchell, J.D., Miller, A.J., Lobova, T.A. 2011. Anacardiaceae. *In*: Kubitzki, K. (ed.). Flowering Plants Eudicots. Berlin, Heidelberg: Springer. 10: 7–50.

Pietrovski, E.F., Rosa, K.A., Facundo, V.A., Rios, K., Marques, M.C.A., Santos, A.R. 2006. Antinociceptive properties of the ethanolic extract and of the triterpene 3β, 6β, 16β-trihidroxilup-20 (29)-ene obtained from the flowers of *Combretum leprosum* in mice. Pharmacol. Biochem. Behavior, 83(1): 90–99. DOI:10.1016/j.pbb.2005.12.010.

Pio Correa, M. 1984. Dicionário das Plantas Úteis do Brasil e das Exóticas Cultivadas, 3rd ed. Imprensa Nacional, Rio de Janeiro.

Pinafo, M.S., Benedetti, P.R., Gaiotte, L.B., Costa, F.G., Schoffen, J.P.F., Fernandes, G.S., Seiva, F.R. 2019. Effects of *Bauhinia forficata* on glycaemia, lipid profile, hepatic glycogen content and oxidative stress in rats exposed to Bisphenol A. Toxicol. Reports, 6: 244–252. DOI:10.1016/j.toxrep.2019.03.001.

Pinto, M.D.S.T., Ribeiro, J.M., Araújo, F.P., Melo, N.F., Fernandes, K.V.S. 2015. Purification and characterization of a peroxidase present in Xilopodium exsudates of umbu plants (*Spondias tuberosa* A.). Afr. J. Biotechnol., 14: 1838–1845. DOI: 10.5897/AJB2015.14521.

Piriz, M.A., Lima, C.A.B., Jardim, V.M.R., Mesquita, M.K., Souza, A.D.Z., Heck, R.M. 2014. Plantas medicinais no processo de cicatrização de feridas: uma revisão de literature. Rev. Bras. Plantas Med., 16: 628–636. http://dx.doi.org/10.1590/1983-084X/12_178.

Porter, R.S., Bode, R.F. 2017. A review of the antiviral properties of black elder (*Sambucus nigra* L.) products. Phytother. Res., 31: 533–554. DOI: 10.1002/ptr.5782.

Rai, A., Das, S., Chamallamudi, M.R., Nandakumar, K., Shetty, R., Gill, M., Sumalatha, S., Devkar, R., Gourishetti, K., Kumar, N. 2018. Evaluation of the aphrodisiac potential of a chemically characterized aqueous extract of *Tamarindus indica* pulp. J. Ethnopharmacol., 210: 118–124. DOI: 10.1016/j.jep.2017.08.016.

Rang, H.P., Dale, M.M., Ritter, J.M., Flower, R.J. 2007. Farmacologia, 6ª ed. Elsevier: Rio de Janeiro.

Rao, U.S.M., Subramanian, S.M. 2009. Biochemical evaluation of antihyperglycemic and antioxidative effects of *Morinda citrifolia* fruit extract studied in streptozotocin-induced diabetic rats. Med. Chem. Res., 18(6): 433–46. DOI:10.1007/s00044-008-9140-1.

Ranaivoson, T., Brinkmann, K., Rakouth, B., Buerkert, A. 2015. Distribution, biomass and local importance of tamarind trees in south-western Madagascar. Glob. Ecol. Conserv., 14–25. DOI: 10.1016/j.gecco.2015.05.004.

Ribeiro, A.R.S., Diniz, P.B., Estevam, C.S., Pinheiro, M.S., Albuquerque-Júnior, R.L., Thomazzi, S.M. 2013. Gastroprotective activity of the ethanol extract from the inner bark of *Caesalpinia pyramidalis* in rats. J. Ethnopharmacol., 147(2): 383–388. DOI:10.1016/j.jep.2013.03.023.

Ribeiro, A.M., Estevinho, B.N., Rocha, F. 2019. Microencapsulation of polyphenols-the specific case of the microencapsulation of *Sambucus Nigra* L. extracts—A review. Trends. Food Sci. Tech., 01–14. DOI: 10.1016/j.tifs.2019.03.011.

Rios, R., Da Silva, H.B.F., Carneiro, N.V.Q., De Oliveira Pires, A., Carneiro, T.C.B., dos Santos Costa, R., Da Silva, T.M. 2017a. *Solanum paniculatum* L. decreases levels of inflammatory cytokines by reducing NFKB, TBET and GATA3 gene expression *in vitro*. J. Ethnopharmacol., 209: 32–40. DOI: 10.1016/j.jep.2017.07.014.

Rios, R., Silva, H.B.F., Carneiro, N.V.Q., Costa, R.S., Carneiro, T.C.B., Marques, C.R., Conceição, A.S. 2017b. Anti-inflammatory activity of Jurubeba (*Solanum paniculatum* L.) through reducing the T-bet and GATA3 gene expression, *in vitro*. J. Allergy Clin. Immunol., 139(2): AB268. DOI: 10.1016/j.jaci.2016.12.865.

Rios, R., Silva, H.B.F., Carneiro, N.V.Q., Pires, A.O., Carneiro, T.C.B., Costa, R.S., Marques, C.R., Machado, M.S.S., Velozo, E.S., Silva, T.M.G., Silva, T.M.S., Conceição, A.S., Alcântara-Neves, N.M., Figueireido, C.A. 2017c. *Solanum paniculatum* L. decreases levels of inflammatory cytokines by reducing NFKB, TBET and GATA3 gene expression in vitro. J. Ehtnopharmacol., 209: 32–40. DOI:10.1016/j.jep.2017.07.014.

Rocha, G.P. 2012. Juazeiro (*Ziziphus joazeiro* Mart.). Disponível em: <http://blog.tocandira.com.br/sobrevivencia/juazeiro-ziziphusjoazeiro-mart/>. Acesso em: 25 dez. 2019.

Rodrigues, L.A., Magalhães, L.S., Pussente, C.G., Crespo, J.M.R.S., Soares, A.F. 2014. Elaboração de um fitoterápico antibiótico a partir da casca de *Anadenanthera macrocarpa*. Rev. Cient. da Faminas., 10(3): 23–39.

Rodrigues, P.S., Souza, M.M., Melo, C.A.F., Pereira, T.N.S., Corrêa, R.X. 2018. Karyotype diversity and 2C DNA content in species of the Caesalpinia group. BMC Genet., 19: 01–25. DOI: 10.1186/s12863-018-0610-2.

Romão, M.O.C., Costa, A.M.D., Terra, F.S., Boriollo, M.F.G., Soares, E.A. 2010. Avaliação da atividade protetora gástrica do extrato de raspas de juá: Rev. Brás clin Med., 8(3): 222–7.

Roque, A.A., Rocha, R.M., Loiola, M.I.B. 2010. Uso e diversidade de plantas medicinais da Caatinga na comunidade rural de Laginhas, município de Caicó, Rio Grande do Norte (nordeste do Brasil). Rev. Bras. Pl. Med., 12: 31–42.

Roschek, J.B., Fink, R.C., McMichael, M.D., Li, D., Alberte, R.S. 2009. Elderberry flavonoids bind to and prevent H1N1 infection *in vitro*. Phytochem., 70: 1255–1261. DOI: 10.1016/j.phytochem.2009.06.003.

Salgado, H.R.N., Neto, F.C., Fernandes, F.H.A., Kogawa, A.C. 2017. Antimicrobial Activity of *Sideroxylon obtusifolium* (Roem. and Schult) TD Penn.(Sapo-taceae). EC Microbiology, 11.6: 250–256.

Sameh, S., Al-Sayed, E., Labib, R.M., Singab, A.N. 2018. Genus *Spondias*: A phytochemical and pharmacological review. evid-based. Compl. Alt., 2018: 01–13. DOI: 10.1155/2018/5382904.

Santana, D.G., Santos, C.A., Santos, A.D., Nogueira, P.C., Thomazzi, S.M., Estevam, C.S., Antoniolli, A.R., Camargo, E.A. 2012. Beneficial effects of the ethanol extract of *Caesalpinia pyramidalis* on the inflammatory response and abdominal hyperalgesia in rats with acute pancreatitis. J. Ethnopharmacol., 142(2): 445–455. DOI:10.1016/j.jep.2012.05.015.

Santos, C.A., Passos, A.M., Andrade, F.C., Camargo, E.A., Estevam, C.S., Santos, M.R., Thomazzi, S.M. 2011. Antinociceptive and anti-inflammatory effects of *Caesalpinia pyramidalis* in rodents. Rev. Bras. Farmacogn., 21(6): 1077–1083. DOI:10.1590/S0102-695X2011005000179.

Santos, C.C., Araújo, S.S., Santos, A.L., Almeida, E.C., Dias, A.S., Damascena, N.P., Lima, A.C. 2014. Evaluation of the toxicity and molluscicidal and larvicidal activities of *Schinopsis brasiliensis* stem bark extract and its fractions. Rev. Bras. Farmacog., 24(3): 298–303. DOI: 10.1016/j.bjp.2014.07.006.

Santos-Filho, F.S., Almeida-Junior, E.B., Lima, P.B., Soares, C.J.R.S. 2015. Checklist of the flora of the restingas of Piauí state, Northeast Brazil. Checklist 11: 1–10. DOI:10.15560/11.2.1598.

Santos, J.S., Marinho, R.R., Ekundi-Valentim, E., Rodrigues, L., Yamamoto, M.H., Teixeira, S.A., Muscara, M.N., Costa, S.K., Thomazzi, S.M. 2013. Beneficial effects of *Anadenanthera colubrina* (Vell.) Brenan extract on the inflammatory and nociceptive responses in rodent models. J. Ethnopharmacol., 148: 218–22. DOI:10.1016/j.jep.2013.04.012.

Santos, P., Benedito, C.P., Alves, T.R., Paiva, E.P., Sousa, E.C., Freires, A.L. 2018. Water stress and temperature on germination and vigor of *Handroanthus impetiginosus* (Mart. ex DC). Rev. Bras. Eng. Agr. Amb., 22(5): 349–354. DOI:10.1590/1807-1929/agriambi.v22n5p349-354.

Santos-Serejo, J.A., Barbosa, A.M.M., Luna, J.V.U. 2009. Tamarindo. *In*: Santos Serejo, J.A., Dantas, J.L.L., Sampaio, C.V., Coelho, Y.S. (eds.). Fruticultura Tropical: espécies regionais e exóticas. 1. ed. Brasília, DF: Embrapa Informação Tecnológica, 437–455.

Saraiva, A.M., Saraiva, C.L., Cordeiro, R.P., Soares, R.R., Xavier, H.S., Caetano, N. 2013. Atividade antimicrobiana e sinérgica das frações das folhas de *Schinopsis brasiliensis* Engl. frente a clones multirresistentes de *Staphylococcus aureus*. Rev. Bras. Pl. Med., 15(2): 199–207. DOI: 10.1590/S1516-05722013000200006.

Shettar, A.K., Sateesh, M.K., Kaliwal, B.B., Vedamurthy, A.B. 2017. *In vitro* antidiabetic activities and GC-MS phytochemical analysis of *Ximenia americana* extracts. S. Afr. J. Bot., 111: 202–211. DOI: 10.1016/j.sajb.2017.03.014.

Sidor, A., Gramza-Michałowska, A. 2015. Advanced research on the antioxidant and health benefit of elderberry (*Sambucus nigra*) in food—a review. J. Funct. Foods, 18: 941–958. DOI: 10.1016/j.jff.2014.07.012.

Silva, F.F.S., Dantas, B. 2017. *Sideroxylon obtusifolium* (Humb. ex Roem. & Schult.) TD Penn. quixabeira. Embrapa Semiárido-Nota Técnica/Nota Científica (ALICE).

Silva, F.R.M.B., Szpoganicz, B., Pizzolatti, M.G., Willrich, M.A.V., De Sousa, E. 2002. Acute effect of *Bauhinia forficata* on serum glucose levels in normal and alloxan-induced diabetic rats. J. Ethnopharmacol., 83(1-2): 33–37. DOI: 10.1016/S0378-8741(02)00193-9.

Silva, J.M.C., Tabarelli, M., Fonseca, M.T., Lins, L.V. 2003. Biodiversidade da caatinga: áreas e ações prioritárias para a conservação. Brasília, DF: Ministério do Meio Ambiente, Universidade Federal de Pernambuco. 382 p. DOI:10.5380/dma.v41i0.49254.

Silva, M.C., Santana, L.A., Mentele, R., Ferreira, R.S., de Miranda, A., Silva-Lucca, R.A., Oliva, M.L. 2012. Purification, primary structure and potential functions of a novel lectin from *Bauhinia forficata* seeds. Process biochem., 47(7): 1049–1059. DOI: 10.1016/j.procbio.2012.03.008.

Silva, P.K., Farias, G.A., Araújo, E.R., Sapucay, M.J.L.C., Colares, P.N.Q., Pedroza, C.M., Rêgo, E.R., Rêgo, M.M. 2009. Caracterização física e química de genótipos de caju. Hortic. Bras., (27): 1551–1555.

Silva, T.M.S., Nascimento, R.J.B., Batista, M.M., Agra, M.F., Camara, C.A. 2007. Brine shrimp bioassay of some species of *Solanum* from Northestern Brazil. Rev. Bras. Farmacog., 17(1): 35–38. DOI: 10.1590/S0102-695X2007000100008.

Siqueira, E.M.S, Félix-Silva, J., De Araújo, L.M., Fernandes, J.M., Cabral, B., Gomes, J.A. 2016a. *Spondias tuberosa* (Anacardiaceae) leaves: profiling phenolic compouds by HPLC-DAD and LC-MS/MS and *in vivo* anti-inflammatory activity. Biomed Chromatogr., 30: 1656–1665. DOI: 10.1002/bmc.3738.

Siqueira, S., Michell, E., Félix-Silva, J., Araújo, L.M.L., Fernandes, J.M., Cabral, B., Freitas Fernandes-Pedrosa, M. 2016b. *Spondias tuberosa* (Anacardiaceae) leaves: profiling phenolic compounds by HPLC-DAD and LC–MS/MS and *in vivo* anti-inflammatory activity. Biomed Chromatogr., 30: 1656–1665. DOI: 10.1002/bmc.3738.

Soares Neto, R.L., Cordeiro, L.S., Loiola, M.I.B. 2014. Flora of Ceará, Brazil: Combretaceae. Rodriguésia, 65(3): 685–700. DOI: 10.1590/2175-7860201465308.

Soraru, S.B., Bandoni, A.L. 1978. Plantas de la medicina popular argentina. Guía ilustrada de las cincuenta plantas indígenas más empleadas. Ed. Albatros, Buenos Aires, 107–109.

Souza, A.B., Souza, L.M.S., Carvalho, J.C.T., Maistro, E.L. 2006. No clastogenic activity of *Caesalpinia ferrea* Mart. (Leguminosae) extract on bone marrow cells of Wistar rats. Genet. Mol., 29: 380–383. DOI: 10.1590/S1415-47572006000200028.

Souza, R.K.D., Silva, M.A.P., Menezes, I.R.A., Ribeiro, D.A., Bezerra, L.R., Souza, M.M.A. 2014. Ethnopharmacology of medicinal plants of carrasco, northeastern Brazil. J. Ethno Pharmacol., 157: 99–104. DOI:10.1016/j.jep.2014.09.001.

Street, R., Prinsloo, G. 2012. Commercially important medicinal plants of South Africa: a review. J. Chem., 2013: 01–17. DOI: 10.1155/2013/205048.

Tamiello-Rosa, C.S., Cantu-Jungles, T.M., Iacomini, M. 2019. Pectins from cashew apple fruit (*Anacardium occidentale*): Extraction and chemical characterization. Carbohydr. Res., 483: 107–152. DOI:10.1016/j.carres.2019.107752.

Tanae, M.M., Lima-Landman, M.T.R., Lima, T.C.M., Souccar, C., Lapa, A.J. 2007. Chemical standardization of the aqueous extract of *Cecropia glaziovii* Sneth endowed with antihypertensive, bronchodilator, antiacid secretion and antidepressant-like activities. Phytomedicine, 14(5): 309–313. DOI:10.1016/j.phymed.2007.03.002.

Tasneen, S., Liu, B., Li, B., Choudhary, M.I., Wang, W. 2018. Molecular pharmacology of Inflammation: medicinal plants as anti-inflammatory agents. Phar. Res., 139: 126–140. DOI:10.1016/j.phrs.2018.11.001.

Tenório, J.A., Dulciana, S., da Silva, T.M., da Silva, T.G., Ramos, C.S. 2016. *Solanum paniculatum* root extract reduces diarrhea in rats. Rev. Bras. Farmacog., 26(3): 375–378. DOI: 10.1016/j.bjp.2016.02.003.

Teske, M., Trentini, A.M.M. 1997. Herbarium: compêndio de fitoterapia. 3. ed. Curitiba.

Tufail, T., Saeed, F., Ain, H.B., Niaz, B., Afzaal, M., Din, A., Suleira, H.A.R. 2019. Cashew nut allergy; immune health challenge. Trends Food Sci. Tec., 86: 209–216. DOI:10.1016/j.tifs.2019.02.014.

Uchôa, A.D., Oliveira, W.F., Pereira, A.P., Silva, A.G., Cordeiro, B.M., Malafaia, C.B., Almeida, C.M.A., Silva, N.H., Albuquerque, J.F.C., Silva, M.S., Correia, M.T.S. 2015. Antioxidant activity and phytochemical profile of *Spondias tuberosa* Arruda leaves extracts. Am. J. Plant Sci., 6: 3038. DOI: 10.4236/ajps.2015.619298.

Uchôa, V.T., Sousa, C.M.M., Carvalho, A.A., Sant'Ana, A.E.G., Chaves, M.H. 2016. Free radical scavenging ability of *Ximenia americana* L. stem bark and leaf extracts. J. Appl. Pharm. Sci., 6: 091–096. DOI: 10.7324/JAPS.2016.60213.

Ueda, T., Yamaguchi, M., Uchimiya, H., Nakano, A. 2001. Ara6, a plant-unique novel type Rab GTPase, functions in the endocytic pathway of *Arabidopsis thaliana.* Embo J, 20: 4730–4741. DOI: 10.1093/emboj/20.17.4730.

Valerino-Díaz, A.B., Gamiotea-Turro, D., Zanatta, A.C., Vilegas, W., Gomes Martins, C.H., De Souza Silva, T., Rastrelli, L., Dos Santos, L.C. 2018. New polyhydroxylated steroidal saponins from *Solanum paniculatum* L. leaf alcohol tincture with antibacterial activity against oral pathogens. J. Agricult. Food Chem., 66(33): 8703–8713. DOI: 10.1021/acs.jafc.8b01262.

Van-Lume, B., Esposito, T., Diniz-Filho, J.A.F., Gagnon, E., Lewis, G.P., Souza, G. 2017. Heterochromatic and cytomolecular diversifcation in the *Caesalpinia* group (Leguminosae): relationships between phylogenetic and cytogeographical data. Perspect. Plant Ecol., 29: 51–63. DOI: 10.1016/j.ppees.2017.11.004.

Vasconcelos, C.F.B., Maranhão, H.M.L., Batista, T.M., Carneiro, E.M., Ferreira, F., Costa, J., Soares, L.A.L., Sá, M.D.C., Souza, T.P., Wanderley, A.G. 2011. Hypoglycaemic activity and molecular mechanisms of *Caesalpinia ferrea* Martius bark extract on streptozotocin-induced diabetes in Wistar rats. J. Ethnopharmacol., 137: 1533–1541. DOI: 10.1016/j.jep.2011.08.059.

Veberic. R., Jakopic J., Stampar F., Schmitzer, V. 2009. European elderberry (*Sambucus nigra* L.) rich in sugars, organic acids, anthocyanins and selected polyphenols. Food Chem., 114: 511–515. DOI: 10.1016/j.foodchem.2008.09.080.

Viau, C.M., Moura, D.J., Facundo, V.A., Saffi, J. 2014. The natural triterpene 3β, 6β, 16β-trihydroxy-lup-20 (29)-ene obtained from the flowers of *Combretum leprosum* induces apoptosis in MCF-7 breast cancer cells. BMC Compl. Alt. Med., 14(1): 280. DOI: 10.1186/1472-6882-14-280.

Viau, C.M., Moura, D.J., Pflüger, P., Facundo, V.A., Saffi, J. 2016. Structural aspects of antioxidant and genotoxic activities of two flavonoids obtained from ethanolic extract of *Combretum leprosum*. Evid-Based Compl. Alt. Med. DOI:10.1155/2016/9849134.

Vieira, E., Galvão, F.C.A., Barros, A.L. 2019. Influence of water limitation on the competitive interaction between two Cerrado species and the invasive grass Brachiaria brizantha cv. Piatã. Plant Physiol. Biochem., 135: 206–214. DOI:10.1016/j.plaphy.2018.12.002.

Vieira, L.M., Sousa, M.S.B., Mancini Filho, J., Lima, A. 2011. Fenólicos totais e capacidade antioxidante *in vitro* de polpas de frutos tropicais. Rev. Bras. Frutic., 33: 888–897. DOI: 10.1590/S0100-29452011005000099.

Vieira, P.M., Santos, S.D.C., Chen-Chen, L. 2010. Assessment of mutagenicity and cytotoxicity of *Solanum paniculatum* L. extracts using *in vivo* micronucleus test in mice. Braz. J. Biol., 70(3): 601–606. DOI:10.1590/S1519-69842010000300017.

Volpato, G.T., Damasceno, D.C., Rudge, M.V.C., Padovani, C.R., Calderon, I.D.M.P. 2008. Effect of *Bauhinia forficata* aqueous extract on the maternal-fetal outcome and oxidative stress biomarkers of streptozotocin-induced diabetic rats. J. Ethnopharmacol., 116(1): 131–137. DOI: 10.1016/j.jep.2007.11.013.

Wei, H., Liu, X.Q., Zhu, J.J., Gao, L.L., Wang, Z.M., Yan, L.H. 2015. A new cassane diterpenoid from the seeds of *Caesalpinia decapetala*. J. Asian Nat. Prod. Res., 18: 1–5. DOI: 10.1080/10286020.2015.1055255.

Weiss, R., Fintelmann, V. 2000. Herbal Medicine. Thieme, Stuttgart.

Xavier, R., Oliveira, J., Mota, E., Silva, G.M. 2018. Avaliação da qualidade de amostras comerciais de chás na cidade de Vitória da Conquista-Bahia. Rev. Fitos., 12(1): 8–17. DOI:10.5935/2446-4775.20180002.

Ximenes, N.C., Júnior, M.A.D.S., Vasconcelos, G.S., Dias, K.C.F., Jucá, M.M., Silva, A.H., Vasconcelos, S.M.M. 2018. Ethanolic extract of *Erythrina velutina* Willd ameliorate schizophrenia-like behavior induced by ketamine in mice. J. Complem. Integr. Med., 16(2). DOI:10.1515/jcim-2018-0038.

Yagi, A. 1997. Isolation and characterization of the glycoprotein fraction with a proliferation-promoting activity on human and hamster cells *in vitro* from *Aloe vera* gel. Planta Med., 63(1): 18–21. DOI:10.1055/s-1997-957595.

Zakay-Rones, Z., Varsano, N., Zlotnik, M., Manor, O., Regev, L., Schlesinger, M., Mumcuoglu, M. 1995. Inhibition of several strains of influenza virus *in vitro* and reduction of symptoms by an elderberry extract (*Sambucus nigra* L.) during an outbreak of influenza B Panama. J. Altern. Complem. Med., 1: 361–369.

Zeraik, M.L., Queiroz, E.F., Marcourt, L., Ciclet, O., Castro-Gamboa, I., Silva, D.H.S., Cuendet, M., da Silva Bolzani, V., Wolfender, J.L. 2016. Antioxidants, quinone reductase inducers and acetylcholinesterase inhibitors from Spondias tuberosa fruits. J. Funct. Foods, 21: 396–405. DOI: 10.1016/j.jff.2015.12.009.

Zielinski, A.A.F., Ávila, S., Ito, V., Nogueira, A., Wosiacki, G., Haminiuk, C.W.I. 2014. The association between chromaticity, phenolics, carotenoids, and *in vitro* antioxidant activity of frozen fruit pulp in Brazil: An application of chemometrics. J. Food Sci., 79: 510–516. DOI: 10.1111/1750-3841.12389.

Chapter 9

Eichhornia crassipes: Shedding Light on its Chemical Composition, Biological Activities and Industrial Uses

Fadia S. Youssef,[1] *Noor H. Aysa*[2] and *Mohamed L. Ashour*[1,*]

Introduction

Eichhornia crassipes (Mart.) Solms, also termed water hyacinth, belongs to family Pontideraceae (Figure 9.1). It is considered as one of the highly productive plants in addition to being the most harmful warm aquatic one all over the globe. Amazon in Brazil is the main origin of this aquatic plant from which it spreads to additional areas in South America and is subsequently transported to tropical and subtropical areas by human action (Tyagi and Agarwal 2017). Being an invasive weed, it triggers many dramatic problems in lakes, rivers and many reservoirs around the globe, particularly in Africa. Water hyacinth has elicited many ecological problems with subsequent social and economic hazards mainly in the area ranging between 40° North to 45° South. These problems are represented by the formation of heavy mats of the plant that completely coat the river, causing blockage of water paths and interference with the transportation of water that badly affects tourism and agriculture field irrigation as well (Aboul-Enein et al. 2014).

Figure 9.1. Photographs showing whole plant (A), flowers (B), leaves (C) and roots (D) of *Eichhornia crassipes*.

[1] Department of Pharmacognosy, Faculty of Pharmacy, Ain Shams University, Abbassia, 11566-Cairo, Egypt.
[2] Clinical and Laboratory Science Department, College of Pharmacy, University of Babylon, Babylon, Iraq.
* Corresponding author: mohamed_ashour@pharm.asu.edu.eg

Additionally, a massive amount of nutrients is required for the invasive growth of *Eichhornia* that seriously affects the growth of other aquatic species. Meanwhile, the decomposition of *E. crassipes* dead bodies evokes the release of many nutrients that badly deteriorates the purity and quality of drinking water, eventually threatening man's health. The condition becomes more seriously exaggerated in summer due to temperature elevation above 35° where the plant grows faster with a consequent fast spread in the rivers causing many hazards owing to prolonged exposure to light and high sun radiation as well (Dandelot et al. 2008).

Although *Eichhornia* is reported as an invasive aquatic plant, it can afford a golden mine of compounds with marked bioactivities owing to its richness with phytosterols as well as many steroids. Thus, more attention has been given recently to get the benefit of *E. crassipes* as a remedy to alleviate several disorders like Alzheimer's, cancer, cardiac diseases as well as aging and diabetes (Aboul-Enein et al. 2014). These diseases are triggered by the elevation of the reactive oxygen species and the oxidative stress that is undoubtedly combated by antioxidants represented by many naturally occurring phytoconstituents. Thus, this chapter highlights the beneficial uses of *Eichhornia crassipes* concerning their biological activities, industrial uses and chemical constituents that are responsible for its activities.

Chemical Composition of *Eichhornia crassipes*

Preliminary phytochemical screening of *Eichhornia crassipes* resulted in the identification of thiol group-containing compounds (Ding et al. 1994), flavonoids, alkaloids, anthraquinones and anthocyanins (Jayanthi et al. 2012) in addition to phenols and saponins, as presented in Figure 9.2 (Gutierrez-Morales et al. 2017).

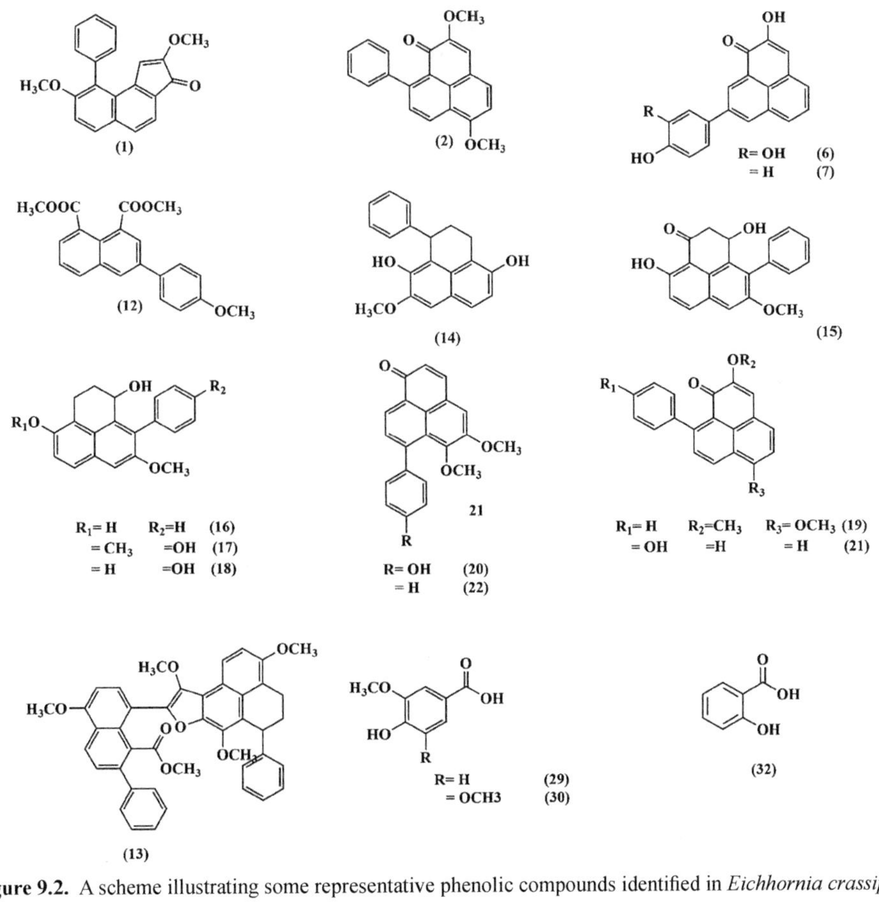

Figure 9.2. A scheme illustrating some representative phenolic compounds identified in *Eichhornia crassipes*.

A thiol group-containing compound acts as a protectant that is commonly adopted in the chemistry of peptides and proteins as well. Thus, in biochemistry, during the assessment of heavy metal content in peptides and proteins, it is usually added in large amounts that differ according to the prepared protein. The presence of thiol-containing compounds is perfectly used to determine the heavy metal bio-concentration by water hyacinth (Ding et al. 1994).

Meanwhile, flavonoids represent a compilation of regular compounds possessing different phenolic groups that naturally occur in many products of natural origin derived from vegetables, stems, barks, and blossoms in addition to wine and tea. Flavonoids and their related compounds perfectly act as a core in the implementation of various restorative, pharmaceutical, therapeutic and nutraceutical applications. It was postulated that the richness of water hyacinth with secondary metabolites as anthocyanins, alkaloids, phenols, flavonoids and anthraquinones in its different extracts and fractions greatly accounts for its pupicidal, larvicidal and mosquito repellant potential (Jayanthi et al. 2012).

Furthermore, a lot of phenolic compounds were identified from *Eichhornia crassipes* as, 2,5-dimethoxy-4-phenyl-benzoindenone **(1)**, phenalene derivatives, and additional permethylated phenalene derivatives were isolated from the ethyl acetate extract represented by 2,6-dimethoxy-9-phenyl-1H-phenalen-1-one **(2)**, 4,9-dimethoxy-7-phenyl-2,3-dihydrophenalen-1olo-O-methyl ether **(3)**, 4,9-dimethoxy-7-(4'-methoxy-phenyl)-2,3-dihydro-phenalen-l-olo-O-methyl ether **(4)**, 4,5-dimethoxy-9-phenyl-2,3-dihydro-phenalen-1-ol-O-methyl ether **(5)**. In addition, 2-hydroxy-8-(3,4-dihydroxyphenyl)-phenalen-1-one **(6)** and 2-Hydroxy-8-(4-hydroxyphenyl)-phenalen-1-one **(7)** were obtained from the acetone extract. Moreover, 4,4",8,8",9,9"-hexamethoxy-1,1"-diphenyl-2,2", 3,3"-tetrahydro-7,7"-bi(1H-phenalene), 4,8,9-trimethoxy-1-phenyl-2,3-dihydro1H-phenalene **(8)**, 6,6",8,8",9,9",4',4'"octamethoxy-1,1"diphenyl-2,2",3,3"-tetrahydro-7,7"-bi(1H-phenalene) **(9)**, 4,8,9-trimethoxy-1-(4'methoxyphenyl)-2,3-dihydro1H-phenalene **(10)** and 4,4",8,8",9,9",4',4'"-octamethoxy-1,1"-diphenyl-2,2",3,3"-tetrahydro-7,7"-bi(1H-phenalene) **(11)**, together with other phenylphenalenes represented by dimethyl 3-(4 -methoxyphenyl) naphthalene-1,8-dicarboxylate **(12)**, methyl 5-methoxy-2-phenyl-8(3,7,10-trimethoxy-6-phenyl-5,6-dihydro-4H-phenaleno[2,1-b]furan-9-yl)-1-naphthoate **(13)**, 2,3-dihydro-4,9-dihydroxy-8-methoxy-1-phenylphenalene **(14)**, 2,3-dihydro-3,9-dihydroxy-5-methoxy4-phenyl-1H-phenalen-1-one **(15)**, 2,3-dihydro-8-methoxy-9-phenyl-1H-phenalene-1,4-diol **(16)**, 2,3-dihydro-4,8-dimethoxy-9-phenyl-1H-phenalen-1-ol **(17)**, 2,3-dihydro-9-(4-hydroxyphenyl)-8-methoxy-1H-phenalene-1,4-diol **(18)**, 2,6-dimethoxy-9-phenyl-1H-phenalen-1-one **(19)**, 7-(4-hydroxyphenyl)-5,6-dimethoxy-1H-phenalen-1-one **(20)**, 2-hydroxy-9-(4-hydroxyphenyl)-1H-phenalen1-one **(21)**, 5,6-dimethoxy-7-phenyl-1H-phenalen-1-one **(22)** were isolated from *Eichhornia crassipe*s (Della Greca et al. 1992, Hölscher and Schneider 2005). It is worth mentioning that phenylphenalenones possessing a parallel phenyl ring in irregular 8-position have been isolated from water hyacinth. The structures were ascertained employing various spectroscopic and spectrometric techniques comprising NMR and MS assessments. The linkage of two atoms of [$^{1-13}$C] phenylalanine indicates its biosynthesis in virtue of 1,2-aryl relocation from the middle of the road of the 9-phenylphenalenone type (Holscher and Schneider 2005).

Many other phenolic compounds such as pyrogallol **(23)**, 4-methylresorcinol **(24)**, catechol **(25)**, 2-methylresorcinol **(26)**, *p*-hydroxybenzoic **(27)**, gentisic **(28)**, vanillic **(29)**, syringic **(30)**, methyl 3-(4-hydroxy-3-methoxyphenyl) prop-2-enoate (methyl ferulate) **(31)**, and salicylic acids **(32)** were detected (Figure 9.2) (Lata and Dubey 2010a).

Eichhornia crassipes is a rich source of flavonoids represented by gossypetin **(33)**, tricin **(34)**, azaleatin **(35)**, chrysoeriol **(36)**, luteolin **(37)**, apigenin **(38)** from the aqueous and petroleum extracts of the rhizome and the shoot that was ascertained using paper chromatography. In addition, orientin **(39)**, kaempferol **(40)**, quercetin **(41)**, isovitexin **(42)** were identified also from the roots and shoots; meanwhile, 4H-pyran4one,2,3-dihydro-3,5-dihydroxy-6-methyl **(43)** was obtained from the ethanol extract (Figure 9.3) (Nyananyo et al. 2007, Vasu et al. 2009, Lata and Dubey 2010a, Jayanthi et al. 2011).

Anthocyanins, represented by (6'"-*O*-(delphinidin 3-*O*-(6'"-*O*-(β-D-glucopyranosyl)-β-D-glucopyranosyl))) (6'"-*O*-(apigenin 7-*O*-(β-D-glucopyranosyl))) malonate **(44),** was isolated from the blue-purple flowers of water hyacinth that was ambiguously determined using various spectroscopic techniques (Toki et al. 1994).

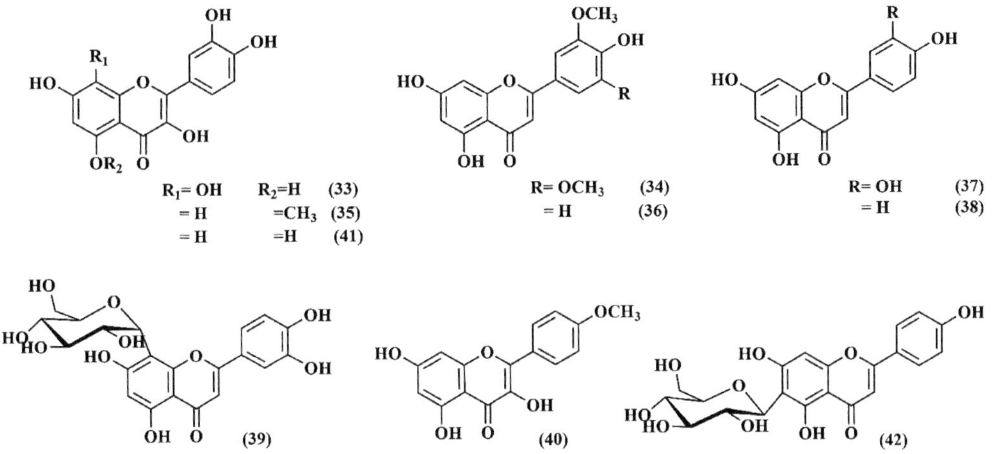

Figure 9.3. A scheme illustrating some representative flavonoids identified in *Eichhornia crassipes*.

Furthermore, the presence of saponins in *E. crassipes* makes it a good candidate with notable biopotency as an antimicrobial agent combating staphylococcal infections, pathogenic MRSA existing in cattle and CoNS bacteria in rabbits. This may be due to the lethal effect of saponins glycosides and their aglycones on the growth and division of bacteria playing an observable role in reinforcing cell immediately (GutierreMorales et al. 2017).

Moreover, mono **(45)** and di-galactosyl diglycerides **(46)** represent the main glycolipids existing in the shoot aqueous and chloroform extracts; meanwhile, phosphatidylethanolamine **(47)**, phosphatidylcholine **(48)** and phosphatidylglycerol **(49)** are the main phospholipids existing in the flowers, roots and leaf stalks (Lalitha et al. 2012). Besides, linoleic **(50)**, palmitic **(51)** and linolenic acids **(52)** represent the main fatty acids that predominate the flowers, leaf stalks, roots and leaves; meanwhile, sucrose, fructose, and other soluble sugars like glucose, arabinose and galactose greatly exist in the leaves in addition to the presence of branched (1-3)-*β*-D-glucan **(53)** and a galactomannan **(54)** (Arayana et al. 1984).

Additionally, terpenoids in *E. crassipes* (leaf aqueous extracts) account for the fragrant behavior of the extract and could be used as natural cures (Kumar et al. 2014). GC/MS analysis of *E. crassipes* leaf revealed the presence of many compounds with major peaks represented by linolenic acid ethyl ester **(55)** (26.26%), palmitic acid ethyl ester **(56)** (12.09%), 9,12-Octadecadienoic acid, ethyl ester **(57)** (3.79%), phytol **(58)** (2.12%), glyceryl linolenate **(59)** (1.35%), *E*-11-hexadecanoic acid ethyl ester **(60)** (1.09%) and stearic acid ethyl ester **(61)** (0.98%), each of which manifested a plethora of biological activities (Tyagi and Agarwal 2017).

Regarding sterols, 6α-hydroxystigmata-4,22-dien-3-one **(62)**, 4α-methyl-5α-ergosta-7,24(28)-diene3β,4β-diol **(63)**, 4α-methyl-5α-ergosta8,14,24(28)-triene-3β,4β-diol **(64)**, were isolated, meanwhile, a mixture of β-sitosterol **(65)**, campesterol **(66)** and stigmasterol **(67)** was detected in the acetone extract (Figure 9.4) (Lata and Dubey 2010a, Lalitha et al. 2012).

Besides, *E. crassipes* is a rich source of alkaloids, as revealed by the qualitative analysis of the extract adopted by thin-layer chromatography. Alkaloids are represented in Figure 9.5. Tomatine **(68)** and cytosine **(69)** predominate in the rhizome and shoot; meanwhile, quinine **(70)**, thebaine **(71)** and codeine **(72)** exist only in the shoot, whereas nicotine is present in the rhizome. Furthermore, 16,17,20,21-tetradehydro-16-(hydroxymethyl)-methyl ester **(73)** and 18,19-secoyohimban-19-oic acid **(74)** are proved to have existed in the *E. crassipes* methanol extract; however, 1H-pyrrole,1-phenyl **(75)** and pipradrol **(76)** were found in the ethanol extract using GC technique (Lata and Dubey 2010b, Muthunarayanan et al. 2011, Shanab et al. 2011).

The root and stem biomass of water hyacinth is a rich source of cellulose xanthogenate as well as the alkali-treated straw that is obtained upon treatment with carbon sulfide and sodium hydroxide as well. Treatment with sodium hydroxide resulted in the removal of hemicellulose and lignin; meanwhile,

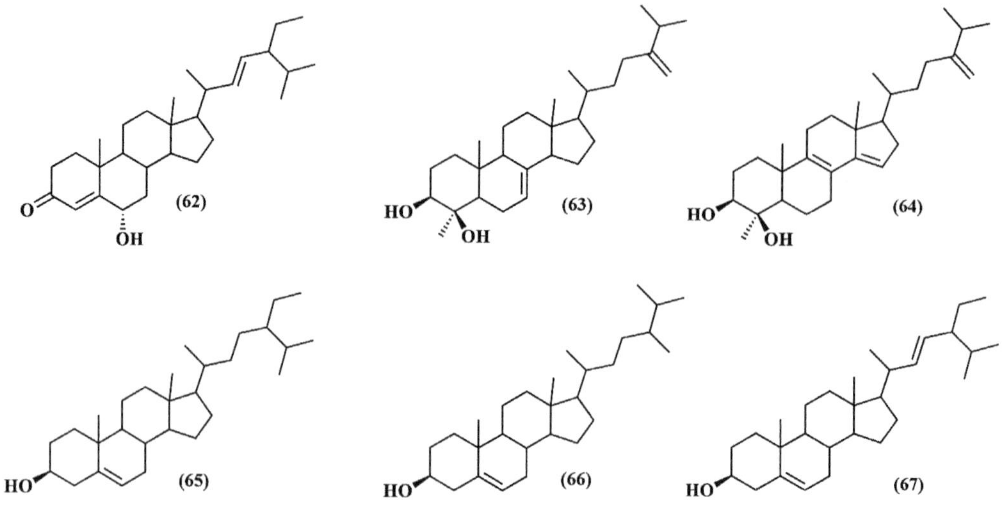

Figure 9.4. A scheme illustrating some representative sterols identified in *Eichhornia crassipes.*

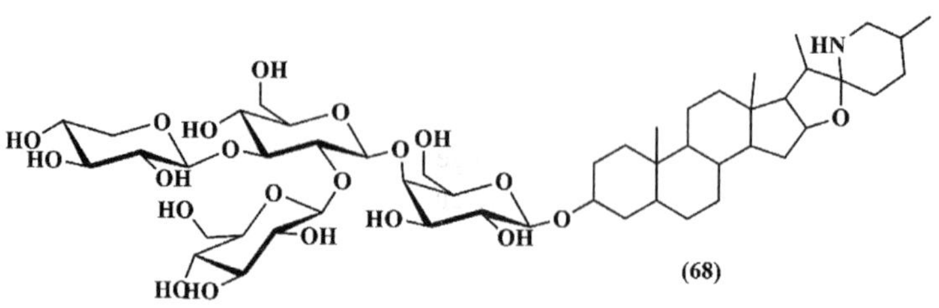

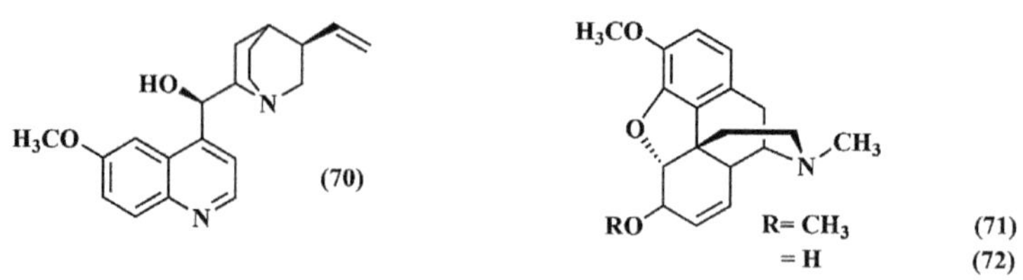

Figure 9.5. A scheme illustrating some representative alkaloids identified in *Eichhornia crassipes.*

carbon sulfide induced the formation of cellulose xanthogenate via the introduction of new functional moieties represented by O–CS–S in addition to C = S (Zhou et al. 2009).

Other compounds belonging to different classes were also detected such as 1,8-naphthalenedicarboxylic acid dimethyl esters, which were obtained from the acid fraction of the ethyl acetate extract represented by 2-phenyl-5-methoxy-1,8-naphthalene dicarboxylic acid dimethyl ester **(77)**, 2-(*p*-methoxyphenyl)-5-methoxy-1,8 naphthalene dicarboxylic acid dimethyl ester **(78)**, 2-phenyl-6methoxy-1,8-naphthalene dicarboxylic acid dimethyl ester **(79)**, 2-phenyl-1,8-naphthalene dicarboxylic acid dimethyl ester linked at C-5 to a phenalenol **(80)** (Della Greca et al. 1992).

Additional compounds represented by 18, 19-Secoyohimban-19-oic acid **(81)**, 16, 17, 20, 21-tetradehydro-16-(hydroxymethyl)-, methyl ester **(82)**, 1,2-benzene dicarboxylic acid, mono-(2-ethylhexyl ester) **(83)**, 1, 2 benzene dicarboxylic acid, diisooctyl ester **(84)**, di amino-di nitro-methyl dioctyl phthalate **(85)**, 1, 2 benzene dicarboxylic acid, dioctyl ester **(86)**, 9-(2,2-Dimethyl propanoilhydrazono)-2,7-bis-[2-{diethylamino)-ethoxy] fluorine **(87)**, 3-Methyl phenyl)-phenyl methanol, 4-(diethylamino)-alpha-[4-(diethylamino) phenyl], isooctyl phthalate **(88)** were also isolated from by *E. crassipes* and showed different antioxidant and anticancer activity in different degrees (Aboul-Enein et al. 2014).

Biological Activities of *Eichhornia crassipes*

Neuro-pharmacological activity

Eichhornia crassipes exerted a potent behavioral neuro-pharmacological potential as evidenced by the notable anti-epileptic, anti-psychosis, anti-depressant, analgesic, anti-anxiety, CNS depressant, sedative, and memory enhancing activities exerted by the ethanol extract of *E. crassipes* leaves, alone or in combination with *Nelumbo nucifera* leaves ethanol extract, in a mouse model approaching that of diazepam, the standard drug (Farheen et al. 2015).

Antioxidant activity

Free radicals are chemical compounds that possess unpaired electrons that make them likely unstable in their outermost energy levels. They may come from external sources like pollution, smoking, infection and ultraviolet radiation. They can also be produced internally within the living cells. Oxidative stress triggered by the hyper-production of free radicals led to the occurrence of life-threatening ailments such as diabetes, neurodegenerative diseases, cancers and cardiovascular disorders. Herbal drugs act as potential sources of bioactive phytoconstituents as phenylpropanoids, phenolic acids, flavonoids, vitamins and tannins, which display free radicals scavenging potential (Chukwurah et al. 2014).

E. crassipes showed a promising antioxidant potential in which the crude extract showed higher activity compared to the isolated compounds from alkaloid and terpenoid classes. The highest antioxidant activity was exhibited by the methanol extract with an IC_{50} value of 74.80 ± 4.5 μg/mL. The highest antioxidant potential of the crude extract may be interpreted by the synergistic activity of all the biological compounds. Additionally, the antioxidant potential was further ascertained by evaluating the scavenging potential of the liquid extracts versus oxygenated free radicals liberated through electrolysis in a phosphate buffer using a platinum electrode. Results of colorimetric assessments done at 515 nm adopting N, N-diethyl-1,4phenylenediamine (DPD) assay, displayed a reduction in the absorbance of the sample, comparable to blank produced via electrolysis of the buffer lacking the extracts, reflecting the existence of antioxidant constituents in water hyacinth. Furthermore, the antioxidant potential was expressed in terms of equivalent glutathione (EG) in which the EG value increased by reducing the dilution factor and changed with varying medium pH, postulating the influence of the medium pH on the antioxidant agent's yields in the extracts (Bodo et al. 2004).

Hepatoprotective activity

However, liver disorders such as hepatocellular carcinoma, liver cirrhosis and fibrosis as well as hepatitis are commonly caused by oxidative stress enhanced by hepatotoxins and xenobiotics, which mediate hepatic inflammation with an ultimate loss of liver integrity and function (Youssef et al. 2016).

E. crassipes represents an effective hepatoprotective agent manifested by its *in vivo* effect on liver markers and in combating oxidative stress as well, where *E. crassipes* co-administration with isoniazid treated rodents exhibited 46% reduction in malondialdehyde level with concomitant elevation in FRAP of plasma. It also counteracted isoniazid elicited effects in serum levels regarding ALP, ALT, AST, creatinine, complete bilirubin and hemoglobin (Kumar et al. 2014).

Anti-cancer activity

Cancer is considered as one of the most prevalent non-communicable diseases worldwide and is recognized as the second most frequent cause of death in the majority of the developed countries and the fourth leading cause of death in the developing nations. It is commonly treated by surgery, chemotherapy as well as radiotherapy. Although chemotherapeutic agents can alleviate symptoms, prolong life or cause a complete cure in some cancer cases, the lack of selectivity is the most deleterious problem facing chemotherapy causing massive damage to both cancer as well as normal cells. Thus, many patients seek alternative and complementary therapies to overcome these undesirable outcomes (Liu et al. 2016).

The crude extract of *E. crassipes* also revealed a notable potency versus MCF-7, HeLa cells, EACC and HepG2 cell lines with an IC_{50} value of 1.2, 1.6, 6.04 and 7.6 μg/mL, respectively, which is comparable to doxorubicin, a standard drug that revealed 0.28 μg/mL for Hela and 0.42 μg/mL for both MCF-7 and HepG2 cell lines (Aboul-Enein et al. 2014).

Antibacterial and antifungal activity

Another health threat facing human civilization is the evolution of antibiotic-resistant microorganisms. Serious infectious diseases caused by various pathogenic microorganisms constitute a massive problem threatening the global health care system. The problem is extremely complicated owing to the emergence of new infectious diseases, the continuous mutation of microorganisms, and the toxicity of the currently used synthetic antibiotics and conventional antimicrobial agents (Alanis 2005).

E. crassipes displayed antibacterial activities against both the Gram-positive bacteria represented by *Bacillus subtilis* and *Streptococcus faecalis*; and Gram-negative ones exemplified by *Escherichia coli* and *Staphylococcus aureus;* meanwhile, development of *Aspergillus flavus* and *Aspergillus niger* were not repressed by either *E. crassipes* rough concentrate nor its five divisions. Interestingly, *Candida albicans* (yeast) was repressed by *E. crassipes* extract and its fractions (Shanab et al. 2010). Concerning staphylococcal contaminations, *E. crassipes* showed a high potency versus MRSA existing in cows, SOSA and CoNS in bunnies (Gutierrez-Morales et al. 2017). An additional study was done on *E. crassipes* shoots and leaves ethanol extracts to evaluate their antibacterial and antifungal effects versus two bacteria, namely, *Rhodospirillum rubrum* and *Micrococcus luteus,* and the two fungi, *Rhodospirillum rubrum* and *Monoscus ruber,* employing the disk diffusion method. *E. crassipes* shoots and leaves ethanol extracts showed a notable behavior towards all the tested organisms comparable to the standard (Thamaraiselvi and Jayanthi 2012).

Larvicidal activity

Insects are capable of transmitting various neglected diseases that severely harm man with negative impacts on his health like filariasis, malaria dengue fever, yellow fever, in addition to many other disorders (Dargahi et al. 2014). Controlling vectors is the most successful method to eradicate the disease; however, most of the eradication processes are mainly dependent on insecticides which are highly toxic as organophosphates that has many side effects on man, animals, in addition to polluting the environment. Many herbs have long been known to control larvae and mosquitoes as well and hence could offer a relatively safe natural alternative (Katade et al. 2006, Rani and Rajasekharreddy 2009).

Water hyacinth displayed an effective larvicidal activity in which the crude root extract 100% affected *Chironomus ramosus* eggs and larvae in addition to the toxic potential of the acetone extract towards two pests *Spodoptera litura*, the tobacco cutworm, and *Achaea janata*, the castor semilooper (Devanand and Rani 2008).

Wound healing and skin whitening activity

Skin represents the main line of defense for the body acting as the crucial barrier versus exogenous pathogenic organisms, thus any destruction caused to the skin may severely threat human survival. It is worth mentioning that serious damage in the function and structure of the skin triggered by the occurrence of wounds causes hazardous effects to the deep underlying tissues causing profound inflammation. In addition, wounds could be subjected to microbial infection that prolongs the time for wound healing (Fonder et al. 2008).

Thus, possessing wound healing efficiency and skin whitening potential, *E. crassipes* could be used in cosmeceutical preparations. Regarding the wound healing potential, the methanol extract of water hyacinth leaves was formulated as an ointment, using 10 and 15% of leaf extract which exerted greater wound contraction potential compared to the control (Ali et al. 2010). Meanwhile, the ethyl acetate extract of water hyacinth in combination with musk and lemon was formulated as cream and revealed 8 to 11% tyrosinase inhibition with skin whitening effect (Jayanthi and Lalitha 2014).

All the biological activities of *Eichhornia crassipes* are demonstrated in Figure 9.6.

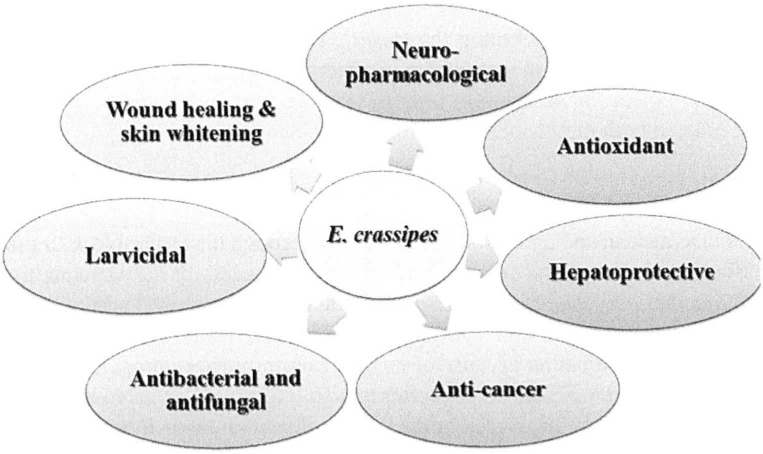

Figure 9.6. A scheme showing the biological activities of *Eichhornia crassipes*.

Industrial Uses of *Eichhornia crassipes*

Treatment of wastewater

E. crassipes is an effective phytoremediation employed for the crucial abolishing of hazardous poisons and risky pollutants from mines wastewater that may be adopted as an effective and natural decision to animate the removal and degradation of agro-current debased wastewater (Petrucio and Esteves 2000a, Jayaweera and Kasturiarachchi 2004, Mahmood et al. 2005, Xia and Ma 2006, Jayaweera et al. 2007a, 2008, Lu et al. 2008, Kularatne et al. 2009, Chen et al. 2010, Valipour et al. 2011, Akinbile and Yusoff 2012, Nesterenko-Malkovskaya et al. 2012, Gwenzi et al. 2014, Lu et al. 2014, Anudechakul et al. 2015, Melignani et al. 2015, Victor et al. 2016, Feng et al. 2017, Mishra and Maiti 2017, Afzal et al. 2018, Agarry et al. 2018, Kaur et al. 2018, Mahunon et al. 2018, Saha et al. 2018, Sricoth et al. 2018, Balasubramanian et al. 2019, Rangel-Peraza et al. 2019, Sayago 2019, Yi et al. 2019).

In addition, *E. crassipes* is perfectly applied for swine wastewater treatment to reduce phosphorous and nitrogen levels in the form of straw. It is also employed to expel fluoroquinolones as a multi-practical initiated carbon (MFAC) (Petrucio and Esteves 2000a, Petrucio and Esteves 2000b, Jayaweera and Kasturiarachchi 2004, Xia and Ma 2006, Jayaweera et al. 2008, Lu et al. 2008, Hadad et al. 2009, Kularatne et al. 2009, Chen et al. 2010, Albano Perez et al. 2011, Akinbile and Yusoff 2012, Fu et al. 2014, Li et al. 2015, Mercado-Borrayo et al. 2015, Sung et al. 2015, Dahroug et al. 2016, Aramburo-Miranda and Ruelas-Ramirez 2017, de Souza et al. 2018, Souza et al. 2018, Rangel-Peraza et al. 2019).

Besides, water hyacinth could effectively remove cyanide blowdown water owing to its tremendous biomass generation. Like CN, it can also be employed to evacuate some poisons generated by paper factories that are responsible for ocean assets eutrophication (Ebel et al. 2007, Saha et al. 2018). *E. crassipes* also acts as an effective bioindicator of polluted water with organic matters as neonicotinoids and endocrine disruptors and thus could act as a perspective for polluted water remediation by alarming natural organic pollutants (De Laet et al. 2019, Eid et al. 2019).

Additionally, *E. crassipes* is considered as a promising bio-sorbent for the expulsion of uranium contaminations and could be also helpful in treating cyanide effluents from little scope gold mines (Bhainsa and D'Souza 2001, Newete et al. 2016, Yi et al. 2016, Srivastava et al. 2019). It also gives an elective solution for the expulsion of co-pollution among antibiotics and heavy metals from domesticated animals and poultry wastewater (Lu et al. 2014, Yan et al. 2019a,b). *E. crassipes* is a viable and economical biomaterial for color expulsion from fluid color arrangements and mechanical effluents (El Zawahry and Kamel 2004, Renganathan et al. 2008, Wanyonyi et al. 2013, El-Zawahry et al. 2016, Mishra and Maiti 2017, Ramanathan et al. 2019, Tabinda et al. 2019).

E. crassipes could be used with minimal effort as a naturally agreeable oil biosorbent in contrast to initiated carbon in applications including the specific partition of Zn(2+) from multi-metal particle arrangements containing Ni(2+) and in wastewater treatment (Gwenzi et al. 2014, Yang et al. 2014). With great effectiveness, it expels fomesafen and mesotrione, two herbicides, quickly from agro-industrial wastewater contaminated with mesotrione and fomesafen (Chen et al. 2019b).

Biogas production

Biogas is considered as a clean and agreeable fuel generated through the anaerobic assimilation of natural squanders such as cow excrement and vegetable squanders. Nowadays, it has become highly significant for local industries as fuel because of its expenses and tidiness. The essential part of the gas is methane, carbon dioxide, hydrogen, nitrogen and hydrogen sulfide; thus, water hyacinth can be utilized as a potential feedstock for biogas creation because of its high carbon-nitrogen proportion. Biogas generation restores bioenergy via anaerobic absorption of water hyacinth followed by its co-assimilation with leafy foods squander and nitrogen to be recovered in the fluid profluent and thus it can be used as an effective fluid manure. Minimal effort biomass might be utilized to create fluids, gases, and biochar in a cost-efficient and ecologically inviting way through pyrolysis or co-pyrolysis in the future (Jayaweera et al. 2007b, Hernandez-Shek et al. 2016, Mishra and Maiti 2017).

The adsorbent of heavy metals

The contamination of aquatic ecosystems with poisonous substantial metals is considered a major natural hazard due to the discharge of these non-biodegradable metals to the biological surrounding without being treated. Various strategies were adopted and considered as an endeavor to get rid of the overwhelming metals from squandered water without reaching the ground. The dried leaves powder of water hyacinth was found effective in evacuating a portion of these substantial metals, comprising lead, copper, cadmium, and chrome, from fluid arrangements using bio-sorption which is influenced by few variable tests factors. Thiol group content that exists in water hyacinth makes it possible to assess the bio-concentration of heavy metal ions as cadmium (Cd), chromium (Cr), copper (Cu), manganese (Mn), nickel (Ni), lead (Pb), uranium (U), and zinc (Zn) and thus water hyacinth serves as a general agent for monitoring the heavy metal pollution of water owing to its richness with thiol group containing compounds (Srivastava et al. 2019). Hence, this species is appropriate for phytoremediation of water, subject to the release of blended modern effluents containing raised Cu fixations (>/=15 mg Cu/L), just like rich local wastewaters (Wolverton and McDonald 1978, Muramoto and Oki 1983, 1984, So et al. 2003, Olivares-Rieumont et al. 2007, Tiwari et al. 2007, Wu et al. 2008, Mishra and Tripathi 2009, Dixit and Dhote 2010, Mahamadi and Nharingo 2010, Sabale et al. 2010, Hadad et al. 2011, Deng et al. 2012, Elfeky et al. 2013, Li et al. 2016, Newete et al. 2016, Tabinda et al. 2018, Cao et al. 2019, Pal et al. 2019). It also utilized a biosorbent to expel watery sulfachloropyridazine (SCP) and different sulfonamides (Liu et al. 2018).

Alternative energy source

The rapid environmental variations and their concomitant effects greatly influence the advances in sustainable international power sources. There is a developing mission to enlarge the stockpile of biomass assets for bioenergy creation. *E. crassipes* acts as a potential alternative energy source since it is available extensively in freshwater, marine, and aquatic ecosystems throughout the world (Sricoth et al. 2018, Venu et al. 2019). It acts as a promising candidate for fuel ethanol production in tropical countries because of their high availability and high biomass yield (Das et al. 2016, Sayago 2019).

Several studies converted water hyacinth to charcoal dust through pyrolysis to act as an available source for producing locally needed fuels and to enhance this quality via changing this charcoal into briquettes with molasses as a fastener to elevate its strength and durability with a subsequent reduction in transportation, handling as well as storage (Carnaje et al. 2018).

Animal feed

E. crassipes was effective for purifying wastewater from an extensive duck ranch during the water hyacinth developing season, as water hyacinth had a great role as duck feed. It also positively affects its quality and increases the eggshell thickness as well (Lu et al. 2008). It upgrades the expulsion of contaminants through their utilization as supplements by their rhizospheric microscopic organisms causing reduction of naphthalene (a polyaromatic hydrocarbon) present in wastewater and wetlands (Nesterenko-Malkovskaya et al. 2012). It can also be used as a natural fertilizer for *Colossoma macropomum* (tambaqui) hatchlings

Table 9.1. Industrial uses of different parts of *E. crassipes* (water hyacinth).

Part used	Feasibility and Industrial uses	References
Leaves	• Accelerates the removal and degradation of agro-industrial wastewater polluted with ethion	(Xia and Ma 2006)
Leaves	• Plants (rough surface) used as cost-effective and environmentally friendly oil sorbents	(Yang et al. 2014)
Leaves	• Used as an immunostimulant for prawn through dietary administration to enhance immune responses and resistance of *M. rosenbergii* against *L. garvieae*	(Chang et al. 2013)
Roots	• Removal of 54% of the initial uranium present within 4 min of contact time	(Bhainsa and D'Souza 2001)
Roots	• Determines the concentration of heavy metals present in surface water such as cadmium (Cd), chromium (Cr), copper (Cu), manganese (Mn), nickel (Ni), lead (Pb), uranium (U), and zinc (Zn)	(Abou-Shanab et al. 2007, Malar et al. 2016, Chen et al. 2019a, Srivastava et al. 2019)
Roots	• To assess the bioconcentration of heavy metal ions using the thiol group content in water hyacinth • To monitor the heavy metal pollution of water	(Ding et al. 1994, Lopes and Piedade 2014)
Roots	• To remove fluoroquinolones (FQs) from contaminated water acting as a multi-functional activated carbon (MFAC)	(Liu et al. 2019)
Roots	• To absorb and adsorb metals (arsenic (As), gold (Au), copper (Cu), iron (Fe), mercury (Hg), manganese (Mn), uranium (U), and zinc (Zn)) within or on the external root surfaces	(Lenka et al. 1990, Newete et al. 2016, Pandey 2016, Eid et al. 2019, Zhang et al. 2019)
Root powder	• Used as a biosorbent to remove aqueous sulfachloropyridazine (SCP) and other sulfonamides	(Liu et al. 2018)
Stem biochar	• Removal of Cd (2+) from aqueous solution	(Zhou et al. 2019)
Plant straw	• To treat swine wastewater sequentially for nitrogen and phosphorus reduction	(Chen et al. 2010, Li et al. 2015)
Plant straw	• To adsorb heavy metals	(Deng et al. 2012)
Roots & Leaves	• For remedy of waters subjected to discharge of mixed industrial effluents containing elevated Cu concentrations and nutrient-rich domestic wastewaters	(Melignani et al. 2015, Mishra and Maiti 2017, Tabinda et al. 2018)
Roots & Leaves	• Used in phytoremediation programs in environments contaminated by arsenic (As) • To produce biogas	(Jayaweera et al. 2007a, de Souza Reis et al. 2019)
Roots & Leaves	• To remove Copper (Cu) pollution in aquaculture ponds poses substantial ecological threats	(Goswami and Das 2018)

Table 9.1 Contd. ...

...Table 9.1 Contd.

Part used	Feasibility and Industrial uses	References
Roots & Leaves	• Serves as an efficient and reliable accumulator for heavy metals	(Wu et al. 2008, Sabale et al. 2010)
Roots, Stems, & Leaves	• Acts as a biomaterial for dye removal from aqueous dye solutions and industrial effluent	(Wanyonyi et al. 2013)
Whole plant	• To purify wastewater from extensive duck ranch during the water hyacinth developing season, as water hyacinth had a great role as duck feed • Affects positively duck quality and increases the eggshell thickness	(Lu et al. 2008)
Whole plant	• To produce biogas • To recover bioenergy • Used as a liquid fertilizer	(Hernandez-Shek et al. 2016)
Whole plant	• To produce fuel ethanol in tropical countries because of their high availability and high biomass yield	(Mishima et al. 2008, Das et al. 2016, Sayago 2019)
Whole plant	• Enhances the removal of pollutants through their consumption as nutrients by rhizospheric bacteria • To reduce naphthalene (a polyaromatic hydrocarbon) present in wastewater and wetlands	(Nesterenko-Malkovskaya et al. 2012)
Whole plant	• To remove cyanide blowdown water due to its high biomass production and tolerance to cyanide (CN)	(Saha et al. 2018)
Whole plant	• To remove different pollutants of paper mill effluent, responsible for the eutrophication of the aquatic resources	(Kumar et al. 2018)
Whole plant	• Used as an alternative fuel source by converting its charcoal into briquettes with molasses as a binder	(Carnaje et al. 2018)
Whole plant	• Used as cellulose source and polyvinyl alcohol showing excellent oil/solvent sorption capacities, super-hydrophobicity, and remarkable reusability	(Yin et al. 2017)
Whole plant	• Used as a bioindicator of water polluted by worrying organic pollutants such as endocrine disruptors and neonicotinoids	(De Laet et al. 2019)
Whole plant	• Used as bio-sorbent for removal of uranium pollutants	(Yi et al. 2016)
Whole plant	• Used in treating cyanide effluents from small-scale gold mines	(Ebel et al. 2007)
Whole plant	• Uptake of most heavy metals	(Muramoto and Oki 1983, Li et al. 2018, Shirinpur-Valadi et al. 2019)
Whole plant	• To treat Cd-polluted waters given its ability to tolerate high Cd concentrations in the media (up to 10 mg Cd/L) and its capacity for uptake and accumulation	(Modenes et al. 2011, Melignani et al. 2019, Shirinpur-Valadi et al. 2019)
Whole plant	• To remove the co-contamination between antibiotics and heavy metals from livestock and poultry wastewater	(Dixit and Dhote 2010, Lu et al. 2014)
Whole plant	• Used as an alternative energy source	(Sricoth et al. 2018, Venu et al. 2019)
Whole plant	• To reduce salinity in natural conditions	(de Lacerda et al. 2019)
Whole plant	• To remove two herbicides (mesotrione and fomesafen) with long degradation cycles in water	(Chen et al. 2019b)
Whole plant	• Used in phytoremediation of the antibiotic sulfadiazine (SDZ)	(Yan et al. 2019a)

Table 9.1 Contd. ...

...Table 9.1 Contd.

Whole plant	• Used as a low-cost biosorbent as an alternative to activated carbon in the selective separation of Zn(2+) from multi-metal ion solutions containing Ni(2+), and water • Used in wastewater treatment	(Gwenzi et al. 2014)
Part used	**Feasibility and Industrial uses**	**References**
Whole plant	• To produce liquids, gases, and biochar in a cost-efficient and environmentally friendly way via pyrolysis or co-pyrolysis in the future	(Afzal et al. 2018)
Macrophyte	• Used as an organic fertilizer of *Colossoma macropomum* (tambaqui) larvae in ponds	(Sipauba-Tavares and Braga 2007)
Macrophytes	• To treat textile effluents rich in COD, BOD, dyes, and heavy metals (Pb, Fe, Cd, Cu) by phytoremediation	(Tabinda et al. 2019)

in lakes. It is very modest and effectively accessible, and might be advantageously used to upgrade fish yield in lakes (Sipauba-Tavares and Braga 2007).
Table 9.1 summarizes different industrial uses of *E. crassipes*.

Conclusion

Thus, it can be concluded that although water hyacinth is reported as an invasive aquatic plant causing many dramatic problems in lakes, rivers and many reservoirs around the globe, particularly in Africa, it can act as an excellent source of compounds with many beneficial pharmacological potentials. In addition, the harvesting of the plant on a large scale greatly elevates its importance and makes it possible to beneficially use its isolated phytochemicals either as they are or after undergoing certain chemical modification. Although many studies were previously conducted and reported the utilization of water hyacinth in bioremediation and energy generation, few studies were conducted on its phytochemicals. Thus, this chapter highlights the economic impact of water hyacinth if perfectly used in virtue of its growth control and in overcoming the problems triggered by the plant on the ecosystem. Thus, planned harvesting of *E. crassipes* should be effectively utilized in the extraction of its phytoconstituents and in pharmaceutical industries in the manufacture of fortified products enriched with the extracts, which could solve the problem of its invasive behavior. Hence, more efforts should be directed to discover more of its chemical and biological effects comprising its anti-infectious potential against life-threatening organisms, like novel coronavirus (COVID-19), and its industrial importance as well.

References

Abou-Shanab, R.A., Angle, J.S., van Berkum, P. 2007. Chromate-tolerant bacteria for enhanced metal uptake by *Eichhornia crassipes* (Mart.). Int. J. Phytoremediation, 9: 91–105.

Aboul-Enein, A.M., Shanab, S.M., Shalaby, E.A., Zahran, M.M., Lightfoot, D.A., El-Shemy, H.A. 2014. Cytotoxic and antioxidant properties of active principals isolated from water hyacinth against four cancer cells lines. BMC Comp. Alt. Med., 14: 397. https://doi.org/10.1186/1472-6882-14-397.

Afzal, I., Ahmad, M.S., Malik, S., Ibrahim, M., Al Ayed, O.S., Qadir, G., Al Doghaither, H., Gull, M. 2018. Thermodynamics and kinetics parameters of *Eichhornia crassipes* biomass bioenergy. Prot. Pept Lett., 25: 187–194.

Agarry, S.E., Oghenejoboh, K.M., Latinwo, G.K., Owabor, C.N. 2018. Biotreatment of petroleum refinery wastewater in vertical surface-flow constructed wetland vegetated with *Eichhornia crassipes*: lab-scale experimental and kinetic modelling. Environ. Technol., 1–21. https://doi.org/10.1080/09593330.2018.1549106.

Akinbile, C.O., Yusoff, M.S. 2012. Assessing water hyacinth (*Eichhornia crassopes*) and lettuce (*Pistia stratiotes*) effectiveness in aquaculture wastewater treatment. Int. J. Phytoremediation, 14: 201–211.

Alanis, A.J. 2005. Resistance to antibiotics: are we in the post-antibiotic era? Arch. Med. Res., 36: 697–705.

Albano Perez, E., Ruiz Tellez, T., Sanchez Guzman, J.M. 2011. Influence of physico-chemical parameters of the aquatic medium on germination of *Eichhornia crassipes* seeds. Plant Biol. (Stuttg), 13: 643–648.

Ali, H., Lata, N., Ahi, J., Ganesh, N. 2010. Evaluation of wound-healing activity of *Eichornia crassipes*: A novel approach. Drug Inv. Today, 2: 212–214.

Anudechakul, C., Vangnai, A.S., Ariyakanon, N. 2015. Removal of chlorpyrifos by water hyacinth (*Eichhornia crassipes*) and the role of a plant-associated bacterium. Int. J. Phytoremediation, 17: 678–685.

Aramburo-Miranda, I.V., Ruelas-Ramirez, E.H. 2017. Desalination of sea water with aquatic lily (*Eichhornia crassipes*). Environ. Sci. Pollut. Res. Int., 24: 25676–25681.

Arayana, G.L., Rao, K.S., Pantulu, A., Thyagarajan, G. 1984. Composition of lipids in roots, stalks, leaves and flowers of *Eichhornia crassipes* (Mart.) Solms. Aquatic Bot., 20: 219–227.

Balasubramanian, U.M., Vaiyazhipalayam Murugaiyan, S., Marimuthu, T. 2019. Enhanced adsorption of Cr(VI), Ni(II) ions from aqueous solution using modified *Eichhornia crassipes* and Lemna minor. Environ. Sci. Pollut. Res. Int., https://doi.org/10.1007/s11356-019-06357-7.

Bhainsa, K.C., D'Souza, S.F. 2001. Uranium(VI) biosorption by dried roots of *Eichhornia crassipes* (water hyacinth). J. Environ. Sci. Health. A Tox. Hazard Subst. Environ. Eng., 36: 1621–1631.

Bodo, R., Azzouz, A., Hausler, R. 2004. Antioxidative activity of water hyacinth components. Plant Sci., 166: 893–899.

Cao, F., Lian, C., Yu, J., Yang, H., Lin, S. 2019. Study on the adsorption performance and competitive mechanism for heavy metal contaminants removal using novel multi-pore activated carbons derived from recyclable long-root *Eichhornia crassipes*. Bioresour. Technol., 276: 211–218.

Carnaje, N.P., Talagon, R.B., Peralta, J.P., Shah, K., Paz-Ferreiro, J. 2018. Development and characterisation of charcoal briquettes from water hyacinth (*Eichhornia crassipes*)-molasses blend. PLoS One, 13: e0207135.

Chang, C.C., Tan, H.C., Cheng, W. 2013. Effects of dietary administration of water hyacinth (*Eichhornia crassipes*) extracts on the immune responses and disease resistance of giant freshwater prawn, *Macrobrachium rosenbergii*. Fish Shellfish Immunol., 35: 92–100.

Chen, X., Chen, X., Wan, X., Weng, B., Huang, Q. 2010. Water hyacinth (*Eichhornia crassipes*) waste as an adsorbent for phosphorus removal from swine wastewater. Bioresour. Technol., 101: 9025–9030.

Chen, X.L., Li, F., Xie, X.J., Li, Z., Chen, L. 2019a. Nanoscale zero-valent iron and chitosan functionalized *Eichhornia crassipes* biochar for efficient hexavalent chromium removal. Int. J. Environ. Res. Public Health, 16: https://doi.org/10.3390/ijerph16173046.

Chen, Z., Huang, L., Song, S., Zhang, Y., Li, Y., Tan, H., Li, X. 2019b. Enhanced disappearance of mesotrione and fomesafen by water hyacinth (*Eichhornia crassipes*) in water. Int. J. Phytoremediation, 21: 583–589.

Chukwurah, P.N., Brisibe, E.A., Osuagwu, A.N., Okoko, T. 2014. Protective capacity of *Artemisia annua* as a potent antioxidant remedy against free radical damage. Asian Pac. J. Trop. Biomed., 4: S92–S98.

Dahroug, Z., Santana, N.F., Pagioro, T.A. 2016. *Eichhornia azurea* decomposition and the bacterial dynamic: an experimental research. Braz. J. Microbiol., 47: 279–286.

Dandelot, S., Robles, C., Pech, N., Cazaubon, A., Verlaque, R. 2008. Allelopathic potential of two invasive alien *Ludwigia* spp. Aquat.Bot., 88: 311–316.

Dargahi, L., Razavi-Azarkhiavi, K., Ramezani, M., Abaee, M.R., Behravan, J. 2014. Insecticidal activity of the essential oil of Thymus transcaspicus against Anopheles stephensi. Asian Pac. J. Trop. Biomed., 4: S589–S591.

Das, A., Ghosh, P., Paul, T., Ghosh, U., Pati, B.R., Mondal, K.C. 2016. Production of bioethanol as useful biofuel through the bioconversion of water hyacinth (*Eichhornia crassipes*). 3 Biotech., 6: 70. https://doi.org/10.1007/s13205-016-0385-y.

de Lacerda, L.P., Lange, L.C., Costa Franca, M.G., Diniz Leao, M.M. 2019. Growth and differential salinity reduction between Portulaca oleracea and *Eichhornia crassipes* in experimental hydroponic units. Environ. Technol., 40: 2267–2275.

De Laet, C., Matringe, T., Petit, E., Grison, C. 2019. *Eichhornia crassipes*: a powerful bio-indicator for water pollution by emerging pollutants. Sci. Rep., 9: 7326. https://doi.org/10.1038/s41598-019-43769-4.

de Souza Reis, I.N.R., Alves de Oliveira, J., Ventrella, M.C., Otoni, W.C., Marinato, C.S., Paiva de Matos, L. 2019. Involvement of glutathione metabolism in *Eichhornia crassipes* tolerance to arsenic. Plant Biol., https://doi.org/10.1111/plb.12988.

Della Greca, M., Lanzetta, R., Molinaro, A., Monaco, P., Previtera, L. 1992. Phenalene metabolites from *Eichhornia crassipes*. Bioorg. & Med. Chem. Lett., 2: 311–314.

Deng, L., Geng, M., Zhu, D., Zhou, W., Langdon, A., Wu, H., Yu, Y., Zhu, Z., Wang, Y. 2012. Effect of chemical and biological degumming on the adsorption of heavy metal by cellulose xanthogenates prepared from *Eichhornia crassipes*. Bioresour. Technol., 107: 41–45.

Devanand, P., Rani, P.U. 2008. Biological potency of certain plant extracts in management of two lepidopteran pests of *Ricinus communis* L. J. Biopest., 1: 170–176.

Ding, X., Jiang, J., Wang, Y., Wang, W., Ru, B. 1994. Bioconcentration of cadmium in water hyacinth (*Eichhornia crassipes*) in relation to thiol group content. Environ. Pollut., 84: 93–96.

Dixit, S., Dhote, S. 2010. Evaluation of uptake rate of heavy metals by *Eichhornia crassipes* and Hydrilla verticillata. Environ. Monit. Assess., 169: 367–374.

Ebel, M., Evangelou, M.W., Schaeffer, A. 2007. Cyanide phytoremediation by water hyacinths (*Eichhornia crassipes*). Chemosphere, 66: 816–823.

Eid, E.M., Shaltout, K.H., Moghanm, F.S., Youssef, M.S.G., El-Mohsnawy, E., Haroun, S.A. 2019. Bioaccumulation and translocation of nine heavy metals by *Eichhornia crassipes* in Nile Delta, Egypt: perspectives for phytoremediation. Int. J. Phytoremediation, 21: 821–830.

El Zawahry, M.M., Kamel, M.M. 2004. Removal of azo and anthraquinone dyes from aqueous solutions by *Eichhornia Crassipes*. Water Res., 38: 2967–2972.

El-Zawahry, M.M., Abdelghaffar, F., Abdelghaffar, R.A., Hassabo, A.G. 2016. Equilibrium and kinetic models on the adsorption of Reactive Black 5 from aqueous solution using *Eichhornia crassipes*/chitosan composite. Carbohydr. Polym., 136: 507–515.

Elfeky, S.A., Imam, H., Alsherbini, A.A. 2013. Bio-absorption of Ni and Cd on *Eichhornia crassipes* root thin film. Environ. Sci. Pollut. Res. Int., 20: 8220–8226.

Farheen, M., Arman, S., Hussaini, R., Kouser, A., Pharm, M. 2015. Phytochemical evaluation and pharmacological screening of ethanolic leaf extracts of *Eichhornia Crassipes* and *Nelumbo nucifera* for neuropharmacological activity in psychoneurosis induced mice. Med.

Feng, W., Xiao, K., Zhou, W., Zhu, D., Zhou, Y., Yuan, Y., Xiao, N., Wan, X., Hua, Y., Zhao, J. 2017. Analysis of utilization technologies for *Eichhornia crassipes* biomass harvested after restoration of wastewater. Bioresour. Technol., 223: 287–295.

Fonder, M.A., Lazarus, G.S., Cowan, D.A., Aronson-Cook, B., Kohli, A.R., Mamelak, A.J. 2008. Treating the chronic wound: A practical approach to the care of nonhealing wounds and wound care dressings. J. Am. Acad. Dermatol., 58: 185–206.

Fu, M., Jiang, L., Li, Y., Yan, G., Zheng, L., Jinping, P. 2014. Identification of gene fragments related to nitrogen deficiency in *Eichhornia crassipes* (Pontederiaceae). Rev. Biol. Trop., 62: 1637–1648.

Goswami, S., Das, S. 2018. *Eichhornia crassipes* mediated copper phytoremediation and its success using catfish bioassay. Chemosphere, 210: 440–448.

Gutierrez-Morales, A., Velazquez-Ordonez, V., Khusro, A., Salem, A.Z.M., Estrada-Zuniga, M.E., Salem, M.Z.M., Valladares-Carranza, B., Burrola-Aguilar, C. 2017. Anti-staphylococcal properties of *Eichhornia crassipes, Pistacia vera*, and *Ziziphus* amole leaf extracts: Isolates from cattle and rabbits. Microb. Pathog., 113: 181–189.

Gwenzi, W., Musarurwa, T., Nyamugafata, P., Chaukura, N., Chaparadza, A., Mbera, S. 2014. Adsorption of Zn(2+) and Ni(2+) in a binary aqueous solution by biosorbents derived from sawdust and water hyacinth (*Eichhornia crassipes*). Water Sci. Technol., 70: 1419–1427.

Hadad, H.R., Maine, M.A., Pinciroli, M., Mufarrege, M.M. 2009. Nickel and phosphorous sorption efficiencies, tissue accumulation kinetics and morphological effects on *Eichhornia crassipes*. Ecotoxicology, 18: 504–513.

Hadad, H.R., Maine, M.A., Mufarrege, M.M., Del Sastre, M.V., Di Luca, G.A. 2011. Bioaccumulation kinetics and toxic effects of Cr, Ni and Zn on *Eichhornia crassipes*. J. Hazard Mater., 190: 1016–1022.

Hernandez-Shek, M.A., Cadavid-Rodriguez, L.S., Bolanos, I.V., Agudelo-Henao, A.C. 2016. Recovering biomethane and nutrients from anaerobic digestion of water hyacinth (*Eichhornia crassipes*) and its co-digestion with fruit and vegetable waste. Water Sci. Technol., 73: 355–361.

Holscher, D., Schneider, B. 2005. The biosynthesis of 8-phenylphenalenones from *Eichhornia crassipes* involves a putative aryl migration step. Phytochemistry, 66: 59–64.

Jayanthi, P., Lalitha, P., Shubashini, K. 2011. Phytochemical investigation of the extracts of *Eichhornia crassipes* and its solvent fractionates. J. Pharm. Res., 4: 1405–1406.

Jayanthi, P., Lalitha, P., Aarthi, N. 2012. Larvicidal and pupicidal activity of extracts and fractionates of *Eichhornia crassipes* (Mart.) Solms against the filarial vector Culex quinquefasciatus Say. Parasitol. Res., 111: 2129–2135.

Jayanthi, P., Lalitha, P. 2014. Formulation development and assessment of skin whitening efficiency of ethyl acetate extract of *Eichhornia crassipes* (Mart.) solms by *in vitro* tyrosinase activity. Int. J. Chem. Tech. Res., 6: 178–182.

Jayaweera, M.W., Kasturiarachchi, J.C. 2004. Removal of nitrogen and phosphorus from industrial wastewaters by phytoremediation using water hyacinth (*Eichhornia crassipes* (Mart.) Solms). Water Sci. Technol., 50: 217–225.

Jayaweera, M.W., Kasturiarachchi, J.C., Kularatne, R.K., Wijeyekoon, S.L. 2007a. Removal of aluminium by constructed wetlands with water hyacinth (*Eichhornia crassipes* (Mart.) Solms) grown under different nutritional conditions. J. Environ. Sci. Health A Tox. Hazard Subst. Environ. Eng., 42: 185–193.

Jayaweera, M.W., Dilhani, J.A., Kularatne, R.K., Wijeyekoon, S.L. 2007b. Biogas production from water hyacinth (*Eichhornia crassipes* (Mart.) Solms) grown under different nitrogen concentrations. J. Environ. Sci. Health A Tox. Hazard Subst. Environ. Eng., 42: 925–932.

Jayaweera, M.W., Kasturiarachchi, J.C., Kularatne, R.K., Wijeyekoon, S.L. 2008. Contribution of water hyacinth (*Eichhornia crassipes* (Mart.) Solms) grown under different nutrient conditions to Fe-removal mechanisms in constructed wetlands. J. Environ. Manage., 87: 450–460.

Katade, S.R., Pawar, P.V., Tungikar, V.B., Tambe, A.S., Kalal, K.M., Wakharkar, R.D., Deshpande, N.R. 2006. Larvicidal activity of Bis (2-ethylhexyl) benzene-1, 2-dicarboxylate from *Sterculia guttata* seeds against two mosquito species. Chem. Biodivers, 3: 49–53.

Kaur, P., Singh, S., Kumar, V., Singh, N., Singh, J. 2018. Effect of rhizobacteria on arsenic uptake by macrophyte *Eichhornia crassipes* (Mart.) Solms. Int. J. Phytoremediation, 20: 114–120.

Kularatne, R.K., Kasturiarachchi, J.C., Manatunge, J.M., Wijeyekoon, S.L. 2009. Mechanisms of manganese removal from wastewaters in constructed wetlands comprising water hyacinth (*Eichhornia crassipes* (Mart.) Solms) grown under different nutrient conditions. Water Environ. Res., 81: 165–172.

Kumar, S., Kumar, R., Dwivedi, A., Pandey, A.K. 2014. *In vitro* antioxidant, antibacterial, and cytotoxic activity and in vivo effect of *Syngonium podophyllum* and *Eichhornia crassipes* leaf extracts on isoniazid induced oxidative stress and hepatic markers. Biomed. Res. Int., 2014: 459452.

Kumar, V., Singh, J., Chopra, A.K. 2018. Assessment of phytokinetic removal of pollutants of paper mill effluent using water hyacinth (*Eichhornia crassipes* [Mart.] Solms). Environ. Technol., 39: 2781–2791.

Lalitha, P., Sripathi, S.K., Jayanthi, P. 2012. Secondary metabolites of *Eichhornia crassipes* (waterhyacinth): a review (1949 to 2011). Nat. Prod. Com., 7: 1934578X1200700939.

Lata, N., Dubey, V. 2010a. Preliminary phytochemical screening of *Eichhornia crassipes*: the world's worst aquatic weed. J. Pharm. Res., 3: 1240–1242.

Lata, N., Dubey, V. 2010b. Quantification and identification of alkaloids of *Eichhornia crassipes*: the world's worst aquatic plant. J. Phar. Res., 3: 1229–1231.

Lenka, M., Panda, K.K., Panda, B.B. 1990. Studies on the ability of water hyacinth (*Eichhornia crassipes*) to bioconcentrate and biomonitor aquatic mercury. Environ. Pollut., 66: 89–99.

Li, Q., Chen, B., Lin, P., Zhou, J., Zhan, J., Shen, Q., Pan, X. 2016. Adsorption of heavy metal from aqueous solution by dehydrated root powder of long-root *Eichhornia crassipes*. Int. J. Phytoremediation, 18: 103–109.

Li, Q., Tang, L., Hu, J., Jiang, M., Shi, X., Zhang, T., Li, Y., Pan, X. 2018. Removal of toxic metals from aqueous solution by biochars derived from long-root *Eichhornia crassipes*. R. Soc. Open Sci., 5: 180966. https://doi.org/10.1098/rsos.180966.

Li, X., Xi, H., Sun, X., Yang, Y., Yang, S., Zhou, Y., Zhou, X., Yang, Y. 2015. Comparative proteomics exploring the molecular mechanism of eutrophic water purification using water hyacinth (*Eichhornia crassipes*). Environ. Sci. Pollut. Res. Int., 22: 8643–8658.

Liu, L., Hu, S., Shen, G., Farooq, U., Zhang, W., Lin, S., Lin, K. 2018. Adsorption dynamics and mechanism of aqueous sulfachloropyridazine and analogues using the root powder of recyclable long-root *Eichhornia crassipes*. Chemosphere, 196: 409–417.

Liu, L., Chen, X., Wang, Z., Lin, S. 2019. Removal of aqueous fluoroquinolones with multi-functional activated carbon (MFAC) derived from recycled long-root *Eichhornia crassipes*: batch and column studies. Environ. Sci. Pollut. Res. Int., 26: 34345–34356.

Liu, N., Yang, H.L., Wang, P., Lu, Y.C., Yang, Y.J., Wang, L. , Lee, S.C. 2016. Functional proteomic analysis revels that the ethanol extract of Annona muricata L. induces liver cancer cell apoptosis through endoplasmic reticulum stress pathway. J. Ethnopharmacol., 189: 210–217.

Lopes, A., Piedade, M.T. 2014. Experimental study on the survival of the water hyacinth (*Eichhornia crassipes* (Mart.) Solms--Pontederiaceae) under different oil doses and times of exposure. Environ. Sci. Pollut. Res. Int., 21: 13503–13511.

Lu, J., Fu, Z., Yin, Z. 2008. Performance of a water hyacinth (*Eichhornia crassipes*) system in the treatment of wastewater from a duck farm and the effects of using water hyacinth as duck feed. J. Environ. Sci. (China), 20: 513–519.

Lu, X., Gao, Y., Luo, J., Yan, S., Rengel, Z., Zhang, Z. 2014. Interaction of veterinary antibiotic tetracyclines and copper on their fates in water and water hyacinth (*Eichhornia crassipes*). J. Hazard Mater., 280: 389–398.

Mahamadi, C., Nharingo, T. 2010. Competitive adsorption of Pb^{2+}, Cd^{2+} and Zn^{2+} ions onto *Eichhornia crassipes* in binary and ternary systems. Bioresour. Technol., 101: 859–864.

Mahmood, Q., Siddiqi, M.R., Islam, E., Azim, M.R., Zheng, P., Hayat, Y. 2005. Anatomical studies on water hyacinth (*Eichhornia crassipes* (Mart.) Solms) under the influence of textile wastewater. J. Zhejiang. Univ. Sci. B, 6: 991–998.

Mahunon, S.E.R., Aina, M.P., Akowanou, A.V.O., Kouassi, E.K., Yao, B.K., Adouby, K., Drogui, P. 2018. Optimization process of organic matter removal from wastewater by using *Eichhornia crassipes*. Environ. Sci. Pollut. Res. Int., 25: 29219–29226.

Malar, S., Shivendra Vikram, S., Jc Favas, P., Perumal, V. 2016. Lead heavy metal toxicity induced changes on growth and antioxidative enzymes level in water hyacinths [*Eichhornia crassipes* (Mart.)]. Bot Stud., 55: 54. https://doi.org/10.1186/s40529-014-0054-6.

Melignani, E., de Cabo, L.I., Faggi, A.M. 2015. Copper uptake by *Eichhornia crassipes* exposed at high level concentrations. Environ. Sci. Pollut. Res. Int., 22: 8307–8315.

Melignani, E., Faggi, A.M. and de Cabo, L.I. 2019. Growth, accumulation and uptake of *Eichhornia crassipes* exposed to high cadmium concentrations. Environ. Sci. Pollut. Res. Int., 26: 22826–22834.

Mercado-Borrayo, B.M., Cram Heydrich, S., Perez, I.R., Hernandez Quiroz, M., De Leon Hill, C.P. 2015. Organophosphorus and organochlorine pesticides bioaccumulation by *Eichhornia crassipes* in irrigation canals in an urban agricultural system. Int. J. Phytoremediation, 17: 701–708.

Mishima, D., Kuniki, M., Sei, K., Soda, S., Ike, M., Fujita, M. 2008. Ethanol production from candidate energy crops: water hyacinth (*Eichhornia crassipes*) and water lettuce (*Pistia stratiotes* L.). Bioresour. Technol., 99: 2495–2500.

Mishra, S., Maiti, A. 2017. The efficiency of *Eichhornia crassipes* in the removal of organic and inorganic pollutants from wastewater: a review. Environ. Sci. Pollut. Res. Int., 24: 7921–7937.

Mishra, V.K., Tripathi, B.D. 2009. Accumulation of chromium and zinc from aqueous solutions using water hyacinth (*Eichhornia crassipes*). J. Hazard Mater. 164: 1059–1063.

Modenes, A.N., Espinoza-Quinones, F.R., Borba, C.E., Trigueros, D.E., Lavarda, F.L., Abugderah, M.M., Kroumov, A.D. 2011. Adsorption of Zn(II) and Cd(II) ions in batch system by using the *Eichhornia crassipes*. Water Sci. Technol., 64: 1857–1863.

Muramoto, S., Oki, Y. 1983. Removal of some heavy metals from polluted water by water hyacinth (*Eichhornia crassipes*). Bull. Environ. Contam Toxicol., 30: 170–177.

Muramoto, S., Oki, Y. 1984. Influence of anionic surface-active agents on the uptake of heavy metals by water hyacinth (*Eichhornia crassipes*). Bull. Environ. Contam. Toxicol., 33: 444–450.

Muthunarayanan, V., Santhiya, M., Swabna, V., Geetha, A. 2011. Phytodegradation of textile dyes by water hyacinth (*Eichhornia crassipes*) from aqueous dye solutions. Int. J. Environ. Sci., 1: 1702–1717.

Nesterenko-Malkovskaya, A., Kirzhner, F., Zimmels, Y., Armon, R. 2012. *Eichhornia crassipes* capability to remove naphthalene from wastewater in the absence of bacteria. Chemosphere, 87: 1186–1191.

Newete, S.W., Erasmus, B.F., Weiersbye, I.M., Byrne, M.J. 2016. Sequestration of precious and pollutant metals in biomass of cultured water hyacinth (*Eichhornia crassipes*). Environ. Sci. Pollut. Res. Int., 23: 20805–20818.

Nyananyo, B., Gijo, A., Ogamba, E. 2007. The physico-chemistry and distribution of water hyacinth (*Eichhornia cressipes*) on the river Nun in the Niger Nelta. J. Appl. Sci. Environ. Manag., 11: doi: 10.4314/jasem.v11i3.55158.

Olivares-Rieumont, S., Lima, L., De la Rosa, D., Graham, D.W., Columbie, I., Santana, J.L., Sanchez, M.J. 2007. Water hyacinths (*Eichhornia crassipes*) as indicators of heavy metal impact of a large landfill on the Almendares River near Havana, Cuba. Bull. Environ. Contam. Toxicol., 79: 583–587.

Pal, R., Kaur, R., Rajwar, D., Narayan Rai, J.P. 2019. Induction of non-protein thiols and phytochelatins by cadmium in *Eichhornia crassipes*. Int. J. Phytoremediation, 21: 790–798.

Pandey, V.C. 2016. Phytoremediation efficiency of *Eichhornia crassipes* in fly ash pond. Int. J. Phytoremediation, 18: 450–452.

Petrucio, M.M., Esteves, F.A. 2000a. Influence of photoperiod on the uptake of nitrogen and phosphorus in the water by *Eichhornia crassipes* and *Salvinia auriculata*. Rev. Bras. Biol., 60: 373–379.

Petrucio, M.M., Esteves, F.A. 2000b. Uptake rates of nitrogen and phosphorus in the water by *Eichhornia crassipes* and *Salvinia auriculata*. Rev. Bras. Biol., 60: 229–236.

Ramanathan, S., Selvin, S.P., Obadiah, A., Durairaj, A., Santhoshkumar, P., Lydia, S., Ramasundaram, S., Vasanthkumar, S. 2019. Synthesis of reduced graphene oxide/ZnO nanocomposites using grape fruit extract and *Eichhornia crassipes* leaf extract and a comparative study of their photocatalytic property in degrading Rhodamine B dye. J. Environ. Health Sci. Eng., 17: 195–207.

Rangel-Peraza, J.G., Mendivil-Garcia, K., Cedillo-Herrera, C.I.G., Rochin-Medina, J.J., Rodriguez-Mata, A.E., Bustos-Terrones, Y.A. 2019. Optimization of organic matter degradation kinetics and nutrient removal on artificial wetlands using *Eichhornia crassipes* and *Typha domingensis*. Environ. Technol., 40: 633–641.

Rani, P.U., Rajasekharreddy, P. 2009. Toxic and antifeedant activities of *Sterculia foetida* (L.) seed crude extract against *Spodoptera litura* (F.) and *Achaea Janata* (L.). J. Biopestic. 2: 161–164.

Renganathan, S., Venkatakrishnan, R., Venkataramana, S., Kumar, M.D., Deepak, S., Miranda, L.R., Velan, M. 2008. Kinetic studies on sorption of basic dye using *Eichhornia crassipes*. J. Environ. Sci. Eng., 50: 249–254.

Sabale, S., Jadhav, V., Jadhav, D., Mohite, B.S., Patil, K.J. 2010. Lake contamination by accumulation of heavy metal ions in *Eichhornia crassipes*: a case study of Rankala Lake, Kolhapur (India). J. Environ. Sci. Eng., 52: 155–156.

Saha, P., Mondal, A., Sarkar, S. 2018. Phytoremediation of cyanide containing steel industrial wastewater by *Eichhornia crassipes*. Int. J. Phytoremediation, 20: 1205–1214.

Sayago, U.F.C. 2019. Design of a sustainable development process between phytoremediation and production of bioethanol with *Eichhornia crassipes*. Environ. Monit Assess. 191: 221. https://doi.org/10.1007/s10661-019-7328-0.

Shanab, S., Ameer, M., Fekry, A., Ghoneim, A., Shalaby, E. 2011. Corrosion resistance of magnesium alloy (AZ31E) as orthopaedic biomaterials in sodium chloride containing antioxidantly active compounds from *Eichhornia crassipes*. Int. J. Electrochem. Sci., 6: 3017–3035.

Shanab, S.M., Shalaby, E.A., Lightfoot, D.A., El-Shemy, H.A. 2010. Allelopathic effects of water hyacinth [*Eichhornia crassipes*]. PLoS One, 5: e13200.

Shirinpur-Valadi, A., Hatamzadeh, A., Sedaghathoor, S. 2019. Study of the accumulation of contaminants by Cyperus alternifolius, Lemna minor, *Eichhornia crassipes*, and Canna x generalis in some contaminated aquatic environments. Environ. Sci. Pollut. Res. Int., 26: 21340–21350.

Sipauba-Tavares, L.H., Braga, F.M. 2007. The feeding activity of *Colossoma macropomum* larvae (tambaqui) in fishponds with water hyacinth (*Eichhornia crassipes*) fertilizer. Braz. J. Biol., 67: 459–466.

So, L.M., Chu, L.M., Wong, P.K. 2003. Microbial enhancement of Cu^{2+} removal capacity of *Eichhornia crassipes* (Mart.). Chemosphere, 52: 1499–1503.

Souza, T.D., Borges, A.C., Matos, A.T., Veloso, R.W., Braga, A.F. 2018. Kinetics of arsenic absorption by the species *Eichhornia crassipes* and Lemna valdiviana under optimized conditions. Chemosphere, 209: 866–874.

Sricoth, T., Meeinkuirt, W., Pichtel, J., Taeprayoon, P., Saengwilai, P. 2018. Synergistic phytoremediation of wastewater by two aquatic plants (*Typha angustifolia* and *Eichhornia crassipes*) and potential as biomass fuel. Environ. Sci. Pollut. Res. Int., 25: 5344–5358.

Srivastava, A., Chahar, V., Sharma, V., Swain, K.K., Hoyler, F., Murthy, G.S., Scherer, U.W., Rupp, H., Knolle, F., Maekawa, M., Schnug, E. 2019. Study of toxic elements in river water and wetland using water hyacinth (*Eichhornia crassipes*) as pollution monitor. Glob Chall., 3: 1800087.

Stuttg de Souza, T.D., Borges, A.C., Teixeira de Matos, A., Veloso, R.W., Braga, A.F. 2018. Optimization of arsenic phytoremediation using *Eichhornia crassipes*. Int. J. Phytoremediation, 20: 1129–1135.

Sung, K., Lee, G.J., Munster, C. 2015. Effects of *Eichhornia crassipes* and *Ceratophyllum demersum* on Soil and water environments and nutrient removal in wetland microcosms. Int. J. Phytoremediation, 17: 936–944.

Tabinda, A.B., Irfan, R., Yasar, A., Iqbal, A., Mahmood, A. 2018. Phytoremediation potential of *Pistia stratiotes* and *Eichhornia crassipes* to remove chromium and copper. Environ. Technol., 1–6. https://doi.org/10.1080/09593330.2018.1540662.

Tabinda, A.B., Arif, R.A., Yasar, A., Baqir, M., Rasheed, R., Mahmood, A., Iqbal, A. 2019. Treatment of textile effluents with Pistia stratiotes, *Eichhornia crassipes* and *Oedogonium* sp. Int. J. Phytoremediation, 21: 939–943.

Thamaraiselvi, P., Jayanthi, P. 2012. Preliminary studies on phytochemicals and antimicrobial activity of solvent extracts of *Eichhornia crassipes* (Mart.) Solms. Asian J. Plant Sci. Res., 2: 115–122.

Tiwari, S., Dixit, S., Verma, N. 2007. An effective means of biofiltration of heavy metal contaminated water bodies using aquatic weed *Eichhornia crassipes*. Environ Monit Assess, 129: 253–256.

Toki, K., Saito, N., Iimura, K., Suzuki, T., Honda, T. 1994. (Delphinidin 3-gentiobiosyl) (apigenin 7-glucosyl) malonate from the flowers of *Eichhornia crassipes*. Phytochemistry, 36: 1181–1183.

Tyagi, T., Agarwal, M. 2017. Ethnomedicine analysis of bioactive constituents in ethanolic leaf extract of *Pistia stratiotes* L. And *Eichhornia Crassipes* (Mart.) Solms By GC-MS. Adv. Biores., 8: 204–211.

Valipour, A., Raman, V.K., Ghole, V.S. 2011. Phytoremediation of domestic wastewater using *Eichhornia crassipes*. J. Environ. Sci. Eng., 53: 183–190.

Vasu, K., Goud, J.V., Suryam, A., Charya, M.S. 2009. Biomolecular and phytochemical analyses of three aquatic angiosperms. Afr. J. Microbiol. Res., 3: 418–421.

Venu, H., Venkataraman, D., Purushothaman, P., Vallapudi, D.R. 2019. *Eichhornia crassipes* biodiesel as a renewable green fuel for diesel engine applications: performance, combustion, and emission characteristics. Environ. Sci. Pollut. Res. Int., 26: 18084–18097.

Victor, K.K., Seka, Y., Norbert, K.K., Sanogo, T.A., Celestin, A.B. 2016. Phytoremediation of wastewater toxicity using water hyacinth (*Eichhornia crassipes*) and water lettuce (*Pistia stratiotes*). Int. J. Phytoremediation, 18: 949–955.

Wanyonyi, W.C., Onyari, J.M., Shiundu, P.M. 2013. Adsorption of methylene blue dye from aqueous solutions using *Eichhornia crassipes*. Bull. Environ. Contam. Toxicol. 91: 362–366.

Wolverton, B.C., McDonald, R.C. 1978. Bioaccumulation and detection of trace levels of cadmium in aquatic systems by *Eichhornia crassipes*. Environ. Health Perspect., 27: 161–164.

Wu, J.S., Ho, T.C., Chien, H.C., Wu, Y.J., Lin, S.M., Juang, R.H. 2008. Characterization of the high molecular weight Cd-binding complex in water hyacinth (*Eichhornia crassipes*) when exposed to Cd. J. Agric. Food Chem., 56: 5806–5812.

Xia, H., Ma, X. 2006. Phytoremediation of ethion by water hyacinth (*Eichhornia crassipes*) from water. Bioresour. Technol., 97: 1050–1054.

Yan, Y., Chen, Y., Xu, X., Zhang, L., Wang, G. 2019a. Effects and removal of the antibiotic sulfadiazine by *Eichhornia crassipes*: Potential use for phytoremediation. Bull. Environ. Contam. Toxicol., 103: 342–347.

Yan, Y., Xu, X., Shi, C., Yan, W., Zhang, L., Wang, G. 2019b. Ecotoxicological effects and accumulation of ciprofloxacin in *Eichhornia crassipes* under hydroponic conditions. Environ. Sci. Pollut. Res. Int., 26: 30348–30355.

Yang, X., Chen, S., Zhang, R. 2014. Utilization of two invasive free-floating aquatic plants (*Pistia stratiotes* and *Eichhornia crassipes*) as sorbents for oil removal. Environ. Sci. Pollut. Res. Int., 21: 781–786.

Yi, Y., Wei, Y., Tsang, P.E., Fang, Z. 2019. Aging effects on the stabilisation and reactivity of iron-based nanoparticles green synthesised using aqueous extracts of *Eichhornia crassipes*. Environ. Sci. Pollut. Res. Int., 26: 28361–28371.

Yi, Z.J., Yao, J., Chen, H.L., Wang, F., Yuan, Z.M., Liu, X. 2016. Uranium biosorption from aqueous solution onto *Eichhornia crassipes*. J. Environ. Radioact., 154: 43–51.

Yin, T., Zhang, X., Liu, X., Wang, C. 2017. Resource recovery of *Eichhornia crassipes* as oil superabsorbent. Mar. Pollut. Bull., 118: 267–274.

Youssef, F., Ashour, M., Sobeh M., El-Beshbishy, H., Singab, A., Wink, M. 2016. *Eremophila maculata*- Isolation of a rare naturally-occurring lignan glycoside and the hepatoprotective activity of the leaf extract. Phytomedicine, 23: 1484–1493.

Zhang, J., Li, Y., Liu, C., Li, F., Zhu, L., Qiu, Z., Xiao, M., Yang, Z., Cai, Y. 2019. Concentration levels, biological enrichment capacities and potential health risk assessment of trace elements in *Eichhornia crassipes* from Honghu Lake, China. Sci. Rep., 9: 24–31.

Zhou, R., Zhang, M., Zhou, J., Wang, J. 2019. Optimization of biochar preparation from the stem of *Eichhornia crassipes* using response surface methodology on adsorption of Cd(2). Sci. Rep., 9: 17538. https://doi.org/10.1038/s41598-019-54105-1.

Zhou, W., Zhu, D., Langdon, A., Li, L., Liao, S., Tan, L. 2009. The structure characterization of cellulose xanthogenate derived from the straw of *Eichhornia crassipes*. Bioresour. Technol., 100: 5366–5369.

Section II
Ethnobotany and Phytotherapy

Chapter 10

Distribution and Use (Ethnobotany and Trade) of *Bergenia ciliata* (Haw.) Sternb. in Nepal

Ripu M. Kunwar

Introduction

The modern facilities of health care are not easily accessible, particularly in the rural areas of Nepal, resulting in heavy reliance of the rural populace on traditional medicines (TM) for primary health care. According to the World Health Organization, over a two-third percentage of the population in developing countries depend on TM for their primary health care (WHO 2002). Because of the high dependency of people on TM, its importance is frequently being hailed (Goleniowski et al. 2006). Its increasing application is also due to its low side effects, low cost, high effectiveness, easy availability of nearby human settlements and application of local resources (Kunwar et al. 2013). There is increasing convergence between TM and modern medicine because both have benefited greatly from natural products (Newman et al. 2003). Since prehistoric times, humans have used natural products, such as plants, animals, microorganisms, and marine organisms, in medicines to alleviate human ailments and diseases. According to fossil records, the human use of plants as medicine may be traced back at least 60,000 years (Fabricant and Farnsworth 2001). Sustainable harvesting is necessary for any medicinal plant but it is generally observed that the high-value medicinal plants such as *Bergenia ciliata, Paris polyphylla, Nardostachys grandiflora,* etc. are overexploited and prematurely harvested (Kunwar et al. 2016, Biswas et al. 2017) because of the accelerated market demand (Pyakurel et al. 2017). *Bergenia ciliata* is one of the high valued medicinal plants of Nepal, widely overexploited for marketing purposes (Olsen 2005).

Bergenia ciliata belongs to the family Saxifragaceae which consists of 30 genera and 580 species. It is one of the six species of genus *Bergenia*. There are two species of *Bergenia* in Nepal that are *B. ciliata* and *B. purpurascens*. *B. ciliata* is distributed in the cold and temperate Himalayas and Central and Eastern Asia between 900 m to 3,600 m (Anon 1970–1988, Kumar and Tyagi 2013) (Figure 10.1). It is an evergreen perennial herb found in hilly areas of Nepal, India, Pakistan, Bhutan, etc. (Venkatadri et al. 2010). It has a long history of use in the Himalayas for curing pulmonary infections, leucorrhea, piles and for dissolving bladder and kidney-stones (Asolkar et al. 2010). In the Ayurveda system of medicine, it is commonly used as tonic, astringent, antiscorbutic, and laxative, and for spleen enlargement, dysuria, and ulcers (Misra 1993, Bagul et al. 2003).

Ethnobotanical Society of Nepal, Kathmandu.
Email: ripukunwar@gmail.com

Figure 10.1. (a) Flowers of *Bergenia ciliata* (@ www.iprsindh.com.pk), (b) *Bergenia ciliata* plant (@ Ripu Kunwar).

Approach and Methods

There are many studies dealing with the different aspects of *Bergenia ciliata* in Nepal but none is accountable for a comprehensive review of taxonomy, distribution and use of *Bergenia ciliata.* India and Pakistan are the countries leading the research on the pharmacology of *Bergenia* (Ahmad et al. 2018). The present review is the first attempt to gather the fragmented information about the taxonomy, distribution, and use (ethnobotany, ethnomedicine and trade) of *B. ciliata* in Nepal. This review also uncovers the gaps in current knowledge that help develop a better guide for sustainable management of *B. ciliata*. Over 100 articles available on Pubmed, Google Scholar and Nepjol, and published books until December 2019 were reviewed for this communication.

Synonyms and Vernacular Names of *Bergenia ciliata*

Bergenia ciliata (Haw.) Sternb. has three synonyms that are *Megasea ciliata* Haw., *Saxifraga ciliata* Royle, and *S. pacumbis* Buch.-Ham. ex D.Don (Akiyama et al. 2012). As *Bergenia ciliata* is one of the popular medicinal plants, it has been given various local names on the basis of local language spoken in different regions. Some of the local names of this plant are illustrated in Table 10.1.

Taxonomy

Bergenia ciliata is an evergreen perennial herb of 5–25 cm height with suborbicular to broadly obovate leaves, which are rounded at the apex and cordate at base. Leaf margins are finely denticulate, fringed with soft hairs. Leaves are alternate, opposite and ex-stipulate. Flowering stems are green to reddish, thick, and sparsely glandular (Chauhan et al. 2012). Flowers 1–20 are showy pinkish-white with ovate petals, lobes acute and denticular near apex, hermaphrodite, calyx oblong 5 adnate to the ovary (more or less) apex rounded, margin denticulate or ciliate; corolla pink, 4 or 5, perigynous and imbricate, stamens indefinite, ovary 4 or 5 and united, fruit capsular or sometimes baccate, seeds numerous, blooms in spring from February to April and fruiting from March to July (Grierson and Long 1987). Seeds 1.5–2 mm (Akiyama et al. 2012).

Distribution

In Nepal, the genus *Bergenia* is distributed from east to west within the altitudinal range of 900 m to 4700 m (Press et al. 2000, Ghimire et al. 2008, DPR 2017). The density of *Bergenia ciliata* is found highest in undisturbed forest, intermediate under restored forest and the lowest under disturbed forest, which indicates forest disturbance may lead to the extinction of this species (Adnan et al. 2014). Detailed

Table 10.1. Local names of *Bergenia ciliata*.

Language	Names	Language	Names
Amchi	*Khadhur*	Magar	*Hangpaat*
Arabian	*Junteyenah*	Malayalam	*Kallurvanchi, Kallurvanni, Kallorvanchi*
Assamese	*Patharkuchi*	Nepalese	*Dhungri ko jara, Pakhanbedha, Pashanbhed, Silparo, Silpu, Simpate, Silphode*
Baitadeli, Doteli	*Vedaite, Silphode, Dhungeful*	Oriya	*Pasanbhedi, Pashanbheda*
Bengali	*Patharchuri*	Pakistan	*Zakham-e-hayat, Bhat-pay*
Chinese	*Honhitao*	Persian	*Gashah*
Danuwar	*Simtadi*	Punjabi	*Kachalu, Pashanbhed*
English	*Rock-foil, Hairy bergenia*	Rai	*Ghayabatam*
German	*Steinbrech*	Sanskrit	*Ashamabhed, Aswamahan, Nagamit, Pasanved, Pashana veda, Samwata, Silaved, Upalbhedi*
Gurung	*Padambet, Pakhanved*	Sherpa	*Chyucha, Chyurpu*
Gujarati	*Pashanbheda*	Sinhalese	*Pahanbeya*
Hindi	*Pakhenbhed*	Tamil	*Sirupilai*
Japanese	*Yukinoshita*	Tamang	*Bregyal*
Kannada	*Alepgaya, Pahanbhedi, Hittaga, Pasanaberu, Hittulaka*	Tibetan	*A-ama-bhe-da*
Kashmir	*Pashanbhed*	Tharu	*Chihuli*
Marathi	*Pashanbheda*	Telugu	*Kondapindi*
Kham	*Gatik mukpo*	Unani	*Mukha*
Kumaoni	*Patharchud*	Urdu	*Zahkm-e-hayat*

distribution and population study of *B. ciliata* in Nepal are yet to be carried out. One study revealed that it is distributed in 16 districts (Kunwar et al. 2011; Figure 10.2).

According to the International Plant Names Index (www.ipni.org), there are 33 species in the genus *Bergenia* Moench (Family: Saxifragaceae) throughout the world, while Tropicos (www.tropicos.org) lists 22 species of *Bergenia* (Table 10.2). Plants of the World Online (www.powo.science.kew.org) illustrates

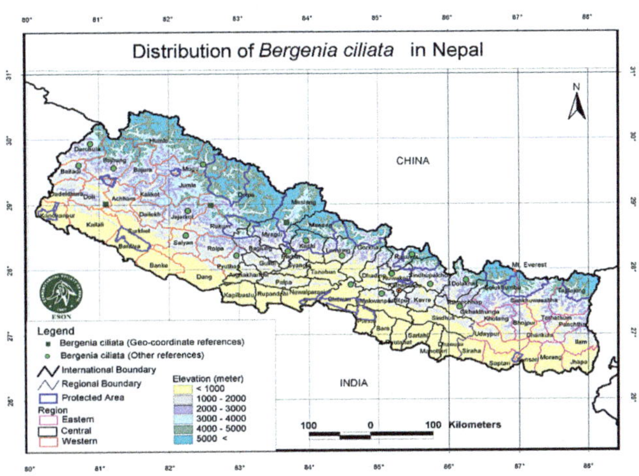

Figure 10.2. Distribution of *B. ciliata* in Nepal (Source: www.eson.org.np).

Table 10.2. Illustration of *Bergenia ciliata* in different databases.

SN	Name	Illustrations				
1	*Bergenia beesiana* hort. ex C.Schneider	ipni				
2	*Bergenia biflora* Moench	ipni	Tropicos			
3	*Bergenia bifolia* Moench		Tropicos			
4	*Bergenia ciliata* (Haw.) Sternb.	ipni	Tropicos	powo	Press et al. (2000)	FoN
5	*Bergenia cordifolia* (Haw.) Sternb.	ipni	Tropicos			
6	*Bergenia coreana* Nakai	ipni	Tropicos			
7	*Bergenia crassifolia* (L.) Fritsch	ipni	Tropicos	powo		
8	*Bergenia delavayi* (Franch.) Engl.	ipni	Tropicos			
9	*Bergenia emeiensis* C.Y.Wu ex J.T.Pan	ipni	Tropicos	powo		
10	*Bergenia fortunei* Stein	ipni		powo		
11	*Bergenia gorbunovii* B.Fedtsch.	ipni				
12	*Bergenia gorbunowii* B.Fedtsch.	ipni	Tropicos			
13	*Bergenia himalaica* Boriss.	ipni	Tropicos			
14	*Bergenia hissarica* Boriss.	ipni	Tropicos	powo		
15	*Bergenia ligulata* (Wall.) Engl.	ipni	Tropicos			
16	*Bergenia media* Engl.	ipni	Tropicos			
17	*Bergenia milesii* Stein	ipni	Tropicos			
18	*Bergenia* × *newryensis* Yeo	ipni				
19	*Bergenia orbicularis* Stein	ipni		powo		
20	*Bergenia ornata* Stein	ipni				
21	*Bergenia pacifica* Kom.	ipni	Tropicos			
22	*Bergenia pacumbis* (Buch.-Ham. ex D.Don) C.Y.Wu & J.T.Pan	ipni	Tropicos	powo		
23	*Bergenia purpurascens* (Hook.f. & Thom.) Engl.	ipni	Tropicos	powo	Press et al. (2000)	FoN
24	*Bergenia schmidtii* (Regel) Silva Tar.	ipni				
25	*Bergenia scopulosa* T.P.Wang	ipni	Tropicos	powo		
26	*Bergenia siphilitica* Raf.	ipni		powo		
27	*Bergenia smithii* Engl.	ipni				
28	*Bergenia spathulata* Nagels ex Guillaumin	ipni				
29	*Bergenia stracheyi* (Hook.f. & Thom.) Engl.	ipni	Tropicos	powo	Press et al. (2000)	
30	*Bergenia thysanodes* (Lindl.) C.Schneider	ipni	Tropicos			
31	*Bergenia tianquanensis* J.T.Pan	ipni	Tropicos	powo		
32	*Bergenia ugamica* V.N.Pavlov	ipni	Tropicos	powo		
33	*Bergenia yunnanensis* hort.	ipni				
34	*Bergenia* × *media* (Haw.) Engl.			powo		
35	*Bergenia* × *schmidtii* (Regel) Silva Tar.	ipni		powo		
36	*Bergenia* × *smithii* (Engl.)			powo		
37	*Bergenia* × *spathulata* Nagels ex Guillumin			powo		

(ipni = International Plant Names Index, powo = Plants of World Online, FoN = Flora of Nepal).

17 species of *Bergenia* (Table 10.2). Wu and Raven (2001) reported seven species of *Bergenia—B. scopulosa, B. crassifolia, B. purpurascens, B. emeiensis, B. pacumbis, B. tianquanensis* and *B. stracheyi—* in Flora of China. Press et al. (2000) reported three species of *Bergenia* in Nepal (Table 10.2). Later on, Akiyama et al. (2012) reported only two species, *Bergenia purpurascens* and *Bergenia ciliata,* in Nepal.

Ethnobotanical and medicinal uses

In Ayurveda, the *Bergenia ciliata* is considered as an astringent, diuretic and used in piles, abdominal mass, urinary calculi, heart disease, urinary problem and wound (Mira 1993, Bagul et al. 2003). The plant is also considered to be useful in dysentery, cancer, lungs problem, fever, diarrhoea, bladder stone, cough, and kidney problems (Bajracharya 1979, Kaul 1997). In folklore, the rhizome of the plant is used for a wide range of purposes such as tonic to medicinal, vegetable and pickle. Its flowers are used in making pickles (Manandhar 2002). It is commonly used for the treatment of various ailments for centuries (Rafi et al. 2017). Leaves and whole plants are also used but less frequently. The paste of rhizome or juice is either given orally or paste, powder or sap is applied on the body. Sometimes, the rhizome is chewed. Among the various ailments cured by its rhizome, the problems associated with urinary, kidney stone and gallstone are frequently alleviated. A total of 52 different types of ailments are treated using *Bergenia ciliata* (Table 10.3).

Table 10.3. Common ethnomedicinal use of *Bergenia ciliata.*

Ailment	Number of citations (n)	Relative citation (n/N)	References
Kidney and gall bladder stones, urinary problems	21	0.13	Ambasta et al. (1992), Adhikari (1997), IUCN (2000), Joshi and Joshi (2001), Manandhar (2002), Islam et al. (2002a), Watanabe et al. (2005), Prajapati et al. (2006), DPR (2007), Pant et al. (2009), Pant and Samant (2010), Samal et al. (2010), Kumari et al. (2012), Singh et al. (2012), Kunwar et al. (2013), Parajuli (2013), Saha and Verma (2013), Singh et al. (2018), Nand and Naithani (2018), Gophane and Khobragade (2019), Verma et al. (2019)
Diarrhea	12	0.07	Shrestha (1988), Pohle (1990), IUCN (2000), Lama et al. (2001), Joshi and Joshi (2001), Manandhar (2002), Watanabe et al. (2005), Bhattarai et al. (2006), DPR (2007), Gewali (2008), Adhikary et al. (2011), Singh et al. (2018)
Fever	11	0.06	Bhattarai (1992), IUCN (2000), Joshi and Joshi (2001), Manandhar (2002), Watanabe et al. (2005), DPR (2007), Kumari et al. (2012), Singh et al. (2012), Shrestha et al. (2016), Singh et al. (2018), Radha and Puri (2019)
Dysentery	7	0.04	Lama et al. (2001), Watanabe et al. (2005), Bhattarai et al. (2006), DPR (2007), Gewali (2008), Kumari et al. (2012), Singh et al. (2018)
Piles	5	0.03	Dey (1998), Watanabe et al. (2005), DPR (2007), Kumari et al. (2012), Singh et al. (2018)
Wounds	5	0.03	Qureshi et al. (2009), Abbasi et al. (2010), Hasan et al. (2013), Shrestha et al. (2016), Singh et al. (2018)
Boils	4	0.02	IUCN (2000), Joshi and Joshi (2001), Manandhar (2002), Kumari et al. (2012)
Cough and cold	4	0.02	Manandhar (2002), Watanabe et al. (2005), Shrestha et al. (2016), Singh et al. (2018)
Cut and burns	4	0.02	Asolkar et al. (2010), Hasan et al. (2013), Shrestha et al. (2016), Singh et al. (2018)
Indigestion, digestion	4	0.02	Lama et al. (2001), Manandhar (2002), Singh et al. (2012), Begum et al. (2014)

Table 10.3 Contd. ...

...Table 10.3 Contd.

Ailment	Number of citations (n)	Relative citation (n/N)	References
Pain in urination, menstruation	4	0.02	Misra (1993), Adhikari (1997), Kumari et al. (2012), Parajuli (2013)
Stomachache	4	0.02	Adhikari (1997), Bhattarai et al. (2006), Gewali (2008), Kunwar et al. (2019)
Aphrodisiac	4	0.02	Ambasta et al. (1992), IUCN (2000), Joshi and Joshi (2001), Singh et al. (2018)
Asthma	3	0.02	Manandhar (2002), Kumari et al. (2012), Singh et al. (2018)
Body pain, bodyache	3	0.02	Manandhar (2002), Shrestha et al. (2016), Nand and Naithani (2018)
Demulcent	3	0.02	Ambasta et al. (1992), IUCN (2000), Joshi and Joshi (2001)
Diuretic	3	0.02	Ambasta et al. (1992), IUCN (2000), Joshi and Joshi (2001)
Ear ache	3	0.02	Manandhar (2002), Asolkar et al. (2010), Hasan et al. (2013)
Heart diseases	3	0.02	Dey (1998), Watanabe et al. (2005), DPR (2007)
Pulmonary infection	3	0.02	Ambasta et al. (1992), IUCN (2000), Joshi and Joshi (2001)
Tonsils	3	0.02	Dutta (2007), Shrestha et al. (2016), Singh et al. (2018)
Ulcer	3	0.02	Qureshi et al. (2009), Gilani et al. (2014), Singh et al. (2018)
Astringent	2	0.01	IUCN (2000), Joshi and Joshi (2001)
Blindness	2	0.01	Bhattarai et al. (2006), Gewali (2008)
Cardiac problems	2	0.01	Kaul (1997), Singh et al. (2018)
Colic, colitis	2	0.01	Manandhar (2002), Singh et al. (2018)
Eye ailment	2	0.01	Dey (1998), Singh et al. (2018)
Ophthalmia	2	0.006	IUCN (2000), Joshi and Joshi (2001)
Post-partum complication	2	0.01	Manandhar (1989), Shrestha et al. (2016)
Round worms	2	0.01	Watanabe et al. (2005), Adhikary et al. (2011)
Tonic	2	0.01	Qureshi et al. (2009), Singh et al. (2012)
Toothache	2	0.01	Dey (1998), Shrestha et al. (2016)
Tumors	2	0.01	Watanabe et al. (2005), DPR (2007)
Anthelmintic	1	0.006	Manandhar (2002)
Bile disorder	1	0.006	Lama et al. (2001)
Carminative	1	0.006	Adhikary et al. (2011)
Constipation	1	0.006	Kumari et al. (2012)
Cutaneous infections	1	0.006	Singh et al. (2018)
Diabetes	1	0.006	Begum et al. (2014)
Fracture	1	0.006	Shrestha et al. (2016)
Gastro-intestinal problem	1	0.006	Singh et al. (2018)
Helmintic	1	0.006	Singh et al. (2018)
Hemorrhoids	1	0.006	Manandhar (2002)
Inflammation	1	0.006	Singh et al. (2018)

Table 10.3 Contd. ...

...Table 10.3 Contd.

Ailment	Number of citations (n)	Relative citation (n/N)	References
Joint Pain	1	0.006	Radha and Puri (2019)
Liver disorder	1	0.006	Lama et al. (2001)
Lungs problem	1	0.006	Watanabe et al. (2005)
Muscular disorder	1	0.006	Qureshi et al. (2009)
Rheumatic	1	0.006	Singh et al. (2018)
Septic pimples on the head of new born baby	1	0.006	Singh et al. (2018)
Skeleto-muscular problem	1	0.006	Bhattarai (2018)
Vomiting	1	0.006	Lama et al. (2001)
N	157		

Phyto-chemical constituents

Several kinds of plant-derived natural products and their derivatives are being used in pharmaceutical industries for the production of various medicines (Agnihotri et al. 2014). Several studies have been carried out to extract many useful substances from *B. ciliata* (Table 10.4). These bioactive compounds are chiefly extracted from rhizome and leaves. Some of the important phytochemical substances in this plant are alkaloid, bergenin, catechin, flavonoid, gallic acid, glycoside, phenolic, tannin, etc. (Dhalwal et al. 2008, Adhikary et al. 2011, Shrestha and Pant 2011, Chauhan et al. 2013). Bhandari et al. (2019) reported 11 different phytochemical constituents in this plant, while Verma et al. (2019) reported 27 constituents. Gyawali and Kim (2012) reported the highest number of phytochemical constituents, 44 to be precise, in this plant (Table 10.4). *Bergenia ciliata* is widely used to treat several ailments of human beings because of the presence of various phytochemical constituents in this plant.

Biological activity

In vitro and *in vivo* study of *Bergenia ciliata* has been carried out by Yousaf et al. (2018) and Dar et al. (2019) to explore the biological activity of *B. ciliata*. Anti-bacterial activity means to destroy the bacterial cells or suppress their growth and control their reproduction ability. The anti-bacterial activity of *Bergenia ciliata* plant is best studied by Sinha et al. (2001a), Pokhrel et al. (2014b), Ahmed et al. (2016), Khan et al. (2018), Yousaf et al. (2018), Haq et al. (2019), Verma et al. (2019).

Bergenia ciliata also contains chemical substances, which are useful for the treatment of cancer. The anti-cancer properties of this plant are illustrated by Qureshi et al. (2009), Gilani et al. (2014), Ahmed et al. (2016), and Dar et al. (2019). Anti-oxidants are substances, which inhibit the oxidation process and prevent or delay cell damage. The source of anti-oxidants may be natural or artificial. The *Bergenia ciliata* plant is rich in anti-oxidant properties (Venkatadri et al. 2010, Uddin et al. 2012, Ahmed et al. 2016, Phull et al. 2016, Khan et al. 2018, Yousaf et al. 2018, Zafar et al. 2019). Because of several useful biological activities of this plant, this is in high demand in these days (DoF 2015, 2016). Table 10.5 shows its various biological activities pertinent to pharmacological actions. Methanolic aqueous extract of *Bergenia ciliata* with its trypsin shows considerable antiviral activity against the influenza virus (Rajbhandari 2001). Acetone extract of roots produced dose-dependent CNS depressant, analgesic, diuretic, and anti-inflammatory effect in acute and sub-acute models of inflammation; the extract showed a decrease in heart rate with the increase in the force of contraction of perfused frog heart (Pandey et al. 1996).

Table 10.4. Chemical constituents of *Bergenia ciliata*.

Compounds Name	**References**
(-)-3-0-galloylcatechin; (-)-3-0-galloylepicatechin	Ruby et al. (2012), Srivastava et al. (2015)
(+)-Afzelechin; Arbutin; Bergenin; Catechin; Gallic acid; Tannic acid	Gupta et al. (2003), Dhalwal et al. (2008), Shrestha and Pant (2011), Chauhan et al. (2012, 2013), Kumar and Tyagi (2013)
[E,Z]-2,4-Decadienal; [E]-4-Hepten-2-one; 2,4-Dimethyl-3-pentanone; 2,4-Hexadienal; 2,4-Nonadienal; 2-Ethyl hexanol; 2-Methyl butanoic acid; 2-Methyl-1-propanol; 2-Nitropropane; 2-Pentanol; 2-Pentanone; 2-Phenylisopropanol; 3-Methyl butanol; 3-Methyl-4-hexen-2-one; 3-pentanol; 5,6-Dihydro-2-pyranone; 5-[2-Propenyl]-1,3-benzodioxole; Acetic acid; Camphor; Decanoic acid; Ethanol; Ethyl acetate; Heptanal; Heptanoic acid; Heptanol; Hexanal; Hexanenitril; Hexanoic acid; Hexanol; Isobutyrophenone; Limonene; Linalool; Methyl cinnamate; Methyl nonanoate; Nonanoic acid; Ocatanol; Octanoic acid; Pentanoic acid; Pentanol; α-Phenylethyl alcohol; α-Terpineol; β-Caryophyllene; β-Phelleandrene; δ-Hexalactone	Gyawali and Kim (2012)
11-0-galloylbergenin; bergenin	Watanabe et al. (2005)
1Monolinoleoylglyceroltrimethylsilyl ether/Monolinolein TMS; 2-Decenal; 4 Cetene/1-Hexadecene; 4Octadecenal; 9 Octadecenoic acid/oleic acid; 9,12,15 Octadecatrienoic acid; 2,3bis [(trimethylsilyl) oxy]propyl ester,(Z,Z,Z)/1-Linolenoyl glycerol; 9 Octadecenoicacid, (2phenyl1, 3dioxolan4yl) methyl ester, cis; Astaxanthin; Azafrin; Bergenin; Beta-eudesmol; Calcitriol; Caryophyllene; Damascenone; Ethyl iso-allocholate; Gallic acid; Hexadecanoic acid/Palmitic acid; Hexanoic acid (Caproic acid); Linalool; n-Pentacosane; Octadecamethyl cyclononasiloxane; Phthalic acid, butyl hex3ylester; Psi psi Carotene; Sitosterol; Tetradecamethyl hexasiloxane; Z10Tetradecen1ol acetate	Verma et al. (2019)
Afzelechin, catechin-3-gallate, saxin	Rastogi and Mehrotra (1970–1979)
Albumen; calcium oxalate; gallic acid; leuco-anthocyanoidic galate; mucilage; starch	Dey (1998)
Albumin; glucose; starch; tannic acid; vitamin	Bhattarai and Ghimire (2006)
Alkaloid; Beta cyanin; Amino acids; Carbohydrates; Flavonoid; Glycoside; Phenolic; Steroids; Tannin; Coumarins	Adhikary et al. (2011), Shankar et al. (2016), Rafi et al. (2017)
Alkaloid; Carbohydrates; Fatty acid; Flavonoid; Glycoside; Phenolic; Protein; Quinone; Reducing Sugar; Saponins, Steroids; Tannin; Terpenoid	Agnihotri et al. (2014), Phull et al. (2016), Zafar et al. (2019), Bhandari et al. (2019)
Alkaloid; Carbohydrates; Flavonoid; Glycoside; Protein; Saponins; Steroids; Tannin	Gophane and Khobragade (2019)
Alkaloid; Catechin; Gallic acid; Gallicin; Glycoside; Phenolic; Saponins; Steroids; Tannin; Terpenoid	Pokhrel et al. (2014a), Ahmed et al. (2016), Yousaf et al. (2018)
Bergenia c-glycoside	GoN (2006)
Cardiac glycoside; Terpenoid, leucocyanidin	Gautam et al. (2007), Ullah et al. (2015)
Leucoanthocyanidin-4-glucoside	Prajapati et al. (2006)
Paashaano lactone	Chandrareddy et al. (1998)
Tannin; wax	Ambasta et al. (1992)

Table 10.5. Biological activity of *Bergenia ciliata*.

Activity/Property	References
Anti-bacterial	Sinha et al. (2001a), Pokhrel et al. (2014b), Ahmed et al. (2016), Khan et al. (2018), Yousaf et al. (2018), Verma et al. (2019), Haq et al. (2019)
Anti-cancer	Qureshi et al. (2009), Gilani et al. (2014), Ahmed et al. (2016), Dar et al. (2019)
Anti-cholinesterase	Zafar et al. (2019)
Anti-diabetic	Bhandari et al. (2008)
Anti-fungal	Islam et al. (2002b), Yousaf et al. (2018)
Anti-haemolytic	Yousaf et al. (2018)
Anti-inflammatory	Kumar et al. (2002)
Anti-malarial	Walter et al. (2013)
Anti-microbial	Khan et al. (2014), Phull et al. (2016)
Anti-neoplastic	Venkatadri et al. (2011)
Anti-oxidant	Venkatadri et al. (2010), Uddin et al. (2012), Ahmed et al. (2016), Phull et al. (2016), Khan et al. (2018), Yousaf et al. (2018), Zafar et al. (2019)
Anti-pyretic	Sinha et al. (2000)
Anti-tussive	Sinha et al. (2001b)
Anti-ulcer	Kakub and Gulfraz (2007), Ruby et al. (2012), Ali et al. (2018)
Anti-urolithiasis	Byahatti et al. (2010), Saha and Verma (2011)
Anti-viral	Rajbhandari (2001), Rajbhandari et al. (2007), Ruby et al. (2012)
Cytotoxic	Phull et al. (2016), Yousaf et al. (2018)

Trade

Quantity of *Bergenia ciliata* collected and traded is not uniform at both spatial and temporal scales (Table 10.6). It is variable with various regions and fiscal years. On analyzing the data from *Hamro Ban* volumes of Department of Forest, the Midwestern region is reported as a reliable source of *Bergenia* rhizome, whereas the production and collection were found least in eastern Nepal (Table 10.6).

Table 10.6. Collection and trade data of *Bergenia ciliata* in Nepal (amount in kg).

FY in BS	AD	Eastern	Central	Western	Mid-western	Far-western
72/73	2017	10	14,009	200	165,302	0
71/72	2016	0	0	0	34,838	27,135
70/71	2015	0	4,000	0	82,955	0
69/70	2014	0	6,085	811	10,220	37,296
68/69	2013	0	0	0	50,023	725
67/68	2012	0	2,100	600	125,575	37,950
66/67	2011	0	0	200	99,269	13,565
65/66	2010	0	0	140	41,381	21,600
64/65	2009	0	15,450	200	207,769	115,330
63/64	2008	4,995	19,400	2,200	102,315	64,400
62/63	2007	0	9,700	8,700	296,285	123,000
61/62	2006	0	375	1,500	7,200	0
60/61	2005	0	101	1,500	48,769	24,106
59/60	2004	0	45	270	185,240	143,270

Source: *Hamro Ban* volumes, Department of Forests.

It is noted that local people started collecting *Bergenia* rhizomes before 2000. A total of 145 tons of *Bergenia* rhizomes were collected in 1997–1998 (Olsen 2005). It was found that the least (9.08 tons) the *Bergenia* rhizomes were collected was in 2006, whereas the highest quantity (437 tons) was collected in 2007 followed by 363 tons in 2018. Overtime, the collected and traded quantity was found to be highly fluctuating and declining (Figure 10.3). The traded quantity was far greater in non-government documents, revealing that the conservation status of *Bergenia* is quite imperiled (Pyakurel 2015). Although the quantity of *Bergenia* rhizome traded highly fluctuates, its price has gradually been increasing for a decade.

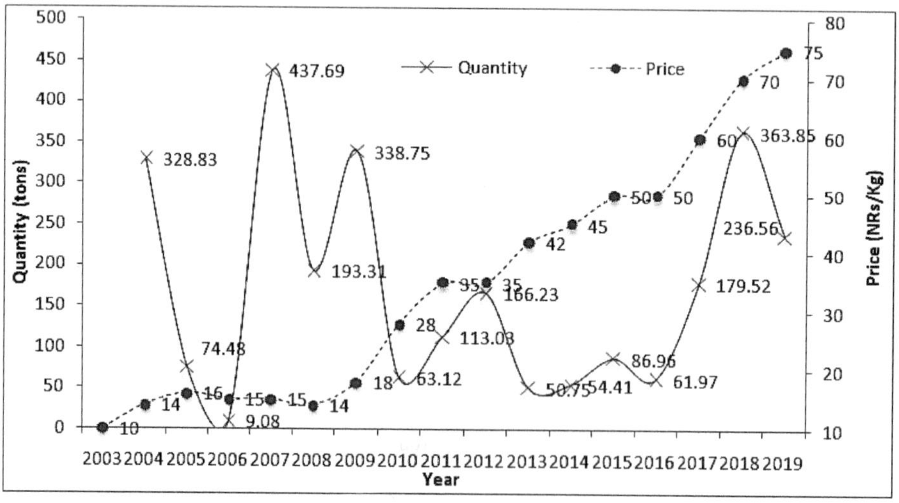

Figure 10.3. Trade and price of *Bergenia ciliata* in Nepal (Source: *Hamro Ban* volumes, JABAN database).

Conservation

As prescribed by the Forest Regulation 1995 and its third amendment 2005, the government royalty of rhizomes of *Bergenia ciliata* is NRs. 2 per kg (GoN 2005). Both *ex situ* and *in situ* conservation of the species are recommended; however, *in situ* conservation is extensively applied. *Ex situ* conservation as cultivation is recently introduced. For sustainable harvesting of this plant, the rhizomes are collected before flowering from the matured individuals; the collection was done by leaving immature rhizomes in soil (Khanal 2003). 10–15 cm long rhizomes should be left inside the soil while collecting (Shrestha and Shrestha 2004). The harvesting period is from September to March (Khanal 2003) and the harvestable amount is 70–80% of the mother stock (Bhattarai and Ghimire 2006). The place where *Bergenia ciliata* was collected should be divided into many blocks and a rotation system of the collection was done every year (Bhattarai and Ghimire 2006).

Conclusions

The present chapter was designed with the aim to compile the discrete information about *Bergenia ciliata* in addition to synthesize its distribution, uses, biological activities and trade in Nepal. It was found that *B. ciliata* is distributed in 16 mid-hill districts of Nepal within 900 m to 3,600 m elevation. It is one of the prioritized plants for the economic development of Nepal. It has been traditionally used among the various communities across the country for treating 52 ailments including kidney and gallstones, fever, diarrhea and dysentery. Almost all parts of the plant are used for curing different ailments; the most frequent part used is rhizome followed by leaf. A total of 44 useful phytochemical compounds were reported from *B. ciliata* plants, which hinted that it is an important resource-base material for traditional medicines and pharmaceutical industries. However, the high-value medicinal plants including *B. ciliata* are declining, which is attributed to the market-driven collection and extensive uses in traditional medicines. Proper management and harvesting of this plant are deemed necessary for sustainable production and use.

References

Abbasi, A.M., M.A. Khan, M. Ahmad, M. Zafar, S. Jahan, S. Sultana. 2010. Ethnopharmacological application of medicinal plants to cure skin diseases and in folk cosmetics among the tribal communities of North-West Frontier Province, Pakistan. J. Ethnopharmacol., 128: 322–325.

Adhikari, S. 1997. Sthaniya Jadibuti dwara Swastha Rakchha (Health Care by Local Herbal Resources). (in Nepali) Himalayan Ayurved Research Institute. Kathmandu, Nepal.

Adhikary, P., Roshan K.C., Kayastha, D., Thapa, D., Shrestha, R., Shrestha, T.M., Gyawali, R. 2011. Phytochemical screening and anti-microbial properties of medicinal plants of Dhunkharka Community, Kavrepalanchowk, Nepal. Int. J. Pharm. Biol. Sci. Arch., 2(6): 1663–1667.

Adnan, M., Tariq, A., Begum, S., Ullah, A., Mussarat, S. 2014. Medicinal plants after forest disturbance, restoration and cultivation in Pakistani Himalaya. Int. J. Agric. Biol., 16: 1006–1010.

Agnihotri, V., Sati, P., Jantwal, A., Pandey, A. 2014. Antimicrobial and antioxidant phytochemicals in leaf extracts of *Bergenia ligulata*: a Himalayan herb of medicinal value. Nat. Prod. Res., 29(11): 1074–1077.

Ahmad, M., Butt, M.A., Zhang, G., Sultana, S., Tariq, A., Zafar, M. 2018. *Bergenia ciliata*: a comprehensive review of its traditional uses, phytochemistry, pharmacology and safety. Biomed. Pharmacother., 97: 708–721.

Ahmed, M., Phul, A.R., Haq, I., Bibi, G., Mazhar, K., Rehman, T., Zia, M., Mirza, B. 2016. Antioxidant, anticancer and antibacterial potential of Zakhme-e-hayat rhizomes crude extract and fractions. Pak. J. Pharm. Sci., 29(3): 895–902.

Akiyama, S., Gornall, R.J., Adhikari, B., Pendry, C.A., Watson, M.F. 2012. Flora of Nepal. Watson, M.F., Akiyama, S., Ikeda, H., Pendry, C.A., Rajbhandari, K.R., Shrestha, K.K. (eds.). Web edition 1. Retrieved on 15.08.2019 from http://data.rbge.org.uk/publications/FloraofNepal/library/Saxifragaceae/1.

Ali, E., Arshad, N., Bukhari, N.I., Tahir, M.N., Zafar, S., Hussain, A., Parveen, S., Qamar, S., Shehzadi, N., Hussain, K. 2018. Linking traditional anti-ulcer use of rhizomes of *Bergenia ciliata* (Haw.) to its *anti-Helicobacter pylori* constituents. Nat. Prod. Res. DOI: 10.1080/14786419.2018.1488711.

Ambasta, S.P., Ramchandran, K., Kashyapa, K., Chand, R. 1992. The Useful Plants of India. Council of Science and Industrial Research (CSIR), New Delhi.

Anon. 1970–1988. The Wealth of India: Raw Materials, Vol. I–XI. CSIR, New Delhi. (Reprinted 1984–89).

Asolkar, L.V., Kakkar, K.K., Chakre, O.J. 2010. Second Supplement to Glossary of Indian Medicinal Plants with Active Principles; 1965–1981. Publication & Information Directorate, Council of Scientific and Industrial Research, New Delhi.

Bagul, M.S., Ravishankara, M., Padh, H., Rajani, M. 2003. Phytochemical evaluation and free radical scavenging properties of rhizome of *Bergenia ciliata* (Haw.) Sternb. forma ligulata Yeo. J. Nat. Remedies, 3(1): 83–89.

Bajracharya, M.B. 1979. Ayurvedic Medicinal Plants and General Treatment. Piyusavarsi Ausadhalaya Mahaboudha, Kathmandu, Nepal, 230 p.

Begum, S., AbdElslam, N.M., Adnan, M., Tariq, A., Yasmin, A., Hameed, R. 2014. Ethnomedicines of highly utilized plants in the temperate himalayan region. Afr. J. Tradit. Complement. Altern. Med., 11(3): 132–142.

Bhandari, L., Bista, B.B., Khanal, C. 2019. Phytochemical, microscopic and standardization of *Bergenia ciliata* for authentication. J. Pl. Res., 17(1): 118–124.

Bhandari, M.R., Anurakkun, N.J., Gao, H., Kawabata, J. 2008. α-Glucosidase and α–amylase inhibitory activities of Nepalese medicinal herb Pakhanbhed (*Bergenia ciliata* Haw.). Food Chem., 106: 247–252.

Bhattarai, K.R., Ghimire, M.D. 2006. Cultivation and Sustainable Harvesting of Commercially Important Medicinal and Aromatic Plants of Nepal. Heritage Research and Development Forum, Nepal, 393 p.

Bhattarai, K.R. 2018. Ethnobotanical study of plants used by Thami community in Ilam District, eastern Nepal. Our Nature, 16(1): 55–67. DOI: http://dx.doi.org/ 10.3126/on.v16i1.22123.

Bhattarai, N.K. 1992. Folk herbal remedies of Sindhupalchok District, Central Nepal. Fitoterapia, 63(2): 145–155.

Bhattarai, S., Chaudhary, R.P., Taylor, R.S.L. 2006. Ethnomedicinal plants used by the peoplc of Manang district, central Nepal. J. Ethnobiol. Ethnomed., 2(1): 41. doi:10.1186/1746-4269-2-41.

Biswas, S., Rawat, M.S., Tantray, F.A., Sharma, S. 2017. Medicinal plants conservation and development areas (MPCDAs)—An initiative towards conservation of medicinal plants. Med. Plants, 9(3): 143–149. DOI: 10.5958/0975-6892.2017.00022.3.

Byahatti, V.V., Pai, K.V., D'Souza, M.G. 2010. Effect of phenolic compounds from *Bergenia ciliata* (Haw.) Sternb. leaves on experimental kidney stones. Anc. Sci. Life, 30(1): 14–17.

Chandrareddy, U.D., Chawla, A.S., Mundkinajeddu, D., Maurya, R., Handa, S.S. 1998. Paashaanolactone from *Bergenia ligulata*. Phytochemistry, 47(5): 907–909.

Chauhan, R., Ruby, K., Dwivedi, J. 2012. *Bergenia ciliata* mine of medicinal properties: a review. Int. J. Pharm. Sci. Rev. Res., 15(2): 20–23.

Chauhan, R., Ruby, K., Dwivedi, J. 2013. Secondary metabolites found in Bergenia species: A compendious review. Int. J. Pharm. Pharm. Sci., 5(1): 9–16.

Dar, K.K., Ali, S., Ejaz, M., Nasreen, S., Ashraf, N., Gillani, S.F., Shafi, N., Safeer, S., Khan, M.A., Andleeb, S., Mughal, T.A. 2019. *In vivo* induction of hepatocellular carcinoma by diethylnitrosoamine and pharmacological intervention in Balb C mice using *Bergenia ciliata* extracts. Braz. J. Biol., 79(4): 629–638. DOI: https://doi.org/10.1590/1519-6984.186565.

Dey, A.C. 1998. Indian Medicinal Plants Used in Ayurvedic Preparation. Bishen Singh Mahendra Pal Singh, Dehra Dun, India.

Dhalwal, K., Shinde, V., Biradar, Y., Mahadik, K. 2008. Simultaneous quantification of bergenin, catechin, and gallic acid from *Bergenia ciliata* and *Bergenia ligulata* by using thin-layer chromatography. J. Food Compos. Anal., 21(6): 496–500.

DoF. 2015. *Hamro Ban*. Annual Report of Department of Forest, Government of Nepal, Ministry of Forest and Soil Conservation (Fiscal year 2071/72, in Nepali).

DoF. 2016. *Hamro Ban*. Annual Report of Department of Forest, Government of Nepal, Ministry of Forest and Soil Conservation. Kathmandu, Nepal.

DPR. 2007. Medicinal Plants of Nepal (Revised), Bull. Department of Plant Resources No. 28, Thapathali Kathmandu, Nepal.

DPR. 2017. Jadibuti tatha gairkashtha ban paidawar sambandhi talim digdarsan, Department of Plant Resources, Thapathali, Kathmandu, Nepal. 2074 (B.S.).

Dutta, IC. 2007. Non-Timber Forest Products of Nepal: Identification, Classification, Ethnic Uses and Cultivation. Hill Side Press, Kathmandu, Nepal, 484p.

Fabricant, D.S., Farnsworth, N.R. 2001. The value of plants used in traditional medicine for drug discovery. Environ. Health Perspect., 109: 69–75. doi: 10.1289/ehp.01109s169.

Gautam, L.N., Belbase, B., Tamrakar, B., Shrestha, S. 2007. Biophytochemical investigation of *B. ciliata*, a rocky Himalayan medicinal plants of Nepal. International Conference on Emerging Issues on Research and Development, JUSAN.

Gewali, M.B. 2008. Aspects of Traditional Medicine in Nepal. Institute of Natural Medicine, University of Toyama. Japan. 175 p.

Ghimire, S.K., Sapkota, I.B., Oli, B.R., Parajuli-Rai, R. 2008. Non-Timber Forest Products of Nepal Himalaya: Database of Some Important Species Found in the Mountain Protected Areas and Surrounding Regions. WWF Nepal, Kathmandu.

Gilani, A.G., Khan, A.M., Qureshi, R.A., Sherwani, S.K., Khan, R.U., Bokhari, T.Z. 2014. Ethno-medicinal treatment of common gastrointestinal disorders by indigenous people in Pakistan. Adv. Biores., 5(1): 42–49. DOI: 10.15515/abr.0976-4585.5.42-49.

Goleniowski, M.E., Bongiovanni, G.A., Bongiovanni, L., Palacio, C.O., Cantero, J.J. 2006. Medicinal plants from the "Sierra de Comechingones", Argentina. J. Ethnopharmacol., 107: 324–341.

GoN. 2005. Forest Regulation: Third Amendment September 2005. Nepal Gazette Vol 3: 55(37). Ministry of Forests and Soil Conservation, Kathmandu, Nepal.

GoN. 2006. Nepal ko Aarthik Bikaska lagi Prathamikata Prapta Jadibutiharu. Department of Plant Resources, Ministry of Forest and Soil Conservation, Thapathali, Kathmandu Nepal, 125 p.

Gophane, S.R., Khobragade, C.N. 2019. Chemical profiling of *Bergenia ciliata* (Haw.) Sternb. and inhibition of key enzymes linked to Primary Hyperoxaluria Type1 (PH1): An *in vitro* and *in silico* Study. IJESRT, 8(1): 263–289. DOI: 10.5281/zenodo.2553038.

Grierson, A.J.C., Long, D.G. 1987. Flora of Bhutan (Vol. 1, Part 3). Royal Botanic Garden, Edinburgh, pp. 492.

Gupta, A.K., Tandon, N., Sharma, M. 2003. Quality Standards of Indian Medicinal Plants. ICMR, New Delhi, India. 1: 95–101.

Gyawali, R., Kim, K.S. 2012. Bioactive volatile compounds of three medicinal plants from Nepal, Kathmandu Univers. J. Sci. Eng. Technol., 8(1): 51–62.

Haq, I., Khurshid, G., Abbasi, A.Z., Nawaz, I., Naqvi, T., Arfan, M., Shah, M.M., Amin, B.A.A., Hussain, J., Ali, M.A., Rehman, S.U., Ahmad, R. 2019. Successful callogenesis from leaf and petiole of *Bergenia ciliata* (Haw.) Sternb. and antibacterial activity of Callus extracts. Pak. J. Bot., 51(5): 1847–1852.

Hasan, M.K., Gatto, P., Jha, P.K. 2013. Traditional uses of wild medicinal plants and their management practices in Nepal-A study in Makawanpur district. Int. J. Med. Arom. Plants, 3(1): 102–112.

Islam, M., Azhar, I., Khan, U., Aslam, M., Ahmad, A., Shahbuddin. 2002a. Bioactivity evaluation of *Berginia ciliata*. Pak. J. Pharm. Sci., 15(1): 15–33.

Islam, M., Azhar, I., Khan, U., Gill, M.A., Din, S., Ahmad, A. 2002b. Antifungal activity evaluation of *Bergenia ciliata*. Pak. J. Pharmacol., 19(2): 1–6

IUCN. 2000. National Register of Medicinal Plants. Kathmandu: IUCN Nepal,. 121 p.

JABAN. 2019. Jadibuti Association of Nepal. Nepalgunj, Nepal.

Joshi, K.K., Joshi, S.D. 2001. Genetic Heritage of Medicinal and Aromatic Plants of Nepal Himalayas, Buddha Academic Publishers and Distributors Pvt. Ltd.

Kakub, G., Gulfraz, M. 2007. Cytoprotective effects of *Bergenia ciliata* Sternb, extract on gastric ulcer in rats. Phytother. Res., 21(12): 1217–20.

Kaul, M.K. 1997. Medicinal Plants of Kashmir and Ladakh: Temperate and Cold Arid Himalaya. Indus Publishing Company, New Delhi.

Khan, A., Jan, J., Khan, A., Jan, F.G., Danish, M. 2018. Evaluation of antioxidant and antimicrobial activities of *Bergenia ciliata* Sternb (Rhizome) crude extract and fractions. Pak. J. Pharm. Sci., 31(1): 31–35.

Khan, N., Abbasi, A.M., Dastagir, G., Nazir, A., Shah, G.M., Shah, M.M., Shah, M.H. 2014. Ethnobotanical and antimicrobial study of some selected medicinal plants used in Khyber Pakhtunkhwa (KPK) as a potential source to cure infectious diseases. BMC Complement. Altern. Med., 14: 122. doi:10.1186/1472-6882-14-122.

Khanal, M. 2003 (2060 BS). Pasanved-Samkshipta Parichaya. Kalpabrikshya, 14(152): 15–18.

Kumar, V., Tyagi, D. 2013 Review on phytochemical, ethnomedical and biological studies of medically useful genus *Bergenia*. Int. J Curr. Microbiol. App. Sci., 2(5): 328–334.

Kumar, V., Shah, G.B., Baheti, J.R., Deshpande, S.S., Parmar, N.S. 2002. Anti-inflammatory activity of aqueous extract of *Bergenia ciliata* rhizomes. JNR, 2(2): 189–190.

Kumari, P., Joshi, G.C., Tewari, L.M. 2012. Biodiversity status, distribution and use pattern of some ethnomedicinal plants. Int. J. Conserv. Sci., 3(4): 309–318.

Kunwar, R.M., Baral, K., Paudel, P., Acharya, R.P., Thapa-Magar, K.B., Cameron, M., Bussmann, R.W. 2016. Land-use and socioeconomic change, medicinal plant selection and biodiversity resilience in Far Western Nepal. PLoS ONE, 11(12): e0167812. https://doi.org/10.1371/journal.pone.0167812.

Kunwar, R.M., Shrestha, K., Malla, S., Acharya, T., Sementelli, A., Kutal, D., Bussmann, R.W. 2019. Relation of medicinal plants, their use patterns and availability in the lower kailash sacred landscape, Nepal. Ethnobot. Res. Appl., 18(6). DOI: http://dx.doi.org/10.17348/era.18.7.1-14.

Kunwar, R.M., Thapa, K.P., Shrestha, R., Shrestha, P.R., Bhattarai, N.K., Tiwari, N.N., Shrestha, K.K. 2011. Medicinal and aromatic plants network (Maps-Net) Nepal: an open-access digital database. Banko Janakari, 21: 48–50.

Kunwar, R.M., Mahat, M., Acharya, R.P., Bussmann, R.W. 2013. Medicinal plants, traditional medicine, markets and management in far-west Nepal. J. Ethnobiol. Ethnomed., 9(1): 24. doi:10.1186/1746-4269-9-24.

Lama, Y.C., Ghimire, S.K., Aumeeruddy-Thomas, Y. 2001. Medicinal Plants of Dolpo: Amchi's Knowledge and Conservation. WWF Nepal Program and People and Plants Initiative, Kathmandu Nepal.

Manandhar, N.P. 1989. Medicinal plants used by Chepang tribes of Makawanpur district, Nepal. Fitoterapia, 60(1): 61–68.

Manandhar, N.P. 2002. Plants and People of Nepal, Timber Press Inc., The Haseltine Building, 133 S.W. Second Avenue, Suite 450, Portland, Oregon 97204, USA.

Misra, B. 1993. *Bhavprakash Nighantu.* (Commented by Chunakar, K.C., Pandey, G.S. (eds.)). Chaukhumba Bharati Academy. Gokul Bhawan, Varanasi, India, 984 p.

Nand, K., Naithani, S. 2018. Ethnobotanical uses of wild medicinal plants by the local community in the Asi Ganga sub-basin, Western Himalaya. J. Complement. Med. Res., 9(1): 34–36.

Newman, D.J., Cragg, G.M., Snader, K.M. 2003. Natural products as sources of new drugs over the period 1981–2002. J. Nat. Prod., 66: 1022–1037. doi: 10.1021/np030096l.

Olsen, C.S. 2005. Quantification of the trade in medicinal and aromatic plants in and from Nepal. Acta Horticulture, 678: 29–35.

Pandey, V.N., Malhotra, S.C., Sharma, D.P. 1996. Pharmacological Investigation of Certain Medicinal Plants and Compound Formulations Used in Ayurveda and Siddha. Central Council for Research in Ayurveda & Siddha, New Delhi.

Pant, S., Samant, S.S. 2010. Ethnobotanical observations in the Mornaula Reserve forest of Kumoun, West Himalaya, India. Ethnobot. Leaflets, 14: 193–217.

Pant, S., Samant, S.S., Arya, S.C. 2009. Diversity and indigenous household remedies of the inhabitants surrounding Mornaula reserve forest in West Himalaya. Indian J. Tradit. Know., 8(4): 606–610.

Parajuli, R.R. 2013. Study on local uses of medicinal plant in Nayabazar, Pyang and Jamuna VDCs of Ilam District. Nepal J. Sci. Technol., 14(1): 57–66.

Phull, A.R., Abbas, Q., Ali, A., Raza, H., Kim, S.J., Zia, M., ul Haq, I. 2016. Antioxidant, cytotoxic and antimicrobial activities of green synthesized silver nanoparticles from crude extract of *Bergenia ciliata.* Future J. Pharm. Sci. doi: 10.1016/j.fjps.2016.03.001.

Pohle, P. 1990. Useful Plants of Manang District. (A contribution to the ethnobotany of Nepal Himalaya). Nepal Research Centre Publication No. 16, ed. Wezler A. Franz Steiner Verlag Wiebaden GMBH, Stuttgart.

Pokhrel, P., Banerjee, J., Dahal, P., Khanal, H., Gupta, A.K., Dey, B.K. 2014b. Phytochemical screening and biological evaluation of different parts of plant *Bergenia ciliata.* J. Pharmacogn. Phytochem., 3(4): 220–224.

Pokhrel, P., Parajuli, R.R., Tiwari, A.K., Banerjee, J.A. 2014a. Short glimpse on promising pharmacological effects of *Bergenia ciliata.* JAPTR Online, 2(1): 1–6.

Prajapati, N.D., Purohit, S.S., Sharma, A.K., Kumar, T. 2006. A Handbook of Medicinal Plants: A Complement Source Book. Agrobios India.

Press, J.R., Shrestha, K.K., Sutton, D.A. 2000. Annotated Checklist of Flowering Plants of Nepal. Natural History Museum, London and Central Department of Botany, Tribhuvan University, Kathmandu, Nepal.

Pyakurel, D. 2015. Identification of Highly Traded and Commercially Important Medicinal Plants of Nepal. Nepal Herbs and Herbal Products Association, Kathmandu, Nepal.

Pyakurel, D., Bhattarai, I., Ghimire, S.K. 2017. Trade and conservation of medicinal and aromatic plants in western Nepal. Bot. Orient. , 11: 27–37.

Qureshi, R.A., Ghufran, M.A. Gilani, S.A., Yousaf, Z., Abbas, G., Batool, A. 2009. Indigenous medicinal plants used by local women in southern Himalayan regions of Pakistan. Pak. J. Bot., 41(1): 19–25.

Radha, Puri, S. 2019. Survey of ethnomedicinal plants used by migratory shepherds in shimla district of Himachal Pradesh. Plant Arch., 19(1): 477–482.

Rafi, S., Kamili, A.N., Ganai, B.A., Mir, M.Y., Parray, J.A. 2017. Phytochemical analysis of *Bergenia ciliata* (Haw.) Sternb. J. Res. Development, 17: 31–34.

Rajbhandari, K.R. 2001. Ethnobotany of Nepal. Ethnobotanical Society of Nepal (ESON), Kathmandu, Nepal, 189 p.

Rajbhandari, M., Mentel, R., Jha, P.K., Chaudhary, R.P., Bhattarai, S., Gewali, M.B., Karmacharya, N., Hipper, M., Lindequist, U. 2007. Antiviral activity of some plants used in nepalese traditional medicine. eCAM, 6(4): 517–522. doi:10.1093/ecam/nem156.

Rastogi, R.P., Mehrotra, B.N. 1970–1979. A Compendium of Medicinal Plants Vol. 2. Central Drugs Research Institute, Lucknow and Publication and Information Directorate, New Delhi, 833 p.

Ruby, K., Chauhan, R., Sharma, S., Dwivedi, J. 2012. Polypharmacological activities of *Bergenia* species. Int. J. Pharm. Sci., 13(1): 100–109.

Saha, S., Verma, R.J. 2011. *Bergenia ciliata* extract prevents ethylene glycol induced histological changes in the kidney. Acta Pol. Pharm., 68(5): 711–715.

Saha, S., Verma, R.J. 2013. Inhibition of calcium oxalate crystallization *in vitro* by an extract of *Bergenia ciliata*. Arab J. Urol., 11(2): 187–192.

Samal, P.K., Dhyani, P.P., Dollo, M. 2010. Indigenous medicinal practices of bhotia tribal community in Indian central Himalaya. Indian J. Tradit. Know., 1: 140–144.

Shankar, K.G., Fleming, A.T., Vidhya, R., Pradhan, N. 2016. Synergistic efficacy of three plant extracts, *Bergenia ciliata*, *Acorus calamus* and *Dioscorea bulbifera* for antimicrobial activity. Int. J. Pharm. Biol. Sci., 7(4): 619–628.

Shrestha, N., Shrestha, S., Koju, L., Shrestha, K.K., Wang, Z. 2016. Medicinal plant diversity and traditional healing practices in Eastern Nepal. J. Ethnopharmacol., 192: 292–301. DOI: http://dx.doi.org/10.1016/j.jep.2016.07.067.

Shrestha, P. 1988. Contribution to the ethnobotany of the Tamang of Kathmandu valley. Ibid., 15(2): 247–266.

Shrestha, U.B., Shrestha, S. 2004 (2061 BS). Major Non-Timber Forest Products of Nepal. Bhundipuran Prakashan, Kathmandu, Nepal, 411 p.

Shrestha, U.K., Pant, B. 2011. Production of bergenin, an active chemical constituent in the callus of *Bergenia ciliata* (Haw.) Sternb. BOTOR, 8: 40–44. doi: http://dx.doi.org/10.3126/botor.v8i0.5557.

Singh, H., Husain, T., Agnihotri, P., Pande, P.C., Iqbal, M. 2012. Biodiversity conservation through traditional beliefs system: A case study from Kumaon Himalayas, India. Int. J. Conserv. Sci., 3(1): 33–40.

Singh, L., Kumar, A., Paul, A. 2018. *Bergenia ciliata*: The medicinal herb of cold desert. Int. J. Chem. Stud., 6(3): 3609–3613.

Sinha, S., Murugesan, T., Maiti, K., Gayen, J.R., Pal, M., Saha, B.P. 2000. Evaluation of antipyretic potential of *Bergenia ciliata* Sternb. rhizome extract. Pharm. Pharmacol. Commun., 6(12): 549–551.

Sinha, S., Murugesan, T., Maiti, K., Gayen, J.R., Pal, B., Pal, M., Saha, B.P. 2001a. Antibacterial activity of *Bergenia ciliata* rhizome. Fitoterapia, 72(5): 550–552.

Sinha, S., Murugesan, T., Pal, B., Pal, M., Saha, B.P. 2001b. Evaluation of anti-tussive activity of *Bergenia ciliata* Sternb. rhizome extract in mice. Phytomedicine, 8(4): 298–301.

Srivastava, N., Srivastava, A., Srivastava, S., Rawat, A.K.S., Khan, A.R. 2015. Simultaneous quantification of bergenin, epicatechin, (+)-catechin, and gallicin in *Bergenia ciliata* using high performance liquid chromatography. J. Liq. Chromatogr. Relat. Technol., 38(12): 1207–1212.

Uddin, G., Rauf, A., Arfan, M., Ali, M., Qaisar, M., Saadiq, M., Atif, M. 2012. Preliminary phytochemical screening and antioxidant activity of *Bergenia ciliata*. Middle East J. Sci. Res., 11(8): 1140–1142.

Ullah, N., Haq, I.U., Mirza, B. 2015. Phytotoxicity evaluation and phytochemical analysis of three medicinally important plants from Pakistan. Toxicol. Ind. Health, 31(5): 389–395.

Venkatadri, R., G. Guha, Rangasamy, A.K. 2011. Anti-neoplastic activities of *Bergenia ciliata* rhizome. J. Pharm. Res., 4(2): 443–445.

Venkatadri, V., Guha, G., Kumar, R.A., Mathew, L. 2010. Evaluation of antioxidant activities of *Bergenia ciliata* rhizome. Rec. Nat. Prod., 4(1): 38–48.

Verma, R., Tapwal, A., Kumar, D., Puri, S. 2019. Assessment of antimicrobial potential and phytochemical profiling of ethnomedicinal plant *Bergenia ciliata* (Haw.) Sternb. in Western Himalaya. J. Microbiol. Biotech. Food. Sci., 9(1): 15–20. doi: 10.15414/jmbfs.2019.9.1.15-20.

Chapter 11

Cenostigma pyramidale: Ethnomedicinal Properties and Perspectives on A Legume Tree Highly Adapted to Semiarid '*Caatinga*' Region

Livia Maria Batista Vilela, Carlos André dos Santos-Silva, Ricardo Salas Roldan Filho, Silvany de Sousa Araújo, José Ribamar Costa Ferreira-Neto, Wilson Dias de Oliveira, Lidiane Lindinalva Barbosa Amorim, Valesca Pandolfi and *Ana Maria Benko-Iseppon**

Introduction

For more than five millennia, the human population has been using plants to treat and cure diseases. The oldest recognized medical record dates back 4,000 years, reporting herbal remedies used to treat a large number of diseases (Morais and Braz-Filho 2007). Plant therapeutic properties have become part of popular culture through observation and experimentation, especially by people with different cultures and ethnicities (Silva et al. 2015a), including those from the northeastern region of Brazil. The ethnobotanical practices of this region were shaped by Amerindian (different tribes), European colonizers, African slaves and their descendants (Benko-Iseppon and Crovella 2010). This region comprises the so-called *Caatinga* vegetation. Within this vegetational domain—more specifically in an archeological site located close to the 'Serra da Capivara National Park' in Piauí—the oldest evidence of human occupation in South America was reported, indicating human presence as early as 20,000 years before Christ (Lahaye et al. 2013).

The term *Caatinga* means "white forest" and was designated by the indigenous Tupi-Guarani population, due to its vegetation lacking leaves and with trunks and branches of gray-white color. The *Caatinga* is the only ecological region entirely restricted to Brazil. It houses 3,150 angiosperms, 6% of them endemic. Such vegetation includes plants that evolved from humid forests along millions of years, being related to species of other surrounding vegetation types (*Cerrado*, *Campos Rupestres, Amazon,* and *Atlantic* forests). This diversity was shaped by a challenging semi-arid climate whose species originated from the mid-Miocene (11–15 million years ago) to the Holocene period (18,000 years ago) (Tabarelli et al. 2018).

Among the plants adapted to the extreme environmental conditions of the *Caatinga*, a plant whose popular name is '*catingueira*' (*Cenostigma pyramidale*) stands out as one of the most important arboreal representatives of this environment. This species is also known by other vernacular names as '*catinga-de-*

Universidade Federal de Pernambuco, Centro de Biociências, Genetics Dept., Plant Genetics and Biotechnology Lab, Av. Prof. Moraes Rego, 1235, 50.670-423 – Recife, PE, Brazil.

* Corresponding author: ana.iseppon@gmail.com

Walter, N.S., Bagai, U., Kalia, S. 2013. Antimalarial activity of *Bergenia ciliata* (Haw.) Sternb. against *Plasmodium berghei*. Parasitol. Res., 112(9): 3123–3128.

Watanabe, T., Rajbhandari, K.R., Malla, K.J., Yahar, S. 2005. A Handbook of Medicinal Plants of Nepal, AYUR SEED, Life Environmental Institute (Ayurseed L.E.I.), Japan.

World Health Organization. 2002. Traditional Medicine Strategy 2002–2005, WHO, Rome, Italy.

Wu, Z., Raven, P.H. 2001. Flora of China. Missouri Botanical Garden Press, 8: 278–280.

Yousaf, S., Kaukab, G., Gul, H., Khalid, N., Kausar, R., Ahmed, H., Ajab, H., Gulfraz, M. 2018. Pharmacological and phytochemical analysis of *Bergenia ciliata* leaf and rhizome extracts. Pak. J. Pharm. Sci., 31(5): 1911–1916.

Zafar, R., Ullah, H., Zahoor, M., Sadiq, A. 2019. Isolation of bioactive compounds from *Bergenia ciliata* (Haw.) Sternb. rhizome and their antioxidant and anticholinesterase activities. BMC Complement. Altern. Med., 19(296). DOI: https://doi.org/10.1186/s12906-019-2679-1.

porco' (meaning 'pork stink') and '*pau-de-rato*' (meaning 'rat stick'), this last name more used in Bahia state (Pio-Correa 1931). These vernacular names sometimes result in confusion, as the name '*catinga-de-porco*' is also attributed to *Croton adenocalyx* Baill. (Euphorbiaceae), *Maytenus gonocladus* Griseb (Celastraceae), and also the related species *Caesalpinia microphylla* Mart. ex G.Don and *C. porcina* Mart. ex Steud, whose bark infusion is reported to be effective against scabies. In turn, the name '*catingueira*' is also used for *Poincianella gardneriana* Benth (L.P. Queiroz), another related species (Pio-Correa 1931).

The use of *C. pyramidale* by the human population very likely precedes the European colonization, possibly because it is one of the most widely dispersed plant species in the *Caatinga*, vegetating in both wetlands and semi-arid. This chapter addresses different aspects of this important biological resource, including its taxonomy, distribution, diversity, uses by human populations, evaluation of compounds and effects, as well as associated genetic and genomic resources. In view of the exacerbated extraction of plant parts from natural resources, management errors and introduction of exotic species to the *Caatinga*, there are concerns about the sustainability of the traditional forms of use of this precious resource, raising demands towards the conservation of this species.

Taxonomy and Distribution of *Cenostigma pyramidale*

The genus *Cenostigma* belongs to the Fabaceae family, subfamily Caesalpinioideae (LPWG 2017). In the actual circumscription, the genus includes seven species endemic to Brazil, where the group is widely distributed, occurring in all phytogeographic domains. Among them, the species *C. pyramidale* stands out as an endemic taxon to the northeastern region of Brazil (Braga 1960, Queiroz 2010).

The species *C. pyramidale* was formerly known as *Poincianella pyramidales* (Tul.) L.P. Queiroz and, in earlier times, as *Caesalpinia pyramidalis* Tul. These names are sometimes still used in scientific publications. The use of these non-valid names brings some difficulties in searching for information about the *catingueira* along the scientific literature. Throughout this chapter, we will use the currently valid name (*Cenostigma pyramidale*), even when citing works where no longer valid names have been cited.

According to Maia (2004), *C. pyramidale* is widely dispersed in the northeastern region of Brazil, comprising eight states (Figure 11.1) including Piauí, Ceará, Rio Grande do Norte, Paraíba, Pernambuco, Alagoas, Sergipe and Bahia (Queiroz 2010, Chaves et al. 2016).

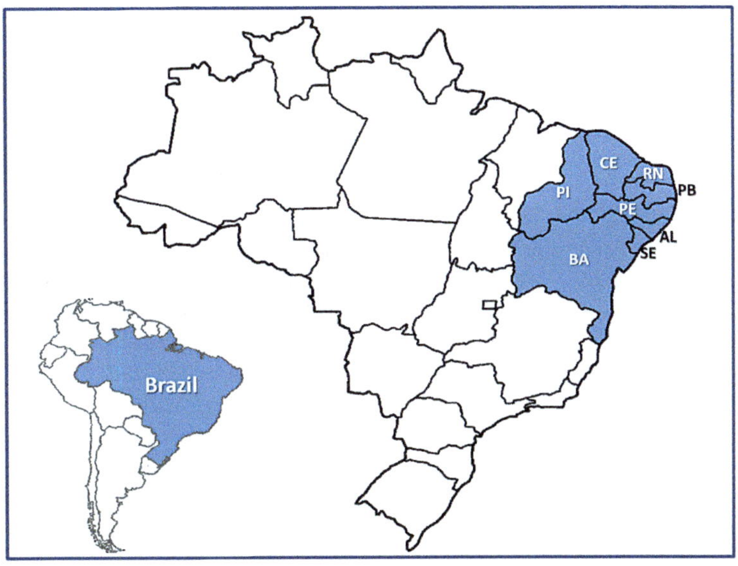

Figure 11.1. Reported distribution of *Cenostigma pyramidale* in the northeastern region of Brazil (Queiroz 2010, Chaves et al. 2016). Abbreviation of the federation states: PI = Piauí, CE = Ceará, RN = Rio Grande do Norte, PB = Paraíba, PE = Pernambuco, AL = Alagoas, SE = Sergipe, and BA = Bahia.

C. pyramidale is generally a medium-sized tree that can reach 4–6 m in height, sometimes reaching up to 12 m and greater heights in favorable conditions. It presents an irregular and open canopy, with light gray or brown bark in adulthood. When old, the heartwood of the trunk completely rots, becoming a shelter for small animals. The leaves are bipinnate with five to 11 leaflets that are sessile, alternate, leathery, oblong and obtuse. It has yellow flowers arranged in racemes with similar length to the leaves. The fruit is a sessile pod that is leathery, flat, dark in color, and can be up to 11 cm long (Matos et al. 2019). The seeds are mature by the end of the dry season and require harvest before autochoric (explosive) seed dispersion. Seeds are light brown, flattened and ovate, with no dormancy. Germination occurs in less than 48 hours once given optimum conditions (Dantas et al. 2008).

The *catingueira* is self-incompatible and seems to have limitations in pollen dispersion since there are high rates of pods developing without seeds (Leite and Machado 2009). Thus, hand cross-pollination increases reproductive success compared to the natural fruit set (Borges et al. 2009). Natural pollination is performed mainly by solitary and polylectic bees, such as *Xylocopa* and *Centris* (Borges et al. 2009). Besides, it inhabits mainly stony ground but grows well in humid lowlands (Lima 1996, Oliveira 2010). These features ensure its successful growth throughout the Brazilian semiarid climate (Fabricante et al. 2009). Our group has observed populations of this species in recent decades. The ability of the *catingueira* to adapt to different soil conditions, rainfall, temperature, and insolation indicate that this species has great rusticity and adaptability, and can be designated as an '*extremophile species*', that is, able to survive contrasting conditions in which most angiosperms would perish.

The *catingueira* is also considered an indicator of the arrival of the coming rainy season because its buds sprout already in the first discrete manifestations of humidity (Alves et al. 2007). Interestingly, a few days after the rain starts, the foliage of this species exudes an overpowering smell, mainly in the reproductive period where its seeds are formed. This characteristic keeps the cattle apart which keeps the plant intact, thus favoring the soil cover. As far as we know, the biochemical nature of this strong odor is still unknown. Later, when its reproductive cycle ends and the leaves dry up, they serve as food for animals (Dantas et al. 2011).

C. pyramidale is used to recover degraded areas, including timber uses for reforestation because it is adapted to xeric and degraded environments, behaving like a pioneer species, also occupying fire-disturbed sites. In addition, it may be an interesting model for studying the mechanisms of tolerance to abiotic factors such as drought and salinity, conditions frequent in their area of occurrence. Recent studies on this species focused on the ecophysiological aspects of successional gradients (Falcão et al. 2015) and on physiological performance in association with arbuscular mycorrhizal fungi (AMF) under hydrated (Frosi et al. 2016a) and drought conditions (Frosi et al. 2016b).

Previous studies report that *C. pyramidale* is tolerant to water deficit (Antunes et al. 2011), high salinity and heat (Matias et al. 2018, Dantas et al. 2019) and also to alternating temperatures (Antunes et al. 2010, Oliveira et al. 2011, Matias et al. 2013, Matias et al. 2014).

Genetic Information on *C. pyramidale* in Public Databases

To date, there are few studies specifically on *C. pyramidale* genetics. A data mining carried out in August 2019 with the keywords '*Cenostigma pyramidale* OR *Caesalpinia pyramidalis* OR *Poincianella pyramidalis* AND Genetics' in the PubMed database—one of the main for biomedical and life sciences journal literature—returned only seven publications (Table 11.1). A thorough analysis of these manuscripts revealed that only three addressed the mined theme, being associated with genetic and cytogenetic diversity analyzes. In addition to the small amount observed, the tools used in the available manuscripts also indicate that genetic studies of this species are incipient: two (Mendes et al. 2014, Belarmino et al. 2017) of the three manuscripts (Table 11.1) reported genetic diversity analyzes with the dominant marker RAPD (Random Amplified Polymorphic DNA). The main advantage of RAPD is that it does not require prior knowledge of DNA sequences of the target species, which is consistent with the scientific *status quo* of *C. pyramidale*, actually lacking a sequenced genome. RAPD primers exhibit small size (generally 10 nucleotides) which increases the annealing possibility. Despite this advantage, their reproducibility

Table 11.1. Informational content on the academic manuscripts on genetics and biomolecules available for *C. pyramidale*, considering different databases. The topics covered in the manuscripts are presented, as well as the number of papers/sequences returned in the general search and those filtered (specific).

Search for	Database	Website	# Retrieved items	# Filtered items	Addressed Subject	References
Scientific Manuscripts	PubMed	https://www.ncbi.nlm.nih.gov/pubmed/	7	3	Genetic diversity | RAPD marker	Mendes et al. (2014)
					Cytogenetics | karyomorphology, and karyotype	Rodrigues et al. (2014)
					Genetic diversity | RAPD marker	Belarmino et al. (2017)
	Google Scholar	https://scholar.google.com/	18	3	Molecular phylogeny | rps16 plastid marker	Gagnon et al. (2013)
					Genetic diversity | RAPD marker	Mendes et al. (2014)
					Cytogenetics | karyotype diversity and 2C DNA content	Rodrigues et al. (2018)
	CAPES Publication Portal	http://www.periodicos.capes.gov.br/	3	none	-	-
	Google	https://www.google.com/	541	none	-	-
Biomolecules	Nucleotide (GenBank)	https://www.ncbi.nlm.nih.gov/nucleotide/	113	11	Phylogenetic informative sequences (chloroplast sequences)	Gagnon et al. (2013)
	Protein (GenBank)	https://www.ncbi.nlm.nih.gov/protein	53	1	Maturase K, partial chloroplast protein sequence	Tomlinson et al. (unpublished)
	SRA (GenBank)	https://www.ncbi.nlm.nih.gov/sra	18	18	Response to salt stress or to inoculation with Arbuscular Mycorrhizal Fungi	unpublished

is low, and the results may differ even between aliquots of the same sample at different dilutions, and especially among laboratories (revised by Nadeem et al. 2018).

The first study was performed by Mendes et al. (2014), who analyzed ten accessions [eight of *C. pyramidale* and two of *Poincianella bracteosa* (Tul.) L.P.Queiroz] from a semiarid zone in the state of Piauí (northeastern Brazil). The similarity coefficient between accessions was low (mean value 0.59) but presented a wide variation (from 0.443 to 0.748). The dendrogram constructed using the UPGMA (Unweighted Pair Group Method with Arithmetic Mean) based on a similarity matrix allowed the separation of the accessions into two major clusters represented by the two distinct species, while the accessions of *C. pyramidale* could be further separated into three subgroups. The second study was carried out by Belarmino et al. (2017) that evaluated the genetic diversity among 20 *C. pyramidale* individuals collected in the state of Paraíba (Brazil). The mean genetic diversity among the genotypes was 63.28%, ranging from 30.9 to 97.7%. The population evaluated in this study showed high genetic diversity, and formed twelve clusters based on the UPGMA analysis. The third article was published by Rodrigues et al. (2014), who analyzed the karyomorphology and karyotype asymmetry of species of the South American *Caesalpinia* (a previous generic epithet of *C. pyramidale*), including *C. pyramidale* individuals collected from Bom Jesus da Lapa (Bahia state, Brazil). It was reported that the mentioned species is diploid and has 24 chromosomes, all metacentric, with a size variation of 49.2% between the largest and the smallest chromosomal pair.

The same search performed in PubMed was made in the Google Scholar database (Table 11.1), returning 18 matches. From these, three actually addressed the genetics of *Catingueira*, two of which (Table 11.1) were absent in the PubMed results. In the first work, Rodrigues et al. (2018) evaluated the karyotypic diversity and the 2C DNA content variation in 10 species of the 'Caesalpinia group', representing six genera: *Paubrasilia, Caesalpinia, Cenostigma, Poincianella, Erythrostemon,* and *Libidibia*. The highest number of $CMA_3^+/DAPI^-$ (chromomycin/4′,6-diamidino-2-phenylindole) terminal bands was observed in *C. pyramidale* that presented four chromosome pairs with GC-rich heterochromatic blocks associated with 45S rDNA sites. The 2C genome size was 1.92 pg; similar to those of related species belonging to the genera *Cenostigma*, *Poincianella,* and *Caesalpinia*. The second publication was carried out by Gagnon et al. (2013) that published a new phylogeny of the 'Caesalpinia group' including *C. pyramidale*, based on rps16 plastid sequences. The species were clustered into the group '*Poncianella B/Cenostigma*', which contains several polytomies and a low posterior probability (55). The authors also proposed that several newly identified clades are candidates for recognition as new genera.

The availability and the amount of DNA and protein sequences in databases can be used as a gauge to reflect the amount of molecular research available on a given species. Few biomolecules (DNA, RNA, and proteins) on *catingueira* are deposited to date (Table 11.1). Data mining in the 'Nucleotide' database (Table 11.1)—which has a collection of sequences from several sources, including GenBank, NCBI RefSeq (Reference Sequence Database), TPA (Third Party Annotation Sequence Database) and PDB (Protein Data Bank)—was carried out adopting the keywords '*Cenostigma pyramidale* OR *Caesalpinia pyramidalis* OR *Poincianella pyramidalis*'. The results retrieved 113 gene sequences and intergenic spacers. However, only 11 nucleotide sequences comprised *C. pyramidale* sequences, which were generated for phylogeny analyses. The rest of them have been sequenced from microorganisms that may be present in the plant rhizosphere (such as *Streptomyces* sp.) or host it as symbiont (such as *Capnodiales* sp. and *Diaporthe* sp.). For proteins, a single sequence was available for the focus species (Table 11.1) regarding a chloroplast protein.

Ethnomedicinal Uses and Applications

Ethnobotanical reports indicate that all *C. pyramidale* tissues (stem, bark, flower, leaf, and root) have been used in folk medicine (Table 11.2). Forms of preparation include decoction (tea), dipping of leaves, infusion, juices, patches, and production of syrups. Trunk bark is the most used plant part, followed by leaves and flowers considering existing reports (Table 11.2).

Therapeutic uses have been reported by communities for a myriad of applications, including diseases of the respiratory tract (as an expectorant, to treat bronchitis, asthma, etc.), and to treat symptoms of

Table 11.2. Reported ethnobotanical indication and medicinal importance of *C. pyramidale*. Legend to abbreviations: NI = not informed. [1]Plants parts: R = root; SB = stem bark; IB = inner bark; LV = leaves; FL = flower; FR = fruit. [2]Category of indications: DIG = digestive system; DEF = defense against bacteria and fungi infections; RES = respiratory system; OTH = whole body, nervous system, etc. Tested and observed biological activities were marked with "*" and "**" to extracts and isolated substances, respectively. Type of study: 1 = Ethnobotanical interviews; 2 = *In vitro* assays; 3 = *In vivo* assays with rats or mice.

Bioactivity/Effect	Plant parts used[1]	Form of use (or*extract/ **bioactive substance)	Category[2]	Type	Reference
Anti-diarrheic/dysentery	R, SB, LV, FL, FR	Tea/Decoction/Infusion	DIG	1	Albuquerque et al. (2007), Agra et al. (2008), Gomes and Bandeira (2012), Macedo et al. (2018)
Anti-inflammatory	SB, FL	Tea/Syrup	OTH	1	Silva et al. (2015a)
Anti-inflammatory	IB	*Ethanolic extract	OTH	3	Moraes et al. (2013), Santana et al. (2012)
Anti-inflammatory	LV	**Agathisflavone	OTH	2	Santos-Souza et al. (2018)
Antimicrobial	R, SB, LV, FL	Tea/Syrup	DEF	1	Agra et al. (2008, Gomes and Bandeira (2012)
Antimicrobial	R, SB, LV, FL, FR	*Methanolic/Ethyl acetate extract	DEF	2	Novais et al. (2003), Saraiva et al. (2012)
Antimicrobial	SB, IB, LV	*Ethanolic extract	DEF	2	Silva et al. (2012), Ribeiro et al. (2013), Chaves et al. (2016)
Antimicrobial	LV	*Aqueous extract	DEF	2	Cruz et al. (2007)
Antinociceptive	IB	*Ethanolic extract	OTH	3	Santana et al. (2012)
Antioxidant	SB, LV	*Ethanolic extract	OTH	2	Silva et al. (2011), Chaves et al. (2016)
Antioxidant	LV	*Methanolic extract	OTH	2	Gomes de Melo et al. (2010)
Antioxidant	LV	**Agathisflavone	OTH	2	Andrade et al. (2018)
Aphrodisiac	SB	Tonic	OTH	1	Agra et al. (2008)
Bronchitis	SB, FL	Syrup/Tea	RES	1	Albuquerque et al. (2007), Agra et al. (2008), Costa et al. (2016)
Colic	R, SB, LV, FL	NI	OTH	1	Albuquerque et al. (2007)
Diabetes	R, SB, LV, FL	NI	OTH	1	Albuquerque et al. (2007)
Fever	R, SB, LV, FL	NI	OTH	1	Albuquerque et al. (2007)
Flatulence	R, SB, LV, FL	NI	DIG	1	Albuquerque et al. (2007)
Gastroprotective	IB	*Ethanolic extract	DIG	3	Ribeiro et al. (2013)
Heartburn	R, SB, LV, FL	NI	DIG	1	Albuquerque et al. (2007)
Hemostatic	SB, FL	Decoction	OTH	1	Cartaxo et al. (2010)
Influenza	SB, FL	Tea/Decoction/Syrup	RES	1	Cartaxo et al. (2010), Silva et al. (2015a), Macedo et al. (2018)
Neurogenesis	LV	**Agathisflavone	OTH	2	Santos-Souza et al. (2018)

Table 11.2 Contd. ...

...Table 11.2 Contd.

Bioactivity/Effect	Plant parts used[1]	Form of use (or*extract/ **bioactive substance)	Category[2]	Type	Reference
Neuroprotective	LV	**Agathisflavone	OTH	2	Santos-Souza et al. (2018)
Respiratory clearance	SB, FL	Decoction	RES	1	Cartaxo et al. (2010), Macedo et al. (2018)
Respiratory clearance	SB, FL	Decoction/Tea/Infusion	RES	1	Cartaxo et al. (2010), Saraiva et al. (2015), Costa et al. (2016)
Stomach pain	SB	Tea	DIG	1	Agra et al. (2008)
Stomach pain	R, SB, LV, FL, FR	Decoction/Infusion/Juice	DIG	1	Albuquerque et al. (2007), Cartaxo et al. (2010), Macedo et al. (2018)
Wound healing	FL, SB	Decoction	OTH	1	Albuquerque et al. (2007), Cartaxo et al. (2010)

influenza. Considering diseases of the digestive trait, the applications include treatment of gastritis, colic, flatulence, diarrhea, stomach heartburn and pain (Table 11.2). The time of treatment may range from four days to up to one month of daily use, mainly for respiratory problems.

Other biological activities have been described using plant extracts, such as antinociceptive, anti-inflammatory, antioxidant, molluscicidal, antimicrobial, anthelmintic and antidiabetic, as well as antiproliferative, among others (Albuquerque et al. 2007, Cruz et al. 2007, Gomes de Melo et al. 2010, Cartaxo et al. 2010, Santos et al. 2012, Oliveira et al. 2016a, b). Despite its therapeutic potential, usage of *C. pyramidale* is still restricted to traditional communities, even considering the existence of some scientifically proven medicinal properties. In this sense, the species has been the target for the identification of several secondary metabolites, with potential applications to treat some diseases of the respiratory and digestive tract besides neuronal disorders (Ribeiro et al. 2013, Santos-Souza et al. 2018).

For instance, Dos Santos Souza et al. (2018) isolated a flavonoid named agathisflavone from the leaves of *catingueira*. This compound presents the ability to improve the neuronal population, being able to interfere in the estrogen signaling, thus stimulating the *in vitro* production of neurons and improving the neuroprotective properties of microglia and astrocytes. It also significantly decreased glutamate-mediated neurotoxicity-mediating the excitotoxicity, a process responsible for the gradual loss of neurons, which can trigger an ischemic stroke in humans (Dos Santos Souza et al. 2018).

Another study with agathisfavone revealed its *in vitro* antioxidant capacity, significantly slowing or preventing lipid peroxidation and inhibiting free radical accumulation, inducing the production of some molecules [as 2,2-diphenyl-1-picrylhydraz (DPPH), 2,20-azinobis (3-ethylbenzthiazoline-6-sulfonic acid; ABTS), 2,20-azobis-2-amidinopropane dihydrochloride (AAPH), hydroxyl (OH) and nitric oxide (NO)]. In this way, it acts by decreasing oxidative stress, which causes damage to the brain and other organs in humans and animals (Andrade et al. 2018).

Other relevant applications refer to the gastroprotective activity of *C. pyramidale* ethanolic extract in mice (Ribeiro et al. 2013). This study reported the inhibition of the development of chemical-induced ulcers and the increased protection of the stomach mucosa after intraduodenal administration in mice (at concentrations of 300 mg/kg of *C. pyramidale* ethanolic extract), when compared to the inhibitors cimetidine (100 mg/kg) and Omeprazole (30 mg/kg).

Most current researches focus on the exploration and discovery of bioactive molecules isolated from *C. pyramidale* (Table 11.3). This is the case of the study carried out by Silva et al. (2012) that analyzed the phytochemical profile and antibacterial activity of the crude ethanol extract of *Catingueira* bark and leaves (as also described by Mendes et al. 2000, Bahia et al. 2005, 2010, Silva et al. 2011, Chaves et al. 2016, Oliveira et al. 2016a, b) under different extract types. The phytochemical composition analyzed by thin-layer chromatography revealed the presence of anthocyanins, sterols, flavonoids, lignans, and terpenoids. In addition, the extract showed antibacterial activity against several Gram-positive bacteria. For example, *C. pyramidale* bark extract presented effective antimicrobial activity against multidrug-resistant *Staphylococcus* strains, demanding studies for the isolation of the compounds responsible for these activities.

Saraiva et al. (2012) conducted a study on the phytochemical profile of *C. pyramidale* methanolic and ethyl acetate extracts of various tissues (leaves, stem bark, roots, flowers, fruits, and seeds). In addition, they tested the antimicrobial activity of the extracts against seventeen *S. aureus* isolates, including Methicillin-resistant strains. The phytochemical profile confirmed the presence of several known antimicrobial agents, including ursolic acid, quercetin, catechin, ellagic acid, sitosterol, flavonoids, proanthocyanidins and gallic acid (Table 11.3).

In another study conducted by Gomes de Melo et al. (2010), the antioxidant capacity and tannin content of plants from the *Caatinga* environment were evaluated, including methanol extracts of *C. pyramidale* leaves. The results revealed an antioxidant activity (IC_{50}) of 42.95 ± 1.77 μg/ml and a tannin content of 8.17% ± 0.64. Given its high antioxidant potential and renewability of *C. pyramidale*'s leaves, they could be a viable resource for obtaining compounds with commercial potential for applications in the chemical, food and pharmaceutical industries (Silva et al. 2011).

Table 11.3. Main phytochemical profile of *C. pyramidale* based on literature reports (Bahia et al. 2005, 2010, Mendes et al. 2000, Oliveira et al. 2016a, b, Santana et al. 2012, Silva et al. 2012, Saraiva et al. 2012, Chaves et al. 2016). Abbreviations: "+" = presence; "–" = absence; "X" = not found in the literature.

Phytochemicals		Root	Stem bark	Leaves	Flower	Fruit
Alkaloids		X	–	–	X	X
Cinnamic derivatives		+	+	+	+	+
Fatty acids derivatives		+	X	X	X	X
Fatty alcohol derivatives		X	X	X	+	X
Terpenes	Monoterpenes	X	+	+	X	X
	Triterpenes and steroids	+	–	+	X	X
	Saponin	X	+	–	X	X
	Sitosterol	+	+	+	+	+
	Ursolic acid	+	–	–	–	–
Polyphenols	Anthocyanins	X	–	+	X	X
	Biflavones	+	+	+	X	X
	Catechin	+	+	–	–	–
	Coumarin	X	–	–	X	X
	Condensed proanthocyanidins	+	+	–	–	–
	Ellagic acid derivatives	+	–	–	+	+
	Flavonoid of aglicones	+	+	+	+	+
	Gallic acid	–	–	–	+	+
	Lignans	X	+	–	X	X
	Quercetin	+	–	–	–	–
	Tannin (Condensed)	X	+	+	X	X
	Tannin (Hydrolysable)	X	–	–	X	X

Bahia et al. (2010) performed a chromatographic analysis of methanolic extracts of *C. pyramidale* leaves. The results revealed the presence of agathisflavone and taxifolin, as well as minor biflavonoids such as loniflavone, amentoflavone, 5'-hydroxyamoflavone and podocarpusflavone A. In addition, the structures of the compounds were established based on nuclear magnetic resonance (NMR) and mass spectrometry (MS) data analysis. The results permitted the proposition of the molecular formula $C_{30}H_{18}O_{10}$ and $C_{30}H_{18}O_{11}$, consequently determining the biflavonoid nature of compounds 1 and 3, respectively. In the same way, it contributed to the structural definition of the resonances of aromatic carbons of B ring in Unity II.

Oliveira et al. (2016) investigated the chemical composition of the methanolic extract of root barks of *C. pyramidale*, resulting in the isolation of 3,3'-dimethylellagic acid and 3,3'-dimethylellagic acid-4'-O-β-D-xyloside. Lupeol, β-sitosterol/stigmasterol, and a mixture of fatty acid methyl ester derivatives were also obtained. In this same study, chromatographic procedures of the methanolic extract of the flowers of this species led to obtaining an unusual mixture of fatty alcohols, β-sitosterol/stigmasterol, α-amyrin, β-amyrin and methyl gallate. This was the first report of 3,3'-dimethylellagic acid, 3,3'-dimethylellagic acid-4'-O-β-D-xyloside and free fatty alcohols in the Fabaceae family.

In a study with the sister genus *Caesalpinia* group, Moreira et al. (2019) determined and quantified the bioactive phenolics (gallic acid, methyl gallate, and ellagic acid) and the biflavonoids (agathisflavone and amentoflavone) in methanolic extracts of the leaves and stem bark of the sister species *Cenostigma macrophyllum*. Besides, antioxidant, acetylcholinesterase inhibition, anti-inflammatory and antinociceptive activities of the standardized hydromethanolic extracts of leaves and stem bark of this species were evaluated. The methanolic extracts showed antioxidant activity (EC_{50} 69.09 and 83.06

μg ml^{-1}), and acetylcholinesterase inhibition (52.23 and 83.36%), also inhibiting the formalin-induced nociception and reducing edema formations at 100 mg kg^{-1} doses. Such results indicate that *catingueira* related species present comparable properties, also deserving bioprospecting evaluations.

Aligning traditional knowledge with scientific evaluation through laboratory testing, *C. pyramidale* represents a promising candidate for the transition from folk medicine to modern therapy based on its phytochemical content and antioxidant activity (Silva et al. 2011).

Cytotoxicity, Mutagenicity, *In Vivo* and *In Vitro Studies*

Natural products are often less toxic than synthetic compounds. Despite this assumption, treatments with medicinal plants—as, for example, teas and infusions—can contain toxic substances that may have mutagenic and cytotoxic effects (Silva et al. 2015b). Toxicity is defined as an inherent property of compounds that cause harmful effects to exposed organisms over a specified time and concentration. The effects caused by these compounds include immobility, mortality, and inhibition of reproduction and reduction of cell growth of the test organism. As highlighted by Araujo et al. (2016), several steps and procedures are involved in the evaluation of possible cytotoxic, genotoxic, mutagenic and even carcinogenic processes of a given compound. Some plant-based phytochemicals, including some flavonoids and aromatic compounds (as estragole, safrole and methyl-eugenol), have mutagenic and carcinogenic potential (Rietjens et al. 2005). Van den Berg et al. (2011) listed compounds of plant origin that have such effects. Monoterpenes, coumarin, polyphenols, alkaloids and anthraquinones figure among them, besides those above mentioned.

Cytotoxicity, genotoxicity, and mutagenicity tests of different plant parts or extracts are mandatory to guarantee the safety of their use by the population. Besides, such studies may bring insights into their mechanisms of action. Despite the widespread use in traditional medicine and the existing scientific studies on the therapeutic activity of *C. pyramidale* (especially the bark), there are few reports on its cellular toxicity (cytotoxicity) or its effects on the genetic material (genotoxicity and mutagenicity) (Meneguetti et al. 2014). Silva et al. (2015b) tested the cytotoxicity of *C. pyramidale* bark infusion at concentrations of 1 and 2 μg/ml using the *Allium cepa* bioassay. The results revealed cytotoxic activity of both concentrations on meristematic cells of *A. cepa* roots, after 24 and 48 h, with an observed inhibition of cell division.

In vivo assays were also performed to test the possible effects of the ethanol extract (at concentrations of 100, 500 and 1000 ppm) of *C. pyramidale* stem bark on *Artemia salina* (Crustacea), *Aedes aegypti* and *Biomphalaria glabrata* eggs (Luna et al. 2005). The authors found discrete toxicological activity (ca.10% mortality) considering the concentrations studied.

Lima et al. (2019) studied the hydroalcoholic extract of *C. pyramidale* leaves by means of the bioassay with *Artemia salina*. The concentrations tested (1000 μg/ml; 750 μg/ml; 500 μg/ml; 250 μg/ml; 100 μg/ml and 50 μg/ml) were nontoxic, and presented a high Average Lethal Dose (LC_{50} = 1442.8433 μg/ml). In another study performing the same test, Oliveira (2010) described the toxicity according to the polarity of the extracts. For this, the authors recovered four phases from the root bark of *C. pyramidale*: hexanic, ethyl acetate, methanolic, and butanolic. The obtained fractions were tested on *A. salina* at concentrations of 25, 50, 100, 150, 200 and 250 μg/ml. After 24 hours incubation, some of the fractions presented cytotoxic activity. The ethyl acetate phase showed higher toxicity followed by the methanolic phase (LC_{50} of 149.12 and 193.98 μg/ml, respectively). In turn, the butanolic phase presented LC_{50} of 228.97 μg/ml, the lowest in relation to the polar phases. For the hexane phase, the LC_{50} was 535.55 μg/ml, considered nontoxic at the tested concentrations. Thus, in ascending order of cytotoxicity, we have the phases: ethyl acetate > methanolic > butanolic > hexanic.

Chaves et al. (2019) evaluated the cytotoxic and genotoxic potential of the *C. pyramidale* hydroalcoholic extract in Swiss mice at concentrations 100, 200, and 500 mg kg^{-1}. The authors found that the highest concentrations were cytotoxic to polychromatic erythrocytes of peripheral blood and bone marrow. Considering the genotoxicity, the extract promoted genotoxic effects exclusively in the concentration of 100 mg kg^{-1} for peripheral blood, whereas no effect was observed in the marrow, demanding a further investigation of the antiproliferative potential of this extract.

Reported Collateral Effects

Evaluating the teratogenic effects of *C. pyramidale* in sheep (*Ovis aries*), Correia et al. (2017) demonstrated that the direct consumption of its leaves by pregnant females might cause congenital malformation, abortion, and perinatal mortality. Therefore, it is imperative to keep the plants out of reach of pregnant sheep throughout the gestational period. Similarly, Reis et al. (2016) and Santos et al. (2018) tested the effects of the consumption of the leaves on pregnant goats (*Capra aegagrus hircus*) and concluded that it causes embryonic losses and congenital malformations. According to these authors, the leaves of *C. pyramidale* have high palatability for these animals and should be considered as one of the leading causes of abortion in both species in the northeastern region of Brazil.

Câmara et al. (2017) confirmed the teratogenic effects of *C. pyramidale* by reproducing congenital malformations in the offspring of rat (*Rattus norvegicus*) female rats that received the plant during pregnancy. In addition to congenital malformations, they found that diets prepared from the leaves of *C. pyramidale* also cause mortality and reduced offspring number, besides lower length and weight of fetuses, also providing increased post-implantation losses and placental damage.

Considering all reports, despite the acceptable taste (palatability) for the tested animals, direct consumption of *C. pyramidale* leaves should be avoided, especially in pregnant females. These results on other mammal species indicate that caution should be exercised in the internal use of fresh leaves or their products (decoction, infusion, juice, tea, etc.) even by humans.

Other Applications

C. pyramidale is highly explored for non-medicinal purposes including animal feeding, recovery of degraded areas, ecological and cultural potential, among others (Salvat et al. 2004, Lucena et al. 2008, Ramos et al. 2008, Santos et al. 2008, Sá et al. 2009, Almeida et al. 2010).

The wood is yellowish-white and, due to high amount of cellulose and lignin (Silva et al. 2009, Matias et al. 2017), it is used to produce wooden fences, in the construction of houses, as well as an excellent source of energy with promising perspectives for the production of alcohol and charcoal (Maia 2004, Silva et al. 2009, Braga 2015). The trunk and branches of *C. pyramidale* are used for fuel even at the immature stage because of their high fiber concentration, low vessel, and parenchyma amount (Silva et al. 2009). Consequently, this species stands out among the most widely used as fuelwood source in the *Caatinga* (Ramos et al. 2008).

C. pyramidale figures among the deciduous *Caatinga* species. That means this tree loses all leaves during the dry season. Total leaf senescence happens when the precipitation inter-pulses intensify as the dry season progresses, characterizing deciduousness (Lima et al. 2018). The fallen leaves remain in the ground serving as food for animals in dry periods (Sampaio and Rodal 2000, Maia 2012, Salvat et al. 2004, Silva et al. 2015c). Although the trees are relatively small (mainly in relation to some other trees in the region), they are better able to withstand coppicing and pollarding, showing a rapid regrowth and lower mortality (Leme and Gasson 2012). Indeed, after application of different harvesting techniques (coppicing, pollarding, or thinning) in the wood production and mortality rates of four *Caatinga* native tree species, *C. pyramidale* showed to be better able to withstand all treatments, with almost no mortality, contributing to long-term fuelwood productivity (Milliken et al. 2018).

Environmental Role and Sustainability

As emphasized, *C. pyramidale* is capable of adapting to different environmental conditions, including poor soils, low rainfall, high temperature and insolation, being classified by our group as an 'extremophile' species. Despite this rusticity, a study realized by Ramos and Albuquerque (2012) showed that the seasonality interferes with patterns of firewood collection. According to authors, more collections are performed in the dry season, and the local vegetation suffers due to uncontrolled collection of whole plants and plant parts, reinforcing the need for the use of species with adequate biomass production by supplying firewood or charcoal with sustainable procedures. We observed the intensive use of *catingueira* firewood as fuel for pottery in the state of Rio Grande do Norte and for the plastering industry in the state of Pernambuco.

The *Caatinga* comprises one of the most fragile and least protected environments in Brazil. In this context, no genetic banks involving seeds or specimens of *C. pyramidale* were found in this review, and there is a single evaluation of its *in vitro* behavior, with possible conservation applications (Silva et al. 2019). Additionally, one of the main threats to native *Caatinga* species is the presence of exotic invasive tree species, with emphasis on *Prosopis juliflora* (Sw.) DC. This species—known in the region by the popular name of 'algaroba'—is also a tree legume (Fabaceae) and drastically affects the native biodiversity and the structure of the invaded communities. A study conducted by Andrade et al. (2009) highlights the high invasiveness and replacement of native tree and shrub species by this invasive species, affecting the entire ecosystem, including *C. pyramidale*. Despite that, the *catingueira* has been ranked among the most tolerant to anthropic and biotic disturbances, being classified among the most recommended for recovery of degraded *Caatinga* areas.

With the exploitation of natural resources, especially for medicinal and economic uses, *Caatinga* species are under anthropic pressure, where native forests are being reduced to small fragments concerning their original areas (Santos et al. 2012). This situation leads to loss of biodiversity of plants typical of the biome and to environmental imbalances, such as ecological changes. Due to its high rusticity and ability to adapt to very contrasting environmental conditions, for example, salinity and drought (Falcão et al. 2015, Frosi et al. 2016a, 2018), allied to its fast-growing and good potential for regrowth (Figueirôa et al. 2008), *C. pyramidale* stands out as a promising plant in the recovery of deforested and degraded areas (Maia 2004, Sales et al. 2019).

Concluding Remarks and Perspectives

The vast majority of studies on *catingueira*'s ethnomedicinal properties are represented by literature in Portuguese that is inaccessible in scientific databases. Most of them regard monographs of undergraduate students of the northeastern region of Brazil. Studies carried out by our group show that such monographs are generally based on the application of questionnaires to the population of a given region, normally small communities. Also, considering reports published in scientific journals and books, like those cited here, it is noteworthy that more than 40% of the references on *catingueira* uses were published in Portuguese. Unlike other medicinal species from Brazil, *catingueira* does not occur naturally in other Brazilian regions, and there are no reports of its cultivation/production outside the northeastern region of Brazil. This fact justifies the few ethnobotanical reports on this species by groups outside its region of occurrence.

In many villages throughout the northeastern of Brazil, plants are the primary source of health care, if not the single one. It is clear that the therapeutic potential of the plants from the *Caatinga* environment has been largely undervalued, being progressively less taught to young people of the most recent generations. Additionally, this knowledge is mostly maintained only by herbal traders (called "erveiros" in Brazil) and remains absent from the world-wide-web environment. Among the existing phytodiversity, the *catingueira* figures as a valuable source of biomolecules, applied to digestive and respiratory diseases, besides antimicrobial properties. Especially, the antioxidant activity of leaves and bark deserve further research as well as the effects of the *C. pyramidale*-derived flavonoid named agathisflavone with *in vitro* neuroprotective effects. Despite these promising properties, dosages and forms of use (especially considering internal use) should be further evaluated, considering potential toxicity, teratogenicity and other collateral effects.

As highlighted here, few genetic information is available on *C. pyramidale*. Our group has just sequenced the genome and the transcriptome of this species under salinity and in association with arbuscular mycorrhizal fungi in different exposition times. Considering its importance and role as an '*extremophile species*' able to grow under diverse environmental situations and also the potential to deliver interesting biomolecules, the obtained data will certainly mean a great advance for the understanding of its genetic, molecular and physiological properties.

References

Agra, M.F., Silva, K.N., Basílio, I.J.L.D., Freitas, P.F., Barbosa-Filho, J.M. 2008. Survey of medicinal plants used in the region Northeast of Brazil. Braz. J. Pharmacog., 18(3): 472–508.

Albuquerque, U.P., de Medeiros, P.M., de Almeida, A.L.S., Monteiro, J.M., de Freitas Lins Neto, E.M., de Melo, J.G., dos Santos, J.P. 2007. Medicinal plants of the caatinga (semi-arid) vegetation of NE Brazil: A quantitative approach. J. Ethnopharmacol., 114: 325–354.

Almeida, C.F.C.B.R., Ramos, M.A., Amorim, E.L.C., Albuquerque, U.P. 2010. A comparison of knowledge about medicinal plants for three rural communities in the semi-arid region of northeast of Brazil. J. Ethnopharmacol., 127: 674–84.

Alves, E.U., Cardoso, E.A., Bruno, R.L.A., Alves, A.U.A., Alves, A.U., Galindo, E.A., Braga Junior, J.M. 2007. Superação da dormência em sementes de *Poincianella pyramidalis* Tul. R. Árvore, 31(3): 405–415.

Andrade, A.W.L., Machado, K.C., Machado, K.C., Figueiredo, D.D.R., David, J.M., Islam, M.T., Uddin, S.J., Shilpi, J.A., Costa, J.P. 2018. *In vitro* antioxidant properties of the biflavonoid agathisflavone. Chem. Cent. J., 12(1): 75.

Andrade, L.A. de, Fabricante, J.R., Oliveira, F.X. de. 2009. Invasão biológica por *Prosopis juliflora* (Sw.) DC.: impactos sobre a diversidade e a estrutura do componente arbustivo-arbóreo da caatinga no estado do Rio Grande do Norte, Brasil. Acta Bot. Brasilica, 23: 935–943.

Antunes, C.G.C., Pelacani, C.R., Ribeiro, R.C., Gomes, H.L.R., Castro, R.D. 2010. Influência do armazenamento na qualidade fisiológica de sementes de *Caesalpinia pyramidalis* Tul. R. Árvore, 34(6): 1001–1008.

Antunes, C.G.C., Pelacani, C.R., Conduru, R.C., Souza, J.V., Souza, C.L.M., Castro, R.D. 2011. Germinação de sementes de *Caesalpinia pyramidalis* Tul. (Catingueira) submetidas a deficiência hídrica. R. Árvore, 35(5): 1007–1015.

Araújo, S.S., Fernandes, T.C., Marin-Morales, M.A., Benko-Iseppon, A.M., Brasileiro-Vidal, A.C. 2016. Mutagenicity, genotoxicity and cytotoxicity assays of medicinal plants: first step for drug development. *In*: Rai, M., Duarte, M.C.T. (eds.). Therapeutic Medicinal Plants—From Lab to the Market. 7: 130–153. CRC Press.

Bahia, M.V., dos Santos, J.B., David, J.P., David, J.M. 2005. Biflavonoids and other phenolics from *Caesalpinia pyramidalis* (Fabaceae). J. Braz. Chem. Soc., 16: 1402–1405.

Bahia, M.V., David, J.P., David, J.M. 2010. Occurrence of biflavones in leaves of *Caesalpinia pyramidalis* specimens. Quím. Nova, 31: 297–1300.

Belarmino, K.S., Rego, M.M., Bruno, R.L.A., Medeiros, G.D.A., Andrade, A.P., Rego, E.R. 2017. Genetic diversity in a *Poincianella pyramidalis* (Tul.) L.P. Queiroz population assessed by RAPD molecular markers. Genet. Mol. Res., 16: gmr16039663.

Benko-Iseppon, A.M., Crovella, S. 2010. Ethnobotanical bioprospection of candidates for potential antimicrobial drugs from Brazilian plants: state of art and perspectives. Curr. Protein Pept. Sci., 11: 189–194.

Benko-Iseppon, A.M., Pinangé, D.S.B., Chang, S.C., Morawetz, W. 2012. Ethnobotanical uses of the native flora from Brazilian North-Eastern region. pp. 84–105. *In*: Rai, M., Rastrelli, L., Marinoff, M., Martinez, J.L., Cordell, G. (eds.). Medicinal Plants: Biodiversity & Drugs. CRC Press, New Hampshire, Reino Unido. ISBN 978-1-5708-000-0. Chapter 4.

Borges, L.A., Sobrinho, M.S., Lopes, A.V. 2009. Phenology, pollination and breeding system of the threatened tree *Caesalpinia echinata* Lam. (Fabaceae), and a review of studies on the reproductive biology in genus. Flora, 204: 111–130.

Braga, R. 1960. Plantas do Nordeste, especialmente do Ceará. 2ª edição. Fortaleza: Imprensa Oficial do Ceará, p. 540.

Câmara, A.C.L., Gadelha, I.C.N., Castro, M.B., Medeiros, R.M.T., Riet-Correa, F., Soto-Blanco, B. 2017. Embryotoxic effects of *Poincianella* (*Caesalpinia*) *pyramidalis* leaves on pregnant rats. J. Vet. Diagn. Invest., 29: 137–142.

Cartaxo, S.L., Almeida Souza, M.M., Albuquerque, U.P. 2010. Medicinal plants with bioprospecting potential used in semi-arid northeastern Brazil. J. Ethnopharmacol., 131: 326–342.

Chaves, T.P., Fernandes, F.H.A., Santana, C.P., Santos, J.S., Medeiros, F.D., Felismino, D.C., Medeiros, A.C.D. 2016. Evaluation of the interaction between the *Poincianella pyramidalis* (Tul.) LP Queiroz extract and antimicrobials using biological and analytical models. Plos One, 11: 1–23.

Chaves, T.P., Medeiros, F.D., Sousa, J.M.C., Silva, L.A.P., Lima, M.A., Coutinho, H.D.M., Medeiros, A.C.D. 2019. Phytochemical characterization and mutagenicity, cytotoxicity, antimicrobial and modulatory activities of *Poincianella pyramidalis* (Tul.) L.P. *Queiroz*. Nat. Prod. Res., 1: 1–7.

Correia, D.A.B., Neto, G.B.M., Diego Lima da S. Gomes, D.L.S., Torres, M.B.A.M. 2017. Malformações congênitas e abortos induzidos experimentalmente pela ingestão de *Poincianella pyramidalis* (Tul.) L.P. Queiroz (catingueira) em ovelhas. Pesq. Vet. Bras., 37: 1430–1436.

Costa, J.C., Marinho, M.G.V. 2016. Etnobotânica de plantas medicinais em duas comunidades do município de Picuí, Paraíba, Brasil. Rev. Bras. Plantas Med., 18(1): 125–134.

Cruz, MC. 2007. Antifungal activity of Brazilian medicinal plants involved in popular treatment of mycoses. J. Ethnopharmacol., 111: 409–412.

Dantas, B.F., Correia, J.D.S., Marinho, L.B., Aragão, C.A. 2008. Biochemical changes during imbibition of *Caesalpinia pyramidalis* Tul. seeds. Rev. Bras. Sementes, 30: 221–227.

Dantas, B.F., Lopes, A.P., Silva, F.F.S., Lúcio, A.A., Batista, P.F., Pires, M.M.L., Aragão, C.A. 2011. Production of seedlings of true catingueira (*Caesalpinia pyramidalis* Tul.) in function of substrates and luminosities. Scientific, 39(1:2): 34–43.

Dantas, B.F., Moura, M.S.B., Pelacani, C.R., Angelotti, F., Taura, T.A., Oliveira, G.M., Bispo, J.S., Matias, J.R., Silva, F.F.S., Pritchard, H.W., Seal, C.E. 2019. Rainfall, not soil temperature, will limit the seed germination of dry forest species with climate change. Oecologia.

Dos Santos Souza, C., Grangeiro, M.S., Lima Pereira, E.P., Dos Santos, C.C, Da Silva, A.B., Sampaio, G.P., Figueiredo, D.D.R., David, J.M., David, J.P., Da Silva, V.D.A., Butt, A.M., Costa, S.L. 2018. Agathisflavone, a flavonoid derived from *Poincianella pyramidalis* (Tul.), enhances neuronal population and protects against glutamate excitotoxicity. Neurotoxicology, 65(3): 85–97.

Fabricante, J.R., Feitosa, S.S., Bezerra, F.T.C., Feitosa, R.C., Xavier, K.R.F. 2009. Análise populacional de *Caesalpinia pyramidalis* Tul. (Fabaceae) na caatinga da região do Seridó nordestino. Rev. Bras. Bioci., 7(3): 285–90.

Falcão, H.M., Medeiros, C.D., Silva, B.L.R., Sampaio, E.V.S.B., Almeida-Cortez, J.S., Santos, M.G. 2015. Phenotypic plasticity and ecophysiological strategies in a tropical dry forest chronosequence: A study case with *Poincianella pyramidalis*. For. Ecol. Manag., 340: 62–69.

Figueirôa, J.M., Araújo, E.L., Pareyn, F.G.C., Cutler, D.F., Gasson, P., Lima, K.C., Santos, V.F. 2008. Variações sazonais na sobrevivência e produção de biomassa de *Caesalpinia pyramidalis* Tul. após o corte raso e implicações para o manejo da espécie. R. Árvore, 32: 1041–1049.

Frosi, G., Barros, V.A., Oliveira, M.T., Cavalcante, U.M., Maia, L.C., Santos, M.G. 2016a. Increase in biomass of two woody species from a seasonal dry tropical forest in association with AMF with different phosphorus levels. Appl. Soil. Ecol., 102: 46–52.

Frosi, G., Barros, V.A., Oliveira, M.T., Santos, M., Ramos, D.G., Maia, L.C., Santos, MG. 2016b. Symbiosis with AMF and leaf Pi supply increases water deficit tolerance of woody species from seasonal dry tropical forest. J. Plant Physiol., 207: 84–93.

Frosi, G., Barros, V.A., Oliveira, M.T., Santos, M., Ramos, D.G., Maia, L.C., Santos, M.G. 2018. Arbuscular mycorrhizal fungi and foliar phosphorus inorganic supply alleviate salt stress effects in physiological attributes, but only arbuscular mycorrhizal fungi increase biomass in woody species of a semiarid environment. Tree Physiol., 38: 25–36.

Gagnon, E., Lewis, G.P., Sotuyo, J.S., Hughes, C.E., Bruneau, A. 2013. A molecular phylogeny of Caesalpinia sensu lato: Increased sampling reveals new insights and more genera than expected. South African J. Bot., 89: 111–127.

Gomes de Melo, J., Araújo, T.A.S., Almeida Castro, V.T.N., Cabral, D.L.V., Rodrigues, M.D., Nascimento, S.C., Amorim, E.L.C., Albuquerque, U.P. 2010. Antiproliferative activity, antioxidant capacity and tannin content in plants of semi-arid northeastern Brazil. Molecules, 15: 8534–8542.

Gomes, T.B.G., Bandeira, F.P.S.F. 2012. Uso e diversidade de plantas medicinais em uma comunidade quilombola no Raso da Catarina, Bahia. Acta Bot. Bras., 26(4): 796–809.

Lahaye, C., Hernandez, M., Boëda, E., Felice, G.D., Guidon, N., Hoeltz, S., Lourdeau, A., Pagli, M., Pessis, A.-M., Rasse, M., Viana, S. 2013. Human occupation in South America by 20,000 BC: the Toca da Tira Peia site, Piauí, Brazil. J. Archaeol. Sci., 40: 2840–2847.

Leite, A.V., Machado, I.C. 2009. Biologia reprodutiva da "catingueira" (*Poincianella pyramidalis* Tul., Leguminosae-Caesalpinioideae), uma espécie endêmica da Caatinga. Rev. Bras. Bot., 32: 55–64.

Leme, C.L.D., Gasson, P. 2012. Anatomical comparison of original and regrowth wood from coppiced and pollarded *Poincianella pyramidalis* trees in the caatinga of Pernambuco, Brazil. IAWA J., 33: 63–72.

Lima, C.R., Bruno, R.D.L.A., Andrade, A.P. de, Pacheco, M.V., Quirino, Z.G.M., Silva, K.D.R.G., Belarmino, K.D.S. 2018. Fenologia de *Poincianella pyramidalis* (Tul.) L.P. Queiroz e sua relação com a distribuição temporal da precipitação pluvial no semiárido brasileiro. Ci. Fl, 28: 1035.

Lima, J.L.S. 1996. Plantas forrageiras da caatinga usos e potencialidades. Embrapa. Petrolina- PE, p. 44.

Lima, M.F.F., de Araujo Silva, J.W.S., Silva, J.K., Moura, A.H.N., Lopes, R.L.F., Cordeiro, B.A., Cordeiro, R.P., Melo, A.F.M. 2019. Avaliação toxicológica através do bioensaio com Artemia salina Leach de espécimes vegetais pertencentes à caatinga/Toxicological evaluation by *Artemia salina* Leach bioassay of caatinga plant specimens. Braz. J. Hea. Rev., 2: 5950–5963.

LPWG [Legume Phylogeny Working Group]. 2017. A new subfamily classification of the Leguminosae based on a taxonomically comprehensive phylogeny. Taxon, 66(1): 44–77.

Lucena, R.F.P., Nascimento, V.T., Araújo, E.D.L., Albuquerque, U.P. 2008. Local uses of native plants in an area of caatinga vegetation (Pernambuco, NE Brazil). Ethnobotan. Res. Applic., 6: 3–13.

Luna, J.S., Santos, A.F., Lima, M.R.F., Omena, M.C., Mendonça, F.A.C., Bieber, L.W., Sant'Ana, A.E.G. 2005. A study of the larvicidal and molluscicidal activities of some medicinal plants from northeast Brazil. J. Ethnopharmacol., 97: 199–206.

Macedo, J.G.F., Menezes, I.R.A., Ribeiro, D.A., Santos, M.O., Mâcedo, D.G., Macêdo, M.J.F., Almeida, B.V., Oliveira, L.G.S., Leite, C.P., Souza, M.M.A. 2018. Analysis of the variability of therapeutic Indications of medicinal species in the Northeast of Brazil: Comparative study. Ev. Based Compl. Alternat; Med. J. ID 6769193, 28 pp.

Maia, G.N. 2004. Catingueira. pp. 159–169. *In*: Maia, G.N. (ed.). Caatinga: Árvores e Arbustos e suas Utilidades. São Paulo: Leitura e Arte.

Maia, G.N. 2012. Caatinga: Árvores e Arbustos e suas Utilidades, 2a ed. Print color Gráfica e Editora, Fortaleza.

Matias, J.R., Conduru, R.C., Oliveira, G.M., Affonso, I.B., Silva, T.B., Costa, D.C.C., Bispo, J.S., Mendes, R.B., Dantas, B.F. 2014. Temperatura limitante à germinação de semente de Catingueira-Verdadeira. Info. Abrates, 24(3): 87.

Matias, J.R., Silva, F.F.S., Dantas, B.F. 2017. Catingueira-verdadeira *Poincianella pyramidalis* [Tul.] L.P. Queiroz. Abrates. Nota Técnica, n° 6. Embrapa Semiárido.

Matias, J.R., Torres, S.B., Freire, J.N.T., Alencar, S.S., Dantas, B.F. 2018. Germinação de sementes de *Cenostigma pyramidalis* sob diferentes temperaturas e salinidade. Info. Abrates, 8(1): 115–118.

Matos, S.S., Melo, A.L., Santos-Silva, J. 2019. Caesalpinioideae e Cercidoideae (Leguminosae) no Parque Estadual Mata da Pimenteira, Semiárido de Pernambuco, Brasil. Rodriguésia 70(1): e01842017.

Mendes, C.C., Bahia, M.V., David, J.M., David, J.P. 2000. Constituents of *Caesalpinia pyramidalis*. Fitoterapia, 71: 205–207.

Mendes, R.F.M., Araujo Neto, R.B., Nascimento, M.P.S.B.C., Lima, P.S.C. 2014. RAPD analysis of the genetic diversity among accessions of Fabaceous forages (*Poincianella* spp.) from the Caatinga. Genet. Mol. Res., 13: 5832–5839.

Meneguetti, D.U.O., Lima, R.A., Silva, J.B., Silva, R.P., de Cassia Pagotto, R., Facundo, V.A. 2014. Análise citotóxica e mutagênica do extrato aquoso de *Maytenus guyanensis* Klotzsch ex Reissek (celastraceae) chichuá (xixuá) amazônico. Cien. Nat., 36: 301–309.

Milliken, W., Gasson, P., Pareyn, F., Sampaio, E.V.S.B., Lee, M., Baracat, A., Araújo, E.L., Cutler, D. 2018. Impact of management regime and frequency on the survival and productivity of four native tree species used for fuelwood and charcoal in the caatinga of northeast Brazil. Biomass Bioen., 116: 18–25.

Moraes, J.P., Pereira, D.S., Matos, A.S., Santana, D.G., Santos, C.A., Estevam, C.S., Fakhouri, R., Junior, W.L., Camargo, E.A. 2013. The ethanol extract of the inner bark of *Caesalpinia pyramidalis* (Tul.) reduces urinary bladder damage during cyclophosphamide-induced cystitis in rats. Sci. World J.

Moreira, B.O., Carvalho, A.L., Alves, C.Q., Morbeck, L.L.B., Cruz, M.P., Yatsuda, R., David, J.M. 2019. Evaluation of anti-inflammatory, antinociceptive and biological activities of *Cenostigma macrophyllum* standardized extracts and determination and quantification of the main metabolites. RSC Adv., 9(70): 41256–41268.

Nadeem, M.A., Nawaz, M.A., Shahid, M.Q., Doğan, Y., Comertpay, G., Yıldız, M., Hatipoğlu, R., Ahmad, F., Alsaleh, A., Labhane, N., Özkan, H., Chung, G., Baloch, F.S. 2018. DNA molecular markers in plant breeding: current status and recent advancements in genomic selection and genome editing. Biotechnol. Biotechnol. Equip., 32: 261–285.

Novais, T.S., Costa, J.F.O., David, J.P.L., David, J.M., Queiroz, L.P., França, F., Giulietti, A.M., Soares, M.B.P., Santos, R.R. 2003. Atividade antibacteriana em alguns extratos de vegetais do semi-árido brasileiro. Rev. Bras. Farmacogn., 13: 5–8.

Oliveira, F.C.S., Barros, R.F.M., Moita Neto, J.M. 2010. Plantas medicinais utilizadas em comunidades rurais de Oeiras, semiárido piauiense. Rev. Bras. Plant. Med., 12(3): 282–301.

Oliveira, J.C.S. 2010. Estudo químico e avaliação biológica do extrato das cascas das raízes de *Caesalpinia pyramidalis* Tul. (Leguminosae). Dissertação. Universidade Federal da Bahia, Programa de Pós-Graduação em Química. 2010.

Oliveira, J.C.S. de, David, J.M., David, J.P. 2016a. Composição química das cascas das raízes e flores de *Poincianella pyramidalis* (Fabaceae). Química Nov., 39: 189–193.

Oliveira, J.C.S.D., David, J.P., David, J.M. 2016b. Biflavonoids from the bark roots of *Poincianella pyramidalis* (Fabaceae). Phytochem. Lett., 16: 18–22.

Oliveira, L.M., Bruno, R.L.A., Silva, K.R.G., ALVES, E.U., Silva, G.Z., Andrade, A.P. 2011. Qualidade fisiológicas de sementes de *Caesalpinia pyramidalis* Tul. durante o armazenamento. Rev. Bras. de Sementes, 33(2): 289–298.

Pio-Corrêa, M. 1931. Dicionário das plantas úteis do Brasil e das exóticas cultivadas. Rio de Janeiro: Ministério da Agricultura. 2: 141–143.

Queiroz, L.P. 2010. New combinations in *Libidibia* (DC.) Schltdl. and *Poincianella* Britton & Rose (Leguminosae, Caesalpinioideae). Neodiversity, 5: 11–12.

Ramos, M.A., Medeiros, P.M., Almeida, A.L.S., Feliciano, A.L.P., Albuquerque, U.P. 2008. Can wood quality justify local preferences for firewood in an area of Caatinga (dryland) vegetation? Biomass Bioen., 32: 503–509.

Ramos, M.A., Albuquerque, U.P. 2012. The domestic use of firewood in rural communities of the Caatinga: How seasonality interferes with patterns of firewood collection. Biomass Bioen., 39: 147–158.

Reis, S.D.S., Oliveira, R.S., Marcelino, S.A.C., Macêdo, J.T.S.A., Riet-Correa, F., Pimentel, L.A., Pedroso, P.M.O. 2016. Congenital malformations and other reproductive losses in goats due to poisoning by *Poincianella pyramidallis* (Tul.) LP. Queiroz (*Caesalpineia pyramidalis* Tul.). Toxicon., 118: 91–94.

Ribeiro, A.R.S., Diniz, P.B.F., Estevam, C.S., Pinheiro, M.S., Albuquerque-Júnior, R.L.C., Thomazzi, S.M. 2013. Gastroprotective activity of the ethanol extract from the inner bark of *Caesalpinia pyramidalis* in rats. J. Ethnopharmacol., 147(2): 383–88.

Rietjens, I.M., Boersma, M.G., Van der Woude, H., Jeurissen, S.M., Schutte, M.E., Alink, G.M. 2005. Flavonoids and alkenylbenzenes: mechanisms of mutagenic action and carcinogenic risk. Mutat. Res., Fundam. Mol. Mech. Mutagen, 574: 124–138.

Rodrigues, P.S., Souza, M.M., Correa, R.X. 2014. Karyomorphology and karyotype asymmetry in the South American Caesalpinia species (Leguminosae and Caesalpinioideae). Genet. Mol. Res., 13: 8278–8293.

Rodrigues, P.S., Souza, M.M., Melo, C.A.F., Pereira, T.N.S., Correa, R.X. 2018. Karyotype diversity and 2C DNA content in species of the Caesalpinia group. BMC Genet., 19: 25.

Sá e Silva, I.M.M., Marangon, L.C., Hanazaki, N., Albuquerque, U.P. 2009). Use and knowledge of fuelwood in three rural caatinga (dryland) communities in NE Brazil. Environ. Dev. Sustain. 11, 833–851.

Sales, F.D.C.V., Bakke, O.A., Bakke, I.A., Souza, B.V., Ferreira, C.D., Bakke, E.A. 2019. How do native trees establish on degraded caatinga sites? J. Exp. Agric. Int., 32: 1–9.

Salvat, A., Antonacci, L., Fortunato, R.H., Suarez, E.Y., Godoy, H.M. 2004. Antimicrobial activity in methanolic extracts of several plant species from northern Argentina. Phytomedicine, 11: 230–234.

Sampaio, E.V.S.B., Rodal, M.J. 2002. Usos das plantas da caatinga. In: Sampaio EVSB, Giulietti AM, Virgínio J, Gamarra-Rojas CFL, Editors. Vegetação e flora da caatinga. Recife, Brazil: APNE – CNIP, p. 176–90.

Santos, J.P., Araújo, E.L., Albuquerque, U.P. 2008. Richness and distribution of useful woody plants in the semi-arid region of northeastern Brazil. J. Arid. Environ., 72: 652–663.

Santos, J.R.S., Lopes, J.R.G., Medeiros, M.A., Campos, E.M., Medeiros, R.M.T., Riet-Correa, F. 2018. Embryonic mortality and abortion in goats caused by ingestion of *Poincianella pyramidalis.* Braz. J. Vet. Res., 38: 1259–1263.

Santos, J.S., Mendes, S.S., Conde, D.C., Delmondez, R.C. 2012. Genetic diversity assessment of *Poincianella pyramidalis* (Tul.) L.P. Queiroz accessions using RAPD markers. Sci. Plena., 8: 1–8.

Saraiva, A.M., Saraiva, C.L., Gonçalves, A.M., Soares, R.R., Mendes, F.O., Cordeiro, R.P., Xavier, H.S., Pisciottano, M.N.C. 2012. Antimicrobial activity and bioautographic study of antistaphylococcal components from *Caesalpinia pyramidalis* Tull. Braz. J. Pharm. Sci., 48(1): 147–54.

Saraiva, M.E., Ulisses, A.V.R.A., Ribeiro, D.A., Oliveira, L.G.S., Macêdo, D.G., Sousa, F.F.S., Menezes, I.R.A., Sampaio, E.V.S.B., Souza, M.M.A. 2015. Plant species as a therapeutic resource in areas of the savanna in the state of Pernambuco, Northeast Brazil. J. Ethnopharmacol., 171(1): 141–53.

Silva, C.G., Marinho, M.G.V., Lucena, M.F.A., Costa, J.G.M. 2015a. Levantamento etnobotânico de plantas medicinais em área de Caatinga na comunidade do Sítio Nazaré, município de Milagres, Ceará, Brasil. Rev. Bras. Plantas Med., 17(1): 133–42.

Silva, C.H.T.P., Peixoto Sobrinho, T.J.S., Castro, V.T.N.A., Lima, D.C.A., Amorim, E.L.C. 2011. Antioxidant Capacity and Phenolic Content of *Caesalpinia pyramidalis* Tul. and *Sapium glandulosum* (L.) Morong from Northeastern Brazil. Molecules, 16: 4728–4739.

Silva, C.H.T.P., Sobrinho, T.J.S.P., Saraiva, A.M, Pisciottano, M.N.C., Amorim, E.L.C. 2012. Phytochemical profile and antibacterial activity of bark and leaves of *Caesalpinia pyramidalis* Tul. and *Sapium glandulosum* (L.) Morong. J. Med. Plants Res., 6(32): 4766–71.

Silva, D.L.S., Oliveira, K.P., Aroeira, L.J.M., Chaves, D. F., Ponciano, M.F., Braga, A.P., Júnior, D.M.L. 2015c. Chemical composition of caatinga potential forages species. Trop. Subtrop. Agroecosyst., 18: 267–272.

Silva, F.D.B., Sales, M.A.G., Sá, O.R.M. Santana, G.M., De Deus, M.S.M., Castro e Sousa, J.M., Ferreira, P.M.P., Peron, A.P. 2015b. Potencial citotóxico, genotóxico e citoprotetor de extratos aquosos de *Caesalpinia pyramidalis* Tul., *Caesalpinia ferrea* Mart. e *Caesalpinia pulcherrima* Sw. Braz. J. BioS., 13: 101–109.

Silva, L.B., Santos, F.A.R., Gasson, P., Cutler, D. 2009. Anatomia e densidade básica da madeira de *Caesalpinia pyramidalis* Tul. (Fabaceae), espécie endêmica da caatinga do Nordeste do Brasil. Acta Bot. Bras., 23: 436–445.

Silva, T., dos, S., Nepomuceno, C.F., Soares, T.L., Santana, J.R.F. de. 2019. *In vitro* conservation of *Poincianella pyramidalis* (Tul.) L.P. Queiroz under minimal growth conditions. Ciênc. Agrotec., 43.

Tabarelli, M., Leal, I.R., Scarano, F.R., Silva, J.M.C. 2018. Caatinga: legado, trajetória e desafios rumo à sustentabilidade. Cien. Cult., 70: 24–29.

Valle, A.L. 2018. Current methodologies in assessing toxicity of natural products. Int. J. Phytocos. Nat. Ingred., 5: 3.

Van den Berg, S.J., Restani, P., Boersma, M.G., Delmulle, L., Rietjens, I.M. 2011. Levels of genotoxic and carcinogenic ingredients in plant food supplements and associated risk assessment. Food Nutr. Sci., 2: 989–1010.

Chapter 12

Traditional Uses, Phytochemicals and Pharmacological Properties of *Chenopodium ambrosioides* L. (*Dysphania ambrosioides*) L. Mosyakin & Clemants

Sabrina Baleixo da Silva,[1,*] *Jhonatas Rodrigues Barbosa,*[2] *Luiza Helena da Silva Martins,*[3] *Mahendra Rai*[4] and *Alessandra Santos Lopes*[1]

Introduction

Chenopodium ambrosioides is a medicinal plant, found in countries with tropical, subtropical, temperate climate and some regions of the Mediterranean and Central America (Sá 2015). It is commonly known as mastruz, tea from Mexico. Traditionally, it is used against hookworm and roundworm (Castellanos 2008, Sérvio et al. 2011, Grassi et al. 2013). From a botanical point of view, it is considered an annual or short-lived perennial herb and its reproduction occurs by seeds (Dembitsky et al. 2008).

For many generations, people from various regions of the world have used extracts of mastruz in popular medicine in the treatment of various diseases, healing action, and fever, regulation of the menstrual cycle, repellent and especially against worms. In ancient cultures of the southern Mexican region, the mastruz was used for both food and medicinal purposes. Even today, in Mexico the plant is used as a form of food seasoning (Picó and Nuez 2000, Lorenzi and Matos 2002).

Ethnopharmacological studies from different countries point to the various properties of *C. ambrosioides* (Castellanos 2008, Napoli 2008, Silva et al. 2009, Gbolade et al. 2010, Chu et al. 2011, Ayoub et al. 2014, Silva 2015). Moreover, popular knowledge among countries is quite similar, especially in the treatment of intestinal infections and inflammations, therefore, this primitive knowledge was fundamental in advancing researches, allowing the identification of chemical compounds with bioactive potential (Grassi et al. 2013). In this context, the leaves of *C. ambrosioides* are among the most commercialized medicinal herbs at fairs and health food stores. The main compounds in the extracts of *C. ambrosioides* are essential oils rich in monoterpenes, hydrocarbons and tannins (Monzote et al. 2014a). Extractions with methanol and ethanol solvents were also used to recover alkaloids, flavonoids

[1] LABIOTEC/FEA (Faculty of Food Engineering) Graduate Program in Food Science and Technology Federal University of Pará Augusto Corrêa Avenue, Guamá, 66.075-900, Belém, Pará, Brazil.

[2] LABEX/FEA (Faculty of Food Engineering) Graduate Program in Food Science and Technology Federal University of Pará Augusto Corrêa Avenue, Guamá, 66.075-900, Belém, Pará, Brazil.

[3] ISPA (Institute of Animal Health and Production) Federal Rural University of Amazonia Avenue Presidente Tancredo Neves, 2501, Terra Firme, 66.077-830, Belém, Pará, Brazil.

[4] Department of Biotechnology, SGB Amravati University Amravati 444602, Maharashtra, India.

* Corresponding author: sabrinabaleixo@gmail.com

and terpenes. Several of these chemical compounds were elucidated, leading to new possibilities for research with *C. ambrosioides*, as tests to assess the bioactive potential, as well as applications in food area (Zohra et al. 2019).

The bioactive potential of various extracts and essential oils from the mastruz has already been evaluated (Cavalli et al. 2004, Nascimento et al. 2006, Prasad et al. 2010, Monzote et al. 2014b, Zohra et al. 2018, Pavela et al. 2018, Pizzolitto et al. 2020, Khezri et al. 2020). Several bioactivities were reported such as antitumor activity in an *in vivo* model with Ehrlich tumor (Nascimento et al. 2006), Allelopathic activity (Jiménez-Osornio et al. 1996), direct action on mitochondria (Monzote et al. 2009), anti-inflammatory, anti-nociceptive and curative effects (Grassi et al. 2013) among others. Further studies may transform the potential into a pharmacological reality in the future, bringing ethnopharmacological knowledge to the level of scientific knowledge and subsequently the production and commercialization of drugs based on extracts of *C. ambrosioides*.

The purpose of this chapter is to show the different aspects of *C. ambrosioides* including the development and evolution of knowledge about botanical, systematic, phytogeographical, ethnopharmacological, phytochemical and biological applications. This chapter was structured starting with the systematic approach and botanical classification, covering topics related to phytogeography and phylogeny of the species. Traditional uses and ethnopharmacological aspects are also reported, with an emphasis on traditional knowledge and its relationship with the foundations of their cultures. The phytochemical approach describes the main chemical compounds and their relationship with the techniques of extraction and geographical distribution. Finally, a description of the main bioactivities is reported. Specific topics are used to emphasize bioactivities that are more prominent.

Systematics

The genus *Chenopodium* described by Linnaeus has more than two hundred native species with different structural, morphological and medicinal properties. *C. ambrosioides* belongs to the Kingdom: Plantae, Division: *Magnoliophyta*, Class: *Magnoliopsida*, Order: *Caryophyllales* and Family: *Amaranthaceae*. Some of these species have high similarity; so before consumption, it is important to be sure of their origin and to know their main characteristics (Winsor 2001).

Among the various spp. of *Chenopodium*, some species are more often used than others, mainly in the form of food, for example *C. quinoa* (Romano et al. 2019) and *C. pallidicaule*. These species are rich in essential amino acids, and thus indispensable for a healthy diet (Ayala 1998, Repo Carrasco et al. 2003, Ishimoto et al. 2010, Farinazzi et al. 2012). *C. ambrosioides* is known worldwide for its medicinal properties; however, in some cultures, the medicinal herb can be found in gastronomic recipes, as well as applied for medicinal purposes (Castellanos 2008, Sá 2015). The study of systematics contributes to a better understanding of the genus.

Phytogeography of *C. ambrosioides*

Phytogeography comprises a set of botanical concepts that involve knowledge about the plants' origin, specific taxa based on its evolutionary history, association possibilities, climatic and physiological aspects of the plants' evolution and adaptation to its natural environment. The phytogeographic study is important because it allows knowing the characteristics of the plant and the influence of different geographical distributions on its development (Zhou and Sun 2016). The genus *C. ambrosioides* is distributed across the globe, while the species *C. ambrosioides* is found frequently in regions of tropical, subtropical climate, such as in America, Africa, Mediterranean regions and central Europe (Bieski et al. 2015, Jesus et al. 2018).

C. ambrosioides is originally from Central America in Mexico, and distributed even in regions with warmer weather. It is considered a weed for its spontaneous growth and its exotic characteristics. The plant is distributed around the world through natural dispersers such as birds, insects, and displacement of sea currents, wind action, in addition to artificial dispersion caused by man. Medium texture, fertile soils with partial moisture are the ones that most contribute to the vegetative development of the plant. As it is a small plant, vegetative development is faster even in adverse conditions (Dembitsky et al. 2008).

Basic Morphology and Phylogenetic Aspects

C. ambrosioides (Figure 12.1) is an annual or short-lived perennial herb with an herbaceous stem, malleable without the presence of lignin, covered by small fluff and reddish-brown color. The leaves are lanceolate, alternate, irregularly toothed, light green or yellowish-green. The secondary flowering branches are found in the leaf axils of the main vegetative branches, whereas the flowers are greenish-yellow, hermaphrodite, and found in the upper part of the leaves.

The seeds are black, light, rich in essential oil and responsible for the reproduction of the plant (Gadano et al. 2006). The plant has a strong smell, not attractive, due to the presence of organic compounds such as aromatic hydrocarbons, tannins and some phenolic, responsible for protecting the plant against herbivorous predators (Winsor 2001).

Phylogenetic researches showed the molecular similarity of the species *C. ambrosioides* with the genus *Dysphania* spp., due to the validation of taxonomic combinations of the different lineages between the genera. Therefore, there were additional transfers and expansion of the circumscription of some species of the genus Chenopodium spp., for *Dysphania* spp. such as *C. melanocarpa*, *C. truncata*, *C. saxatilis*, *C. andícola* and *C. bothioide* (Mosyakin 1993, 1996, Clemants and Mosyakin 2003, Mosyakin and Clemants 2008, Fuentes-Bazan et al. 2012).

Thus, *C. ambrosioid* can also be descri*bed as Dysphania ambrosioides* (L.) Mosyakin & Clemants (sin: *C. ambrosioides* L.). The circumscription of the name is recent; therefore, few researches are found in the literature with the term *Dysphania ambrosioides*.

Figure 12.1. *Chenopodium ambrosioides* (L.) by Trivellato et al. (2013), with permission from the Journal of Ethnopharmacology.

Traditional Knowledge and Ethnopharmacology

Traditional knowledge is an important form of subsistence for human civilization, from ancient times to the present. Traditional knowledge involves specific knowledge and techniques for hunting, agriculture, traditional medicine, among others, which were developed empirically by communities and are perpetuated throughout the generations. The knowledge about medicinal plants is of fundamental importance in the advancement of science, as it enables the discovery of plants with bioactivity. The knowledge accumulated for generations can be used in the production of new drugs, cosmetics and ingredients for the food industry. In this context, it is important to address the ethnopharmacological study of medicinal plants, as it enables the advance of new, renewable and sustainable technologies (Heywood 2011).

C. ambrosioides is known worldwide for its different uses and applications, mainly pharmacological. It has different popular names according to their physical characteristics and functions. Because of its therapeutic effects, the World Health Organization (WHO) considers it as one of the most used medicinal plants in the world (Lorenzi and Matos 2002), proven in studies of *in vitro* and *in vivo* tests. It is among the 71 medicinal plants listed by the Brazilian Ministry of Health for interest in research and pharmacological applications (Brasil 2006).

The Pharmacological Uses of *C. ambrosioides*

It had its first account in the ancient civilization of the pre-Columbian Mayans of Yucatan, a region in the current South of Mexico. The plant known as lucumxiu (worm plant) was used to eliminate intestinal worms; with the passing of the years, it gained other names such as: ambrosia from Mexico, tea from Mexico, Santa Maria herb and tea from worms. This plant resource has influenced the development of traditions and forms of use over time. In Mexico, unlike other countries, in addition to using *C. ambrosioides* in popular medicine, its consumption was included in the gastronomical culture of ancient Precolombian base, in the form of condiments for the preparation of snacks such as cheese quesadilla with spazote (Picó and Nuez 2000).

The different uses of the mastruz influence different cultures across the globe. Different cultures have specialized in the most varied applications for the mastruz; among these cultures, we can observe specialties in common, indicating that the application of the mast for pharmacological activities has broad ethnopharmacological support (Picó and Nuez 2000).

In Brazil, the plant is known mainly as the mastruz and erva-de-santa-maria, and has an extensive distribution throughout the country because of its favorable biomes. Religious groups, rural and indigenous population and quilombola communities have used it, through the consumption of grass mixed with milk or infusion of leaves (Pinto et al. 2006). According to traditional knowledge, the plant acts mainly in the treatment of worms, against skin disorders, edema, digestive processes, gastritis, menstrual cramps, and stomachaches among others (Di Stasi and Hiruma-Lima 2002).

In Europe in the 18th century, between 1715 and 1779, the Native Americans, due to its property of eliminating intestinal worms, highlighting its anthelmintic property (Kliks 1985), knew the plant. In western Spain and northeastern Portugal, the leaves known as Spanish tea are dried and traditionally prepared by infusion, and are used to assist in the treatment of digestive disorders, treatment for eczema, febrifuge, anthelmintic and expectorant (Muntane 1994).

Asian countries are known culturally for the intense use of medicinal plants for herbal purposes. *C. ambrosioides* is traditionally recognized in China and Western India by the popular name of wormseed, Mexican tea, and epazote. Consumed in the form of tea, it is traditionally used against worms and parasites, intestinal diseases and to treat rheumatism (Yang and Tang 1998, Prasad et al. 2010).

C. ambrosioides has a high potential for pharmacological advancement, since the plant can be easily grown in different biomes and is inserted in different cultures all over the world. Furthermore, the ethnopharmacological knowledge of the mastruz shows that societies separated by time, space and different cultures such as language, traditions, cuisine, and religion have similar popular knowledge of traditional medicine, antimicrobial, antifungal, antirheumatic, antioxidant and anti-inflammatory and expectorant that were perpetuated for generations and have scientific proof based on detailed studies of their phytochemical profile (Sérvio et al. 2011, Grassi et al. 2013).

Phytochemical Profile and Value-added Compounds

The phytochemical profile of *C. ambrosioides* was explored by various analytical techniques such as gas chromatography, liquid chromatography, and arrangements of these technologies with different detectors. In recent years, it was demonstrated that essential oil and hydroalcoholic extracts have different chemical constituents. In addition, due to the geographical distribution of the species around the world, some chemical compounds can be found more frequently (Chu et al. 2011).

Cavalli et al. (2004) demonstrated, by GC, GC-MS and ^{13}C-NMR spectroscopy, that the essential oil from *C. ambrosioides* of Madagascar has several chemical compounds; among them, ascaridole was found as the main constituent. In addition, other compounds such as isoascaridole, p-cymene, α-terpinene and limonene are also present. Chu et al. (2011) and Monzote et al. (2014) reported similar results. However, Harraz et al. (2014) extracted the essential oil of *C. ambrosioides* by hydrodistillation and analyzed it by GC/MS. The authors identified 12 volatile constituents. The main constituent identified was o-cymene, followed by α-terpinene and monoterpenes. As can be seen through the results of the study, the regional difference and the extraction method interfere in the composition result and in the

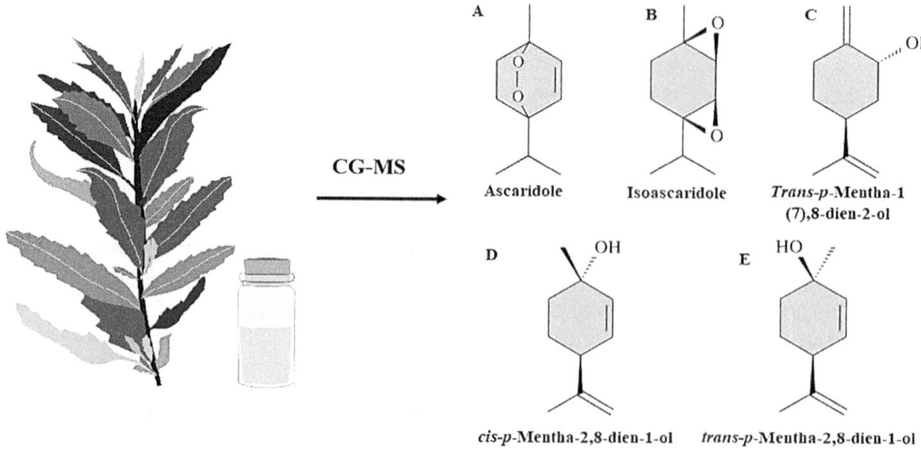

Figure 12.2. Graphic summary shows the chemical structures of the main compounds obtained from the essential oil of *Chenopodium ambrosioides* (L.).

main components obtained from *C. ambrosioides*. Figure 12.2 shows a summary of the extraction of essential oils and the chemical structures of the main chemical components obtained from the essential oil of *C. ambrosioides*.

Adejumo et al. (2011) obtained methanol extracts of *C. ambrosioides* by solid/liquid partition for seven days. Various compounds were found such as alkaloids, tannins, saponins, and anthraquinones. Shah and Khan (2017) isolated and quantified β-sitosterol, octadecanoic acid, scopoletin, and 1-piperoylpiperidine from *C. ambrosioides*.

Hou et al. (2017), using an ethyl alcohol extract, isolated and identified polyol monoterpenes. Two new polyol monoterpenes were identified: 4-hydroxy-4 (α or β)-isopropyl-2-methyl-2cyclohexen-1-one. In addition to these, other terpenes have been identified such as 1-methyl-4β-isopropyl-1-cyclohexene4α, 5α, 6α-triol, (1S, 2S, 3R, 4S)-1-methyl-4-(propan-2-yl) cyclohexane-1,2,3,4-tetrol, (1R, 2S, 3S, 4S)-1,2,3,4-tetrahydroxy-p-menthane, (1R, 2S)-3-pmenthen-1,2-diol, (1R, 4S)-p-menth-2-en-1-ol and 1,4-dihydroxyp-menth-2-ene.

In the most recent work, Zohra et al. (2018), using 18 solvents with different polarities, optimized the extraction of total phenolic and flavonoid compounds. HPLC analysis exhibited the presence of rutin, myricetin, and quercetin. The authors consider that the difference in solvent polarity is essential for the selection of phenolic and flavonoid compounds. Figure 12.3 shows a summary of the extraction

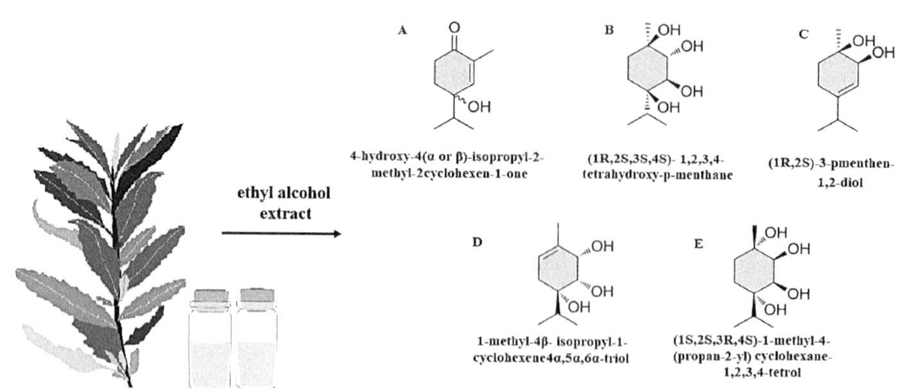

Figure 12.3. Graphic summary shows the chemical structures of the main compounds obtained in from the ethanol extract of *Chenopodium ambrosioides* (L.).

of secondary compounds and the chemical structures of the main chemical components obtained from *C. ambrosioides*.

Pharmacological Properties

The presence of organic compounds, such as essential oils and phenolic compounds present in *C. ambrosioides*, gives the plant some medicinal properties. Studies show that *C. ambrosioides* extracts have activity against Leishmania (Monzote et al. 2014), *in vitro* antitumor activity and *in vivo* model against Ehrlich tumor (Nascimento et al. 2006), allelopathic activity (Jiménez-Osornio et al. 1996), direct action on mitochondria (Monzote et al. 2009), nematicides (Bai et al. 2011), insecticidal activity (Chu et al. 2011), antifungals (Prasad et al. 2010), anti-inflammatories, antinociceptives and healing effects (Grassi et al. 2013).

C. ambrosioides as a Potential Source of Antimicrobial, Anti-aflatoxigenic and Antioxidant Activity

The use of natural compounds obtained from many resources such as plants is considered an alternative, promising and ecological approach for the control of pests such as fungi, bacteria, algae, and viruses. Some substances, such as essential oils (OEs), can act promptly, breaking the growth cycle of some microorganisms and preventing the production of toxic chemical compounds such as aflatoxins. In addition, the presence of organic compounds such as alkaloids and flavonoids linked to fatty acids, produce important biological effects as antioxidant activities (Pizzolitto et al. 2020). As shown in the summary (Figure 12.4), essential oils (OEs) can act to break the cycle of growth and production of enzymes in microorganisms.

The antimicrobial effect of essential oils (OEs) were tracked in a wide range of microorganisms, as reported in several recent studies (Gavahian et al. 2020, Zhang et al. 2020, Siddique et al. 2020). Several mechanisms were proposed; however, some relevant aspects were not properly understood. Essential oils (OEs) are a complex mixture of organic compounds, whose action is hard to be understood only through studies with isolated substances. The essential oils (OEs) have molecular complexes that are difficult to isolate and synthesize, and therefore the separate study of these compounds is not feasible (Khezri et al. 2020).

The antimicrobial action of essential oils were addressed in recent reports (Debonne et al. 2018, Tariq et al. 2019, Kharbach et al. 2020). These authors consider that the antimicrobial action may be related to the ability of essential oils to penetrate the cell membrane leading to inhibitory activities and important pathways of microbial metabolisms. The nature of innumerable phenolic compounds present in essential

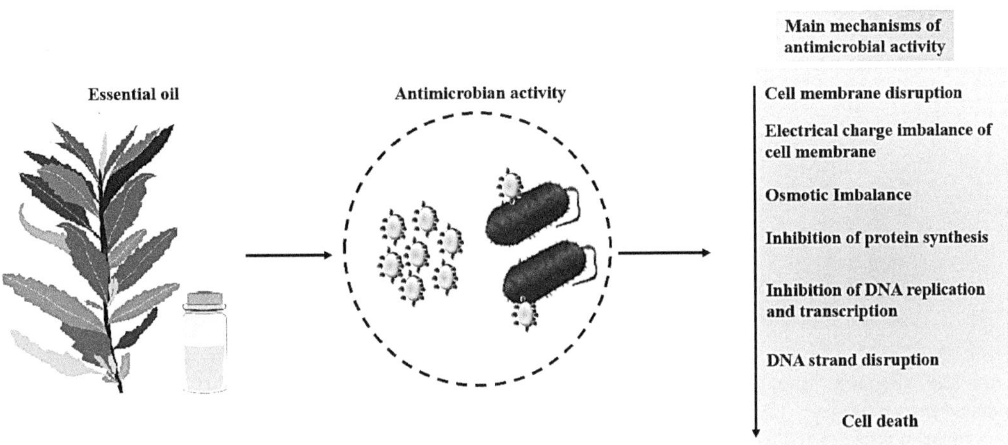

Figure 12.4. Graphic summary of the action of essential oils (OEs) with antimicrobial activity.

oils also acts as disturbers of the electrical components of the cell membrane, leading to the rupture and leakage of cytoplasm compounds. In addition, these compounds can interfere with the energy generation system (ATP), essential in the production of energy responsible for all metabolic routes. The action of these compounds lead to the arrest of the cell cycle, the shutdown of defense systems, inactivation of the mechanisms of enzymatic production and cell death (Calvo-Irabien 2018).

Kumar et al. (2007) extracted essential oil from the leaves of *C. ambrosioides* and evaluated the potential for antifungal, antiaflatoxigenic and antioxidant activity. The results show that 100 μg/ml of essential oil inhibited the mycelial growth of various fungi such as *Aspergillus niger*, *Aspergillus fumigatus*, *Botryodiplodia theobromae*, *Fusarium oxysporum*, *Sclerotium rolfsii*, *Macrophomina phaseolin*a, *Cladosporium* cladosporioides and helminth. In addition, the extract has antioxidant activity in models such as ABTS (2, 2'-azino-bis (3-ethylbenzothiazoline-6-sulfonic acid)). The inhibition of aflatoxin B1 production by the aflatoxigenic strain of *Aspergillus flavus* was also observed.

Zerrifi et al. (2020), while studying essential oil from several Moroccan plants, also studied *C. ambrosioides* and the anti-cyanobacterial action of its essential oil. Several chemical compounds were identified in the oil, such as 5-3-carene, β-cymene, 3,4-diepoxy-β-mentane and 1,4-epoxy-β-menth-2-ene. The essential oil inhibited the growth of the single-celled cyanobacterium *Microcystis aeruginosa*. The authors concluded that *C. ambrosioides* is a botanical platform, ecologically appropriate to be used as an alternative antimicrobial agent.

Larvicidal and Repellent Potential of *C. ambrosioides* Essential Oil

Many researchers (Khader et al. 2018, Agyare et al. 2018, Chaachouay et al. 2019, Verma and Singh 2020) explored the larvicidal potential of plants of ethnopharmacological relevance. Recently, several studies with different plants (Benelli et al. 2020, Jesser et al. 2020, Baccari et al. 2020) demonstrated that extracts and essential oil of *C. ambrosioides* have larvicidal activity against various disease-transmitting insects. The potential of plant extracts as larvicides was the subject of study for some years. The ethnopharmacological relevance of some plants attracted the attention of researchers, who explored the details of popular culture and discovered new organic compounds with industrial potential. Figure 12.5 explains the larvicidal potential of *C. ambrosioides* extracts.

In this scenario, the *C. ambrosioides* was also explored for its larvicidal potential. Bigoga et al. (2013) demonstrated that concentrations of 200 ppm and 300 ppm increase the mortality of Anopheles gambiae mosquito larvae. In addition, the main compounds found in essential oils were alpha-terpinene and para-cymene. The traditional use of *C. ambrosioides* oil to repel insects was demonstrated, although further studies are needed to improve the potential for oil release in a controlled manner.

In the study described by Pavela et al. (2018), several secondary organic compounds present in the essential oil of the species *Dysphania ambrosioides* were extracted and characterized by gas chromatography and mass spectrometry. In addition, the larvicidal potential was evaluated in relation

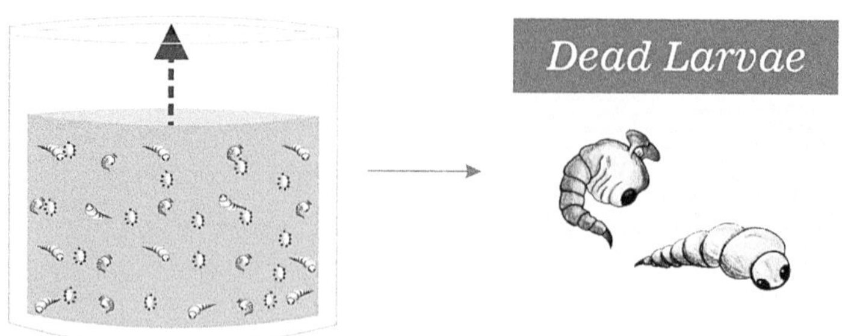

Figure 12.5. Graphical summary of the larvicidal potential of *Chenopodium ambrosioides* (L.) extracts. In step (1), *Chenopodium ambrosioides* (L.) essential oil is obtained. In step (2), the essential oil rich in compounds such as terpenes and hydrocarbons is used to kill mosquito larvae. In step (3), due to the cytotoxic effect of the essential oil, the mosquito larvae are killed.

to the filariasis vector, *Culex quinquefasciatus*. The results showed that the essential oil has high concentrations of monoterpene peroxide ascaridol and β-chimena, with potential larvicide, mainly by activating the inhibition system of the enzyme acetylcholinesterase.

Antimalarial Potential of *C. ambrosioides* Leaves

Malaria is a parasitic disease, transmitted by more than 120 species of *Plasmodium* spp., spread over more than 91 tropical countries. Of the 120 species of *Plasmodium*, only 6 are recognized for infecting humans (Ashley et al. 2014). In recent years, the number of malaria cases has been varied, with epidemiological outbreaks in some places and setbacks in others, as shown in recent studies (Kim et al. 2019, Okunlola and Oyeyemi 2019, Eijk et al. 2019, Kibret et al. 2019, Kagaya et al. 2019).

From an epidemiological point of view, malaria is transmitted by the female mosquito *Anopheles* spp., distributed in regions of tropical and subtropical climate. The progress and proliferation of the disease depend on environmental suitability, such as altitude, climate, vegetation, as well as public sanitation, poverty, natural disasters, and wars. In addition, the movement of contaminated people to other places helps in the spread of the parasite (O'Brien et al. 2015).

Different plants were used for the extraction of bioactive compounds, in a joint effort by researchers from different areas to identify new therapeutic alternatives for the treatment of malaria. Thus, in the study by Cysne et al. (2016), crude hydro alcoholic extract (HCE) from the leaves of *C. ambrosioides* was used in biological tests as therapeutic alternatives for the treatment of malaria. Using surface plasmon resonance (SPR), the molecular affinity between the total proteins of erythrocytes infected with *Plasmodium falciparum* and the crude hydro alcoholic extract was evaluated. In addition, in an *in vivo* study, the antimalarial potential was evaluated, considering the survival rate and parasitemia. The

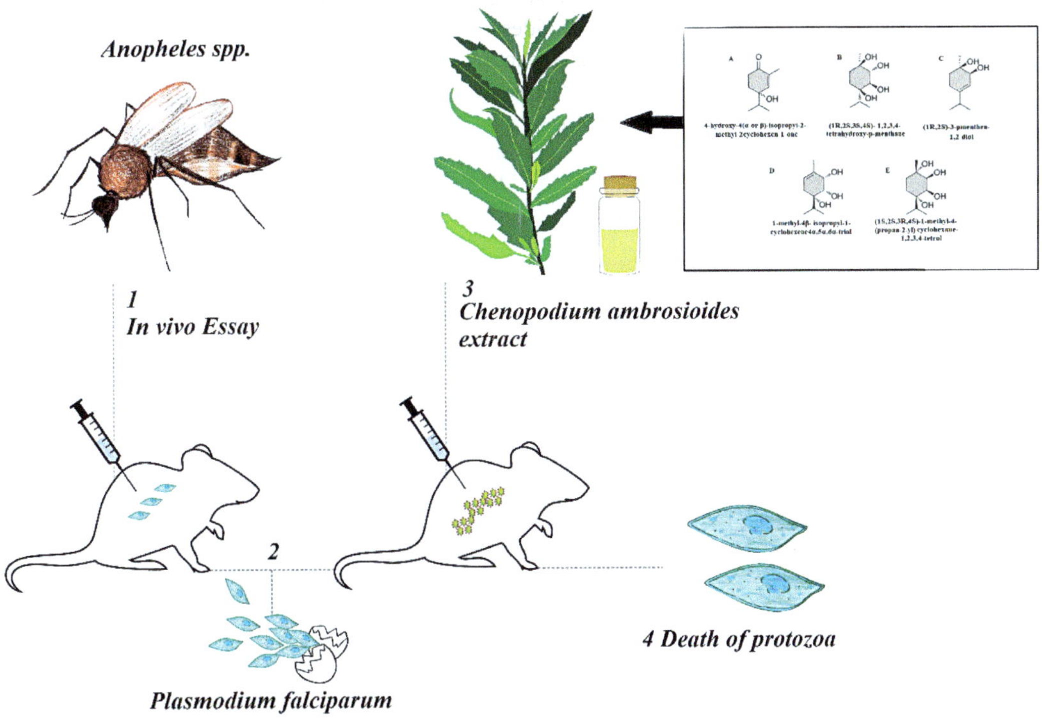

Figure 12.6. Summary of the antimalarial activity of *Chenopodium ambrosioides* (L.) extracts. In step (1), the protozoan *Plasmodium falciparum* is applied to rats, in the *in vivo* activity model. In step (2), the protozoan breaks the spores and multiplies. In step (3), *Chenopodium ambrosioides* (L.) extract is applied to the rats, to assess the antimalarial effect. In step (4), the chemical compounds present in the extract inhibit the metabolic development of the protozoan leading to death of protozoa.

results show that the crude hydro alcoholic extract can bind to the total proteins of infected erythrocytes and inhibit the growth of parasites in the *in vitro* assay, while *in vivo* treatment increased the rate of cell survival and decreased parasitemia in infected animals. *C. ambrosioides* extracts have antimalarial activity (Figure 12.6).

Conclusion

C. ambrosioides or *Dysphania ambrosioides* (L.) Mosyakin & Clemants is a medicinal plant with a social impact, which is found in several places around the world. In the present chapter, the discussion on the classification and systematics of the species shows that from a phylogenetic point of view, the species can be correlated with other genera.

The phytochemical study of the species revealed a complex set of chemical structures, such as monoterpenes, alkaloids, flavonoids and hydrocarbons. Finally, several bioactivities of *C. ambrosioides* were reported, with an emphasis on antimicrobial, repellent and antimalarial activities. The potential of the mastruz goes beyond the line of pharmacological research, as in the future new studies could be carried out to explore extracts for encapsulation, impregnation, and production of nanocomposites and nanoparticles.

Acknowledgement

All authors acknowledge the institutions Federal University of Pará (UFPA), Federal Rural University of Amazonia (UFRA), and SGB Amravati University for supporting research and for the access to didactic materials and scientific papers. We also acknowledge the designer Wendy Gonçalves Francês for the work on illustrations.

References

Adejumo, O.E., Owa-Agbanah, I.S., Kolapo, A.L., Ayoola, M.D. 2011. Phytochemical and antisickling activities of *Entandrophragma utile*, *Chenopodium ambrosioides* and *Petiveria alliacea*. J. Med. Plants Res., 5: 1531–1535.

Agyare, C., Spiegler, V., Asase, A., Scholz, M., Hempel, G., Hensel, A. 2018. An ethnopharmacological survey of medicinal plants traditionally used for cancer treatment in the Ashanti region, Ghana. J. Ethnopharmacol., 212: 137–152.

Ashley, E.A., Dhorda, M., Fairhurst, R.M., Amaratunga, C., Lim, P., Suon, S., Sopha, C. 2014. Spread of artemisinin resistance in *Plasmodium falciparum* malaria. N. Engl. J. Med., 371: 411–423.

Ayala, G. 1998. Consumption of quinoa (*Chenopodium quinoa*), kiwicha (*Amaranthus caudatus*) and tarwi (*Lupinus mutabilis*) and strategies to promote its consumption. Technical Meeting and Formulation Workshop of the Regional Project on Human Production and Nutrition based on Andean Crops. Arequipa, 1: 115–122.

Ayoub, S., Hamed, A.E., Osman, S. 2014. Estudo etnofarmacológico de um remédio herbal usado na medicina popular sudanesa para tratamento da doença de herpes zoster. Am. J. Res. Commun., 2: 1–11.

Baccari, W., Znati, M., Zardi-Bergaoui, A., Chaieb, I., Flamini, G., Ascrizzi, R., Jannet, H.B. 2020. Composition and insecticide potential against *Tribolium castaneum* of the fractionated essential oil from the flowers of the Tunisian endemic plant *Ferula tunetana* Pomel ex Batt. Ind Crops Prod., 143: 1–7.

Bai, C.Q., Liu, Z.L., Liu, Q.Z. 2011. Nematicidal constituents from the essential oil of *Chenopodium ambrosioides* aerial parts. J. Chem., 8: 143–148.

Benelli, G., Pavela, R., Rakotosaona, R., Nzekoue, F.K., Canale, A., Nicoletti, M., Maggi, F. 2020. Insecticidal and mosquito repellent efficacy of the essential oils from stem bark and wood of *Hazomalania voyroni*. J. Ethnopharmacol., 248: 1–19.

Bieski, I.G.C., Leonti, M., Arnason, J.T., Ferrier, J., Rapinski, M., Violante, I.M.P., Silva, D.R. 2015. Ethnobotanical study of medicinal plants by population of valley of Juruena region, legal Amazon, Mato Grosso, Brazil. J. Ethnopharmacol., 173: 383–423.

Bigoga, J.D., Saahkem, P.A., Ndindeng, S.A., Ngondi, J.L., Nyegue, M., Oben, J.E., Leke, R.G. 2013. Larvicidal and repellent potential of *Chenopodium ambrosioides* Linn essential oil against *Anopheles gambiae* Giles (Diptera: Culicidae). Open Entomol. J., 7: 16–22.

Brazil. 2006. Ministry of Health. Phytotherapy at SUS and the Medicinal Plant Research Program at the Medicines Center. Brasília: Ministry of Health.

Calvo-Irabien, L.M. 2018. Native Mexican aromatic flora and essential oils: Current research status, gaps in knowledge and agro-industrial potential. Ind. Crops Prod., 111: 807–822.

Castellanos, J.R.G. 2008. Epazote *Chenopodium ambrosioides*. Revisión a sus características morfológicas, actividad farmacológica, y biogénesis de su principal principio activo, ascaridol. Bol. Latinoam. Caribe Plantas Med. Aromát., 7: 3–9.

Cavalli, J.F., Tomi, F., Bernardini, A.F., Casanova, J. 2004. Combined analysis of the essential oil of *Chenopodium ambrosioides* by GC, GC-MS and 13C-NMR spectroscopy: quantitative determination of ascaridole, a heat-sensitive compound. Phyto Anal. Tec., 15: 275–279.

Chaachouay, N., Benkhnigue, O., Fadli, M., El Ibaoui, H., Zidane, L. 2019. Ethnobotanical and ethnopharmacological studies of medicinal and aromatic plants used in the treatment of metabolic diseases in the Moroccan Rif. Heliyon., 5: 1–9.

Chu, S.S., Hu, J.F., Liu, Z.L. 2011. Composition of essential oil of Chinese *Chenopodium ambrosioides* and insecticidal activity against maize weevil, *Sitophilus zeamais*. Pest Manag. Sci., 67: 714–718.

Clemants, S.E., Mosyakin, S.L. 2003. Flora of North America Editorial Committee. eds. *Dysphania*. North Am. Flora, 4: 267–275.

Cysne, D.N., Fortes, T.S., Reis, A.S., de Paulo Ribeiro, B., dos, S., Ferreira, A., do Amaral, F.M.M., Nascimento, F.R.F. 2016. Antimalarial potential of leaves of *Chenopodium ambrosioides* L. Parasitol. Res., 115: 4327–4334.

Debonne, E., Van Bockstaele, F., Samapundo, S., Eeckhout, M., Devlieghere, F. 2018. The use of essential oils as natural antifungal preservatives in bread products. J. Essent. Oil Res., 30: 309–318.

Dembitsky, V., Shkrob, I., Hanus, L.O. 2008. Ascaridole and related peroxides from the genus *Chenopodium*. Biomed. Pap. Med. Fac. Univ. Palacky Olomouc., 152: 29–33.

Di Stasi, L.C., Hiruma-Lima, C.A. 2002. Plantas medicinais na Amazônia e na Mata Atlântica. Editora Unesp., 2: 323–330.

Eijk, V.A.M., Sutton, P.L., Ramanathapuram, L., Sullivan, S.A., Kanagaraj, D., Priya, G.S.L., Wassmer, S.C. 2019. The burden of submicroscopic and asymptomatic malaria in India revealed from epidemiology studies at three varied transmission sites in India. Sci. Rep., 9: 1–11.

Farinazzi-Machado, F.M.V., Barbalho, S.M., Oshiiwa, M., Goulart, R., Pessan Junior, O. 2012. Use of cereal bars with quinoa (*Chenopodium quinoa* W.) to reduce risk factors related to cardiovascular diseases. Food Sci. Technol., 32: 239–244.

Fuentes-Bazan, S., Mansion, G., Borsch, T. 2012. Towards a species level tree of the globally diverse genus *Chenopodium* (*Chenopodiaceae*). Mol. Phylogenet. Evol., 621: 359–374.

Gadano, A.B., Gurni, A.A., Carballo, M.A. 2006. Argentine folk medicine: genotoxic effects of *Chenopodiaceae* family. J. Ethnopharmacol., 103: 246–251.

Gavahian, M., Chu, Y.H., Lorenzo, J.M., Mousavi Khaneghah, A., Barba, F.J. 2020. Essential oils as natural preservatives for bakery products: Understanding the mechanisms of action, recent findings, and applications. Crit. Rev. Food Sci. Nutr., 60: 310–321.

Gbolade, A.A., Tira-Picos, V., Nogueria, J.M.F. 2010. Chemical constituents of *Chenopodium ambrosioides* L. var anthelminticum herb essential oil from Nigeria. Chem. Nat. Comp., 45: 654–655.

Grassi, T.L., Malheiros, A.M.S.C., da Silva Buss, Z., Monguilhott, E.D., Frode, T.S., de Souza, M.M. 2013. From popular use to pharmacological validation: a study of the anti-inflammatory, anti-nociceptive and healing effects of *Chenopodium ambrosioides* extract. J. Ethnopharmacol., 145: 127–138.

Harraz, F.M., Hammoda, H.M., El Ghazouly, M.G., Farag, M.A., El-Aswad, A.F., Bassam, S.M. 2014. Chemical composition, antimicrobial and insecticidal activities of the essential oils of Conyza linifolia and *Chenopodium ambrosioides*. Nat. Prod. Res., 29: 879–882.

Heywood, V.H. 2011. Ethnopharmacology, food production, nutrition and biodiversity conservation: towards a sustainable future for indigenous peoples. J. Ethnopharmacol., 137: 1–15.

Hou, S.Q., Li, Y.H., Huang, X.Z., Li, R., Lu, H., Tian, K., Li, Y.K. 2017. Polyol monoterpenes isolated from *Chenopodium ambrosioides*. Nat. Prod. Res., 31: 2467–2472.

Jesser, E., Lorenzetti, A.S., Yeguerman, C., Murray, A.P., Domini, C., Werdin-González, J.O. 2020. Ultrasound assisted formation of essential oil nanoemulsions: Emerging alternative for *Culex pipiens Say* (*Diptera: Culicidae*) and *Plodia interpunctella hubner* (*Lepidoptera: Pyralidae*) management. Ultrason. Sonochem., 61: 1–33.

Jesus, R.S., Piana, M., Freitas, R.B., Brum, T.F., Alves, C.F., Belke, B.V., Bianchini, B.V. 2018. *In vitro* antimicrobial and antimycobacterial activity and HPLC-DAD screening of phenolics from *Chenopodium ambrosioides* L. Braz. J. Microbiol., 49: 296–302.

Jiménez-Osornio, F.J., Kumamoto, J., Wasser, C. 1996. Allelopathic activity of *Chenopodium ambrosioides* L. Biochem. Syst. Ecol., 24: 195–205.

Kagaya, W., Gitaka, J., Chan, C.W., Kongere, J., Idris, Z.M., Deng, C., Kaneko, A. 2019. Malaria resurgence after significant reduction by mass drug administration on Ngodhe Island, Kenya. Sci. Rep., 9: 1–11.

Kharbach, M., Marmouzi, I., El Jemli, M., Bouklouze, A., Vander Heyden, Y. 2019. Recent advances in untargeted and targeted approaches applied in herbal-extracts and essential-oils fingerprinting—A review. J. Pharmaceut. Biomed., 177: 1–52.

Khader, S.Z.A., Ahmed, S.S.Z., Sathyan, J., Mahboob, M.R., Venkatesh, K.P., Ramesh, K. 2018. A comparative study on larvicidal potential of selected medicinal plants over green synthesized silver nano particles. Egypt J. Basic Appl. Sci., 5: 54–62.

Khezri, K., Farahpou, M.R., Rad, S.M. 2020. Efficacy of Mentha pulegium essential oil encapsulated into nanostructured lipid carriers as an *in vitro* antibacterial and infected wound healing agent. Colloids Surf. A Physicochem. Eng. Asp., 59: 1–9.

Kibret, S., Ryder, D., Wilson, G.G., Kumar, L. 2019. Modeling reservoir management for malaria control in ethiopia. Sci. Rep., 9: 1–11.

Kim, Y., Ratnam, J.V., Doi, T., Morioka, Y., Behera, S., Tsuzuki, A., Imai, C.C. 2019. Malaria predictions based on seasonal climate forecasts in South Africa: A time series distributed lag nonlinear model. Sci. Rep., 9: 1–10.

Kliks, M.M. 1985. Studies on the traditional herbal anthelmintic *Chenopodium ambrosioides* L.: ethnopharmacological evaluation and clinical field trials. Soc. Sci. Med., 21: 879–886.

Kumar, R., Mishra, A.K., Dubey, N.K., Tripathi, Y.B. 2007. Evaluation of *Chenopodium ambrosioides* oil as a potential source of antifungal, antiaflatoxigenic and antioxidant activity. Int. J. Food Microbiol., 115: 159–164.

Lorenzi, H., Matos, F.J.A. 2002. Plantas medicinais no Brasil: nativas e exóticas. 2.ed. Nova Odessa—SP: Instituto Plantarum de Estudos da Flora 542.

Monzote, L., Stamberg, W., Staniek, K., Gille, L. 2009. Toxic effects of carvacrol, caryophyllene oxide, and ascaridole from essential oil of *Chenopodium ambrosioides* on mitochondria. Toxicol. Appl. Pharm., 240: 337–347.

Monzote, L., Pastor, J., Scull, R., Gille, L. 2014. Antileishmanial activity of essential oil from *Chenopodium ambrosioides* and its main components against experimental cutaneous leishmaniasis in BALB/c mice. Phytomedicine, 21: 1048–1052.

Monzote, L., García, M., Pastor, J., Gil, L., Scull, R., Maes, L., Gille, L. 2014. Essential oil from *Chenopodium ambrosioides* and main components: activity against Leishmania, their mitochondria and other microorganisms. Exp. Parasitol., 136: 20–26.

Mosyakin, S.L. 1993. Um esboço de um sistema para *Chenopodium* L. (espécies da Europa, Norte e Ásia Central). Ukr. Bot. Zh, 50: 71–77.

Mosyakin, S.L., Clemants, S.E. 1996. New infrageneric taxa and combinations in *Chenopodium* L. (*Chenopodiaceae*). Novon, 6: 398–403.

Mosyakin, S.L., Clemants, S.E. 2008. Further transfers of glandular-pubescent species from *Chenopodium* subg. Ambrosia to *Dysphania* (*Chenopodiaceae*). J. Bot. Res. Inst. Texas, 2: 425–431.

Muntané, J. 1994. Tresor de la saviesa popular de les herbes, remeis i creences de Cerdanya del temps antic. Institut d'estudis ceretans, Puigcerda, 1: 476.

Napoli, M. 2008. The plants, rituals and spells that 'cured' helminthiasis in Sicily. J. Ethnobiol. Ethnomed., 4: 21.

Nascimento, F.R., Cruz, G.V., Pereira, P.V.S., Maciel, M.C., Silva, L.A., Azevedo, A.P.S., Guerra, R.N. 2006. Ascitic and solid Ehrlich tumor inhibition by *Chenopodium ambrosioides* L. treatment. Life Sci., 78: 2650–2653.

O'Brien, S.F., Delage, G., Seed, C.R., Pillonel, J., Fabra, C.C., Davison, K., Leiby, D.A. 2015. The epidemiology of imported malaria and transfusion policy in 5 nonendemic countries. Transfus. Med. Rev., 29: 162–171.

Okunlola, O.A., Oyeyemi, O.T. 2019. Spatio-temporal analysis of association between incidence of malaria and environmental predictors of malaria transmission in Nigeria. Sci. Rep., 9: 1–11.

Pavela, R., Maggi, F., Lupidi, G., Mbuntcha, H., Woguem, V., Womeni, H.M., Benelli, G. 2018. Clausena anisata and *Dysphania ambrosioides* essential oils: from ethno-medicine to modern uses as effective insecticides. Environ. Sci. Pollut., 25: 10493–10503.

Picó, B., Nuez, F. 2000. Minor crops of Mesoamerica in early sources II. Herbs used as condiments. Genet. Resour. Crop Ev., 47: 541–52.

Pinto, E.D.P.P., Amorozo, M.C.D.M., Furlan, A. 2006. Popular knowledge about medicinal plants in rural communities in the Atlantic Forest-Itacaré, BA, Brazil. Acta Bot. Bras., 20: 751–762.

Pizzolitto, R.P., Jacquat, A.G., Usseglio, V.L., Achimón, F., Cuello, A.E., Zygadlo, J.A., Dambolena, J.S. 2020. Quantitative-structure-activity relationship study to predict the antifungal activity of essential oils against Fusarium verticillioides. Food Control., 108: 1–8.

Prasad, C.S., Shukla, R., Kumar, A., Dubey, N.K. 2010. *In vitro* and *in vivo* antifungal activity of essential oils of *Cymbopogon martini* and *Chenopodium ambrosioides* and their synergism against dermatophytes. Mycoses, 53: 123–129.

Repo-Carrasco, R., Espinoza, C., Jacobsen, S.E. 2003. Nutritional value and use of the Andean crops quinoa (*Chenopodium quinoa*) and kañiwa (*Chenopodium pallidicaule*). Food Rev. Int., 19: 179–189.

Romano, N.G., Ureta, M.M., Guerrero Sánchez, M., Gómez-Zavaglia, A. 2019. Nutritional and technological properties of a quinoa (*Chenopodium quinoa* Willd.). Food Res., 1: 1–11.

Sá, R.D., Soares, L.A.L., Randau, K.P. 2015. Karina Perrelli. Óleo essencial de *Chenopodium ambrosioides* L.: estado da arte. J. Appl. Pharm. Sci., 36: 267–276.

Sérvio, E.M.L., de Araújo, K.S., da Silva Nascimento, L.R., da Costa, C.L.S., Mendes, L.M.S., Maia Filho, A.L.M., Santos, Í.M.S.P. 2011. Cicatrização de feridas com a utilização do extrato de *Chenopodium ambrosioides* mastruz e cobertura secundária estéril de gaze em ratos. ConScientiae Saúde, 10: 441–448.

Shah, H., Khan, A.A. 2017. Phytochemical characterisation of an important medicinal plant, *Chenopodium ambrosioides* Linn. Nat. Prod. Res., 31: 2321–2324.

Shimoto, E.Y., Monteiro, M.P. 2010. Quinoa (*Chenopodium quinoa willd quinoa*) (*chenopodium quinoa willd*) as functional food. Rev. Aten. Saú., 8: 1–6.

Siddique, S., Parveen, Z., Mazhar, S. 2020. Chemical composition, antibacterial and antioxidant activities of essential oils from leaves of three *Melaleuca* species of Pakistani flora. Arabian J. Chem., 13: 67–74.

Silva, D.M., Magalhães, V.C., Macedo, T.P., Almassy Jr, A.A., da Silva, F. 2009. Levantamento etnofarmacológico em comunidades rurais do Recôncavo da Bahia/BA. Rev. Bras. Agroec., 4: 1–5.

Silva, L.E., de Quadros, D.A., Neto, A.J.M. 2015. Estudo etnobotânico e etnofarmacológico de plantas medicinais utilizadas na região de Matinhos-PR. Ciên. Nat., 37: 266–276.

Tariq, S., Wani, S., Rasool, W., Bhat, M.A., Prabhakar, A., Shalla, A.H., Rather, M.A. 2019. A comprehensive review of the antibacterial, antifungal and antiviral potential of essential oils and their chemical constituents against drug-resistant microbial pathogens. Microb. Pathog., 134: 1–20.

Trivellato Grassi, L., Malheiros, A., Meyre-Silva, C., da Silva Buss, Z., Monguilhott, E.D., Fröde, T.S., de Souza, M.M. 2013. From popular use to pharmacological validation: a study of the anti-inflammatory, anti-nociceptive and healing effects of *Chenopodium ambrosioides* extract. J. Ethnopharmacol., 145(1): 127–138.

Verma, A.k., Singh, S. 2020. Phytochemical analysis and *in vitro* cytostatic potential of ethnopharmacological important medicinal plants. Toxicol. Rep., 7: 443–452.

Winsor, M.P. 2001. Cain on Linnaeus: the scientist-historian as unanalysed entity. Studies in History and Philosophy of Science Part C: Stud. Hist. Philos. Biol. Biom. Sci., 32: 239–254.

Yang, R.Z., Tang, C.S. 1988. Plants used for pest control in China: a literature review. Econ. Bot., 42: 376–406.

Zerrifi, S.E.A., Kasrati, A., Tazart, Z., El Khalloufi, F., Abbad, A., Oudra, B., Vasconcelos, V. 2020. Essential oils from Moroccan plants as promising ecofriendly tools to control toxic cyanobacteria blooms. Ind. Crops Prod., 143: 1–8.

Zhang, Y., Wei, J., Chen, H., Song, Z., Guo, H., Yuan, Y., Yue, T. 2020. Antibacterial activity of essential oils against Stenotrophomonas maltophilia and the effect of citral on cell membrane. LWT, 117: 1–8.

Zhou, Z., Sun, H. 2016. Wu Zhengyi and his contributions to plant taxonomy and phytogeography. Plant Divers, 38: 259–261.

Zohra, T., Ovais, M., Khalil, A.T., Qasim, M., Ayaz, M., Shinwari, Z.K. 2019. Extraction optimization, total phenolic, flavonoid contents, HPLC-DAD analysis and diverse pharmacological evaluations of *Dysphania ambrosioides* (L.) Mosyakin & Clemants. Nat. Prod. Res., 33: 136–142.

Chapter 13
Apiaceae Plants Growing in the East
Centuries of Healing Traditions and Culture

Sherweit El-Ahmady,[1,*] *Nehal Ibrahim*,[1] *Nermeen Farag*[2] and *Sara Gabr*[3]

Introduction

Family Apiaceae (Umbelliferae) is considered one of the most important families of flowering plants, which consists of approximately 3780 species in 434 genera (Sayed-Ahmad et al. 2017). Members of family Apiaceae are distributed worldwide, mostly in the northern temperate regions and are cultivated to be used for food such as vegetables, spices, and herbs or primarily for medicinal purposes. The most common characteristics of Apiaceae species are the aromatic herbaceous nature, simple or compound umbel inflorescences and indehiscent fruits (schizocarps) containing schizogenous oil ducts providing oils with distinctive flavours (Aćimović 2017). The most common vegetable crops of this family are carrot and celery, while other members are used for their fragrant leaves, for example, parsley, coriander (cilantro) and dill. The fruits, commonly and mistakenly called seeds, of anise, sweet fennel, caraway, coriander, and cumin are also used for their aromatic character, hence they are usually dried and ground to make popular culinary spices. Several plants of this family have been used since antiquity to treat various illnesses related to digestion, reproduction, and respiration. This is probably due to the variety of phytochemicals which members of this family are known for: flavonoids, coumarins, terpenoids, triterpenoids, polyacetylenes, in addition to essential oils. In turn, several studies have reported potential biological activities such as antimicrobial, anti-inflammatory, analgesic, anti-tumor, antioxidant, diuretic, gastrointestinal and anti-obesity properties (Sayed-Ahmad et al. 2017).

In this chapter, we present nine commonly used members of family Apiaceae that are popular and native to the Middle-East region and Asia, for various purposes associated to their fragrant properties, medicinal constituents and nutrients (carrots, anise, fennel, parsley, celery, cumin, and coriander); others solely for their bioactive constituents (Ajwain, *Ammi visnaga* and *Ammi majus*). Traditional uses for each plant will be presented followed by reported biological activities, then phytochemical content mostly associated with the pharmacological activity of each species will be discussed (Figures 13.1, 13.2, 13.3a, 13.3b).

[1] Department of Pharmacognosy, Faculty of Pharmacy, Ain Shams University, Cairo, Egypt.
[2] Department of Pharmacognosy, Faculty of Pharmacy, Cairo University, Cairo, Egypt.
[3] Department of Pharmacognosy, Faculty of Pharmacy, Heliopolis University, Cairo, Egypt.
* Corresponding author

Figure 13.1. Apiaceae Umbels (Image by andreas 160578 from Pixabay).

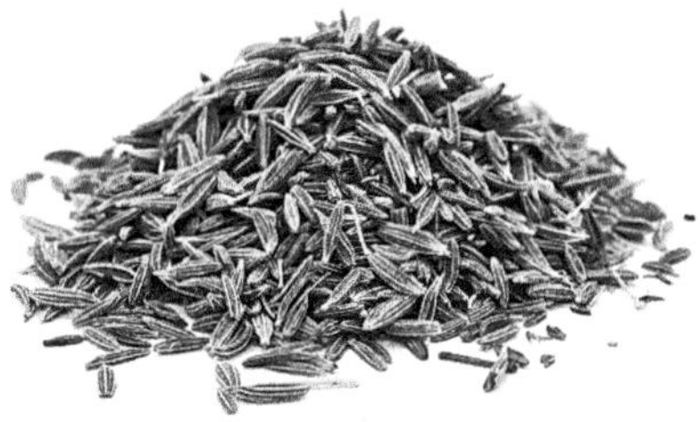

Figure 13.2. Caraway fruit (Images purchased from Shutterstock).

Figure 13.3. Dill leaves (a); **dill fruit** (b) (Image (a) by Wanjae Lee from Pixabay; Images (b) purchased from Shutterstock).

Ajwain (*Trachyspermum ammi*)

Trachyspermum ammi L., also known as Ajwain or Bishop's weed, is a prominent member of the Apiaceae family. It is a highly valued medicinally important plant native to Egypt and is cultivated in Iraq, Iran, Afghanistan, Pakistan, and India. It is also widely grown in arid and semi-arid regions where soils contain high levels of salts. Ajwain is a slightly branched annual herb, growing 60–90 cm tall with a striated stem and compound umbel inflorescence, each containing up to 16 flowers. The grayish-brown, ovoid, aromatic fruit is 2 mm long and 1.7 mm wide and is harvested from February to March (Bairwa et al. 2012).

Traditional uses

The Ajwain fruits are edible and usually used as a spice in India, the Middle East and Asia, and some parts of America. The seeds possess remarkable digestive and antiseptic properties and are used in traditional medicine, primarily to control bowel disorders such as indigestion, flatulence, colic, and diarrhea. They are also used as a stimulant, stomachic, carminative, aromatic, antispasmodic, antihypertensive, antiseptic, antiparasitic, antiscorbutic, antihistamine, vermicide, emmenagogue, sialagogue, and anti-inflammatory (Anwar et al. 2016). Ajwain oil contains around 50% thymol, which is a strong germicide, antibacterial, antispasmodic, and fungicide. The diuretic properties of *T. ammi* seeds have been widely reported for use in some drug formulations for kidney stone treatments (Kaur et al. 2009) (Figure 13.4).

Figure 13.4. Ajwain fruit (Images purchased from Shutterstock).

Biological activities

Antioxidant activity

The essential oil exhibited a high degree of FRAP, a good DPPH radical scavenging activity and a moderate H_2O_2 radical scavenging activity making it a high bioresource of antioxidant (Chatterjee et al. 2013). Extracts of fresh and frozen ajwain leaves were analyzed *in vitro* using free radical scavenging DPPH assay. Frozen ajwain leaves exhibited significantly higher antioxidant activity than the fresh ones due to higher lutein content in frozen leaves (Raza et al. 2015). In another recent study, ajwain seed essential oil, its fractions, and extracts were analysed by the DPPH, hydroxyl, nitric oxide radical scavenging methods and FRAP assay. All the tested components showed significant antioxidant activity where the methanol, dichloromethane extracts of ajwain seeds showed the highest scavenging potential in all four methods due to its rich phenolic and flavonoid content (Dhaiwal et al. 2017).

Antidiabetic activity

Ajwain oil effectively inhibited the key enzymes α-amylase and α-glucosidase, thereby decreasing the postprandial hyperglycemia and proving its ability to promote glucose uptake in the L6 myotubes (Aneesa et al. 2019).

Antitussive and bronchodilatory effects

In a study to evaluate the antitussive effect of aqueous and macerated extract of ajwain seeds, a significant reduction of cough number was observed (Vickers 2017). Ajwain extract and essential oil inhibited the effect of both Histamine (H1) receptors isolated from guinea-pig tracheal chains due to the presence of carvacrol. In a clinical trial on asthmatic patients, the bronchodilatory effect of decocted extract of ajwain was examined and according to the results, the extract had a relative bronchodilatory effect on asthmatic airways compared to the effect of theophylline (Zarshenas et al. 2013).

Diuretic and anti-lithiasis activity

Ajwain was attributed to have diuretic and anti-lithiasis activity in traditional medicinal reports. A human study was performed where the seeds were decocted in milk and given orally to volunteers suffering from urinary stone for a nine days period. The results were reported satisfactory against a pure ca-oxalate stone (Zarshenas et al. 2013).

Gastrointestinal effect

The use of ajwain showed the increased secretion of gastric acids, bile acids and activity of digestive enzymes and also reduced the food transient time, proving its traditional usage as a digestive stimulant for digestive purposes. In another study, ajwain reinforced the pancreatic lipase and amylase effectiveness supporting the fact that it has a digestive stimulant activity (Zarshenas et al. 2013). Animal pretreated with ethanolic extract of ajwain fruits showed significant protection by reducing ulcerative lesions when compared with a control group of animals (Platel and Srinivasan 2001).

Anti-inflammatory activity

The anti-inflammatory potential was evaluated using carrageenan-induced rat paw oedema model and a cotton pellet induced granuloma model. Aspirin (ASA) (150 mg/kg) and anti-inflammatory drug phenyl butazone(PBZ) (150 mg/kg) were used as standard positive controls. Accordingly, both total alcoholic extract and total aqueous extract possessed a significant *in vivo* anti-inflammatory effect (Thangam and Dhananjayan 2003).

Hypolipidimic activity

In vivo studies on albino rats were assessed for the antihyperlipidemic effect of ajwain seeds. It proved to be effective on the lipid profile decreasing the total cholesterol, LDL-cholesterol, triglycerides and total lipids. Moreover, organic extract of seeds reduced the atherogenic index and increased the level of HDL-cholesterol in albino rabbits (Meena et al. 2018).

Antimicrobial activity

Ajwain seed extract was reported to possess fungicidal action against *Rhizoctonia solani*, a causative agent of sheath blight of rice (Ansari 1995). The oil exhibited remarkable antibacterial activity against *Staphylococcus aureus, Escherichia coli, Salmonella typhi, Shigella dysenteriae* and *Vibrio cholera.* Extracts prepared in different solvents exhibited variable activity against *E. coli, P. aeruginosa, S. typhi* and *S. aureus* (Patel et al. 2008), suggesting their centuries-old usage in the treatment of gastrointestinal disorders. This historical use of ajwain seeds to cure various gastrointestinal disorders has also been scientifically proved in another study carried out by Kaur and Arora (Kaur and Arora 2009), wherein aqueous and organic extracts of ajwain seeds have also shown their antibacterial effect. Methanolic and hexane extract of ajwain seeds have been lately validated for its antimicrobial activity for the previously mentioned reports (Sharma et al. 2018).

Chemistry

Essential oil

Ajwain is known for its brownish essential oil, which is responsible for its odor and taste. The chemical composition of ajwain seed essential oil is influenced by different factors including pressure, temperature, and modifier volume and extraction time. Extraction based on the supercritical fluid extraction yielded only three components, including γ-terpinene (14.2%), p-cymene (23.1%) and thymol (62.0%), which constituted more than 99% of the oil. It was considered to be the most selective method for the extraction of thymol. However, in hydrodistilled oil, there were eight components including thymol (49.0%), γ-terpinene (30.8%), p-cymene (15.7%), β-pinene (2.1%), myrcene (0.8%) and limonene (0.7%) (Zarshenas et al. 2013).

Miscellaneous

Alcoholic extracts contain a highly hygroscopic saponin. From the fruits, a yellow, crystalline flavone and a steroid-like substance have been isolated; it also contains 6-O-β-glucopyranosyloxythymol glucoside and yields 25% oleoresin containing 12% volatile oil (thymol, γ-terpinene, para-cymene, and α- and β-pinene) (Bairwa et al. 2012).

Ammi Species

Ammi visnaga L. (*Visnaga daucoides* Gaertn), commonly known as the toothpick plant and "khella or khella baladi" in Arabic, is a short annual or biennial plant indigenous to the Mediterranean region of North Africa, Asia, and Europe (Chevalier 1996). *Ammi majus* L. is indigenous to Egypt growing in the Nile Valley, but is also found in some Mediterranean countries, West Africa and parts of Iran. It is commonly known as Bishop's weed and "khella shaitani" in Arabic. Even though these herbs are the lesser known members of family Apiaceae, nevertheless, their medicinal properties that go back to ancient times make them appreciated wonders of this plant family.

Traditional use

Ammi visnaga L. has been used in folk medicine in the Middle East since ancient times. A decoction of the fruit was used for the treatment of renal colic by the ancient Egyptians, in the treatment of kidney inflammation in Iraq and Palestine, and urolithiasis and prostatic pain in Algeria. The herb extract or plant powder has also been traditionally used for the treatment of abdominal cramps, mild anginal symptoms, as an emmenagogue, diuretic, and has also been prescribed for the amelioration of headaches, vertigo, and diabetes. It was a popular herb used in the treatment of vitiligo and psoriasis. It is generally used to dilate bronchial, urinary, and blood vessels without affecting blood pressure and hence is used for bronchial asthma, urinary calculi and kidney stones. It also has been used as an antidiabetic in Palestine and Morocco (Khalil et al. 2020, Miara et al. 2019). The fruits of *Ammi majus* were used traditionally for the treatment of skin disorders, psoriasis and vitiligo and was also used as an emmenagogue, as a diuretic, and for treatment of leprosy, kidney stones, and urinary tract infections (Al-Snafi 2013) (Figure 13.5).

Figure 13.5. ***Ammi visnaga* dried umbel** (Images purchased from Shutterstock).

Biological activities

Kidney diseases

As mentioned previously, the popularity of this herb goes back to ancient times where it was widely used for renal colic and other urinary tract problems. *A. visnaga* has been shown to be effective in the treatment of nephrolithiasis and uremia and its use in the treatment of various kidney disorders was associated with khellin and visnagin. These compounds have been shown to prevent hyperoxaluria by inhibiting the oxalate formation by increasing the urinary pH and citrate concentration, hence protecting the renal epithelial cell (Nirumand et al. 2018, Vanachayangkul et al. 2010).

Cardiovascular effects

A. visnaga has been used in the treatment of angina pectoris by being an effective smooth muscle relaxant acting on peripheral and coronary vasodilation without affecting blood pressure. This property is associated with the two main γ-pyrones, khellin and visnagin having calcium antagonistic activity causing the vasodilating activities, as well as the pyranocoumarin, visnadin which has specifically been shown to possess both peripheral and coronary vasodilator activities. It may also interfere with other sites involved in vascular smooth muscle contraction and hence the treatment of angina pectoris. The administration of khellin, either orally or by intramuscular injection, showed promising results in treating angina pectoris and is favored in the case of coronary thrombosis. It was also found that samidin and khellol glycoside induced a positive inotropic effect on the heart (Duarte et al. 1995, 2000, Campos-Toimil et al. 2002, Venugopala et al. 2013).

Antidiabetic activity

A decoction of *A. visnaga* fruit was shown to reduce blood glucose levels by 51% in normoglycemic rats, compared to an oral hypoglycaemic agent (tolbutamide) and an aqueous extract showed a significant hypoglycemic effect when given to both normal and streptozotocin-diabetic rats (Hassan et al. 2002).

Vitiligo treatment

Many studies have been conducted on the effectiveness of khellin, both orally and topically, for the treatment of vitiligo. The treatment with a gel formulation of khellin based upon a water/2-propanolpropylene glycol (khellin-WPG) system combined with ultraviolet A (UVA) was found to considerably improve the clinical outcome of patients by facilitating the availability of the drug in the skin. This treatment was shown to be safe for both short and long-term treatments. The treatment was shown to be highly effective when used on 74 patients suffering from vitiligo, and did not cause any side effects. Khellin was encapsulated in L-phenylalanine stabilized phosphatidylcholine liposomes in combination with ultraviolet A/ultraviolet B (UVA/UVB) light therapy (KPLUV) and was shown to possess a potent melanin biosynthesis inhibitory effects when investigated using a B16 melanoma cell line (Hofer et al. 2001, De Leeuw et al. 2003, 2011).

As for *A. majus*, there are older reports of its use in the treatment of vitiligo, psoriasis, and hypopigmentation tinea versicolor. Clinical studies conducted using topical or oral administration or a combination of both showed promising results in numerous studies; the constituents associated with these properties were concluded to be furanocoumarins, specifically xanthotoxins (Al-Snafi et al. 2013, El-Mofty et al. 1995, De Berker 1997, Parsad et al. 1998).

Anti-inflammatory and analgesic activity

Visnagin was found to be the compound responsible for the anti-inflammatory effect of *A. visnaga*; it caused a decrease in mRNA expression and the release of certain inflammation factors like TNF-α, IL-1β, and IFNγ. It was also shown to reduce LPS-induced IL-6 and MCP-1 mRNA level, and in that regard, the inhibition of transcription factors such as AP-1 and NF-κB (Lee et al. 2010). *A. visnaga* was one of the plants shown to have analgesic potential in a recent study conducted on various plants in Saudi Arabia (Gouda et al. 2014). *Ammi majus* coumarins were shown to have anti-inflammatory properties by the carrageenan induced rat paw edema method (Selim and Ouf 2012).

Antioxidant and cytotoxic activity

The antioxidant activity of *A. visnaga* and its association to the flavonoid content was investigated in a study conducted on the aerial parts of the plant where results showed an equivalent antioxidant activity (IC_{50} of 8.77 ± 0.2 μg/mL) compared to the standard antioxidant rutin (IC_{50} of 3.01 ± 0.2 μg/mL) (Bencheraiet et al. 2011). Regarding the plant's cytotoxic activity, three studies investigated the effect of khellin and visnagin against different cell lines. In one study, khellin showed mild to moderate activity when tested against the hepatocarcinoma cell line (HepG2). In another study, the ethanolic extract of *A. visnaga* produced inhibitory effects on both Hela (cervical cancer) and MCF7 cell lines. In a recent comparative study on the effect of each of khellin and visnagin against four human cell lines, Hela, Hep-G2, HCT 116, and MCF7, the results revealed good cytotoxic activity of both compounds against the Hep-G2 cell line (Abdel-Monem et al. 2014, Beltagy et al. 2015, Pakfetrat et al. 2015).

Antimicrobial activity

Different extracts of *A. visnaga* have shown antimicrobial activity related to various organisms including *Candida*, *Mycobacterium tuberculosis*, *Streptococcus mutans*, *Streptococcus salivarius*, and *Streptococcus sanguis* as well as Gram-negative bacteria *Escherichia coli* and *Klebsiella pneumoniae*. Certain components, i.e., khellin, xanthotoxin, and bergapten, were found to inhibit aflatoxin production of *Aspergillus flavus* and other toxigenic fungi (Mahmoud 1999, Ghareeb et al. 2011, Semyari et al. 2011). Several studies have also focused on the antimicrobial effects of the essential oils of the plant, showing their effectiveness against *Escherichia coli*, *Pseudomonas aeruginosa*, and *Klebsiella pneumoniae* strains (Satrani et al. 2004, Rasooli et al. 2007, Sellami et al. 2011). Furocoumarins of *A. majus* have been shown to possess bactericidal, fungicidal, antiviral activities (Al-Snafi 2013), insecticidal, larvicidal, moluscicidal, nematicidal, ovicidal, viricidal and herbicidal activities.

Other biological activities

Visnadin was shown to have the potential for inhibiting hair loss. A lotion for the scalp composed of visnadin, in addition to other components, led to an increase in arterial and arteriolar sphygmic activity in the subpapillary plexus, leading to an improvement in local microcirculatory flow (Khalil et al. 2019). *A. visnaga* has been established as a bronchodilator for its smooth muscle relaxant effect and hence its use in the treatment of asthma (Al-Snafi 2016). Studies have also shown the potential of khellin and visnagin as promising insecticides and could be used for the development of new botanical acaricides (Maleck et al. 2013, Pavela 2015). Furocoumarins of *A. majus* have been shown to possess insecticidal, larvicidal, moluscicidal, nematicidal, ovicidal and herbicidal activities (Al-Snafi 2013). It is worth noting that there are reports of contraindications on both *Ammi* species; *A. majus* is contraindicated in children under the age of 12 years and in pregnancy, nursing, tuberculosis, liver and kidney diseases and autoimmune diseases. There are also some reports of sensitivity to xanthotoxin (psoralens) and warnings in diseases associated with photosensitivity, cataract, and invasive squamous cell cancer. Also, the exposure to sun or other sources of ultraviolet light should be avoided during treatment with *A. visnaga* and its components to minimize photosensitivity (Al-Snafi 2013).

Chemistry

The chemical constituents of *A. visnaga* are various and have been reported through numerous studies over the years. The main bioactive compounds include γ-pyrones, coumarins flavonoids, and essential oils.

γ-Pyrones and Coumarins

The main active constituents of *A. visnaga* are the γ-Pyrones and coumarins which make this plant unique among all other Apiaceae members. The former γ-Pyrones are furanochromone derivatives of which the major ones are Khellin and visnagin. Others include 4-norvisnagin, khellinol, visamminol, ammiol, and khellol as well as 5,7-dihydroxy-2-methyl-γ-pyrone-7-O-glucoside and pimolin (III), and khellinin, khellinone, and visnaginone. Another major class is the coumarins, which are divided into two sub-groups, namely, pyranocoumarins and furanocoumarins. The former is isolated from the fruits and named cis-khellactone-30-β-d-glucopyranoside, in addition to visnadin, samidin, and dihydrosamidin, while furanocoumarins are present in smaller amounts and include xanthotoxin, ammoidin, bergapten, and psoralene (Abou-Mustafa et al. 1990, Elgamal et al. 1998, Guenaydin and Beyazit 2004, Hashim et al. 2014, Sonnenberg et al. 1995, Zgorka et al. 1998).

The major constituents of *A. majus* are furanocoumarins, which include xanthotoxins (methoxsalen, 8-methoxypsoralen, ammoidin), imperatorin (ammidin) and bergaptens (heraclin, majudin, 5-methoxypsoralen), marmesin, isoimperatorin, heraclenin, isopimpinellin, 6-hydroxy-7-methoxy-4 methyl coumarin and 6-hydroxy-7-methoxy coumarin as well as nonfurocoumarin and umbelliprenin (Al-Snafi 2013). Recently, newly identified coumarins with cytotoxic properties were isolated (El-Sharkawy and Selim 2018).

Phenolic compounds

The main phenolics identified in *A. visnaga* belong to the flavonoids class. These include flavonoid aglycones identified as quercetin, kaempferol, rhamnocitrin, and rhamnetin, and rhamnazin. Flavonoid glycosides include quercetin-3-glucoside, kaempferol-3-glucoside, and isorhamnetin 3-β-d-glucoside, rhamnetin-3-O-glucoside, isorhamnetin-3-O-glucoside, rhamnazin-3-glucoside, isorhamnetin-7-O-glucoside, and quercetin-7,3,3'-O-triglucoside. *A. visnaga* also constitutes flavonoidal sulfates as quercetin 3-sulfate, rhamnocitrin 3-sulfate, rhamnetin, and isorhamnetin-3-sulfate. Flavones are represented by Apigenin, Luteolin, and Chrysoeriol, while the isolflavones identified include daidzin, genistin, sissotrin, isoformononetin, formononetin, prunetin, biochaninA, coumestrol, daidzein, and 6,7,40-trihydroxyisoflavone (Bencheraiet 2011, Elgamal et al. 1998). Flavonoids reportedly identified in fruits of *A. majus* include quercetin and kaempferol and glycosides of quercetin and luteolin (Abdul-Jalil et al. 2010).

Essential oils and fatty acids

The essential oils of *A. visnaga* were mainly extracted from the umbels and fruits and constituted nonterpenes, monoterpenes, diterpenes and sesquiterpenes. Oxygenated or hydrocarbon monoterpenes comprised most of the identified oil constituents, where linalool and thymol represented the highest percentage of the constituents. Other identified monoterpenes include α-thujene, α-pinène, β-pinene, and β-myrcene. Other classes of chemical compounds identified in much smaller amounts include sterols like β-sitosterol and β-sitosterol-glucoside, in addition to fatty acids such as palmitic, palmitoleic, stearic, petroselinic, linoleic, linolinic, arachidic, and tetracosanoic acids (Khadri et al. 2011, Khalfallah et al. 2011, Keddad et al. 2016, Sellami et al. 2011).

In the case of *A. majus*, the essential oil extracted from the fruits constituted dipiperitone, unsaturated cyclic terpeniole and a mixture of furocoumarins. Also, fatty acids were identified in the plant oil including methyl ester of linoleic acid, methyl ester of oleic acid, palmitic acid and linolenic acids. Other fatty acids included hexanoic acid, caprylic acid, capric acid, lauric acid, myristic acid, pentadecanoic acid, palmitic acid, margaric acid, stearic acid, elaidic acid, arachidic acid, behenic acid, tricosnoic acid and tetracosanoic acid (Hussain et al. 2012).

Alkaloids

Two alkaloids of the furoquinoline-type were isolated from *Ammi majus* L.; they are identified as 4-hydro-7-hydroxy-8-methoxyfuroquinoline and 4-hydro-7-hydroxy-8-prenyloxyfuroquinoline (Mohammed and El-Sharkawy 2017).

Anise (*Pimpinella anisum*)

Pimpinella anisum L. is commonly known as "anise". The synonyms used for this plant include *Anisum odoratum* Raf., *Anisum officinale* DC., *Anisum officinarum* Moench, *Anisum vulgare* Gaertn., *Apium anisum* (L.) *Crantz, Carum anisum* (L.) Baill., *Ptychotis vargasiana* DC., *Selinum anisum* (L.) E.H.L. Krause, *Seseli gilliesii* Hook. & Arn., *Sison anisum* (L.) Spreng., and *Tragium anisum* (L.) Link. (Rocha and Fernandes 2016, Sun et al. 2019).

Anise (*Pimpinella anisum*) is an herbaceous annual plant that grows spontaneously in the Mediterranean region and is often cultivated in order to obtain its fruits, which are primarily the plant parts used for essential oil extraction, although extraction of roots can also be performed (Rocha and Fernandes 2016, Sun et al. 2019). The small size and form of its dried ripe fruits cause them to be mistakenly referred to as "seeds," with the raw plant material often described as "aniseed" (Rocha and Fernandes 2016).

Traditional uses

P. anisum is one of the most widely used plants for the treatment of various ailments and as a spice since ancient times. Decoctions and infusions prepared from aerial parts of *P. anisum* have been used in the treatment of renal colic, gastrointestinal colic, and upper respiratory tract ailments in Egypt (AbouZid and Mohamed 2011). In Turkish folk medicine, the aerial parts have traditionally been used as carminative, expectorant, sedative, antidepressant, antiseptic, insecticidal, antiviral, antispasmodic, nematocidal, mutagenic, analgesic, antifungal, antibacterial, diuretic, estrogenic, antimalarial, and pectoral stimulant (Kuruuzum-Uz et al. 2010), while the seeds have been used as analgesic, anti-inflammatory, appetizing, hypnotic, antispasmodic, expectorant, antibacterial, and hepato-protective (Tepe and Tepe 2015). Hot water extracts of the seeds have been used also in traditional Beirut (Lebanon) folk medicine for their diuretic and laxative effect, expectorant and anti-spasmodic action, and their ability to ease intestinal colic and flatulence (Kreydiyyeh et al. 2003). In traditional Iranian medicine, the fruits have been used in the treatment of epilepsy (due to its anticonvulsant activity) (Abdollahi Fard and Shojaii 2013). The whole plant of *P. anisum* in Palestine is used as a herbal self-therapy agent (Sawalha et al. 2008) (Figure 13.6).

Figure 13.6. Anise fruit (Images by Gokalp Iscan from Pixabay).

Biological activities

Gastrointestinal effect

Aqueous suspension of anise significantly inhibits gastric mucosal damage induced by necrotizing agents and indomethacin (Al Mofleh et al. 2007). Hydroalcoholic extract of the aerial part of *P. anisum* reduced the contractions induced by acetylcholine in the rat anococcygeus smooth muscle. In addition to its antispasmodic effect, the extract also relaxed the preparations pre-contracted with acetylcholine. The action displayed by the extract was mainly dependent on the activation of the NO-cGMP pathway (Tirapelli et al. 2007).

Antidiabetic activity

The oil of *P. anisum* was shown to significantly enhance glucose absorption from the rat jejunum and increase the Na+–K+ ATPase activity in a jejunal homogenate in a dose dependent manner. Accordingly, *P. anisum* oil increased glucose absorption by increasing the activity of the Na+–K+ ATPase and consequently the sodium gradient needed for the sugar transport (Tepe and Tepe 2015).

Antioxidant activity

Several studies were done to test the antioxidant potential of *Pimpenella anisum* for being rich with phenolics content (Nickavar and Abolhasani 2009). Water and ethanolic extracts of aniseeds showed higher antioxidant activity, reducing power, DPPH radical and superoxide anion scavenging, hydrogen peroxide scavenging, and metal chelating activities when compared to BHA, BHT, and α-tocopherol, where water extract exhibited greater antioxidant capacity than ethanolic extract. The antioxidant potential of essential oil and oleoresins from anise seeds showed highest antioxidant activity, even higher than synthetic antioxidants like butylated hydroxyanisole (BHA), and butylated hydroxytoluene (BHT) (Shojaii and Abdollahi Fard 2012, Tepe and Tepe 2015).

Anti-inflammatory and analgesic activities

Fixed oil of *P. anisum* showed anti-inflammatory activity as strong as indomethacin using carrageenan-induced rat paw edema test and remarkable results were obtained (Tepe and Tepe 2015). Topical application of methanolic extract of *Pimpinella anisum* stimulated wound healing activity in diabetic rats. Wound size, inflammation and cellularity were significantly reduced, and the re-epithelialization rate, collagen content, fibroblastic response were enhanced. More scar tissue in the early stages of wound healing was also produced and in the later stages, *P. anisum* significantly decreased oxidative stress, scar tissue formation, and improved tissue maturity and remodeling (Hashemnia et al. 2019).

P. anisum exhibited significant analgesic activity versus benzoquinone induced writing and in thermal tests (Twaij et al. 1987). The essential oil also showed significant analgesic effect similar to morphine and aspirin (Tas 2009) and an analgesic effect comparable to that of 100 mg/kg aspirin and 10 mg/kg morphine (Tas et al. 2006).

Bronchodilatory activity

Ethanolic extract of *P. anisum* exhibited an inhibitory effect on calcium channels which may contribute to its bronchodilator effect on guinea pig tracheal chains, while the water extract and essential oil exhibited a relaxant effect on target tissue due to the potassium channel opening effect by the plant. It was also claimed that the relaxant effect of this plant is not due to an inhibitory effect of histamine (H1) or stimulatory effect of 2-adrenergic receptors, but due to inhibitory effect on muscarinic receptors (Tepe and Tepe 2015).

CNS effects

The fruit oil of *P. anisum* was found to have an effect on neuronal excitability and neuroprotective effects. The oil changed the firing pattern and increased the frequency of firing and decreased hyperpolarization potential. Moreover, the oil inhibited the production of dark neurons in different regions of the brain in epileptic rats resulting in a neuroprotective effect (Janahmadi et al. 2008, Karimzadeh et al. 2012).

In a study, forced swimming test (FST) and tail suspension test (TST) were used to determine the antidepressant effects of *P. anisum* in comparison to fluoxetine (a serotonin reuptake inhibitor). The results of this study revealed that aqueous and ethanolic extracts of *P. anisum* fruits reduced immobility time and increased the swimming time of mice in a dose-dependent manner. The serotonergic neurotransmission may be the key role in the antidepressant-like activity of *P. anisum* extracts where the inhibition of 5-HT reuptake, as well as inhibition of Monoamine oxidase by the major compound anethole, increases 5-HT availability in synaptic clefts, which could desensitize the 5-HT receptors involved in depression (Shahamat et al. 2016).

The use of aniseed essential oil has shown to change the effect of drugs that act in central nervous system. The analgesic effect of codeine is dependent on its biotransformation (O-demethylation) into morphine by liver isoenzyme CYP2D6. Examining the influence of EO on codeine, the treatment with aniseed EO exhibited statistically significant prolongation in response latencies up to the 50th minute when compared with the control group. The reason for the increased analgesic effect of codeine could be due to a metabolic interaction as a consequence of CYP2D6 induction (Samojlik et al. 2012). An alteration in the pentobarbital metabolism was also observed in mice groups pretreated with anise essential oil where the sleeping time decreased significantly when compared to the control group reflecting the change in pentobarbital induced sleeping time. This also supports that trans-anethole induced microsomal enzymes and increased metabolism of pentobarbital in the liver.

Gynecological effect

Aqueous plant extract at a concentration range of 10–100 g/ml stimulated osteoblastic cell differentiation and exhibited an anti-estrogenic effect on breast cancer cells without proliferative effects on cervical adenocarcinoma cells (Kassi et al. 2004). In a double-blind clinical trial, anise extract reduced hot flash frequency and intensity in postmenopausal women (Nahidi et al. 2008).

Antimicrobal activity

The essential oil and several types of extracts of the aerial parts, fruits and seeds of *P. anisum* exhibited remarkable activities against a broad spectrum of Gram-negative and Gram-positive bacteria, and fungi (Tepe and Tepe 2015). In a study, the antibacterial activities of the aqueous and 50% (v/v) methanol extracts of *Pimpinella anisum* fruits were tested against four pathogenic bacteria *Staphylococcus aureus, Streptococcus pyogenes, Escherchia coli,* and *Klebsiella pneumoniae* by disc diffusion method. The results showed a fair antibacterial activity against all of the tested bacteria where the aqueous extract was found to be more effective than methanolic extract (Akhtar et al. 2008). Antimicrobial effects of water and ethanolic extracts of aniseed were effective against *Candida albicans* (Gülçın et al. 2003).

The essential oil extracted from the aerial part of Iranian *P. anisum* had an antimicrobial effect on *E. coli* (Ehsani and Mahmoudi 2012). Methanolic extract of the aerial part of *P. anisum* from Jordan increased the inhibitory effect of chloramphenicol, neomycin, doxycycline, cephalexin and nalidixic acid against multidrug resistant *E. coli* (Darwish and Aburjai 2010). In another study, essential oil and methanol extract of *P. anisum* seeds were evaluated for their single and combined antibacterial activities against *Staphylococcus aureus, Bacillus cereus, Escherichia coli, Proteus vulgaris, Proteus mirabilis, Salmonella typhi, Salmonella typhimurium, Klebsiella pneumoniae* and *Pseudomonas aeruginosa* with *Thymus vulgaris* (Al-Bayati 2008). Combinations between two essential oils resulted in high inhibitory activities against most of the pathogenic bacteria and mostly on *Proteus vulgaris* (Al-Bayati 2008).

Antifungal activities of fluid extract and essential oil of anise fruits were tested on seven species of yeasts and four species of dermatophytes by the means of diffusion method with cylinders and the broth dilution method. The aniseed extract showed anti-mycotic activity against *Candida albicans, C. parapsilosis, C. tropicalis, C. pseudotropicalis,* and *C. krusei*, and the largest inhibition zone was observed in *C. albicans*. It also showed inhibitory effect against dermatophyte species (*Trichophyton rubrum, T. mentagrophytes, Microsporum canis,* and *M. gypseum*), and the largest inhibitory zone was detected for *T. mentagrophytes* (Kosalec et al. 2005). The essential oil, however, exhibited stronger antifungal activities than its extract against yeasts and dermatophytes. Antifungal activity of anise essential oil was also reported against *Aternaria alternata, Aspergillus niger,* and *Aspergillus parasiticus*, and the most affected fungus was *A. parasiticus* (Özcan and Chalchat 2006).

Chemistry

Anise is well known for essential oil, which gives it the characteristics of odor and aroma. Although the major component of anise oil is trans-anethole (75–90%), other constituents include coumarins (umbelliferone, umbelliprenine, bergapten, and scopoletin), lipids (fatty acids, betaarmyrin, stigmasterol, and its salts), and flavonoids (flavonol, flavone, glycosides, rutin, isoorientin).

Phenolic compounds

Quantitative analysis of total phenolic compounds using HPLC indicated that the anise ethyl acetate extract contained more total phenolics (10.18 mg/g for Tunisian anise seeds (TAS) and 7.68 mg/g for Egyptian anise seeds (EAS)) than ethanol extract (7.44 mg/g for TAS and 5.73 mg/g for EAS). The phenolic compositions of TAS and EAS revealed that ethyl acetate extracts showed higher proportions of naringin, chlorogenic acid, and rosmarinic acid. However, ethanol extracts were richer in larcitrin, rosmarinic acid and cirsimartin. Six hydroxycinnamic acid derivatives and ten flavones derivatives, mainly luteolin and apigenin derivatives, were detected by HPLC and quercetin 3-glucuronide, rutin, luteolin 7-glucoside, isoorientin, isovitexin apigenin 7-glucoside and luteolin glycoside were isolated (Bettaieb Rebey et al. 2018).

Essential oil

Anethole is the major component of the aniseed oil where it forms 88.5%–96.9% of the total oil when oil is extracted by hydrodistillation or supercritical fluid (Rocha and Fernandes 2016). Other studies have demonstrated the presence of eugenol trans-anethole, methyl chavicol, anisaldehyde, estragole, coumarins, scopoletin, umbelliferone, estrols, terpene hydrocarbons, polyenes, and polyacetylenes (Gülçın et al. 2003). Other compounds that were found with a concentration higher than 0.06% were (E)-methyleugenol, α-cuparene, α-himachalene, β-bisabolene, p-anisaldehyde, and cis-anethole (Özcan and Chalchat 2006).

Fatty acids

Regarding the monounsaturated fatty acid (MUFA) proportion, it was shown to be predominant for Tunisian aniseeds and Egyptian aniseeds at 67.65% and 56.87%, respectively, where petroselinic acid was the most prevalent in seed oil at 46.60% and 38.40% for the Tunisian and Egyptian variety, respectively. Aniseed oil was also characterized by an important level of linoleic acid. Moreover, the amount of saturated fatty acids (SFA) in these oils was considerably low represented mainly by palmitic acid. It is worth noting that petroselinic acid is the major fatty acid component in Apiaceae seed oil rather than oleic acid.

The Carrot (*Daucus carota*)

There are about 28 species of Daucus, a herbaceous genus of the family Apiaceae of which the best-known species and one of the most relevant crops is the cultivated carrot. Carrots rank among the top 10 vegetable crops in the United States, China and Russia (Dawid et al. 2015). *D. carota* is a complex, very variable species comprising wild and cultivated carrots, resulting in a confused taxonomy. The complex is subdivided into 13 subspecies, 12 for wild taxa and one for the cultivated taxon (subsp. *sativus* (Hoffm.) Arc.). However, for cultivated carrot it is better to classify directly at cultivar level below the species level. In its wild state, it is a tiny, bitter root with little appeal as a food, but years of human cultivation and domestication has made it an extremely versatile vegetable, appearing in several colors, shapes, and sizes (Brothwell and Brothwell 1969, Stolarczyk and Janick 2011). Six wild species are grown in Egypt among which the two varieties *Daucus carota* var. *sativus* and var. *boissieri* are widely cultivated for the edible roots (Boulos 1999, Khalil et al. 2015).

Wild carrot is sometimes referred to as the subspecies *carota*, whereas the carotene-rich, the orange-rooted cultivated carrot is *D. carota* subsp. *sativus* (Hoffm.) Arcang. It differs from its wild ancestor, which has whitish/ivory colored roots, by being a shorter plant having larger, more palatable, less woody-textured edible taproots, normally orange, but different colors such as yellow-white or blackish-purple in other varieties and cultivars. Carrots are eaten both raw and cooked, in sweet and savory dishes and it is known for its high beta-carotene content, which the body converts to Vitamin A. It also forms a major ingredient in the food processing industry, and is a significant constituent of cosmetic products. The leaves are also consumed in salads and the seeds made into herbal tea (Stolarczyk and Janick 2011).

To unravel the history of the carrot, we have to go back a very long way. The carrot dates back about 5,000 years ago when the root was found to be growing in Middle Asia around Afghanistan and adjacent regions of Russia, Iran, India and Pakistan as the greatest diversity of wild types was found there (Vavilov et al. 1992). Drawings found in temples from Egypt in 2000 BC show a purple plant, which some Egyptologists believe to be a purple carrot (anthocyanin-containing). The first evidence of carrot being cultivated as a food crop was in the Iranian Plateau and Persia in the 10th century (Brothwell and Brothwell 1969). Throughout the centuries, Arab merchants travelled the trade routes of Arabia, Asia and Africa bringing home to their villages the seeds of the purple carrot. These carrots spread east and west from Iran and northern Arabia to be known across the Middle East including the Mediterranean area, North Africa, Western Europe and China by the mid-15th century (Simon 2000). During these years, the vegetable appeared in a variety of colors ranging from purple to white, pale yellow, red, green/yellow and black, but not the common orange that we see today (Brothwell and Brothwell 1969, Rubatzky et al. 1999, Stolarczyk and Janick 2011). In the late 1500s, Dutch agricultural scientists used a mutant yellow carrot seed to develop a stable, large, straight, sweet, orange carrot like the ones we eat today. These beta carotene-containing orange cultivars were developed to become the dominant species across the world and formed the basis for modern cultivars of ssp. *sativus* (Banga 1963).

Traditional uses

The species *Daucus carota* L. is recognized worldwide due to its roots, widely used for both food and medicinal purposes (da Silva Dias 2014). The record of the use of carrots in herbalism dates back to the 10th century, with mentions in the Old English Herbarium and the Leech Book of Bald (the oldest surviving medical textbook in England) indicating the use of the root as an emmenagogue as well as a treatment for smallpox and cough (http://www.carrotmuseum.co.uk/herbalists.html). Worldwide, both root and seed have documented historical uses, mainly to promote menstruation or as a diuretic. The essential oil was used as a flavor for liqueurs and perfumes. Seeds were used as a carminative, diuretic, stimulant, emmenagogue, and were used for chronic dysentery, kidney ailments, worms, as an aphrodisiac and for uterine pain. Roots were used as an infusion for threadworm, as a diuretic and for eliminating uric acid; carrots also belonged in the diet of gout-prone people (Reed 1976, Duke 1983). The ethnobotanical uses of its roots also included applications in the treatment of tumors, cancerous ulcers and wounds, mammary carcinoma, skin cancer, nephrosis and as a nephroprotective agent (Duke 1983, Duke et al. 2002, Kumarasamy et al. 2005). *D. carota* was used by the Ancient Egyptians as a stimulant, carminative, diuretic, anthelmintic and as a decoction for infantile diarrhea (Boulos 1999, Van Wyk and Wink 2004). It has been also used in the Lebanese traditional medicine to protect against gastric ulcer, treatment for cancer, diabetes and muscle back pain, to enhance liver function and boost immunity (Zgheib et al. 2014). Different parts of the carrot are used in Indian traditional medicine for the treatment of a broad spectrum of ailments including kidney dysfunction, asthma, dropsy, inflammation, leprosy, worm troubles, a local stimulant for ulcers, to regulate the functions of the stomach and intestines and it has been long considered to be as good for liver (Kirtikar and Baasu 1933). Most of the literature does not clearly identify the taxonomic status of the plant material used, so it has not been clearly possible to discriminate between results obtained with "wild carrot" and cultivated forms of the plant. Most chemical and pharmacological work on carrot has undoubtedly been conducted using cultivated material, but the information is not available to establish the degree to which this may differ from "wild carrot" seed in terms of chemistry and pharmacological activity (Figures 13.7a, b).

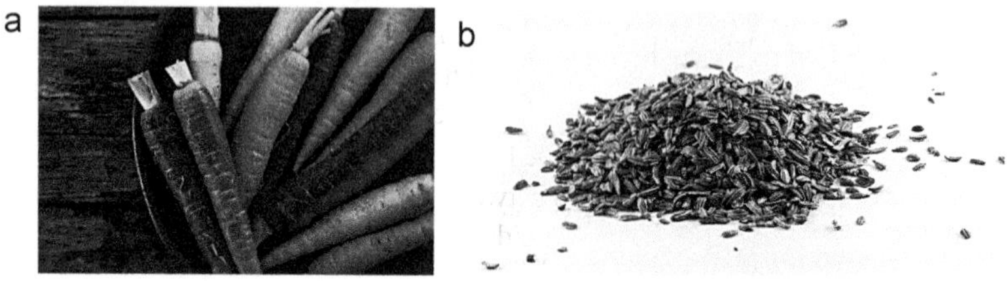

Figure 13.7. ***Daucus carota*** **varieties** (a); **fruit** (b) (Images purchased from Shutterstock).

Biological activities

Activity on the gastrointestinal tract

Many studies supported the traditional use of *D. carota* for gastric protection. The antisecretory, gastroprotective and antacid capacity of 50% ethanol extract from *D. carota* roots (EDC) was studied using experimental rats. The EDC significantly reduced gastric lesions, the volume of gastric content and the total acidity. Histopathological studies further confirmed the effects of EDC by inhibiting congestion, edema, hemorrhage, and necrosis in the gastric mucosa (Chandra et al. 2015). Additionally, the gastroprotective potential of the fresh juice and aqueous extracts of *D. carota* roots were further observed as they significantly decreased free acidity, total acidity and ulcer index, while they increased the pH and the mucus content against induced gastric ulceration in rats (Khatib et al. 2010, Jiin et al. 2014). Also, the aqueous and methanolic extracts of *D. carota* umbels exhibited significant protection against ethanol-induced gastric ulcer in rats (Wehbe et al. 2009). The observed gastroprotective effect of *D. carota* could be mainly attributed to the presence of flavonoids and other bioactive compounds as carotenoids by reducing gastric acidity and by protecting against free radical induced damage produced by the damaged mucosal cells (Gálvez et al. 2001, Amarowicz et al. 2004).

Nephroprotective and hepatoprotective effects

Daucus carota showed a significant nephroprotective effect in a dose-dependent manner by ameliorating the induced nephrotoxicity and thus authenticates its ethnomedicinal use. The renoprotective activity of *D. carota* root extracts (petroleum ether, ethanolic, fractional methanolic and methanolic extracts) was studied in renal ischemia-reperfusion injury in rats. The significantly decreased activity of superoxide dismutase, catalase, creatinine, uric acid, urea and reduced glutathione enzymes was significantly less in the group treated with *D. carota* root extracts, while the significant decrease in the high malondialdehyde levels suggested that *D. carota* extracts probably exerted renoprotective activity by the free radical scavenging activity (Mital et al. 2011, Afzal et al. 2013, Sodimbaku et al. 2016).

The hepatoprotective and antioxidant activity of methanolic extract of *D. carota* seeds was studied in experimental rats. A significant decrease in liver enzymes, serum glutamic pyruvic transaminase

(SGPT), serum glutamic-oxaloacetic transaminase (SGOT) and alkaline phosphatase (ALP) in blood serum with significant increase in superoxide dismutase (SOD), catalase (CAT), glutathione reductase (GRD), glutathione peroxidase (GPX), glutathione-S-transferase (GST) and lipid peroxidation (LPO) was estimated in all drug-treated groups (Singh et al. 2012). The extracts significantly lowered the serum levels of SGOT, SGPT, ALP, lactate dehydrogenase, sorbitol and glutamate dehydrogenase elevated by CCl4-induction in mice (Bishayee et al. 1995). The hepatoprotective effect of kaempferol isolated from *D. carota* leaves was tested in paracetamol induced liver damage of albino rats. In a dose dependent manner, oral treatment with kaempferol reversed all the serum and liver parameters (Jain et al. 2012).

Antidiabetic activity

The dichloromethane (DCM) extract of carrot roots stimulated insulin-dependent glucose uptake (GU) in adipocytes in a dose-dependent manner. The isolated polyacetylenes, falcarinol and falcarindiol significantly stimulated basal and/or insulin-dependent GU in 3T3-L1 adipocytes, porcine myotube cell cultures and peroxisome proliferator-activated receptor (PPAR) γ-mediated transactivation (El-Houri et al. 2015), while the administration of *D. carota* seed extracts in type-1 diabetic rats significantly decreased the elevated serum levels of glucose, cholesterol, triglycerides, creatinine, LDL, VLDL, aspartate aminotransferase, urea and alanine aminotransferase that are related to renal dysfunction and metabolic disturbance in diabetic rats (Pouraboli and Ranjbar 2010, 2012, 2015). Another research (Coyne et al. 2005) demonstrated a significant association between vitamin A-rich carotenoids and diabetes status suggesting that carrot and vitamin A-rich carotenoids might help diabetics to manage their condition. Additionally, the efficacy of purple carrot juice against metabolic syndrome was investigated. Purple carrot diet was supplemented in a high-carbohydrate, high-fat diet-fed rat model. Interestingly, there was a reduction in impaired glucose tolerance, endothelial function and abdominal fat deposits. The authors concluded that the anthocyanins of the carrot juice were responsible for the beneficial effect (Poudyal et al. 2010).

Cardiovascular effects

Daucus carota extracts caused a dose-dependent decrease in systolic and diastolic arterial blood pressure in normotensive anesthetized rats, a concentration-dependent decrease in force and rate of atrial contractions in spontaneously beating guinea-pig paired atria and inhibition of K+-induced contractions in rabbit thoracic aorta (Gilani et al. 1994, 2000). Aqueous extract of *D. carota* tubers was investigated for inotropic and cardioprotective effects by measuring various biochemical parameters. Cardiac tonicity was estimated by evaluating Na+K+-ATPase, Mg^{2+}-ATPase and Ca^{2+}-ATPase levels in the heart. The levels of Na+K+-ATPase and Mg^{2+}-ATPase were decreased and that of Ca^{2+}-ATPase were increased significantly. Cardioprotection was assessed by estimating the altered serum levels of aspartate transaminase, alanine transaminase, lipid peroxidase, and lactate dehydrogenase levels and cardiac total protein, lipid peroxidase, and lactate dehydrogenase, which were restored significantly by the administration of the extract (Muralidharan et al. 2008).

The associations between fruits and vegetables of different colors and 10-years coronary heart disease (CHD) incidence were examined and concluded that "a higher intake of deep orange fruit and vegetables, and especially carrots, may protect against CHD" (Griep et al. 2011).

Antioxidant and anticancer activities

Like many other colored vegetables, carrot is one of nature's antioxidant reservoirs. Carotenoids, polyphenols and vitamins present in carrot function as free radical scavengers (Robards et al. 1999, Sharm et al. 2012). Carrots of different colors displayed a high variation in antioxidant properties (Alasalvar et al. 2005). The results consistently indicated that among different carrot colors, purple manifested the highest antioxidant capacity due to its higher phenolic content (Leja et al. 2013). Additionally, the antioxidant properties of these compounds exhibited anti-carcinogenic and anti-aging properties, contributed to reducing the risk of cardiovascular diseases and protected against Alzheimer's disease (Zaman et al. 1992, Faulks and Southon 2001, Dias 2012).

The flavones isolated from the methanol extract of *D. carota* seeds were studied for antioxidant effects where luteolin showed the highest degree of free radical scavenging activity (Kumarasamy et al. 2005), while another study showed that the antioxidant features of the black carrot extracts appeared significantly higher than that of orange carrots (Algarra et al. 2014).

Treatment of leukemia cell lines with carrot juice extract induced apoptosis and caused cell cycle arrest (Zaini et al. 2011), while the effects of five fractions from carrot juice extract (CJE) [three polyacetylenes (falcarinol, falcarindiol and falcarindiol-3-acetate) and two carotenoids (beta-carotene and lutein)] were studied in human lymphoid leukemia cell lines. Treatment of all three lymphoid leukemia cell lines with the fraction from carrot extracts containing polyacetylenes and carotenoids showed that they were significantly cytotoxic. All the AML cell lines tested were sensitive to the extract, mainly by inducing apoptotic cell death (Tawil et al. 2015). These studies support the suggested use of carrots as a potential treatment for leukaemia in traditional medicine.

The polyphenolics of black carrot, mainly its anthocyanins, were effective for the risk reduction of different types of cancer and exerted a defense role against certain ailments, including diabetes, cardiovascular and oxidative stress (Andersen and Jordheim 2006, Poudyal et al. 2010, Ekinci et al. 2016). This was further supported by a study that showed the use of lyophilized powder of aqueous black carrot anthocyanins at doses ranging from 0.0 to 2.0 mg/mL, which inhibited the growth of HT-29 and HL-60 cancer cells (Jing et al. 2008, Akhtar 2017). Ethanol extracts of black carrot anthocyanins were used in the treatment of human colon, breast, and prostate cancers and were found to have antioxidant and anti-proliferative properties against different cancer cell lines (Netzl et al. 2007, Wang and Stoner 2008).

Additionally, *D. carota* pentane/diethyl ether fraction (50:50) caused a pronounced decrease in cancer cell motility and invasion in the investigated lung, skin, breast and glioblastoma cancer cell lines (Zgheib et al. 2014), while the anticancer activity of the pentane fraction of the *D. carota* oil extract was evaluated against human colon adenocarcinoma cell lines (HT-29 and Caco-2). Treatment of cells with various concentrations of F1 or F2 fractions produced a dose-dependent suppression of cell proliferation and induction of apoptotic cell death (Shebaby et al. 2015).

Analgesic and anti-inflammatory activities

The anti-inflammatory activity of *D. carota* has been authenticated on different inflammation models. The volatile oil of *D. carota* seeds resulted in inhibition of carrageenan-induced rat paw edema and writhing as well as led to the decrease in nitric oxide production in LPS-stimulated macrophages, indicating its anti-inflammatory and analgesic activities (Porchezhian et al. 2000, Alves-Silva et al. 2016). Additionally, carrot seed extracts had the potential to be consumed for preventing arthritic, gout and other COX-II enzyme-mediated inflammatory pain, mainly via 2,4,5-trimethoxybenzaldehyde, oleic acid and trans-asarone constituents (Momin et al. 2003). Moreover, the ethanolic extract of *D. carota* exhibited anti-inflammatory and analgesic effects via attenuation of the writhing responses, inhibition of carrageenan, histamine and serotonin-induced paw edema as well as formaldehyde-induced arthritis (Mani et al. 2006, Arya et al. 2011). The anti-inflammatory effect of aqueous extracts of *D. carota* root in acetic acid-induced experimental colitis has been demonstrated by inhibiting the release of oxido-inflammatory mediators such as myeloperoxidase and nitric oxide (Patil et al. 2012).

Effects on the reproductive system

The seeds of wild carrot have a long history of use relating to fertility, especially as an anti-fertility agent. Carrot seeds have been described as an abortifacient, emmenagogue and contraceptive throughout European history, with contemporary reports from India and the United States. These documents also indicate its use by women for over 2,000 years as a mean to control fertility (Riddle 1997, Tobyn et al. 2011, Maurya et al. 2004, Bennett and Schuler 2011). Scientific *in vivo* and *ex vivo* studies suggest that several modes of action may contribute to the anti-fertility effect, including an effect on the oestrous cycle and anti-progestogenic activity (Jansen and Wohlmuth 2014).

Animal studies of the anti-fertility activity of *D. carota* began in the 1970s, with scientists attempting to demonstrate efficacy and understand the actions of *D. carota* extract in rodents. The anti-fertility action of *D. carota* seed successively extracted with petroleum ether, 95% ethanol and water were tested on early pregnancy in albino rats. At 100 and 500 mg/kg body weight (BW), the ethanolic residue prevented implantation in 40% and 67% of animals, respectively. The corresponding figure was 60% and 40% for the aqueous residue and petroleum ether extract, respectively. The abortifacient activity of the extract was also noted in some animals (Garg and Garg 1971, Garg et al. 1978). In follow-up research, the effects of different chromatographic fractions of DC seed on fertility in albino rats were investigated. Results showed that pregnancy was inhibited most effectively by chloroform and methanol fractions of the petroleum ether extract with complete anti-implantation activity (Garg and Mathur 1972, Kamboj and Hawan 1982).

Petroleum ether extract of carrot seeds administered subcutaneously on days 7–13 of pregnancy caused abortion in most rats, and uterine and ovarian weight was considerably reduced with profuse vaginal bleeding within three days (Kaliwal et al. 1984). The oil of *D. carota* seed had 40% anti-implantation activity when administered 1–7 days post-coitally in rats at 500 mg/kg BW (Kamboj and Hawan 1982). The volatile oil was also found to terminate pregnancy in mice and rats when terpenoid fraction was administered subcutaneously. The peripheral plasma concentration of progesterone in pregnant rats decreased significantly 24 and 48 hours after subcutaneous injection, suggesting the anti-fertility activity may be related to the inhibition of progesterone (Chu et al. 1985).

Effect on intraocular pressure

In a study on rabbits, topical application of *Daucus carota* seeds extracts in normotensive rabbits resulted in diminished mean intraocular pressure (IOP). In experimentally elevated IOP, topical application of the extract with a concentration of 0.6% yielded 29.39% decrease of IOP in water loaded animals which was comparable to the efficacy of Pilocarpine and 30.27% decrease of IOP in steroid pretreated rabbits, which was significantly higher than that of Pilocarpine (Agarwal et al. 2008).

Antimicrobial activity

The *in vitro* antimicrobial activity of essential oils of *D. carota* seeds was evaluated, using the disk-diffusion method, against one Gram-positive (*Staphylococcus aureus*) and two Gram-negative bacteria (*Escherichia coli* and *Salmonella typhimurium*), and a pathogenic yeast (*Candida albicans*). All tested essential oils exhibited antibacterial and antifungal activities against the assayed microorganisms (Rokbeni et al. 2013). The antimicrobial effect of wild *Daucus carota* seed extracts (70% and 40% ethanol) was examined against Gram-positive (*S. aureus, Staphylococcus hyicus*—isolated from the soil, *Micrococcus luteus*—isolated from soil, criptogamic culture of *Bacillus subtilis*), Gram-negative (*P. aeruginosa, E. coli* and *Salmonella abony* and *Acinetobacter johnsonii*—isolated from the environment, *Moellerella wisconsensis*—isolated from the environment) and fungi (*C. albicans, Candida utilis, Saccharomyces cerevisiae* and *Aspergillus brasiliensis*). The extracts were more active against Gram-positive bacteria and fungi than against the strains of Gram-negative bacteria (Pavlyuk et al. 2015).

The antimicrobial activity of the essential oils of the flowering and mature umbels, with seeds of wild *D. carota* L. subsp. *carota* from two different sites in Tunisia, were assayed by using the broth dilution method on *E. coli* and *S. aureus*, and clinical strains of *C. albicans* and *C. tropicalis*. The MIC values obtained were all > 2.5% (v/v) (Marzouki et al. 2010). The antimicrobial activity of the essential oil of *D. carota* subsp. *carota* from Portugal was evaluated against several Gram-positive and Gram-negative bacteria, yeasts, dermatophytes, and *Aspergillus* strains. The results showed a significant activity towards Gram-positive bacteria, *Cryptococcus neoformans*, and dermatophytes (Alves-Silva et al. 2016).

Activity on CNS

The effects of the ethanolic extract of *D. carota* seeds on cognitive functions, total serum cholesterol levels, and brain cholinesterase activity has been studied in mice. The extract showed significant improvements in the memory scores and reduced the brain acetylcholinesterase activity and cholesterol levels of young and aged mice. It also reversed the amnesia induced by scopolamine and diazepam (Vasudevan and Parle 2006). The seeds contain choline and can inhibit brain cholinesterase activity, and may elevate the brain acetylcholine levels via increased synthesis of acetylcholine. Furthermore, the antidepressant activity of ethanol root extract of *D. carota* was comparable to that of standard drugs such as Fluoxetine in different animal models (Babu et al. 2014).

Chemistry

A variety of bioactive compounds from different chemical classes have been identified in *D. carota* including phenolics, carotenoids, polyacetylenes, ascorbic acid, essential oil and others.

Phenolic compounds

Phenolic compounds are present in high concentrations in the root periderm tissues of carrots as the concentrations of phenolic compounds in different carrot root tissues decrease from the peel (periderm) to the xylem. However, the concentration depends on the cultivar, the extraction method, the manner in which the results are expressed, and the post-harvest and processing circumstances (Zhang and Hamauzu 2004, Alasalvar et al. 2005). The fresh purple genotypes have 2.9 times higher phenolic concentrations (102 ± 3.8 mg/100 g) than orange genotypes (34.8 ± 1.9 mg/100 g) (Alasalvar et al. 2005).

Carrot roots contain hydroxycinnamic acids and derivatives with chlorogenic acid as the main hydroxycinnamic acid identified in different carrot tissues, accounting for 42.2% to 61.8% of total phenolics, with the highest concentration being 54.1 mg/100 g in purple carrots (Zhang and Hamauzu 2004, Sharm et al. 2012). Carrots are also rich in phenolic acids, such as *p*-hydroxybenzoic, *p*-coumaroylquinic and caffeoylquinic acids, as well as in anthocyanins (Alasalvar et al. 2001, Gonçalves et al. 2013). Black carrots have relatively higher flavonoids contents as compared to red and orange cultivars. Quercetin, luteolin, kaempferol, myricetin and apigenin have been identified as major flavonoids present in carrots and the seed extracts of wild carrot (Bahorun et al. 2004, Marinova et al. 2005, Pavlyuk et al. 2015). Additionally, carrots of different colors displayed a high variation in antioxidant properties (Gajewski et al. 2007). The results consistently indicated that among different carrot colors, purple exhibited the highest antioxidant capacity due to its higher phenolic compound concentration (Leja et al. 2013).

Anthocyanins are mainly present in purple and black carrots. Cyanidin, delphinidin, petunidin, peonidin, malvidin and pelargonidin anthocyanins have been reported in black carrot (Sun et al. 2009, Arscott and Tanumihardjo 2010, Baranski et al. 2012). Anthocyanins available in black carrots have been extensively investigated and reviewed in the literature (Simon 2000, Kammerer et al. 2003, Surles et al. 2004). These water-soluble pigments are mainly responsible for imparting red, purple, and blue colors to many fruits, vegetables, flowers, and grains and have been widely used as natural colorants in food and beverage industries (Reed 2002, Mazza 2007). Five major anthocyanin pigments, two nonacylated cyanidins (cyanidin 3-xylosylglucosylgalactoside and cyanidin 3-xylosylgalactoside) and three cyanidins acylated with sinapic acid (cyanidin 3-sinapoylxylosylglucosylgalactoside, cyanidin 3-feruloylxylosylglucosylgalactoside(ferulicacid)andcyanidin3-*p*-coumaroylxylosylglucosylgalactoside (*p*-coumaric acid)) have been reported in black carrots (Alasalvar et al. 2001, Algarra et al. 2014).

Carotenoids

Carotenoids are named after carrot because carrot accumulates an enormous number of carotenoids in its roots. The different colors of carrots reveal their various concentrations of phytochemicals. Carotenoids give yellow, orange, and red carrots their colors. There are two types of carotenoids present in carrot, namely, carotenes and xanthophylls (Ahmad et al. 2019). The major carotenoids in carrot roots are β-carotene (75%); α-carotene (23%) and lutein (1.9%), lycopene, and zeaxanthin (Søltoft et al. 2011). Orange carrots contain high quantities of beta-carotene. Yellow carrots contain low quantities of beta-carotene, but higher levels of lutein, which may protect from age-related macular degeneration and be beneficial for eyesight. Red carrots contain lycopene—a potent antioxidant with potential anti-cancer activity—in concentrations similar to that of tomatoes. Red carrots also contain moderate levels of alpha- and beta-carotene and lutein. The white variety has low levels of these phytochemicals, but contains high levels of potassium. The dietary intake of carotenoids has been related to the protection of DNA, proteins, and lipids from oxidative damage, as well as to the maintenance of the normal function of normal skin, normal mucosal membranes, and normal vision (Søltoft et al. 2011, Mustafa et al. 2012, Sommer and

Vyas 2012, Elvira-Torales et al. 2019). Carotenoids have been also linked with the enhancement of the immune system and decreased risk of degenerative diseases such as cancer and cardiovascular disease (Faulks and Southon 2001).

Essential oil

The chemical composition of the steam-distilled carrot is complex and variable. Most published analyses have been of seed oil distilled from cultivated plants or from various subspecies of *D. carota.* Steam distilled carrot seed oil typically contains significant amounts of the sesquiterpene alcohol carotol and the monoterpenoid ester geranyl acetate, with highly variable amounts of the monoterpenoids *α*-pinene, sabinene and linalool, and of the sesquiterpenoids *β*-bisabolene and daucol (Lawrence 1992, 1999).

The chemical composition of the essential oils from the fruits of *Daucus carota* var. *sativus* (yellow carrot) and var. *boissieri* (red carrot) cultivated in Egypt was determined using GC/FID and GC/MS. A total of 29 and 32 compounds were identified accounting for 96.58 and 96.72% of the total detected components in the hydrodistilled yellow and red carrot oils, respectively. Yellow carrot fruit essential oil had a lower percentage of monoterpenes containing mainly *β*-pinene and *α*-limonene and in addition to oxygenated monoterpenes represented mainly by linalool (0.34%). Red carrot fruit essential oil had a higher percentage of monoterpenes containing mainly *β*-pinene and *p*-cymene and didn't contain any oxygenated monoterpenes. On the other hand, sesquiterpenes in yellow carrot fruit essential oil constituted mainly carotol, *β*-bisabolene, *trans*-*α*-bergamotene, germacrene and *α*-curcumene. As for red carrot fruit essential oil, carotol, *β*-bisabolene, *trans*-*β*-caryophyllene and *trans*-*α*-asarone were the main components (Khalil et al. 2015).

A total of 48 and 46 compounds were identified in the essential oil from leaves and seeds of wild carrot. The seeds essential oil was predominantly composed of oxygenated monoterpens (66.08%) and oxygenated sesquiterpenes (16.41%). The main components were geranyl acetate (52.45%), cedrone S (14.04%), and asarone (11.39%). The oil from leaves is mainly composed of monoterpene hydrocarbons (64.59%) and sesquiterpene hydrocarbons (22.18%) with *α*-pinene (27.44%), sabinene (25.34%) and germacrene D (16.33%) as its main components (Ksouri et al. 2015). While ninety-one compounds were identified in the essential oil of carrot leaves (*Daucus carota* L. subsp. *sativus*) from Iran, the main classes were monoterpenes (30.0%), sesquiterpenes (27.8%) and phenyl propanes (26.4%). The major constituents were *trans*-anethole (23.5%) and myrcene (14.5%) (Mojaba et al. 2008).

The analysis of the cultivated carrot seed and edible seed oil in Turkey showed that they contained 34 and 14 components, respectively. The major constituents of seed essential oil were carotol (66.78%), daucene (8.74%), (*Z*,*Z*)-*α*-farnesene (5.86%), germacrene D (2.34%), *trans*-*α*-bergamotene (2.41%), *β*-selinene (2.20%), *β*-bisabolene (1.90%), bicyclogermacrene (1.87%) and *β*-caryophyllene (1.10%). The fatty acid composition of the oil showed that it contained (mg/100 g): palmitic: 10.01 ± 0.13, palmitoleic: 0.64 ± 0.02, stearic: 2.41 ± 0.06, oleic: 0.17 ± 0.01, linoleic: 11.82 ± 1.17, petroselinic: 59.35 ± 3.81, vaccenic: 0.55 ± 0.01 and arachidic: 0.81 ± 0.03 (Özcan and Chalchat 2007). The essential oil of *Daucus carota* ssp. *maximus* fruits revealed the presence of at least 19 components. The fruit essential oil consisted chiefly of phenylpropanoids and sesquiterpene hydrocarbons, accompanied by relatively much smaller amounts of monoterpenes and oxygenated sesquiterpenes.

The essential oils of *Daucus carota* seeds from ten wild populations spread over northern Tunisia were characterized by a predominance of sesquiterpene hydrocarbons in most samples. The main volatile compounds identified were *β*-bisabolene, sabinene, geranyl acetate and elemicin (Rokbeni et al. 2013). Phytochemical investigation of the fruits of *Daucus carota* resulted in the isolation of a new sesquiterpene named as daucucarotol (Fu et al. 2009).

Polyacetylenes

Twelve polyacetylenes were isolated from carrot, where falcarinol, falcarindiol and falcarindiol-3-acetate are the essential ones predominately found in the roots. The other nine polyacetylenes isolated are: (E)-isofalcarinolone, falcarindiol-8-acetate, 1,2-dihydrofalcarindiol-3-acetate, (E)-falcarindiolone-8-acetate, (E)-falcarindiolone-9-acetate, 1,2-dihydrofalcarindiol, (E)-1-methoxyfalcarindiolone-8-acetate, (E)-1-methoxy-falcarindiolone-9-acetate, and panaxydiol (Schmiech et al. 2009, Dawid et al. 2015).

Their distribution has been shown to be linked to the cultivars and the concentration of polyacetylenes found, maybe 10–20 times more in wild type of carrots than domesticated carrots (Pferschy-Wenzig et al. 2009, Dawid et al. 2015).

In vitro biological studies on plant extracts containing falcarinol-type polyacetylenes showed they have anti-cancer and anti-inflammatory actions. Polyacetylenes are extremely cytotoxic against several cancer cell lines and have revealed antifungal and anti-inflammatory properties (Kreutzmann et al. 2008, Baranska et al. 2013). It was suggested that the hydroxyl group (–OH) at C3 may account for these activities and that synergistic interaction between bioactive polyacetylenes may be important for their bioactivity (Purup et al. 2009). It was found that C17-polyacetylenes inhibit the breast cancer resistance protein BCRP/ABCG2 when used as a multidrug resistance reversal agent (Tan et al. 2014). Falcarinol and falcarindiol may be used as antidiabetic agents in the treatment of diabetes due to their ability to arouse basal or insulin-dependent glucose absorption in adipocytes and porcine myotube cell cultures based on different doses (El-Houri et al. 2015). Additionally, polyacetylenes of purple carrots are involved in anti-inflammatory bioactivity in humans with falcarinol being the most bioactive polyacetylene (Metzger et al. 2008).

Celery (*Apium graveolens*)

Apium graveolens Linn (Celery), known as karafs in Arabic, is a common plant of Arab, Islamic and Ayurvedic traditional medicine systems. This annual or biennial herb has originated from the Middle East and the Mediterranean basin (Li et al. 2018), and is currently cultivated worldwide for its leaves, fruits, bulbous roots, volatile oil and oleoresin. Three major taxonomic morphotypes of celery are currently used: *Apium graveolens* var. *secalinum*, known as leaf celery or smallage type, mainly used in the Mediterranean region, *Apium graveolens* var. *dulce*, known as celery, stalk celery or blanched celery and *Apium graveolens* var. *rapaceum*, known as celeriac or celery root (Malhotra 2012).

Traditional uses

In various folk medicine systems, celery is primarily used as diuretic, urinary tract antiseptic, a sedative in sleep disturbances, aphrodisiac and emmenagogue (Al-Asmari et al. 2017). Other traditional uses of celery include antiarthritic, anthelmintic, antispasmodic for griping pains, carminative and laxative (Fazal and Singla 2012). In Ayurveda, celery is prescribed for cold, flu, maldigestion, water retention, arthritis and disorders of liver and spleen (Fazal and Singla 2012). However, the excessive intake of celery is discouraged in kidney inflammation and during pregnancy (Kooti et al. 2015).

Figure 13.8. Celery stem and leaves (Image by Ptanpm from Pixabay).

Biological activities

Effect on fertility

In vivo studies validated the positive effects of celery on fertility. Celery elicited a protective role on the testicular tissues against the antiepileptic agent sodium valproate. In addition, celery improved spermatogenesis and amplified the sperm profile (Hamza and Amin 2007, Hardani et al. 2015). It is noteworthy to mention that chronic use of high doses of *A. graveolens* can exert adverse effects on fertility (Kooti et al. 2017).

Effect on the urinary tract

Conforming to its traditional use as a diuretic plant, celery accentuates urinary calcium excretion and prevents calcium deposition in renal parenchyma which can help prevent and treat nephrocalcinosis (Al Jawad et al. 2011).

Antioxidant activity

Several *in vitro* and *in vivo* models demonstrated the powerful antioxidant capacity of celery juice, aqueous, ethanol, methanol and butanol extracts of leaves, fruits and roots, besides the essential oil. This was manifested by interesting DPPH radical scavenging capacity and an increase in ferric reducing antioxidant power (FRAP), cupric ion reducing antioxidant capacity (CUPRAC), total antioxidant capacity (TAOC), glutathione content and antioxidant enzymes in liver homogenate (Jung et al. 2011, Kooti and Daraei 2017).

Antidiabetic and hypolipidemic effects

The fruit extract of *A. graveolens* showed amelioration in serum level of insulin and HDL in streptozotocin-induced diabetic rats. On the histopathological level, celery exhibited protective effects against atrophy, necrosis and inflammation of the pancreatic tissue (Tashakori-Sabzevar et al. 2016, Yusni et al. 2018). The hydroalcoholic extract of *A. graveolens* ameliorated the lipid profile in rats fed with high-fat diet by reducing triglycerides, cholesterol and LDL levels (Kooti et al. 2014). The hypolipidemic effects of celery were suggested to be mediated by inhibition of cholesterol hepatic biosynthesis in addition to impairment of lipid absorption from the intestine and increasing fecal bile excretion (Al-Snafi 2014).

Anti-inflammatory and antiarthritic effects

Celery extracts have been proved to reduce the carrageenan-induced paw edema and inflammation (Ramezani et al. 2009, Powanda and Rainsford 2011). In hyperuricemic rats, a decrease in serum uric acid and xanthine oxidase levels was achieved by treatment with celery fruit aqueous extract and oil. In acute gouty arthritic rats, the same treatment successfully alleviated joints swelling and inflammatory infiltration, besides the decrease in interleukin (IL)-1β and TNF-α levels (El-Rahman and Abd–Elhak 2015, Li et al. 2019). The antigout and antiarthritic potential of celery can be explained in light of its powerful antioxidant and anti-inflammatory properties.

Effects on the gastrointestinal tract

Like most of the Apiaceae, celery has been proved to possess a spasmolytic effect on griping pains. This antispasmodic effect was corroborated on rat ileum and was attributed to the presence of apigenin (Branković et al. 2015).

Moreover, celery aqueous and alcoholic extracts exhibited significant protective and antiulcer effects on gastric mucosa against alcohol-, indomethacin- or strong alkali-induced ulcers comparable to omeprazole (Al-Howiriny et al. 2010).

Effects on the cardiovascular system

The leaf extract of *A. graveolens* decreased systolic blood pressure and lipid parameters and increased the heart rate in fructose-induced and deoxycorticosterone acetate-induced hypertensive rats (Moghadam et al. 2013, Dianat et al. 2015). Interestingly, no effect on blood pressure or heart rate was observed in normotensive rats. One of the major components of celery fruit oil, 3-n-butylphthalide, displayed antihypertensive and vasorelaxant activity (Sowbhagya 2014). Accordingly, celery can be considered as a potential herbal treatment for chronic hypertension in humans.

Other biological effects

The neuroprotective effects of 3-n-butylphthalide from celery have been illustrated and proved to be mediated by its angiogenic and antioxidant properties in addition to its role in impairing β-amyloid metabolism (Liang et al. 2015). The later plays an important role in the pathogenesis of Alzheimer's and Parkinson's diseases (Niu et al. 2019). Celery is a promising herbal drug for the treatment of vascular dementia. The same compound exhibited significant antiplatelet aggregation and antithrombotic properties which can inhibit several ischemic cardiovascular and cerebrovascular disorders (Peng et al. 2004).

Celery fruit oil demonstrated powerful larvicidal and repellent activity against *Aedes aegypti*, known as a vector for many pathogens. This represents a biodegradable and eco-friendly solution in mosquito intervention measures (Kumar et al. 2014).

Chemistry

Volatile oil

A. graveolens is characterized by a peculiar fragrance attributed to its volatile constituents. Celery leaves display the highest yield of volatile oil which is mainly composed of α- and β-pinene, myrcene, limonene, 1-dodecanol, 9-octadecen-12-ynoic acid methyl ester and tetradecence-1-ol acetate (Malhotra 2012, Nagella et al. 2012). On the other hand, the fruit volatile oil contains mainly limonene, phthalides as 3-n-butyl phthalide, sedanenolide and sedanolide in addition to α- and β-selinene, apiol, sedanonic anhydride and caryophyllene (Hassanen et al. 2015). The fruit oil is more enriched in sesquiterpenoids than leaf oil, while the leaf oil contains more monoterpenoids (Sowbhagya 2014). Phthalides are responsible for most of the medicinal effects of celery including anti-inflammatory, hypotensive and insecticidal activities (Sowbhagya 2014).

Phenolic compounds

The phenolic acids identified in celery are caffeic and chlorogenic acids. In addition, a wide range of flavonoids has been isolated from celery fruits including apigenin, luteolin, kaempferol, graveobioside A and B and quercetin whose biological activities extend from antioxidant and anti-inflammatory to anticancer, vasorelaxant and cardioprotective effects (Rathee et al. 2009, Tapas et al. 2008, Zhang et al. 2000, Ziyan et al. 2007). Various furo- and 7-hydroxycoumarins have been also isolated from celery fruits such as bergapten, 8-hydroxyl 5-methoxypsoralen known for their antipsoriatic activity, isopimpinellin, isoimperatorin, apiumoside, apiumetin, apigravrin, osthenol and celereoside (Kooti et al. 2015, Al-Asmari et al. 2017).

Miscellaneous compounds

In addition to flavonoid glycosides, other glycosides were isolated from celery fruit. These include the sesquiterpenoid glucosides celeriosides and the phthalide glycosides celephtalides (Kitajima et al. 2003).

The chief fatty acids of *A. graveolens* fixed oil were oleic, palmitic, petroselinic, petriselaidic, stearic and myristic acids (Malhotra 2012).

Coriander (*Coriandrum sativum*)

Coriandrum sativum L. was one of the earliest spices used by mankind. Coriander, also called cilantro or Chinese parsley, is an annual herb that reaches 0.20–1.40 m high at flowering (Diederichsen 1996). It probably originated from the Eastern-Mediterranean area and is currently cultivated all over the world (Small 2006, Wei et al. 2019). The plant's name is derived from the Greek name of bug "Korion" because of the characteristic odor of the green herb which is similar to that of bugs (Diederichsen 1996). This odor is due to different aldehydic components of the essential oil that disappear upon ripening leaving linalool as the major component of coriander essential oil (Msaada et al. 2007). This is why the odor of ripe fruits (schizocarpic cremocarp) is very different from the green plant (Parthasarathy et al. 2008). It is worth noting that the genus *Coriandrum* constitutes the cultivated species *Coriandrum sativum* (coriander) and the wild species *Coriandrum tordylium*. The mature fruits, incorrectly termed seeds, are largely used all over the world for flavoring sweets, beverages and tobacco products and in perfumery, besides their culinary uses, especially in Middle-Eastern cuisine where both the fruits and the herbaceous aerial parts are widely used.

Traditional uses

The use of coriander has been long-standing in the Mediterranean area and the best have been imported from Egypt (Panagiotakopulu et al. 1995, Diederichsen 1996), where its traditional use dates back to the pharaonic era. Coriander is named in Ebers papyrus (1550 BC) and remains of coriander mericarps have been found in the tomb of Tut-Ankh-Amon (Germer 1989, de Vartavan 1990) and other Egyptian tombs which show that coriander had a mythological significance. In the ancient Egyptian culture and medicine, coriander was considered to have cooling, stimulant, carminative and digestive properties. It was prescribed in the form of tea for digestive and urinary complaints including cystitis (Aboelsoud 2010). In addition, it has been noted for several purposes: as a preservative for meat, a poison for intestinal parasites, an insect repellent, an additive in flour and other food and medicine with magical properties (Rackham et al. 1947, Panagiotakopulu et al. 1995). Herein, we focus on the biological activities that have been scientifically confirmed by pharmacological studies (Figures 13.9a, b).

Figure 13.9. Coriander leaves (cilantro) (a); **fruit** (b) (Image (a) by restyledliving from Pixabay; Image (b) by Gokalp Iscan from Pixabay).

Biological activities

Activity on the gastrointestinal tract

The most popular traditional use of coriander in the Middle-East is for the relief of gastrointestinal complaints such as anorexia, indigestion, flatulence, griping and abdominal spasm. This use was supported by the finding that coriander stimulates gastric acid secretion via the cholinergic pathway (Vasudevan et al. 2000). Interestingly, coriander fruit hydro-methanolic crude extract demonstrated the spasmogenic effect, followed by a spasmolytic effect at higher doses on rabbit jejunum (Jabeen et al. 2009). Fractionation of the crude extract revealed that the cholinergic spasmogenic activity of coriander was separated in its aqueous fraction, while the organic fractions showed a dose-dependent spasmolytic effect. This spasmolytic activity of coriander has been proved to be mediated through Ca^{2+} channel blockade (Jabeen et al. 2009). Flavonoids as quercetin, apigenin and naringenin and EO constituents as linalool are reported to exert an antispasmodic effect (Gharzouli and Holzer 2004, Kundu et al. 2014). This explains the coriander's relaxing effect on hyperactive gut disorders such as abdominal colic and diarrhea.

Activity on the urinary system

The traditional use of coriander as a diuretic agent and to treat urinary tract infection is reputed in the oriental culture (Abu-Rabia 2005). The diuretic effect of coriander fruit hydro-methanolic extract was validated by an increase in urine output (Jabeen et al. 2009), glomerular filtration rate and increased urinary loss of electrolytes, like furosemide (Aissaoui et al. 2008).

Antidiabetic activity

Coriander has been used traditionally to manage diabetes (Otoom et al. 2006, Tahraoui et al. 2007). The antihyperglycemic effect of coriander fruits has been authenticated in streptozotocin-diabetic mice, in obese–hyperglycemic–hyperlipidemic rats and clinically in type-2 diabetic patients (Aissaoui et al. 2011, Swanston-Flatt et al. 1990, Waheed et al. 2006). Also, the methanolic extract of coriander leaves demonstrated a significant decrease in blood glucose levels (Mazhar and Mazumder 2013). The antihyperglycemic effect of coriander was found to be associated with stimulation of insulin release from clonal B-cell line, enhancement of glucose uptake and metabolism by muscles (insulin-like action) (Gray and Flatt 1999) and competitive inhibition of α-glucosidase (Brindis et al. 2014).

Antihyperlipidemic activity

The *in vivo* antihyperlipidemic effect of fresh leaves of *Coriandrum sativum* was manifested by reduced levels of serum LDL-cholesterol and triglycerides, increased level of HDL-cholesterol, increased fecal excretion of cholesterol and phospholipids and less cholesterol deposits in the aorta of high cholesterolemic animals (Kousar et al. 2011, Joshi et al. 2012). Among the proposed mechanisms is the reduction of cholesterol biosynthesis through the inhibition of 3-hydroxy-3-methylglutaryl CoA reductase (HMG-CoAR) enzyme, the enzyme that catalyzes the rate-limiting step in cholesterol biosynthesis (Cho et al. 2011).

Antioxidant activity

Coriander essential oil (EO) and extracts have been proposed as natural antioxidants that can extend the shelf-life of foods and cosmetics. Furthermore, the pathogenesis of many diseases is related to abnormally elevated levels of reactive oxygen species (ROS). Such diseases include diabetes, atherosclerosis, inflammation, dementia and cancer. The antioxidant power of coriander can reduce oxidative stress, participate in the body's own antioxidant system and hence fight these diseases. Coriander antioxidant potential certainly contributes, at least in part, to many of its biological effects. Commercial coriander EO and linalool exhibited good radical scavenging activities and inhibited lipid peroxidation (Duarte et al. 2016). The fixed oil and its fractions were even better than nigella oil in radical scavenging activity. This activity is positively correlated with the total content of polyunsaturated fatty acids (PUFA) (Ramadan et al. 2003). Interestingly, coriander fruit aqueous extracts proved to be stronger than ascorbic acid in the inhibition of lipid peroxide and hydroxyl radicals. This may be attributed to the glycosides of terpenoids and flavonoids (Satyanarayana et al. 2004). Moreover, the methanolic extracts of the freeze-dried cilantro leave demonstrated significant DPPH radical-scavenging and ferrous ion-chelating activities (Wong and Kitts 2006), while ethanol extracts exhibited ferric-reducing antioxidant power (FRAP), *in vitro* anti-lipid-peroxidation and DNA protective activity (Harsha and Anilakumar 2014).

Among different solvents, the ethyl acetate extract showed the highest phenolic content, FRAP and DPPH scavenging activity (Tang et al. 2013). It should be noted that coriander whole crude extracts are more powerful antioxidants than their fractions, a fact that can be explained by the synergism among various constituents. It is noteworthy that thermal processing doesn't significantly affect the chemical composition and antioxidant profile of coriander fruit extracts (Zhang et al. 2015).

Analgesic and anti-inflammatory activities

The analgesic effect of different extracts of coriander fruit and aerial parts has been established against acute and chronic pain models in various nociceptive tests using heat and chemical stimuli (Taherian et al. 2012, Kazempor 2015). Coriander antinociceptive effect is suggested to be mediated, at least partially, by the activation of the opioid system (Taherian et al. 2012).

The use of coriander in the traditional treatment of arthritis and other inflammatory conditions has been authenticated in exhaustive animal and clinical studies. In this context, the role of a traditional formulation containing coriander in carrageenan-induced rat paw edema and in human rheumatoid arthritis patients has been studied revealing an enhancement in pain tolerance and in the mobility of the affected joints (Thabrew et al. 2003). Biochemically, the hydroalcoholic extract of coriander fruit reduced the levels of serum inflammatory mediators such as interleukin-6 and interleukin-1β, reduced the expression of the inducible nitric oxide synthase (iNOS), cyclooxygenase-2 (COX-2), and decreased the production of NO and prostaglandin E2 (PGE2) in rat models (Wu et al. 2010, Nair et al. 2013). Interestingly, comparable anti-inflammatory activities have been observed for raw and roasted coriander fruits (Zhang et al. 2015). However, the oral intake of coriander produces very little or no reliable antinociceptive or anti-inflammatory activities in mice. The intraperitoneal route is much preferred (Begnami et al. 2018).

Activity on CNS

Coriander has been recommended for the relief of anxiety and insomnia in several traditional medicine systems (Duke 2002). Pharmacological studies conducted by Emamghoreishi et al. validated this beneficial attribute and proved that the aqueous extract of coriander fruit has anxiolytic, sedative and hypnotic effects (Emamghoreishi et al. 2005, Emamghoreishi and Heidari-Hamedani 2006). The anxiolytic activity of coriander is ascribed to linalool, which is well known for its sedative and relaxant properties through various mechanisms including inhibition of glutamate release, modulation of monoamine neurotransmitters and stimulation of $GABA_A$ receptors (Cheng et al. 2015, Prachayasittikul et al. 2018).

Antimicrobial activity

The coriander oil and different extracts of leaves and fruits displayed varying levels of inhibitory and germicidal effects against wide variety of pathogenic microorganisms including *Streptococcus pyogenes*, methicillin-resistant *Staphylococcus aureus*, *Listeria monocytogenes*, *Enterococcus faecalis*, *Klebsiella pneumoniae*, *Proteus mirabilis*, *Pseudomonas pseudoalcaligenes*, *Bacillus cereus*, *B. subtilis*, *Flavobacterium indologenes*, *Escherichia coli*, *Providencia alcalifaciens*, *Pseudomonas aeruginosa*, *P. fluorescens*, *Salmonella typhimurium*, *S. typhi*, *Yersinia enterocolitica*, *Shigella dysenteriae*, *Acinetobacter lwoffi* and *Acinetobacter baumannii* (Alves et al. 2016, Casetti et al. 2012, Yildiz 2016). Coriander fruit essential oil demonstrated better antimicrobial activity than its main components (Alves et al. 2016). The oil can also have a synergistic effect with conventional antibiotics (Duarte et al. 2012). Coriander EO and its major component linalool displayed anti-campylobacter activity. They can also inhibit biofilm formation and disrupt the quorum sensing (QS) system even at sub-MICs (Duarte et al. 2016).

The EO of aerial parts also shows an interesting antifungal activity against *Microsporum canis* and oral *Candida albicans*. The major antifungal constituents are 2-Hexen-1-ol, 3-hexen-1-ol and cyclohexane. Interestingly, a combination of these compounds exhibited a synergistic effect (Furletti et al. 2011). The oil and linalool also reduce aflatoxin B1 (AFB1) synthesis by *Aspergillus flavus* (Lasram et al. 2019). The antibacterial, antifungal, antiaflatoxinogenic antioxidant and lipid peroxidation inhibitory effects of coriander make it an excellent candidate in food and cosmetic preservation, supporting its traditional use as meat, flour and food additive to prevent food-borne diseases and food spoilage (Darughe et al. 2012, Khan et al. 2013).

Chemistry

A variety of bioactive compounds from different chemical classes have been isolated from coriander including essential oils, fixed oils, polyphenols, sterols and others.

Polyphenols

Many phenolic compounds from different classes have been identified in coriander seeds and leaves, including phenolic acids (caffeic, ferulic, gallic, chlorogenic, *o*-coumaric, *p*-coumaric, rosmarinic and salicylic acids), flavonoids (quercetin, rutin, luteolin, kaempferol, naringin, apigenin, hyperoside and catechin) and coumarins (esculin, esculetin, scopoletin, 4-hydroxycoumarin and umbelliferone).

The total phenolic and total flavonoid contents of coriander seeds and leaves from different geographical origins have been extensively studied together with their corresponding antioxidant potential. The variation was due to genotype, geographical origin and, most importantly, the extraction solvent. Generally, the total flavonoid content was significantly higher in leaves than in fruits or seeds. While the vegetative parts demonstrate the presence of good amounts of quercetin and its derivatives (quercetin-3-O-rutinoside), fruits displayed mostly phenolic acids, especially caffeic acid and its derivatives (caffeoyl N-tryptophan hexoside) (Barros et al. 2012).

Investigation of methanolic extracts of coriander seeds from different geographical origins revealed that Syrian coriander possesses higher total phenolic content than Tunisian and Egyptian cultivars (all in the range 0.94–1.09 mg GAE/g dry weight). Similarly, the highest total flavonoid and total condensed tannin contents of seeds methanolic extracts have been observed in the Syrian coriander (Msaada et al.

2017). Interestingly, the methanolic seed extract of a Canadian cultivar showed a total phenolic content of 15.16 mg GAE/g dry weight (Sriti et al. 2011). In addition, anthocyanins have been detected in leaves of Jordanian wild coriander (Ereifej et al. 2015). They are also found in *in vitro* grown coriander samples (Rahimi et al. 2013).

Carotenoids

Coriander fresh leaves are considered a good source of carotenoids, especially β-carotene, which accumulates in mature leaves in larger quantities than the seedling stage (Divya et al. 2012). Five fractions have been isolated from coriander leaf ether extracts, containing β-carotene, β-cryptoxanthin epoxide, lutein-5,6-epoxide, violaxanthin and neoxanthin. However, each of them displayed less antioxidant potential than the crude extract suggesting synergistic effects among these compounds (Guerra et al. 2005).

Essential oil

Volatile constituents remain the most widely used and studied components of *C. sativum* due to their relevance to most of coriander biological activities. However, it should be noted that essential oil composition is affected by several factors including the cultivar, geographical origin, growing conditions, harvesting time, post-harvest processing, storage duration and conditions and, most significantly, plant organ and developmental stage. All organs produce oil to different degrees: fruits, seeds, leaves, stems, flowers and roots. (E)-2-Decenal is the main constituent of coriander leaves oil, while stems contain 1-ethenyl-cyclododecanol and phytol (Chung et al. 2012, Wei et al. 2019). Linalool is the major component of mature fruit oil accounting for 25–87.5% of the oil (Msaada et al. 2009). Linalool content below 55% is usually associated with coriander native to Asia. Aldehydes constitute collectively about 49.22% of coriander roots oil and the major compound is 2-dodecenal (41.82%). Flowers' oil contains linalool and citronellol as main constituents (Punetha et al. 2018).

Fruits are the most widely used part of coriander with approximately in addition to the oleoresin which is produced on a relatively small scale (Carrubba et al. 2002). Coriander oleoresin normally contains about 90% fixed oil and 5% volatile oil (Parthasarathy et al. 2008). Two varieties of coriander exist: var. *microcarpum,* which is characterized by small fruits, and var. *vulgare,* which is large-fruited (Laribi et al. 2015). European, Caucasian and Russian coriander grown in cool, wet conditions are usually small-fruited with an oil content greater than 0.5%(Diederichsen 1996, Wei et al. 2019). Some Russian cultivars display the highest OIL yield (Parthasarathy et al. 2008). Indian and North-African corianders produce larger fruits and are considered inferior to the European types with less oil content. In addition, linalool content is higher in small-fruited cultivars than the large-fruited ones (Telci et al. 2006).

The method of extraction can influence oil composition. Among different preparation methods, hydrodistillation affords the highest linalool content of coriander fruit oil (835.2 mg/g extract) followed by subcritical water extraction. The least linalool content was obtained by Soxhlet extraction (52.5 mg/g extract). Similarly, hydrodistillation provides the highest content of γ-terpinene and (+)-limonene (Pavlić et al. 2015).

Variations in oil composition were observed during different stages of plant growth. At the seedling stage, the oil composition of leaves and stems was almost the same with decanal, (E)-2-decenal, (E)-2-undecenal, 2-dodecenal, and (E)-2-tetradecenal as major constituents. These aliphatic aldehydes, present in the vegetative parts and unripe fruits, impart a characteristic fetid odor to the seedling and the green herb, an odor that is appreciated in Asian cuisine (Kohara et al. 2006). Upon ripening, fruits gain a more agreeable aroma due to the presence of large amounts of monoterpene alcohol, linalool (Parthasarathy et al. 2008). In fact, the oil profile of coriander fruits changes considerably during maturation. A sharp decrease of monoterpene esters is accompanied by a concomitant rise in monoterpene alcohols. Geranyl acetate, for example, drops from 46.3% to less than 0.9% and linalool increases from 10.9% to 87.5% as the fruits transform from the immature stage to the final stage of maturity. Other constituents of fruit oil include neryl acetate, γ-terpinene, α-pinene, *p*-cymene, limonene, camphor and phellandrene (Msaada et al. 2007).

Fatty acids and sterols

Significant accumulation of fatty acids (FAs) in coriander seeds and fruits has been noted and the abundance of unsaturated FA has drawn considerable attention to coriander oil. Seeds produce 19.24 to 28.4% fixed oil (based on dry weight) (Sriti et al. 2010) and more than 95% of seeds FA are unsaturated (Sriti et al. 2009). As in the case of essential oils, the FA profile depends on the organ, maturation stage and genotype. However, some FAs predominate in most of coriander oils. Petroselinic acid (C18:1n-12), linoleic acid (C18:2n-6), oleic acid (C18:1n-9) and palmitic acid (C16:0) represent the major FAs in coriander oil. Other FAs include stearic, α-linolenic and arachidic acids. During maturation, there is a significant decrease in saturated and polyunsaturated FAs in favor of monounsaturated FAs, but organically cultivated fruits showed higher contents of PUFA (Nguyen et al. 2015).

As in the case of essential oils, the small-fruited coriander (*C. sativum*, var. *microcarpum*) produces more oil. However, the content of petroselinic acid is higher in the large-fruited variety (*C. sativum*, var. *vulgare*) (Kiralan et al. 2009). Soil salinity does not encourage FA accumulation, especially the unsaturated FAs (Neffati and Marzouk 2008). On the other hand, the FA profile of coriander leaves was quite different. A high content of PUFA was observed with α-linolenic acid constituting 39.4–41.1% of total FAs. Other identified FAs in coriander leaves include linoleic, heptadecenoic and palmitic acids (Neffati and Marzouk 2008).

Coriander seeds are particularly rich in sterols with a total sterol content ranging from 36.93 to 51.86 mg/g seed oil followed by fruits. β-sitosterol is the main sterol in coriander fruits and is especially concentrated in the pericarp where it accounts for 49.4% of total sterols. Stigmasterol is found mainly in seed oil (Sriti et al. 2009). Sterols have received considerable attention due to their inhibitory effect on dietary cholesterol absorption which can be beneficial in hyperlipidemic patients.

Vitamin E

Tocols (tocopherols and tocotrienols) are termed collectively vitamin E. Coriander seeds and fruits represent a good source of tocols with a total tocol content of 26.4 mg/100 g seed oil. Seeds are rich in γ-tocotrienol (19.5 mg/100 g oil), followed by α-tocotrienol (5.7 mg/100 g oil), while α-tocopherol is concentrated mainly in the pericarp (Sriti et al. 2010). Tocols are powerful antioxidants and due to their lipophilic character, they are specifically important in inhibiting lipid peroxidation and scavenging lipid hydroperoxide radicals.

Cumin (*Cuminium cyminum*)

Genus Cuminum encompasses four species: *Cuminum borszczowii*, *C. setifolium*, *C. sudanense* and *C. cyminum*, which is widely used as a spice. *Cuminum cyminum* L., commonly known as cumin, is a small herbaceous annual plant native to the Eastern Mediterranean regions. Cumin fruits have been used as spice and medicine since antiquity, especially in North African and Middle-Eastern areas, and continue to be popularly used till today.

Traditional uses

The medicinal uses of cumin have been immensely widespread in diverse folk medicine systems in the Middle-East and its beneficial effects have been described for gastrointestinal, gynecological, and respiratory disorders. The uses vary from the external application as antiseptic and anti-inflammatory to the management of diabetes and hypertension (Leporatti and Ghedira 2009). In indigenous Saudi-Arabian folk medicine, ground cumin fruits are applied externally to scorpion stings (Al-Yahya 1986). Herein, we discuss the biological activities of cumin in the light of established therapeutic effects (Figure 13.10).

Figure 13.10. Cumin fruit (Images purchased from Shutterstock).

Biological activities

Effects on the gastrointestinal tract

The usage of cumin for gastrointestinal disorders is among the most prominent ethnomedicinal practices in the Eastern-Mediterranean region where cumin is considered as carminative, antispasmodic and eupeptic agent used in the treatment of mild digestive complaints including flatulence, bloating, colic, dyspepsia and diarrhea.

Screening of the antispasmodic potential of seven Apiaceae plants revealed that *Cuminum cyminum* induces dose-dependent relaxation in guinea pig ileum (Saini et al. 2014). This was confirmed in Albino rats where the aqueous extract of cumin displayed a significant decrease in the frequency of diarrhea induced with castor oil, as well as a reduction in intestinal fluid secretion and intestinal propulsion (Sahoo et al. 2014). Furthermore, the continuous dietary intake of cumin (1.25%) significantly enhanced the activity of pancreatic trypsin, chymotrypsin, and amylase (Platel and Srinivasan 2000a, Milan et al. 2008), as well as intestinal maltase (Platel and Srinivasan 1996). A significant stimulatory effect was also observed on bile flow (Platel and Srinivasan 2000b). This supports the use of cumin in home remedies to aid digestion.

Antidiabetic activity

C. cyminum is able to reduce blood glucose levels without initiating hypoglycaemia or β-cell burn out. This antidiabetic activity was validated in the petroleum ether fraction of *C. cyminum* distillate through *in vitro* and *in vivo* studies. The hypoglycaemic effect was ascribed to cuminal and cuminol in the essential oil. The latter compounds were identified as potent glucose-dependent insulinotropic agents with 3.34- and 3.85-fold increase in insulin secretion, for cuminal and cuminol, respectively, compared to the 11.8 mM glucose control. This action was mediated by blocking the ATP-sensitive K^+ channels and increasing the intracellular Ca^{2+} concentration in cultured rat pancreatic cells. In addition, both compounds exert antioxidant protective effects on insulin-secreting β cells (Patil et al. 2013). Furthermore, cuminal has been proved to inhibit aldose reductase and α-glucosidase enzymes with IC_{50} values of 0.00085 mg/mL and 0.5 mg/mL, respectively (Lee 2005). Aldose reductase is a key enzyme in the onset of most of the diabetic complications. On the other hand, inhibition of α-glucosidase can delay carbohydrate digestion and decrease the absorption rate of glucose (Kalita et al. 2018).

Antihyperlipidemic activity

The petroleum ether fraction of cumin essential oil improved the lipid profile in streptozotocin-induced diabetic rats. In clinical trials, the oil, as well as cumin powder, exhibited the same effect in overweight, pre-diabetic and type-2 diabetic subjects with a significant decrease in triglycerides and LDL-cholesterol levels and an increase in HDL-cholesterol level (Zare et al. 2014, Keihan et al. 2016). Cumin oil can be regarded as an adjuvant treatment to improve the metabolic status in pre-diabetics (Jafari et al. 2018).

Gynecological and pregnancy-related activities

In traditional Tunisian medicine, the decoction of unripe cumin fruits is considered abortive and galactagogue (Leporatti and Ghedira 2009). Cumin is an ingredient in the Indian herbal preparation "Payapro", proved to have a significant galactagogue effect (Bhatt et al. 2009). Interestingly, cumin fruit's extract markedly enhanced milk production by 13% in cumin-fed goats. An increase in the content of monounsaturated FA and PUFA was observed in the milk of supplemented animals (Miri et al. 2013).

In addition to its galactopoietic properties, cumin is especially rich in minerals. Among spices and herbs, cumin fruits contain the highest amount of iron: 66.36 mg/100 g, representing about 5 times the daily requirement of iron (Narain and Ilango 2015, Saini et al. 2016). This may also account for cumin traditional use during lactation. However, the presence of phytates, iron-binding phenolic compounds (tannins) and calcium in cumin can decrease iron bioavailability due to their iron absorption-inhibitory effects (Saini et al. 2016), knowing that cumin fruits contain substantial amounts of calcium (765–1080 mg/100 g) (Tangkanakul et al. 2009, Trailokya et al. 2017). Moreover, cumin possesses an osteoprotective effect comparable to estradiol. The methanolic extract of cumin fruits showed a significant anti-osteoporotic effect in ovariectomized rats with a reduction in calcium urinary excretion and an increase in calcium content and mechanical strength of bones (Shirke et al. 2008). More investigation is required to assess the possibility of using cumin in treating postmenopausal osteoporosis.

Antioxidant activity and anticancer

Oxidative stress can be traced in most chronic diseases such as diabetes, hyperlipidemia, atherosclerosis and rheumatoid arthritis and this can result in the production of harmful free radicals and reactive oxygen species (ROS). The fruits of cumin contain considerable amounts of phenolics and thus offer an excellent antioxidant source. Moreover, the EO demonstrated higher antioxidant activity compared to the synthetic antioxidants BHT and BHA (Allahghadri et al. 2010). Constituents of cumin EO are reported to exert strong antioxidant and chelating effects and to protect against oxidative liver damage, especially cuminal and *p*-cymene (Babri et al. 2015). In addition, the acetone extract of cumin flowers is a powerful DPPH radical scavenger, lipid peroxidation inhibitor and reducing agent, with IC_{50} values of 4, 32, and 8 µg/mL, respectively (Bettaieb et al. 2010). The antioxidant profile of cumin has been suggested as the underlying mechanism for its multifaceted pharmacological effects such as antidiabetic, antihyperlipidemic, antistress, neuroprotective and antimicrobial, as outlined in the previous and succeeding sections.

Dietary supplementation with cumin was reported to inhibit the induction of various types of carcinoma including gastric squamous cell carcinomas, colonic cancer, forestomach tumorigenesis, uterine cervix tumorigenesis and hepatomas in mice and rats. Cumin was suggested to modulate carcinogen metabolism via metabolizing phase I and phase II enzymes (Nalini et al. 1998, Gagandeep et al. 2003).

Anti-inflammatory and analgesic activities

Cumin was an important healing agent in ancient Egyptian medicine and religion. Cumin, powdered and mixed with wheat flour and water, was used externally to alleviate the pain of arthritic joints (Aboelsoud 2010). An anti-inflammatory effect of cumin essential oil was described by suppressing different features of the lipopolysaccharide-induced inflammation pathway in cultured macrophages. Cumin fruits EO, in which the most abundant constituent was cuminal (48.77%), significantly decreased the expression of the pro-inflammatory cytokines: interleukin 1 and 6 (IL-1, IL-6), inducible nitric oxide synthase (iNOS) and cyclooxygenase (COX-2) (Wei et al. 2015). Furthermore, activity-guided screening of the *C. cyminum* crude extract resulted in the identification of cuminal as a competitive inhibitor of lipoxygenase, a key enzyme in the inflammatory pathway. It was suggested that a coordinate bond between the Fe^{2+} in the lipoxygenase active site and cuminal may play a role in enzyme inhibition (Tomy et al. 2014).

The oil, as well as the aqueous and alcoholic extracts of *C. cyminum*, demonstrate significant and dose-dependent analgesic effect in chronic and inflammatory pain models in rats and mice (using chemical and thermal stimuli) (Sayyah et al. 2002b, Bhat et al. 2014, Khichi et al. 2019). The analgesic and anti-inflammatory attributes of cumin can validate its external use in the case of scorpion stings.

CNS effects

Cumin extract demonstrated antistress and memory-enhancing activities in scopolamine-induced amnesic rats (Koppula and Choi 2011). Cuminal, the major constituent of *C. cyminum* oil, was found to inhibit the fibrillation of α-synuclein, which is a pathogenesis hallmark of neurodegenerative diseases, especially Parkinson's disease (Morshedi et al. 2015), suggesting a protective role of cumin against neurodegenerative diseases. Cumin oil showed sedative as well as anticonvulsant effects against the maximal electroshock- or pentylenetetrazole-induced seizures in mice through $GABA_A$ receptor-mediated inhibitory neurotransmission in mice (Sayyah et al. 2002a).

Antimicrobial activity

The broad-spectrum antimicrobial activity of cumin has been authenticated on different bacterial and fungal strains. This can explain why cumin was an essential ingredient in the mummification process in ancient Egypt (Counsell 2006) and support its external use as a disinfectant for skin disorders in the form of a poultice.

Cumin oil demonstrated a potent bactericidal effect against the Gram (–) bacterial strains *E. coli, Bordetella bronchiseptica, Salmonella typhi, Klebsiella pneumoniae, Vibrio cholerae, Helicobacter pylori, Shigella dysenteriae, Neisseria gonorrhoeae* and a moderate-weak bactericidal effect against the Gram (+) strains *Staphylococcus aureus, Listeria monocytogenes* and the acid-fast *Mycobacterium tuberculosis* (Chaudhary et al. 2014a, Khalil et al. 2018).

Furthermore, cumin oil displayed marked antifungal activity against *Candida albicans* strains identified in vaginal samples of women with vulvovaginal candidiasis (Minooeianhaghighi Sepehrian et al. 2017). The oil also showed marked toxicity against a broad spectrum of foodborne fungal isolates from stored wheat and chickpeas as well as aflatoxin inhibitory effect (Kedia et al. 2014). In addition, the oil of Moroccan cumin demonstrated strong antifungal activity against six yeast strains, belonging to four species (Petretto et al. 2018). Among the different constituents of cumin EO, cuminal is reported to possess good antibacterial activity against multidrug-resistant strains of *Salmonella* spp., followed by moderate activities against *E. coli*, *Staphylococcus aureus* and *Bacillus licheniformis* strains (Naveed et al. 2013).

Insecticidal activity

Cumin is regarded as a potentially useful source of green pesticides. Interesting pesticidal and insecticidal activities of cumin oil were obtained on *Myzus persicae* (the peach-potato aphid) and *Spodoptera littoralis* (the tobacco cutworm) and on two insect vectors, *Musca domestica* (the common housefly) and *Culex quinquefasciatus* (the lymphatic filariasis and Zika virus vector) (Benelli et al. 2018, Phumee et al. 2019). The oil was found safe on the beneficial organisms *Eisenia fetida* (the earthworm) and *Harmonia axyridis* (aphid predator).

Chemistry

C. cyminum contains different phytochemical constituents, including volatile oils, fatty acids, phenolic compounds and proteins.

Phenolic compounds

The total phenolic content of cumin fruits hydroalcoholic extracts varied widely according to the method of extraction and ranged from 290.29 to 1159.54 mg GAE/100 g dry weight in ultrasound-assisted extraction and microwave-assisted extraction, respectively. A variety of phenolic compounds including phenolic acids, flavonoids and phenolic diterpenes were identified in cumin fruits. This is associated with strong antioxidant capacity manifested by 88.43% inhibition of DPPH and 140.32 mmol Trolox equivalent/100 g dry weight in FRAP assay (Gallo et al. 2010).

The phenolic acids identified in cumin include salicylic acid, gallic acid, cinnamic acid and p-hydroxybenzoic acid and the flavonoids include rutin, quercetin, luteolin and apigenin. Cumin fruits were reported to contain an atypical nitrogen-containing flavonoid that suppresses the expression of inducible nitric oxide synthase and cyclooxygenase-2 enzymes revealing an interesting anti-inflammatory activity (Kang et al. 2019).

Quercetin was the main phenolic compound in cumin roots (26%), while in the stems and leaves, *p*-coumaric, rosmarinic, trans-2-dihydrocinnamic acids and resorcinol were predominant. Vanillic acid was the chief compound in cumin flowers (51%) (Bettaieb et al. 2010). Rosmarinic acid was the major phenolic acid in the unripe fruits (Rebey et al. 2014). In addition, moderate levels of genistein (exceeding 10 mg/kg) have been detected in cumin (Clarke et al. 2004). This phytoestrogen together with quercetin can account, at least in part, for the estrogenic and anti-osteoporotic potential of cumin.

Essential oil

Cumin fruits are rich sources of essential oil with a yield ranging from 1–5% (Singh et al. 2017). Of this, 40–65% is cuminaldehyde (Parthasarathy et al. 2008). The essential oil from eight Moroccan *C. cyminum* populations revealed common fingerprint with cuminal (cuminaldehyde), β-pinene, *p*-cymene, γ-terpinene, and α-terpinen-7-al as major oil components (Petretto et al. 2018). However, Indian cumin revealed a slightly different oil profile with *p*-Mentha-1,4-dien-7-ol as major constituent and less cuminal content (Dubey et al. 2018). The essential oil yield and composition of cumin fruits have been investigated during different stages of maturation. The levels of α-pinene, β-pinene, α-phellandrene, α-terpinene, and γ-terpinene were found to decrease, while the content of *p*-cymene, α-terpineol, cuminal, and safranal increased during maturation. While the oil yield of cumin fruits recorded its maximum yield in the mature stage, the total phenolic content and antioxidant potential was highest in the premature stage (Moghaddam et al. 2015).

Roasting of cumin fruits results in the formation of other aromatic compounds, the substituted pyrazines including 2-ethoxy-3-isopropylpyrazine, 2-methoxy-3-sec-butylpyrazine, and 2-methoxy-3-methylpyrazine (Li and Jiang 2004). Less essential oil yields were recorded in roots (0.03%), stem and leaves (0.1%), and flowers (1.7%) and the major components were bornyl acetate (23%), α-terpinene (34%), and γ-terpinene (51%) in roots, stems and leaves, and flowers, respectively. Recently, a new method was proposed for the extraction of oil from cumin. Based on natural deep eutectic solvents pretreatment coupled with microwave hydrodistillation, this method offered a doubling of extraction yield compared with microwave hydrodistillation (2.2% w/w versus 1.1% w/w, respectively). Furthermore, premium quality oil with 58 volatile components was obtained, compared with 45 volatile components identified in the microwave hydrodistillation method (Zhao et al. 2019). Similarly, cryogenic grinding of cumin fruits resulted in better retention of flavor and a significant increase in cuminal content (Sharma et al. 2016). As discussed above, cuminal mediates many of cumin's therapeutic benefits.

Other constituents

Cumin fruits contain about 10% fixed oil. The oil is composed mainly of petroselinic acid (55.9% of total fatty acids) which is of great nutritional importance, palmitic acid (23.82%) and linoleic acid (12.40%) as major constituents (Rebey et al. 2012). Sesquiterpenoid glucosides, namely cuminoside A and B, besides alkyl glycosides and steroidal glycosidic esters, have been identified in the methanolic extract of cumin fruits (Takayanagi et al. 2003, Chaudhary et al. 2014b).

Fennel (*Foeniculum vulgare*)

Genus Foeniculum comprises 3 species: *Foeniculum scoparium, F. subinodorum* and the commonly cultivated *F. vulgare*. *F vulgare* Mill. is a biennial medicinal and aromatic plant; it originated in the southern Mediterranean region and has been well known as a spice and medicinal plant since the time of the ancient Egyptians. It is known as Fennel and has more than 100 names (Badgujar et al. 2014). It is a hardy, perennial umbelliferous herb with yellow flowers and feathery leaves that grow up to 40 cm long, while the fruit is a dry seed, 4–10 mm long, elliptical, slightly curved and obtuse at the ends with five prominent ridges. There are two distinguished, two sub-species of fennel, *piperitum* and *vulgare*: sub-species *piperitum* has bitter seeds, while sub-species *vulgare* has sweet seeds that are used as flavouring agents due to their characteristic anise odour (Díaz-Maroto et al. 2006).

Traditional uses

Most of the plant parts of fennel have been used in traditional medicine for a wide range of diseases in the Middle East. In Jordan, an infusion of the leaves and seeds is used for abdominal pains, flatulence and irritable colon (Alzweiri et al. 2011). At present, the ripe fruit of the plant is still widely utilized in

Arabian folk medicine systems as a diuretic, stimulant, appetizer, digestive, an infantile febrifuge and to treat various other ailments (Tanira et al. 1996). In Turkey, leaves are made into a paste and used for abdominal pains, while the aerial parts are inhaled into the eyes in case of itching and blurry eyes (Bulut and Tuzlaci 2013). In northern Iran, an infusion of the seeds, stem and leaves is used for stomachache and as a carminative (Ghorbani 2005). Herein, we focus on the biological activities that have been scientifically confirmed by pharmacological studies (Figure 13.11).

Figure 13.11. Fennel fruit (Images by Gokalp Iscan from Pixabay).

Biological activities

The indigenous uses of *F. vulgare* have been supported by a number of biological studies and the extracts, as well as isolated compounds, have been evaluated for several activities.

Gastrointestinal effect

The most popular traditional use of fennel in the Middle-East is for the relief of gastrointestinal complaints such as irritable colon, flatulence, abdominal pain and constipation. The aqueous extract of *F. vulgare* showed a remarkable anti-ulcerogenic effect against ethanol-induced gastric lesions in rats (Birdane et al. 2007). An alcoholic extract of the fruits of *F. vulgare* possesses antispasmodic activity, which inhibits the acetylcholine and histamine-induced guinea pig ileal contractions *in vitro*. The essential oil, which is obtained from the fruits, inhibited oxytocin and prostaglandin (Forster et al. 1980). The essential oil was shown to regulate the motility of smooth muscles of the intestine, while, at the same time, reducing intestinal gas (Chakŭrski et al. 1981).

Antidiabetic activity

Ingestion of the essential oil of the seed corrected the hyperglycemia in diabetic rats and the activity of serum glutathione peroxidase as well as improving the pathological changes noticed in their kidney and pancreas. Trans-anethole was reported to be the major constituent of fennel volatile oil that has the ability to enhance insulin secretion, improve glucose tolerance and inhibit the inflammation pathway of kappa B (Badgujar et al. 2014, Al-Okbi et al. 2018).

Hypolipidemic activity

The aqueous extract of fennel was shown to cause significant reduction of plasma lipid levels, that is, cholesterol, triglycerides, LDL cholesterol, and apolipoprotein-B decreased by 40%, 23%, 61%, and 61%, respectively, and increase in HDL-cholesterol and apolipoprotein A1 by 85% and 58%, respectively (Oulmouden et al. 2011).

Antioxidant activity and anticancer activity

The essential oil and extracts have been known as an excellent source of natural antioxidants and contributed to the daily antioxidant diet. In a study, essential oils of the fruits of three organically grown cultivars of Egyptian fennel (*F. vulgare* var. *azoricum, F. vulgare* var. *dulce* and *F. vulgare* var. *vulgare*) were reported to possess antioxidant activity due to its high phenolic and flavonoid content (Shahat et al. 2011). The methanolic extract of the fruits showed good antitumor activity at a concentration of 200 μg/mL against B16F10 melanoma cell line (Pradhan et al. 2008). A significant increase in the malondialdehyde levels and a significant decrease in the catalase activity and glutathione content in liver and tumor tissue were observed in mice bearing Ehrlich ascites carcinoma (Mohamad et al. 2011).

Anti-Inflammatory and analgesic activities

Oral administration of methanol extract of *F. vulgare* fruit given to rats and mice exhibited inhibitory effects against acute and subacute inflammatory diseases. For acute inflammation, the methanol extract exhibited 69% inhibition of paw edema induced by carrageenan injection as compared to the control group of animals, which inhibited 70% of ear-edema induced by arachidonic acid in mice as well as increased the level of serum transaminase, aspartate aminotransferase (AST), and alanine aminotransferase (ALT) on inflammation induced by formaldehyde (Choi and Hwang 2004). Hexane, methylene chloride, ethyl acetate, and methanolic extract of the aerial part of *F. vulgare* showed remarkable antinociceptive activity against acetic acid-induced writhing in mice (Nassar et al. 2010).

Estrogenic and galactogenic properties

Oral administration of the acetone extract of fruits and seeds exhibited estrogen-like activity in male and female rats (Mallni et al. 1985). The oil was reported to exhibit estrogenic activity in promoting menstruation and alleviate the symptoms of female climacteric, and increase libido (Albert-Puleo 1980). The oil also revealed a significant inhibitory effect against prostaglandin E2 and significantly reduced the frequency of uterine contraction induced by prostaglandin E2 (Ostad et al. 2001). Anethole, which is structurally similar to dopamine, was found to help stimulate milk production in pregnant women (Badgujar et al. 2014).

Antimicrobial activities

Foeniculum vulgare has been used as an ethnic remedy for the cure of numerous infectious disorders of bacterial, fungal, viral, and mycobacterial origin (Badgujar et al. 2014). The aqueous and methanolic extract of the dried seeds was shown to have a broad antibacterial activity and was individually tested against a number of pathogenic bacteria by the disc diffusion method. This was due to the presence of compounds such as linoleic acid, undecanal, 1, 3-benzenediol, oleic acid and 2, 4-undecadienal in the extract. It is reported that fennel seed extract also contains 5-hydroxy-furanocoumarin, which contributed to the antimicrobial property of this plant (Khan 2017). In another *in vitro* study, aqueous and alcoholic seed extracts of *F. vulgare* exhibited inhibitory effect against *Alternaria alternata, Mucor rouxii*, and *Aspergillus flavus* (Thakur et al. 2013). Twenty compounds were isolated from the hexane extract and tested against one sensitive and three MDR strains of *Mycobacterium tuberculosis* using the Alamar Blue microassay (Esquivel-Ferriño et al. 2012).

In a recent study, essential oil exhibited a potential for the control of multidrug-resistant *Acinetobacter baumannii* infections. Ethanol and water extracts of *F. vulgare* have shown activity against *Campylobacter jejuni* and *Helicobacter pylori* (Rather et al. 2016). In another study the essential oil, as well as the ethanolic and methanolic fruit extracts, were shown to have antibacterial effect against *Bacillus cereus, B. megaterium, B. pumilus, B. subtilis, Escherichia coli, Klebsiella pneumonia, Micrococcus luteus, Pseudomonas putida, P. syringae*, and *Candida albicans*. Essential oil of *F. vulgare* had significant antimicrobial activities against some of the microorganisms as compared to the methanolic and ethanolic extracts (Gulfraz et al. 2008). Dichloromethane extract and essential oil from *F. vulgare* showed good potential as an antifungal agent against *C. albicans*. The essential oil also displayed a strong antiviral activity against *HSV-1* (Orhan et al. 2012).

Chemistry

A variety of bioactive compounds of different classes have been isolated from *F. vulgare* including volatile components, fatty acids, phenolic compounds, and hydrocarbons from different parts of the plant.

Polyphenols

The most prevalent flavonoids reported in fennel include quercetin-3-glucuronide, isoquercitrin, quercetin-3-arabinoside, kaempferol-3-glucuronide, kaempferol-3-arabinoside, isorhamnetin glucoside, Quercetin-3-O-galactoside, kaempferol-3-O-rutinoside, and kaempferol-3-O-glucoside (Kunzemann and Herrmann 1977, Badgujar et al. 2014). Flavonoids like isorhamnetin3-O-α-rhamnoside, quercetin, and kaempferol were also isolated from the ethyl acetate extract, whereas quercetin 3-O-rutinoside, kaempferol 3-*O*-rutinoside, and quercetin 3-O-β-glucoside were isolated from the methanol extract (Nassar et al. 2010). Diglucoside stilbene trimers and benzoisofuranone derivatives have also been isolated from *F. vulgare* fruit, together with cis-miyabenol C, trans-miyabenol C, trans-resveratrol-3-*O*-b-D-glucopyranoside, sinapyl glucoside, syringin-4-*O*-b-glucoside, oleanolic acid, 7a-hydroxycampesterol, (3b,5a,8a,22E) 5,8-epidioxy-ergosta-6,22-dien-3-ol, and 2,3-dihydropropylheptadec-5-onoate (Badgujar et al. 2014).

Essential oil

F. vulgare is well known for the characteristic anise odor of its essential oil. The essential oil composition exhibits considerable chemo diversity depending upon the method of extraction and geographical origin (Díaz-Maroto et al. 2006). Moreover, the essential oil content and composition vary during the different maturation stages where it was reported to decrease upon fruit maturation (Telci et al. 2009).

The major components of *F. vulgare* seed essential oil have been reported to be trans-anethole, fenchone, estragol (methyl chavicol), and a-phellandrene. It has been also reported to contain more than 87 volatile compounds (Badgujar et al. 2014). A total of 37 volatile compounds were identified from the pentane extracts of leaves, stems, and seeds of *F. vulgare* using gas chromatography (GC) and gas chromatography-mass spectrometry (GC-MS) techniques. In the supercritical CO_2 (SCCO 2) seed extracts of fennel, a total of 28 compounds were identified with major compounds being *trans*-anethole (68.6–75.0%), fenchone (8.40–14.7%), and methylchavicol (5.09–9.10%), whereas only 19 compounds were detected from hydrodistilled oil of fennel.

The hexane extracts of fennel were analyzed by GC-MS and 78 compounds were identified from these extracts; the major compounds were identified as 1,3-benzenediol, 1-methoxycyclohexene, o-cymene, sorbic acid, 2-hydroxy-3-methyl-2-cyclopenten-1-one, estragole, limonene-10-ol, and 3-methyl-2-cyclopenten-1-one (Esquivel-Ferriño et al. 2012). GC–MS analysis of Egyptian fennel essential oil identified 27 constituents, representing 89.24% of the total oil. Estragole (51.04%), limonene (11.45%), l-fenchone (8.19%) and trans-anethole (3.62%) were the major constituents of the essential oil tested (Ahmed et al. 2019).

Fatty acids

The content of unsaturated fatty acids ranges from 66% to 80% and predominates saturated fatty acids (Barros et al. 2010). The dominant fatty acid was petroselinic and oleic acid at 75.0–82.8%, followed by linoleic acid, palmitic, stearic and myristic acid. About 21 fatty acids were identified and quantified from fennel including caproic acid, caprylic acid, capric acid, undecanoic acid, lauric acid, myristic acid, myristoleic acid, pentadecanoic acid, palmitic acid, heptadecanoic acid, stearic acid, oleic acid, linoleic acid, α-linolenic acid, and arachidic acid, (Badgujar et al. 2014). This is also reported for the aerial parts and fruits of *F. vulgare* subsp. *piperitum* collected from North-Western Mediterranean coastal strip near El-Salloum, Egypt (Al-Snafi 2018).

Parsley (*Petroselinum* Species)

Petroselinum (parsley) is a genus of two species of flowering plants: *Petroselinum crispum* (garden parsley) and *P. segetum* (corn parsley). The widely cultivated *P. crispum* (parsley or garden parsley) is an important culinary herb, widely used as a flavoring agent and as a vegetable. It is native to the central Mediterranean region (Cyprus, southern Italy, Greece, Portugal, Spain, Malta, Morocco, Algeria, and Tunisia), but has naturalized elsewhere in Europe and planted throughout the world due to its usage in the food industry and pharmaceutical, cosmetic and perfume manufacturing. There are two main cultivated varieties of parsley plants used as herbs in recipes: *P. crispum* var. neapolitanum, and *P. crispum* var. *crispum*, the latter is also known as curled-leaf parsley (French parsley), while *P. crispum* var. neapolitanum is flat-leafed (Italian) parsley. Flat-leaf Italian parsley is more closely related to the wild parsley species that was first grown in the Mediterranean. Compared to curly parsley, flat-leafed parsley is generally harder, has a stronger flavor and is easier to grow. However, some people prefer the curly-leaf variety because of its decorative appearance in garnishing. Both parsley types taste very similar and both offer similar health benefits (Herbst 2001).

Traditional uses

In ancient times, parsley was not only used for culinary and medical purposes, it was subjected to superstitious beliefs by the ancient Greeks and Romans. The ancient Greeks mainly used parsley as a form of decoration for funeral wreaths and they made crowns of parsleys to honor the winners of sports games (Tucker and DeBaggio 2009). The Romans used parsley to mask the alcohol breath and for deodorizing the corpse. Parsley was primarily used for its medicinal properties long before being introduced as a spice or food. It was probably first commonly eaten in Europe in the Middle Ages. The flat-leafed variety was not initially eaten because it was easily confused with false-parsley (a noxious weed). However, the curly-leafed variety was soon commonly used as a garnish in a wide variety of dishes (Yanardağ et al. 2003, Grieve 2014). It is commercially sold in both fresh and dried forms (Rayment 2016).

Myriad of pharmacological activities of parsley herb and parsley essential oil have been mentioned in traditional and folklore medicines as carminative, gastro tonic, stomachic, diuretic, natural detox remedies, as an emmenagogue, abortifacient, antiseptic of urinary tract, anti-urolithiasis, and also for hypertension, cardiac disease, diabetes and for various dermal diseases (Duke 1985, Tyler 1993, Bisset 1994, Anderson et al. 1997, Robbers and Tyler 1999). In Chinese traditional medicine, the leaf is used as a food flavoring (Wong and Kitts 2006). In the Persian traditional medicine, *Petroselinum crispum* seeds have been also used for gastrointestinal disorder, inflammation, halitosis, kidney stone, and amenorrhoea (Moazedi et al. 2007, Tonkaboni et al. 2007, Behtash et al. 2008, Aghili et al. 2009), while leaves are mainly employed as a food flavor and used for gastrointestinal disorders and for skin problems like dermatitis, alphosis and macula and also for treating hemorrhoids and kidney stones. The leaves also possess abortifacient activity and are useful in skin disease, hypertension, hyperlipidemia, hepatotoxic, diabetes, cardiac diseases, eczema, amenorrhoea, dysmenorrhea, urinary tract infection, hyperuricaemia, prostatitis, halitosis, anaemia, constipation and baldness in ethnomedicine of other countries like Italy, Spain, Morocco, Serbia and Turkey (Oztürk et al. 1991, Merzouki et al. 1997, Ziyyat et al. 1997, Jouad et al. 2001, Eddouks et al. 2002, Bolkent et al. 2004, Benítez et al. 2010, Montesano et al. 2012, Aljanaby 2013, Savikin et al. 2013). In Indian traditional medicine, the leaf is used as a diuretic for the management of bladder disorders and fluid retention, for treating premenstrual syndrome and menopausal hot flashes; it also exhibited antispasmodic, emmenagogue, carminative, expectorant, antiseptic, and anti-inflammatory properties (Khare 2008) (Figures 13.12a and 13.12b).

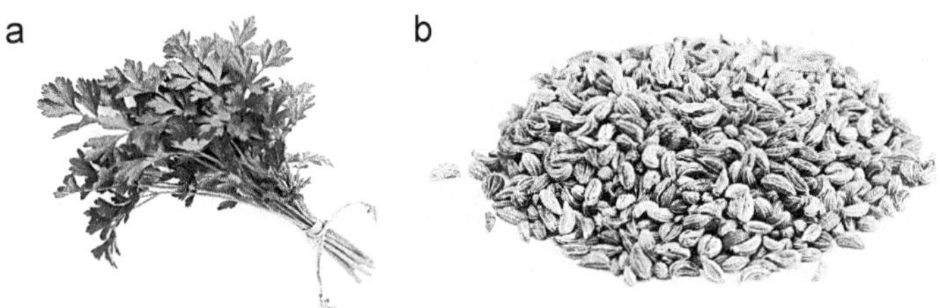

Figure 13.12. Parsley leaves (a); **parsley fruit** (b) (purchased from Shutterstock).

Biological activities

Antioxidant activity

The intake of a fortified diet with *Petroselinum crispum* by 14 people for one week caused a significant increase in antioxidant enzymes compared with their levels in the basic diet. Apigenin was suggested as the main constituent responsible for the observed differences (Nielsen et al. 1999). Different extracts from *P. crispum* leaves and stems exhibited antioxidant properties in various *in vitro* models (Fejes et al. 2000, Popović et al. 2007, Vora et al. 2009). The antioxidant activity of essential oils extracted from parsley flowers was measured using the 2,2,1-diphenyl-1-picrylhydrazyl (DPPH) radical scavenging activity and potassium ferricyanide–ferric chloride assay and resulted in DPPH radical inhibiting at 64.28% and ferric reducing power of 0.93 mmol/L Trolox (Marín et al. 2016). The antioxidant activity of parsley fermented juice (rich in nitrite and polyphenols) obtained from its fresh roots was assessed by measuring its ability to reduce DPPH synthetic radical, reducing the power of Fe^{3+} ion and the capacity of chelating transition metal ions. The ability of parsley juice to reduce DPPH synthetic radical was 79.45%, while the capacity to reduce Fe^{3+} was 0.758 ± 0.14 and to chelate Fe^{2+} ion was 23.64% (Papuc et al. 2016). The aqueous extracts of *P. crispum* leaves and its isolated flavonoids, quercetin and kaempferol increased the total antioxidant capacity and decreased malondialdehyde concentration in hyperuricemic rats (Haidari et al. 2011). Leaf and stem of aqueous and methanol extracts of *P. crispum* have been identified to possess antioxidant activity *in vitro* via the DPPH radical-scavenging, ion-chelating, and hydroxyl radical assays. Essential oil from seeds of *P. crispum* exhibited antioxidant activity using *β*-carotene bleaching, DPPH free radical scavenging and Fe^{2+} metal-chelating assays with apiol and myristicin as being the main two constituents responsible for the obtained antioxidant activity (Zhang et al. 2006). The antioxidant activity of *P. crispum* ethyl acetate fraction derived from the methanolic extract exhibited the highest total phenolic and total flavonoids contents. Concurrently, it demonstrated the most potent DPPH and ABTS (2, 2′-azino-bis (3-ethylbenzthiazoline-6-sulphonic acid)) activities and the highest total antioxidant capacity (El-Sayed et al. 2018).

By fighting free radical damage, or oxidative stress—that mainly contributes to almost every age-related disease, including cancer, heart disease and neurodegenerative diseases-antioxidants exhibit anti-carcinogenic and anti-aging properties, contributing to reducing the risk of cardiovascular diseases and against Alzheimer's disease and thus authenticates the above-mentioned ethnomedicinal uses of parsley.

Antidiabetic activity

The aqueous extract of *P. crispum* leaves decreased blood glucose level and exhibited significant hepatoprotective effect in diabetic rats (Bolkent et al. 2004, Ozsoy-Sacan et al. 2006). Additionally, the antioxidant activity *P. crispum* was identified to increase lipid peroxidation and decrease glutathione levels in hyperglycemia-induced heart and aorta oxidative damage in rats via its antioxidant activity in the heart and aorta tissue (Sener et al. 2003). The *o*-toluidine and 2-thiobarbituric acid assays showed that the aqueous extract of *P. crispum* leaves prevented an increase in blood glucose level in rats (Tunali et al. 1999). Blood glucose levels of experimental rats administered *P. crispum* extract were significantly reduced in streptozotocin-induced diabetic rats using the *o*-toluidine assay (Yanardağ et al. 2003). Fetuses of the diabetic mothers which were treated with parsley extract demonstrated somewhat normal morphological development and minimized the observed drastic changes in the diabetic rats and their fetuses (Abd Rabou and Eid 2017).

Analgesic and spasmolytic activity

The ethanol seed extract of *P. crispum* showed a significant dose-dependent analgesic activity by reducing KCl and $CaCl_2$-induced contractions on isolated rat ileum via blocking voltage-gated calcium channels (Moazedi et al. 2007). Additionally, the ethanol seed extract of *P. crispum* induced relaxation of isolated ilea from adult male Wistar rat induced by 60 mM potassium chloride (KCl) in the presence of two antagonists of α- and *β*-adrenoceptors (Damabi et al. 2010). Aqueous and ethanol leaf extracts of *P. crispum* also demonstrated antispasmodic activity on acetylcholine-, formalin- and acetic acid-induced contractions of mice and rat isolated ileum in dose-dependent manner (Eidi et al. 2009).

Gastrointestinal effects

The *P. crispum* effects on gastrointestinal disorders have been authenticated on different rat models. *P. crispum* leaves exhibited beneficial effects on different peptic ulcer models in rats via its antisecretory and cytoprotective activities using the cold-restraint ulcer (CRU) technique (Al-Howiriny et al. 2003). Aqueous extract from *P. crispum* seeds demonstrated laxative activity in the rat by reducing the sodium and water absorption and by enhancing Na-K-2Cl transporter activity in the rat colon (Kreydiyyeh et al. 2001).

Effects on genitourinary system

Methanol extract from *P. crispum* aerial part showed proliferative activity in estrogen-sensitive breast cancer cell line (MCF-7) equal to soybean isoflavone glycosides and the activity was mainly related to its phenolic content: 6"-acetylapiin, aglicones; apigenin, diosmetin, and kaempferol. Furthermore, methanol extract restored the uterus weight in ovariectomized mice when orally administered for consecutive 7 days where apiin and apigenin were suggested as being responsible for the observed activity (Yoshikawa et al. 2000). *P. crispum* oil demonstrated significant protective activity against zearalenone-induced reproductive toxicity and significantly improved testosterone level, sperm count and sperm motility and inhibited germ cells chromosomal aberrations in matured male mice (Abdel-Wahhab et al. 2006). The ethanol seed extract of *P. crispum* (5 mg/kg) reduced the dysfunction in rat kidney caused by prostodin-induced abortion via immunohistochemical and immunofluorescent staining and biochemical analysis (Rezazad and Farokhi 2014).

Cardiovascular effects

The hypotensive and negative inotropic and chronotropic activity of *P. crispum* leaves resulted from the decreased mean blood pressure and the rate and amplitude of contraction of isolated rat atria (Brankovic et al. 2008). *P. crispum* crude aqueous extract exhibited antiplatelet activity in experimental animals on platelet aggregation *in vitro* and *ex vivo*, and on bleeding time *in vivo*. The crude extract with aglycone flavonoids—keampferol, apigenin and cosmosiin—as the main active compounds significantly inhibited platelet aggregation *ex vivo* and prolonged bleeding time without changes in the amount of platelet (Gadi et al. 2009, 2012). In addition, apigenin and cosmosiin flavonoids isolated from aqueous extracts of *P. crispum* leaves exhibited strongly *in vitro* antiplatelet aggregation activity. While the aqueous *P. crispum* extract showed no inhibition on clotting activity when compared with the control, it exhibited strong antiplatelet aggregation activity (Chaves et al. 2011). In a recent study, the antihypertensive activity of the aqueous extract of parsley aerial parts was evaluated using *in vivo* and *in vitro* studies. Parsley extract was able to decrease the systolic, diastolic, mean arterial blood pressure in normotensive and hypertensive rats mainly through vasodilatory properties via an endothelium-independent pathway including both voltage-operated and receptor-operated calcium channel (Ajebli and Eddouks 2019).

Antimicrobial activity

The extracts from *P. crispum*'s different organs and their isolated compounds demonstrated antibacterial and antifungal activities against different strains like *Lactobacillus plantarum*, *Leuconostoc mesenteroides*, *E. coli*, and *B. subtilis* (Kim et al. 1998, Holton and Basset 2005, Wong and Kitts 2006). Furocoumarins isolated from aqueous extracts of *P. crispum* leaves, psoralen, 8-methoxypsoralen, 5-methoxypsoralen, oxypeucedanin and isopimpinellin, exhibited inhibitory activity against *Escherichia coli, Listeria monocytogenes, Erwinia carotovora*, and *Listeria innocua* (Manderfield et al. 1997). It was also reported that ethanol seed extract of *P. crispum* exhibited antibacterial activity against *Brucella melitensis* (Seyyednejad et al. 2008). Hot and cold water extract from *P. crispum* leaves demonstrated antibacterial activity against *P. aeruginosa*, *S. aureus* and *S. pyogenes* isolated from a patient with burn infection (Aljanaby 2013). Methanol extract of *P. crispum* leaves demonstrated antimicrobial activity on *B. subtilis*, *P. aeruginosa*, *S. epidermidis*, *S. aureus* and *S. cerevisiae in vitro* and coumarins were suggested as the main constituents responsible for the observed activity (Ojala et al. 2000). Additionally, all parsley extracts, along with the essential oil and their major constituents, myristicin and parsley-apiole, significantly inhibited the radial growth of the phytopathogenic fungus *C. acutatum* (Pineda et al.

2018). Furthermore, parsley essential oil and its major compounds apiol, myristicin, and β-phellandrene showed antibacterial and antifungal activities against all tested bacteria foodborne diseases, mainly, *S. aureus*, *Listeria monocytogenes*, *Salmonella enterica*, *Penicillium ochrochloron* and *Trichoderma viride* (Linde et al. 2016).

Chemistry

Flavonoids

Flavonoids are dominant compounds of this plant (Pápay et al. 2012). Flavonoids including Apigenin, luteolin, chrysoeriol, quercetin and isorhamnetin were detected in cell suspension cultures of *P. crispum* (Hempel et al. 1999). Flavonoids apigenin, cosmosiin, oxypeucedanin hydrate, apiin, quercetin and kaempferol were detected from *P. crispum* leaf extract (Chaves et al. 2011, Gadi et al. 2012, Stan et al. 2012). Parsley is considered one of the major sources of flavones like luteolin and apigenin (Panche et al. 2016). A flavone glycoside, 6-acetylapiin and petroside, its monoterpene glucoside, the furanocoumarin cnidilin and the flavone glycosides [diosmetin 7-*O*-β-d-glucopyranoside, and kaempferol 3-*O*-β-d-glucopyranoside] were isolated from the methanol aerial part of *P. crispum* (Yoshikawa et al. 2000). However, apiin and luteolin are the main flavonoids reported in *P. crispum* (Fejes et al. 1998, Nielsen et al. 1999, Fejes et al. 2000). Apiose and d-glucose are the sugars that mostly contribute to the structure of flavonoid glycosides (Farzaei et al. 2013).

Essential oil components

P. crispum roots, leaves and seeds contain essential oils, where the latter is considered as its main source (Bruneton 1999). Myristicin and apiol are the two main constituents of *P. crispum* essential oil and they are responsible for its antioxidant activity (Zhang et al. 2006). α-Pinene, sabinene, β-pinene, ρ-cymene, limonene, β-phellandrene, γ-terpinene, 1-allyl-2,3,4,5-tetra-methoxy-benzene, eugenol, and elemicin have also been isolated from seeds of *P. crispum* (Wagner and Bladt 1996, Zhang et al. 2006). Additionally, *P. crispum* seeds have been found to contain fatty acid and petroselinic acid (Guieta et al. 2003). Essential oil obtained from the leaves of *P. crispum* have been revealed to contain sesquiterpene hydrocarbons; β-caryophyllene, γ-elemene and β-elemene, aldehydes; phenylacetaldehyde and benzaldehyde, monoterpene hydrocarbons; β-pinene, sabinene, 3-carene, camphene, α-thujene, myrcene, monoterpene alcohols; α-terpineol and 2-(ρ-Tolyl) propan-2-ol, aromatic compounds; toluene and m-xylene, sesquiterpene hydrocarbons; α-cubebene, α-copaene and β-bisabolene (MacLeod et al. 1985), while *P. crispum* roots have been found to contain two C17 polyacetylenic alcohols [heptadeca-1,9(*Z*)-diene-4,6-diyn-3-ol (14) and heptadeca-1,9(*Z*)-diene-4,6-diyn-3,8-diol] (Nitz et al. 1990, Christensen and Brandt 2006). The triacylglycerol, tripetroselinin, has been isolated from the seeds of *P. crispum* (Destaillats et al. 2009). Among three cultivars (Italian Giant, Italian Plain, and Local) of parsley, the highest essential oil content was obtained from Italian Giant was obtained from Italian Giant with B-myrcene . β-myrcene, 1,3,8-*p*-menthatriene, β-phellandrene, and myristcin were the major compounds in all three cultivars (Alharbi et al. 2019).

Coumarins

Oxypeucedanin has been identified as a major constituent in parsley. Oxypeucedanin is reported to be mainly responsible for contact photodermatitis induced by this plant. Additionally, other photoactive furocoumarins including isoimperatorin, psoralen, isopimpinellin, 8-methoxypsoralen, 5-methoxypsoralen, and imperatorin have been isolated from the leaves and roots of *P. crispum* (Chaudhary et al. 1986).

Conclusions

The increase in literature, scientific reports and ongoing research interest in medicine from nature comes as no wonder as both the medical industry and consumers strive to find the optimum medication with the least side effects. Reported beneficial pharmacological effects and positive clinical studies provide a solid foundation for the validation of folk medicine and the traditional use of common plants used for hundreds of years. This, in turn, encourages scientists to pursue further investigation and validation. Members of family Apiaceae are no exception with their vast chemical constituents and pharmacological efficacy pertaining to historically-based folk medicine. Essential oils and crude extracts of anise, fennel, coriander, celery, parsley, ajwain and cumin have shown tremendous common therapeutic potential as antimicrobials, antioxidants, and anti-inflammatory, in addition to individual remedial uniqueness as antilipidemic, as in the case of coriander and cumin. Specific compounds isolated from other members, as coumarins from *Ammi* species and monoterpenes from essential oils, have also served as the basis for medicinally significant therapeutics worldwide; carotenoids in carrots and coumarins in ammi species are associated cardiovascular health. Although safety concerns, as well as rigid quality and efficacy assessment for such herbal products, remain a global objective, nature will remain the primary source, origin and inspiration for most medicine.

References

Abdel-Monem, A., Maged, A., Zeinab, E., Fawkia, A. 2014. Biological activity of the Egyptian medicinal plants: Part 4 Cytotoxicity of 50 Egyptian plants and spices against hepatocellular carcinoma. Am. J. Ethnomed., 1: 56–63.

Abdul-Jalil, T.Z., Saour, K., Nasser, A. 2010. Phytochemical study of some flavonoids present in the fruits of two Ammi L. species wildly grown in Iraq. Iraqi J. Pharm. Sci., 19(1): 48–57.

Abdulmanea, K., Prokudina, E.A., Lanková, P., Vaníˇcková, L., Koblovská, R., Zelený, V., Lapčík, O. 2012. Immunochemical and HPLC identification of isoflavonoids in the Apiaceae family. Biochem. Syst. Ecol., 45: 237–243.

Abou-Mustafa, E.A., Saleh, N.A., Elgamal, M.H., Shalaby, N.M., Duddeck, H. 1990. A further contribution to the gamma-pyrone constituents of Ammi visnaga fruits. Planta Med., 56: 134.

Abdel-Wahhab, M.A., Abbes, S., Salah-Abbes, J., Hassan, A., Oueslati, R. 2006. Parsley oil protects against Zearalenone-induced alteration in reproductive function in male mice. Toxicol. Lett., 164: S266.

Abdollahi Fard, M., Shojaii, A. 2013. Efficacy of Iranian traditional medicine in the treatment of epilepsy. BioMed. Res. Int.

Abd Rabou, M.A., Eid, F.A. 2017. Possible protective role of parsley extract on the diabetic pregnant rats and their fetuses. Pak. J. Biol. Sci., 20(11): 552–562.

Aboelsoud, N.H. 2010. Herbal medicine in ancient Egypt. J. Med. Plants Res., 4: 082–086.

AbouZid, S.F., Mohamed, A.A. 2011. Survey on medicinal plants and spices used in Beni-Sueif, Upper Egypt. Journal of Ethnobiol. and Ethnomed., 7: 18.

Aćimović, M.G. 2017. Nutraceutical potential of Apiaceae. pp. 1–31. *In*: Mérillon, J.-M., Ramawat, K.G. (eds.). Bioactive Molecules in Food. Springer International Publishing, Cham.

Afzal, M., Kazmi, I., Kaur, R., Ahmad, A., Pravez, M., Anwar, F. 2013. Comparison of protective and curative potential of *Daucus carota* root extract on renal ischemia reperfusion injury in rats. Pharm. Biol., 51: 856–862.

Agarwal, R., Gupta, S.K., Srivastava, S., Agrawal, S.S., Saxena, R. 2008. Lowering of intraocular pressure by topical application of *Daucus carota* seed extract in rabbits. Indian J. Exp. Biol., 46: 541–546.

Aghili, M.H., Makhzan-al-Advia, Rahimi, R., Shams Ardekani, M.R., Farjadmand, F. 2009. Tehran: Tehran Univ. Med. Sci., 329–330.

Ahmad, T., Cawood, M., Iqbal, Q., Ariño, A., Batool, A., Tariq, R.M.S., Azam, M., Akhtar, S. 2019. Phytochemicals in *Daucus carota* and their health benefits—review article. Foods, 8: 424–445.

Ahmed, A.F., Shi, M., Liu, C., Kang, W. 2019. Comparative analysis of antioxidant activities of essential oils and extracts of fennel (*Foeniculum vulgare* Mill.) seeds from Egypt and China. Food Sci. and Human Wellness, 8: 67–72.

Aissaoui, A., El-Hilaly, J., Israili, Z.H., Lyoussi, B. 2008. Acute diuretic effect of continuous intravenous infusion of an aqueous extract of *Coriandrum sativum* L. in anesthetized rats. J. Ethnopharmacol., 115: 89–95.

Ajebli, M., Eddouks, M. 2019. Antihypertensive activity of *Petroselinum crispum* through inhibition of vascular calcium channels in rats. J. Ethnopharmacol.

Aissaoui, A., Zizi, S., Israili, Z.H., Lyoussi, B. 2011. Hypoglycemic and hypolipidemic effects of *Coriandrum sativum* L. in Meriones shawi rats. J. Ethnopharmacol., 137: 652–661.

Akhtar, A., Deshmukh, A., Bhonsle, A., Kshirsagar, P., Kolekar, M. 2008. *In vitro* antibacterial activity of *Pimpinella anisum* fruit extracts against some pathogenic bacteria. Vet. World, 1: 272–274.

Akhtar, S., Rauf, A., Imran, M., Qamar, M., Riaz, M., Mubarak, M.S. 2017. Black carrot (*Daucus carota* L.), dietary and health promoting perspectives of its polyphenols: A review. Trends Food Sci. Technol., 66: 36–47.

Alasalvar, C., Al-Farsi, M., Quantick, P.C., Shahidi, F., Wiktorowicz, R. 2005. Effect of chill storage and modified atmosphere packaging (MAP) on antioxidant activity, anthocyanins, carotenoids, phenolics and sensory quality of ready-to-eat shredded orange and purple carrots. Food Chem., 89: 69–76.

Al-Asmari, A.K., Athar, M.T., Kadasah, S.G. 2017. An updated phytopharmacological review on medicinal plant of Arab region: *Apium graveolens* linn. Pharmacog. Rev., 11: 13.

Al-Bayati, F.A. 2008. Synergistic antibacterial activity between *Thymus vulgaris* and *Pimpinella anisum* essential oils and methanol extracts. J. Ethnopharmacol., 116: 403–406.

Albert-Puleo, M. 1980. Fennel and anise as estrogenic agents. J. Ethnopharmacol., 2: 337–344.

Algarra, M., Fernandes, A., Mateus, N., Freitas, V., Esteves da Silva, J.C.G., Casado, J. 2014. Anthocyanin profile and antioxidant capacity of black carrots (*Daucus carota* L. spp. *sativus* var. atrorubens Af.) from Cuevas Bajas, Spain. J. Food Compos. Anal., 33: 71–76.

Alharbi, M., Mahmoud, A.A., Astatkie, T., Said-Al Ahl, H. 2019. Growth and essential oil composition responses of parsley cultivars to phosphorus fertilization and harvest date. J. Plant Nutr., 42(18): 2395–2405.

Al-Howiriny, T., Al-Sohaibani, M., El-Tahir, K., Rafatullah, S. 2003. Prevention of experimentally-induced gastric ulcers in rats by an ethanolic extract of "Parsley" *Petroselinum crispum*. Am. J. Chinese Med., 31(5): 699–711.

Al-Howiriny, T., Alsheikh, A., Alqasoumi, S., Al-Yahya, M., ElTahir, K., Rafatullah, S. 2010. Gastric antiulcer, antisecretory and cytoprotective properties of celery (*Apium graveolens*) in rats. Pharmaceutical Biol., 48: 786–793.

Aljanaby, A.A.J.J. 2013. Antibacterial activity of an aqueous extract of *Petroselinum crispum* leaves against pathogenic bacteria isolated from patients with burns infections in Al-najaf Governorate, Iraq. Res. Chem. Intermed., 39(8): 3709–3714.

Al Jawad, F.H., Al Razzuqi, R.A., Al Jeboori, A.A. 2011. *Apium graveolens* accentuates urinary Ca^{2+} excretions in experimental model of nephrocalcinosis. Int. J. of Green Pharm. (IJGP), 5.

Allahghadri, T., Rasooli, I., Owlia, P., Nadooshan, M.J., Ghazanfari, T., Taghizadeh, M., Astaneh, S.D.A. 2010. Antimicrobial property, antioxidant capacity, and cytotoxicity of essential oil from cumin produced in Iran. J. Food Sci., 75: H54–H61.

Al Mofleh, I.A., Alhaider, A.A., Mossa, J.S., Al-Soohaibani, M.O., Rafatullah, S. 2007. Aqueous suspension of anise "*Pimpinella anisum*" protects rats against chemically induced gastric ulcers. World J. Gastroenterol. WJG, 13: 1112.

Al-Okbi, S.Y., Hussein, A.M., Elbakry, H.F., Fouda, K.A., Mahmoud, K.F., Hassan, M.E. 2018. Health benefits of fennel, rosemary volatile oils and their nano-forms in dyslipidemic rat model. Pakistan J. Biol. Sci.: PJBS, 21: 348–358.

Al-Snafi A. 2013. Chemical constituents and pharmacological activities of *Ammi majus* and *Ammi visnaga*. A review. Int. J. Pharm. & Ind. Res., 03(03): 257–265.

Al-Snafi, A.E. 2014. The pharmacology of *Apium graveolens*—A review. Int. J. Pharmaceutical Res. Scholars, 3: 671–677.

Al-Snafi, A.E. 2016. A review of medicinal plants with Broncho-Dilatory effect-Part 1. Sch. Acad. J. Pharm., 5: 297–304.

Al-Snafi, A.E. 2018. The chemical constituents and pharmacological effects of *Foeniculum vulgare*—A review. IOSR J. Pharmacy, 8: 81–96.

Alves, S., Duarte, A., Sousa, S., Domingues, F.C. 2016. Study of the major essential oil compounds of *Coriandrum sativum* against Acinetobacter baumannii and the effect of linalool on adhesion, biofilms and quorum sensing. Biofouling, 32: 155–165.

Alves-Silva, J.M., Zuzarte, m., Gonçalves, M.J., Cavaleiro, C., Cruz, M.T., Cardoso, S.M., Salgueiro, L. 2016. New claims for wild carrot (*Daucus carota* subsp. *carota*) essential oil. Evid.-Based Compl. Altern. Med. http://dx.doi.org/10.1155/2016/9045196.

Al-Yahya, M. 1986. Phytochemical studies of the plants used in traditional medicine of Saudi Arabia. Fitoterapia.

Alzweiri, M., Al Sarhan, A., Mansi, K., Hudaib, M., Aburjai, T. 2011. Ethnopharmacological survey of medicinal herbs in Jordan, the Northern Badia region. J. Ethnopharmacol., 137: 27–35.

Amarowicz, R., Pegg, R.B., Rahimi-Moghaddam, P., Barl, B., Weil, J.A. 2004. Free-radical scavenging capacity and antioxidant activity of selected plant species from the Canadian prairies. Food Chem., 84: 551–62.

Anderson, L.A., Newall, C.A., Phillipson, J.D. 1997. Herbal Medicine. A Guide for Health-care Professionals. London: The Pharmaceutical Press, 203–204.

Andersen, O.M., Jordheim, M. 2006. The anthocyanins. pp. 471–552. *In*: Andersen, O.M., Jordheim, M. (eds.). Flavonoids, Chemistry, Biochemistry and Applications. CRC Press, London.

Aneesa, N.N., Anitha, R., Varghese, S. 2019. Antidiabetic activity of ajwain oil in different *in vitro* models. J. Pharmacy Bioallied Sci., 11: 142.

Anwar, S., Ahmed, N., Habibatni, S., Abusamra, Y. 2016. Ajwain (*Trachyspermum ammi* L.) Oils, Essential Oils in Food Preservation, Flavor and Safety. Elsevier, pp. 181–192.

Arscott, S.A., Tanumihardjo, S.A. 2010. Carrots of many colors provide basic nutrition and bioavailable phytochemicals acting as a functional food. Compr. Rev. Food Sci. F, 9: 223–239.

Arya, V., Gupta, V.K., Kaur, R. 2011. A review on plants having anti-arthritic potential. Int. J. Pharm. Sci. Rev. Res., 7: 131–136.

Babri, R.A., Khokhar, I., Mahmud, S. 2015. Antioxidant potential and iron chelating activity of some bioactive compounds. J. Chem. Soc. Pakistan, 37.

Babu, P.N., Nagaraju, B., Yamini, K., Dhananjaneyulu, M., Venkateswarlu, K., Mubina, M. 2014. Evaluation of antidepressant activity of ethanolic extract of *Daucus carota* in mice. J. Pharm. Sci. & Res., 6: 73–77.

Badgujar, S.B., Patel, V.V., Bandivdekar, A.H. 2014. *Foeniculum vulgare* Mill: a review of its botany, phytochemistry, pharmacology, contemporary application, and toxicology. BioMed Research International.

Bahorun, T., Luximon-Ramma, A., Crozier, A., Aruoma, O.L. 2004. Total phenol, flavonoid, proanthocyanidin and vitamin C levels and antioxidant activities of Mauritian vegetables. J. Sci. Food Agric, 84: 1553–1561.

Bairwa, R., Sodha, R., Rajawat, B. 2012. Trachyspermum ammi. Pharmacog. Rev., 6: 56.

Banga, O. 1963. Main types of the western carotene carrot and their origin. Zwolle [Netherlands] W. E. J. Tjeenk Willink.

Baranska, M., Roman, M., Dobrowolski, J.C., Schulz, H., Baranski, R. 2013. Recent advances in Raman analysis of plants: alkaloids, carotenoids, and polyacetylenes. Curr. Anal. Chem., 9: 108–127.

Baranski, R., Allender, C., Klimek-Chodacka, M. 2012. Towards better tasting and more nutritious carrots, carotenoid and sugar content variation in carrot genetic resources. Food Res. Int., 47: 182–187.

Barros, L., Carvalho, A.M., Ferreira, I.C. 2010. The nutritional composition of fennel (*Foeniculum vulgare*): Shoots, leaves, stems and inflorescences. LWT-Food Sci. Technol., 43: 814–818.

Barros, L., Dueñas, M., Dias, M.I., Sousa, M.J., Santos-Buelga, C., Ferreira, I.C. 2012. Phenolic profiles of *in vivo* and *in vitro* grown *Coriandrum sativum* L. Food Chem., 132: 841–848.

Begnami, A.F., Spindola, H.M., Ruiz, A.L.T.G., de Carvalho, J.E., Groppo, F.C., Rehder, V.L.G. 2018. Antinociceptive and anti-edema properties of the ethyl acetate fraction obtained from extracts of *Coriandrum sativum* Linn. leaves. Biomed. Pharmacotherapy, 103: 1617–1622.

Behtash, N., Kargarzadeh, F., Shafaroudi, H. 2008. Analgesic effects of seed extract from *Petroselinum crispum* (Tagetes minuta) in animal models. Toxicol Lett., 180(Suppl 5): S127–S128.

Beltagy, A.M., Beltagy, D.M. 2015. Chemical composition of *Ammi visnaga* L. and new cytotoxic activity of its constituents Khellin and Visnagin. J. Pharma. Sci. Res., 7: 285–291.

Bencheraiet, R., Kherrab, H., Kabouche, A., Kabouche, Z., Jay, M. 2011. Flavonols and antioxidant activity of *Ammi visnaga* L. (Apiaceae). Rec. Nat. Prod., 5: 52–55.

Benelli, G., Pavela, R., Petrelli, R., Cappellacci, L., Canale, A., Senthil-Nathan, S., Maggi, F. 2018. Not just popular spices! Essential oils from *Cuminum cyminum* and *Pimpinella anisum* are toxic to insect pests and vectors without affecting non-target invertebrates. Ind. Crops Prod., 124: 236–243.

Benítez, G., González-Tejero, M.R., Molero-Mesa, J. 2010. Pharmaceutical ethnobotany in the western part of Granada province (southern Spain): Ethnopharmacological synthesis. J. Ethnopharmacol., 129(1): 87–105.

Bennett, R., Schuler, M. 2011. Wild Carrot Study—Final Summary. Retrieved on November 22nd, 2019 from http://robinrosebennett.com/articles/wild-carrot-study-final-summary-august-2011/.

Bettaieb Rebey, I., Bourgou, S., Aidi Wannes, W., Hamrouni Selami, I., Saidani Tounsi, M., Marzouk, B., Fauconnier, M.-L., Ksouri, R. 2018. Comparative assessment of phytochemical profiles and antioxidant properties of Tunisian and Egyptian anise (*Pimpinella anisum* L.) seeds. Plant Biosystems—An International Journal Dealing with all Aspects of Plant Biology, 152: 971–978.

Bettaieb, I., Bourgou, S., Wannes, W.A., Hamrouni, I., Limam, F., Marzouk, B. 2010. Essential oils, phenolics, and antioxidant activities of different parts of cumin (*Cuminum cyminum* L.). J. Agri. Food Chem., 58: 10410–10418.

Bhat, S.P., Rizvi, W., Kumar, A. 2014. Effect of *Cuminum cyminum* L. seed extracts on pain and inflammation. J. Nat. Remedies, 14: 186–192.

Bhatt, N., Singh, M., Ali, A. 2009. Effect of feeding herbal preparations on milk yield and rumen parameters in lactating crossbred cows. Int. J. Agric. Biol., 11: 721–726.

Birdane, F.M., Cemek, M., Birdane, Y.O., Gülçin, İ., Büyükokuroğlu, M.E. 2007. Beneficial effects of *Foeniculum vulgare* on ethanol-induced acute gastric mucosal injury in rats. World J. Gastroenterology: WJG, 13: 607.

Bishayee, A., Sarkar, A., Chatterjee, M. 1995. Hepatoprotective activity of carrot (*Daucus carota* L.) against carbon tetrachloride intoxication in mouse liver. J. Ethnopharmacol., 47: 69–74.

Bisset, N.G. 1994. Herbal Drugs and Phytotherapeuticals. Medpharm Scientific Publishers, CRC Press, Boca Raton, Ann Arbor, London Tokyo, 369–370.

Bolkent, S., Yanardag, R., Ozsoy-Sacan, O., Karabulut-Bulan, O. 2004. Effects of parsley (*Petroselinum crispum*) on the liver of diabetic rats: a morphological and biochemical study. Phytother. Res., 18(12): 996–999.

Boulos, L. 1999. Flora of Egypt. Cairo, Egypt: Al Hadara Pub.

Boskabady, M., Ramazani-Assari, M. 2005. Possible mechanism for the relaxant effect of *Pimpinella anisum* on guinea pig tracheal chains. Pharmaceutical Biol., 42: 621–625.

Brankovic, S., Djosev, S., Kitic, D., Radenkovic, M., Veljkovic, S., Nesic, M., Pavlovic, D. 2008. Hypotensive and negative chronotropic and inotropic effects of the aqueous and ethanol extract from parsley leaves. J. Clin. Lipidol., 2(5)(Suppl 1): S191–S408.

Branković, S., Gočmanac-Ignjatović, M., Kostić, M., Veljković, M., Miladinović, B., Milutinovic, M., Radenković, M. 2015. Spasmolytic activity of the aqueous and ethanol celery leaves (*Apium graveolens* L.) extracts on the contraction of isolated rat ileum. Acta medica Medianae, 54: 11–16.

Brindis, F., González-Andrade, M., González-Trujano, M., Estrada-Soto, S., Villalobos-Molina, R. 2014. Postprandial glycaemia and inhibition of α-glucosidase activity by aqueous extract from *Coriandrum sativum*. Nat. Prod. Res., 28: 2021–2025.

Bruneton, J. 1999. Pharmacognosy, Phytochemistry Medicinal Plants, Second Edition Intercept Ltd., London, pp. 519–520.

Brothwell, D., Brothwell, P. 1969. Food in Antiquity. The Johns Hopkins University Press, Baltimore, pp. 111–112.

Bulut, G., Tuzlaci, E. 2013. An ethnobotanical study of medicinal plants in Turgutlu (Manisa—Turkey). J. Ethnopharmacol., 149: 633–647.

Campos-Toimil, M., Orallo, F., Santana, L., Uriarte, E. 2002. Synthesis and vasorelaxant activity of new coumarin and furocoumarin derivatives. Bioorg. Med. Chem. Lett., 12, 783–786.

Carrubba, A., la Torre, R., Prima, A.D., Saiano, F., Alonzo, G. 2002. Statistical analyses on the essential oil of Italian coriander (*Coriandrum sativum* L.) fruits of different ages and origins. J. Essential Oil Res., 14: 389–396.

Casetti, F., Bartelke, S., Biehler, K., Augustin, M., Schempp, C., Frank, U. 2012. Antimicrobial activity against bacteria with dermatological relevance and skin tolerance of the essential oil from *Coriandrum sativum* L. fruits. Phytotherapy Res., 26: 420–424.

Chakŭrski, I., Matev, M., Koĭchev, A., Angelova, I., Stefanov, G. 1981. Treatment of chronic colitis with an herbal combination of *Taraxacum officinale*, *Hypericum perforatum*, *Melissa officinalis*, *Calendula officinalis* and *Foeniculum vulgare*. Vutreshni bolesti, 20: 51–54.

Chandra, P., Kishore, K., Ghosh, A.K. 2015. Assessment of antisecretory, gastroprotective, and *in vitro* antacid potential of *Daucus carota* in experimental rats. Osong Public Health Res. Perspect., 6: 329–335.

Chatterjee, S., Goswami, N., Kothari, N. 2013. Evaluation of antioxidant activity of essential oil from Ajwain (*Trachyspermum ammi*) seeds. Int. J. of Green Pharmacy (IJGP), 7.

Chaudhary, S.K., Ceska, O., Têtu, C., Warrington, P.J., Ashwood-Smith, M.J., Poulton, G.A. 1986. Oxypeucedanin, a major furocoumarin in Parsley, *P. crispum*. Planta Med., 52(6): 462–464.

Chaudhary, N., Husain, S.S., Ali, M. 2014a. Chemical composition and antimicrobial activity of volatile oil of the seeds of *Cuminum cyminum* L. World J. of Pharmacy and Pharmaceutical Sci., 3: 1428–1441.

Chaudhary, N., Husain, S.S., Ali, M. 2014b. New phenolic, triterpenic and steroidal constituents from the fruits of *Cuminum cyminum* L. J. Pharmacog. Phytochem., 3: 149–154.

Chaves, D.S., Frattani, F.S., Assafim, M., de Almeida, A.P., de Zingali, R.B., Costa, S.S. 2011. Phenolic chemical composition of *P. crispum* extract and its effect on haemostasis. Nat. Prod. Commun., 6(7): 961–964.

Chen, Y.C., Shen, S.C., Lee, W.R., Hou, W.C., Yang, L.L., Lee, T.J. 2001. Inhibition of nitric oxide synthase inhibitors and lipopolysaccharide induced inducible NOS and cyclooxygenase-2 gene expressions by rutin, quercetin, and quercetin pentaacetate in RAW 264.7 macrophages. J. Cellular Biochem., 82: 537–548.

Cheng, B.-H., Sheen, L.-Y., Chang, S.-T. 2015. Evaluation of anxiolytic potency of essential oil and S-(+)-linalool from *Cinnamomum osmophloeum* ct. linalool leaves in mice. J. Traditional and Complementary Med., 5: 27–34.

Chevallier, A. 1996. The Encyclopedia of Medicinal Plants. Dorling Kindersley: London, UK.

Cho, K.-J., Han, S.H., Kim, B.Y., Hwang, S.-G., Park, K.-K., Yang, K.-H., Chung, A.-S. 2000. Chlorophyllin suppression of lipopolysaccharide-induced nitric oxide production in RAW 264.7 cells. Toxicol. Applied Pharmacol., 166: 120–127.

Cho, S.-Y., Jun, H.-j., Lee, J.H., Jia, Y., Kim, K.H., Lee, S.-J. 2011. Linalool reduces the expression of 3-hydroxy-3-methylglutaryl CoA reductase via sterol regulatory element binding protein-2-and ubiquitin-dependent mechanisms. FEBS Letters, 585: 3289–3296.

Choi, E.-M., Hwang, J.-K. 2004. Antiinflammatory, analgesic and antioxidant activities of the fruit of *Foeniculum vulgare*. Fitoterapia, 75: 557–565.

Christensena, L.P., Brandtb, K. 2006. Bioactive polyacetylenes in food plants of the Apiaceae family: occurrence, bioactivity and analysis. J. Pharm. Biomed. Anal., 41(3): 683–693.

Chu, Y.H., Zhou, M.H., Li, Q., Bao, Y.M. 1985. Antifertility effect of volatile oil of *Daucus carota* seeds. Reprod. Contracept., 5: 37–40.

Chung, I.-M., Ahmad, A., Kim, S.-J., Naik, P.M., Nagella, P. 2012. Composition of the essential oil constituents from leaves and stems of Korean *Coriandrum sativum* and their immunotoxicity activity on the Aedes aegypti L. Immunopharmacol. and Immunotoxicol., 34: 152–156.

Clarke, D., Barnes, K., Lloyd, A. 2004. Determination of unusual soya and non-soya phytoestrogen sources in beer, fish products and other foods. Food Additives and Contaminants, 21: 949–962.

Counsell, D.J. 2006. Intoxicants in Ancient Egypt: The Application of Modern Forensic Analytical Techniques to Ancient Artefacts and Mummified Remains in the Evaluation of Drug Use by an Ancient Society: A Historical and Scientific Investigation. The University of Manchester (United Kingdom).

Coyne, T., Ibiebele, T.I, Baade, P.D., Dobson, A., McClintock, C., Dunn, S., Leonard, D., Shaw, J. 2005. Diabetes mellitus and serum carotenoids: findings of a population-based study in Queensland, Australia. Am. J. Clin. Nutr., 82: 685–693.

Czepa, A., Hofmann, T. 2003. Structural and sensory characterization of compounds contributing to the bitter off-taste of carrots (*Daucus carota* L.) and carrot puree. J. Agric. Food Chem., 51: 3865–3873.

Czepa, A., Hofmann, T. 2004. Quantitative studies and sensory analyses on the influence of cultivar, spatial tissue distribution, and industrial processing on the bitter off-taste of carrots (*Daucus carota* L.) and carrot products. J. Agric. Food Chem., 52: 4508–4514.

Damabi, N.M., Moazedi, A., Seyyednejad, S. 2010. The role of α- and β-adrenergic receptors in the spasmolytic effects on rat ileum of *Petroselinum crispum*. Asian Pac. J. Trop. Med., 3(11): 866–870.

Dhaiwal, K., Chahal, K.K., Kataria, D., Kumar, A. 2017. Gas chromatography-mass spectrometry analysis and *in vitro* antioxidant potential of ajwain seed (Trachyspermum ammi L.) essential oil and its extracts. J. Food Biochem., 41: e12364.

da Silva Dias, J.C. 2014. Nutritional and health benefits of carrots and their seed extracts. Food Nutr. Sci., 5: 2147–2156.

Darughe, F., Barzegar, M., Sahari, M. 2012. Antioxidant and antifungal activity of Coriander (*Coriandrum sativum* L.) essential oil in cake. Int. Food Res. J., 19: 1253–1260.

Darwish, R.M., Aburjai, T.A. 2010. Effect of ethnomedicinal plants used in folklore medicine in Jordan as antibiotic resistant inhibitors on *Escherichia coli*. BMC Complementary Alternative Med., 10: 9.

Dawid, C., Dunemann, F., Schwab, W., Nothnagel, T., Hofmann, T. 2015. Bioactive C 17-polyacetylenes in carrots (*Daucus carota* L.): Current knowledge and future perspectives. J. Agric. Food Chem., 63: 9211–9222.

De Berker, D.A. 1997. Comparison of psoralen-UVB and psoralen UVA photochemotherapy in the treatment of psoriasis. J. American Acad. Dermatol., 36: 577–581.

De Leeuw, J., Van der Beek, N., Maierhofer, G., Neugebauer, W.D. 2003. A case study to evaluate the treatment of vitiligo with khellin encapsulated in L-phenylalanin stabilized phosphatidylcholine liposomes in combination with ultraviolet light therapy. Eur. J. Dermatol., 13: 474–477.

De Leeuw, J., Assen, Y.J., van der Beek, N., Bjerring, P., Martino Neumann, H.A. 2011. Treatment of vitiligo with khellin liposomes, ultraviolet light and blister roof transplantation. J. Eur. Acad. Dermatol. Venereol., 25: 74–81.

Deepa, B., Anuradha, C.V. 2011. Linalool, a plant derived monoterpene alcohol, rescues kidney from diabetes-induced nephropathic changes via blood glucose reduction. Diabetologia Croatica, 40.

Destaillats, F., Keskitalo, M., Arul, J., Angers, P. 2009. Triacylglycerols of Apiaceae seed oils: Composition and regiodistribution of fatty acids. Eur. J. Lipid Sci. Technol., 111: 164–169.

de Vartavan, C. 1990. Contaminated plant-foods from the tomb of Tutankhamun: A new interpretive system. J. Archaeological Sci., 17: 473–494.

Dianat, M., Veisi, A., Ahangarpour, A., Moghaddam, H.F. 2015. The effect of hydro-alcoholic celery (*Apium graveolens*) leaf extract on cardiovascular parameters and lipid profile in animal model of hypertension induced by fructose. Avicenna J. Phytomed., 5: 203.

Dias, J.S. 2012. Nutritional quality and health benefits of veg: a review. Food Nutr. Sci., 3: 1354–74.

Díaz-Maroto, M.C., Pérez-Coello, M.S., Esteban, J., Sanz, J. 2006. Comparison of the volatile composition of wild fennel samples (*Foeniculum vulgare* Mill.) from Central Spain. J. Agric. and Food Chem., 54: 6814–6818.

Diederichsen, A. 1996. Coriander (*Coriandrum sativum* L.); promoting the conservation and use of underutilized and neglected crops. 3. Institute of Plant Genetics and Crop Plant Research, Gatersleben/International Plant Genetic Resources Institute, Rome, 1–82.

Divya, P., Puthusseri, B., Neelwarne, B. 2012. Carotenoid content, its stability during drying and the antioxidant activity of commercial coriander (*Coriandrum sativum* L.) varieties. Food Res. Int., 45: 342–350.

Duarte, J., Perez-Vizcaino, F., Torres, A.I., Zarzuelo, A., Jimenez, J., Tamargo, J. 1995. Vasodilator effects of visnagin in isolated rat vascular smooth muscle. Eur. J. Pharm., 286: 115–122.

Duarte, J., Torres, A.I., Zarzuelo, A. 2000. Cardiovascular effects of Visnagin on rats. Planta Med., 66: 35–39.

Duarte, A., Ferreira, S., Silva, F., Domingues, F. 2012. Synergistic activity of coriander oil and conventional antibiotics against *Acinetobacter baumannii*. Phytomed., 19: 236–238.

Duarte, A., Luís, Â., Oleastro, M., Domingues, F.C. 2016. Antioxidant properties of coriander essential oil and linalool and their potential to control *Campylobacter* spp. Food Control, 61: 115–122.

Dubey, P., Saxena, S., Lal, G., Kant, K., Mishra, B., Agarwal, D., John, S., Sharma, L. 2018. Essential oil constituents and unsaturated fatty acids in Indian *Cuminum cyminum* L. seed oil under varying agro climatic environments. Int. J. Seed Spices, 8: 7–14.

Duke, J.A. 1983. Handbook of Energy Crops. Purdue University, Center for New Crops & Plants Products.

Duke, J.A. 1985. Handbook of Medicinal Herbs. Boca Raton: CRC Press, 356–357.

Duke, J.A. 2002. Handbook of Medicinal Herbs. CRC Press.

Eddouks, M., Maghrani, M., Lemhadri, A., Ouahidi, M.L., Jouad, H. 2002. Ethnopharmacological survey of medicinal plants used for the treatment of diabetes mellitus, hypertension and cardiac diseases in the south-east region of Morocco (Tafilalet). J. Ethnopharmacol., 82(2-3): 97–103.

Ehsani, A., Mahmoudi, R. 2012. Phytochemical properties and hygienic effects of Allium ascalonicum and Pimpinella anisum essential oils in Iranian white brined cheese. J. Essent. Oil Bearing Plants, 15: 1013–1020.

Eidi, A., Eidi, M., Badiei, L. 2009. Antinociceptive effects of ethanolic extract of parsley (*Petroselinum crispum* L.) leaves in mice. Med. Sci. J. Islamic Azad Univ., 19(3): Pe181–Pe186.

Ekinci, F.Y., Baser, G.M., Özcan, E., Üstündağ, Ö.G., Korachi, M., Sofu, A., Blumberg, J.B., Chen, C.-Y.O. 2016. Characterization of chemical, biological, and antiproliferative properties of fermented black carrot juice, shalgam. Eur. Food Res. Technol., 242: 1355–1368.

Elgamal, M.H.A., Shalaby, N.M.M., El-Hagrassy, A.M., Toth, G., Simon, A., Duddeck, H. 1998. A further contribution to some gamma-pyrone constituents of Ammi visnaga. Fitoterapia, 69: 549–550.

El-Houri, R.B., Kotowska, D., Christensen, K.B., Bhattacharya, S., Oksbjerg, N., Wolber, G., Kristiansen, K., Christensen, L.P. 2015. Polyacetylenes from carrots (*Daucus carota*) improve glucose uptake *in vitro* in adipocytes and myotubes. Food Funct., 6: 2135–2144.

El-Mofty, A.M., El-Sawalhy, H., El-Mofty, M. 1995. Photochemotherapy in the treatment of post tinea versicolor hypopigmentation. Med. J. Cairo. Univ., 61(4): 632–637.

El-Rahman, H., Abd–Elhak, N. 2015. Xanthine oxidase inhibitory activity and antigout of celery leek parsley and molokhia. Adv. Biochem., 3: 40–50.

El-Sayed, M.M., Metwally, N.H., Ibrahim, I.A., Abdel-Hady, H., Abdel-Wahab, B.S.A. 2018. Antioxidant activity, total phenolic and flavonoid contents of *Petroselinum crispum* Mill. J. Appl. Life Sci. Int., 19(2): 1–7.

El-Sharkawy, E., Selim, Y. 2018. Three new coumarin types from aerial parts of Ammi majus L. and their cytotoxic activity. Zeitschrift fur Naturforschung. C, J. Biosciences, 73: 1–7.

Elvira-Torales, L.I., García-Alonso, J., Periago-Castón, M.J. 2019. Nutritional importance of carotenoids and their effect on liver health: A review. Antioxidants, 8: 229.

Emamghoreishi, M., Khasaki, M., Aazam, M.F. 2005. Coriandrum sativum: evaluation of its anxiolytic effect in the elevated plus-maze. J. Ethnopharmacol., 96: 365–370.

Emamghoreishi, M., Heidari-Hamedani, G. 2006. Sedative-hypnotic activity of extracts and essential oil of coriander seeds. Iranian J. Med. Sci., 31: 22–27.

Ereifej, K.I., Feng, H., Rababah, T., Almajwal, A., Alu'datt, M., Gammoh, S.I., Oweis, L.I. 2015. Chemical composition, phenolics, anthocyanins concentration and antioxidant activity of ten wild edible plants. Food and Nutrition Sci., 6: 581.

Esquivel-Ferriño, P.C., Favela-Hernández, J.M.J., Garza-González, E., Waksman, N., Ríos, M.Y., Camacho-Corona, M.d.R. 2012. Antimycobacterial activity of constituents from *Foeniculum vulgare* var. dulce grown in Mexico. Molecules, 17: 8471–8482.

Faulks, R.M., Southon, S. 2001. Carotenoids, metabolism and disease. *In*: Handbook of Nutraceuticals and Functional Foods. CRC Press, Florida, p. 9.

Fazal, S.S., Singla, R.K. 2012. Review on the pharmacognostical & pharmacological characterization of *Apium graveolens* Linn. Indo Global J. of Pharmaceutical Sci., 2: 36–42.

Fejes, S., Kéry, A., Blázovics, A., Lugasi, A., Lemberkovics, E., Petri, G., Szöke, E. 1998. Investigation of the *in vitro* antioxidant effect of *Petroselinum crispum* (Mill.) Nym. ex A.W. Hill. Acta Pharm. Hung., 68(3): 150–156.

Fejes, S.Z., Blázovics, A., Lemberkovics, E., Petri, G., Sz"oke, E., Kéry, A. 2000. Free radical scavenging and membrane protective effects of methanol extracts from *Anthriscus cerefolium* L. (Hoffm.) and *Petroselinum crispum* (Mill.) nym. ex A.W. Hill. Phytother. Res., 14(5): 362–365.

Forster, H., Niklas, H., Lutz, S. 1980. Antispasmodic effects of some medicinal plants. Planta Med., 40: 309–319.

Fu, H.-W., Zhang, L., Yi, T., Tian, J.-K. 2009. A new sesquiterpene from the fruits of *Daucus carota* L. Molecules, 14: 2862–2867.

Furletti, V., Teixeira, I., Obando-Pereda, G., Mardegan, R., Sartoratto, A., Figueira, G., Duarte, R., Rehder, V., Duarte, M., Höfling, J. 2011. Action of *Coriandrum sativum* L. essential oil upon oral Candida albicans biofilm formation. Evidence-Based Complementary and Alternative Med.

Gadi, D., Bnouham, M., Aziz, M., Ziyyat, A., Legssyer, A., Legran, C., Lafeve, F.F., Mekhfi, H. 2009. Parsley extract inhibits *in vitro* and *ex vivo* platelet aggregation and prolongs bleeding time in rats. J. Ethnopharmacol., 125(1): 170–174.

Gadi, D., Bnouham, M., Aziz, M., Ziyyat, A., Legssyer, A., Bruel, A., Berrabah, M., Legrand, C., Fauvel- Lafeve, F., Mekhfi, H. 2012. Flavonoids purified from parsley inhibit human blood platelet aggregation and adhesion to collagen under flow. J. Complement. Integr. Med., 9: 19.

Gagandeep, S., Dhanalakshmi, S., Mendiz, E., Rao, A.R., Kale, R.K. 2003. Chemopreventive effects of *Cuminum cyminum* in chemically induced forestomach and uterine cervix tumors in murine model systems. Nutr. and cancer, 47: 171–180.

Gajewski, M., Szymczak, P., Elkner, K., Dąbrowska, A., Kret, A., Danilcenko, H. 2007. Some aspects of nutritive and biological value of carrot cultivars with orange, yellow and purple-coloured roots. Veg. Crop. Res. Bull, 67: 149–161.

Gallo, M., Ferracane, R., Graziani, G., Ritieni, A., Fogliano, V. 2010. Microwave assisted extraction of phenolic compounds from four different spices. Molecules, 15: 6365–6374.

Gálvez, J., Sánchez De Medina, F., Jiménez, J., Zarzuelo, A. 2001. Effects of flavonoids on gastrointestinal disorders. Studies in Nat. Prod. Chem., 25: 607–49.

Garg, S., Garg, G. 1971. Antifertility screening of plants. Part VII Effect of five indigenous plants on early pregnancy in albino rats. Indian J. Med. Res., 59: 302–306.

Garg, S., Mathur, V. 1972. Effect of chromatographic fractions of *Daucus carota* Linn (seeds) on fertility in female albino rats. J. Reprod. Fertil., 31: 143–145.

Garg, S., Mathur, V.S., Chaudhury, R.R. 1978. Screening of Indian plants for antifertility activity. Indian. J. Exp. Biol., 16: 1077–1079.

Gastón, M.S., Cid, M.P., Vázquez, A.M., Decarlini, M.F., Demmel, G.I., Rossi, L.I., Aimar, M.L., Salvatierra, N.A. 2016. Sedative effect of central administration of *Coriandrum sativum* essential oil and its major component linalool in neonatal chicks. Pharmaceutical Biol., 54: 1954–1961.

Germer, R. 1989. Die pflanzenmaterialien aus dem grab des Tutanchamun. Gerstenberg.

Ghanem, M.T., Radwan, H.M., Mahdy, E.-S.M., Elkholy, Y.M., Hassanein, H.D., Shahat, A.A. 2012. Phenolic compounds from *Foeniculum vulgare* (Subsp. Piperitum) (Apiaceae) herb and evaluation of hepatoprotective antioxidant activity. Pharmacognosy Res., 4: 104.

Ghareeb, A.M., Zedan, T.H., Gharb, L.A. 2011. Antibacterial and antifungal activities of *Ammi visnaga* extracts against pathogenic microorganisms. Iraqi J. Sci., 52: 30–36.

Gharzouli, K., Holzer, P. 2004. Inhibition of guinea pig intestinal peristalsis by the flavonoids quercetin, naringenin, apigenin and genistein. Pharmacol., 70: 5.

Ghorbani, A. 2005. Studies on pharmaceutical ethnobotany in the region of Turkmen Sahra, north of Iran: (Part 1): General results. J. Ethnopharmacol., 102: 58–68.

Gilani, A., Shaheen, F., Saeed, S.A. 1994. Cardiovascular action of *Daucus carota*. Arch. Pharm. Res., 17: 150–153.

Gonçalves, E.M., Pinheiro, J., Abreu, M., Brandão, T.R.S., Silva, C.L.M. 2013. Carrot (*Daucus carota* L.) peroxidase inactivation, phenolic content and physical changes kinetics due to blanching. J. Food Eng., 97: 574–581.

Gouda, Y.G., Hamam, F.S., AlRemawi, M.M., Abdallah, M.A., AlAbbadi, M.A., AlShehary, S.M., AlGohary, R.T., Mohamed, K.M. 2014. Study of analgesia caused by some commonly used herbs in the kingdom of Saudi Arabia. Ann. Biol. Sci., 2: 48–55.

Gray, A.M., Flatt, P.R. 1999. Insulin-releasing and insulin-like activity of the traditional anti-diabetic plant *Coriandrum sativum* (coriander). British J. Nutr., 81: 203–209.

Griep, L.M., Verschuren, W.M., Kromhout, D., Ocké, M.C., Geleijnse, J.M. 2011. Colors of fruit and vegetables and 10-year incidence of CHD. British J. Nutr., 106: 1562–1569.

Grieve, M. 2014. https://www.botanical.com/botanical/mgmh/p/parsle09.html.

Guenaydin, K., Beyazit, N. 2004. The chemical investigations on the ripe fruits of *Ammi visnaga* (Lam.) Lamarck growing in Turkey. Nat. Prod. Res., 18: 169–175.

Guerra, N.B., de Almeida Melo, E., Mancini Filho, J. 2005. Antioxidant compounds from coriander (*Coriandrum sativum* L.) etheric extract. J. Food Comp. and Analysis, 18: 193–199.

Guieta, S., Robinsa, R.J., Leesb, M., Billaulta, I. 2003. Phytochemistry quantitative 2 H NMR analysis of deuterium distribution in petroselinic acid isolated from parsley seed. Phytochem., 64(1): 227–233.

Gülçın, İ., Oktay, M., Kıreçcı, E., Küfrevıoğlu, Ö.İ. 2003. Screening of antioxidant and antimicrobial activities of anise (*Pimpinella anisum* L.) seed extracts. Food Chem., 83: 371–382.

Gulfraz, M., Mehmood, S., Minhas, N., Jabeen, N., Kausar, R., Jabeen, K., Arshad, G. 2008. Composition and antimicrobial properties of essential oil of *Foeniculum vulgare*. African J. of Biotechnol., 7.

Haidari, F., Ali, S., Mohammad, M., Mahboob, S. 2011. Effects of parsley (*Petroselinum crispum*) and its flavonol constituents, kaempferol and quercetin, on serum uric acid levels, biomarkers of oxidative stress and liver xanthine oxidoreductase activity in oxonate-induced hyperuricemic rats. Iran. J. Pharm. Res., 10: 811–819.

Hamza, A.A., Amin, A. 2007. *Apium graveolens* modulates sodium valproate-induced reproductive toxicity in rats. Journal of Experimental Zoology Part A: Ecological Genetics and Physiol., 307: 199–206.

Hardani, A., Afzalzadeh, M.R., Amirzargar, A., Mansouri, E., Meamar, Z. 2015. Effects of aqueous extract of celery (*Apium graveolens* L.) leaves on spermatogenesis in healthy male rats. Avicenna J. Phytomed., 5: 113.

Harsha, S.N., Anilakumar, K.R. 2014. *In vitro* free radical scavenging and DNA damage protective property of *Coriandrum sativum* L. leaves extract. J. Food Sci. Technolo., 51: 1533–1539.

Hashemnia, M., Nikousefat, Z., Mohammadalipour, A., Zangeneh, M.-M., Zangeneh, A. 2019. Wound healing activity of *Pimpinella anisum* methanolic extract in streptozotocin-induced diabetic rats. J. Wound Care, 28: S26–S36.

Hashim, S., Jan, A., Marwat, K.B., Khan, M.A. 2014. Phytochemistry and medicinal properties of *Ammi visnaga* (Apiacae). Pak. J. Bot., 46: 861–867.

Hassan, J., Mhamed, M., Mohammed, E. 2002. Hypoglycemic effect of aqueous extract of *Ammi visnaga* in normal and streptozotocin-induced diabetic rats. J. Herb. Pharma., 2: 19–29.

Hassanen, N.H., Eissa, A., Hafez, S., Mosa, E.A. 2015. Antioxidant and antimicrobial activity of celery (*Apium graveolens*) and coriander (*Coriandrum sativum*) herb and seed essential oils. Int. J. Curr. Microbiol. App. Sci., 4: 284–296.

Hempel, J., Pforte, H., Raab, B., Engst, W., Bo, H., Jacobasch, G. 1999. Flavonols and flavones of parsley cell suspension culture change the antioxidative capacity of plasma in rats. Nahrung, 43(3): 201–204.

Herbst, S.T. 2001. The New Food Lover's Companion: Comprehensive Definitions of Nearly 6,000 Food, Drink, and Culinary Terms. Barron's Cooking Guide. Hauppauge, NY: Barron's Educational Series.

Hofer, A., Kerl, H., Wolf, P. 2001. Long-term results in the treatment of vitiligo with oral khellin plus UVA. Eur. J. Dermatol., 11: 225–229.

Holton, J., Basset, C. 2005. Bactericidal and anti-adhesive properties of culinary and medicinal plants against *Helicobacter pylori*. World J. Gastroenterol., 11(47): 7499–7507.

Hussain, I., Khan, S., Khan, M.I., Ur Rehma, I., Ahmed, M. 2012. Investigation of fatty Acid composition of *Ammi majus* seed oil by gas chromatography mass spectrometry. J. Chin. Chem. Soc., 59: 1–4.

Jabeen, Q., Bashir, S., Lyoussi, B., Gilani, A.H. 2009. Coriander fruit exhibits gut modulatory, blood pressure lowering and diuretic activities. J. Ethnopharmacol., 122: 123–130.

Jafari, T., Mahmoodnia, L., Tahmasebi, P., Memarzadeh, M.R., Sedehi, M., Beigi, M., Fallah, A.A. 2018. Effect of cumin (*Cuminum cyminum*) essential oil supplementation on metabolic profile and serum leptin in pre-diabetic subjects: A randomized double-blind placebo-controlled clinical trial. J. Func. Foods, 47: 416–422.

Jain, P.K., Khurana, N., Pounikar, Y., Patil, S., Gajbhiye, A. 2012. Hepatoprotective effect of carrot (*Daucus carota* L.) on paracetamol intoxicated rats. IJPPT, 1: 17–22.

Janahmadi, M., Farajnia, S., Vatanparast, J., Abbasipour, H., Kamalinejad, M. 2008. The fruit essential oil of *Pimpinella anisum* L. (Umblliferae) induces neuronal hyperexcitability in snail partly through attenuation of after-hyperpolarization. J. Ethnopharmacol., 120: 360–365.

Jansen, G.C., Wohlmuth, H. 2014. Carrot seed for contraception: a review. Aust. J. Herb. Med., 26: 10–17.

Jiin, W.H., Hidayat, E.M., Lukman, K. 2014. Gastroprotective effect of carrot (*Daucus carota* L.) juice in rat models. Althea Med. J., 1: 35–39.

Jing, P., Bomser, J.A., Schwartz, S.J., He, J., Magnuson, B.A., Giusti, M.M. 2008. Structure-function relationships of anthocyanins from various anthocyanin-rich extracts on the inhibition of colon cancer cell growth. J. Agric. Food Chem., 56: 9391–9398.

Joshi, S.C., Sharma, N., Sharma, P. 2012. Antioxidant and lipid lowering effects of *Coriandrum sativum* in cholesterol fed rabbits. Int. J. Pharmacy and Pharmaceutical Sci., 4: 231–234.

Jouad, H., Haloui, M., Rhiouani, H., El Hilaly, J., Eddouks, M. 2001. Ethnobotanical survey of medicinal plants used for the treatment of diabetes, cardiac and renal diseases in the North centre region of Morocco (Fez-Boulemane). J. Ethnopharmacol., 77(2-3): 175–182.

Jung, W., Chung, I., Kim, S., Kim, M., Ahmad, A., Praveen, N. 2011. *In vitro* antioxidant activity, total phenolics and flavonoids from celery (*Apium graveolens*) leaves. J. Med. Plants Res., 5: 7022–7030.

Kalita, D., Holm, D.G., LaBarbera, D.V., Petrash, J.M., Jayanty, S.S. 2018. Inhibition of α-glucosidase, α-amylase, and aldose reductase by potato polyphenolic compounds. PloS One, 13: e0191025.

Kaliwal, B.B., Ahamed, R.N., Rao, M.A. 1984. Dose and temporal effect of carrot seed (*Daucus carota*) extract on pregnancy in albino rats. Comp. Physiol. Ecol., 9: 173–177.

Kamboj, V., Hawan, B. 1982. Research on plants for fertility regulation in India. J. Ethnopharmacol., 6: 191–226.

Kammerer, D., Carle, R., Schieber, A. 2003. Detection of peonidin and pelargonidin glycosides in black carrots (*Daucus carota* ssp. *sativus* var. atrorubens Alef.) by high performance liquid chromatography/electrospray ionization mass spectrometry. Rapid Commun. Mass Sp., 17: 2407–2412.

Kang, N., Yuan, R., Huang, L., Liu, Z., Huang, D., Huang, L., Gao, H., Liu, Y., Xu, Q.-M., Yang, S. 2019. Atypical nitrogen-containing flavonoid in the fruits of cumin (*Cuminum cyminum* L.) with anti-inflammatory activity. J. Agric. Food Chem., 67: 8339–8347.

Karimzadeh, F., Hosseini, M., Mangeng, D., Alavi, H., Hassanzadeh, G.R., Bayat, M., Jafarian, M., Kazemi, H., Gorji, A. 2012. Anticonvulsant and neuroprotective effects of *Pimpinella anisum* in rat brain. BMC Complementary and Alternative Med., 12: 76.

Kassi, E., Papoutsi, Z., Fokialakis, N., Messari, I., Mitakou, S., Moutsatsou, P. 2004. Greek plant extracts exhibit selective estrogen receptor modulator (SERM)-like properties. J. Agric. Food Chem., 52: 6956–6961.

Kaur, G.J., Arora, D.S. 2009. Antibacterial and phytochemical screening of *Anethum graveolens*, *Foeniculum vulgare* and *Trachyspermum ammi*. BMC Complementary and Alternative Med., 9: 30.

Kaur, T., Bijarnia, R.K., Singla, S.K., Tandon, C. 2009. *In vivo* efficacy of *Trachyspermum ammi* anticalcifying protein in urolithiatic rat model. J. Ethnopharmacol., 126: 459–462.

Kazempor, S.F. 2015. The analgesic effects of different extracts of aerial parts of *Coriandrum Sativum* in mice. Int. J. Biomed. Sci. IJBS, 11: 23.

Keddad, A., Baaliouamer, A., Hazzit, M. 2016. Chemical composition and antioxidant activity of essential oils from umbels of algerian *Ammi visnaga* (L.). J. Oil Bearing. Plants, 19: 1243–1250.

Kedia, A., Prakash, B., Mishra, P.K., Dubey, N.K. 2014. Antifungal and antiaflatoxigenic properties of *Cuminum cyminum* (L.) seed essential oil and its efficacy as a preservative in stored commodities. Int. J. Food Microbiol., 168-169: 1–7.

Keihan, G.S., Gharib, M.H., Momeni, A., Hemati, Z., Sedighin, R. 2016. A comparison between the effect of *cuminum cyminum* and vitamin E on the level of leptin, paraoxonase 1, hbA1c and oxidized LDL in diabetic patients. Int. J. Mol. Cellular Med., 5: 229.

Khadhri, A., ElMokni, R., Mguis, K., Ouerfelli, I., Araujo, M. 2011. Variability of two essential oils of *Ammi visnaga* (L.) Lam. a traditional Tunisian medicinal plant. J. Med. Plants Res., 5: 5079–5082.

Khalfallah, A., Labed, A., Semra, Z., Al kaki, B., Kabouche, A., Touzani, R., Kabouche, Z. 2011. Antibacterial activity and chemical composition of the essential oil of *Ammi visnaga* L. (Apiaceae) from Constantine, Algeria. Int. J. Med. Aromat. Plants, 1: 302–305.

Khalil, N., Ashour, M.L., Singab, A.N., Salama, O. 2015. Chemical composition and biological activity of the essential oils obtained from yellow and red carrot fruits cultivated in Egypt. J. Pharm. Biol. Sci., 10: 13–19.

Khalil, N., Ashour, M., Fikry, S., Singab, A.N., Salama, O. 2018. Chemical composition and antimicrobial activity of the essential oils of selected Apiaceous fruits. Future J. Pharmaceutical Sci., 4: 88–92.

Khalil, N., Bishr, M., Desouky, S., Salama, O. 2020. Ammi visnaga L., a potential medicinal plant: a review. Molecules, 25.

Khan, D.A., Hassan, F., Ullah, H., Karim, S., Baseer, A., Abid, M.A., Ubaid, M., Khan, S.A., Murtaza, G. 2013. Antibacterial activity of *Phyllantus emblica*, *Coriandrum sativum*, *Culinaris medic*, *Lawsonia alba* and *Cucumis sativus*. Acta Pol. Pharm. Drug Res., 70: 855–860.

Khan, N. 2017. *In vitro* antibacterial activity of *Foeniculum vulgare* seed extract. Agrotechnol., 6: 2.

Khare, C.P. 2008. Indian Medicinal Plants: An Illustrated Dictionary. Springer Science & Business Media.

Khichi, K.K., Ali, A., Gehlot, A., Rathore, R. 2019. An experimental study of *Cuminum cyminum* for its analgesic properties in albino rat. Int. J. Sci. Res., 8.

Kim, O.M., Kim, M.K., Lee, S.O., Lee, K.R., Kim, S.D. 1998. Antimicrobial effect of ethanol extracts from spices against *Lactobacillus plantarum* and *Leuconostoc mesenteroides* isolated from kimchi. J. Korean Soc. Food Sci. Nutr., 27(3): 455–460.

Kiralan, M., Calikoglu, E., Ipek, A., Bayrak, A., Gurbuz, B. 2009. Fatty acid and volatile oil composition of different coriander (*Coriandrum sativum*) registered varieties cultivated in Turkey. Chem. Natural Compounds, 45: 100–102.

Kirtikar, K.R., Basu, B.D. 1933. Indian Medicinal Plants, Vol. II. Dehradun (Uttanchal): International Book Distributors. pp. 1229–1231.

Kitajima, J., Ishikawa, T., Satoh, M. 2003. Polar constituents of celery seed. Phytochem., 64: 1003–1011.

Kohara, K., Sakamoto, Y., Hasegawa, H., Kozuka, H., Sakamoto, K., Hayata, Y. 2006. Fluctuations in volatile compounds in leaves, stems, and fruits of growing coriander (*Coriandrum sativum* L.) plants. J. Jap. Society for Horticultural Sci., 75: 267–269.

Koppula, S., Choi, D.K. 2011. *Cuminum cyminum* extract attenuates scopolamine-induced memory loss and stress-induced urinary biochemical changes in rats: a noninvasive biochemical approach. Pharmaceutical Biol., 49: 702–708.

Kosalec, I., Pepeljnjak, S., Kuštrak, D. 2005. Antifungal activity of fluid extract and essential oil from anise fruits (*Pimpinella anisum* L., Apiaceae). Acta Pharmaceutica, 55: 377–385.

Kooti, W., Ghasemiboroon, M., Asadi-Samani, M., Ahangarpoor, A., Noori Ahmad Abadi, M., Afrisham, R., Dashti, N. 2014. The effects of hydro-alcoholic extract of celery on lipid profile of rats fed a high fat diet. Advances in Environmental Biol., 8: 325–330.

Kooti, W., Ali-Akbari, S., Asadi-Samani, M., Ghadery, H., Ashtary-Larky, D. 2015. A review on medicinal plant of *Apium graveolens*. Advanced Herbal Med., 1: 48–59.

Kooti, W., Daraei, N. 2017. A review of the antioxidant activity of celery (*Apium graveolens* L). J. Evidence-based Complementary & Alternative Med., 22: 1029–1034.

Kooti, W., Moradi, M., Peyro, K., Sharghi, M., Alamiri, F., Azami, M., Firoozbakht, M., Ghafourian, M. 2017. The effect of celery (*Apium graveolens* L.) on fertility: A systematic review. J. Complementary and Integrative Med., 15.

Kousar, S., Jahan, N., Khalil-ur-Rehman, Nosheen, S. 2011. Antilipidemic activity of *Coriandrum sativum*. Biosci. Res., 8: 8–14.

Kreutzmann, S., Christensen, L.P., Edelenbos, M. 2008. Investigation of bitterness in carrots (*Daucus carota* L.) based on quantitative chemical and sensory analyses. LWT Food Sci. Technol., 41: 193–205.

Kreydiyyeh, S.I., Usta, J., Kaouk, I., Al-Sadi, R. 2001. The mechanism underlying the laxative properties of Parsley extract. Phytomed., 8(5): 382–388.

Kreydiyyeh, S.I., Usta, J., Knio, K., Markossian, S., Dagher, S. 2003. Aniseed oil increases glucose absorption and reduces urine output in the rat. Life Sci., 74: 663–673.

Ksouri, A., Dob, T., Belkebir, A., Krimat, S., Chelghoum, C. 2015. Chemical composition and antioxidant activity of the essential oil and the methanol extract of Algerian wild carrot *Daucus carota* L. ssp. *carota* (L.) Thell. J. Mater Environ. Sci., 6: 784–791.

Kumar, S., Mishra, M., Wahab, N., Warikoo, R. 2014. Larvicidal, repellent, and irritant potential of the seed-derived essential oil of Apium graveolens against dengue vector, *Aedes aegypti* L. (Diptera: Culicidae). Frontiers in Public Health, 2.

Kumarasamy, Y., Nahar, L., Byres, M., Delazar, A., Sarker, S.D. 2005. The assessment of biological activities associated with the major constituents of the methanol extract of 'wild carrot' (*Daucus carota* L.) seeds. J. Herb Pharmacother., 5: 61–72.

Kundu, S., Shabir, H., Basir, S.F., Khan, L.A. 2014. Inhibition of As (III) and Hg (II) caused aortic hypercontraction by eugenol, linalool and carvone. J. Smooth Muscle Res., 50: 93–102.

Kunzemann, J., Herrmann, K. 1977. Isolation and identification of flavon (ol)-O-glycosides in caraway (*Carum carvi* L.), fennel (*Foeniculum vulgare* Mill.), anise (*Pimpinella anisum* L.), and coriander (*Coriandrum sativum* L.), and of flavon-C-glycosides in anise. I. Phenolics of spices (author's transl). Zeitschrift fur Lebensmittel-untersuchung und-Forschung, 164: 194–200.

Kuruuzum-Uz, A., Guvenalp, Z., Yuzbasioglu, M., Ozbek, H., Kazaz, C., Demirezer, L. 2010. Flavonoids from *Pimpinella kotschyana*. Planta Med., 76: P274.

Laribi, B., Kouki, K., M'Hamdi, M., Bettaieb, T. 2015. Coriander (*Coriandrum sativum* L.) and its bioactive constituents. Fitoterapia, 103: 9–26.

Lasram, S., Zemni, H., Hamdi, Z., Chenenaoui, S., Houissa, H., Tounsi, M.S., Ghorbel, A. 2019. Antifungal and antiaflatoxinogenic activities of *Carum carvi* L., *Coriandrum sativum* L. seed essential oils and their major terpene component against *Aspergillus flavus*. Ind. Crops Prod., 134: 11–18.

Lawrence, B.M. 1992. Progress in essential oils. Perfumer & Flavorist, 17: 71–72.

Lawrence, B.M. 1999. Progress in essential oils. Perfumer & Flavorist, 24: 7–8.

Lee, H.-S. 2005. Cuminaldehyde: Aldose reductase and α-glucosidase inhibitor derived from *Cuminum cyminum* L. seeds. J. Agric. Food Chem., 53: 2446–2450.

Lee, J., Jung, J., Park, S., Sim, Y., Kim, S., Ha, T., Suh, H. 2010. Anti-inflammatory effect of visnagin in lipopolysaccharide-stimulated BV-2 microglial cells. Arch. Pharma. Res., 33: 1843–1850.

Leja, M., Kamińska, I., Kramer, M., Maksylewicz-Kaul, A., Kammerer, D., Carle, R., Baranski, R. 2013. The content of phenolic compounds and radical scavenging activity varies with carrot origin and root color. Plant. Foods Hum. Nutr., 68: 163–170.

Leporatti, M.L., Ghedira, K. 2009. Comparative analysis of medicinal plants used in traditional medicine in Italy and Tunisia. J. Ethnobiol. and Ethnomed., 5: 31–31.

Li, M.-Y., Hou, X.-L., Wang, F., Tan, G.-F., Xu, Z.-S., Xiong, A.-S. 2018. Advances in the research of celery, an important Apiaceae vegetable crop. Critical Reviews in Biotechnology, 38: 172–183.

Li, R., Jiang, Z.T. 2004. Chemical composition of the essential oil of *Cuminum cyminum* L. from China. Flavour and Frag. J., 19: 311–313.

Li, S., Li, L., Yan, H., Jiang, X., Hu, W., Han, N., Wang, D. 2019. Anti-gouty arthritis and anti-hyperuricemia properties of celery seed extracts in rodent models. Mol. Med. Rep., 20: 4623–4633.

Liang, W., Wing-Shing Cheung, D., Chai Tang, H., Ping Lam, W., Zhang, L., Chung Leung, P., Tai-Wai Yew, D. 2015. Chapter 89—Effects of 3-n-Butylphthalide from celery on vascular dementia. pp. 963–971. *In*: Martin, C.R., Preedy, V.R. (eds.). Diet and Nutrition in Dementia and Cognitive Decline. Academic Press, San Diego.

Linde, G.A., Gazim, Z.C., Cardoso, B.K., Jorge, L.F., Tešević, V., Glamoćlija, J., Soković, M., Colauto, N.B. 2016. Antifungal and antibacterial activities of *Petroselinum crispum* essential oil. Genet. Mol. Res., 15(3).

MacLeod, A.J., Snyder, C.H., Subramanian, G. 1985. Volatile aroma constituents of parsley leaves. Phytochem., 24(11): 2623–2627.

Mahmoud, A.L.E. 1999. Inhibition of growth and aflatoxin biosynthesis of Aspergillus flavus by extracts of some Egyptian plants. Lett. Appl. Microbiol., 29: 334–336.

Maleck, M., Dos Santos, F.C.C., Serdeiro, M.T., Guimaraes, A.E., Ferreira, B., Gunaydin, K., de Almeida, A.P. 2013. Khellin: a furanochromone with toxicity against *Oncopeltus fasciatus* (Hemiptera) and *Aedes aegypti* (Diptera). J. Nat. Pharma., 4: 32–36.

Malhotra, S.K. 2012. 12—Celery. pp. 249–267. *In*: Peter, K.V. (ed.). Handbook of Herbs and Spices (Second Edition). Woodhead Publishing.

Mallni, T., Vanithakumari, G., Devi, N., Fiango, V. 1985. Effect of foeniculuai vulgare. Mill seed extract on the genital organs of male and female rats. Indian J. Physiol. Pharmacol., 29: 21–26.

Mani, V., Gunnam, K.K., Parle, M. 2006. Antinociceptive and anti-Inflammatory properties of *Daucus carota* seeds extract. J. Health Sci., 52: 598–606.

Manderfield, M.M., Schafer, H.W., Davidson, P.M., Zottola, E.A. 1997. Isolation and identification of antimicrobial furocoumarins from parsley. J. Food Prot., 60: 72–77.

Marín, I., Sayas-Barberá, E., Viuda-Martos, M., Navarro, C., Sendra, E. 2016. Chemical composition, antioxidant and antimicrobial activity of essential oils from organic fennel, parsley, and lavender from Spain. Foods, 5(18): 1–10.

Marinova, D., Ribarova, F., Atanassova, M. 2005. Total phenolics and total flavonoids in Bulgarian fruits and vegetables. J. Chem. Technol. Metall., 40: 255–260.

Marzouki, H., Khaldi, A., Falconieri, D., Piras, A., Marongiu, B., Molicotti, P., Zanetti, S. 2010. Essential oils of *Daucus carota* subsp. *carota* of Tunisia obtained by supercritical carbon dioxide extraction. Nat. Prod. Commun., 5: 1955–1958.

Maurya, R., Srivastava, S., Kulshreshta, D.K., Gupta, C.M. 2004. Traditional remedies for fertility regulation. Curr. Med. Chem., 11: 1431–1450.

Mazhar, J., Mazumder, A. 2013. Evaluation of antidiabetic activity of methanolic leaf extract of *Coriandrum sativum* in alloxan induced diabetic rats. Res. J. Pharm. Biol. Chem. Sci., 4: 500–507.

Mazza, G. 2007. Anthocyanins and heart health. Ann. Ist Super Sanita., 43: 369–374.

Meena, S., Lal, G., Meena, R. 2018. Multidimensional therapeutic uses of ajwain (*Trachyspermum ammi* L.). Int. J. Seed Spices, 8: 1–5.

Merzouki, A., Ed-Derfoufi, F., El-Aallall, A., Molero-Mesa, J. 1997. Wild medicinal plants used by local Bouhmed population (Morocco). Fitoterapia, 68(5): 444–460.

Metzger, B.T., Barnes, D.M., Reed, J.D. 2008. Purple carrot (*Daucus carota* L.) polyacetylenes decrease lipopolysaccharide-induced expression of inflammatory proteins in macrophage and endothelial cells. J. Agric. Food Chem., 56: 3554–3560.

Miara, M., Bendif, H., Rebbas, K., Rabah, B., Hammou, M., Maggi, F. 2019. Medicinal plants and their traditional uses in the highland region of Bordj Bou Arreridj (Northeast Algeria). J. Herb. Med., 16: 100262.

Milan, K.M., Dholakia, H., Tiku, P.K., Vishveshwaraiah, P. 2008. Enhancement of digestive enzymatic activity by cumin (*Cuminum cyminum* L.) and role of spent cumin as a bionutrient. Food Chem., 110: 678–683.

Miri, V.H., Tyagi, A.K., Ebrahimi, S.H., Mohini, M. 2013. Effect of cumin (*Cuminum cyminum*) seed extract on milk fatty acid profile and methane emission in lactating goat. Small Ruminant Res., 113: 66–72.

Mital, P.R., Laxman, J.P., Rameshvar, K.P. 2011. Protective effect of *Daucus carota* root extract against ischemia reperfusion injury in rats. Pharmacol., 1: 432–439.

Moazedi, A.A., Mirzaie, D.N., Seyyednejad, S.M., Zadkarami, M.R., Amirzargar, A. 2007. Spasmolytic effect of *Petroselinum crispum* (Parsley) on rat's ileum at different calcium chloride concentrations. Pak J. Biol. Sci., 10(22): 4036–4042.

Moghaddam, M., Miran, S.N.K., Pirbalouti, A.G., Mehdizadeh, L., Ghaderi, Y. 2015. Variation in essential oil composition and antioxidant activity of cumin (*Cuminum cyminum* L.) fruits during stages of maturity. Ind. Crops Prod., 70: 163–169.

Moghadam, M.H., Imenshahidi, M., Mohajeri, S.A. 2013. Antihypertensive effect of celery seed on rat blood pressure in chronic administration. J. Med. Food, 16: 558–563.

Mohamad, R.H., El-Bastawesy, A.M., Abdel-Monem, M.G., Noor, A.M., Al-Mehdar, H.A.R., Sharawy, S.M., El-Merzabani, M.M. 2011. Antioxidant and anticarcinogenic effects of methanolic extract and volatile oil of fennel seeds (*Foeniculum vulgare*). J. Med. Food, 14: 986–1001.

Mohammed, M.M., El-Sharkawy, E.R. 2017. Cytotoxic new furoquinoline alkaloid isolated from *Ammi majus* L. growing in Egypt. Nat. Prod. Res., 31: 645–652.

Mojaba, F., Hamedi, A., Nickavar, B., Javidnia, K. 2008. Hydrodistilled volatile constituents of the leaves of *Daucus carota* L. subsp. *sativus* (Hoffman.) Arcang. (Apiaceae) from Iran. J. Essent. Oil Bearing Plants, 11: 271–277.

Momin, R.A., De Witt, D.L., Nair, M.G. 2003. Inhibition of cyclooxygenase (COX) enzymes by compounds from *Daucus carota* L. seeds. Phytother. Res., 17: 976–9.

Montesano, V., Negro, D., Sarli, G., De Lisi, A., Laghetti, G., Hammer, A. 2012. Notes about the uses of plants by one of the last healers in the Basilicata Region (South Italy). J. Ethnobiol. Ethnomed., 8: 15.

Morshedi, D., Aliakbari, F., Tayaranian-Marvian, A., Fassihi, A., Pan-Montojo, F., Perez-Sanchez, H. 2015. Cuminaldehyde as the major component of *Cuminum cyminum*, a natural aldehyde with inhibitory effect on alpha-synuclein fibrillation and cytotoxicity. J. Food Sci., 80: H2336–2345.

Msaada, K., Hosni, K., Taarit, M.B., Chahed, T., Kchouk, M.E., Marzouk, B. 2007. Changes on essential oil composition of coriander (*Coriandrum sativum* L.) fruits during three stages of maturity. Food Chem., 102: 1131–1134.

Msaada, K., Taarit, M.B., Hosni, K., Hammami, M., Marzouk, B. 2009. Regional and maturational effects on essential oils yields and composition of coriander (*Coriandrum sativum* L.) fruits. Scientia Horticulturae, 122: 116–124.

Msaada, K., Jemia, M.B., Salem, N., Bachrouch, O., Sriti, J., Tammar, S., Bettaieb, I., Jabri, I., Kefi, S., Limam, F. 2017. Antioxidant activity of methanolic extracts from three coriander (*Coriandrum sativum* L.) fruit varieties. Arabian J. Chem., 10: S3176–S3183.

Muralidharan, P., Balamurugan, G., Pavan, K. 2008. Inotropic and cardioprotective effects of *Daucus carota* Linn. on isoproterenol-induced myocardial infarction. Bangladesh J. Pharmacol., 3: 74–79.

Mustafa, A., Trevino, L.M., Turner, C. 2012. Pressurized hot ethanol extraction of carotenoids from carrot by-products. Molecules, 17: 1809–1818.

Nagella, P., Ahmad, A., Kim, S.-J., Chung, I.-M. 2012. Chemical composition, antioxidant activity and larvicidal effects of essential oil from leaves of *Apium graveolens*. Immunopharmacol. Immunotoxicol., 34: 205–209.

Nahidi, F., Taherpoor, M., Mojab, F., Majd, H. 2008. Effect of Anise extract on hot flush of menopause. Pajoohandeh J., 13: 167–173.

Nair, V., Singh, S., Gupta, Y.K. 2013. Anti-granuloma activity of *Coriandrum sativum* in experimental models. J. Ayurveda Integrative Med., 4: 13.

Nalini, N., Sabitha, K., Viswanathan, P., Menon, V. 1998. Influence of spices on the bacterial (enzyme) activity in experimental colon cancer. J. Ethnopharmacol., 62: 15–24.

Narain, R., Ilango, V. 2015. Analysis of iron content of selected vegetarian food items in Dubai, UAE. Int. J. Sci., Environment and Technol., 4: 543–552.

Nassar, M.I., Aboutabl, E.-s.A., Makled, Y.A., El-Khrisy, E.-D., Osman, A.F. 2010. Secondary metabolites and pharmacology of *Foeniculum vulgare* Mill. subsp. Piperitum. Revista latinoamericana de química, 38: 103–112.

Naveed, R., Hussain, I., Tawab, A., Tariq, M., Rahman, M., Hameed, S., Mahmood, M.S., Siddique, A.B., Iqbal, M. 2013. Antimicrobial activity of the bioactive components of essential oils from Pakistani spices against Salmonella and other multi-drug resistant bacteria. BMC Complement Altern. Med., 13: 265.

Nayeem, K., Godad, A., Hashilkar, N., Joshi, R.K. 2010. Gastroprotective activity of the aqueous extract from the roots of *Daucus carota* L. in rats. Int. J. Res. Ayurveda Pharm., 1: 112–119.

Neffati, M., Marzouk, B. 2008. Changes in essential oil and fatty acid composition in coriander (*Coriandrum sativum* L.) leaves under saline conditions. Ind. Crops and Prod., 28: 137–142.

Netzel, M., Netzel, G., Kammerer, D.R., Schieber, A., Carle, R., Simons, L., Bitsch, I., Bitsch, R., Konczak, I. 2007. Cancer cell antiproliferation activity and metabolism of black carrot anthocyanins. Innov. Food Sci. Emerg. Technol., 8: 365–372.

Nguyen, Q.-H., Talou, T., Cerny, M., Evon, P., Merah, O. 2015. Oil and fatty acid accumulation during coriander (*Coriandrum sativum* L.) fruit ripening under organic cultivation. The Crop J., 13: 366–369.

Nickavar, B., Abolhasani, F.A.-S. 2009. Screening of antioxidant properties of seven Umbelliferae fruits from Iran. Pakistan J. Pharmaceutical Sci., 22.

Nielsen, S.E., Young, J.F., Daneshvar, B., Lauridsen, S.T., Knuthsen, P., Sandström, B., Dragsted, L.O. 1999. Effect of parsley (*Petroselinum crispum*) intake on urinary apigenin excretion, blood antioxidant enzymes and biomarkers for oxidative stress in human subjects. Br. J. Nutr., 81(6): 447–455.

Nirumand, M.C., Hajialyani, M., Rahimi, R., Farzaei, M.H., Zingue, S., Nabavi, S.M., Bishayee, A. 2018. Dietary plants for the prevention and management of kidney stones: preclinical and clinical evidence and molecular mechanisms. Int. J. Mol. Sci., 19: 765.

Nitz, S., Spraul, M.H., Drawert, F. 1990. C17 polyacetylenic alcohols as the major constituents in roots of *Petroselinum crispum* Mill. ssp. *tuberosum*. J. Agric. Food Chem., 38(7): 1445–1447.

Niu, F., Sharma, A., Feng, L., Ozkizilcik, A., Muresanu, D.F., Lafuente, J.V., Tian, Z.R., Nozari, A., Sharma, H.S. 2019. Chapter 3—Nanowired delivery of DL-3-n-butylphthalide induces superior neuroprotection in concussive head injury. pp. 89–118. *In*: Sharma, A., Sharma, H.S. (eds.). Progress in Brain Research. Elsevier.

Ojala, T., Remes, S., Haansuu, P., Vuorela, H., Hiltunen, R., Haahtela, K., Vuorela, P. 2000. Antimicrobial activity of some coumarin containing herbal plants growing in Finland. J. Ethnopharmacol., 73(1): 299–305.

Orhan, İ.E., ÖZÇELİK, B., Kartal, M., Kan, Y. 2012. Antimicrobial and antiviral effects of essential oils from selected Umbelliferae and Labiatae plants and individual essential oil components. Turkish J. Biol., 36: 239–246.

Ostad, S., Soodi, M., Shariffzadeh, M., Khorshidi, N., Marzban, H. 2001. The effect of fennel essential oil on uterine contraction as a model for dysmenorrhea, pharmacology and toxicology study. J. Ethnopharmacol., 76: 299–304.

Otoom, S., Al-Safi, S., Kerem, Z., Alkofahi, A. 2006. The use of medicinal herbs by diabetic Jordanian patients. J. Herbal Pharmacotherapy, 6: 31–41.

Oulmouden, F., Saïle, R., Gnaoui, N., Benomar, H., Lkhider, M., Amrani, S. 2011. Hypolipidemic and anti-atherogenic effect of aqueous extract of fennel (*Foeniculum Vulgare*) extract in an experimental model of atherosclerosis induced by triton WR-1339. European J. Sci. Res., 52: 91–99.

Özcan, M.M., Chalchat, J.C. 2006. Chemical composition and antifungal effect of anise (*Pimpinella anisum* L.) fruit oil at ripening stage. Annals of Microbiol., 56: 353–358.

Özcan, M.M., Chalchat, J.C. 2007. Chemical composition of carrot seeds (*Daucus carota* L.) cultivated in Turkey: characterization of the seed oil and essential oil. Grasas Y Aceites, 58: 359–365.

Ozsoy-Sacan, O., Yanardag, R., Orak, H., Ozgey, Y., Yarat, A., Tunali, T. 2006. Effects of parsley (*Petroselinum crispum*) extract versus glibornuride on the liver of streptozotocin-induced diabetic rats. J. Ethnopharmacol., 104(1-2): 175–181.

Oztürk, Y., Baser, C.H.K., Aydın, S. 1991. Hepatoprotective (antihepatotoxic) plants in Turkey. pp. 40–50. Baser, K.H.C. (ed.). Proceedings of the 9th Symposium on Plant Drugs. Eskisehir: Turkey.

Pakfetrat, H., Nemati, N., Shiravi, A. 2015. Cytotoxicity effects of Ammi visnaga extract on Hela and MCF7 cancer cell lines. J. Anim. Bio., 7: 25–33.

Panagiotakopulu, E., Buckland, P.C., Day, P.M., Doumas, C. 1995. Natural insecticides and insect repellents in antiquity: a review of the evidence. J. Archaeological Sci., 22: 705–710.

Panche, A.N., Diwan, A.D., Achandra, S.R. 2016. Flavonoids: An overview. J. Nutr. Sci.

Pápay, Z.E., Kósa, A., Boldizsár, I., Ruszkai, A., Balogh, E., Klebovich, I. 2012. Pharmaceutical and formulation aspects of *P. crispum* extract. Acta Pharm. Hung., 82(1): 3–14.

Papuc, C., Predescu, C., Nicorescu, V., Stefan, G., Nicorescu, S. 2016. Antioxidant properties of a parsley (*Petroselinum crispum*) juice rich in polyphenols and nitrites. 1st International Multidisciplinary Conference on Nutraceuticals and Functional Foods Current Research in Nutrition and Food Science, 4(SI. 2): 114–118.

Parejo, I., Jauregui, O., Sánchez-Rabaneda, F., Viladomat, F., Bastida, J., Codina, C. 2004. Separation and characterization of phenolic compounds in fennel (*Foeniculum vulgare*) using liquid chromatography-negative electrospray ionization tandem mass spectrometry. J. Agric. Food Chem., 52: 3679–3687.

Parsad, D., Saini, R., Verma, N. 1998. Combination of PUVAsol and topical calcipotriol in vitiligo. Dermatol., 197: 167–170.

Parthasarathy, V.A., Chempakam, B., Zachariah, T.J. 2008. Chemistry of Spices. Cabi.

Patel, J.D., Patel, D.K., Shrivastava, A., Kumar, V. 2008. Screening of plant extracts used in traditional antidiarrhoeal medicines against pathogenic *Escherichia coli*. Sci. World, 6: 63–67.

Patil, M.V.K., Kandhare, A., Bhise, S. 2012. Anti-inflammatory effect of *Daucus carota* root on experimental colitis in rats. Int. J. Pharm. Pharm. Sci., 4: 337–343.

Patil, S.B., Takalikar, S.S., Joglekar, M.M., Haldavnekar, V.S., Arvindekar, A.U. 2013. Insulinotropic and β-cell protective action of cuminaldehyde, cuminol and an inhibitor isolated from *Cuminum cyminum* in streptozotocin-induced diabetic rats. British J. Nutr., 110: 1434–1443.

Pavela, R. 2015. Acaricidal properties of extracts and major furanochromenes from the seeds of *Ammi visnaga* Linn. against Tetranychus urticae Koch. Ind. Crop. Prod., 67: 108–113.

Pavlić, B., Vidović, S., Vladić, J., Radosavljević, R., Zeković, Z. 2015. Isolation of coriander (*Coriandrum sativum* L.) essential oil by green extractions versus traditional techniques. The J. Supercritical Fluids, 99: 23–28.

Pavlyuk, I., Stadnytska, N., Jasicka-Misiak, I., Górka, B.M., Wieczorek, P., Novikov, V. 2015. A study of the chemical composition and biological activity of extracts from wild carrot (*Daucus carota* L.) seeds waste. Res. J. Pharm. Biol. Chem. Sci., 6: 603–611.

Peana, A.T., Paolo, S.D., Chessa, M.L., Moretti, M.D., Serra, G., Pippia, P. 2003. (−)-Linalool produces antinociception in two experimental models of pain. European J. Pharmacol., 460: 37–41.

Peng, Y., Zeng, X., Feng, Y., Wang, X. 2004. Antiplatelet and antithrombotic activity of L-3-n-butylphthalide in rats. J. Cardiovascular Pharmacol., 43: 876–881.

Petretto, G.L., Fancello, F., Bakhy, K., Faiz, C.A., Sibawayh, Z., Chessa, M., Zara, S., Sanna, M.L., Maldini, M., Rourke, J.P., Pintore, G. 2018. Chemical composition and antimicrobial activity of essential oils from *Cuminum cyminum* L. collected in different areas of Morocco. Food Biosci., 22: 50–58.

Pferschy-Wenzig, E.-M., Getzinger, V., Kunert, O., Woelkart, K., Zahrl, J., Bauer, R. 2009. Determination of falcarinol in carrot (*Daucus carota* L.) genotypes using liquid chromatography/mass spectrometry. Food Chem., 114: 1083–1090.

Phumee, A., Chompoosri, J., Intayot, P., Boonserm, R., Boonyasuppayakorn, S., Buathong, R., Thavara, U., Tawatsin, A., Joyjinda, Y., Wacharapluesadee, S. 2019. Vertical transmission of Zika virus in *Culex quinquefasciatus* Say and *Aedes aegypti* (L.) mosquitoes. Scientific Rep., 9: 5257.

Pineda, R., Vizcaíno, S., García, C.M., Gil, J.H., Durango, D. 2018. Antifungal activity of extracts, essential oil and constituents from *Petroselinum crispum* against *Colletotrichum acutatum*. Revista Facultad Nacional de Agronomía, 71(3): 8563–8572.

Platel, K., Srinivasan, K. 1996. Influence of dietary spices or their active principles on digestive enzymes of small intestinal mucosa in rats. Int. J. Food Sci. Nut., 47: 55–59.

Platel, K., Srinivasan, K. 2000a. Influence of dietary spices and their active principles on pancreatic digestive enzymes in albino rats. Food/Nahrung, 44: 42–46.

Platel, K., Srinivasan, K. 2000b. Stimulatory influence of select spices on bile secretion in rats. Nutr. Res., 20: 1493–1503.

Platel, K., Srinivasan, K. 2001. Studies on the influence of dietary spices on food transit time in experimental rats. Nutr. Res., 21: 1309–1314.

Popović, M., Kaurinović, B., Jakovljević, V., Mimica-Dukic, N., Bursać, M. 2007. Effect of parsley (*Petroselinum crispum* (Mill.) Nym. ex A.W. Hill, Apiaceae) extracts on some biochemical parameters of oxidative stress in mice treated with CCl(4). Phytother. Res., 21(8): 717–723.

Porchezhian, E., Ansari, S.H., Ali, m. 2000. Analgesic and anti-inflammatory activity of volatile oil from *Daucus carota* Linn seeds. Indian J. Nat. Prod. Resour., 16: 24–26.

Poudyal, H., Panchal S, Brown L. 2010. Comparison of purple carrot juice and β-carotene in a high-carbohydrate, high-fat diet-fed rat model of the metabolic syndrome. Br. J. Nutr., 104: 1322–1332.

Pouraboli, I., Ranjbar, B. 2012. Effect of *Daucus carota* seeds extract on serum levels of glucose, lipids and lipoproteins in type I diabetic male rats. Iran J. Biol., 5: 679–687.

Pouraboli, I., Ranjbar, B. 2015. The effect of *Daucus carota* seeds extract on lipid profile, LFT and kidney function indicators in streptozocin-induced diabetic rats. Int. J. Plant Sci. and Ecol., 3: 84–87.

Powanda, M., Rainsford, K. 2011. A toxicological investigation of a celery seed extract having anti-inflammatory activity. Inflammopharmacol., 19: 227–233.

Prachayasittikul, V., Prachayasittikul, S., Ruchirawat, S., Prachayasittikul, V. 2018. Coriander (*Coriandrum sativum*): A promising functional food toward the well-being. Food Res. Int., 105: 305–323.

Pradhan, M., Sribhuwaneswari, S., Karthikeyan, D., Minz, S., Sure, P., Chandu, A.N., Mishra, U., Kamalakannan, K., Saravanankumar, A., Sivakumar, T. 2008. *In-vitro* cytoprotection activity of *Foeniculum vulgare* and *Helicteres isora* in cultured human blood lymphocytes and antitumour activity against B16F10 melanoma cell line. Res. J. Pharmacy and Technol., 1: 450–452.

Punetha, D., Tewari, G., Pande, C. 2018. Compositional variability in inflorescence essential oil of *Coriandrum sativum* from North India. J. of Essential Oil Res., 30: 113–119.

Purup, S., Larsen, E., Christensen, L.P. 2009. Differential effects of falcarinol and related aliphatic C17-polyacetylenes on intestinal cell proliferation. J. Agric. Food Chem., 57: 8290–8296.

Rackham, H., Jones, W.H.S., Eichholz, D. 1947. Pliny: Natural History. Harvard University Press.

Rahimi, A.R., Babaei, S., Mashayekhi, K., Rokhzadi, A., Amini, S. 2013. Anthocyanin content of coriander (*Coriandrum sativum* L.) leaves as affected by salicylic acid and nutrients application. IJB, 3: 141–145.

Ramadan, M.F., Kroh, L.W., Mörsel, J.-T. 2003. Radical scavenging activity of black cumin (*Nigella sativa* L.), coriander (*Coriandrum sativum* L.), and Niger (*Guizotia abyssinica* Cass.) crude seed oils and oil fractions. J. Agric. Food Chem., 51: 6961–6969.

Ramezani, M., Nasri, S., Yassa, N. 2009. Antinociceptive and anti-inflammatory effects of isolated fractions from *Apium graveolens* seeds in mice. Pharmaceutical Biol., 47: 740–743.

Ranjbar, B., Pouraboli, I., Mehrabani, M.S., Dabiri, S., Javadi, A. 2010. Effect of the methanolic extract of *Daucus carota* seeds on carbohydrate metabolism and morphology of pancrease in type I diabetic male rats. Physiol. Pharmacol., 1: 85–93.

Rasooli, I., Taghizadeh, M., Astaneh, S.D.A., Rezaei, M.B., Jaimand, K. 2007. Phytobiological properties of *Ammi visnaga* L. and *Lavandula angustifolia* Mill. essential oils. Int. J. Essent. Oil Ther., 1: 72–78.

Rathee, P., Chaudhary, H., Rathee, S., Rathee, D., Kumar, V., Kohli, K. 2009. Mechanism of action of flavonoids as anti-inflammatory agents: a review. Inflammation & Allergy-Drug Targets (Formerly Current Drug Targets-Inflammation & Allergy), 8: 229–235.

Rather, M.A., Dar, B.A., Sofi, S.N., Bhat, B.A., Qurishi, M.A. 2016. Foeniculum vulgare: A comprehensive review of its traditional use, phytochemistry, pharmacology, and safety. Arabian J. Chem., 9: S1574–S1583.

Raza, M., Shukla, A.K., Fatima, T., Ali, S. 2015. Comparative study of antioxidant activity of polyphenols isolated from frozen and fresh leaves of *Trachyspermum ammi* (Ajwain). J. Pharmacognosy and Phytochem., 3: 122–124.

Rebey, I.B., Jabri-Karoui, I., Hamrouni-Sellami, I., Bourgou, S., Limam, F., Marzouk, B. 2012. Effect of drought on the biochemical composition and antioxidant activities of cumin (*Cuminum cyminum* L.) seeds. Ind. Crops Prod., 36: 238–245.

Rebey, I.B., Kefi, S., Bourgou, S., Ouerghemmi, I., Ksouri, R., Tounsi, M.S., Marzouk, B. 2014. Ripening stage and extraction method effects on physical properties, polyphenol composition and antioxidant activities of cumin (*Cuminum cyminum* L.) seeds. Plant Foods for Human Nutr., 69: 358–364.

Reed, C.F. 1976. Information summaries on 1000 economic plants. Typescripts Submitted to the USDA 1976.

Reed, J. 2002. Cranberry flavonoids, atherosclerosis, and cardiovascular health. Crit. Rev. Food Sci. Nutr., 42: 301–316.

Rezazad, M., Farokhi, F. 2014. Protective effect of *Petroselinum crispum* extract in abortion using prostadin—induced renal dysfunction in female rats. Avicenna J. Phytomed., 4(5): 312–319.

Riddle, J. 1997. Eve's Herbs: A History of Contraception and Abortion in the West. Cambridge, Massachusetts: Harvard University Press.

Robards, K., Prenzler, P.D., Tucker, G., Swatsitang, P., Glover, W. 1999. Phenolic compounds and their role in oxidative processes in fruits. Food Chem., 66: 401–436.

Robbers, J.E., Tyler, V.E. 1999. Tylers' Herbs of Choice: The Therapeutic Use of Phytochemicals. New York: Haworth Herbal Press, p. 92.

Rocha, L., Fernandes, C.P. 2016. Chapter 22—Aniseed (*Pimpinella anisum*, Apiaceae) oils. pp. 209–213. *In*: Preedy, V.R. (ed.). Essential Oils in Food Preservation, Flavor and Safety. Academic Press, San Diego.

Rokbeni, N., M'rabet, Y., Dziri, S., Chaabane, H., Jemli, M., Fernandez, X., Boulila, A. 2013. Variation of the chemical composition and antimicrobial activity of the essential oils of natural populations of Tunisian *Daucus carota* L. (Apiaceae). Chem. Biodiversity, 10: 2278–2290.

Rubatzky, V.E., Quiros, C., Simon, P. 1999. Carrots and Related Vegetable Umbelliferae. CABI, New York.

Sahoo, H.B., Sahoo, S.K., Sarangi, S.P., Sagar, R., Kori, M.L. 2014. Anti-diarrhoeal investigation from aqueous extract of *Cuminum cyminum* Linn. Seed in Albino rats. Pharmacognosy Res., 6: 204–209.

Saini, N., Singh, G., Nagori, B. 2014. Antispasmodic potential of umbelliferae plants a functional approach. Int. J. Pharmaceutical & Biol. Arch., 5.

Saini, R.K., Nile, S.H., Keum, Y.-S. 2016. Food science and technology for management of iron deficiency in humans: A review. Trends in Food Sci. Technol., 53: 13–22.

Samojlik, I., Mijatović, V., Petković, S., Škrbić, B., Božin, B. 2012. The influence of essential oil of aniseed (*Pimpinella anisum*, L.) on drug effects on the central nervous system. Fitoterapia, 83: 1466–1473.

Satrani, B., Farah, A., Fechtal, M., Talbi, M., Bouamrani, M.L. 2004. Chemical composition and antibacterial and antimycotic activities of the essential oil of *Ammi visnaga* (L.) Lam. from Morocco. Acta Bot. Gall., 151: 65–71.

Satyanarayana, S., Sushruta, K., Sarma, G.S., Srinivas, N., Raju, G.V.S. 2004. Antioxidant activity of the aqueous extracts of spicy food additives—evaluation and comparison with ascorbic acid in *in vitro* systems. J. Herbal Pharmacotherapy, 4: 1–10.

Savikin, K., Zdunic, G., Menkovic, N., Zivković, J., Cujić, N., Tereščenko, M., Bigović, D. 2013. Ethnobotanical study on traditional use of medicinal plants in South-Western Serbia, Zlatibor district. J. Ethnopharmacol., 146(3): 803–810.

Sawalha, A.F., Sweileh, W.M., Sa'ed, H.Z., Jabi, S.W. 2008. Self-therapy practices among university students in Palestine: focus on herbal remedies. Complementary Therapies in Med., 16: 343–349.

Sayed-Ahmad, B., Talou, T., Saad, Z., Hijazi, A., Merah, O. 2017. The Apiaceae: Ethnomedicinal family as source for industrial uses. Ind. Crops Prod., 109: 661–671.

Sayyah, M., Mahboubi, A., Kamalinejad, M. 2002a. Anticonvulsant effect of the fruit essential oil of *Cuminum cyminum* in mice. Pharmaceutical Biol., 40: 478–480.

Sayyah, M., Peirovi, A., Kamalinejad, M. 2002b. Anti-nociceptive effect of the fruit essential oil of *Cuminum cyminum* L. in rat. Iranian Biomed. J., 6: 141–145.

Schmiech, L., Alayrac, C., Witulski, B., Hofmann, T. 2009. Structure determination of bisacetylenic oxylipins in carrots (*Daucus carota* L.) and enantioselective synthesis of falcarindiol. J. Agric. Food Chem., 57: 11030–11040.

Selim, Y.A., Ouf, N.H. 2012. Anti-inflammatory new coumarin from the *Ammi majus* L. Med. Chem. Letters, 2: 1–4.

Sellami, H., Flamini, G., Cioni, P., Smiti, S. 2011. Composition of the essential oils in various organs at different developmental stages of *Ammi visnaga* (L.) Lam. from Tunisia. Chem. Biodiversity, 8: 1990–2004.

Semyari, H., Owlia, P., Farhadi, S., Tabrizi, S. 2011. Evaluation of antimicrobial effect of "*Ammi visnaga*" against oral streptococci. J. Microbiol. Antimicrob., 3: 126–129.

Sener, G.K., Sacan, O., Yanardag, R., Ayanoglu-Du, G.L. 2003. Effects of parsley (*Petroselinum crispum*) on the aorta and heart of stz-induced diabetic rats. Plant Foods Hum. Nutr., 58(3): 1–7.
Seyyednejad, S.M., Maleki, S., Mirzaei Damab, N., Motamedi, H. 2008. Antibacterial activity of prunus mahaleb and parsley (*Petroselinum crispum*) against some pathogen. Asian J. Biol. Sci., 1(1): 51–55.
Shahamat, Z., Abbasi-Maleki, S., Motamed, S.M. 2016. Evaluation of antidepressant-like effects of aqueous and ethanolic extracts of *Pimpinella anisum* fruit in mice. Avicenna J. Phytomed., 6: 322.
Shahat, A., Ibrahim, A., Hendawy, S., Omer, E., Hammouda, F., Abdel-Rahman, F., Saleh, M. 2011. Chemical composition, antimicrobial and antioxidant activities of essential oils from organically cultivated fennel cultivars. Molecules, 16: 1366–1377.
Sharma, K.D., Karki, S., Thakur, N.S., Attri, S. 2012. Chemical composition, functional properties and processing of carrot—A review. J. Food Sci. Technol., 49: 22–32.
Sharma, L., Agarwal, D., Saxena, S., Kumar, H., Kumar, M., Verma, J., Singh, B. 2018. Antibacterial and antifungal activity of ajwain (*Trachyspermum ammi*) in different solvent. J. Pharmacognosy and Phytochem., 7: 2672–2674.
Sharma, L.K., Agarwal, D., Rathore, S.S., Malhotra, S.K., Saxena, S.N. 2016. Effect of cryogenic grinding on volatile and fatty oil constituents of cumin (*Cuminum cyminum* L.) genotypes. J. Food Sci. Technol., 53: 2827–2834.
Shebaby, W.N., Bodman-Smith, K.B., Mansour, A., Mroueh, M., Taleb, R.I., El-Sibai, M., Daher, C.F. 2015. *Daucus carota* pentane-based fractions suppress proliferation and induce apoptosis in human colon adenocarcinoma HT-29 cells by inhibiting the MAPK and PI3K pathways. J. Med. Food, 18: 745–752.
Shirke, S.S., Jadhav, S.R., Jagtap, A.G. 2008. Methanolic extract of *Cuminum cyminum* inhibits ovariectomy-induced bone loss in rats. Experimental Biol. and Med., 233: 1403–1410.
Shojaii, A., Abdollahi Fard, M. 2012. Review of pharmacological properties and chemical constituents of *Pimpinella anisum*. ISRN Pharmaceutics.
Simon, P.W. 2000. Domestication, historical development, and modern breeding of carrot. Plant Breed. Rev., 19: 147–190.
Singh, K., Singh, N., Chandy, A., Manigauha, A. 2012. *In vivo* antioxidant and hepatoprotective activity of methanolic extracts of *Daucus carota* seeds in experimental animals. Asian Pac. J. Trop. Biomed., 2: 385–388.
Singh, R.P., Gangadharappa, H., Mruthunjaya, K. 2017. *Cuminum cyminum*—A popular spice: An updated review. Pharmacognosy J., 9.
Small, E. 2006. Culinary herbs. NRC Research Press.
Sodimbaku, V., Pujari, L., Mullangi, R., Marri, S. 2016. Carrot (*Daucus carota* L.): Nephroprotective against gentamicin-induced nephrotoxicity in rats. Indian J. Pharmacol., 48: 122–127.
Søltoft, M., Bysted, A., Madsen, K.H., Mark, A.B., Bügel, S.G., Nielsen, J., Knuthsen, P. 2011. Effects of organic and conventional growth systems on the content of carotenoids in carrot roots, and on intake and plasma status of carotenoids in humans. J. Sci. Food Agric., 91: 767–775.
Sommer, A., Vyas, K.S. 2012. A global clinical view on vitamin A and carotenoids. Am. J. Clin. Nutr., 96: 1204S–1206S.
Sonnenberg, H., Kaloga, M., Eisenbach, N., Froemming, K.K. 1995. Isolation and characterization of an angular-type dihydropyranocoumarine glycoside from the fruits of *Ammi visnaga* (L.) Lam. (Apiaceae). Z. Fur Nat., 50: 729–731.
Sowbhagya, H. 2014. Chemistry, technology, and nutraceutical functions of celery (*Apium graveolens* L.): an overview. Critical Rev. Food Sci. Nutr., 54: 389–398.
Sriti, J., Talou, T., Wannes, W.A., Cerny, M., Marzouk, B. 2009. Essential oil, fatty acid and sterol composition of Tunisian coriander fruit different parts. J. Sci. Food Agric., 89: 1659–1664.
Sriti, J., Wannes, W.A., Talou, T., Mhamdi, B., Hamdaoui, G., Marzouk, B. 2010. Lipid, fatty acid and tocol distribution of coriander fruit's different parts. Ind. Crops Prod., 31: 294–300.
Sriti, J., Talou, T., Faye, M., Vilarem, G., Marzouk, B. 2011. Oil extraction from coriander fruits by extrusion and comparison with solvent extraction processes. Ind. Crops Prods., 33: 659–664.
Stan, M., Soran, M.L., Varodi, C., Lung, I. 2012. Extraction and identification of flavonoids from parsley extracts by HPLC analysis. AIP Conference Proceedings, 1425: 50.
Stolarczyk, J., Janick, J. 2011. Carrot: History and iconography. Chronica Horticulturae, 51: 13–18.
Sun, W., Shahrajabian, M.H., Cheng, Q. 2019. Anise (*Pimpinella anisum* L.), a dominant spice and traditional medicinal herb for both food and medicinal purposes. Cogent Biology, 5: 1673688.
Surles, R.L., Weng, N., Simon, P.W., Tanumihardjo, S.A. 2004. Carotenoid profiles and consumer sensory evaluation of specialty carrots (*Daucus carota*) of various colors. J. Agr. Food Chem., 52: 3417–3421.
Swanston-Flatt, S., Day, C., Bailey, C., Flatt, P. 1990. Traditional plant treatments for diabetes. Studies in normal and streptozotocin diabetic mice. Diabetologia, 33: 462–464.
Tabanca, N., Khan, S.I., Bedir, E., Annavarapu, S., Willett, K., Khan, I.A., Kirimer, N., Baser, K.H.C. 2004. Estrogenic activity of isolated compounds and essential oils of *Pimpinella species* from Turkey, evaluated using a recombinant yeast screen. Planta Med., 70: 728–735.
Taherian, A.A., Vafaei, A.A., Ameri, J. 2012. Opiate system mediate the antinociceptive effects of *Coriandrum sativum* in mice. Iranian J. Pharmaceutical Res.: IJPR, 11: 679.

Tahraoui, A., El-Hilaly, J., Israili, Z., Lyoussi, B. 2007. Ethnopharmacological survey of plants used in the traditional treatment of hypertension and diabetes in south-eastern Morocco (Errachidia province). J. Ethnopharmacol., 110: 105–117.

Takayanagi, T., Ishikawa, T., Kitajima, J. 2003. Sesquiterpene lactone glucosides and alkyl glycosides from the fruit of cumin. Phytochem., 63: 479–484.

Tan, K.W., Killeen, D.P., Li, Y., Paxton, J.W., Birch, N.P., Scheepens, A. 2014. Molecular and cellular pharmacology Dietary polyacetylenes of the falcarinol type are inhibitors of breast cancer resistance protein (BCRP/ABCG2). Eur. J. Pharmacol., 723: 346–352.

Tang, E.L., Rajarajeswaran, J., Fung, S.Y., Kanthimathi, M. 2013. Antioxidant activity of *Coriandrum sativum* and protection against DNA damage and cancer cell migration. BMC Complementary and Alternative Med., 13: 347.

Tangkanakul, P., Auttaviboonkul, P., Niyomwit, B., Lowvitoon, N., Charoenthamawat, P., Trakoontivakorn, G. 2009. Antioxidant capacity, total phenolic content and nutritional composition of Asian foods after thermal processing. Int. Food Res. J., 16: 571–580.

Tanira, M., Shah, A., Mohsin, A., Ageel, A., Qureshi, S. 1996. Pharmacological and toxicological investigations on *Foeniculum vulgare* dried fruit extract in experimental animals. Phytotherapy Res., 10: 33–36.

Tas, A., Ozbek, H., Atasoy, N., Enes Altug, M., Ceylan, E. 2006. Evaluation of analgesic and anti inflammatory activity of *Pimpinella anisum* fixed oil extract. Indian Vet. J., 83: 840–843.

Tas, A. 2009. Analgesic effect of *Pimpinella anisum* L. essential oil extract in mice. Indian Vet. J., 86: 145–147.

Tawil, M., Bekdash, A., Mroueh, M., Daher, C.F., Abi-Habib, R.J. 2015. Wild carrot oil extract is selectively cytotoxic to human acute myeloid leukemia cells. Asian Pac. J. Cancer Prev., 16: 761–767.

Tapas, A.R., Sakarkar, D., Kakde, R. 2008. Flavonoids as nutraceuticals: a review. Trop. J. Pharmaceutical Res., 7: 1089–1099.

Tashakori-Sabzevar, F., Ramezani, M., Hosseinzadeh, H., Parizadeh, S.M.R., Movassaghi, A.R., Ghorbani, A., Mohajeri, S.A. 2016. Protective and hypoglycemic effects of celery seed on streptozotocin-induced diabetic rats: experimental and histopathological evaluation. Acta Diabetologica, 53: 609–619.

Telci, I., Toncer, O.G., Sahbaz, N. 2006. Yield, essential oil content and composition of *Coriandrum sativum* varieties (var. vulgare Alef and var. microcarpum DC.) grown in two different locations. J. Essential Oil Res., 18: 189–193.

Telci, I., Demirtas, I., Sahin, A. 2009. Variation in plant properties and essential oil composition of sweet fennel (*Foeniculum vulgare* Mill.) fruits during stages of maturity. Ind. Crops Prod., 30: 126–130.

Tepe, A.S., Tepe, B. 2015. Traditional use, biological activity potential and toxicity of *Pimpinella* species. Ind. Crops Prod., 69: 153–166.

Thabrew, M.I., Dharmasiri, M., Senaratne, L. 2003. Anti-inflammatory and analgesic activity in the polyherbal formulation Maharasnadhi Quathar. J. Ethnopharmacol., 85: 261–267.

Thakur, N., Sareen, N., Shama, B., Jagota, K. 2013. Studies on *in vitro* antifungal activity of *Foeniculum vulgare* Mill. against spoilage fungi. Global J. Bio-Science and BioTechnol., 2: 427–430.

Thangam, C., Dhananjayan, R. 2003. Antiinflammatory potential of the seeds of *Carum copticum* Linn. Indian J. Pharmacol., 35: 388–391.

Tirapelli, C.R., de Andrade, C.R., Cassano, A.O., De Souza, F.A., Ambrosio, S.R., da Costa, F.B., de Oliveira, A.M. 2007. Antispasmodic and relaxant effects of the hidroalcoholic extract of *Pimpinella anisum* (Apiaceae) on rat anococcygeus smooth muscle. J. Ethnopharmacol., 110: 23–29.

Tobyn, G., Denham, A., Whitelegg, M. 2011. The Western Herbal Tradition: 2000 Years of Medicinal Plant Knowledge. Edinburgh: Churchill Livingstone.

Tomy, M.J., Dileep, K.V., Prasanth, S., Preethidan, D.S., Sabu, A., Sadasivan, C., Haridas, M. 2014. Cuminaldehyde as a lipoxygenase inhibitor: *in vitro* and *in silico* validation. Applied Biochem. and Biotechnol., 174: 388–397.

Tonkaboni, M.M., Tohfeh-al-Momenin, Rahimi, R., Shams Ardekani, M.R., Farjadmand, F. 2007. Editors. Tehran: Shahid Beheshti University of Medical Sciences, 129.

Trailokya, A., Srivastava, A., Bhole, M., Zalte, N. 2017. Calcium and calcium salts. J. The Association of Physicians of India, 65: 100–103.

Tucker, A.O., DeBaggio, O. 2009. Encyclopedia of Herbs. Timber Press, Portland, Oregon, pp. 393–394.

Tunali, T., Yarat, A., Yanardag, R., elik, F.O., Zsoy, O., Ergenekon, G., Emekli, N.O. 1999. Effect of parsley (*Petroselinum crispum*) on the skin of STZ induced diabetic rats. Phytother. Res., 13(2): 138–141.

Twaij, H.A., Elisha, E.E., Khalid, R.M., Paul, N.J. 1987. Analgesic studies on some Iraqi medicinal plants. Int. J. of Crude Drug Res., 25: 251–254.

Tyler, V.E. 1993. The Honest Herbalist. 3rd ed. New York, London, Norwood: Pharmaceutical Products Press, 235–236.

Van Wyk, B.E., Wink, M. 2004. Medicinal Plants of the World: An Illustrated Scientific Guide to Important Medicinal Plants and Their Uses. Portland, Timber Press.

Vanachayangkul, P., Byer, K., Khan, S., Butterweck, V. 2010. An aqueous extract of *Ammi visnaga* fruits and its constituents khellin and visnagin prevent cell damage caused by oxalate in renal epithelial cells. Phytomed., 17: 653–658.

Vasudevan, K., Vembar, S., Veeraraghavan, K., Haranath, P. 2000. Influence of intragastric perfusion of aqueous spice extracts on acid secretion in anesthetized albino rats. Indian J. Gastroenterology: Official Journal of the Indian Society of Gastroenterology, 19: 53–56.

Vasudevan, M., Parle, M. 2006. Pharmacological evidence for the potential of *Daucus carota* in the management of cognitive dysfunctions. Biol. Pharm. Bull., 29: 1154–1161.

Vavilov, N.I., Vavylov, M.I., Vavílov, N.Í., Dorofeev, V.F. 1992. Origin and Geography of Cultivated Plants. University Press, Cambridge.

Venugopala, K.N., Rashmi, V., Odhav, B. 2013. Review on natural coumarin lead compounds for their pharmacological activity. BioMed. Res. Inter., 1–14.

Vickers, N.J. 2017. Animal communication: When I'm calling you, Will you answer too? Current Biology 27: R713–R715.

Vora, S.R., Patil, R.B., Pillai, M.M. 2009. Protective effects of *Petroselinum crispum* (Mill) Nyman ex A. W. Hillleaf extract on D-galactose-induced oxidative stress in mouse brain. Indian J. Exp. Biol., 47(5): 338–342.

Wagner, H., Bladt, S. 1996. Plant Drug Analysis. Springer-Verlag, Berlin-Heidelberg, pp. 154–175.

Waheed, A., Miana, G., Ahmad, S., Khan, M.A. 2006. Clinical investigation of hypoglycemic effect of *Coriandrum sativum* in type-2 (NIDDM) diabetic patients. Pakistan J. Pharmacol., 23: 7–11.

Wang, L.S., Stoner, G.D. 2008. Anthocyanins and their role in cancer prevention. Cancer Lett., 269: 281–290.

Wehbe, K., Mroueh, M., Daher, C.F. 2009. The potential role of *Daucus carota* aqueous and methanolic extracts on inflammation and gastric ulcers in rats. J. Complement. Integr. Med., 6: 1–16.

Wei, J., Zhang, X., Bi, Y., Miao, R., Zhang, Z., Su, H. 2015. Anti-inflammatory effects of cumin essential oil by blocking JNK, ERK, and NF-kappaB signaling pathways in LPS-stimulated RAW 264.7 cells. Evid Based Complement Alternat. Med., 474509.

Wei, J.-N., Liu, Z.-H., Zhao, Y.-P., Zhao, L.-L., Xue, T.-K., Lan, Q.-K. 2019. Phytochemical and bioactive profile of *Coriandrum sativum* L. Food Chem., 260–267.

Wong, P.Y., Kitts, D.D. 2006. Studies on the dual antioxidant and antibacterial properties of parsley (*Petroselinum crispum*) and cilantro (*Coriandrum sativum*) extracts. Food Chem.. 97(3): 505–515.

Wu, T.T., Tsai, C.W., Yao, H.T., Lii, C.K., Chen, H.W., Wu, Y.L., Chen, P.Y., Liu, K.L. 2010. Suppressive effects of extracts from the aerial part of *Coriandrum sativum* L. on LPS-induced inflammatory responses in murine RAW 264.7 macrophages. J. Sci. Food and Agric., 90: 1846–1854.

Yanardağ, R., Bolkent, S., Tabakoglu-Oguz, A., Ozsoy-Sacan, O. 2003. Effects of *Petroselinum crispum* extract on pancreatic B cells and blood glucose of streptozotocin-induced diabetic rats. Biol. Pharm. Bull., 26(8): 1206–1210.

Yildiz, H. 2016. Chemical composition, antimicrobial, and antioxidant activities of essential oil and ethanol extract of *Coriandrum sativum* L. leaves from Turkey. Int. J. Food Properties, 19: 1593–1603.

Yoshikawa, M., Uemura, T., Shimoda, H., Kishi, A., Kawahara, Y., Matsuda, H. 2000. Medicinal foodstuffs. XVIII. Phytoestrogens from the aerial part of *P. crispum* MIll. (Parsley) and structures of 6-acetylapiin and a new monoterpene glycoside, petroside. Chem. Pharm. Bull., 48(7): 1039–1044.

Yusni, Y., Zufry, H., Meutia, F., Sucipto, K.W. 2018. The effects of celery leaf (Apium graveolens L.) treatment on blood glucose and insulin levels in elderly pre-diabetics. Saudi Med. J., 39: 154.

Zaini, R., Clench, M.R., Le Maitre, C.L. 2011. Bioactive chemicals from carrot (*Daucus carota*) juice extracts for the treatment of leukemia. J. Med. Food, 14: 1303–1312.

Zaini, R.G., Brandt, K., Clench, M.R., Le Maitre, C.L. 2012. Effects of bioactive compounds from carrots (*Daucus carota* L.), polyacetylenes, beta-carotene and lutein on human lymphoid leukaemia cells. Anticancer Agents Med. Chem., 12: 640–652.

Zaman, Z., Roche, S., Fielden, P., Frost, P.G., Niriella, D.C., Cayley, A.C. 1992. Plasma concentration of vitamin A and E and carotenoids in Alzheimer's disease. Age Ageing, 21: 91–96.

Zare, R., Heshmati, F., Fallahzadeh, H., Nadjarzadeh, A. 2014. Effect of cumin powder on body composition and lipid profile in overweight and obese women. Complementary Therapies in Clinical Practice, 20: 297–301.

Zarshenas, M.M., Moein, M., Samani, S.M., Petramfar, P. 2013. An overview on ajwain (*Trachyspermum ammi*) pharmacological effects; modern and traditional. J. Natural Remedies, 14: 98–105.

Zgheib, P., Daher, C.F., Mroueh, M., Nasrallah, A., Taleb, R.I., El-Sibai, M. 2014. *Daucus carota* pentane/diethyl ether fraction inhibits motility and reduces invasion of cancer cells. Chemotherapy, 60: 302–309.

Zgorka, G., Dragan, T., Glowniak, K., Basiura, E. 1998. Determination of furanochromones and pyranocoumarins in drugs and Ammi visnaga fruits by combined solid-phase extraction-high-performance liquid chromatography and thin-layer chromatography-high-performance liquid chromatography. J. Chromatogr. A, 797: 305–309.

Zhao, Y., Wang, P., Zheng, W., Yu, G., Li, Z., She, Y., Lee, M. 2019. Three-stage microwave extraction of cumin (*Cuminum cyminum* L.) Seed essential oil with natural deep eutectic solvents. Ind. Crops Prod., 140: 111660.

Zhang, C.-R., Dissanayake, A.A., Kevseroğlu, K., Nair, M.G. 2015. Evaluation of coriander spice as a functional food by using *in vitro* bioassays. Food Chem., 167: 24–29.

Zhang, D., Hamauzu, Y. 2004. Phenolic compounds and their antioxidant properties in different tissues of carrots (*Daucus carota* L.). J. Food Agric. Environ., 2: 95–100.

Zhang, H., Chen, F., Wang, X., Yao, H. 2006. Evaluation of antioxidant activity of parsley (*Petroselinum crispum*) essential oil and identification of its antioxidant constituents. Food Res. Int., 39(8): 833–839.

Zhang, Y.-H., Park, Y.-S., Kim, T.-J., Fang, L.-H., Ahn, H.-Y., Hong, J., Kim, Y., Lee, C.-K., Yun, Y.-P. 2000. Endothelium-dependent vasorelaxant and antiproliferative effects of apigenin. General Pharmacology: The Vascular System, 35: 341–347.

Ziyan, L., Yongmei, Z., Nan, Z., Ning, T., Baolin, L. 2007. Evaluation of the anti-inflammatory activity of luteolin in experimental animal models. Planta Med., 73: 221–226.

Ziyyat, A., Legssyer, A., Mekhfi, H., Dassouli, A., Serhrouchni, M., Benjelloun, W. 1997. Phytotherapy of hypertension and diabetes in oriental Morocco. J. Ethnopharmacol., 58(1): 45–54.

Chapter 14

Ritualistic and Medicinal Plants from Marajó-PA Island

Paulo Wender Portal Gomes,[1,*,#] *Luiza Helena da Silva Martins,*[2] *Paulo Weslem Portal Gomes,*[3,#] *Emilli Roberta Sousa Pereira,*[4] *Abraão de Jesus Barbosa Muribeca,*[1] *Andrea Komesu*[5] and *Mahendra Rai*[6]

Introduction

The Amazon region is known worldwide because of its biodiversity with 25,000 to 30,000 species of endemic plants in the place; the region also hosts several cultures formed by native, quilombola and mestizo people. The history of the use of medicinal plants in the Amazon is very ancient and involves the influence and interaction between many people (Native, European, African, etc.). The exchanges of knowledge of these people end up reverberating nowadays in the popular traditions of the use of medicinal plants in the Brazilian Amazon (Monteiro et al. 2012).

The Brazilian Amazon is formed by the states of Acre, Amazonas, Amapá, Roraima, Rondônia, Pará and a part of Maranhão, Tocantins and Mato Grosso. In the State of Pará, the Marajó Island is located, with the geographical coordinates of 0°58'S49°34'O, the mouth of the Amazon River and the largest sea-river island in the world. The climate of Marajó is of the tropical rainy type, whose annual rainfall regime delimits two well-defined seasons, one dry and the other rainy. The average annual temperature is 26°C with a high relative humidity that can be higher than 80%. The thermal amplitude is very weak and the days have the same duration as the nights. This island, as well as the entire Amazon region, has a great biological and cultural diversity, being, therefore, an area of interest in the research of plants with different potentials (Monteiro et al. 2012).

The Marajó Island is composed by 12 cities and is divided into two micro-regions: to the east is the Microregion of Campos, which comprises the municipalities of Cachoeira do Arari, Chaves, Muaná,

[1] Institute of Exaction and Natural Sciences (ICEN) Federal University of Pará (UFPA) R. Augusto Corrêa, 01 - Guamá, 66075-110 Belém, PA, Brazil.

[2] Institute of Animal Health and Production (ISPA) Federal Rural University of Amazonia (UFRA) Av Presidente Tancredo Neves N° 2501; Terra Firme, 66.077-830, Belem, Pará, Brazil.

[3] Department of Plant Biology, University of Campinas (UNICAMP) Cidade Universitária Zeferino Vaz, Barão Geraldo, 13083-970, Campinas, SP, Brazil.

[4] Institute of Health Sciences, Faculty of Nutrition (FANUT) Federal University of Pará (UFPA) R. Augusto Corrêa, 01 - Guamá, 66075-110 Belém, PA, Brazil.

[5] Department of Marine Sciences (DCMar) Federal University of São Paulo (UNIFESP) R. Silva Jardim, n° 136, Vila Matias, 11015-020, Santos, SP, Brazil.

[6] Biotechnology Department SGB Amravati University Amravati-444 602, Maharashtra, India.

* Corresponding author: wenderufpa@hotmail.com

The contributions of authors were the same in the chapter.

Ponta de Pedras, Salvaterra, Santa Cruz do Arari and Soure. On the west side of the island is located the Microregion of Furos, with the municipalities of Afuá, Anajás, Breves, Curralinho, and São Sebastião da Boa Vista (Cristina and Borges 2012).

This Island has a history marked by the presence of native people, travelers and foreign explorers, who have made decisive contributions to local development and its culture. Combined with biological diversity, the cultural wealth produced and lived by the Marajoara people gives the place a special tone that intensifies from the aura of mystery and mysticism that surrounds the place (Cristina and Borges 2012).

Filho (2014) clarifies that the tradition of the shamans is a mark of the strength of the native tradition in Marajó. Even with the incorporation of elements from European and African cultures, the matrix of thought of the native people remains strongly inherited from their native ancestors.

The Brazilian Council of Phytotherapy (Conbrafito) considers phytotherapy as "the treatment of diseases through bioactive plants". It is the oldest form of cure, presented and based on plant or plant-derived drugs, in the most varied forms of preparation according to traditional folk experiences and knowledge or modern scientific methods (Conbrafitro 2020).

The traditional populations anywhere in the world preserve knowledge from years passed on by their ancestors. The knowledge includes the use of plants in their daily lives, mainly in healing diseases, since most of these communities are removed from large centers, and there was and is a need for treatment through plants. Moreover, all of this knowledge is transferred from generation to generation through oral reports (Rodrigues and Carvalho 2001).

The Marajó Island gathers knowledge from its colonization and before it, such as the knowledge of the native people, predominant in this region in the pre-colonial period. With the arrival of farmers, people were brought in who were subjected to slavery, and with the escapes, they formed resistance communities, called quilombos (de Souza Lages et al. 2019).

The quilombola and native communities are present to this day in Marajó region, perpetuating the traditional knowledge derived from their ancestors about phytotherapy together with their beliefs (umbanda, candomblé and shamanism) (Cordovil 2012).

This chapter aims to deal with the medicinal and ritualistic plants used on Marajó Island and their chemical effects involved, in order to relate the pharmacological actions of plants and their effects on healing through phenomenological effects on native people's beliefs in the use of plants in their religious rituals, which occur in rituals of the native communities.

Phytotherapy

Plants with biologically active principles are the oldest resources that the human species use in health treatment. A survey carried out in the *Science Direct* database between 1996–2019 and illustrated in Figure 14.1, we can see a growth in publications in the world regarding the term phytotherapy. In this

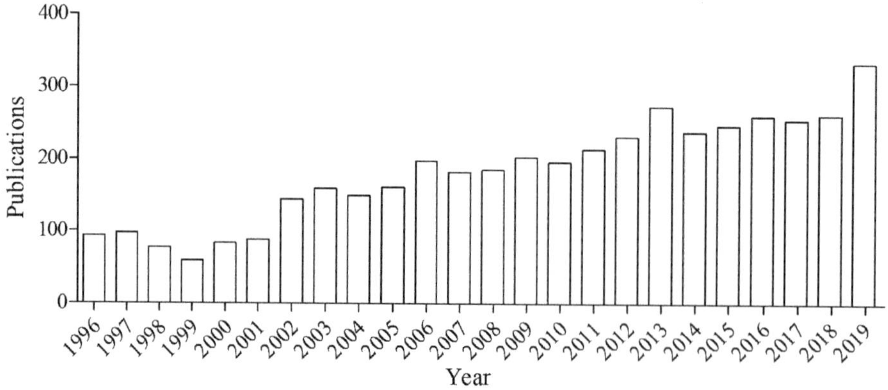

Figure 14.1. Publications from 1996–2019 with the term phytotherapy. Source: http://www.sciencedirect.com/.

context, researches about the bioactive compounds present in ritualist plants composition are a great deal of potential in their medicinal use.

Given the analysis of the literature of the last twenty-four years of research, phytotherapy is defined as the science that studies the healing processes, general indications, contraindications and related processes involved in the use of plants or plant products for medicinal purposes (Heinrich et al. 2017).

In the world of phytotherapy, many terms are used by scientific society. Table 14.1 gathers the most important, varied terms and peripheral definitions according to the purposes of the products and by-products of drugs with therapeutic activity according to Brazilian legislation.

Table 14.1. Most common definitions in phytotherapy.

Terms	Definitions	Ref.
Medicinal plant	Vegetable species, cultivated or not, administered by any route or form, which exerts therapeutic action	(Brasil 2018)
Vegetable drug	Medicinal plant or its parts, after collection, stabilization and drying processes, which may be intact, streaked, crushed or pulverized	(Brasil 2018)
Herbal medicine	Product obtained from a medicinal plant, or its derivatives, except isolated substances, for prophylactic, curative or palliative purposes	(Brasil 2016)
Traditional herbal medicine	The one obtained with the exclusive use of active vegetable raw materials, whose safety is based on traditional usage and which is characterized by reproducibility and constant quality	(Brasil 2013)
Medicinal infusion	Prepared from medicinal plants—by infusion, decoction or maceration in water—and used for therapeutic purposes	(Brasil 2018)
Vegetable derivative	Product obtained from natural medicinal plant or vegetable drug, which contains substances responsible for therapeutic action and can be presented in the form of alcohol, wax, exudate, extract, fixed oil, volatile oil, tincture and others	(Brasil 2018)
Manipulated herbal medicine	Herbal medicine prepared in manipulation pharmacies authorized by health surveillance	(Brasil 2018)

Phytotherapy on Marajó Island

The Marajó Island, positioned in the northeast of the State of Pará (0°58'S49°34'O), is part of a fluvio-marine complex of 49,606 km^2 bounded by the Amazonas River (West), Pará River (South), Tocantins-Marajó Bay (East), and Atlantic Ocean (North). This area is located in a region of the tropical climate, with a mean annual temperature of 28°C and precipitation of 2,500 to 3,000 mm.year^{-1}, 90% being concentrated between January and July (Souza 2013). The population of Marajó emerged from a native population (Indians) and missionaries, travelers and explorers from different parts of Europe, who made decisive contributions to local development. The island has a biological diversity and cultural richness produced by the Marajoara man and gives the place a special tone that intensifies the aura of mystery and mysticism that surrounds the region (Marajó and Barão 1992, La Condamine 2000, Agassiz 2000, Wallace 2004, Baena 2004).

Plants have been used to treat diseases in the Marajó Island since the pre-colonial period in Brazil, where the Indians lived and later with people who were enslaved from the African continent (Miranda 2011). The literature (Hammer and Johns 1993) reports four medicinal plants (*Carapa guianensis*, *Elephantopus scaber*, *Piper umbellatum*, *Stachytarpheta cayenensis*) used by the Caboclo emdo Marajó communities as antiphlogistic, anti-arthritic, antibacterial, anti-inflammatory, diuretic, analgesic and anti-inflammatory antitumor, and antifungal.

In the Marajoara region, it is common to use traditional knowledge, passed down through generations of natives and slaves to the present day. Some people, depending on their position in society, tribe or community, called doctors, shamans, healers, etc., that make use of plants to obtain a therapeutic benefit or, simply, to improve health. This behavior became a common activity in the municipalities of Marajó island (Monteiro et al. 2011).

The task of phytotherapy in the region is also to study the interactions established between natural medicine and a "belief" of a certain group. In this sense, it is important to understand what medicinal and ritualistic plant means. According to the OMS, medicinal plants are those that, introduced or placed in contact with a human or animal organism, perform a pharmacological activity even if harmful to the organism, such as the death of buffalo in farms in Marajó as a result of ingestion of toxic plants (WHO 1998).

Currently, there is no universal definition for ritualistic plants and these can be understood as those used in religious rituals, Umbanda, candomblé, and shamanism capable of assisting in energetic, astral and physical restoration, in addition to enhancing our mediumistic tools through the energy of various herbs (Albuquerque 1997). One example is the use of the plant popularly known as broom (*Scoparia dulcis*), used to "bless" newborn children who acquired "*quebranto*", characterized according to popular belief as a harmful influence on people, like bad energies caused by people envy.

Use of medicinal plants on Marajó Island

In Brazil, more specifically in the Amazon region, where the Marajó Island is located, there is a great potential for the development of phytotherapy, given that, as it is considered to be the greatest plant diversity on the planet, in corroboration, it holds a wide socio-diversity of traditional knowledge regarding the use of medicinal plants, mostly native and quilombola. The market and industry of herbal medicines in the Brazilian state are governed by regulations and laws established by the Ministry of Health, more specifically, by the Health Surveillance Agency (ANVISA), which maintains the quality control of drugs and drug by-products derived from plants with "curative" activities against diseases (Brasil 2011).

The use of medicinal plants, both in-home medicine and in pharmacopoeia, has ancient origins, associated with the beginning of medicine and based on information transmitted through generations. Over the centuries, the knowledge about the healing potential of plants has formed the basis for the therapy of different diseases (Ferreira 2018).

Plants for therapeutic purposes have evolved, from the simplest methods of preparation to the industrial manufacture used by modern man. Plants can cause beneficial activities in the human body if they are administered in the form of infusion or juice, or as a pure substance isolated, or even transformed into pills and ointments, and they show an efficient result in the recovery of health (Lorenzi and Matos 2008, Ferreira 2018).

Religions that Use Ritualistic and Medicinal Plants in Marajó

Umbanda and Candomblé

Afro-Brazilian religions in their ritual plants have two well-defined roles: first, they have a sacral role, linked to a mythical thought, in which axé (vitalizing force of the deities) is deposited; secondly, the functional role, that provides the ideal conditions for contact with the sacred through the trance of possession in order to achieve the healing throughout plants. However, these trances do not result only from the use of these plants, but from other factors specific to the rituals (Camargo 2015, Ferreira 2018).

For Candomblé members, health happens in three dimensions: mental health, body health, and spiritual health. In this context, it is noteworthy that there is a strong heritage of African culture on popular medicine in Brazil, especially with regard to the use of medicinal plants, a resource widely used in the early colonization which continues to this day, consisting of a field of valuable knowledge (Mota and Trad 2011, Paz et al. 2015).

Shamanism

According to Maués (1990) and Albuquerque and Faro (2015), Caboclo shamanism is defined as a set of shamanistic beliefs and practices present in practically the entire Amazonian territory, in which varying degrees of elements of indigenous, Afro-Brazilian and Catholic are included.

It is important to note that there are differences between caboclo and native shamanism. The first results the relationship between different cultural and religious traditions (mainly native, Catholic,

Christian and African) brought about by the process of Brazilian colonization. In addition, the second refers to practices and beliefs restricted to indigenous villages and ethnicities (Maués 1990, Albuquerque and Faro 2015).

The shamanism, according to Lima (2002), is a cult that originated from the indigenous people, passed on to the caboclos and which today, in our civilization, "is the last mark of a cult in danger of extinction" (Lima 2002, Albuquerque and Faro 2015), as it would be increasingly diffused among elements of other religions. She claims that her knowledge and healing practices would be the survival of a cult originally indigenous to the Marajó tribes. According to Lima (2002), the shamanism exercised by her consists of the "marijuana shamanism", which differs from the caboclo shamanism that comes from Maranhão.

The "pajé", who would correspond to the shaman, would be that person who has the gift of curing natural and unnatural diseases with the help of the enchanted. The enchanted or caruanas, as they are also called, would be magical beings that live at the bottom of rivers and forests, therefore being holders of power and wisdom (Albuquerque and Faro 2015).

Practitioners defend the importance of healing through plants as they claim that before doctors and scientists, the healers and midwives were responsible for people's health. The difficulty of moving to a large urban center, such as Belem, in search of specialized treatment makes many people seek the shaman or curator, instead of the doctor (Albuquerque and Faro 2015).

In their research, Pombo and Fares (2018) found that practitioners of Marajó shamanism are attentive to the natural pharmacopeia, to the medicinal herbs that exist on the spot, showing and proving that the plants have healing power.

Legislation for Ritualistic and Medicinal Plants

The toxicity of medicines prepared with plants may seem trivial when compared to conventional treatments; however, it is a serious public health problem. Medicinal plants may have adverse reactions by their constituents because of interactions with other medications or foods, or even related to the characteristics of patients, like sex, age, and others. Diagnostic errors, incorrect identification of plant species and use other than the traditional way would be dangerous, with risks of an overdose, inefficient therapeutic effectiveness and adverse reactions (Balbino and Dias 2010). In addition, the use of these products may compromise the effectiveness of conventional treatments, as they reduce or enhance their effect (Balbino and Dias 2010).

Legislation for medicinal plants

The therapeutic application of natural products in Brazil presents a different history of the pharmaceutical industry and of medical practice in general, applied to synthetic medicine, about the quality control process. Natural products are also explored in a parallel market, with little guarantee of effectiveness and safety. There is no resolution on the subject by the Federal Council of Medicine (Lopes et al. 2019).

The legislation about medicinal plants took place in 1990s, which introduced in Brazil the possibility of expanding the area of research in the internal production and commercialization of medicines of plant origin. In addition to the growth of international trade in these products and the participation of researchers in government, agency demanded the establishment of legislation for quality control and new criteria for the registration of products with governmental control over regulatory body (Fernandes 2004).

The current legislation in the Brazilian State for drugs and related by-products highlights that for the registration and control of synthetic products, it must be based on specific criteria, referenced in international models that establish preventive measures and statistical analysis in the evaluation of quality of industrially produced drugs (Anvisa 2015).

In the 1990s, initially, the analytical control was limited to representative samples as a guarantee of the quality of the entire pharmaceutical production. A complete control of all production steps was carried out, which included technical elements and materials necessary for production (facilities, documentation, procedures, training and training of human resources) (Fernandes 2004).

For the establishment of records of products of plant origin, the National Form of Phytotherapics in RDC n° 60 says that the study group linked to the Ministry of Health, which incorporated the Association of Pharmaceutical Advisors for Industry (AFAI) and the Association of National Pharmaceutical Laboratories (ALANAC), prepared this. The legislation for the control of herbal products in the country also meets the need to carry out the requested tests, favoring the approximation of renowned universities with the industry since it neither has the structure nor can pay to carry them out. In addition. Anvisa (2011) highlighted the role of the university, which is fundamental to the prospects of herbal medicine, as it has already trained human resources and has the necessary knowledge to meet the demand.

Legislation for ritualistic plants

In Brazil, products of plant mixtures are used in baths, variable according to the sector affiliation entity chosen from those traditionally incorporated into popular medical material, and therefore, there is a very old therapeutic tradition, which cannot always be confirmed pharmacology because they are plants that have not yet been studied (Silva et al. 2001). Thus, there is no specific legislation on the use and production of these plant products in rituals (Carvalho et al. 2012).

Brazilian Health System (SUS) and Phytotherapy

In Brazil, the National Program for Medicinal and Phytotherapic Plants (PPNPMF) was created with the main purpose of ensuring safe access and scientifically proven use of medicinal and phytotherapy plants, following the perspective of the sustainable use of biodiversity, the advancement of the production chain and the national industry (Borges 2010).

One of the guiding principles of the PPNPMF (Brasil 2016) is the expansion of therapeutic options and the improvement of health care for users of the Unified Health System (SUS). Another important policy that establishes Phytotherapy in SUS is the Complementary and Integrative Practices Policy, that is, the patient can receive comprehensive care with corroboration of the most varied forms of treatment of the disease, whether by synthetic drugs, herbal medicines or treatments based on wisdom popular as well as in medicine. Currently, the supply of herbal medicines is offered in 1,108 municipalities, according to reports from the Health Information System for Primary Care (SISAB), available on the health information platform in primary units.

Currently, SUS provides twelve herbal medicines through states and cities; these are indicated for gynecological treatment, gastroenterology, bone injuries such as arthritis and osteoarthritis, among others reported in the National List of Essential Medicines – RENAME (Brasil 2017).

It is worth noting that only health professionals, following the regulations of each council, carry out the prescription of herbal medicines and medicinal plants: nutritionist (CFN 2013), pharmacist (CFF 2013), nurse (COFEN 2010), dental surgeon (CFO 2008), doctor (CFM 2009), psychologist-acupuncturist, physiotherapist-acupuncturist, holistic therapist and neurologist (Panizza 2010).

Morphology and Secondary Metabolites of Plants

In the current scenario about medicinal plants, the word "Metabolome" is used to describe all the small molecules that make up a cell, and this science, belonging to the class of "omics", has a very important correlation with biology, that is, with genomics (DNA), transcriptomic (RNA) and proteomics. All this was only possible as mass spectrometry emerged in studies aimed at identifying and quantifying the metabolome, including studies with secondary metabolites of medicinal plants (Wang et al. 2019).

According to Thirumurugan et al. (2018), in the aforementioned perspective, secondary metabolites are defined as a set of products or by-products synthesized and stored in plants, for protection against insect attacks, as well as for adaptation to the environment. Figure 14.2 describes the main pathways of the production of secondary metabolites in plants.

Terpenes are produced in the Pyruvate +3 molecules of phosphoglyceraldehyde and mevalonic acid (HMG-CoA reductase). According to Dewick (2009), these compounds form a diversified group, that is, a family of metabolites derived from C_5 isoprene, which links at the two molecular ends. Some common

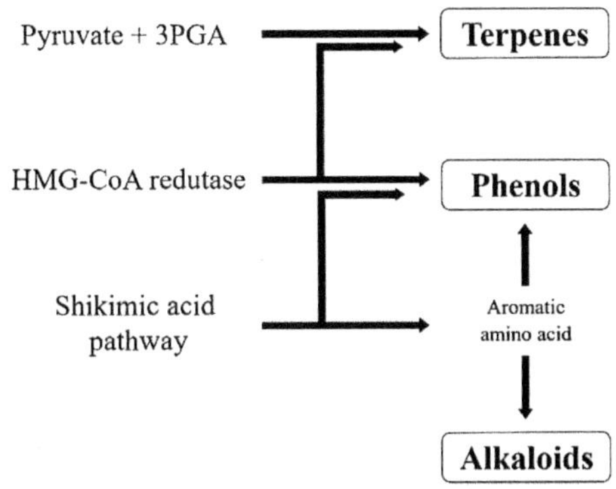

Figure 14.2. Main pathways of production of secondary metabolites in plants. Source: authors.

structures have carbon skeleton derived from $(C_5)_n$ and are classified as hemiterpenes (C_5), monoterpenes (C_{10}), sesquiterpenes (C_{15}), diterpenes (C_{20}), esteresterpenes (C_{25}), triterpenes (C_{30}) and tetraterpenes (C_{40}). The precursor is isopentenyl pyrophosphate (IPP), which can be obtained by two different routes (Figure 14.3).

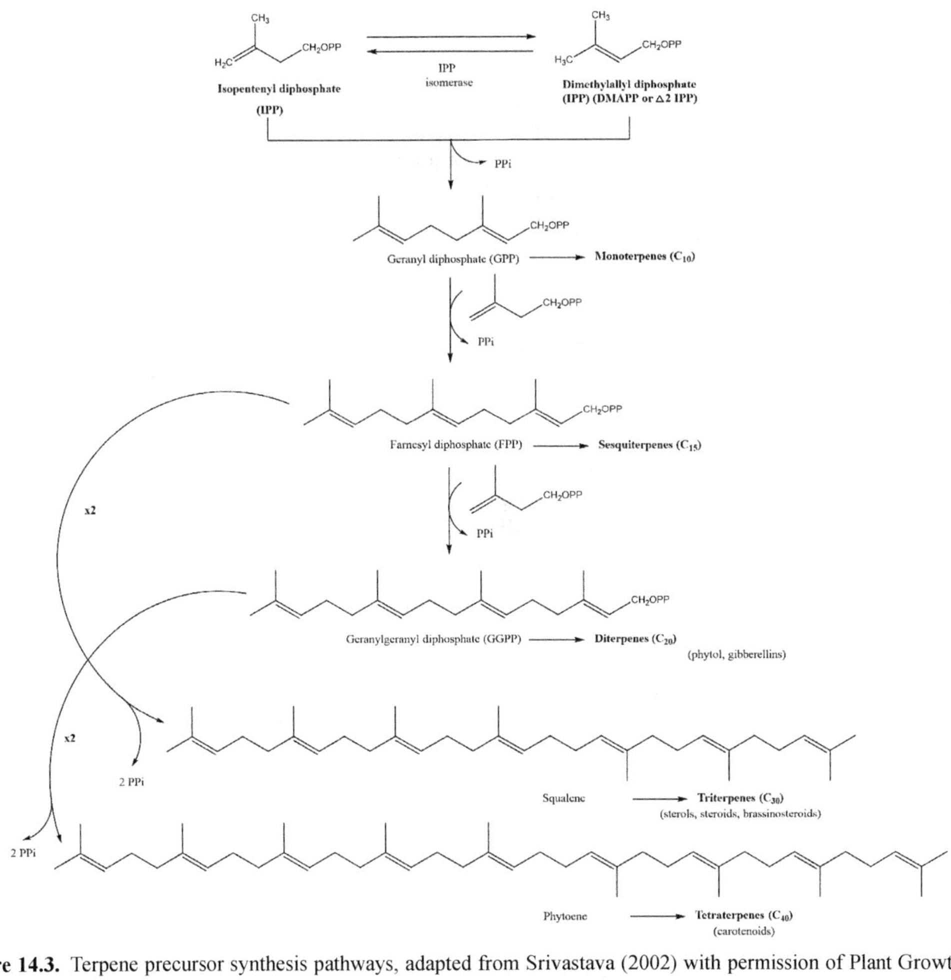

Figure 14.3. Terpene precursor synthesis pathways, adapted from Srivastava (2002) with permission of Plant Growth and Development.

The biosynthetic mevalonic acid route (HMG-CoA reductase) produces phenolic compounds, as well as by the shikimic acid route. The route of the shikimic acid (Figure 14.4) begins with condensation between phosphoenolpyruvate (PEP) and D-erythrose-4-phosphate to generate 7-carbon acid, 3-deoxy-D-arabino-heptulosonic acid-7-phosphate (DAHP), and the whole process is catalyzed by enzyme DAHP synthase (DAHPS) (Kougan et al. 2013).

Alkaloids are compounds that are present in plant tissues as salts (tartaric, acetic, oxalic, citric, malic and lactic acids), esters (for example, atropine, scopolamine, cocaine, aconitine), or combined with tannins or sugars (for example, the glycoalkaloids of the species Solanum) (Svendsen 1983, Kar 2007).

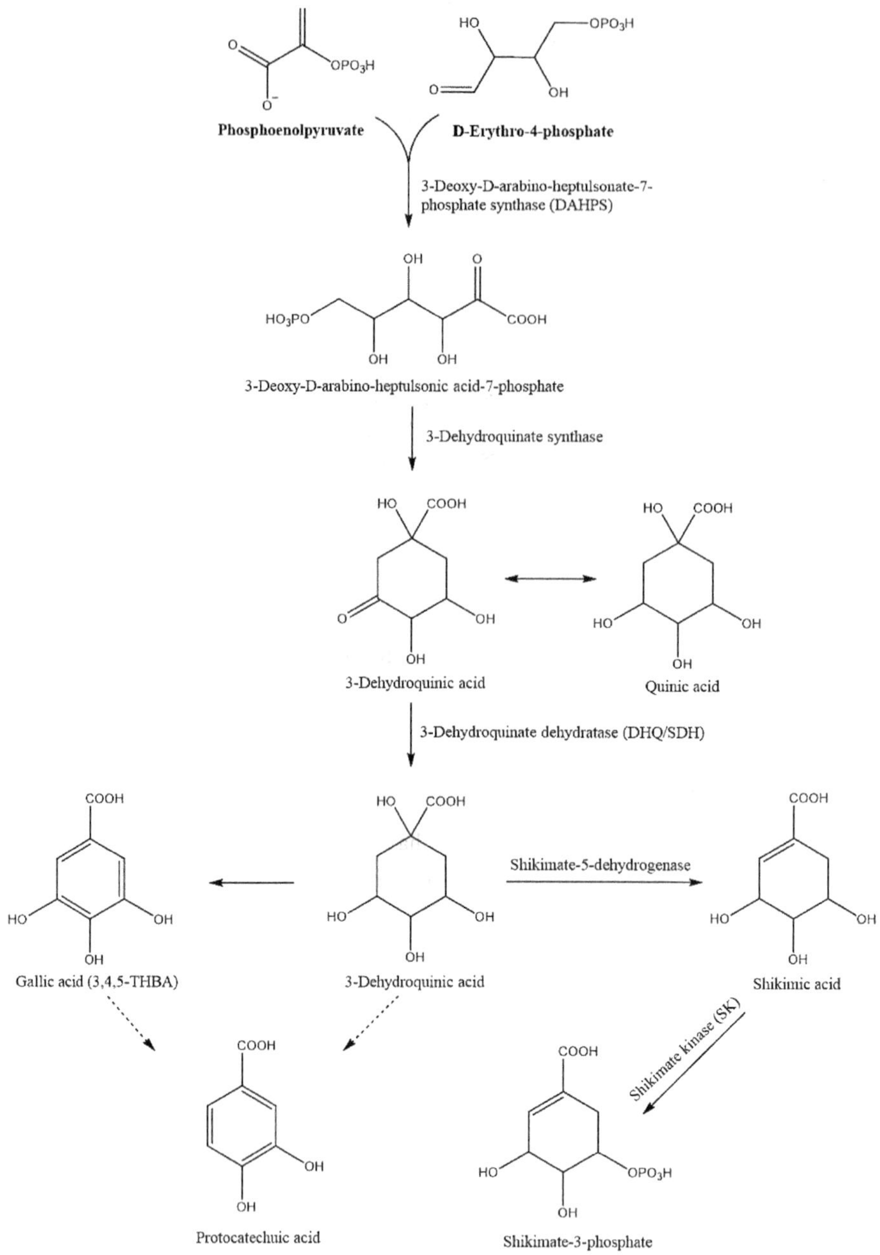

Figure 14.4. Biosynthesis of shikimic acid and some phenolic acids, adapted from Kougan et al. (2013) with permission of Medicinal Plant Research in Africa.

In the biosynthesis of alkaloids, amino acids participate as precursor molecules and undergo a series of transformations, among which, the most common are oxidation, reduction, dehydration, decarboxylation or aldolic condensation reactions. Teles et al. (2019) highlighted two very specific reactions in alkaloid biosynthesis, Mannich condensation or Schiff's base formation (Figure 14.5).

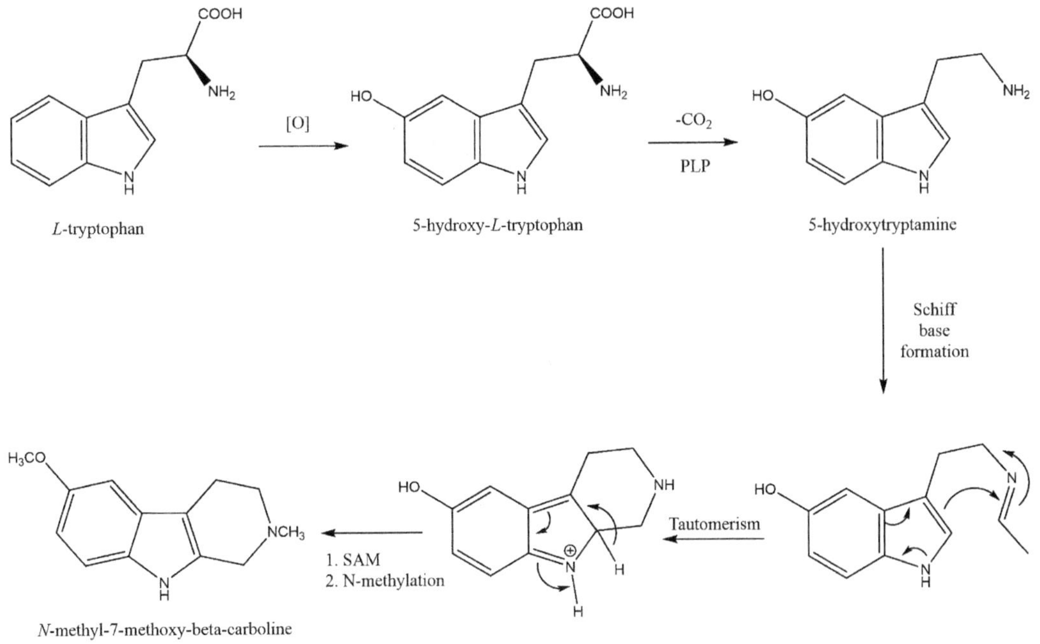

Figure 14.5. Reactions for the production of specific alkaloids, adapted from Teles et al. (2019) with the permission of The Alkaloids: Chemistry and Biology.

The Use of Medicinal Plants: Empirical and Scientific Knowledge

Most used parts of medicinal plants

The use of plants for medicinal purposes dates back to the origin of human civilization, that is, since ancient times plants have been indispensable sources of prevention and treatment of various diseases in humans and other animals (Rahman et al. 2019).

Since the different parts of medicinal plants may have several medicinal properties (Sanchita 2018), the use of leaves, flowers, fruits, seeds, roots, stems, exudation or even the whole plant for a particular disease is justified..

Recent studies on the use of medicinal plants have reported the use of the leaf as the part of the plant most used for medicinal purposes, followed by the bark of the stems, in addition to the fruits, seeds, and roots, which are often prepared by infusion. As an example, we can mention the ethnobotanical study of medicinal plants in the community of Caruarú located in the north of the Amazon, with 38.66% of the citations referring to the use of leaves from 50 species, followed by barks (30.66%) and the root (12%) (Mesquita and Tavares-Martins 2018).

The frequent use of leaves to treat diseases was reported for other communities in the Amazon, such as the Mamangal community (dos Santos et al. 2019), rural community Moacir Lucena (Guerra et al. 2010), coastal Amazon community in Marudá (Coelho-Ferreira 2009) and in the marine extractive reserve community of Source (Rocha et al. 2017), in addition to urban backyards (Palheta et al. 2017), fairs and markets (Linhares et al. 2014).

Probably, the leaves are the most used parts due to their availability in the plant during the year and due to the feasibility of the collection (Guerra et al. 2010), but mainly due to the highest concentrations of metabolites with biological activity for curing diseases, such as high concentrations of flavonoids, polyphenols, and carotenoids and other products (Heinze et al. 2018, Harbaum et al. 2008) as secondary metabolites have natural pharmacologically active products with antimalarial, antidiabetic, protective liver, anti-ulcer, anti-inflammatory and antimicrobial activities, which are fundamental for human well-being (Sanchita 2018), especially from the poorest people, that have no access to synthetic drugs.

Secondary metabolites in medicinal plants

Plants are living chemical factories for various secondary metabolite biosynthesis pathways, among which, some form the basis for many industrial pharmaceutical drugs, such as alkaloids, terpenoids and phenylpropanoids (Sanchita 2018). On the other hand, such bioactive compounds can improve human health through food, such as herbal remedies derived from medicinal plants (Li et al. 2020).

There are about 100,000 secondary metabolites present in the plant kingdom confined to specific taxonomic groups (Li et al. 2020); among these, nitrogen compounds (cyanogenic glycosides, alkaloids, and glycosines), phenolic compounds (flavonoids and phenylpropanoids) and terpenes predominate (Isoprenoids) and play major roles (Fang et al. 2011). The metabolites are widely distributed in different cells, tissues and organs of medicinal plants (Li et al. 2020); however, the metabolites stored in different cells, tissues and organs can present different biological activities according to the stages of development of the plant (Bartwal et al. 2013).

Before using a plant for medicinal purposes, it is essential to consider the season of the year in which the plant is collected and the habitat in which it grows. Secondary metabolite pathways and their regulation are highly susceptible to environmental variations, where different external pressures alter the expression of genes involved in biosynthesis pathways (Borges et al. 2017, Sanchita 2018). Stress conditions in a given habitat lead to an increase and accumulation of secondary metabolites in plants, as the genes involved in secondary metabolic pathways are affected in the presence of different stresses (Debnath et al. 2011). As a result, the synthesis and accumulation of secondary metabolites in plants are strictly controlled by external factors, such as drought or flood, high intensity of light and temperature, poorly drained soil or contaminated by pesticides; the stress of the plants' cells (Ma and Zhang 2010, Verma and Shukla 2015).

Elementary Ritualistic Dogmas in the Preparation of Medicinal Herbs

Several studies dealing with the origin, nuances and singularities of the knowledge of medicinal plants, such as the work of Santos (2000) in "Popular Traditions of the Use of Medicinal Plants in the Amazon", may be enough to convince us that this knowledge is unilaterally linked to transmission conducted over several generations, who are already used to providing their resources for curing illnesses.

In addition, it can be said that it is in this environment that those people acclaimed as "holders" of traditional knowledge are forged, that is, during the experience acquired in the appreciation of the ills that affect their patients; hence, they expect to always find greater experience among older people (Lima et al. 2012).

Thus, there were several types of research about the relationship between the preparations of a remedy reported by ethnomedicine practitioners with the main steps that must be mimicked in the preclinical sphere, in order to evaluate a reported therapeutic action (Oliveira et al. 2012).

In this way, we can understand that the ethnopharmacological study has the main function of compiling the information declared by the users of medicinal flora for chemical and pharmacological studies (Elisabetsky 2003).

Well-oriented methodological strategies restricted to pharmacological efforts can greatly diverge traditional and scientific knowledge (Etkin and Elisabetsky 2005). We can then admit that the scope of these studies must go beyond the eagerness to report, understand and validate some particularity that may be elementary to medicalized therapeutic practices (Heinrich 2014).

We also emphasize that if this is not taken care of, it will often not be possible to establish a striking parameter between the pharmacological response and the therapeutic use of a popular nature. In this sense, Menendez-Baceta et al. (2015) point out that although many pharmacological works are built on the legitimate nature of popular knowledge, many studies reveal that some species of plants traditionally used for therapeutic purposes do not reflect the expected pharmacological activity.

It can be said that this problem often occurs because traditional use needs to be understood and explained beyond the chemical composition of medicines. That is, all dogmas applied from the collection of the plant to the preparation of the medicine must be considered. In this context, many sociolinguistic peculiarities can be the main obstacle to the complete understanding of the process, which factually ends up being neglected when it is not possible to establish the conventional logic (Martins et al. 2005).

It can be considered that this occurrence is parallel both to the linguistic uniqueness of a certain group and to the marked increase in dogmas inherited from the anthropological profile of local practitioners. Reyes-García (2010) points out that it is common for human beings to seek to group and categorize types of animals and plants in names in ways that express a wide or short differentiation. This directly reflects the equivalence between established taxonomies and those of popular origin.

Thus, regarding the extent of the therapeutic activity of many plants, there is still something unknown and divergent from universal cognition. We also highlight that many of the probable active and specific properties of plants may depend on how they fit into the classification system of people (Arruda Camargo 2006).

According to Frazão-Moreira (2016), treatment processes are permeated in the relationship between healer and patient, cultural expressions, perceptions of nature and rituals and symbols. That is why they must be taken together in order to understand the procedural effectiveness of "emic".

In some visits to communities acclaimed as quilombola remnants, already carried out by the members of this work, it was possible to attest that the preparation of many remedies is directly related to local medical concepts, that is, the way of preparation is part of an intrinsic pathological categorization. In this sense, many diseases classified as "diseases of the soul or spirit", and often the preparation and dosage guidelines, do not show consistency with conventional reality (Arruda Camargo 2006).

In this context, many notes provided under an immaterial core of a strictly "supernatural" nature, as already described by Pagani et al. (2017) and Quiroz et al. (2016), are based on behaviors that are designs of nature itself, such as the stars "sun" and "moon", which according to Afonso (2006) are extremely important in African and descendant cultures. Others linked to repetitions established are based on an ordinal number, such as the number 7, for example, which has a wide meaning in umbanda according to Souza (2008).

Hence, the efforts are being made to understand the importance of medicinal herbs based on traditional knowledge, for the development of alternative herbal medicines to treat various diseases, which have been included in the promotion of the National Program for Medicinal Plants and Phytotherapic (Brazil 2018).

In this way, it is suggested that verbal language is the main bias between healer and healer, and it is in this resonant stage that the dosage designations of treatment are established, from preparation to use (Cunha Lima et al. 2012).

Studies of Ramzan et al. (2017) with individuals of Pakistani descent, Coelho-Ferreira (2009) with a community on the Amazon coast and Hajdu and Hhmann (2012) with an indigenous reserve in Bolivia make it possible to notice a marked identity about the use of plants with the cultures from which they originate, where they describe that the preparation of traditional remedies is a very cautious phase and as practitioners, outline the steps unilaterally and invariably.

Elisabetsky (2003) recommends that one should strip away a prejudiced critical view, in order to interpret that the preparation must be followed under a strictly imperative nature. It can be said that it is a safeguard under the belief that if it is not done in the way it was learned, the fear of inefficiency predominates.

However, Djordjevic (2017) warns that many problems can be avoided in the context related to quality that can overshadow the potential health benefits, that is, the elaboration of new herbal products should be taken in a strictly particular way, considering factors like biogenetics.

Conclusion

The medicinal and ritualistic plants used on Marajó Island and their chemical effects involved were evaluated in this chapter. Native communities perpetuate the traditional knowledge derived from their ancestors about phytotherapy together with their beliefs (umbanda, candomblé and shamanism). The use of plants for medicinal purposes has been an indispensable way of prevention and treatment of various diseases in humans. Medicinal plants, such as *Carapa guianensis, Elephantopus scaber, Piper umbellatum,* and *Stachytarpheta cayenensis*, are used by the Marajó communities as antiphlogistic, anti-arthritic, antibacterial, anti-inflammatory, diuretic, analgesic and anti-inflammatory, antitumor, antifungal agents. Brazil has the greatest plant diversity on the planet and a wide socio-diversity of traditional knowledge regarding the use of medicinal plants. Therefore, there is a great potential for the development of phytotherapy in the Marajo Island in Amazon region.

Acknowledgment

All authors acknowledge the institutions the Federal University of Pará, Federal Rural University of Amazonia, Federal University of Sao Paulo, University of Campinas and SGB Amravati University for supporting research and for the access to didactic materials and scientific papers.

References

Abe, F,, Donnelly, D.M.X., Moretti, C., Polonsky, J. 1985. Isoflavanoid constituents from Dalbergia monetaria. Phytochemistry, 24: 1071–1076.

Adeyemi, O.O., Okpo, P.A., Ogunti, O.O. 2002. Analgesic and anti-inflammatory effects of the aqueous extract of leaves of Persea americana Mill (Lauraceae). Fitoterapia, 73: 375–380.

Afonso, G. 2006. Relações Afro-indígenas. Scientific American Brasil, Edição especial Etnoastronomia, 14: 72–79.

Agassiz, L. 2000. Viagem ao Brasil 1865–1866. Brasília: Senado Federal/Conselho Editorial.

Agência Nacional de Vigilância Sanitária – Anvisa. Resolução RDC nº 60, de 10 de novembro de 2011. Aprova o Formulário de Fitoterápicos da Farmacopéia Brasileira, primeira edição e dá outras providências. Diário Oficial União. 11 nov 2011.

Agostini-Costa, T.S., Lima, A., Lima, M.V. 2003. Determinação de taninos em pedúnculo de caju: método da vanilina versus método do butanol ácido. Quím Nova, 26: 763–765.

Albuquerque, M.B.B., Faro, M.C.S. 2015. Saberes de cura: um estudo sobre Pajelança Cabocla e mulheres pajés da Amazônia. Revis. Bras. de Hist., 5(13).

Albuquerque, U.P. 1997. Etnobotânica: uma aproximação teórica e epistemológica. Revista Brasileira de Farmácia, 78(3): 60–64.

Andrade, E.H., Zoghbi, M.G., Maia, J.G. 2011. Volatiles from the leaves and flowers of Carapa guianensis Aubl. J Essent. Oil, 13: 436–438.

Arruda Camargo, M.T.L. 2006. Os poderes das plantas sagradas numa abordagem etnofarmacobotânica. Rev. Mus. Nac. Anrqueol. Etnolo., 15-16: 395–410.

Baena, A.L.M. 2004. Ensaio corográfico sobre a província do Pará. Brasília: Senado Federal/Conselho Editorial.

Balbino, E.E., Dias, M.F. 2010. Farmacovigilância: um passo em direção ao uso racional de plantas medicinais e fitoterápicos. Rev. Bras. Farmacogn., 20(6): 992–1000.

Bartwal, A., Mall, R., Lohani, P., Guru, S.K., Arora, S. 2013. Role of secondary metabolites and brassinosteroids in plant defense against environmental stresses. J. Plant Growth Regul., 32(1): 216–232.

Borges, C.V., Minatel, I.O., Gomez-Gomez, H.A., Lima, G.P.P. 2017. Medicinal plants: influence of environmental factors on the content of secondary metabolites. Medicinal Plants and Environmental Challenges, 259–277.

Borges, T.L. 2010. Ministério da Saúde. Secretaria de Ciência, Tecnologia e Insumos Estratégicos. Departamento de Assistência Farmacêutica e Insumos Estratégicos. Relação Nacional de Medicamentos Essenciais.

Brasil. 2011. Resolução da Diretoria Colegiada - RDC Nº 60, de 10 de novembro de 2011. Diário Oficial [da] República Federativa do Brasil, Poder executivo, Brasília.

Brasil. 2013. Agência Nacional de Vigilância Sanitária. Resolução RDC n. 13, 14 de março de 2013: Dispõe sobre as Boas Práticas de Fabricação de Produtos Tradicionais Fitoterápicos.

Brasil. 2015. Anvisa. Agência Nacional de Vigilância Sanitária. RDC Nº 53, de 4 de Dezembro de 2015. Estabelece parâmetros para a notificação, identificação e qualificação de medicamentos e dá outras providências. Diário Oficial da União. Brasília.

Brasil. 2016. Ministério da Saúde. Secretaria de Ciência, Tecnologia e Insumos Estratégicos. Departamento de Assistência Farmacêutica. Política e Programa Nacional de Plantas Medicinais e Fitoterápicos. Brasília: Ministério da Saúde, pp. 148–149.

Brasil. 2017. Ministério da Saúde. Secretaria de Ciência, Tecnologia e Insumos Estratégicos. Departamento de Assistência Farmacêutica e Insumos Estratégicos. Relação Nacional de Medicamentos Essenciais: RENAME. Brasilia: Ministerio da Saude, Brasília.

Brasil. 2018. Agência Nacional de Vigilância Sanitária – Anvisa. Formulário de Fitoterápicos Farmacopeia Brasileira, 1 ed..

Brasil. 2018. Ministério da Saúde. Secretaria-Executiva. Secretaria de Atenção à Saúde. Glossário temático: práticas integrativas e complementares em saúde. Brasília: Ministério da Saúde, pp. 89–89.

Brito, A.R.M.S., Cota, R.H.S., Nunes, D.S. 1997. Gastric antiulcerogenic effects of Dalbergia monetaria L. in rats. Phytother. Res., 11: 314–316.

Camargo, A. 2015. Rituais com ervas: banhos, defumações e benzimentos. 4. ed. São Paulo: Livre Expressão, 2015. 216 p.

Carvalho, A.C.B., Branco, P.F., Fernandes, L.A. 2012. Regulação Brasileira em Plantas Medicinais e Fitoterápicos. Rev. Fitos. Jan-Mar, 7(1): 5–16.

Carvalho, M.G., Albuquerque, L.A., Alves, C.C.F., Cascon, V. 2008. Cornoside and other constituents from the latex of Parahancornia amapa (Hub.) Ducke (Apocynaceae) a medicinal plant in Northern Brazil. Rev. Bras. Farmacogn., 18: 667–669.

Castro, A.H.F., Paiva, R., Alvarenga, A.A., Vitor, S.M.M. 2009. Calogênese e teores de fenóis e taninos totais em barbatimão [Stryphnodendron adstringens (Mart.) Coville]. Ciênc. agrotec, 33: 385–390.

Chaves, M.H., Citó, A.M.G.L., Lopes, J.A.D., Costa, D.A., Oliveira, C.A.A., Costa, A.F., Brito Júnior, F.E.M. 2010. Fenóis totais, atividade antioxidante e constituintes químicos de extratos de Anacardium occidentale L., Anacardiaceae. Rev. Bras, Farmacogn., 20: 106–112.

Coelho-Ferreira, M. 2009. Medicinal knowledge and plant utilization in Amazonian coastal community of Marudá, Pará State (Brasil). J. Ethnopharmacol., 126: 159–175.

Colpo, E., Vilanova, C.D.D.A., Pereira, R.P., Reetz, L.G.B., Oliveira, L., Farias, I.L.G., Boligon, A.A., Athayde, M.L., Rocha, J.B.T. 2014. Antioxidant effects of Phyllanthus niruri tea on healthy subjects. Asian Pacific J. Trop. Med., 7: 113–118.

Conselho Brasileiro de Fitoterapia. http://www.conbrafito.org.br/index.php?option=com_content&view=article&id=45&It (accessed March 22, 2020).

Cordovil, D. 2012. Afro-Brazilian religions and audiovisual narratives in Amazonia: research tools. Vibrant, Virtual Braz. Anthr., 9(2): 513–534.

Couto, A.G., Kassuya, C.A.L., Calixto, J.B., Petrovick, P.R. 2013. Anti-inflammatory, antiallodynic effects and quantitative analysis of gallic acid in spray dried powders from Phyllanthus niruri leaves, stems, roots and whole plant. Rev. Bras. Farmacogn., 23: 124–131.

Cristina, D.D.O.K., Borges, L.C. 2012. Pajelança, meio ambiente e cotidiano: interação dos pajés com a natureza–cachoeira do Arari/PA. https://ler.letras.up.pt/uploads/ficheiros/10336.pdf (accessed March 22, 2020).

Cunha Lima, S.T., Rodrigues, E.D., Alves, C., Merrigan, T.L., Melo, T., Guedes, M.L.S., Nascimento, A.F., Toralles, M.B. 2012. The use of medicinal plants by an indigenous Pataxó community in NE Brazil. Rev. Bras. Pl. Med., 14(1): 84–91.

de Souza Lages, L.T., Cardoso, L.F.C., Schmitz, H. 2019. Redes de movimentos sociais em comunidade quilombola de Salvaterra, Marajó, Pará. Estud. soc. agric., 27(2): 328–351.

Debnath, M., Pandey, M., Bisen, P.S. 2011. An omics approach to understand the plant abiotic stress. OMICS, 15: 739–762.

Dewick, P.M. 2009. Medicinal Natural Products: A Biosynthetic Approach. John Wiley and Sons, United Kingdom, 3ª ed.

Djordjevic, S.M. 2017. From medicinal plant raw material to herbal remedies. Aromatic and Medicinal Plants: Back to Nature, 269–288.

Dos Santos, E.Q., Costa, J.F.S., Pereira, M.G.S., Costa, J.M., Sousa, R.L. 2019. Etnobotânica da flora medicinal de quintais na comunidade Mamangal, Igarapé-Miri, PA. Sci. Plena, 15: 1–11.

Elisabetsky, E. 2003. Etnofarmacologia. Ciência e Cultura, 55(3): 35–36.

Etkin, N.L., Elisabetsky, E. 2005. Seeking a transdisciplinary and culturally germane science: The future of ethnopharmacology. J. Ethnopharmacol., 100(1-2): 23–26.

Fang, X., Yang, C.M.A.Q., Yang, L., Chen, X. 2011. Genomics grand for diversified plant secondary metabolites. Plant Divers. Resour., 33: 53–64.

Faria, W.C.S., Bett, S.C., Santos, C.G.B., Brasil, A.S., Gauto, R.F., Beserra, A.M.S.S., Oliveira, A.P. 2014. Caracterização físico-química e análise fitoquímica preliminar do fruto noni (Morinda citrofolia L.) produzido na cidade de Cuiabá – MT. Rev. Bras. Tecn. Agroind., 8: 1208–1215.

Fernandes, T.M. 2004. Plantas medicinais: memória da ciência no Brasil. Editora Fiocruz.

Ferreira, M.E.A. 2018. Plantas medicinais utilizadas em rituais de religiões de matriz afro-brasileira: estudo de caso umbanda. Undergraduate thesis, Unesc.

Filho, Florêncio Almeida Vaz. 2014. A crença nos encantados entre os indígenas do baixo Rio Tapajós. Rev. Leetra, 4(1): 84–91.

Frazão-Moreira, A. 2016. The symbolic efficacy of medicinal plants: practices, knowledge, and religious beliefs amongst the Nalu healers of Guinea-Bissau. J. Ethnobiol. Ethnomed., 12(1): 24.
Gramosa, N.V., Silveira, E.R. 2005. Volatile constituents of Copaifera langsdorffii from the Brazilian Northeast. J. Essent. Oil Res., 17: 130–132.
Guerra, A.M.N.M., Pessoa, M.F., Souza, C.S., Maracajá, P.B. 2010. Utilização de plantas medicinais pela comunidade rural Moacir Lucena, Apodi-RN. Biosci. J., 26(3): 442–450.
Hajdu, Z., Hohmann, J. 2012. An ethnopharmacological survey of the traditional medicine utilized in the community of Porvenir, Bajo Paraguá Indian Reservation, Bolivia. J. Ethnopharmacol., 139(3): 838–857.
Hammer, M.L.A., Johns, E.A. 1993. Tapping an Amazonian plethora: four medicinal plants of Marajo Island, Para (Brazil). J. Ethnopharmacol., 40(1): 53–75.
Harbaum, B., Hubbermann, E.M., Zhu, Z., Schwarz, K. 2008. Free and bound phenolic compounds in leaves of pak choi (Brassica campestris L. ssp. chinensis var. communis) and Chinese leaf mustard (Brassica juncea Coss). Food Chem., 110(4): 838–846.
Heinrich, M. 2014. Ethnopharmacology: quo vadis? Challenges for the future. Rev. Bras. Farmacogn, 24(2): 99–102.
Heinrich, M., Barnes, J., Prieto-Garcia, J., Gibbons, S., Williansin, E. 2017. Fundamentals of Pharmacognosy and Phytotherapy E-Book. Elsevier Health Sciences.
Heinze, M., Hanschen, F.S., Wiesner-Reinhold, M., Baldermann, S., Gräfe, J., Schreiner, M., Neugart, S. 2018. Effects of developmental stages and reduced UVB and low UV conditions on plant secondary metabolite profiles in pak choi (Brassica rapa subsp. chinensis). J. Agric. Food Chem., 66: 1678–1692.
Kang, J., Xie, C., Li, Z., Nagarajan, S., Schauss, A.G., Wu, T., Wu, X. 2011. Flavonoids from açai (Euterpe oleracea Mart.) pulp and their antioxidant and anti-inflammatory activities. Food Chem., 128: 152–157.
Kar, A. 2007. Pharmacognosy and pharmacobiotechnology. New Age International (P) Ltd., Publishers, New Delhi, 396(398,464,465).
Kim, H., Moon, J.Y., Kim, H., Lee, D.S., Cho, M., Choi, H.K., Kim, Y.S., Mosaddik, A., Cho, S.K. 2010. Antioxidant and antiproliferative activities of mango (Mangifera indica L.) flesh and peel. Food Chem., 121: 429–436.
Kougan, G.B., Tabopda, T., Kuete, V., Verpoorte, R. 2013. Simple phenols, phenolic acids, and related esters from the medicinal plants of Africa. pp. 225–249. *In*: Victor Kuete (ed.). Medicinal Plant Research in Africa. Elsevier.
La Condamine, C.M. 2005. Viagem na América meridional descendo o rio das Amazonas. Senatus, Brasília, 4(1): 85–86.
Li, Y., Kong, D., Fu, Y., Sussman, M.R., Wu, H. 2020. The effect of developmental and environmental factors on secondary metabolites in medicinal plants. Plant. Physiol. Biochem., 148: 80–89.
Lima, J.C.S., Martins, D.T.O., Souza, Jr P.T. 1998. Experimental evaluation of stembark of Stryphnodendron adstringens (Martius) Coville for anti-inflammatory activity. Phytother. Res., 12: 218–220.
Lima, S.C.S., Arruda, G.O., Renovato, R.D., Alvarenga, M.R.M. 2012. Representações e usos de plantas medicinais por homens idosos. Rev. Latino-Am. Enfermagem, 20(4): 778–786.
Lima, Z. 2002. O Mundo Místico dos Caruanas da Ilha do Marajó. 6 ed. Belém: Cejup.
Linhares, J.F.P., Hortegal, E.V., Rodrigues, M.I.A., Silva, P.S.S. 2014. Etnobotânica das principais plantas medicinais comercializadas em feiras e mercados de São Luís, Estado do Maranhão, Brasil. Rev. Pan-Amaz Saude, 5(3): 39–46.
Lopes, C.M.C., Lima, S.M.R.R., Veiga, E.C.A., Soares-Jr, J.M. 2019. Phytotherapeutic medicines: reality or myth? Rev. Ass. Med. Bras., 65: 292–294.
Lorenzi, H., Matos, F.J.D.A. 2008. Plantas medicinais no Brasil: nativas e exóticas. 2ª edição. São Paulo: Instituto Plantarum.
Luo, F., Fu, Y., Xiang, Y., Yan, S., Hu, G., Huang, X., Huang, G., Sun, C., Li, X., Chen, K. 2014. Identification and quantification of gallotannins in mango (Mangifera indica L.) kernel and peel and their antiproliferative activities. J. Funct. Foods, 8: 282–291.
Ma, Z.Q., Zhang, S.S. 2010. Light intensity affects growth, photosynthetic capability, and total flavonoid accumulation of Anoectochilus plants. HortScience, 45: 863–867.
Marajó, J.C.G.A., Barão, A. 1992. As regiões amazônicas: estudos chorographicos dos Estados do Gram Pará e Amazonas. Belém: SECULT.
Markom, M., Hasan, M., Daud, W.M.W., Singh, H., Jahim, J.M. 2007. Extraction of hydrolysable tannins from Phyllanthus niruri Linn. Effects of solvents and extraction methods. Sep. Purif. Technol., 52: 487–496.
Martins, A.G., Rosário, D.L., Barros, M.N., Jardim, M.A.G. 2005. Levantamento etnobotânico de plantas medicinais, alimentares e toxicas da Ilha do Combu, Município de Belém, Estado do Pará. Rev. Bras. Farmacogn., 86: 31–30.
Matthew, S., Jain, A.K., James, M., Matthew, C., Bhowmik, D. 2013. Analgesic and anti-inflammatory activity of Kalanchoe pinnata (Lam.) Pers. J. Med. Plants Stud., 1: 24–28.
Maués, R.H. 1990. A Ilha Encantada: medicina e xamanismo numa comunidade e de pescadores. Belém: UFPA.
Melo, A.F.M., Santos, E.J.V., Souza, L.F.C., Carvalho, A.A.T., Pereira, M.S.V., Higino, J.S. 2006. Atividade antimicrobiana *in vitro* do extrato de Anacardium occidentale L. sobre espécies de Streotococcus. Rev. Bras. Farmacogn., 16: 202–205.
Menendez-Baceta, G., Aceituno-Mata, L., Reyes-García, V., Tardío, J., Salpeteur, M., Pardo-De-Santayana, M. 2015. The importance of cultural factors in the distribution of medicinal plant knowledge: a case study in four Basque regions. J. Ethnopharmacol., 161: 116–127.

Menon, N., Sparks, J., Omoruyi, F.O. 2016. Oxidative stress parameters and erythrocyte membrane adenosine triphosphatase activities in streptozotocin-induced diabetic rats administered aqueous preparation of Kalanchoe Pinnata leaves. Pharmacogn. Res., 8: 85–88.

Mesquita, U., Tavares-Martins, A.C.C. 2018 Etnobotánica de plantas medicinales en la comunidad de Caruarú, Isla del Mosqueiro, Belém-PA, Brasil. Boletín Latinoamericano y Del Caribe de Plantas Medicinales y Aromáticas, 17(2): 130–159.

Miranda, C.A.C. 2011. A arte de curar nos tempos da colônia: limites e espaços da cura. Ed. UFPE.

Mohan, C.G., Deepak, M., Viswanatha, G.L., Savinay, G., Hanumantharaju, V., Rajendra, C.E., Halemani, P.D. 2013. Antioxidant and anti-inflammatory activity of leaf extracts and fractions of Mangifera indica. Asian Pac. J. Trop. Med., 6: 311–314.

Monteiro, M.V.B., Bevilaqua, C.M.L., Palha, M.D.C., Braga, R.R., Schwanke, K., Rodrigues, S.T., Lameira, A.O. 2011. Ethnoveterinary knowledge of the inhabitants of Marajó island, Eastern Amazonia, Brazil. Acta Amazônica, 41(2): 233–242.

Monteiro, M.V.B., Rodrigues, S.T., Camurça-Vasconcelos, A.L.F. 2012. Plantas medicinais utilizadas na medicina etnoveterinária praticada na ilha do Marajó. Embrapa Amazônia Oriental, ISSN 1983-0513.

Mota, C.S., Trad, L.A.B. 2011. A gente vive pra cuidar da população: estratégias de cuidado e sentidos para a saúde, doença e cura em terreiros de candomblé. Saude e Soc., 20(2): 325–337.

Nascimento, A.L.C.D. 2018. Ciência do sagrado na Amazônia. Encontros entre a tradição e modernidade nas práticas de pajelanças e religiões afro-brasileiras. PhD Thesis, UFPA.

Oliveira, S.G.D., Moura, F.R.R., Demarco, F.F., Nascente, O.S., Del Pino, F.A.B., Lund, R.G. 2012. An ethnomedicinal survey on phytotherapy with professionals and patients from care units in the Brazilian unified health system. J. Ethnopharmacol., 140(2): 428–437.

Pagani, E., Santos, J.F.L., Rodrigues, E. 2017. Culture-bound syndromes of a Brazilian Amazon Riverine population: Tentative correspondence between traditional and conventional medicine terms and possible ethnopharmacological implications. J. Ethnopharmacol., 203: 80–89.

Paiva, L.A.F., Gurgel, L.A., Silva, R.M., Tome, A.R., Gramos, N.V., Silveira, E.R., Santos, F.A., Rao, V.S.N. 2003. Anti-inflammatory effect of kaurenoic acid, a diterpene from Copaifera langsdorffii on acetic acid-induced colitis in rats. Vasc. Pharmacol., 39: 303–307.

Palheta, I.C., Tavares-Martins, A.C.C., Lucas, F.C.A., Jardim, M.A.G. 2017. Ethnobotanical study of medicinal plants in urban home gardens in the city of Abaetetuba, Pará state, Brazil. Bol. latinoam. Caribe plantas med. aromát., 16: 206–262.

Panizza, S.T. 2010. Como prescrever ou recomendar plantas medicinais. São Luís, MA: CONBRAFITO.

Paz, C.E., Lemos, I.C.S., Monteiro, Á.B., de Araújo Delmondes, G., Fernandes, G.P., Coutinho, H.D.M., de Menezes, I.R.A. 2015. Plantas medicinais no candomblé como elemento de resistência cultural e cuidado a saúde. Rev. Cuba. de Plantas Medicinales., 20(1).

Penido, C., Costa, K.A., Pennaforte, R.J., Costa, M.F.S., Pereira, J.F.G., Siani, A.C., Henriques, M.G.M.O. 2005. Anti-allergic effects of natural tetranortriterpenoids isolated from Carapa guianensis Aublet on allergen-induced vascular permeability and hyperalgesia. J. Inflamm. Res., 54: 295–303.

Pino, J.A., Correa, M.T. 2003. Chemical composition of the essential oil from Annatto (Bixa orellana L.) seeds. J. Essent. Oil Res., 15: 66–67.

Pombo, D.P., Fares, J.A. 2018. Vozes femininas no universo marajoara: experiências de vida em situações de cura. Agricultura Familiar: Pesquisa, Formação e Desenvolvimento, 12(1): 95–110.

Pozo-Insfran, D.D., Brenes, C.H., Talcott, S. 2004. Phytochemical composition and pigment stability of açai (Euterpe olaracea Mart.). J. Agric. Food Chem., 52: 1539–1545.

Prado, I.M., Prado, G.H.C., Prado, J.M., Meireles, M.A.A. 2013. Supercritical CO_2 and low-pressure solvent extraction of mango (Mangifera indica) leaves: Global yield, extraction kinetics, chemical composition and cost of manufacturing. Food Bioprod. Proc., 91: 656–664.

Queiroz, Y.S., Ishimoto, E.Y., Bastos, D.H.M., Sampaio, G.R., Torres, E.A.F.S. 2009. Garlic (Allium sativum L.) and ready-to-eat garlic products: *in vitro* antioxidant activity. Food Chem., 115: 371–374.

Quiroz, D., Sosef, M., Van Andel, T. 2016. Why ritual plant use has ethnopharmacological relevance. J. Ethnopharmacol., 188: 48–56.

Rahman, I.U., Afzal, A., Iqbal, Z., Ijaz, F., Ali, N., Shah, S., Ullah, S., Bussmann, R.W. 2019. Historical perspectives of ethnobotany. Clin. Dermatol., 37: 382–388.

Ramzan, S., Soelberg, J., Jäger, A.K., Cantarero-Arévalo, L. 2017. Traditional medicine among people of Pakistani descent in the capital region of Copenhagen. J. Ethnopharmacol., 196: 267–280.

Reyes-García, V. 2010. The relevance of traditional knowledge systems for ethnopharmacological research: theoretical and methodological contributions. J. Ethnobiol. Ethnomed., 6(1): 32.

Rocha, T.T., Tavares-Martins, A.C.C., Lucas, F.C.A. 2017. Traditional populations in environmentally protected areas: an ethnobotanical study in the Source Marine Extractive Reserve of Brazil. Boletín Latinoamericano y Del Caribe de Plantas Medicinales y Aromáticas, 16: 410–427.

Rodrigues, V.E.G., Carvalho, D.A. 2001. Levantamento etnobotânico de plantas medicinais do domínio cerrado na região do Alto Rio Grande, Minas Gerais. Cienc. Agrotec., 25: 102–123.

Rodríguez-Carpena, J.G., Morcuende, D., Andrade, M.J., Kylli, P., Estévez, M. 2011. Avocado (Persea americana Mill.) phenolics, *in vitro* antioxidant and antimicrobial activities, and inhibition of lipid and protein oxidation in porcine patties. J. Agric. Food Chem., 59: 5625–5635.

Sanchita, A.S. 2018. Gene expression analysis in medicinal plants under abiotic stress conditions. Plant Metabolites and Regulation under Environmental Stress, 407–414.

Santos, A.C.F., Aguiar-Dias, A.C.A., Amarante, C.B., Coelho-Ferreira, M. 2013. Estruturas secretoras da lâmina foliar de amapá amargo (Parahancornia fasciculata, Apocynaceae): histoquímica e doseamento de flavonoides. Acta Amazon., 43: 407–414.

Santos, A.O., Ueda-Nakamura, T., Dias Filho, B.P., Veiga Junior, V.F., Pinto, A.C., Nakamura, C.V. 2008. Antimicrobial activity of Brazilian copaiba oils obtained from different species of the Copaifera genus. Mem. Inst. Oswaldo Cruz, 103: 277–281.

Santos, F.S.D. 2000. Tradições populares de uso de plantas medicinais na Amazônia. História, Ciências, Saúde-Manguinhos, 6: 919–939.

Silva, S.R., Buitrón, X., Oliveira, L.H., Martins, M.V.M. 2001. Plantas medicinais do Brasil: aspectos gerais sobre legislação e comércio. Quito, Equador: TRAFFIC América do Sul.

Souza, L.E.A.L. 2008. O espiritismo, a magia e as sete linhas da Umbanda. Editora do Conhecimento.

Souza, L.S.B., Rossetti, D.F., elias, V.R., Prado, R.L. 2013. Neotectonics in Marajó Island, State of Pará (Brazil,) revealed by vertical electric sounding integrated with remote sensing and geological data. An. Acad. Bras. Cienc., 85.1: 73–86.

Souza, T.M., Severi, J.Á., Silva, V.Y.A., Santos, E., Pietro, R.C.L.R. 2007. Bioprospecção de atividade antioxidante e antimicrobiana da casca de Stryphnodendron adstringens (Mart.) Coville (Leguminosae-Mimosoidae). Rev. Ciênc Farm Básica Apl., 28: 221–226.

Srivastava, L.M. 2002. Plant Growth and Development: Hormones and Environment. Elsevier.

Svendsen, A.B., Verpoorte, R. 1983. Chromatography of Alkaloids. Part A: Thin-layer Chromatography, vol. 23A. Amsterdam-Oxford-New York: Elsevier Scientific Publishing Company.

Teles, M.M.R.S., Pinheiro, A.A.V., Dias, C.D.S., Tavares, J.F., Barbosa, Filho J.M., Da Cunha, E.V.L. 2019. Alkaloids of the Lauraceae. The Alkaloids, 147.

Thirumurugan, D., Cholarajan, A., Raja, S.S.S., Vijayakumar, R. 2018. An introductory chapter: secondary metabolites. Second Metab—Sources Appl., 1–21.

Vallespi, R.M.C., Morales, M.L.A.F., Garcia, C.L., Torralba, M.P., Gutierrez, D.S.M. 2013. Quimica Bioorganica Y Productos Naturales. Uned-Universidad Nacion. 354 p.

Verma, N., Shukla, S. 2015. Impact of various factors responsible for fluctuation in plant secondary metabolites. J. Appl. Res. Med. Aromat. Plants, 2(4): 105–113.

Wallace, A.R. 2004. Viagens pelo Amazonas e Rio Negro. Brasília: Senado Federal, Conselho Editorial.

Wang, X.J., Ren, J.L., Zhang, A.H., Sun, H., Yan, G.L., Han, Y., Liu, L. 2019. Novel applications of mass spectrometry-based metabolomics in herbal medicines and its active ingredients: Current evidence. Mass Spectrom. Rev. doi:10.1002/mas.21589.

World Health Organization (WHO). 1998. Regulatory Situation of Herbal Medicines: A Worldwide Review (No. WHO/TRM/98.1). Geneva: World Health Organization.

Chapter 15

Ameliorating Insulin Signalling Pathway by Phytotherapy

Deegendra Khadka

Introduction

Diabetes mellitus (DM) is an endocrine metabolic syndrome that is determined by elevated blood glucose level (hyperglycemia), poor secretion of insulin or miserable sensitivity of cells to the action of insulin (Fatemeh et al. 2017). Three clinical forms of diabetes have been recognized: type 1, type 2 and gestational diabetes. Type 1 diabetes, previously known as insulin-dependent diabetes mellitus (IDDM) or juvenile-onset diabetes, is dominant in 5–10% of all diagnosed cases of diabetes (You and Henneberg 2016). Autoimmune disorders, environmental and genetic factors are known to be responsible for this form of diabetes. Type 2 diabetes was formerly known as non-insulin dependent diabetes mellitus (NIDDM) or adult-onset diabetes, which is the most common form prevailing in 90–95% of diabetic patients. NIDDM is directly linked with obesity, afflicted glucose tolerance, family history of diabetes, prior history of gestational diabetes, physical inactivity, cardiovascular diseases, hypertension and race/ ethnicity (Vijan 2010). Among pregnant women, about 2–5% suffer from gestational diabetes mellitus (GDM) but in the most cases, GDM disappears following delivery (Eman 2015). To date, the exact cause of diabetes to a concerned person is unpredictable. Many reasons including specific hereditary disorders, drugs, malnutrition, surgery, infections, and different ailments are the principal culprit in developing diabetes (Singh and Singh 2008).

The data published in 2000 showed that 2.8% of the total population was the victim of diabetes, which is expected to be 4.4% in 2030 (Wild et al. 2004). According to the World Health Organization (WHO), diabetes will be the seventh leading cause of death in 2030 (Wild et al. 2004). WHO reports published in 2004 unveiled that the occurrence of type II diabetes affects more than 170 million individuals worldwide and it is also expected that the total number of people suffering from diabetes will reach 239 million by the year 2020 (Hoffman 1998, Vats et al. 2002).

Diabetes couldn't be cured completely but blood glucose level can be controlled by injecting insulin for type 1 diabetes mellitus and through oral medication in case of type 2 diabetes mellitus. But, some of the oral anti-diabetic drugs such as sulfonylureas, biguanides and glinides (Krentz and Bailey 2005) have a number of adverse effects. Hence, exploring antidiabetic agent which would be more effective and protective is a challenging job.

Findings related to phytotherapy indicate that there have been many endeavours to use herbal medicines in the treatment of diabetes (Kooti et al. 2016, Ota and Ulrih 2017). Researchers have been working to apply natural products in diabetes management by the mechanisms on inhibition of α-glycosidase and α-amylase, the effects on glucose uptake and glucose transporters, the enhancement of insulin secretion

Faculty of Science, Molecular Biotechnology Laboratory, Nepal Academy of Science and Technology, Khumaltar, Lalitpur.
Emails: deegendra.khadka@nast.gov.np; deegendrakhadka@gmail.com

and pancreatic β-cell proliferation and inhibition of protein tyrosine phosphatase 1B (Rios et al. 2015). Currently available therapeutic alternatives include several drugs that have been consumed orally such as sulfonylureas (metformin), biguanides (glybenclamide), troglitazone (pioglitazone) and inhibitors of DPP-4 (gliptins), SGLT2 (gliflozin) and α-glucosidase (acarbose) (American Diabetes Association 2017). However, the use of these synthetic drugs may have terrible side effects like gastrointestinal disorders and hepatotoxicity (Marín-Peñalver et al. 2016) and thus diabetic patients feel secure to choose natural options (Chang et al. 2007). It has been reported that at least 1,200 species of medicinal plants are used in traditional medicine for their anti-diabetic properties, of which 450 plants have been studied to explore their potential (Medagama and Bandara 2014, Baharvand-Ahmadi et al. 2016, Verma et al. 2018).

The present paper reviews selected medicinal plants that have shown experimental or clinical anti-diabetic activity and that have been used in traditional systems of medicine. Only those medicinal plants showing remarkable activities have been included.

A Glance at Molecular Level for Type 2 Diabetes Mellitus (T2DM)

The insulin receptor is a transmembrane receptor that is activated by insulin molecules (White and Kahn 1994). Two alpha subunits and two beta subunits make up the insulin receptor. The alpha subunits are involved in ligand binding, while the two intracellular tyrosine kinase beta-subunits participate in transducing the signal into the cell. The alpha subunits are linked by disulfide bonds to the beta subunits (Gustafson et al. 1999). Activation of the insulin receptor (IR) after ligand binding is a multi-step process involving structural changes in both the ligand and the receptor. The binding of insulin to IR results in autophosphorylation of the receptor on a number of crucial tyrosine residues. This causes activation of the insulin receptor tyrosine kinase, followed by phosphorylation of various insulin receptor substrate (IRS) proteins to propagate the insulin-signalling event further downstream and mediate various biological effects. The phosphorylation on tyrosine residues in IR and IRS proteins develops docking sites for other enzymes and effector molecules containing SH2 or phosphotyrosine-binding (PTB) domains to propagate insulin signal (Taha and Klip 1999). A schematic view of the role of PTP1B in metabolic insulin signalling pathway is depicted in Figure 15.1.

Insulin signalling is known to be attenuated through dephosphorylation by PTPases. Among those, PTP1B is a key player that negatively regulates insulin signalling pathway. Several *in vivo* and *in vitro* experiments revealed the relationship between PTP1B and insulin receptors (Seely et al. 1996, Ahmad et al. 1995). Insulin stimulates cells to take up glucose and store it as glycogen and to inhibit hepatic glucose production. The effect of PTP1B on the metabolic arm of the insulin signalling pathway is thought to be mediated upstream of phosphatidylinositol 3-kinase (PI3K) (Suzuki et al. 2001), and is capable of interacting with and removing tyrosine phosphates from both IR and IRS proteins (Xie et al. 2003). The schematic view of the role of PTP1B in the metabolic insulin signalling pathway is represented in Figure 15.1.

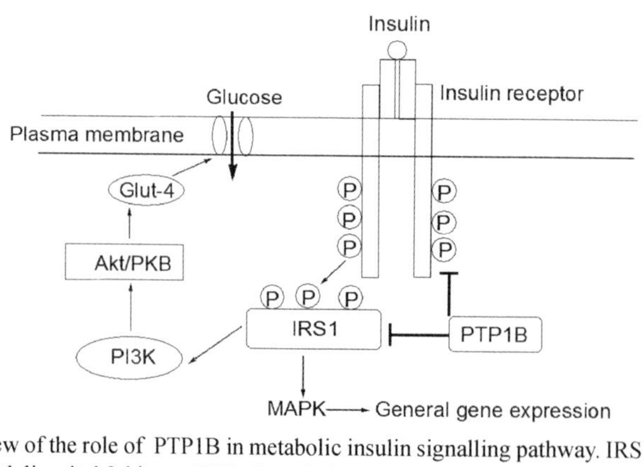

Figure 15.1. Schematic view of the role of PTP1B in metabolic insulin signalling pathway. IRS1: Insulin receptor substrate 1; PI3K: Phosphatidylinositol 3-kinase; PKB: Protein kinase B; MAPK: Mitogen-activated protein kinase.

At the cellular level, insulin resistance is manifested either as reduced capacity or reduced sensitivity to promote glucose transport and metabolism (Goldstein et al. 1998). The mice lacking PTP1B gene have enhanced insulin sensitivity and are resistant to diet-induced obesity (Elchebly et al. 1999). In addition, several studies have revealed that increasing PTP1B expression results in insulin-resistivity which is associated with obesity. The insulin resistance observed in type 2 diabetes and obesity is due to the disequilibrium in enzyme activity between IR and PTPases (Tonks et al. 1988). As a fascinating phosphatase for the treatment of obesity and type 2 diabetes, PTP1B has been a subject of keen interest to the researchers for numerous screenings and development of specific inhibitors.

Antidiabetic Property and Wild Plants

Since the prehistoric time, plants are considered as an effective source of drug molecules and some of the available drugs are derived directly or indirectly from the plant. Phytotherapy is an interestingly growing arena because of having less or no side effects (Ayyanar et al. 2008). Treatments of diabetes by naturally derived agents have promising effects and free from serious consequences. The herbal drug molecules assist in protecting the β-cells which are responsible for insulin secretion during the diabetic condition and keep the blood glucose level in the normal range (Liu et al. 2007).

Based on ethnobotanical information, nearly 800 plant species may contain anti-diabetic property, of which *Momordica charantia*, *Pterocarpus marsupium* and *Trigonella foenum-greacum* have been recommended for diabetes management (Ponnusamy et al. 2011). Medicinal plants exhibited anti-diabetic activities only in the situations when the plants can rejuvenate the pancreatic tissues function by ameliorating insulin secretion or suppressing glucose absorption in intestinal parts or alleviating insulin-dependent metabolic activities (Patel et al. 2012). More than 300 plant species possess hypoglycemic activity; however, discovering more potent antidiabetic drugs from plants based origin is still captivating. Plants containing a remarkable percentage of glycosides, alkaloids, terpenoids, flavonoids, carotenoids, etc. would possess effective anti-diabetic activities (Malaviya et al. 2010).

Highly Effective Anti-diabetic Medicine Plants

Hypoglycemic activities of the medicinal plants have been evaluated by using a different type of experimental techniques. The species described below are included in the list of effective anti-diabetic plants because of their significant *in vitro* and *in vivo* (mouse and human) experimental outputs. In this chapter, plant extracts showing low IC_{50} value and which are efficacious in maintaining blood glucose levels are reviewed. According to Gover et al. (2002), plants exhibited good anti-diabetic activity when they contain a high amount of alkaloids, glycosides, galactomannan guns, polysaccharides, peptidoglycans, hypoglycans, guanidine, steroids, carbohydrates, glycopeptides, terpenoids, amino acids and inorganic ions.

Achyranthes aspera L.

The plant is commonly known as chaff flower or prickly chaff flower in English and is famous for Dattiwan in Nepali. *A. aspera* belonged to the family Amaranthaceae which is distributed mainly in the tropical region of Nepal and India. It is an erect, annual or perennial herb having 1–2 m height, mostly with a woody base. Morphologically, stems are angular, ribbed, simple or branched from the base with tinged purple colour. The plants bloom in the summer period. The flowering part is 8–30 cm long accommodated with several single, white or red flowers of 3–7 mm in width (Verma 2016). The whole plant parts (seeds, roots and shoots) are important because of their use in medicinal purposes. Chemical analysis revealed that the plant contains carbohydrates, protein, glycosides, alkaloids, tannins, saponins, flavonoids, and lignin as major constituents (Ghimire et al. 2015). The chemical structure of the major ingredients is depicted in Figure 15.2.

Several articles have been published describing the effectiveness of *A. aspera* for the management of diabetes but statistical data were available only in a few papers. Srinivasulu et al. (2016) carried out *in vitro* assay to inhibit α-amylase and *in vitro* glucose uptake in isolated psoas muscle and adipose tissue

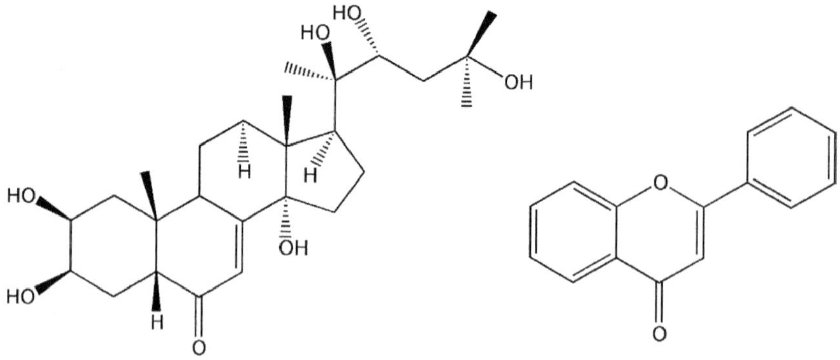

Saponins **Flavonoids**

Figure 15.2. Saponins and Flavonoids are some of the important bioactive compounds of *Achyranthes aspera.*

of 180 g weight male Wistar rats. The inhibition of 0.5% mercuric chloride was arbitrarily fixed as 100% inhibition. Methanolic extract of *A. aspera* leaf (5–25 mg/mL) showed 29.75–71.97% amylase inhibitory activity. *In vitro* studies were carried out in both psoas muscle and adipose tissue in different sets, i.e., tissue alone, with insulin (25 mU/L and 50 mU/L), plant extracts (50, 100 and 150 mg/mL) and plant extracts along with insulin (Srinivasula et al. 2016).

Allium sativum L.

A bulbous plant growing up to 1 m high is popular globally by its pungent flavour as a seasoning or condiment used in spice. *A. sativum* is famous for its English name garlic belonging to the family Amaryllidaceae. The plant consists of a bulb growing to 25–70 cm with bisexual flowers. The species is originally from Asia but is cultivated almost all over the world like China, North Africa (Egypt), Europe and Mexico (Mikaili et al. 2013). Studies were carried out to analyse the chemical composition and found that sulphur compounds such as allicin, alliin are major constituents along with diallyl disulphide (DDS), S-allylcysteine (SAC) and diallyl trisulfide (DTS), (E)-ajoene and other organosulphides as minor components which play crucial pharmacological roles (Lanzotti et al. 2012). Some of the pharmacologically active constituents are shown in Figure 15.3.

Hypoglycemic action of *A. sativum* could be possibly due to an increase in the pancreatic secretion of insulin from β-cells, release of bound insulin or enhancement of insulin sensitivity. It has been previously suggested that garlic can enhance serum insulin by effectively combining with compounds like cysteine which would spare insulin from SH group reactions that are a common cause of insulin inactivation. According to Jelodar et al. (2005) and Bahaman and Yousefian (2009), IC_{50} value of methanolic extract was calculated by inhibiting α-amylase enzyme to establish antidiabetic potentials. IC_{50} of the extract has been reported as 17.95 mg/mL. Oral administration of garlic juice to alloxan-induced diabetic rabbits at a dose of 250 mg/kg produced 60% of the activity of tolbutamide (Mathew and Augusti 1973) which is one of the hypoglycemic medicine. Kiesewetter et al. (1991) experimented oral administration of garlic powder (800 mg/day) to 120 diabetic patients in a month, which revealed the eye-opening consequences of lowering the average blood glucose by 11.6%. Jain and Vyas (1975) commenced an experiment to unveil the effectiveness of garlic extracts for diabetes management to the streptozotocin (STZ) induced diabetic rats whose serum glucose 200% higher than normal rats. The experiment was carried out by daily treating of STZ-induced diabetic rats with an extract of raw garlic (500 mg/kg intraperitoneally) for seven weeks, which showed significantly lowered serum glucose. Compared to control diabetic rats, garlic-treated rats had 57% less serum glucose. Garlic may act as an anti-diabetic agent by increasing either the pancreatic secretion of insulin from the β-cells or the release of bound insulin.

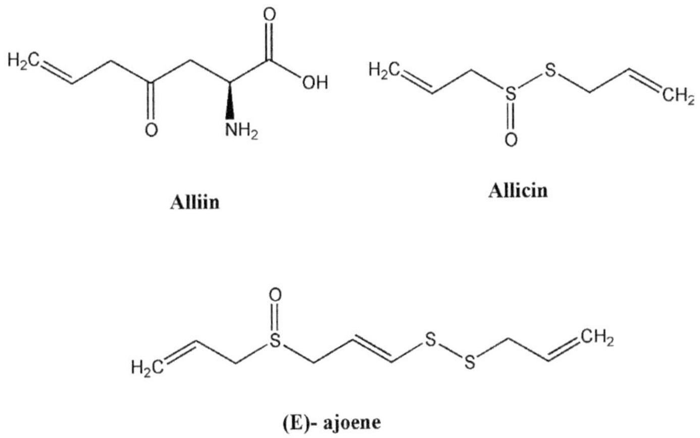

Figure 15.3. Major chemical ingredients Alliin, Allicin and (E)-ajoene found in *Allium sativum*.

Aloe vera (L.) Burm.f.

Aloe vera is a stemless plant of the family Asphodelaceae gaining height up to 60–100 cm. The leaves are thick and fleshy, adorning, green to grey-green, and in some varieties beautified with white flecks on both upper and lower stem surfaces. The leaf border is notched and contains small white teeth. The flower is pendulous with a yellow tubular corolla of 2–3 cm long. Originally, it was grown in the Arabian Peninsula but nowadays the plant is grown around the world in tropical or arid climate (Dehgolan et al. 2012). Chemical compounds found on the leaf are polysaccharides, sugars, minerals, proteins, lipids and phenolic compounds. The studies found that *A. vera* is integrated by 200 potentially active constituents like vitamins, enzymes, minerals, sugars, lignin, saponins, salicylic acids, emodin, aloin and amino acids which are crucial for the multifunctional activity (Kar and Bera 2018). Some of the chemical compounds of *A. vera* are pictured in Figure 15.4.

The plant is useful for many consumer products like beverages, skin lotion, cosmetics or ointments for minor burns and sunburns. The use of the plants as a medicine for the remedy of several ailments including diabetes mellitus is widely known. To prove the plant's effect, *in vivo* test was carried out on the alloxan-induced diabetic rats by intraperitoneal injection of alloxan monohydrate at a dose of 150 mg/kg. Comparing the effects caused by pure leaf extract at a dose of 100, 200, 400 mg/kg with standard drug metformin at a dose 50 mg/kg while 2mL distilled water was taken as a control. After five weeks of the experiment revealed shocking results having no statically significant difference ($P < 0.0001$) at doses of 200 mg/kg and 400 mg/kg (Manjunath et al. 2016). Another research group (Chaudhary et al. 2014) introduced *A. vera* gel powder with 100 mg and 200 mg to diabetic patients for three months along with supplements. It was observed that there was a significant ($p \leq 0.01$) reduction in fasting blood glucose level by 11.4% and 15.4% and postprandial glucose level decreased 18.5% and 27.8% by using 100 mg and 200 mg powder, respectively. With the intervention of *A. vera,* significant reduction was observed in blood glucose of diabetic patients.

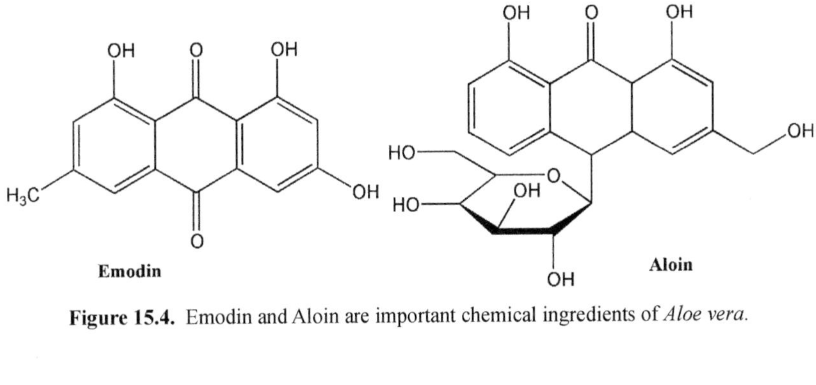

Figure 15.4. Emodin and Aloin are important chemical ingredients of *Aloe vera*.

Andrographis paniculata (Burm f.) Nees

A. paniculata is an annual herb of family Acanthaceae. The plant is commonly known as creat or green chireta. It grows up to 50 cm tall. Stems are 4-angled and glabrous. The leaf blade is ovate-lanceolated or narrowly elliptic with 1.5–7 × 1–2.5 cm. Both surfaces of the leaves are glabrous and abaxially pale green (Shu et al. 2011a). It is distributed in tropical Asian countries but native to India and Sri Lanka. The major chemical constituents are diterpenoids, flavonoids polyphenols, andrograpanin, neoandrographolide and andrographolide (Tan et al. 2016). Representative chemical constituents are shown in Figure 15.5.

Subramania et al. (2008) had immense passion to discover alternative medicine for type 2 diabetes. During the experiment, the research team used ethanolic extract for both *in vitro* and *in vivo* studies. The extract showed appreciable α-glucosidase inhibitory effect in a concentration-dependent manner (IC_{50} 17.2 ± 0.15 mg/mL) and a weak α-amylase inhibitory activity (IC_{50} 50.9 ± 0.17 mg/mL). The *in vivo* studies demonstrated that *A. paniculata* extracts significantly ($P < 0.05$) reduced peak blood glucose and area under curve in diabetic rats when challenged with oral administration of starch and sucrose. Hence, α-glucosidase inhibition may be one of the mechanisms for the extract to exert anti-diabetic activity and indicates that the extract can be considered as a potential candidate. Erindyah et al. (2013) made an effervescent tablet containing the solution of *A. paniculata* and *E. uniflora* and tolbutamide to justify hypoglycemic effect. On the treatment of effervescent solution and tolbutamide (a prescribed medicine) to the glucose loaded rats, the blood glucose level showed significant differences ($p \leq 0.05$), while comparing with the negative control or placebo group. The solution of effervescent tablets decreased blood glucose levels by 13.1% than the negative control (solution of sodium CMC 0.5%). Decreasing blood glucose level of the test solution compared to the placebo by 16.57% proved that extracts contain potential anti-diabetic ingredients.

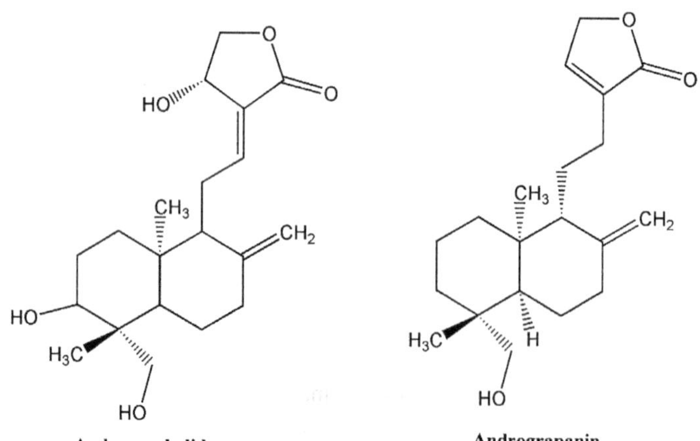

Figure 15.5. Active chemical components, namely, Adrographolide and Andrograpanin of *Andrographis paniculata.*

Artemisia herba-alba Asso

The common name of *A. herba-alba* is white herb because its stems and leaves are white and fluffy. The plant belongs to family Asteraceae and can grow to 20–40 cm high. Leaves produce aromatic fragrance and are blanketed with tiny glandular hair that reflect sunlight imparting a grayish colour to the plant. Sterile shoots possess leaves with grey colour, petiolated and ovate to orbiculared margin. The flowering heads are stalkless, oblong and tapering at the base. Each flowering head consists of 2–5 yellowish bisexual flowers (Pottier-Alapetite 1979). The aerial parts contain essential oil, mainly the presence of α-thujone, 1, 8-cineole, α-chrysanthenone, camphor, β-thujone and piperitenone (Rekkab et al. 2016). Some bioactive compounds are presented in Figure 15.6.

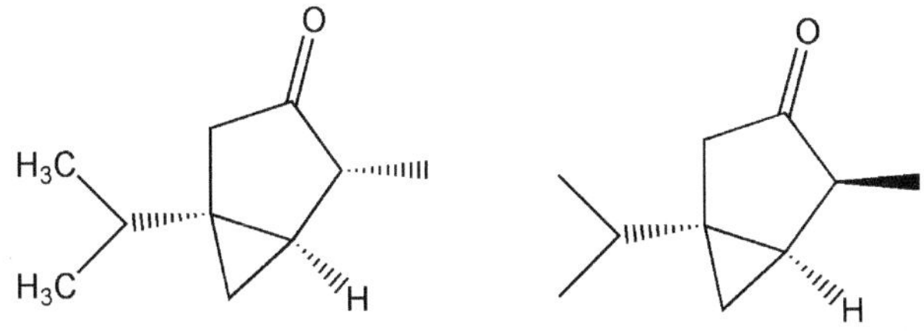

Figure 15.6. α-thujone and β-thujone are some bioactive compounds of *Artemisia herba-alba.*

The species is applied as antiseptic and antispasmodic. *A. herba-alba* is a robust candidate for hypoglycemic activity (Nagwa et al. 2012). The 70% ethyl alcohol extract and mucilage of 70% ethyl extract inhibited the activity of α-amylase by 11% and 2%, respectively. Oral administration (0.39 g/kg b. w.) of the aqueous solution to alloxan-induced diabetic rats exhibited a significant reduction in blood glucose content. The same effect was also seen after the administration of repaglinide (1 mg/kg b. w.) and regular insulin (0.1 IU/kg b. w.). It was demonstrated that the anti-diabetic effect of the plant was similar to that of repaglinide and regular insulin. These results suggested that the plant possesses anti-diabetic property and were able to ameliorate biochemical damages in alloxan-induced diabetic rats (Taskein et al. 2006). The aqueous extracts have been tested on male albino rats, which demonstrated that administration in a dose of 100 mg/kg of body weight for 60 days induced a very significant decrease in glucose level with a significant ($p < 0.01$) increase in total serum cholesterol level, triglycerides, phospholipids, serum aspartate aminotransferase and serum alanine aminotransferase (Mohamed et al. 2010, Kataibeh and Daradka 2007).

Azadirachta indica (A. Juss)

A. indica, usually known as neem in Nepal, belongs to the family Meliaceae. It is a medium-sized tree, reaching 15–30 m high. The leaves are alternate, petiolated, clustered at the end of the branches, dark glossy green at maturity, bearing 10–20 leaflets. The flowers are numerous, fragrant, white and borne in large clusters. The size of fruit is 1–2 cm long drupes, and are smooth. Unripe fruits produce green white milky juice but change into yellow to brown at maturity (Orwa et al. 2009, Puri 1999). Phytochemical analysis revealed that the major chemical constituents are limonoid, nimbocinol, azadirachtin, azadirone, epoxyazadiradione, azadiradione, gedunin, 17-hydroxyazadiradione and cholesterol (Siddiqui et al. 1985). Azadirachtin is one of the principal bioactive compounds (see Figure 15.7).

Anti-fungal, anti-viral, anti-bacterial and other properties of neem increase its popularity for use in cosmetics, fertilizer and medicine. Dinesh Kumar et al. (2010) tested an ethanolic extract in a concentration of 10–100 μg/mL and calculated IC_{50} value against α-amylase enzyme. The amazing result was obtained showing IC_{50} value 62.99 ± 1.20 μg/mL when compared with acarbose (IC_{50} value 83.33 ± 0.75 μg/mL). The outcomes after *in vitro* experiment revealed that *A. indica* might be effective in lowering postprandial hyperglycemia.

Bauhinia racemosa (Lam.)

B. racemosa is a shrub between 3–5 m high which belongs to Fabaceae family and is commonly called Bidi leaf tree. Young branches are zigzag and glabrous. Leaf-blade is ovate-cordate to cordate in shape. The inflorescence is a raceme containing 3–15 flowers. Flower buds are acutely tapered and end in 5 linear calyx teeth. Petals are white, obovate-elliptic (Zong 2010). The plant contains interesting bio-constituents including flavonoids, tannins, methyl gallate, gallic, kaempferol, quercetin, astragalin, querection 3–O–α–rhamnoside, kaempferol 3–O–β–glucoside and Rutin (Rasheda and Butnariub 2014). Some of the chemical ingredients found on *B. racemosa* are presented in Figure 15.8.

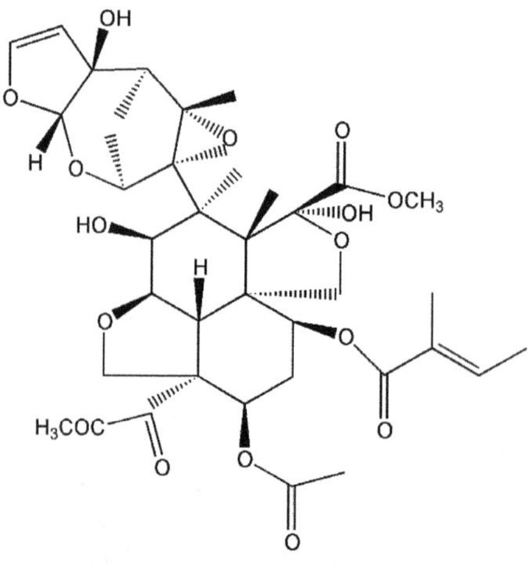

Azadirachtin

Figure 15.7. Azadirachtin is a bioactive compound of *Azadirachta indica*.

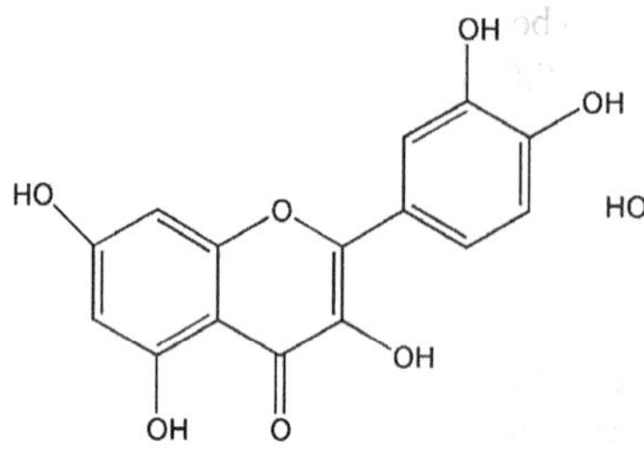

Quercetin

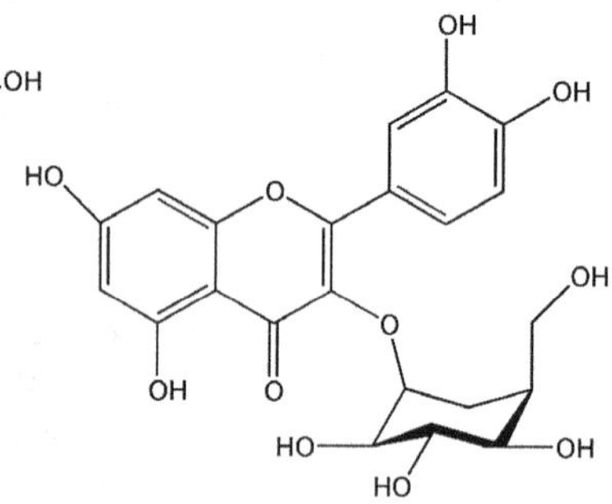

Astragalin

Figure 15.8. Quercetin and Astragalin are some building blocks of *Bauhinia racemosa*.

Both hexane and petroleum ether extracts were used for the experiment. Diabetes was induced by intraperitoneal injection of STZ 45 mg/kg. Animals were divided into five groups, namely, control group, diabetic group, reference group (glibenclamide) and two test groups, BR-1 (250 mg/kg extract) and BR-2 (500 mg/kg). Each group contains six animals. Daily dosing was performed for 28 days and water consumption was observed regularly. The main finding was that the petroleum ether extract of leaves prevented a rise in blood glucose levels in STZ induced diabetic animals. The extract exhibited the concentration-dependent inhibitory effect with an IC_{50} value of 660.26 µg/mL (Vivek et al. 2017). Due to high potential, the plant may be an effective drug for the treatment of diabetes and related complications like obesity and dyslipidemia.

Caesalpinia bonducella (L.) Roxb

C. bonducella belongs to Fabaceae, and is commonly known as grey nicker. It is widely distributed in the tropical and subtropical regions of Asia and the Caribbean. The plant is a shrub, possessing branches in finely gray downy armed with hooked and hard prickles. The leaves are bipinnate with 30–60 cm long having short prickly petioles. The seeds are hard and shiny green coloured, and 1–2 cm long (Alvin Francis et al. 2011). Phytochemical analysis revealed that plants contain alkaloid, phenol, flavonoid, tannin, caesalpinianone, bonducellin and lignin. The nutritional profile of *C. bonducella* describes the presence of carbohydrate, proteins, fat, and fibre (Manikandaselvi et al. 2016). Both caesalpinianone and bonducellin are illustrative compounds of the plant shown in Figure 15.9.

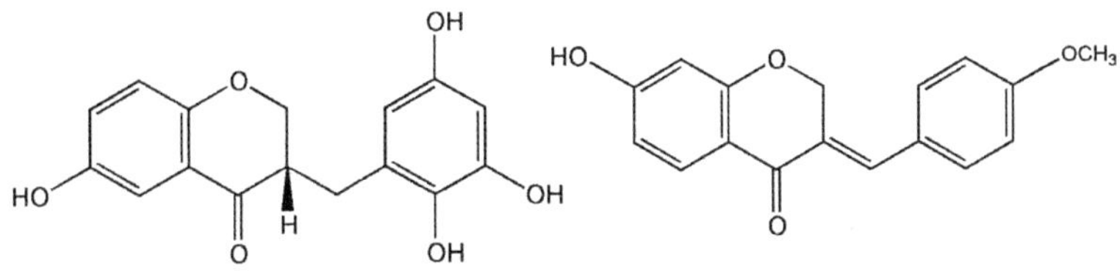

Caesalpinianone **Bonducellin**

Figure 15.9. Some effective compounds—Caesalpinianone and Bonducellin found in *Caesalpinia bonducella*.

All plant parts have medicinal properties. According to Subbiah et al. (2019), shell extract revealed a remarkable effect on diabetes management. The α-amylase activity can be measured *in vitro* by hydrolysis of starch in the presence of α-amylase enzyme. Glycomet GP2 was used as the standard drug (80.97%) at 1000 μg/mL. The maximum activity was at a concentration of 1000 μg/mL and shell showed a better anti-diabetic activity (78.947%). It is also observed that the anti-diabetic activity of the shell extract is close to the standard value. *In vivo* experiment on male Wister albino rats was performed by taking the ethanolic extract. Blood glucose and plasma insulin measurements were done after 14 days from the treatment of the extract at a dose of 500 mg/kg/day and compared with Glibenclamide treated rats as a standard. The results showed Glibenclamide to be more effective in decreasing the postprandial blood glucose (PPBG) level than the seed extract ($p < 0.01$). The results suggested that seed extracted in ethanol possesses anti-diabetic activity against experimentally-induced type 2 diabetes (Widhiantara et al. 2018).

Carthamus tinctorius L.

Carthamus tinctorius belongs to family Asteraceae, and is an annual herb that grows 30–150 cm tall. The plant is called safflower in English and is mainly cultivated for vegetable oil extracted from the seeds. Each flower possesses unique globular flower heads in yellow, orange, or red colour. The individual branch bears one to five flower heads containing 15–20 seeds per head (Anjani 2010). The chemical groups found on *C. tinctorius* are oils, proteins, minerals, phenolics, flavonoids, alkaloids, lignans, serotonin, steroids, polysaccharides, quinochalcone *C*-glycosides and cartormin predominantly (Ali 2015). Figure 15.10 contains a few chemical compounds of *C. tinctorius*.

The plant is used to treat dysmenorrheal, amenorrhea, postpartum, abdominal pain, trauma, etc. Many publications have described significant results of the plant for the treatment of T2DM. The inhibitory activity of hexane extract against PTP1B enzyme exhibited IC_{50} value of 36.1 μg/mL while comparing positive control with IC_{50} value of 5.9 μg/mL (Li et al. 2014). Isolated component of serotonin derivatives showed IC_{50} value against α-glucosidase as 47.2 μm/mL and 99.8 μm/mL, respectively, while that of the reference drugs Acarbose and 1-deoxynojirimycin was estimated as 907.5 μm/mL and 278.0 μm/mL, respectively (Al-Esmail et al. 2019). *In vivo* experiment by Pramesha et al. (2009) introduced albino rats for observation. The rats were distinguished into five groups, viz. group 1 of normal rats, group 2 as diabetic control (untreated) and the remaining three groups served as diabetic induced. Standard drug (glibenclamide) was given to group III while groups IV and V were treated with two different doses of extract, viz., 20 mg/kg and 30 mg/kg b. w. Oral administration of the methanolic extract at two different doses produced significant ($P < 0.01$) decrease in blood glucose in 5th hour and 7th hour onwards for 30 mg/kg and 20 mg/kg, respectively. An additional experiment on alloxan-induced diabetic rabbits was performed by feeding extract at doses of 200 and 300 mg/kg body weight for 30 days, which indicated that extract-treated rabbit had significantly low blood glucose level and high insulin level than diabetic control group (Khan et al. 2014).

Cinnamomum zeylanicum Linn.

C. zeylanicum is a small evergreen tree belonging to the family Lauraceae, and its bark is commonly used as a spice. The plant is evergreen and gains a height above 915 cm distributed in tropical region.

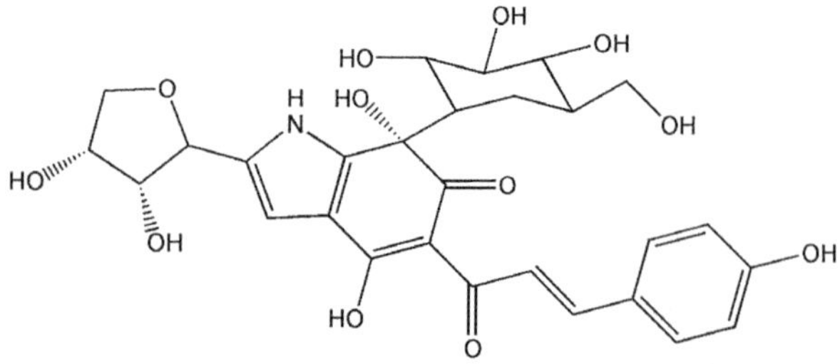

Cartormin

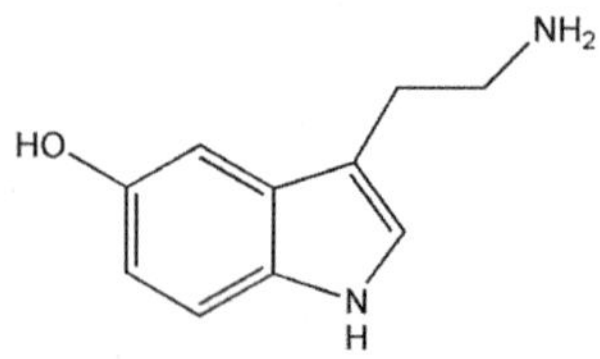

Serotonin

Figure 15.10. Some of the bioactive compounds—Cartormin and Serotonin found in *Carthamus tinctorius.*

Leaves are well veined and ovate in dark green colour. Flowers are small, yellowish to white coloured with aromatic fragrance. The fruit is oval having dark purple colour. The older tree bark is brittle, rough and brownish which is used as an organic spice (Baruah and Nath 2007). The most important compounds found on *C. zeylanicum* are cinnamaldehyde, eugenol and trans-cinnamaldehyde, procyanidins and catechins (Rao and Gan 2014). Cinnamaldehyde and eugenol as representative chemicals are pictured in Figure 15.11.

The plant is traditionally used for many centuries for the treatment of diabetes and other conditions. Effectiveness of *C. zeylanicum* for the management of diabetes is known by half of the inhibitory activity against α-glucosidase and α-amylase enzymes. The plant extracts exhibited a high hypoglycemic effect by inhibition of α-glucosidase and α-amylase with IC_{50} ranging from 0.5 to 8.7 and 37.1 to 52.5 μg/mL, respectively. Obtained results support the traditional use of a number of the analysed species (Saheli et al. 2013). *C. zeylanicum* oil (CTO) and cinnamaldehyde were orally administered to streptozotocin (STZ)-induced diabetic rats to study its effect in both acute and chronic anti-hyperglycemic models. The body weight, oral glucose tolerance test and biochemical parameters, viz., glucose level, insulin level, liver glycogen content, glycosylated hemoglobin, total plasma cholesterol and triglyceride parameters were estimated for all treated groups and compared against a diabetic control group. CTO (100 mg/kg and 200 mg/kg), cinnamaldehyde (20 mg/kg) and glibenclamide (0.6 mg/kg) in respective groups of diabetic animals administered for 28 days reduced the blood glucose level in streptozotocin-induced diabetic rats. There was a significant increase in body weight, liver glycogen content, plasma insulin level and decrease in the blood glucose, glycosylated hemoglobin, and total plasma cholesterol in test groups as compared to the control group. The results of CTO and cinnamaldehyde were found comparable with standard drug glibenclamide (Kumar et al. 2012).

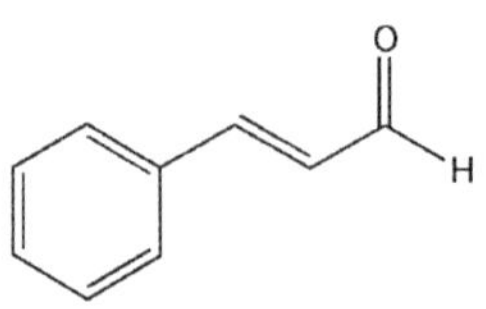

Cinnamaldehyde

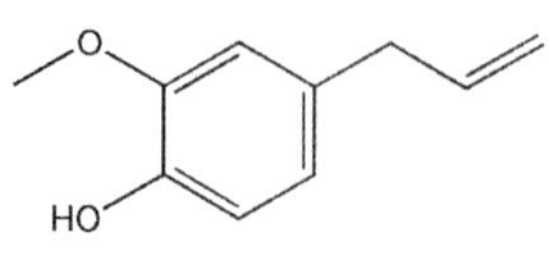

Eugenol

Figure 15.11. Cinnamaldehyde and Eugenol are few bioactive compounds of *Cinnamomum zeylancium.*

Ferula asafoetida H. Karst.

The plant is a perennial herb growing up to 1.5 m, belonging to family Umbelliferae. It is also known as asafoetida, and has a pungent smell, lending it the trivial name of stinking gum. Morphologically, leaves contain spacious sheathing petioles. The height of the flowering stem is 2.5–3 m long and hallow. The pale greenish-yellow coloured flower is in large compound umbels. The shape of the fruits are oval, flatform, thin, reddish brown in colour and contains a milky juice. Generally, this plant is rich in resin, endogeneous gum, and volatile oil. The resin portion is a mixture of sinotannols 'A' and 'B', ferulic acid, ferulsinaic acid and umbelliferone, while volatile oil portion contains organosulfide compounds, such as 2-butyl-propenyl-disulfide, diallyl sulfide, diallyl disulfide and dimethyl trisulfide (Mahendra and Bisht 2012). Figure 15.12 shows a few chemical compounds contained on *F. asafoetida.*

F. asafoetida contains α-glucosidase inhibition property to treat T2DM. *Ex vivo* assay was done by Yarizade et al. (2017), taking an ethanolic extract against the enzyme dipeptidyl peptidase-4 (DPP-IV) and α-glucosidase. The results established that the extract had inhibitory activity on DPP-IV enzyme of 24.5% and the aqueous extract had the highest inhibitory activity of 28% with α-glucosidase. The animal experiments of alloxan-induced diabetic rats with pancreatic beta-cell damage revealed jaw-dropping consequences when the extract was introduced. The extract induced a significant decrease in blood glucose and an increase in serum insulin. Data unveiled that glucose level of the extract feeding alloxan-induced diabetic rats was 10.28 ± 0.85 mmol/L $p < 0.05$ and control group rats was 3.27 ± 0.25 mmol/L $p < 0.05$. Extract feeding diabetic rats secreted insulin significantly higher of the value 0.48 ± 0.05 ng/mL while diabetic rats without extract feeding showed 0.33 ± 0.06 ng/mL (Abuzaiton 2010). The result proved that extract treatment triggers a therapeutic protective effect in diabetes by preserving pancreatic beta cells' integrity, which supports traditional usage for diminishing diabetic complications.

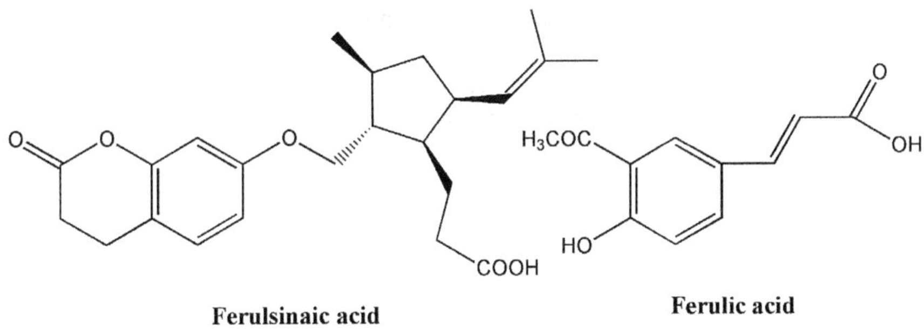

Figure 15.12. Some bioactive compounds of *Ferula asafoetida*—Ferulsinaic acid and Ferulic acid.

Gymnema sylvestre (Retz.) Schult.

G. sylvestre is a perennial woody vine native to tropical Asia, and a member of the family Apocynaceae. Common names are Gymnema in English and gurmar in Hindi, which means sugar destroyer. It is a large climber having elliptic, acuminate and glabrous leaves. The flowers are small, axillary and lateral umbel like cymes, pedicels are long, calyx-lobes ovate, obtuse, pubescent, while corolla is pale yellow, campanulate, valvate (Potawale et al. 2008). The major active components are a group of oleanane type triterpenoid saponins known as Gymnemic acid, along with resins, albumin, carbohydrates, tartaric acid, formic acid, butyric acid, anthraquinone derivatives, inositol alkaloids, lignin and cellulose (Potawale et al. 2008). Gymnemic acid is one of the major ingredients (Figure 15.13).

To confirm the anti-diabetic property of *G. sylvestre, ex vivo* assay was carried out by Ibrahim et al. (2017). Enzymatic assay of the methanolic extract against α-amylase and α-glucosidase measures therapeutic value. The extract demonstrated significant inhibition of α-glucosidase having IC_{50} 182.26 ± 1.05 μg/mL in comparison with standard Acarbose of IC_{50} 189.52 ± 0.46 μg/mL and was more potent than the Acarbose on α-amylase inhibition with IC_{50} of 195.3 ± 4.40 μg/mL and 200.05 ± 7.16 μg/mL,

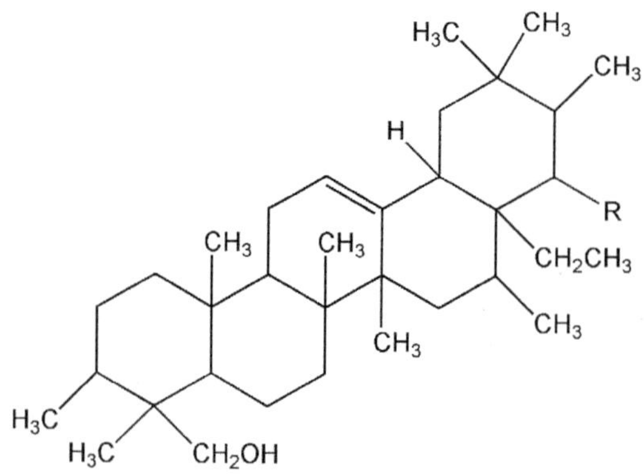

Gymnemic acid

Figure 15.13. Gymnemic acid is the principal chemical ingredient found in *Gymnema sylvestre*.

respectively. The findings indicate the potentiality of consuming extract as a natural remedy for the management of T2DM.

Gongronema latifolium Benth.

G. latifolium, a member of the Apocynaceae family, is a climber with woody hollow glabrous stems growing to 3 m tall. The common name is amaranth globe. The stem is hollow, soft and hairy and contains latex. The leaves are green in colour, simple, opposite and occasionally whorled and exstipulate. The flowers are greenish, bisexual, small, regular and fragrant. The pedicel is 2–4 mm in length with elliptical to rounded and hairy at the apex (Balogun et al. 2016). Several chemical compounds are found on the plants but alkaloid, flavonoid, saponins, proteins, carbohydrates, resins, tannins, reducing sugars polyphenols, triterpenoid and sapogenin are main ingredients (Balogun et al. 2016). Some chemical compounds found on *G. latifolium* are sketched in Figure 15.14.

Ethanolic extract was tested in alloxan-induced diabetic rats to know hypoglycemic activity. Calculating fasting blood glucose (FBG) level at a different time interval (16, 32, 48 and 54 hr) showed the FBG level decreased by 20.25 ± 0.92% at 54 hr but the standard drug glimepiride reduced FBG level by 33.76 ± 0.99%. The extract caused a progressive reduction in FBG concentrations but induced no significant ($p > 0.005$) reduction in FBG concentration between 32 and 48 hr (Ibegbulem and Chikezie 2013).

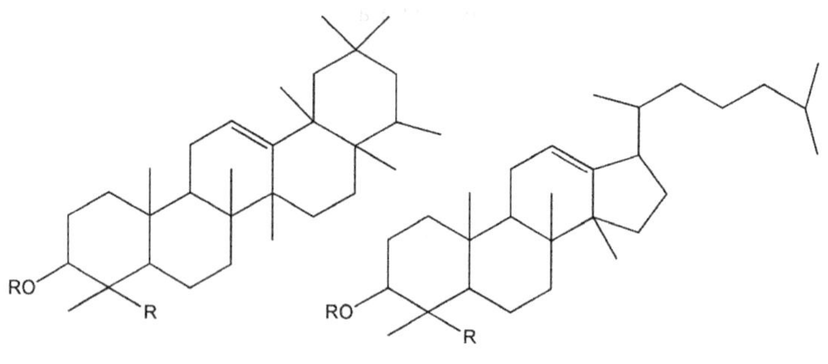

Triterpenoid **Sapogenin**

Figure 15.14. *Gongronema latifolium* contains Triterpenoid and Sapogenin as some of the building blocks.

Mangifera indica L.

Mangifera indica is a member of family Anacardiaceae, commonly called mango in English. It is a large fruit tree growing to a height of 30 m and distributed throughout South-East Asia and the east coast of Africa. The leaves are simple, alternate 16–30 cm long × 3–7 cm broad on flowering branches and up to 50 cm long on sterile branches. The flowers are fragrant, pentameric, greenish-white or pinkish, very small and densely born long and pyramidal panicles. The fruit is large, fleshy, with variable size, shape, colour and taste, and has a characteristic beak at the proximal end (Litz 2009). Ripe fruit contains a major source of mangferin, mangocoumarin, sugars (glucose, fructose, and sucrose) and other carbohydrates such as starch and pectins (Bello-Pérez et al. 2007). The leaves are enriched by sesquiterpenes, while fruit peel oil consists of a very high amount of monoterpenes (D[zbreve]amić et al. 2010). In Figure 15.15, few compounds are presented.

The anti-hyperglycemic activity was proved by an enzymatic reaction. Dinesh Kumar et al. (2010) initiated an *in vitro* assay by using ethanolic extract of stem bark against α-amylase. Taking extract concentration at 10–100 μg/mL range and acarbose as the standard sample showed IC_{50} value of 37.86 ± 0.32 μg/mL of the sample and the standard had IC_{50} value 83.33 ± 0.75 μg/mL. The results showed that *M. indica* might be effective in lowering postprandial hyperglycemia. Additional data (Ganogpichayagrai et al. 2017) was achieved by calculating the inhibitory activity of ethanolic leaf extract against α-glucosidase. The IC_{50} value of the extract obtained 50 μg/mL while comparing with IC_{50} of acarbose, which was 11.9285 μg/mL.

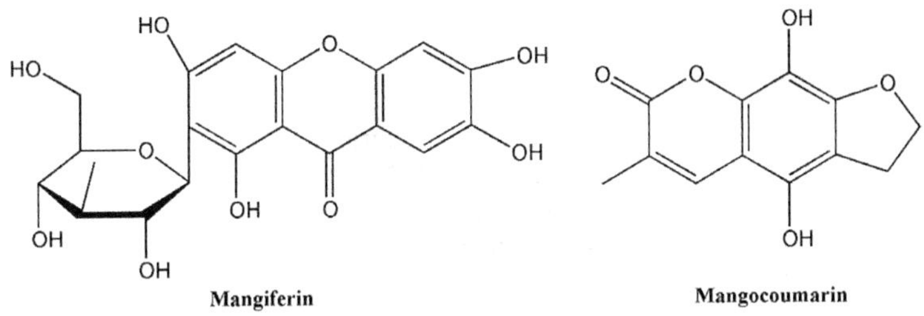

Figure 15.15. Some of the bioactive compounds—Mangiferin and Mangocoumarin found in *Mangifera indica.*

Momordica charantia L.

The species is a tropical and subtropical vine of the family Cucurbitaceae, growing up to 5 cm in length. Commonly known as "bitter gourd", it is popular for its effectiveness to manage T2DM. It is a perennial herb with unbranched tendrils. Plants are dioecious or monoecious. Male flowers are solitary or in a raceme; pedicel often with a large bract calyx tube and corolla yellow or white in color, rotate or broadly campanulate and 5-lobed. In female flowers, the ovary is oblong or fusiform; style elongate; stigmas 3, undivided or 2-lobed; ovules numerous and horizontal (Shu et al. 2011b). Plants contain active phytochemical compounds, such as triterpenes, proteins, momorcharin steroids, fatty acid, lauric, myriaatic, palmitic, stearic and linolaic. Charantin is a typical cucurbitane-type triterpenoid found on *M. charantia* (Goo et al. 2016). Some of the crucial chemical compounds are exhibited in Figure 15.16.

Ethanolic extract was used in *in vitro* experiment against α-amylase and α-glucosidase to calculate half inhibitory concentration. The results were pretty shocking due to having high hypoglycemic properties. *M. charantia* extract contained IC_{50} 56.86 ± 1.12 μg/mL against α-amylase and IC_{50} 88.19 ± 0.74 μg/mL against α-glucosidase, in comparison to Acarbose having IC_{50} values of 93.07 ± 1.49 μg/mL and 77.25 ± 1.20 μg/mL, respectively (Guder 2016). The extract acted as a competitive inhibitor for α-amylase and α-glucosidase (Poovitha and Parani 2016) by oppressing 66% and 69%, respectively, with IC_{50} of 0.26 mg/mL for α-amylase and 0.29 mg/mL for α-glucosidase. Furthermore, STZ-induced diabetic rats were used for supporting the evidence. Oral administration of juice to the diabetic rats at a dose of 10 mL/kg/day for 14 days showed that the juice caused a significant reduction of serum glucose

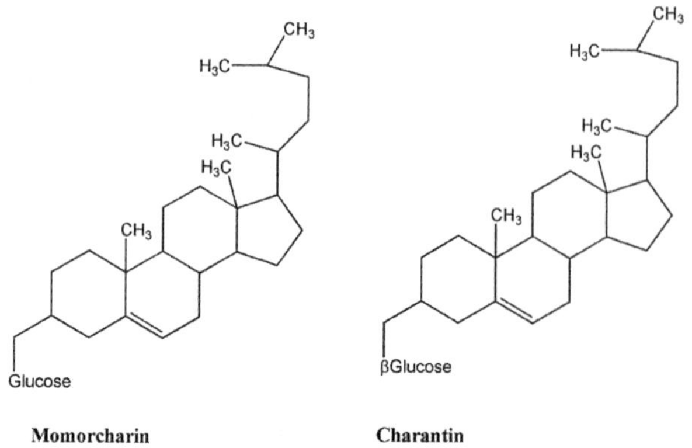

Figure 15.16. Some of the bioactive compounds—Momorcharin and Charantin found in *Momordica charantia*.

(135.99 ± 6.27 mg/mL) and a remarkable increase of serum insulin by 3.41 ± 0.08 mg/mL (Mahmoud et al. 2017). Therefore, *M. charantia* would be one of the best natural treatments for diabetes.

Nigella sativa L.

The common name of *N. sativa* is black caraway. It is an annual flowering plant in the family Ranunculaceae. The plant attains a height of 20–30 cm, having finely divided linear leaves. Pale blue and white-coloured flowers contain five to ten petals. The fruit is a large and inflated capsule, composed of three to seven united follicles, each containing numerous seeds which are used as a spice (Andreas 2005). The plant contains oil, proteins, thymoquinone, dithymoquinone alkaloid, saponin and essential oil. The oil is unsaturated fatty acids, which include arachidonic, eicosadienoic, linoleic, linolenic, oleic, almitoleic, palmitic, stearic and myristic acid as well as p-cymene. The seed contains two different forms of alkaloids: isoquinoline alkaloid that includes nigellicimine, nigellicimine n-oxide and pyrazol alkaloid that includes nigellidine and nigellicine (Forouzanfar et al. 2014). Figure 15.17 contains some few chemical compounds found in *N. sativa.*

Anti-diabetic activity of the plant was measured using ethyl acetate extract to calculate IC_{50} value with α-glucosidase. The extract showed IC_{50} value 3.25 ± 0.05 μg/mL and that of standard Acarbose was 2.35 ± 0.03 μg/mL, which was similar to the standard drug (Amutha and Godavari 2015). The inhibition against the same enzyme by methanolic extract (Salehi et al. 2013) was reflected by IC_{50} of 0.51 ± 0.04 mg/mL in comparison with standard Acarbose and Thymoquinone, which was 0.53 ± 0.06 mg/mL and 0.65 ± 0.05 mg/mL, respectively. Having similar IC_{50} for both extract and the standard medicine, the plant would be one of the effective natural medicines for T2DM.

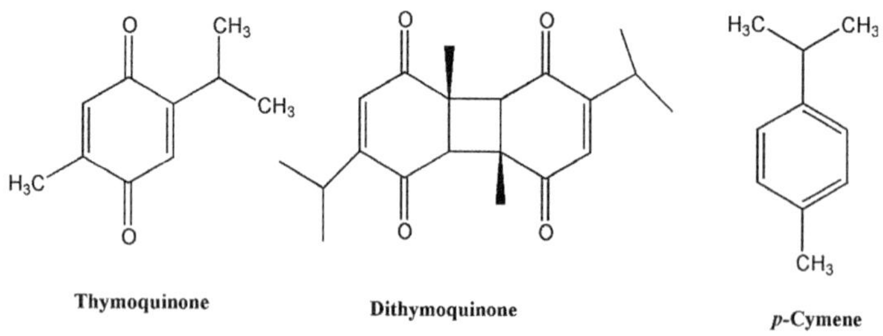

Figure 15.17. Thymoquinone, Dithymoquinone and p-Cymene are the representative chemical ingredients of *Nigella sativa*.

Ocimum tenuiflorum L.

The plant is an aromatic perennial herb in the family Lamiaceae, and is commonly called holy basil or Tulasi. It has been domesticated for its religious and traditional medicinal importance and essential oil. It is extensively consumed as an herbal tea. The plant is erect, many-branched subshrub 30–60 cm with hairy stems. The colour of the leaves is green purple and simple petioled with an ovate, up to 5 cm long blade, and usually has a slightly toothed margin. The flowers are purple, located in close whorls on elongate racemes (Warrier 1995). The major phytochemical constituents are oleanolicacid, ursolic acid, rosmarinic acid, apigenin and essential oils. Eugenol β-elemene, β-caryophyllene and germacrene are the essential oils but terpenes are present in major quantity (Sundaram et al. 2012). Ursolic acid and apigenin representing chemical compounds are shown in Figure 15.18.

Chloroform and aqueous extract exhibited a remarkable enzymatic reaction to show the potential against diabetes. The extracts reflected inhibitory activity against porcine α-amylase, murine pancreatic and liver glucosidases. Chloroform extract possessed more potent inhibitory activity having IC_{50} 1.76 μg/mL against murine pancreatic glucosidases. The inhibitory activity of aqueous extract with IC_{50} values of 1.55 and 9.86 μg/mL was observed for porcine α-amylase and murine liver glucosidase, respectively, which is significantly more than Acarbose (Bhat et al. 2011).

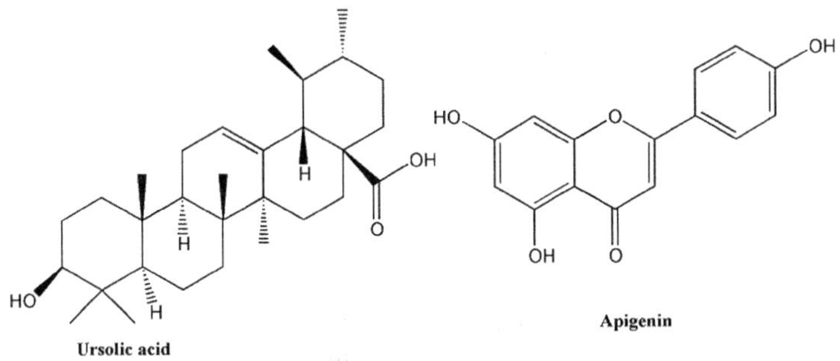

Figure 15.18. Some of the building blocks—Ursolic acid and Apigenin found in *Ocimum tenuiflorum*.

Pterocarpus marsupium Roxb.

P. marsupium, a member of family Fabaceae, is also called Malabar kino or vijayasar. The species is a medium to large-sized deciduous tree that can grow up to 30 m, having wavy margined compound leaves, with 3–7 oblong leaflets that are 3–5 inches long. The species possesses fragrant yellow flowers with large panicles. The seed pods are orbicular, winged and flat (Badkhane et al. 2010). The plant is the building block of pterosupin, pterostilbene, isoliquiritigenin, liquiritigenin, epicatechin, kinotannic acid, kinoin, kino red, marsupinol and beta-edesmol (Badkhane et al. 2010). Some of the building blocks' structure is depicted in Figure 15.19.

Dried gun powder of *P. marsupium* was extracted in ethanol. Ethanolic fraction showed significant inhibitory activity against α-amylase and α-glucosidase enzymes with IC_{50} values of 5.16 μg/mL and 1.06 μg/mL, respectively (Gulati et al. 2012). Having low IC_{50} indicates that the plant would be useful for anti-diabetic medicine.

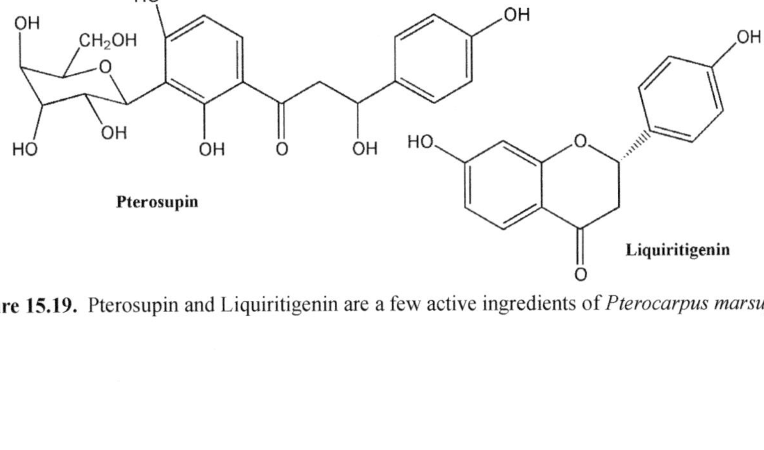

Figure 15.19. Pterosupin and Liquiritigenin are a few active ingredients of *Pterocarpus marsupium*.

Salvia officinalis L.

The species is a member of mint family Lamiaceae, which is a perennial green subshrub with a woody stem. Commonly, the plant is known as culinary saga. It has a long history of medicinal and culinary usage. The plant is about 60 cm tall. Leaves are oblong, ranging in size up to 6.4 cm long by 2.5 cm wide with grey-green, rugose on the upper side and nearly white underneath due to the many short soft hairs. The flower is white, pink or purple (Clebsch and Barner 2003). The plant is characterized by the presence of major constituents as ellagic acid, camphor, α-thujone, apigenin 1, 8-cineole, viridiflorol, β-thujone and β-caryophyllene (Khedher et al. 2017). A few bioactive constituents are shown in Figure 15.20.

Some findings described aqueous leaf extract of *S. officinalis* as containing anti-hyperglycemic property (Salah et al. 2016, Behradmanesh et al. 2013). Introducing plant extract intraperitoneally at a dosage of 300 mg/kg for five weeks to the twenty-eight adult male alloxan-induced albino rats of 7–9 weeks old weighing (240 ± 10 gm) showed very highly significant elevation in mean of blood glucose levels of diabetic animals as compared with control group and extract administration group (Salah et al. 2016). Clinical trial on 80 patients of type II diabetes was performed (Behradmanesh et al. 2013) by feeding extract to the patients for agglomerating strong evidence for anti-diabetic nature. The results demonstrated that fasting blood glucose (FBG) in the drug group was 25 mg/dL less than in the control group; however, the difference was not significant ($p > 0.05$). Two-hour postprandial blood glucose (2 hpp) had a significant difference in the 12th week of the study ($p < 0.05$).

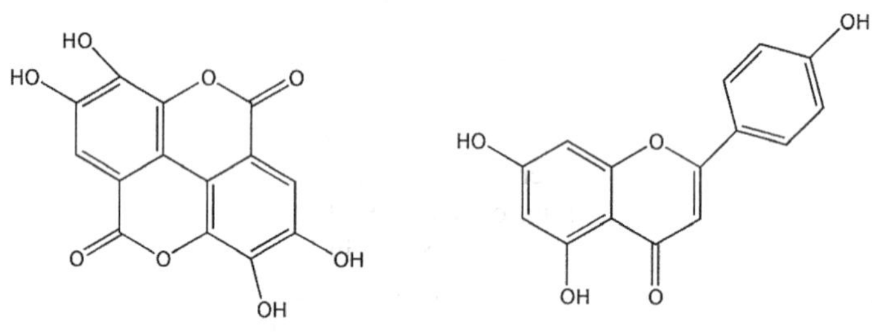

Figure 15.20. Some of the active components—Ellagic acid and Apigenin found in *Salvia offininalis*.

Syzygium cumini (L.) Skeels

S. cumini, commonly known as black plum, is an evergreen tropical tree in the flowering plant family Myrtaceae. A slow-growing species can attain heights of 30 m and stay alive above a century. Leaves have a turpentine smell and are opposite, 5–25 cm long and 2.5–10 cm wide. Flowers are fragrant and appear in clusters with a funnel-shaped calyx and 4–5 united petals. Fruit is mostly found in clusters of just a few or 10–40, is round or oblong, often curved and turns from green to light-magenta (Bijauliya et al. 2018). Chemical analysis revealed that plants contain predominantly kaempferol 7-O-methylether, myristic acid, γ-sitosterol anthocyanins pelargonidin-3-O-glucoside, cyanidin-3-O-malonyl glucoside, malic acid and ferulic acid (Bijauliya et al. 2018). Some chemicals found in *S. cumini* are drawn in Figure 15.21.

The seeds are used in various healing systems like Ayurveda, Unani and Chinese medicine. The whole plant parts contain hypoglycemic activity. *In vitro* assay against porcine pancreatic α-amylase (PPA) was done by separating different portions of a fruit and each was extracted in methanol (Gajera et al. 2017). The kernel fraction of fruit demonstrated lowest IC_{50} (8.3 μg/mL) value followed by seed (12.9 μg/mL), seed coat (50.8 μg/mL) and pulp (270 μg/mL). The seed and kernel inhibited PPA at much lower concentrations than standard Acarbose (24.7 μg/mL). Percentage inhibitions of α-glucosidase activity by different concentrations of four extracts, namely hexane, ethylacetate, methanol and water, were calculated against α-glucosidase (Handunge et al. 2018). The inhibitory effects were rapidly increased with increasing solvent polarity. IC_{50} value of hexane, ethylacetate, methanol and water

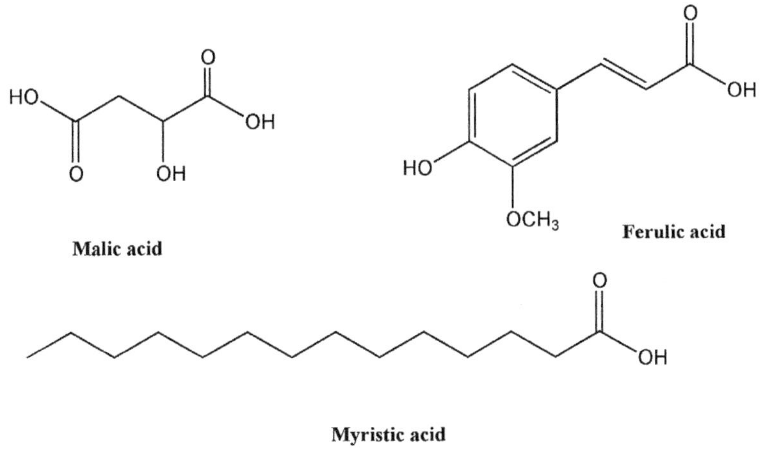

Figure 15.21. Some of the active components—Malic acid, Ferulic acid and Myristic acid found in *Syzygium cumini*.

was demonstrated as 7.89, 3.11, 0.86 and 0.69 μg/mL, respectively, as compared to standard Acarbose having IC_{50} value 210.54 μg/mL. Cold and hot water extracts exhibiting concentration-dependent human pancreatic amylase and half of the inhibitory activity were 42.1 and 4.1 μg/mL, respectively.

Symplocos cochinchinensis (Lour.) S. Moore

The species is an evergreen tree, grows up to 35 m, and belongs to the family Symplocaceae. It has powerful resistance against poisonous crater gases and acid soil conditions. The plant has branchlets, and is glabrous and pubescent with elliptic, or obovate-elliptic, and 6–27 × 0.9–10 cm leaves. Each flower contains calyx lobes 1–3 mm and corolla 3–5 mm in length (Fan 1996). The species is enriched with terpens (arjunolic acid, malic acid and Asiatic acids), L-arbinose, D-galactose and their derivatives (Matta et al. 2017). Some bioactive ingredients are pictured in Figure 15.22.

Diabetes is a complex disease. *In vitro* assay against several targets connected to diabetes such as α-glucosidas inhibition, glucose uptake, adipogenic potential, pancreatic β-cell proliferation, inhibition of protein glycation, protein tyrosine phosphatase 1B (PTP1B) and dipeptidyl peptidase-IV (DPP-IV) was performed from the ethanolic extract of the bark (SCE) and its fractions (hexane, dichloromethane, ethyl acetate and 90% ethanol). While observing the activities, SCE exhibited comparatively better activity like α-glucosidase inhibition (IC_{50} 82.0762 mg/mL), insulin-dependent glucose uptake (3 fold increase) in L6 myotubes, pancreatic beta-cell regeneration in RIN-m5F (3.5 fold increase), reduced triglyceride accumulation (22% decrease) in 3T3L1 cells, and protection from hyperglycemia-induced generation of reactive oxygen species in HepG2 cells (59.57% decrease) with moderate antiglycation and

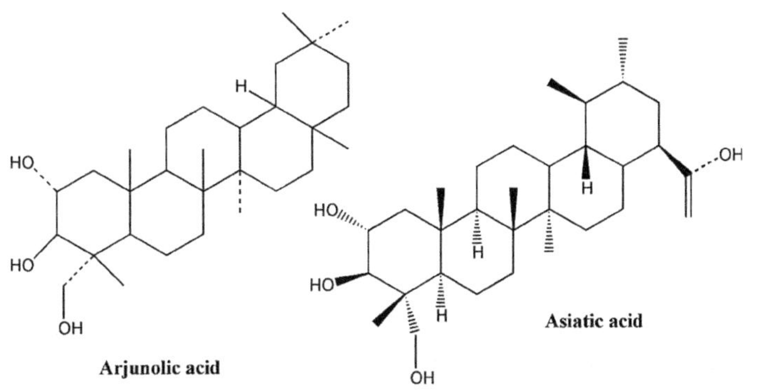

Figure 15.22. Arjunolic acid and Asiatic acid are some active components of *Symplocos cochinchinensis*.

PTP1B inhibition. So, it is concluded that plant mediates anti-diabetic activity mainly via α-glucosidase inhibition, and improves insulin sensitivity with moderate antiglycation (Kalathookunnel et al. 2014).

Swertia chirata Buch.-Ham. ex Wall

S. chirata, usually known as chirata, is an erect, annual, branched herb, up to 1.5 m high. The species belongs to the family Gentianaceae. The plant has an erect and approximately 2–3 ft long stem in orange-brown or purplish colour. Leaves are lanceolate, in opposite pairs and sessile. Flowers are small, numerous, tetramerous, large leafy panicles, green-yellow and tinged with purple and green or white hair (Kumar and Staden 2016). *S. chirata* is incorporated with wide varieties of biologically active compounds such as gentiopicrin, xanthones and their derivatives, lignans, alkaloids, flavonoids, terpenoids, chiratin, palmitic acid, oleic acid and stearic acid. The first isolated dimeric xanthone was chiratanin. The major phytoconstituents swertiamarin, mangiferin, swerchirin, sweroside, amaroswerin and gentiopicrin assist in improving biological activities (Kumar and Staden 2016). Some major components are shown in Figure 15.23.

The plant is famous for indigenous knowledge to treat piles, skin diseases, ulcers and diabetes. Many papers are published regarding anti-diabetic activity of *S. chirata* (Singh et al. 2012, Chandrasekhar et al. 1990). Methanolic extract of the aerial parts showed a pretty substantial blood sugar lowering effect in fasted, fed, glucose loaded and tolbutamide pre-treated albino rat model. The ED_{50} for 40% blood sugar lowering in male albino rats (bodyweight 140–165 g) is 23.1 mg/kg/oral (Singh et al. 2012). It was also found that the ethanolic extract potentially lowered blood sugar level in fasted, glucose fed and tolbutamide pre-treated animals when the animals were administrated with 250 mg/kg plant extract. Chandrasekhar et al. (1990) reported that the hexane fraction (250 mg/kg, b. w. orally for 28 days) lowered blood sugar of albino rats with increased glycogen content of liver and insulin released from pancreatic ß-cells. Swerchirin (50 mg/kg, b.w. orally) isolated from crude extract showed a strong potential to lower blood glucose (up to 60%) after seven hours post-treatment.

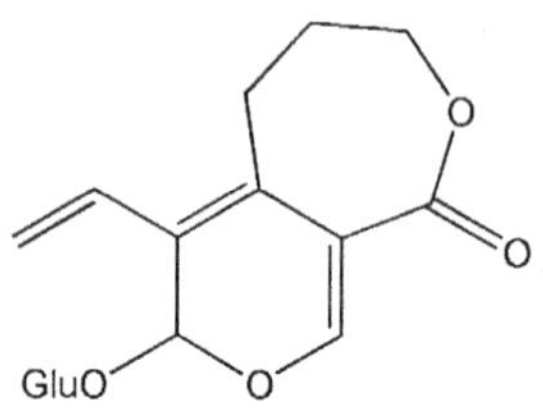

Gentiopicrin

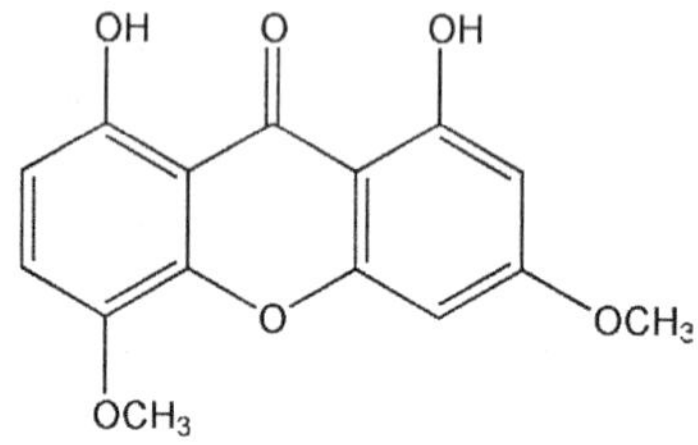

Swerchirin

Figure 15.23. Some of the active components—Gentiopicrin and Swerchirin found in *Swertia chirata*.

Terminalia chebula Retz.

T. chebula, commonly known as black or chebulic myrobalan, is a species in the family *Combretaceae*. The plant is a medium to large deciduous tree growing up to 30 m tall. The leaves are oval, alternate to sub-opposite. Monoecious flowers in dull white to yellow coloured possess a strong unpleasant odour. The fruits are smooth ellipsoid to ovoid drupes, yellow to orange-brown with a single angled stone (Zi 2007). The fruit is drupe and blackish with five longitudinal ridges. The major chemical component found on *T. chebula* is tannin (32%). Plants also contain gallic acid, chebulic acid, punicalagin, chebulanin, neochebulinic, ellagic acid, chebulegic acid, chebulinic acid, flavonol, glycosides, triterpenoids, coumarin conjugated with gallic acid called chebulin and phenolic compounds (Chattopadhyay and Bhattacharyya 2007). Some acids representing major chemical blocks are shown in Figure 15.24.

T. chebula is beneficial for kidney and liver dysfunctions. Dutta and Kalita (2018) used methanolic extract to show the hypoglycemic character by both *in vitro* and *in vivo* experiments. The crude leave extract demonstrated 100% α-glucosidase inhibition with IC_{50} 0.956 ± 0.342 mg/mL compared to the

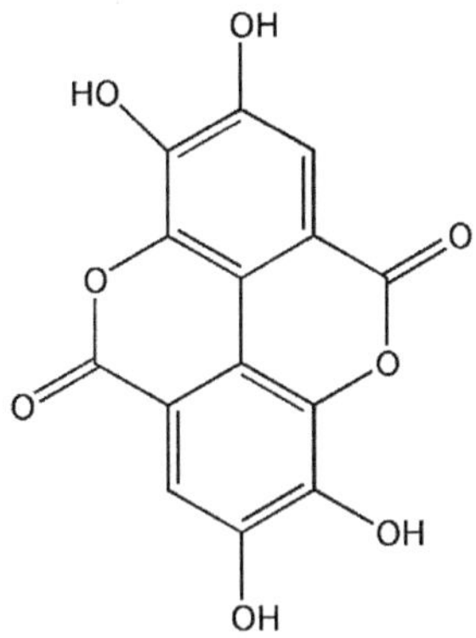

Ellagic acid

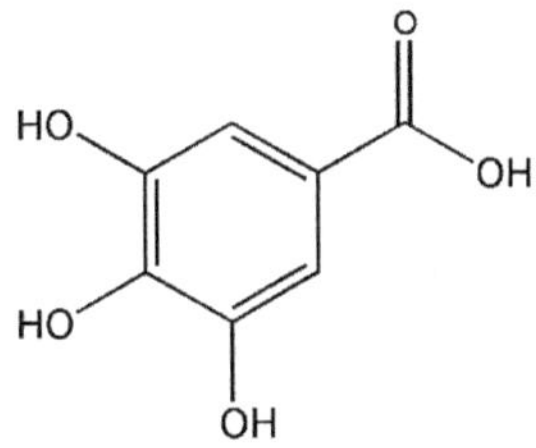

Gallic acid

Figure 15.24. Some of the chemical ingredients of *Terminalia chebula*—Ellagic acid and Gallic acid.

standard drug Acarbose showing IC_{50} 01.50 ± 0.072 mg/mL. Oral administration of the active fraction to diabetic rats loaded with maltose significantly ($P < 0.05$) retarded the postprandial spike of blood glucose level compared to standard drug acarbose. Calculating percentage inhibition by Bansodel and Salalkar (2016) using petroleum ether extract showed significant α-amylase inhibitory activity (51.22%) as compared to the standard drug (65.99%). The studies provided *in vitro* and *in vivo* rational evidence of leaves to retard postprandial hyperglycemia.

Tinospora cordifolia (Willd.) Miers

The plant is an herbaceous vine of the family Menispermaceae. Heart-leaved moonseed, gaduchi and giloya are the common names of *T. cordifolia*. The species is a large, deciduous, extensively-spreading, climbing shrub with several elongated twining branches. Leaves are simple, alternate, exstipulate, roundish and pulvinate, Flowers are unisexual, small, greenish-yellow on axillary and terminal racemes. Especially male flowers are found in clusters, while female flowers are usually solitary. Fruits are of scarlet or orange colour, aggregated in clusters of one to three, ovoid smooth drupelets on thick stalks with subterminal style scars (Sinha et al. 2004). A myriad of biologically active phytoconstituents including alkaloids, tannins, cardiac glycosides, lignin, flavonoids, saponins and steroids exhibited anti-diabetic activities (Reddy and Reddy 2015). In Figure 15.25, some chemical constituents are drawn.

To be proven potential anti-hyperglycemic agents of *T. cordifolia* leaves against alpha amylase enzyme. Shareef et al. (2014) had extracted leaves in petroleum ether, chloroform, ethyl acetate and methanol solvents. The percentage inhibition of α-amylase by the extracts was studied in a concentration range of 10–640 μg/mL. Of the four extracts, ethyl acetate and methanol were comparatively effective than in petroleum ether and chloroform in inhibiting α-amylase. The IC_{50} of petroleum ether and chloroform extract was 100 and 120 μg/mL, respectively, and methanol extracts were 20 μg/mL. The ethylacetate and methanol exhibited a maximum inhibition of 98% at 50 μg/mL concentration. Dinesh Kumar et al. (2010) had performed *in vitro* assay by taking ethanolic extract against α-amylase at concentrations 10–100 μg/mL, which exhibited minimum alpha-amylase inhibitory effects from 12.32 ± 0.79 to 36.41 ± 0.39 μg/mL with an IC_{50} value 138.80 ± 0.16 μg/mL. Dichloromethane (DCM) extract demonstrated

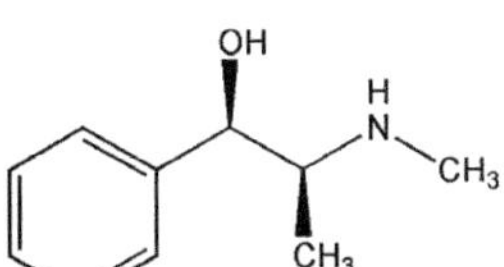

Alkaloid

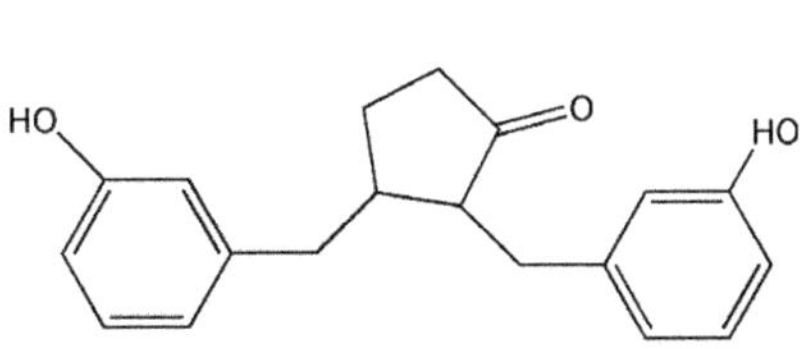

Lignan

Figure 15.25. Alkaloid and Lignan are some active components of *Tinospora cordifolia*.

100% inhibition of the α-glucosidase (Chougale et al. 2009), but in the case of, salivary amylase and pancreatic amylase were 75% and 83%, respectively. Introducing maltose load of 2 mg/g along with 0.3 mg/g b. w. of the DCM stem extract to the normal and diabetic rats, the hypoglycemic activity was raised by 50 and 58%, respectively, as compared to the controls. The extract happened to inhibit α-glycosidase in a non-competitive way.

Trigonella foenum-graecum L.

T. foenum-graecum is famous with its English name fenugreek. The species is an annual plant in the family Fabaceae. The plant is erect or prostrate, straight or profusely branched with 20–130 cm height, and possesses greenish or pinkish coloured stem. Leaves are simple and trifoliate, distinctly petiolate, stipulate with oval leaf lamina or orbicular with an entire margin. Flowers are yellow when young but white at maturity (Mehrafarin et al. 2011). The biological and pharmacological actions of fenugreek are attributed to the variety of its constituents, namely steroids, N-compounds, polyphenolic substances, volatile constituents, amino carbohydrates, proteins, alkaloids, mainly trigonelline, and choline. Gentianine, carpaine, flavonoids, quercetin, kaempeferol and saponins are also major components of the plant (Mehrafarin et al. 2011). Kaempferoal and trigonelline are building blocks which are shown in Figure 15.26.

Fenugreek is consumed as a dietary supplement, animal feed, food additive and in traditional medicine for increasing breast milk. Methanolic extract of seed was supplied for α-glucosidase inhibition assay. The inhibitory activity of the extract at 100 μg/mL was 68% ($p < 0.05$) with IC_{50} value of 57.25 μg/mL when compared to the standard Acarbose of 94% with IC_{50} values of 42.78 μg/mL. The maximum percentage of DPP-4 inhibition of the extract at 320 μg/mL is 77.84% ($p < 0.01$) with IC_{50} value of 52.26 μg/mL, in comparison with standard Vildagliptin which is 80.15% with IC_{50} value of 22.98 μg/mL (Inbaraj and Muniappan 2018). *In vivo* experiment (Mekap et al. 2015) was carried out to reveal hypoglycemic potential. During the experiment, plants (400 mg/kg) were treated to alloxan-treated rats for 10 days showing significantly ($p < 0.001$) attenuated blood glucose level (141.30 ± 11.56 mg/dL) of

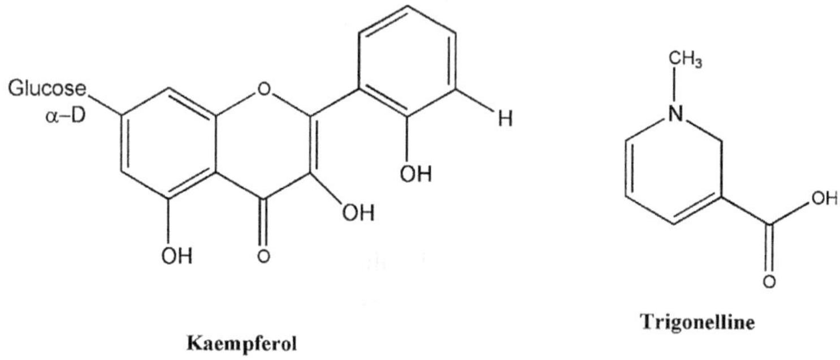

Figure 15.26. Some of the building blocks—Kaempferol and Trigonelline found in *Trigonella foenum-graecum*.

tested animals when compared to alloxan control (293.16 ± 7.32 mg/dL). The data obtained from the study suggested that leaf extracts are effective for T2DM treatment.

Urtica dioica L.

U. dioica, often known as stinging nettle, is an herbaceous perennial plant in the family Urticaceae. The plant grows up to 1–2 m tall. The stem is erect in shape and green in colour, quadrangular with lacunar collenchyma found at each corner. The dorsal surface of the leaves are dark green but ventral side is little pale (Testai et al. 2002). Stinging trichomes cover both stems and leaves. The small dioecious flowers are either male or female in separate inflorescences, brown to greenish in colour and occur as racemes in the axils of the upper leaves (Ahmed and Parasuraman 2014). Flavonoids, tannins, volatile compounds, oxalic acid, tartaric acid, polysaccharides, isolectins, sterols, terpenes, protein, vitamins and minerals are among the main chemical constituents. Leaves containing fluid enriched in histamine, acetylcholine and serotonin are attributed to the burning nature of *U. dioica* (Joshi et al. 2014). Figure 15.27 shows few chemical compounds incorporated in *U. dioica*.

The herb has been used extensively to treat disorders of the kidneys, urinary tract and gastrointestinal tract. Anti-diabetic effects of aqueous extract and its safety were observed on alloxan-induced diabetic mice. The plant extract was administered orally at doses of 25 mg/kg, 100 mg/kg, 200 mg/kg and 300 mg/kg body weight. Toxicity was determined at a dose of 1000 mg/kg body weight. The results indicated that the plant extracts exhibited insulin-mimetic anti-diabetic activity. Evaluation for toxicity had shown that a dose of 1000 mg/kg b. w. preserved the integrity of liver, kidney and lipid profiles for

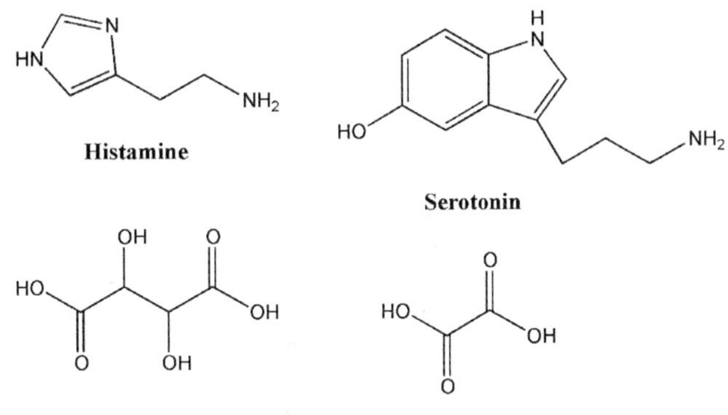

Histamine **Serotonin** **Tartaric acid** **Oxalic acid**

Figure 15.27. Some of the biologically active compounds—Histamin, Serotonin, Tartaric acid and Oxalic acid found in *Urtica dioica*.

biochemical markers. In conclusion, this study confirmed that extract at the dose of 50 mg/kg, 100 mg/kg, 200 mg/kg and 300 mg/kg body weights possess anti-diabetic activity. It is also safe while using a dose of 1000 mg/kg b. w. (Mukundi et al. 2017).

Conclusion

Phytotherapy for diabetes management has been adopted globally since a long time. Plants are well known for natural anti-oxidants and anti-diabetic compounds such as flavonoids, tannins, phenolic, and alkaloids that motivate pancreatic tissues to raise the insulin secretion or decrease the intestinal absorption of glucose. Anti-diabetic herbal medicines show lesser side effects and offer cost-effective management of diabetes through nutrient supplementation. This chapter has shown anti-diabetic properties of selected plant species having potent *in vitro* and *in vivo* efficacies. A large number of plants screened for their anti-diabetic effects have yielded certain interesting leads as mentioned above. Among them, *Allium sativum, Aloe vera, Andrographis paniculata, Azadirachta indica, Gymnema sylvestre, Momordica charantia, Nigella sativa, Syzygium cumini, Ocimum tenuiflorum, Terminalia chebula, Tinospora cordifolia, Trigonella foenum-graecum* and *Urtica dioica* showed promising anti-diabetic activity. Till date, no plant-based drug has reached clinical stage to substitute or reduce the need for the currently-available synthetic drugs, which is the main point to be considered for future research.

References

Abu-zaiton, A.S. 2010. Anti-diabetic activity of *Ferula asafoetida* extracts in normal and alloxan-induced diabetic rats. Pakistan J. Biol. Sci., 13(2): 97–100.

Ahmad, F., Li, P.M., Meyerovitch, J., Goldstein, B.J. 1995. Osmotic loading of neutralizing antibodies demonstrates a role for protein-tyrosine phosphatase 1B in negative regulation of the insulin action pathway. J. Biol. Chem., 270: 20503–20508.

Ahmed, M.K.K., Parasuraman, S. 2014. *Urtica dioica* L. (Urticaceae): a stinging nettle. Syst. Rev. Pharm., 5: 6–8.

Ali Esmail, A.S., Majid, W.J., Talab, T.A., Al-Battat, H.A. 2019. Medicinal plants with anti-diabetic effects—An overview (Part 1). J. Pharma., 9(3): 09–46.

Ali, E.Al-S. 2015. The chemical constituents and pharmacological importance of *Carthamus tinctorius*—An overview. J. Pharm. Biol., 5(3): 143–166.

Alvin Francis, S.L, Lok, W.F., Ang, K.Y., Chong, P.X., Ng, S., Teo, T.K., Yee, C.K., Hugh, Y., Tan, T.W. 2011. The status of *Caesalpinia bonducella* in Singapore. 4: 43–48.

American Diabetes Association. 2017. Pharmacologic approaches to glycemic treatment. Diabetes Care. J. Clin. Appl. Res. Edu., 40(1): S64–S76.

Amutha, K., Godavari, A. 2015. Anti-diabetic and phytochemical screening of *Nigella sativa*. J. Chem. Pharma. Res., 7(8): 432–436.

Andreas, H. 2005. The oldest evidence of *Nigella sativa* L. (Ranunculaceae) and its possible introduction to central Europe. Veg. Hist. Archaeobot., 14(4): 562–570.

Anjani, K. 2010. Morphological description of *Carthamus tinctorius* x *C. oxyacantha* hybrid. Electron. J. Plant Breed., 1(5): 1351–1353.

Attanayake, A.P., Jayatilaka, K.A.P.W., Panthirana, C., Mudduwa, L.K.B. 2015. Antihyperglycemic activity of *Coccinia grandis* (L.) Voigt in streptozotocin-induced diabetic rats. Indian J. Trad. Knowl., 14(3): 376–381.

Ayyanar, M., Sankarasivaraman, K., Gnacimuthu, S.I. 2008. Traditional herbal medicines used for the treatment of diabetes among two major tribal groups in South Tamil Nadu, India. Ethnobot. Leaflets, 12: 276–280.

Badkhane, Y., Yadav, A.S., Sharma, A.K., Raghuwanshi, D.K., Uikey, S.K., Mir, F.A., Lone, S.A., Murab, T. 2010. *Pterocarpus marsupium* Roxb—Biological activities and medicinal properties. Int. J. of Adv. Pharma. Sci., 1: 350–357.

Baharvand-Ahmadi, B., Bahmani, M., Tajeddini, P., Naghdi, N., Rafieian-Kopaei, M. 2016. An ethno-medicinal study of medicinal plants used for the treatment of diabetes. J. Nephropathol., 5(1): 44–50.

Bahman, N., Yousefian, N. 2009. Inhibitory effects of six Allium species on α-amylase enzyme activity. Iranian J. Pharma. Res., 8(1): 53–57.

Balogun, M.E., Besong, E.E., Obimma, J.N., Mbamalu, O.S., Djobissie, S.F.A. 2016. *Gongronema Latifolium*: A phytochemical nutritional and pharmacological review. J. Phys. Pharm. Adv., 6(1): 811–824.

Bansode, T.S., Salalkar, B.K. 2016. Strategies in the design of anti-diabetic drugs from *Terminalia Chebula* Retz. using *in-silico* and *in-vitro* approach. Micro Medicine., 4: 60–67.

Baruah, A., Nath, S.C. 2007. Systematics and diversities of Cinnamomum species used as "cinnamon" spice in North-East India. J. Econ. Tax. Bot., 31(4): 872–887.

Behradmanesh, S., Derees, F., Kopaei, M.R. 2013. Effect of *Salvia officinalis* on diabetic patients. J. Renal Inj. Prev., 2(2): 51–54.

Bello-Pérez, L.A., García-Suárez, F.J., Agama-Acevedo, E. 2007. Mango carbohydrates. Food., 1(1): 36–40.

Bhat, M., Zinjarde, S.S., Bhargava, S.Y., Kumar, A.R., Joshi, B.N. 2011. Hypoglycemic activity of *Swertia chirayita* (Roxb ex Flem) Karst. Evid Based Complement Alternat. Med., Article ID 810207: 1–6.

Bijauliya, K.R., Alok, S., Sabharwal, M., Chanchal, D.K. 2018. *Syzygium cumini* (Linn.)—An overview on morphology, cultivation, traditional uses and pharmacology. Int. J. Pharm. Sci. Res., 9(9): 3608–3620.

Chandrasekar, B., Bajpai, M.B., Mukherjee, S.K. 1990. Hypoglycemic activity of *Swertia chirayita* (Roxb ex Flem) Karst. Indian J. Exp. Biol., 28: 616–618.

Chang, H.Y., Wallis, M., Tiralongo, E. 2017. Use of complementary and alternative medicine among people living with diabetes: literature review. J. Adv. Nurs., 58(4): 307–319.

Chattopadhyay, R.R., Bhattacharyya, S.K. 2007. Plant Review *Terminalia chebula*. Pharmacognos. Rev., 23: 145–150.

Choudhary, M., Kochhar, A., Sangha, J. 2014. Hypoglycemic and hypolipidemic effect of *Aloe vera* L. non-insulin dependent diabetics. J. Food Sci. Technol., 51(1): 90–96.

Chougale, A.D., Ghadyale, V.A., Panaskarand, S.N., Arvindekar, A.U. 2009. Alpha glucosidase inhibition by stem extract of *Tinospora cordifolia*. J. Enzyme Inhib. Med. Chem., 24(4): 998–1001.

Clebsch, B., Barner, C.D. 2003. The New Book of Salvias. Timber Press, p. 216.

Dehgolan, R.R., Sarvestani, Z.T., Rezazadeh, S.A., Dolatabadian, A. 2012. Morphological and physiological characters of *Aloe vera* subjected to saline water irrigation. J. Herbs Spices Med. Plants, 18(3): 222–230.

Dinesh Kumar, B., Mitra, A., Manjunatha, M. 2010. A comparative study of alpha amylase inhibitory activities of common anti-diabetic plants at Kharagpur 1 block. Int. J. Green Pharm., 115–121.

Dutta, J., Kalita, M.C. 2018. Anti-hyperglycemic evaluation of *Terminalia chebula* leaves. Int. J. Pharm. Pharm. Sci., 10(11): 43–48.

D[zbreve]amić, M.A., Marin, P.D., Gbolade, A.A., Ristić, M.S. 2010. Chemical composition of *Mangifera indica* essential oil from Nigeria. J. Essent. Oil Res., 22(2): 123–125.

Elchebly, M., Payette, P., Michaliszyn, E., Cromlish, W., Collins, S., Loy, A.L., Normandin, D., Cheng, A., Himms-Hagen, J., Chan, C.-C., Ramachandran, C., Gresser, M.J., Tremblay, M.L., Kennedy, B.P. 1999. Increased insulin sensitivity and obesity resistance in mice lacking the protein tyrosine phosphatase-1B gene. Science, 283: 1544–1548.

Eman, M.A. 2015. Gestational diabetes mellitus. Saudi Med J., 36(4): 399–406.

Erindyah, R.W., Nurwaini, S., Wahyuningtyas, N., Mohandani, I.P., Hapsoro, S., Aji, W. 2013. Effervescent tablet of Sambiloto (*Andrographis paniculata* [Burm.F.] Ness) and Dewandaru (*Eugenia uniflora* L.) extract: Formulation, anti-diabetic and anti-oxidant properties. Int. J. Pharm. Teach. Pract., 4(2): 689–69.

Fan, Y.N.S. 1996. *Symplocos cochinchinensis* (Loureiro) S. Moore. J. Bot., 52: 148. 1914. Fl China. 15: 235–252.

Fatemeh Farzaei, D.V.M., Morovati, D.R., PharmD, F.F., PharmD, M.H.F. 2017. A mechanistic review on medicinal plants used for diabetes mellitus in traditional persian medicine. Evid Based Complement Alternat. Med., 22(4): 944–955.

Forouzanfar, F.B., Bazzaz, S.F., Hosseinzadeh, H. 2014. Black cumin (*Nigella sativa*) and its constituent (thymoquinone): A review on anti-microbial effects. Iran J. Basic Med. Sci., 17: 929–938.

Gajera, H.P., Gevariya, S.N., Hirpara, D.G., Patel, S.V., Golakiya, B.A. 2017. Anti-diabetic and anti-oxidant functionality associated with phenolic constituents from fruit parts of indigenous black jamun (*Syzygium cumini* L.) landraces. J. Food Sci. Technol., 54(10): 3180–3191.

Ganogpichayagrai, A., Palanuvej, C., Ruangrungsi, N. 2017. Anti-diabetic and anti-cancer activities of *Mangifera indica* cv. Okrong leaves. J. Adv. Pharm. Technol. Res., 8(1): 19–24.

Ghimire, K., Banerjee, J., Gupta, A.K., Dahal, P. 2015. Phytochemical constituents and pharmacological uses of medicinal plant *Achyranthes aspera*: A review. World J. Pharm. Res., 4(1): 470–489.

Goldstein, B.J., Ahmad, F., Ding, W., Li, P.M., Zhang, W.R. 1998. Regulation of the insulin signalling pathway by cellular protein-tyrosine phosphatases. Mol. Cell Biochem., 182: 91–99.

Goo, K.S., Ashari, S., Basuki, N., Sugiharto, A.N. 2016. The bitter gourd *Momordica charantia* L.: Morphological aspects, charantin and vitamin C vontents. J. Agr. Vet. Sci., 9(10)1: 76–81.

Grover, J.K., Yadav, S., Vats, V. 2002. Medicinal plants of India with anti-diabetic potential. J. Ethnopharmacol., 81(1): 81–100.

Güder, A. 2016. Influence of total anthocyanins from bitter melon (*Momordica charantia* Linn.) as antidiabetic and radical scavenging agents. Iran J. Pharma. Res., 15(1): 301–309.

Gulati, V., Harding, I.H., Palombo, E.A. 2012. Enzyme inhibitory and antioxidant activities of traditional medicinal plants: Potential application in the management of hyperglycemia. Complement Altern. Med., 77(12): 1–9.

Gustafson, T.A., Moodie, S.A., Lavan, B.E. 1999. The insulin receptor and metabolic signaling. Rev. Physiol. Biochem. Pharmacol., 137: 71–190.

Handunge, K.I.P., Poongunran, J., Premadasa, W.K.V.K., Jayasinghe, J.A.V.P. 2018. *In-vitro* antiglycation and hypoglycaemic effects of *Syzygium cumini* leaf extracts. J. Natn. Sci. Foundation Sri Lanka, 46(3): 281–291.

Hoffman, D. 1998. The complete illustrated holistic herbal. UK: Element Books, Shaftesbury, Dorset, UK, 228.

Ibegbulem, C.O., Chikezie, P.C. 2013. Hypoglycemic properties of ethanolic extracts of *Gongronema latifolium, Aloe perryi, Viscum album* and *Allium sativum* administered to alloxan-induced diabetic albino rats (*Rattus norvegicus*). Pharmacognosy Comm., 3(2): 12–16.

Ibrahim, A., Babandi, A., Tijjani, A.A., Murtala, Y., Yakasai, H.M., Shehu, D., Babagana, K., Umar, I.A. 2017. *In-vitro* anti-oxidant and anti-diabetic potential of *Gymnema Sylvestre* methanol leaf extract. Europ. Sci. J., 13(36): 218–238.

Inbaraj, S.D., Muniappan, M. 2018. Effect of *Trigonella foenum graceum* on alpha glucosidase and dipeptidyl peptidase-IV inhibitory activity—An *in-vitro* study. Asian J. Pharm. Clin. Res., 11(10): 513–516.

Islam, S.Md., Parvin, S., Uddin, Md.N., Mazid, Md.A. 2014. Anti-diabetic and anti-oxidant activities of decoctions of *Coccinia grandis* Linn. and *Centella asiatica* (L.) on alloxan-induced diabetic rats. Bangladesh Pharma. J., 17(1): 86–91.

Jain, R.C., Vyas, C.R. 1975. Hypoglycemic action of onion and garlic. Am. J. Clin. Nutr., 28: 684–685.

Jelodar, G.A., Maleki, M., Motadayen, M.H., Sirus, S. 2005. Effect of fenugreek, onion and garlic on blood glucose and histopathology of pancreas of alloxan-induced diabetic rats. Indian J. Med. Sci., 59: 64–69.

Joshi, B.C., Mukhija, M., Kalia, A.N. 2014. Pharmacognostical review of *Urtica dioica* L. Int. J. Green Pharm., 8: 201–209.

Kar, S.K., Bera, T.K. 2018. Phytochemical constituents of *Aloe vera* and their multifunctional properties: A comprehensive review. Int. J. Pharm. Sci. Res., 9(4): 1416–1423.

Kalathookunnel, A.A., Riya, M.P., Mishra, A., Anilkumar, K.S., Chandrakanth, C.K., Tamrakar, A.K., Srivastava, A.K., Raghu, K.G. 2014. Anti-diabetic property of *Symplocos cochinchinensisis* mediated by inhibition of alpha glucosidase and enhanced insulin sensitivity. Plos One, 9(9): 1–13.

Khan, R.A., Rizwaniand, G.H., Feroz, Z. 2014. Effect of *Carthamus tinctorius* (Safflower) on fasting blood glucose and insulin levels in alloxan-induced diabetic rabbits. Pak. J. Pharm. Sci., 27(2): 377–380.

Khataibeh, M.H., Daradka, H. 2007. Antiandrogenic activity of *Artemisia herba-alba* in male rats, with emphasis on biochemical parameters. Asian J. Chem., 19(4): 2595–2602.

Khedher, M.R.B, Khedher, S.B., Chaieb, I., Tounsi, S., Hammami, M. 2017. Chemical composition and biological activities of *Salvia officinalis* essential oil from Tunisia. Exp. and Clin. Sci., 16: 160–173.

Kiesewetter, H., Jung, F.F., Pindur, G., Jung, E.M., Mrowietz, C., Wenzel, E. 1991. Effect of garlic on thrombocyte aggregation, microcirculation and other risk factors. Int. J. Clin. Pharmacol. Ther. Toxicol., 29(4): 151–155.

Kooti, W., Farokhipour, M., Asadzadeh, Z., Ashtary-Larky, D., Asadi-Samani, M. 2016. The role of medicinal plants in the treatment of diabetes: A systematic review. Electron. Physician., 8: 1832–1842.

Krentz, A.J., Bailey, C.J. 2005. Oral anti-diabetic agents. Drugs, 65: 385–411.

Kumar, S., Vasudeva, N., Sharma, S. 2012. GC-MS analysis and screening of antidiabetic, antioxidant and hypolipidemic potential of *Cinnamomum tamala* oil in streptozotocin induced diabetes mellitus in rats. Cardiovasc. Diabetol., 11(95): 1–11.

Kumar, V., Staden, J.V. 2016. A review of *Swertia chirayita* (Gentianaceae) as a traditional medicinal plant. Front Pharmacol., 6: 1–14.

Lanzotti, V., Barile, E., Antignani, V., Bonanomi, G., Scala, F. 2012. Anti-fungal saponins from bulbs of garlic, *Allium sativum* L. var. Voghiera. Phytochemistry, 78: 126–34.

Li, L., Wang, Q., Yang, Y., Wu, G., Xin, X., Aisa, H.A. 2012. Chemical components and anti-diabetic activity of essential oils obtained by hydrodistillation and three solvent extraction methods from *Carthamus tinctorius* L. Acta Chromatographica, 24(4): 653–665.

Litz, R.E. 2009. The Mango: Botany, Production and Uses. Second Edition, Cabi Series, CABI, Wallingford, UK.

Liu, I.M., Tzeng, T.F., Liou, S.S., Lan, T.W. 2007. Improvement of insulin sensitivity in obese Zucker rats by myricetin extracted from *Abelmoschus moschatus*. Planta Med., 73: 1054–1060.

Malviya, N., Jain, S., Malviya, S. 2010. Antidiabetic potential of medicinal plants. Acta Pol. Pharm., 67(2): 113–118.

Mahendra, P., Bisht, S. 2012. *Ferula asafoetida*: Traditional uses and pharmacological activity. Pharmacogn. Rev., 6(12): 141–146.

Manikandaselvi, S., Vadivel, V., Brindha, P. 2016. Studies on nutraceutical properties of *Caesalpinia bonducella* L.: An indian traditional medicinal plant. Res. J. Med. Plant., 10(2): 127–139.

Manjunath, K., Prakash, G.B., Subash, K.R., Tadvi, N.A., Manikanta, M., Umamaheswara, R.K. 2016. Effect of *Aloe vera* leaf extract on blood glucose levels in alloxan- induced diabetic rats. Natl. J. Physiol. Pharma. Pharmacol., 6(5): 471–474.

Marín-Peñalver, J.J., Martín-Timón, I., Sevillano-Collantes, C., del Cañizo-Gómez, F.J. 2016. Update on the treatment of type 2 diabetes mellitus. World J. Diabetes, 7(17): 354–395.

Mathew, P.T., Augusti, K.T. 1973. Studies on the effect of allicin (diallyl disulphide-oxide) on alloxan diabetes I. Hypoglycemic action and enhancement of serum insulin effect and glycogen synthesis. Indian J. Biochem. Biophys., 10: 209–212.

Matta, D., Nanda, H., Mahalingam, G. 2017. Phytopharmaceutical potentials of *Prosopis laevigata, Symplocos cochinchinensis* and *Nymphaea alba*: A review. Asian J. Pharm. Clin. Res., 10(10): 63–68.

Medagama, A.B., Bandara, R. 2014. The use of complementary and alternative medicines (CAMs) in the treatment of diabetes mellitus: Is continued use safe and effective? Nutr. J., 13: 1–9.

Mehrafarin, A., Rezazadeh, Sh., Naghdi, H.B., Noormohammadi, Gh., Zand, E., Qaderi, A. 2011. A review on biology, cultivation and biotechnology of Fenugreek (*Trigonella foenum-graecum* L.) as a valuable medicinal plant and multipurpose. J. Med. Plants, 10(37): 1–20.

Mekap, S.K., Panda, A.K., Panda, P.K., Pani, S.R. 2015. Anti-hyperglycemic and lipid lowering activity of *Trigonella foenum-graecum* (Linn.) leaves. Pharma. Biol. Eval., 2(1): 18–24.

Mikaili, P., Maadirad, S., Moloudizargari, M., Aghajanshakeri, S., Sarahroodi, S. 2013. Therapeutic uses and pharmacological properties of garlic, shallot and their biologically active compounds. Iran J. Basic Med. Sci., 16: 1031–1048.

Mohamed, A.H., El-Sayed, M.A., Hegazy, M.E., Helaly, S.E., Esmail, A.M., Mohamed, N.S. 2010. Chemical constituents and biological activities of *Artemisia herba-alba.* Rec. Nat. Prod., 4(1): 1–25.

Mukundi, J., Mwaniki, N.E.N., Piero, N.M., Murugi, N.J., Kelvin, J.K., Yusuf, A.A., John, M.K., Alex, N.K., Daniel, A.S., Peter, G.K., Alice, M.N. 2017. Potential anti-diabetic effects and safety of aqueous extracts of *Urtica dioica* collected from Narok County, Kenya, Mwangi. Pharma. Anal. Acta, 5(8): 1–8.

Mona, F.M., El Zahraa, F., El Ashry, Z., El Maraghy, N.N., Fahmy, A. 2017. Studies on the anti-diabetic activities of *Momordica charantia* fruit juice in streptozotocin-induced diabetic rats. Pharma. Biol., 55(1): 758–765.

Nagwa, E.A., Ahmed, A.S., El-Khayat, Z., Shaffie, N.M., El-Aziz, A.M.A. 2012. Hypoglycemic activity of *Artemisia herba-alba* (Asso.) used in Egyptian traditional medicine as hypoglycemic remedy. J. Appl. Pharma. Sci., 2(3): 30–39.

Orwa, C., Mutua, A., Kindt, R., Jamnadass, R., Anthony, S. 2009. Agroforesty Database: a tree reference and selection guide version 4.0. World Agroforestry Centre, Kenya.

Ota, A., Ulrih, N.P. 2017. An overview of herbal products and secondary metabolites used for management of type two diabetes. Front Pharmacol., 8: 1–14.

Paramesha, M., Ramesh, C.K., Krishna, V., Parvathi, K.M.M., Kuppast, I.J. 2016. Anti-hyperglycemic activity of methanolic extract of *Carthamus tinctorius* L., Annigere-2. Asian J. Exp. Sci., 23(3): 01–06.

Patel, D.K., Prasad, S.K., Kumar, R., Hemalatha, S. 2012. An overview on anti-diabetic medicinal plants having insulin mimetic property. Asian Pac. J. Trop. Biomed., 2(4): 320–330.

Poovitha, S., Parani, M. 2016. *In-vitro* and *in-vivo* α-amylase and α-glycosidase inhibiting activities of the protein extracts from two varieties of bitter gourd (*Momordica charantia* L.). Complement Altern. Med., 16(1): 185.

Ponnusamy, S., Ravindran, R., Zinjarde, S., Bhargava, S., Kumar, A.R. 2011. Evaluation of traditional Indian anti-diabetic medicinal plants for human pancreatic amylase inhibitory effect *in vitro.* Evid Based Complement Altern. Med. Article ID 515647: 1–10.

Potawale, S.E., Shinde, V.M., Anandi, L., Borade, S., Dhalawat, H., Deshmukh, R.S. 2008. *Gymnema sylvestre:* A comprehensive review. Pharmacologyonline, 2: 144–157.

Pottier, A.G. 1979. "Flore de la Tunisie: part 2. Dicotyledones, Gamopetales". Tunis, Ministry of Higher Education and Scientific Research (in French), 2: 1012–3.

Puri, H.S. 1999. Neem: The Divine Tree (*Azadirachta indica*). Harwood Academic Publishers.

Rasheda, K., Butnariub, M. 2014. Anti-microbial and anti-oxidant activities of *Bauhinia racemosa* Lam. and chemical content. Iranian J. Pharma. Res., 13(3): 1073–1080.

Rao, P.V., Gan, S.H. 2014. Cinnamon: A multifaceted medicinal plant. Evid Based Complement Alternat. Med. Article ID 642942: 1–13.

Reddy, N.M., Reddy, R. 2015. *Tinospora cordifolia* chemical constituents and medicinal properties: A review. Sch. Acad. J. Pharm., 4(8): 364–369.

Rekkab, S., Abaza, I., Chibani, S., Kabouche, A., Kabouche, Z. 2016. Chemical composition of the essential oil of aerial parts of *Artemisia herba-alba* Asso. from Oum El-Bouaghi (Algeria) and chemotaxonomic survey. J. Mater. Environ. Sci., 7(12): 4383–4390.

Rios, J.L., Francini, F., Schinella, G.R. 2015. Natural products for the treatment of type 2 diabetes mellitus. Planta Med., 81: 975–994.

Ross, I.A. 2005. *Ferula assafoetida*. Med. Plants World, 3: 223–234.

Salah, M.M.A.C., Hussein, M.S., Rana, I.M., Khalid, L.B. 2016. Effect of *Salvia officinalis* L. (Sage) aqueous extract on liver and testicular function of diabetic albino male rats. J. Babylon Univ/Pure Appl. Sci., 24(2): 390–399.

Salehi, P., Asghari, B., Esmaeili, M.A., Dehghan, H., Ghazi, I. 2013. α-Glucosidase and α-amylase inhibitory effect and anti-oxidant activity of ten plant extracts traditionally used in Iran for diabetes. J. Med. Plants Res., 7(6): 257–266.

Seely, B.L., Staubs, P.A., Reichart, D.R., Berhanu, P., Milarski, K.L., Saltiel, A.R., Kusari, J., Olefsky, J.M. 1996. Protein tyrosine phosphatase 1B interacts with the activated insulin receptor. Diabetes, 45: 1379–1385.

Shareef, M.I., Reddy, P.J.M., Gopinath, S.M., Dayananda, K.S., Mandal, A., Purushotham, K.M. 2014. *In-vitro* α-amylase inhibitory activity of the leaves of *Tinospora cordifolia*. Intl. J. Innov. Res. Sci. Eng. Tech., 3(3): 10091–10096.

Shu, C.X.L., Jiaqi, H., Chi, H.C., Yunfei, D., Daniel, T.F. 2011a. *Andrographis wallich* ex Nees in Wallich, Pl. Asiat. Rar., 3: 77, 116. 1832. Fl China. 19: 473–474.

Shu, K.G., Ming, L.A., Jeffrey, C. 2011b. *Momordica* Linnaeus, Sp. Pl., 2: 1009. 1753. Fl China. 19: 28–30.

Siddiqui, S., Faizi, S., Siddiqui, B.S. 1985. Studies on the chemical constituents of *Azadirachta indica* A. Juss (Meliaceae). Part VII [1]: 922–924.

Singh, A.R., Singh, S.A. 2008. Diseases of poverty and lifestyle, well-being and human development. Mens Sana Monogr., 6(1): 187–225.

Singh, R.L., Singh, P., Agarwal, A. 2012. Chemical constituents and bio-pharmacological activities of *Swertia chirata*: A review. Natural Prod. Indian J., 8(6): 238–247.

Sinha, K., Mishra, N.P., Singh, J., Khanuja, S.P.S. 2004. *Tinospora cordifolia* (Guduchi), a reservoir plant for therapeutic applications: A review. Indian J. Tradit. Know., 3(3): 257–70.

Sobhi, W., Stevigny, C., Duez, P., Calderon, B.B., Atmani, D., Benboubetra, M. 2016. Effect of lipid extracts of *Nigella sativa* L. seeds on the liver ATP reduction and alpha-glucosidase inhibition. Pak. J. Pharm. Sci., 29(1): 111–117.

Srinivasulu, N., Mallaiah, P., Sudhakar, G., SasiBhusana Rao, B., Sarala Kumari, D. 2016. Alpha amylase inhibitory activity and *in-vitro* glucose uptake in psoas muscle and adipose tissue of male Wister rats of leaf methanolic extract of *Achyranthes aspera*. J. Pharm. Phytochem., 5(3): 176–180.

Subramanian, R., ZainiAsmawi, M., Sadikun, A. 2008. *In-vitro* α-glucosidase and α-amylase enzyme inhibitory effects of *Andrographis paniculata* extract and andrographolide. Acta Biochimica Polinica, 55: 391–398.

Sundaram, R.S., Ramanathan, M., Rajesh, R., Satheesh, B., Saravanan, D. 2012. LC-MS quantification of rosmarinic acid and ursolic acid in the Ocimum sanctum linn. leaf extract (Holy Basil, Tulsi). J. Liq. Chromatogr. Related Tech., 35(5): 634.

Suzuki, T., Hiroki, A., Watanabe, T., Yamashita, T., Takei, I., Umezawa, K.J. 2001. Potentiation of insulin-related signal transduction by a novel protein-tyrosine phosphatase inhibitor, Et-3, 4-dephostatin, on cultured 3T3-L1 adipocytes. J. Biol. Chem., 276: 27511–27518.

Tan, M.C.S., Oyong, G.G., Shen, C.C., Ragasa, C.Y. 2016. Chemical composition of *Andrographis paniculata* (Burm.f.) Nees. Res. J. Pharm. Biol. Chem. Sci., 7(6): 2405–2408.

Taha, C., Klip, A. 1999. The insulin signaling pathway. J. Memb. Biol., 169: 1–12.

Taskein, D., Atasever, M., Adiguzel, G., Keles, M., Tastekin, A. 2006. Hypoglycemic effect of *Artemisia herba-alba* in experimental hyperglycemic rats. Bull Vet. Pulawy., 50: 235–238.

Testai, L., Chericoni, S., Calderone, V., Nencioni, G., Nieri, P., Morell, I. 2002. Cardiovascular effects of *Urtica dioica* L. (Urticaceae) roots extracts: *in-vitro* and *in-vivo* pharmacological studies. J. Ethnopharmacol., 81: 105–109.

Tonks, N.K., Diltz, C.D., Fischer, E.H. 1988. Characterization of the major protein-tyrosine-phosphatases of human placenta. J. Biol. Chem., 263: 6722–6730.

Vats, V., Grover, J.K., Rathi, S.S. 2002. Evaluation of anti-hyperglycemic and hypoglycemic effect of *Trigonella foenum-graecum* Linn., *Ocimum sanctum* Linn. and *Pterocarpus marsupium* Linn. in normal and alloxanized diabetic rats. J. Ethnopharmacol., 79: 95–100.

Verma, S. 2016. Review study on *Achyranthes aspera* (Amaranthaceae)—A valuable medicinal herb. J. Med. Plants Stud., 4(3): 06–07.

Verma, S., Gupta, M., Popli, H., Aggarwal, G. 2018. Diabetes mellitus treatment using herbal drugs. Intl. J. Phytomed., 10(1): 1–10.

Vigasini, S., Nagaraja, P., Narayan, P., Nagendra, H.G.R. 2019. Evaluation of pharmacological properties of *Caesalpinia bonducella* seed and shell extract. Pharmacogn. J., 11(1): 150–154.

Vijan, S. 2010. In the clinic type 2 diabetes. Ann. Intern. Med., 152(5): 1–15.

Vivek, K., Rathore, K., Jain, P., Ahmed, Z. 2017. Biological activity of *Bauhinia racemosa* against diabetes and interlinked disorders like obesity and hyperlipidemia. Clin. Phytosci., 3(7): 1–9.

Widhiantara, I.G., Arunngam, P., Siswanto, F.M. 2018. Ethanolic extract of *Caesalpinia bonducella* f. seed ameliorates diabetes phenotype of streptozotocin-nicotinamide-induced type 2 diabetes rat. Biomed. Pharmacol. J., 11(2): 1127–1133.

White, M.F., Kahn, C.R. 1994. The insulin signalling system. J. Biol. Chem., 269: 1–4.

Wild, S., Roglic, G., Green, A., Sicree, R., King, H. 2004. Global prevalence of diabetes estimates for the year 2000 and projections for 2030. Diabetes Care, 27(5): 1047–1053.

Xie, L., Lee, S.Y., Andersen, J.N., Waters, S., Shen, K., Guo, X.L., Moller, N.P., Olefsky, J.M., Lawrence, D.S., Zhang, Z.Y. 2003. Cellular effects of small molecule PTP1B inhibitors on insulin signalling. Biochem., 42: 12792–12804.

Yarizade, A., Kumleh, H.H., Niazi, A. 2017. *In-vitro* anti-diabetic effects of *Ferula asa-foetida* extracts through dipeptidy peptidase IV and α-glycosidase inhibitory activity. Asian J. Pharma. and Clin. Res., 10(5): 357–360.

You, W.P., Henneberg, M. 2016. Type 1 diabetes prevalence increasing globally and regionally: the role of natural selection and life expectancy at birth. BMJ Open Diabetes Res. Care, 4, e000161.

Warrier, P.K. 1995. Indian Medicinal Plants. Orient Longman, p. 168.

Zi, H. 2007. *Terminalia chebula* Retzius, Observ. Bot., 5: 31. 1788. Fl China. 13: 310–314.

Zong, Z.H.Y.T.J. 2010. *Bauhinia racemosa* Lamarck, Encycl., 1: 390. 1785. Fl China. 10: 6–21.

Chapter 16

Croton zehntneri: Essential Oils and Inclusion Complex

Lorenna Fonseca,[1,*] *Mahendra Rai*[2] and *Sidney Gonçalo de Lima*[1,*]

Introduction

Essential oils (EOs) are volatile, aromatic, clear and rarely colored, liposoluble and have a complex chemical composition. They are liquids at room temperature and have been used since antiquity as flavorings, in food preservation, medicine and perfumery. The development of knowledge about essential oils, however, was delayed by the advent of antibiotics in the middle of the 19th century and was recently renewed because of the misuse of antibiotics resulting in the selection of more resistant microorganisms. Thus, it was necessary to look for new alternatives; among them, essential oils have great potential, compared to synthetic antibiotics, as they are natural and environmentally sustainable compounds. In general, they are safe due to less toxicity (Edris 2007, Cook and Lanaras 2016, Zhai et al. 2018, de Matos et al. 2019).

The term "essential oil" was first used in the 16th century by Paracelsus Von Hohenheim, who named the effective component of a drug "Quinta essential". Essential oils are complex mixtures of volatile compounds produced by living organisms and isolated by physical means from an entire or part of a plant of known taxonomic origin. The International Standards Organization (ISO) (ISO/D1S9235.2) defines essential oil as a product obtained from distillation with water or steam or by mechanical processing of citrus peels or by pressing natural materials (Edris 2007, Zhai et al. 2018, Souza et al. 2018).

EOs can be easily isolated from the plant material and are a complex mixture of terpenes and aromatic compounds that compose the characteristic aroma of each plant and its pharmacological properties. Terpenes are organic compounds formed from C_5 isoprene units, being the main constituents of essential oils. Its structure consists of different numbers of isoprene units $(C_5)_n$ and appears as linear or cyclic. They can be C_{10} monoterpenes, C_{15} sesquiterpenes, C_{20} diterpenes, C_{25} sesterterpenes, C_{30} triterpenes, among others. The functionalization of the hydrocarbon skeleton with various functional groups leads to a variety of compounds, such as alcohols, aldehydes or ketones, and are commonly known as terpenoids (Dewick 2009, Ceborska 2018).

EOs can be biosynthesized by all plant organs, that is, buds, flowers, leaves, stems, branches, seeds, fruits, roots, wood or bark and are stored in secretory cells, ducts, epidermal cells or glandular trichomes. The Lamiaceae and Apiaceae families are particularly rich in aromatic plants. Examples include coniferous wood, balm, citrus, coriander, eucalyptus, lemongrass, cloves, cumin, mint, rosemary and thyme, among others (Bakkali et al. 2008, Cook and Lanaras 2016, Zhai et al. 2018, Gil et al. 2019).

[1] Postgraduate Program in Chemistry, Chemistry Department, Natural Sciences Center, Federal University of Piaui, Teresina-PI-64049-550, Brazil.

[2] Department of Biotechnology, SGB Amravati University, Amravati-444602, Maharashtra.

* Corresponding authors: fonseca.lorenna@gmail.com; sidney@ufpi.edu.br

Plants synthesize essential oils as secondary metabolites and with bioactive properties. They play roles in the defense mechanisms of plants against herbivores, fungal, viral and bacterial infections; in attracting insects to favor the dispersion of pollens and seeds, or to repel undesirable ones; in communication and signaling with other plants; and in allopathy, reducing competition from neighboring plant species (Bakkali et al. 2008, Cook and Lanaras 2016, Gil et al. 2019).

It is possible to have several different species producing an essential oil with the same chemical composition, for example, carvacrol in *Origanum vulgare* (Mastro et al. 2017), *Origanum onites* (Ozkan et al. 2010) and *Thymbra capitata* (Palmeira-De-Oliveira et al. 2012). On the other hand, plants of the same species, for example, *Mentha spicata,* can have completely different aromas: mint (menthol and carvone), wild mint (piperitone and piperitenone oxide) and lavender (linalool and linalyl acetate) (Zhao et al. 2013). These are known as chemotypes, that is, plants of the same species having compounds from different biosynthetic pathways (Cook and Lanaras 2016).

It is known that there are compounds with similar biological activities in phylogenetically related plants, which can contribute to the search for new substances with great economic potential (Zamora et al. 2018). However, the chemical composition and content of essential oil in plant material may vary with the season, the bioclimatic zone, growing conditions and stress, as well as due to the conditions of harvesting, processing and storage of plant material, terpenoids and other constituents which are generally volatile, thermosensitive and easily oxidizable and hydrolysable (Cook and Lanaras 2016, de Matos et al. 2019).

The return to use in medicine, cosmetics and the food industry for natural products in everyday life has stimulated interest in different aspects of essential oils. This interest is reflected in the sharp increase in research, with publications on essential oils in the last twenty years focusing mainly not only on the composition and biological properties of essential oil, but also on isolation methodologies, species variation and others. The yield and quality of essential oil can be optimized based on the growth of production technology, which is always looking for more efficient, fast and economical processes to increase quality (composition, color and texture), extraction yield and better storage conditions to extend the shelf life of products (Cook and Lanaras 2016, Gil et al. 2019).

Many extraction methods have been developed to extract essential oils, including microwave by gravity, microwave extraction without solvents and with solvents, extraction with subcritical water, ultrasound extraction, supercritical CO_2 extraction, Soxhlet extraction, among other techniques and traditional extraction methods, such as steam distillation and hydrodistillation. Therefore, the chemical profile of essential oils obtained from the same plant material may differ in the quantity and stereochemistry of the molecules extracted according to the extraction technique (Bakkali et al. 2008, Gao et al. 2017, Carvalho et al. 2018). Other factors that can contribute to the varying quality, quantity and composition are the climate, soil composition, plant organ used for the extraction, age and stage of the vegetative cycle. Thus, to obtain essential oils of constant composition, they must be extracted under the same conditions of the same part of the plant that grows in the same soil, under the same climate and harvested in the same season (Masotti et al. 2003, Angioni et al. 2006, Bakkali et al. 2008).

The complex composition of EOs carries a wide spectrum of possible applications in many fields, for example, in the food, beverage and pharmaceutical industry, in medicine, perfumery, cosmetics, health products, veterinary science, alternative medicine (aromatherapy) and agricultural products. EOs have gained significant interest in the pharmaceutical field (Cook and Lanaras 2016, de Matos et al. 2019). Edris (2007) reviewed the literature on the use of EOs in therapies and listed studies on various pharmacological applications, such as chemotherapy, in the treatment of cardiovascular diseases, as an antidiabetic and many others. EOs have many bioactive compounds that, in addition to their fragrance and flavor, also have a wide range of biological activities and properties. For example, they are antibacterial, antifungal, antioxidant, insecticide, insect repellent, acaricide, larvicide, anthelmintic, anti-inflammatory, cytotoxic, antibiotic, anticarcinogen, analgesic and local anesthetic (Edris 2007, Bakkali et al. 2008).

In this chapter, we have carried out a literature review about the *Croton* genera and the variations in the chemical compositions and biological activities of the essential oils of *Croton zehntneri* according to the geographic regions. Here, we have also highlighted the concept of essential oil and inclusion complex with cyclodextrin.

Inclusion Complex

In order to expand the industrial application of EOs, it is necessary to improve its physical-chemical characteristics such as aqueous solubility and thermal and oxidative instability, which can lead to the degradation and loss of its bioactive compounds by volatilization. The formation of an inclusion complex (IC) with cyclodextrins (CDs) is a way to control degradation and volatilization and has the advantage of increasing bioefficiency due to the ability to enhance absorption through membranes and biological barriers. Due to these characteristics, cyclodextrins are of great interest in pharmaceutical chemistry, agriculture, cosmetics, food, medication administration and industries (Barman et al. 2018, Rocha et al. 2018, de Matos et al. 2019).

Isolated substances with low aqueous solubility generally have less *in vivo* biological activity than those with higher solubility, as well as low availability and incomplete release of active molecules. Thus, compounds that are poorly soluble in water have, in general, low absorption, making bioavailability difficult, often limiting their application. Among the techniques used to increase aqueous solubility and oral bioavailability, the use of inclusion complexes with CDs stands out. Many advantages of complexing with cyclodextrins have been reported in the literature, including increased solubility and bioavailability, improved stability, masking flavor and odor, reducing volatility and increasing thermal stability, in addition to enabling the preparation of a controlled release system (Barman et al. 2018, Hadian et al. 2018, Cavalcante et al. 2019).

During the production of foods containing volatile and aromatic compounds, evaporation and exposure to oxygen often alter their organoleptic characteristics. Thus, the complexation of volatile or unstable food additives using CDs has been shown to prolong the shelf life of the product, improving the stability of these additives which become more resistant to food processing steps, presenting the potential to retain aromatic materials in multiple components food systems (Hadian et al. 2018). In addition, inclusion complexes can increase stability and protect complexed compounds from external conditions, such as heat, pH or light and extend the duration of their effects and are commonly used for chiral discrimination of compounds (Barman et al. 2018, Ceborska 2018).

CDs are also known as cycloamidos, cyclomaltoses and *Schardinger* dextrins and are obtained from the enzymatic degradation of starch (polymer made up of glucose units) from the intramolecular transglycosylation reaction by the action of the cyclodextrin-glycosyl transferase (CGTase) enzyme, produced by some microorganisms. The preparation of CDs is widely described in multiple research works and in several patents (Del Valle 2004, Cavalcante et al. 2019).

CGTases enzymes promote the hydrolysis of certain glycosidic bonds in starch, resulting in mixtures of cyclodextrins and cyclization of oligosaccharide fragments. Among the microorganisms that produce these enzymes include *Bacillus macerans*, *B. megaterium*, *B. megatterium*, *B. circulans*, *B. subtilis B. stearothermophilus*, *B. alcalofilico*, *Micrococcus varians*, *M. luteus* and *Klebsiella pneumoniae*. Moreover, the relative amounts of each type of CD depend on the type of microorganism that produces the enzyme and the reaction conditions (Veiga et al. 2006).

CDs are cyclic oligosaccharides, containing D-glucopyranoside (glucose) units that are linked by α-1,4 glycosidic bonds; the most popular are α-CD, β-CD and γ-CD, each with 6, 7 and 8 glucose units, respectively. All are crystalline, non-hygroscopic and industrially produced. CDs containing less than 6 glucose units do not exist for stoichiometric reasons. On the other hand, CDs composed of more than 8 glucose units have been reported; however, these are produced with extremely low yield and their complexing properties are weak (Veiga et al. 2006, Barman et al. 2018, Ceborska 2018).

Due to the lack of free rotation on the α-1,4 glycosidic bonds and the shape of the chair presented by the glucose units, cyclodextrins are not perfectly cylindrical molecules, but toroidal or conical, with a hydrophobic cavity and a surface hydrophilic external layer, which allows CDs to form inclusion complexes (CI) with guest molecules of different sizes and low polarity. The hydrophilic character of the outer surface is due to the presence of primary and secondary hydroxyls facing the outer surface. The hydrophobic character of the internal cavity of CDs is formed by oxygen atoms that are involved in glycosidic bonds between glucose units (Barman et al. 2018, Rocha et al. 2018).

Based on this structure, the primary hydroxyl groups are located on the narrow side of the cone shape, while the secondary hydroxyl groups are located on the widest edge. The hydrophobic cavity of cyclodextrin is capable of trapping compounds with low polarity, forming stable ICs through interactions, such as hydrogen bonds and Van der Waals forces. As a result, CDs can develop a complex with hydrophobic and amphiphilic compounds that have compatible dimensions, changing the physicochemical properties of these compounds, such as aqueous solubility, stability and bioavailability (Barman et al. 2018, Rocha et al. 2018, Cavalcante et al. 2019).

Among natural CDs, β-CD is the most used, its cavity is suitable for guests with a molecular mass between 100 and 400 g mol^{-1}, the molecular mass range of most molecules of interest currently. In addition, the acquisition cost of β-CD is lower than other CDs and its greater commercial availability leads to its frequent use. Hadian et al. (2018) reported the use of β-CD for the inclusion of geraniol due to its low biotoxicity and high biocompatibility, solubility and bioavailability of poorly soluble compounds in the oral release of the drug. Currently, it is the most common cyclodextrin in pharmaceutical formulations and, therefore, probably the best studied in humans (Del Valle 2004, Hadian et al. 2018, Cavalcante et al. 2019).

The intramolecular hydrogen bonds established between the C_2-OH groups of a glucose unit and the C_3-OH groups of an adjacent glucose unit stabilize the larger ring giving rise to a rigid structure, which in turn is the cause of the reduced aqueous solubility; therefore, β-CD is the CD most likely to precipitate in the form of crystals and is the most commonly used because of its high efficiency of complexation, cavity with adequate dimensions and low cost (Veiga et al. 2006, Zhu et al. 2007, Li et al. 2015, De Souza and Morgon 2016).

CDs are also known as typical host molecules and complexed molecules are called guest molecules. This association results in guest-host complexes, also called the inclusion complex (IC). Inclusion complexes are, therefore, structures composed of two or more molecules, in which the host molecule includes all or part of the guest molecule, but there is no formation or breaking of covalent bonds, only interaction between the molecules; in addition, guest molecules can dissociate from the CD establishing a dynamic balance between free and dissociated molecules (Del Valle 2004, Veiga et al. 2006).

Complexation is a dimensional adjustment between the host cavity and the guest molecule. In aqueous solution, the main driving force for the formation of IC is the repulsion forces between the internal cavity of the host molecule and the aqueous solution, and the guest molecule tends to lodge itself in an energetically favorable way inside the CD cavity. In addition, the guest molecule must have polarity, size and compatible shape in order to obtain an apolar-apolar interaction and decrease the tension of the CD ring, resulting in a more stable and lower energy state, according to Figure 16.1 (Del Valle 2004, Veiga et al. 2006, Cavalcante et al. 2019).

This process is energetically favorable and contributes to increase the stability of the complex, since it promotes a favorable change in enthalpy, an increase in input and a reduction in the total energy of the system. Other forces also contribute to the formation and stabilization of complexes, such as Van der Waals interactions (dipole-dipole connections and London dispersion forces), hydrogen connections (between molecules in use and hydroxyl groups of CDs), hydrophobic interactions and aesthetic effects (Del Valle 2004, Veiga et al. 2006).

The minimum condition required for the formation of inclusion complexes is the geometric compatibility between the host and host molecules, that is, the guest molecule must penetrate and adjust totally or partially in the hydrophobic cavity of the CD. Not only stereochemical factors related to the

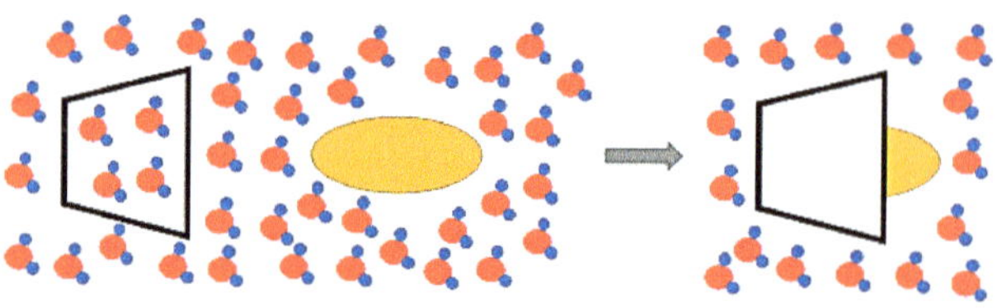

Figure 16.1. Representation of the formation of the inclusion complex (guest/host) in aqueous solution.

shape and size of the host molecules, but also the polarity of these molecules, condition the inclusion in the cavity of the CDs (Veiga et al. 2006).

The choice of the preparation method for the inclusion complex can affect the physico-chemical properties and physiological performance of drug delivery systems (Adeoye et al. 2018). Many methods are described in the literature for the preparation of solid inclusion complexes (compound/CD) and the choice of a method depends on the properties of the guest molecule, such as its availability and cost. Some examples are co-evaporation, co-precipitation, spray drying, lyophilization, grinding, and supercritical fluid technology (Da Cunha Filho and Sá-Barreto 2007, Jantarat et al. 2014, Abarca et al. 2015, Hadian et al. 2018). Many of these processes require very high energy support, organic solvents or both (Rudrangi et al. 2015).

In many cases, the techniques for preparing organic nanostructured systems contain one or more steps, such as heating, solvent evaporation and high-pressure homogenization, in which the guest molecules are subjected to conditions that can cause degradation and loss of active content through volatilization (de Matos et al. 2019).

In this context, CDs are one of the molecules most used for inclusion, protection against oxidation or other processes. In addition, CD is an easy, safe and low-cost alternative capable of improving bioavailability and stability, and controlled release of active compounds (Wenz 2012, Zhang and Ma 2013). Therefore, the development of alternatives in order to improve the physical-chemical properties of the compounds present in essential oils can be considered an important field to produce new products.

Croton genus

Euphorbiaceae is one of the most diverse and complex families of Angiosperms with approximately 6300 species and 245 genera dispersed mainly in tropical regions of the globe. In Brazil, it is among the 10 most representative Angiosperm families, comprising 63 genera and 945 species in all biomes, of which 629 species and 11 genera are endemic. Among the most diverse genres, *Croton*, *Manihot*, *Dalechampia* and *Acalypha* can be highlighted (Wurdack and Davis 2009, Orlandini and Lima 2014, Nascimento et al. 2017).

Croton L. is the second largest genus of the Euphorbiaceae family and the most studied in Brazil and abroad, with about 1200 species, and has a pantropical distribution, with approximately 712 species on the American continent, and in Brazil, 350 are species widely distributed in the more diverse environments, with emphasis on the Cerrado, Caatinga and Campos Rupestres (Craveiro et al. 1978, Sodré and da Silva 2015, Alves et al. 2019). In the Northeast region of Brazil, *Croton* species are popularly known as canopies or quince trees. They are characterized by the presence of latex in their stems, produce essential oils with pleasant odors and are used for the most diverse purposes, including for medicinal use (Craveiro et al. 1978, Oliveira-Tintino et al. 2018).

This genus has a wide variety of biological activities, including anti-inflammatory, antinoceptive, gastroprotective and healing properties. Many species of *Croton* are used in folk medicine for several purposes; *Croton campestris*, for example, is used for general pain, influenza, constipation, bronchitis, pneumonia, asthma, inflammation, among many others. The *Croton nepetaefolius* is widely used as a sedative and antispasmodic agent, and is also used against stomach and intestinal pain (Matias et al. 2010, da Silva and Sales 2014, Soares et al. 2017, Oliveira-Tintino et al. 2018).

Various biological activities, including anti-inflammatory—*Croton campestris* (Oliveira-Tintino et al. 2018), antifungal—*Croton tricolor* (Miranda et al. 2019), larvicide—*Croton nepetaefolius* (Soares et al. 2017), vasorrelaxant activity—*Croton argyrophylloides* (de França-Neto et al. 2012) and gastroprotective—*Croton zehntneri* (Coelho-de-Souza et al. 2013), have been reported for the essential oil of several species of this genus.

Croton zehntneri Pax et Hoffm

It is a plant native to the Northeast of Brazil, popularly known as "canela de cunhã", "canelinha", "canela-brava". It is a deciduous plant; during the dry period (August to October), it loses all foliage and in the rainy period (January to May), it acquires the leaves again (Da Silva-Alves et al. 2015, Coelho-de-Souza et al. 2019a). The leaves and aerial parts of *C. zehntneri* are rich in essential oils and have a strong and pleasant odor reminiscent of anise and cloves (Craveiro et al. 1978, Albuquerque et al. 1995).

Due to the great medicinal and economic importance of plants and their wide distribution in the regions of the country, the Brazilian government produced the National List of Medicinal Plants of Interest to the Unified Health System (RENISUS), a list of plant species already used in traditional medicine that have the potential to generate products that may be relevant to the Unified Health System of Brazil (SUS). The *C. zehntneri* was included in this list due to its pharmacological properties and its possible use in the development of new pharmaceutical products, including medicines (phytotherapy) and adjuvants (Ministério da Saúde 2016).

Here, we have carried out a literature review of the variations in the chemical compositions and biological activities of the essential oils of *Croton zehntneri* according to the geographic regions (Table 16.1).

Table 16.1. Studies of *C. zehntneri* essential oils obtained from different geographic regions.

Main constituents	Parts used	Study	Plant collection	References
estragole (61%)	leaves	Identification of the chemical composition of essential oil of *Croton zehntneri*	Araripe, Ceará, Brazil	(Craveiro et al. 1978)
estragole (56%) methyl-eugenol (34%)	aerial parts			
methyl-eugenol (34%) methyl-isoeugenol (26%)	stalk			
estragole (52%) anethole (32%)	leaves	Identification of the chemical composition of the leaves' essential oil collected from different locations	Tianguá, Ceará, Brazil/1979	(Craveiro et al. 1981)
eugenol (47%) geraniol (23%)			Quixadá, Ceará, Brazil/1977	
-	aerial parts	Depressive effects on the nervous system of rats	Oeiras, Piauí, Brazil	(Batatinha et al. 1995)
estragole (59%) anethole (27%)	leaves	Relaxing effects on the skeletal muscles of rats	-	(Albuquerque et al. 1995)
estragole (57%) anethole (28%)	aerial parts	Effects on intestinal smooth muscle *in vitro*	Tianguá, Ceará, Brazil	(Coelho-de-Souza et al. 1997)
estragole (57%) anethole (28%)	leaves	Specific modulatory effects on isolated smooth muscle contractility	-	(Coelho-de-Souza et al. 1998)
methyl-eugenol (24%) estragole (10%)	aerial parts	Study of central nervous system disorders to assess anxiety and depression	Oeiras, Piauí, Brazil	(Lazarini et al. 2000)
anethole (85%) estragole (5%)	leaves	Evaluation of antinociceptive activity	Viçosa, Minas Gerais, Brazil/1998	(Oliveira et al. 2001)
anethole (85 to 88%)	leaves	Kinetic and solubility study of extracts obtained using pressurized CO_2	Serra do Ibiapaba, Ceará, Brazil	(Sousa et al. 2005)
anethole (31 to 47%) anisaldehyde (19 to 27%)	aerial parts	Antioxidant activity of *Croton* species collected at two different times	Viçosa, Ceará, Brazil/2002	(De Morais et al. 2006)

Table 16.1 Contd. ...

...*Table 16.1 Contd.*

Main constituents	Parts used	Study	Plant collection	References
trans-anethole (94%)	leaves and stalk	Larvicidal activity against *Aedes aegypti*	Viçosa, Ceará, Brazil	(De Lima et al. 2006)
estragole (46%) anethole (42%)	aerial parts	Study of the role of sensory nerves sensitive to capsaicin in mediating cardiovascular effects in anesthetized rats	Viçosa, Ceará, Brazil/2003	(De Siqueira et al. 2006)
anethole (74%) eucalyptol (4%)	leaves	Larvicidal activity against *Aedes aegypti*	Tianguá, Ceará, Brazil/2004	(Santos et al. 2007)
anethole (36%) *p*-anisaldehyde (16%)	stalk			
anethole (90%)	inflorescences			
estragole (72%) anethole (14%)	leaves	Antifungal activity and toxicity assessment	Viçosa, Ceará, Brazil	(Fontenelle et al. 2008)
estragole (77%) eucalyptol (7%)	leaves	Antibacterial activity and toxicity assessment against *Artemia salina*	Salitre, Ceará, Brazil/2006	(Da Costa et al. 2008)
anethole (56%) estragole (28%)	leaves	Effects of essential oil on hepatotoxicity induced by paracetamol in rats	Viçosa, Ceará, Brazil/2004	(Lima et al. 2008)
trans-anethole (94%)	leaves	Antibacterial activity and antibiotic modulation	Crato, Ceará, Brazil/2008	(Rodrigues et al. 2009)
anethole (89%), eugenol (84%), estragole (90%)	leaves	Evaluation of the cholinesterase inhibition effect of three *Croton zehntneri* chemotypes	Ubajara e Croatá da Serra, Ceará, Brazil/2008	(Santos et al. 2010)
anethole (70%) eugenol (49%) z-methyl-isoeugenol (53%)	stalks			
anethole (34%) z-methyl-isoeugenol (81%) e-methyl-isoeugenol (91%)	stalks			
estragole (77%) eucalyptol (7%)	leaves	Study of modulation in antibiotic activity	Crato, Ceará, Brazil/2008	(Coutinho et al. 2010)
anethole (47%)	leaves	Larvicidal activity of alginate spheres coated with cashew gum doped with essential oil of *Croton zehntneri*	Fortaleza, Ceará, Brazil	(De Paula et al. 2010)
estragole (77%) eucalyptol (7%)	aerial parts	Evaluation of cytotoxicity and antimalarial activity of EOs	Salitre, Ceará, Brazil/2007	(Mota et al. 2012)
trans-anethole (85%) estragole (4,8%)	leaves	Evaluation of OE activity in the inflammatory and remodeling phases of the wound healing process	Viçosa, Ceará, Brazil/1998	(Malveira Cavalcanti et al. 2012)

Table 16.1 Contd. ...

...Table 16.1 Contd.

Main constituents	Parts used	Study	Plant collection	References
estragole (46%) anethole (42%)	aerial parts	Investigation of cardiovascular effects in hypertensive rats	Viçosa do Ceará, Ceará, Brazil/2003	(Bezerra de Siqueira et al. 2013)
trans-anethole (86%) estragole (5%)	leaves	Investigation of gastroprotective action in rats with gastric lesion	Viçosa, Ceará, Brazil/1998	(Coelho-de-Souza et al. 2013)
trans-anethole (91%)	aerial parts	Study of insecticidal activity against various stages of life of the *Aedes aegypti*	Viçosa, Ceará, Brazil	(De Lima et al. 2013)
estragole (57%) anethole (28%)	leaves	Evaluation of the relaxing effects on the cavernous body of rats	-	(Cabral et al. 2014)
estragole (99%)	leaves	Preparation and characterization of the inclusion complex of essential oil with β-cyclodextrin	Simões, Piauí, Brazil/2011	(Aguiar et al. 2014)
estragole (72%) *trans*-anethole (23%)	stalks	Antioxidant and antimicrobial activity	Goiás, Bahia, Brazil/2011	(Donati et al. 2014)
estragole (93%)	leaves			
anethole (86%) estragole (5%)	aerial parts	Study of the effects on nervous excitability and sciatic nerve in rats	Viçosa do Ceará, Ceará, Brazil/1988	(Da Silva-Alves et al. 2015)
estragole (85%) espatulenol (6%)	leaves and stalks	Antimicrobial activity (bacteria and fungi) and cytotoxicity against tumor cell lines	Simões, Piauí, Brazil/2011	(Andrade et al. 2015)
anethole (86%) estragole (5%)	leaves	Antioxidant activity to reverse the damage to the respiratory system caused by asthma	Viçosa, Ceará, Brazil	(Serra et al. 2018)
trans-anethole (86%) estragole (5%)	leaves	Study of toxicity after oral treatment in rats	Viçosa do Ceará, Ceará, Brazil	(Coelho-de-Souza et al. 2019)
estragole	leaves	Inclusion complex of EO and β-CD prepared by co-precipitation and spray drying	Simões, Piauí, Brazil	(Fonseca et al. 2019)

The first reports in the literature of use in popular culture indicate the infusion of leaves or extracts of "canelinha" as a drink instead of coffee, in perfumes and as sweeteners in food and drinks. In medicine, infusions are used as a remedy for stomach pain and insomnia (Craveiro et al. 1978), to treat "nervous disorders" such as irritability, anxiety and seizures (Bernardi et al. 1991), sedative effect and to relieve gastrointestinal function disorders (Albuquerque et al. 1995).

The published research on essential oil of *C. zehntneri* until the end of the 1990s concentrated on the identification of the chemical composition (Craveiro et al. 1978), and influences on the central nervous system (Batatinha et al. 1995) and the muscular system (Albuquerque et al. 1995, Coelho-de-Souza et al. 1997, 1998).

From the year 2000, publications related to the essential oil of *C. zehntneri* aimed at biological activities, among them, on the central nervous system (Lazarini et al. 2000), antinociceptive activity (Oliveira et al. 2001), antioxidant activity (De Morais et al. 2006, Donati et al. 2014, Serra et al. 2018), larvicidal activity against *Aedes aegypti* (De Lima et al. 2006, Santos et al. 2007), antimicrobial activity

(Da Costa et al. 2008, Fontenelle et al. 2008, Rodrigues et al. 2009, Coutinho et al. 2010, Andrade et al. 2015) and antimalarial activity (Mota et al. 2012).

Aguiar et al. (2014) reported for the first time the preparation and characterization of the inclusion complex of essential oil of *C. zehntneri*. The essential oil (which was collected in Simões, Piauí) showed presence of estragole as the major compound (99%), and was complexed with β-cyclodextrin, by the co-precipitation method. The complex was characterized by infrared absorption spectroscopy (IR) and differential scanning calorimetry (DSC) (Aguiar et al. 2014).

Fonseca et al. (2019) prepared the inclusion complex of *Croton zehntneri* essential oil with β-cyclodextrin by two methods, spray drying and co-precipitation, and was characterized by vibrational spectroscopy (Raman spectroscopy and FT-IR); the complexation efficiency was determined by gas chromatography coupled to mass spectrometry (GC-MS). It was observed that co-precipitation allowed greater complexation efficiencies when compared to the spray drying (Fonseca et al. 2019).

From the data in Table 16.1, it is possible to observe that the species has aromas that vary between specimens collected in different regions of Northeast Brazil, mainly in the State of Ceará. Such variations in the chemical composition depend on the region of cultivation, method of extraction, analysis, and the part of the plant used to obtain essential oils. Thus, there are four chemical types for this species: anethole—for specimens collected in Tianguá (CE) and Viçosa (CE); eugenol—for those collected in Quixadá (CE); methyl-eugenol—for those collected in Oeiras (PI); estragole—Simões (PI) and Salitre (CE) (Aguiar et al. 2014). The chemical structures of these compounds are shown in Figure 16.2. In addition to the four compounds characteristic of the essential oils of *C. zehntneri*, it is also possible to identify other major compounds such as geraniol, anisaldehyde, eucalyptol and spathulenol.

Estragole, also known as methyl chavicol (Figure 16.2), is a phenylpropanoid and natural constituent of several essential oils from aromatic plants such as basil (*Ocimum basilicum* L., Lamiaceae), tarragon (*Artemisia dracunculus* L., *Asteraceae*) and fennel (*Pimpinella anisum* L., Apiaceae) (Lewinsohn et al. 2000, Zeller et al. 2009). In addition, estragole has been used as a flavoring agent in non-alcoholic beverages, as a condiment and in medicinal products, and has insecticidal, antiviral, antimicrobial, acaricidal, muscle relaxant and anticonvulsant properties (Bagamboula et al. 2004, Chang et al. 2009, Zeller et al. 2009, Obolskiy et al. 2011, Ponte et al. 2012, Cabral et al. 2014, Diao et al. 2014, Rodrigues et al. 2016).

Anethole (Figure 16.2) is found in spices and herbs, such as anise, fennel, lemongrass, coriander, etc., and is the main component of the essential oils of *Illicium verum* and *Foeniculum vulgare* (Domiciano et al. 2013, Diao et al. 2014). Studies show that anethole has antimicrobial, anti-inflammatory and antioxidant properties (Kubo et al. 2008, Huang et al. 2010, Sá et al. 2017), and is thus widely used as a flavoring agent in sweets and alcoholic beverages; in addition, it is capable of inhibiting the growth of pathogens, and can be used as a natural preservative (Zafeiropoulou et al. 2010, Kfoury et al. 2014, 2016a, b, Gharib et al. 2017).

Eugenol, as shown in Figure 16.2, is a phenylpropanoid derivative known for its characteristic flavor and aroma and for its antimicrobial and antioxidant properties, and is used as an agent of food flavor and fragrance. In addition, eugenol is a food additive approved by the U.S. Food and Drug Administration, with an LD_{50} of 3000 mg kg^{-1} for oral rats (Gong et al. 2016, Piletti et al. 2017). It is present in the essential oil of several plants, including *Ocimum tenuiflorum* (Raina et al. 2013), *Ocimum basilicum*

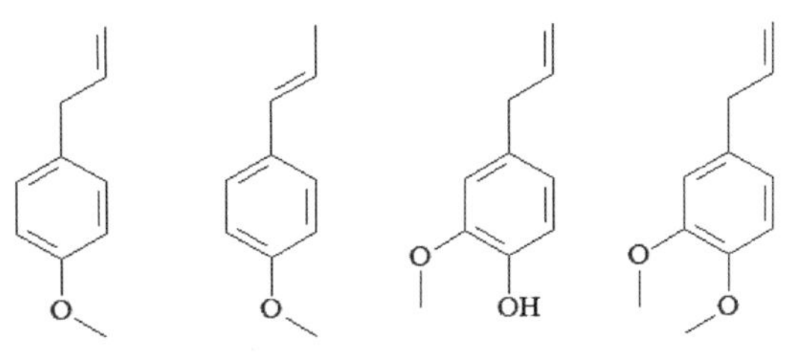

estragole anethole eugenol methyl-eugenol

Figure 16.2. Chemical structures of the main chemical compounds of *Croton zehntneri* essential oils.

(Bhuvaneshwari et al. 2016) and *Syzygium aromaticum* (Sebaaly et al. 2015); it can also be found in nutmeg, cinnamon, among others (Kim et al. 1997).

Methyl-Eugenol (Figure 16.2) is a phenylpropanoid derived from eugenol. The conversion of eugenol to methyl-eugenol is carried out by the enzyme eugenol-O-methyltransferase. The level of activity of this type of enzyme in different plant chemotypes is one of the main factors that affect the distinct composition of phenylpropanoids in EOs. It is present in more than 450 species of plants; among them, it is present in the essential oil of *Ocimum basilicum* L. (Fischer et al. 2011), *Melaleuca alternifolia* and *Melaleuca linariifolia* (Shepherd et al. 2017). Methyl-eugenol has the property of attracting wild male flies of the *Bactrocera* and *Dacus* genera and since the 1950s, it has been used for the monitoring and control of species, attracting and trapping male flies (Reyes-Hernández et al. 2019).

Conclusion

Essential oils (EOs) are volatile, aromatic, clear and rarely colored, liposoluble and have a complex chemical composition. The complex composition of EOs carries a wide spectrum of possible applications in many fields, for example, in the food, beverage and pharmaceutical industry, in medicine, perfumery, cosmetics, health products, veterinary science, alternative medicine (aromatherapy) and produces agricultural products. In order to expand the industrial application of EOs, it is necessary to improve its physical-chemical characteristics such as aqueous solubility and thermal and oxidative instability, which can lead to the degradation and loss of its bioactive compounds by volatilization. The formation of an inclusion complex with cyclodextrins is a way to control degradation and volatilization and has the advantage of increasing bioefficiency. The *Croton zehntneri* leaves have a rich essential oil content and a strong and pleasant odor reminiscent of anise and cloves. After reviewing the variations in chemical compositions of the essential oils of *Croton zehntneri*, according to the geographical region of plant collection, four main compounds were observed: estragole, anethole, eugenol and methyl-eugenol. And the essential oil has many activities, such as antinociceptive activity, antioxidant activity, larvicidal activity against *Aedes aegypti*, antimicrobial activity and antimalarial activity. The essential oil (collected in Simões, Piauí, Brazil) presented estragole as the major compound and was complexed with β-cyclodextrin, by different methods, was characterized by vibrational spectroscopy and the complexation efficiency was determined by gas chromatography coupled to mass spectrometry.

References

Abarca, R.L., Rodríguez, F.J., Guarda, A., Galotto, M.J., Bruna, J.E. 2015. Characterization of beta-cyclodextrin inclusion complexes containing an essential oil component. Food Chem., 196: 968–975.

Adeoye, O., Costa, C., Casimiro, T., Aguiar-ricardo, A., Cabral-marques, H. 2018. Preparation of ibuprofen/hydroxypropyl-γ-cyclodextrin inclusion complexes using supercritical CO_2-assisted spray drying. Supercrit. Fluids, 133: 479–485.

Aguiar, U.N., De Limaa, S.G., Rochaa, M.S., De Freitasa, R.M., Oliveiraa, T.M., Silvab, R.M., Moura, L.C.B., De Almeidab, L.T.G. 2014. Preparação e caracterização do complexo de inclusão do óleo essencial de *Croton zehntneri* com b-ciclodextrina. Quim. Nova, 37: 50–55.

Albuquerque, A.A.C., Sorenson, A.L., Leal-Cardoso, J. 1995. Effects of essential oil of *Croton zehntneri*, and of anethole and estragole on skeletal muscles. J. Ethnopharmacol., 49: 41–49.

Alves, C.S.C., Oliveira, A.P., Oliveira-Junior, R.G., Santos, R.F., Santos, A.D.C., Tavares, G.F., Almeida, J.R.G.S. 2019. Caracterização Físico-Química de Complexos de Inclusão Contendo os Monoterpenos Cânfora e 1,8-Cineol, Constituintes Majoritários dos Óleos Essenciais de *Croton conduplicatus*, em β-Ciclodextrina. Rev. Virtual Química, 11: 401–410.

Andrade, T.C.B., Lima, S.G.D., Freitas, R.M., Rocha, M.S., Islam, T., Silva, T.G.D., Militão, G.C.G. 2015. Isolation, characterization and evaluation of antimicrobial and cytotoxic activity of estragole, obtained from the essential oil of *Croton zehntneri* (euphorbiaceae). An. Acad. Bras. Cienc., 87: 173–182.

Angioni, A., Barra, A., Coroneo, V., Dessi, S., Cabras, P. 2006. Chemical composition, seasonal variability, and antifungal activity of *Lavandula stoechas* L. ssp. stoechas essential oils from stem/leaves and flowers. J. Agric. Food Chem., 54: 4364–4370.

Bagamboula, C.F., Uyttendaele, M., Debevere, J. 2004. Inhibitory effect of thyme and basil essential oils, carvacrol, thymol, estragol, linalool and p-cymene towards *Shigella sonnei* and *S. flexneri*. Food Microbiol., 21: 33–42.

Bakkali, F., Averbeck, S., Averbeck, D., Idaomar, M. 2008. Biological effects of essential oils—A review. Food Chem. Toxicol., 46: 446–475.

Barman, S., Barman, B.K., Roy, M.N. 2018. Preparation, characterization and binding behaviors of host-guest inclusion complexes of metoclopramide hydrochloride with a- and b-cyclodextrin molecules. J. Mol. Struct., 1155: 503–512.

Başer, K.., Buchbauer, G. 2010. Handbook of Essential Oils: Science, Technology, and Applications. Taylor & Francis Group.

Batatinha, M.J.M., de Souza-Spinosa, H., Bernardi, M.M. 1995. *Croton zehntneri*: possible central nervous system effects of the essential oil in rodents. J. Ethnopharmacol., 45: 53–57.

Bernardi, M.M., De Souza-Spinosa, H., Batatinha, M.J.M., Giorgi, R. 1991. *Croton zehntneri*: possible central nervous system effects in rodents. J. Ethnopharmacol., 33: 285–287.

Bezerra de Siqueira, R.J., Duarte, G.P., Magalhães, P.J.C., Lahlou, S. 2013. Cardiovascular effects of the essential oil of *Croton zehntneri* leaves in DOCA-salt hypertensive, conscious rats. Nat. Prod. Commun., 8.

Bhuvaneshwari, K., Gokulanathan, A., Jayanthi, M., Govindasamy, V., Milella, L., Lee, S., Yang, D.C., Girija, S. 2016. Can *Ocimum basilicum* L. and *Ocimum tenuiflorum* L. *in vitro* culture be a potential source of secondary metabolites? Food Chem., 194: 55–60.

Cabral, P.H.B., De Morais Campos, R., Fonteles, M.C., Santos, C.F., Leal Cardoso, J.H., Nascimento, N.R.F. Do, 2014. Effects of the essential oil of *Croton zehntneri* and its major components, anethole and estragole, on the rat corpora cavernosa. Life Sci., 112: 74–81.

Cal, K. 2006. Skin penetration of terpenes from essential oils and topical vehicles. Planta Med., 72: 311–316.

Carvalho, M.T., Bermamasco, R., Gomes, R.G. 2018. Métodos de extração de compostos bioativos: aproveitamento de subprodutos na agroindústria. Rev. UNINGÁ Rev., 33: 66–84.

Cavalcante, A., do N., Feitosa, C.M., da Silva Santos, F.P., de Sousa, A.P.R., dos Santos Sousa, R., de Souza, A.A., Pinto, B.F., Araújo, C.M., Rashed, K. 2019. Elaboration and characterization of the inclusion complex between β-cyclodextrin and the anticholinesterase 2-oleyl-1,3-dipalmitoyl-glycerol extracted from the seeds of Platonia insignis MART. J. Mol. Struct., 1177: 286–301.

Ceborska, M. 2018. Structural investigation of solid state host/guest complexes of native cyclodextrins with monoterpenes and their simple derivatives. J. Mol. Struct., 1165: 62–70.

Chang, C.L., Cho, I.K., Li, Q.X. 2009. Insecticidal activity of basil oil, trans-anethole, estragole, and linalool to adult fruit flies of *Ceratitis capitata, Bactrocera dorsalis*, and *Bactrocera cucurbitae*. J. Econ. Entomol., 102: 203–209.

Coelho-de-Souza, A.N., Barata, E.L., Magalhães, P.J.C., Lima, C.C., Leal-Cardoso, J.H. 1997. Effects of the essential oil of *Croton zehntneri*, and its constituent estragole on intestinal smooth muscle. Phyther. Res., 11: 299–304.

Coelho-de-Souza, A.N., Criddle, D.N., Leal-Cardoso, J.H. 1998. Selective modulatory effects of the essential oil of *Croton zehntneri* on isolated smooth muscle preparations of the guinea-pig. Phyther. Res., 12: 189–194.

Coelho-de-Souza, A.N., Lahlou, S., Barreto, J.E.F., Yum, M.E.M., Oliveira, A.C., Oliveira, H.D., Celedônio, N.R., Feitosa, R.G.F., Duarte, G.P., Santos, C.F., De Albuquerque, A.A.C., Leal-Cardoso, J.H. 2013. Essential oil of *Croton zehntneri* and its major constituent anethole display gastroprotective effect by increasing the surface mucous layer. Fundam. Clin. Pharmacol., 27: 288–298.

Coelho-de-Souza, A.N., Rocha, M.V.A.P., Oliveira, K.A., Vasconcelos, Y.A.G., Santos, E.C., Silva-Alves, K.S., Diniz, L.R.L., Ferreira-da-Silva, F.W., Oliveira, A.C., Ponte, E.L., Evangelista, J.S.A.M., Assreuy, A.M.S., Leal-Cardoso, J.H. 2019. Volatile oil of *Croton zehntneri* per oral sub-acute treatment offers small toxicity: perspective of therapeutic use. Brazilian J. Pharmacogn., 29: 228–233.

Cook, C.M., Lanaras, T. 2016. Essential Oils: Isolation, Production and Uses. 1st ed, Encyclopedia of Food and Health. Elsevier Ltd.

Coutinho, H.D.M., Matias, E.F.F., Santos, K.K.A., Tintino, S.R., Souza, C.E.S., Guedes, G.M.M., Santos, F.A.D., Costa, J.G.M., Falcão-Silva, V.S., Siqueira-Júnior, J.P. 2010. Enhancement of the norfloxacin antibiotic activity by gaseous contact with the essential oil of *Croton zehntneri*. J. Young Pharm., 2: 362–364.

Craveiro, A.A., Andrade, C.H.S., Matos, F.J.A., Alencar, J.W. de. 1978. Anise-like flavor of *Croton zehntneri* Pax et Hoffm. J. Agric. Food Chem., 26: 772–773.

Craveiro, A.A., Rodrigues, A.S., Andrade, C.H.S., Matos, F.J.A., Alencar, J.W., Machado, M.I.L. 1981. Volatile constituents of brazilian euphorbiaceae. genus *Croton*. J. Nat. Prod., 44: 602–608.

Da Costa, J.G.M., Rodrigues, F.F.G., Angélico, E.C., Pereira, C.K.B., De Souza, E.O., Caldas, G.F.R., Silva, M.R., Santos, N.K.A., Mota, M.L., Dos Santos, P.F. 2008. Composição química e avaliação da atividade antibacteriana e toxicidade do óleo essencial de *Croton zehntneri* (variedade estragol). Brazilian J. Pharmacogn., 18: 583–586.

Da Cunha Filho, M.S.S., Sá-Barreto, L.C.L. 2007. Utilização de ciclodextrinas na formação de complexos de inclusão de interesse farmacêutico. Rev. Ciencias Farm. Basica e Apl., 28: 1–9.

Da Silva-Alves, K.S., Ferreira-Da-Silva, F.W., Coelho-De-Souza, A.N., Albuquerque, A.A.C., Do Vale, O.C., Leal-Cardoso, J.H. 2015. Essential oil of *Croton zehntneri* and its main constituent anethole block excitability of rat peripheral nerve. Planta Med., 81: 292–297.

de França-Neto, A., Cardoso-Teixeira, A.C., Medeiros, T.C., do Socorro Quinto-Farias, M., de Souza Sampaio, C.M., Coelho-de-Souza, A.N., Lahlou, S., Leal-Cardoso, J.H. 2012. Essential oil of *Croton Argyrophylloides*: Toxicological aspects and vasorelaxant activity in rats. Nat. Prod. Commun., 7: 1934578X1200701.

De Lima, G.P.G., De Souza, T.M., De Paula Freire, G., Farias, D.F., Cunha, A.P., Ricardo, N.M.P.S., De Morais, S.M., Carvalho, A.F.U. 2013. Further insecticidal activities of essential oils from *Lippia sidoides* and *Croton* species against *Aedes aegypti* L. Parasitol. Res., 112: 1953–1958.

De Lima, M.G.A., Maia, I.C.C., De Sousa, B.D., De Morais, S.M., Freitas, S.M. 2006. Effect of stalk and leaf extracts from euphorbiaceae species on *Aedes aegypti* (Diptera, Culicidae) larvae. Rev. Inst. Med. Trop. Sao Paulo, 48: 211–214.

de Matos, S.P., Lucca, L.G., Koester, L.S. 2019. Essential oils in nanostructured systems: Challenges in preparation and analytical methods. Talanta, 195: 204–214.

De Morais, S.M., Catunda, F.E.A., Da Silva, A.R.A., Neto, J.S.M., Rondina, D., Leal Cardoso, J.H. 2006. Atividade antioxidante de óleos essenciais de espécies de *Croton* do nordeste do Brasil. Quim. Nova, 29: 907–910.

De Paula, H.C.B., De Oliveira, E.F., Abreu, F.O.M.S., De Paula, R.C.M., De Morais, S.M., Forte, M.M.C. 2010. Esferas (Beads) de Alginato como Agente Encapsulante de Óleo de *Croton Zehntneri* Pax et Hoffm. Polimeros, 20: 112–120.

De Siqueira, R.J.B., Leal-Cardoso, J.H., Couture, R., Lahlou, S. 2006. Role of capsaicin-sensitive sensory nerves in mediation of the cardiovascular effects of the essential oil of *Croton zehntneri* leaves in anaesthetized rats. Clin. Exp. Pharmacol. Physiol., 33: 238–247.

de Sousa Barros, A., de Morais, S.M., Ferreira, P.A.T., Vieira, Í.G.P., Craveiro, A.A., dos Santos Fontenelle, R.O., de Menezes, J.E.S.A., da Silva, F.W.F., de Sousa, H.A. 2015. Chemical composition and functional properties of essential oils from *Mentha* species. Ind. Crops Prod., 76: 557–564.

De Souza, A.R., Morgon, N.H. 2016. Interação entre o Dímero de Apocinina e Ciclodextrina Souza,. Rev. Virtual Quim., 8: 417–424.

Del Valle, E.M.M. 2004. Cyclodextrins and their uses: A review. Process Biochem., 39: 1033–1046.

Dewick, P.M. 2009. Medicinal Natural Products Medicinal: A Biosynthetic Approach. John Wiley & Sons.

Diao, W.R., Hu, Q.P., Zhang, H., Xu, J.G. 2014. Chemical composition, antibacterial activity and mechanism of action of essential oil from seeds of fennel (*Foeniculum vulgare* Mill.). Food Control, 35: 109–116.

Domiciano, T.P., Dalalio, M.M.H.D.O., Silva, E.L., Ritter, A.M.V., Estevão-Silva, C.F., Ramos, F.S., Caparroz-Assef, S.M., Cuman, R.K.N., Bersani-Amado, C.A. 2013. Inhibitory effect of anethole in nonimmune acute inflammation. Naunyn. Schmiedebergs. Arch. Pharmacol., 386: 331–338.

Donati, M., Mondin, A., Chen, Z., Miranda, F.M., Do Nascimento, B.B., Schirato, G., Pastore, P., Froldi, G. 2014. Radical scavenging and antimicrobial activities of *Croton zehntneri*, *Pterodon emarginatus* and *Schinopsis brasiliensis* essential oils and their major constituents: Estragole, trans-anethole, β-caryophyllene and myrcene. Nat. Prod. Res., 29: 939–946.

Edris, A.E. 2007. Pharmaceutical and therapeutic Potentials of essential oils and their individual volatile constituents: a review. Phyther. Res., 21: 308–323.

Fischer, R., Nitzan, N., Chaimovitsh, D., Rubin, B., Dudai, N. 2011. Variation in essential oil composition within individual leaves of sweet basil (*Ocimum basilicum* L.) is more affected by leaf position than by leaf age. J. Agric. Food Chem., 59: 4913–4922.

Fonseca, L., Rocha, M., dos S., Brito, L.C.F., Sousa, E.S., Reinaldo, F.S., Pereira, F.M.M., Santos, F.E.P., de Lima, S.G. 2019. Characterization of inclusion complex of *Croton zehntneri* essential oil and β-Cyclodextrin prepared by spray drying and freeze drying. Rev. Virtual Química, 11: 529–542.

Fontenelle, R.O.S., Morais, S.M., Brito, E.H.S., Brilhante, R.S.N., Cordeiro, R.A., Nascimento, N.R.F., Kerntopf, M.R., Sidrim, J.J.C., Rocha, M.F.G. 2008. Antifungal activity of essential oils of *Croton* species from the Brazilian Caatinga biome. J. Appl. Microbiol., 104: 1383–1390.

Gao, X., Lv, S., Wu, Y., Li, J., Zhang, W., Meng, W., Wang, C., Meng, Q. 2017. Volatile components of essential oils extracted from Puerh ripe tea by different extraction methods. Int. J. Food Prop., 20: S240–S253.

Gharib, R., Auezova, L., Charcosset, C., Greige-Gerges, H. 2017. Drug-in-cyclodextrin-in-liposomes as a carrier system for volatile essential oil components: Application to anethole. Food Chem., 218: 365–371.

Gil, N.D., Segura, M.A., Soriano, G., Neri, J., López, N., Mas, E., Carlos, B., Varela, V., Ballesteros, M.F. 2019. Dissolution and recovery of waste expanded polystyrene using alternative essential oils. Fuel, 239: 611–616.

Gong, L., Li, T., Chen, F., Duan, X., Yuan, Y., Zhang, D., Jiang, Y. 2016. An inclusion complex of eugenol into β-cyclodextrin: Preparation, and physicochemical and antifungal characterization. Food Chem., 196: 324–330.

Guimarães, L.G.D.L., Cardoso, M., das G., Zacaroni, L.M., Morais, A.R. De. 2008. Influência da luz e da temperatura sobre a oxidação do óleo essencial de capim-limão (*Cymbopogon citratus* (D.C.) STAPF). Quim. Nova, 31: 1476–1480.

Hadian, Z., Maleki, M., Abdi, K., Atyabi, F., Mohammadi, A., Khaksar, R. 2018. Preparation and characterization of nanoparticle β-Cyclodextrin:Geraniol inclusion complexes. Iran. J. Pharm. Res., 17: 39–51.

Huang, Y., Zhao, J., Zhou, L., Wang, J., Gong, Y., Chen, X., Guo, Z., Wang, Q., Jiang, W. 2010. Antifungal activity of the essential oil of illicium verum fruit and its main component trans-anethole. Molecules, 15: 7558–7569.

Ipek, E., Zeytinoglu, H., Okay, S., Tuylu, B.A., Kurkcuoglu, M., Baser, K.H.C. 2005. Genotoxicity and antigenotoxicity of *Origanum* oil and carvacrol evaluated by *Ames Salmonella*/microsomal test. Food Chem., 93: 551–556.

Jantarat, C., Sirathanarun, P., Ratapongsai, S., Watcharaka, P., Sunyapong, S., Wadu, A. 2014. Curcumin-hydroxypropyl-β-cyclodextrin inclusion preparation methods effect of common solvent evaporation, freeze drying, dan ph shift on solubility and stability of *Curcumin*. Trop. J. Pharm. Res., 13: 1215–1223.

Kfoury, M., Auezova, L., Greige-Gerges, H., Ruellan, S., Fourmentin, S. 2014. Cyclodextrin, an efficient tool for trans-anethole encapsulation: Chromatographic, spectroscopic, thermal and structural studies. Food Chem., 164: 454–461.

Kfoury, M., Auezova, L., Greige-Gerges, H., Fourmentin, S. 2015. Promising applications of cyclodextrins in food: Improvement of essential oils retention, controlled release and antiradical activity. Carbohydr. Polym., 131: 264–272.

Kfoury, M., Auezova, L., Greige-Gerges, H., Larsen, K.L., Fourmentin, S. 2016a. Release studies of trans-anethole from β-cyclodextrin solid inclusion complexes by Multiple Headspace Extraction. Carbohydr. Polym., 151: 1245–1250.

Kfoury, M., Lounès-Hadj Sahraoui, A., Bourdon, N., Laruelle, F., Fontaine, J., Auezova, L., Greige-Gerges, H., Fourmentin, S. 2016b. Solubility, photostability and antifungal activity of phenylpropanoids encapsulated in cyclodextrins. Food Chem., 196: 518–525.

Kim, H.M., Lee, E.H., Kim, C.Y., Chung, J.G., Kim, S.H., Lim, J.P., Shin, T.Y. 1997. Antianaphylactic properties of eugenol. Pharmacol. Res., 36: 475–480.

Kubo, I., Fujita, K., Nihei, K. 2008. Antimicrobial activity of anethole and related compounds from aniseed. J. Sci. of Food Agric., 242–247.

Lazarini, C.A., Uema, A.H., Brandão, G.M.S., Guimarães, A.P.C., Bernardi, M.M. 2000. *Croton zehntneri* essential oil: Effects on behavioral models related to depression and anxiety. Phytomedicine, 7: 477–481.

Lewinsohn, E., Ziv-raz, I., Dudai, N., Tadmor, Y., Lastochkin, E., Larkov, O., Chaimovitsh, D., Ravid, U., Putievsky, E., Pichersky, E., Shoham, Y. 2000. Biosynthesis of estragole and methyl-eugenol in sweet basil Ocimum. Plant Sci., 160: 27–35.

Li, W., Liu, X., Yang, Q., Zhang, N., Du, Y., Zhu, H. 2015. Preparation and characterization of inclusion complex of benzyl isothiocyanate extracted from papaya seed with β-cyclodextrin. Food Chem., 184: 99–104.

Lima, F.C., Sousa, D.F., Ferreira, J.M., Lima, R.C.P., Tome, A.R., Cardoso, J.H.L., Queiroz, M.G.R., Campos, A.R. 2008. *Croton zehntneri* essential oil prevents acetaminophen-induced acute hepatotoxicity in mice. Rec. Nat. Prod., 2: 135–140.

Malveira Cavalcanti, J., Henrique Leal-Cardoso, J., Leite Diniz, L.R., Gomes Portella, V., Oliveira Costa, C., Barreto Medeiros Linard, C.F., Alves, K., De Paula Rocha, M.V.A., Calado Lima, C., Marilande Cecatto, V., Coelho-De-Souza, A.N. 2012. The essential oil of *Croton zehntneri* and trans-anethole improves cutaneous wound healing. J. Ethnopharmacol., 144: 240–247.

Masotti, V., Juteau, F., Bessière, J.M., Viano, J. 2003. Seasonal and phenological variations of the essential oil from the narrow endemic species Artemisia molinieri and its biological activities. J. Agric. Food Chem., 51: 7115–7121.

Mastro, G. De, Tarraf, W., Verdini, L., Brunetti, G., Ruta, C. 2017. Essential oil diversity of *Origanum vulgare* L. populations from Southern Italy. Food Chem., 235: 1–6.

Matias, E.F.F., Santos, K.K.A., Costa, J.G.M., Coutinho, H.D.M. 2010. Light-enhanced antibiotic activity of Brazilian medical plants (*Croton campestris* A, *Ocimum gratissimum* L. and *Cordia verbenaceae* DC). Asian Biomed., 4: 183–186.

Ministério da Saúde. 2016. Política e Programa Nacional de Plantas Medicinais e Fitoterápicos. Biblioteca Virtual em Saúde do Ministério da Saúde.

Miranda, F.M., Braga do Nascimento Junior, B., Aguiar, R.M., Pereira, R.S., De Oliveira Teixeira, A., De Oliveira, D.M., Lima, E. de O., Oigman, S.S., de Rezende, C.M., Froldi, G. 2019. Promising antifungal activity of *Croton tricolor* stem essential oil against Candida yeasts. J. Essent. Oil Res., 31: 223–227.

Mota, M., Pohlit, A., Rocha e Silva, L., Rocha, H., de Andrade Neto, V., Costa, L., Galberto da Costa, J., Lobo, L. 2012. *In vitro* and *in vivo* antimalarial activity of essential oils and chemical components from three medicinal plants found in Northeastern Brazil. Planta Med., 78: 658–664.

Nascimento, M.R., Guilherme Sousa da Silva, Gonçalo Mendes da Conceição. 2017. Euphorbiaceae: checklist de uma coleção botânica, Maranhão, Brasil. Enciclopédia Biosf., 14: 1353.

Obolskiy, D., Pischel, I., Feistel, B., Glotov, N., Heinrich, M. 2011. *Artemisia dracunculus* L. (tarragon): A critical review of its traditional use, chemical composition, pharmacology, and safety. J. Agric. Food Chem., 59: 11367–11384.

Oliveira-Tintino, C.D. de M., Pessoa, R.T., Fernandes, M.N.M., Alcântara, I.S., da Silva, B.A.F., de Oliveira, M.R.C., Martins, A.O.B.P.B., da Silva, M. do S., Tintino, S.R., Rodrigues, F.F.G., da Costa, J.G.M., de Lima, S.G., Kerntopf, M.R., da Silva, T.G., de Menezes, I.R.A. 2018. Anti-inflammatory and anti-edematogenic action of the *Croton campestris* A. St.-Hil (Euphorbiaceae) essential oil and the compound β-caryophyllene in *in vivo* models. Phytomedicine, 41: 82–95.

Oliveira, A.C., Leal-Cardoso, J.H., Santos, C.F., Morais, S.M., Coelho-de-Souza, A.N. 2001. Antinociceptive effects of the essential oil of *Croton zehntneri* in mice. Brazilian J. Med. Biol. Res., 34: 1471–1474.

Orlandini, P., Lima, L.R. de. 2014. Sinopse do gênero *Manihot* Mill. (Euphorbiaceae) no Estado de São Paulo, Brasil. Hoehnea 41: 51–60.

Ozkan, G., Baydar, H., Erbas, S. 2010. The influence of harvest time on essential oil composition, phenolic constituents and antioxidant properties of Turkish oregano (*Origanum onites* L.). J. Sci. Food Agric., 90: 205–209.

Palmeira-De-Oliveira, A., Gaspar, C., Palmeira-De-Oliveira, R., Silva-Dias, A., Salgueiro, L., Cavaleiro, C., Pina-Vaz, C., Martinez-De-Oliveira, J., Queiroz, J.A., Rodrigues, A.G. 2012. The anti-*Candida* activity of *Thymbra capitata* essential oil: Effect upon pre-formed biofilm. J. Ethnopharmacol., 140: 379–383.

Piletti, R., Bugiereck, A.M., Pereira, A.T., Gussati, E., Dal Magro, J., Mello, J.M.M., Dalcanton, F., Ternus, R.Z., Soares, C., Riella, H.G., Fiori, M.A. 2017. Microencapsulation of eugenol molecules by β-cyclodextrine as a thermal protection method of antibacterial action. Mater. Sci. Eng. C, 75: 259–271.

Ponte, E.L., Sousa, P.L., Rocha, M.V.A.P., Soares, P.M.G., Coelho-de-Souza, A.N., Leal-Cardoso, J.H., Assreuy, A.M.S. 2012. Comparative study of the anti-edematogenic effects of anethole and estragole. Pharmacol. Reports, 64: 984–990.

Raina, A.P., Kumar, A., Dutta, M. 2013. Chemical characterization of aroma compounds in essential oil isolated from "Holy Basil" (*Ocimum tenuiflorum* L.) grown in India. Genet. Resour. Crop Evol., 60: 1727–1735.

Reyes-Hernández, M., Thimmappa, R., Abraham, S., Pagadala Damodaram, K.J., Pérez-Staples, D. 2019. Methyl eugenol effects on *Bactrocera* dorsalis male total body protein, reproductive organs and ejaculate. J. Appl. Entomol., 143: 177–186.

Rocha, M.S., De Lima, S.G., Bartolomeu C. Viana, José Galberto Martins Costa, Francisco E.P. Santos. 2018. Characterization of the inclusion complex of the essential oil of *Lantana camara* L. and β-cyclodextrin by vibrational spectroscopy, GC–MS, and X-ray diffraction. J. Incl. Phenom. Macrocycl. Chem., 91: 95–104.

Rodrigues, F.F.G., Costa, J.G.M., Coutinho, H.D.M. 2009. Synergy effects of the antibiotics gentamicin and the essential oil of *Croton zehntneri*. Phytomedicine, 16: 1052–1055.

Rodrigues, L.B., Oliveira Brito Pereira Bezerra Martins, A., Cesário, F.R.A.S., Ferreira e Castro, F., de Albuquerque, T.R., Martins Fernandes, M.N., Fernandes da Silva, B.A., Quintans Júnior, L.J., da Costa, J.G.M., Melo Coutinho, H.D., Barbosa, R., Alencar de Menezes, I.R. 2016. Anti-inflammatory and antiedematogenic activity of the *Ocimum basilicum* essential oil and its main compound estragole: *In vivo* mouse models. Chem. Biol. Interact., 257: 14–25.

Rudrangi, S.R.S., Bhomia, R., Trivedi, V., Vine, G.J., Mitchell, J.C., Alexander, B.D., Wicks, S.R. 2015. Influence of the preparation method on the physicochemical properties of indomethacin and methyl-β-cyclodextrin complexes. Int. J. Pharm., 479: 381–390.

Sá, N.A.R., Araújo, V.R., Correia, H.H.V., Ferreira, A.C.A., Guerreiro, D.D., Sampaio, A.M., Escobar, E., Santos, F.W., Moura, A.A., Lôbo, C.H., Ceccatto, V.M., Campello, C.C., Rodrigues, A.P.R., Leal-Cardoso, J.H., Figueiredo, J.R. 2017. Anethole improves the *in vitro* development of isolated caprine secondary follicles. Theriogenology, 89: 226–234.

Santos, H.S., Santiago, G.M.P., de Oliveira, J.P.P., Arriaga, A.M.C., Marques, D.D., Lemos, T.L.G. 2007. Chemical Composition and larvicidal activity against *Aedes aegypti* of essential oils from *Croton zehntneri*. Nat. Prod. Commun., 2: 1233–1236.

Santos, H.S., Furtado, E.F.F., Bertini, L.M., Bandeira, P.N., Albuquerque, M.R.J.R., Menezes, J.E.S.A., Trevisan, M.T.S., Lemos, T.L. 2010. Chemical composition and cholinesterase inhibition of essential oils of three chemotypes from *Croton zehntneri*. Rev. Latinoam. Quim., 45–51.

Sebaaly, C., Jraij, A., Fessi, H., Charcosset, C., Greige-Gerges, H. 2015. Preparation and characterization of clove essential oil-loaded liposomes. Food Chem., 178: 52–62.

Serra, D.S., Gomes, M.D.M., Cavalcante, F.S.Á., Leal-Cardoso, J.H. 2018. Essential oil of *Croton zehntneri* attenuates lung injury in the OVA-induced asthma model. J. Asthma, 56: 1–10.

Shepherd, M., Savins, D., Dowell, A., Morrow, S., Allen, G., Southwell, I. 2017. Ecotype variation of methyl eugenol content in tea tree (*Melaleuca alternifolia* and *Melaleuca linariifolia*). Chem. Biodivers., 14.

Soares, B.V., Cardoso, A.C.F., Campos, R.R., Gonçalves, B.B., Santos, G.G., Chaves, F.C.M., Chagas, E.C., Tavares-Dias, M. 2017. Antiparasitic, physiological and histological effects of the essential oil of *Lippia origanoides* (Verbenaceae) in native freshwater fish *Colossoma macropomum*. Aquaculture, 469: 72–78.

Sodré, R.C., da Silva, J.M., Sales, M.F. De. 2014. *Croton* L. (Euphorbiaceae) no Parque Estadual da Serra Dourada, Goiás, Brasil. Rodriguesia, 65: 221–234.

Sodré, R.C., da Silva, M.J. 2015. O gênero Croton L. (Euphorbiaceae s. s. – Crotonoideae) na Floresta Nacional da Silvânia, Goiás, Brasil. Iheringia Série Botânica, 70: 89–104.

Sousa, E.M.B.D., Martínez, J., Chiavone-Filho, O., Rosa, P.T.V., Domingos, T., Meireles, M.A.A. 2005. Extraction of volatile oil from *Croton zehntneri* Pax et Hoff with pressurized CO_2: Solubility, composition and kinetics. J. Food Eng., 69: 325–333.

Souza, A.V.V. de, Dos Santos, U.S., de Sá Carvalho, J.R., Barbosa, B.D.R., Canuto, K.M., Rodrigues, T.H.S. 2018. Chemical composition of essential oil of leaves from *Lippia schaueriana* Mart. collected in the caatinga area. Molecules, 23.

Veiga F., Percorelli C., Ribeiro L. 2006. As ciclodextrinas em tecnologia farmacêutica. Coimbra: Editora Minerva Coimbra.

Wenz, G. 2012. Influence of intramolecular hydrogen bonds on the binding potential of methylated β-cyclodextrin derivatives. Beilstein J. Org. Chem., 8: 1890–1895.

Wurdack, K.J., Davis, C.C. 2009. Malpighiales phylogenetics: Gaining ground on one of the most recalcitrant clades in the angiosperm tree of life. Am. J. Bot., 96: 1551–1570.

Yi, F., Sun, J., Bao, X., Ma, B., Sun, M. 2019. Influence of molecular distillation on antioxidant and antimicrobial activities of rose essential oils. LWT - Food Sci. Technol., 102: 310–316.

Zafeiropoulou, T., Evageliou, V., Gardeli, C., Yanniotis, S., Komaitis, M. 2010. Retention of trans-anethole by gelatine and starch matrices. Food Chem., 123: 364–368.

Zamora, C.M.P., Torres, C.A., Nuñez, M.B. 2018. Antimicrobial activity and chemical composition of essential oils from Verbenaceae species growing in South America. Molecules, 23.

Zeller, A., Horst, K., Rychlik, M. 2009. Study of the metabolism of estragole in humans consuming fennel tea. Chem. Res. Toxicol., 22: 1929–1937.

Zhai, H., Liu, H., Wang, S., Wu, J., Kluenter, A.M. 2018. Potential of essential oils for poultry and pigs. Anim. Nutr., 4: 179–186.

Zhang, J., Ma, P.X. 2013. Cyclodextrin-based supramolecular systems for drug delivery: Recent progress and future perspective. Adv. Drug Deliv. Rev., 65: 1215–1233.

Zhao, D., Xu, Y.W., Yang, G.L., Husaini, A.M., Wu, W. 2013. Variation of essential oil of *Mentha haplocalyx* Briq. and *Mentha spicata* L. from China. Ind. Crops Prod., 42: 251–260.

Zhu, X.-L., Wang, H.-B., Chen, Q., Yang, W.-C., Yang, G.-F. 2007. Preparation and characterization of inclusion complex of iprodione and beta-cyclodextrin to improve fungicidal activity. J. Agric. Food Chem., 55: 3535–9.

Chapter 17

Emerging Medicinal Values of Kiwifruit (*Actinidia* Lindl.)

Mira Dhakal[1] and *Shandesh Bhattarai*[2,*]

Introduction

The fruit is a seed-bearing structure created from the ovary through the development of one or more flowers. The gynoecium of the flower forms all or part of the fruit (Esau 1977). The angiosperms disseminate seeds from the fruits. Mostly, fruit plants are woody and perennial (Mauseth 2003, Schlegel 2003). Fruits of angiosperms are variously classified as simple or compound, depending upon whether they are developed from single flower or inflorescence, dry or fleshy, or dehiscent or indehiscent, and so on (Naik 1984). According to Singh (2004), simple fruits, aggregate fruits, and composite/multiple fruits are the three main groups of fruits.

Fruits in which part or all of the pericarp (fruit wall) is fleshy at maturity are simple fleshy fruits. Simple fruits can be either dry or fleshy and result from the ripening of a simple or compound ovary in a flower with only one pistil (Schlegel 2003, Gupta 2007, www.rkv.rgukt.in). Aggregate fruits are produced from single flowers that have multiple carpels that are not joined together, i.e., each pistil contains one carpel. Each pistil forms a fruitlet, and collectively the fruits are called an etaerio (Gupta 2007, www.rkv.rgukt.in). Multiple fruits are formed from a cluster of flowers/inflorescence and each flower produces a fruit, but these mature into a single mass (Schlegel 2003). The evolutionary progress appears to be from fruitlets developed from apocarpous gynoecium to simple fruits to compound or multiple fruits (Naik 1984). It may be proper to regard the Magnoliaceous fruit as the most primitive type, at least among modern members of angiosperms (Ozenda 1949, Canright 1955, Melville 1962, Skipworth and Philipson 1966, Naik 1984).

Humans and animals have become dependent on fruits as a source of food (Lewis 2002), and its consumption has a reduced risk of chronic diseases (Singh 1995). Fruit provides vitamins, proteins, minerals, carbohydrates, fats, energy, organic acids, and enzymes as well as contains several medicinal values. Vitamin A is essential for growth and reproduction which helps in resisting infections, increases longevity and decreases senility (Singh 1995). Vitamin B_1 (Thiamin) is essential for good appetite, digestion, growth, fertility, lactation and normal functioning of nervous tissue. Vitamin B_2 or G (Riboflavin) is important for the growth and health of the skin and respiration in poorly vascularized

[1] Horticulture Research Division, Nepal Agriculture Research Council, Lalitpur, Nepal.
[2] Faculty of Science, Nepal Academy of Science and Technology, Lalitpur, Nepal.
* Corresponding author: shandeshbhattarai@gmail.com

tissues such as the cornea (Singh 1995). Vitamin B_6 deficiency may result from various conditions such as peripheral neuropathy, a pellagra-like syndrome, anemia, and seizures. Vitamin C (Ascorbic acid) is a dietary essential for human beings. Folic acid and B_{12} deficiency may lead to anaemia (Singh 1995). Proteins are important for body growth, contains essential amino acids, and are necessary for the formation and maintenance of body tissues. Similarly, carbohydrates and fats provide energy to the human body (Singh 1995). Ten minerals are essential for the growth and development of the human body and fruits are a rich source of minerals. Of these, calcium, phosphorus, and iron are most essential and found in varying concentrations in different fruits (Singh 1995). Fruits are a good source of energy because many of them contain digestible form of carbohydrates in the form of sugar or starch or both (Singh 1995). A few fruits are also a good source of enzymes that are helpful in metabolic activities, leading to proper digestion of food. Almost all fruits have medicinal value. Many fruits are good sources of fibre which helps to relieve constipation, lowers cholesterol and blood glucose levels, and helps to lose weight (Singh 1995).

Various vitamins and minerals are called antioxidants, which protect cells from damage caused by free radicals (Singh 1995). The potassium in fruit can reduce the risk of heart disease and stroke and also reduce the risk of developing kidney stones and help to decrease bone loss. Women of childbearing age who may become pregnant and those in the first trimester of pregnancy need adequate folate (Benzie and Choi 2014). The consumption of fruit helps to maintain optimum health due to health-promoting phytochemicals (Benzie and Choi 2014). Kiwifruit is the most nutrient-rich of the top 26 fruits consumed in the world today, and thus attracts the attention of the people globally. It also has the highest density of any fruit for vitamin C and magnesium (Lal et al. 2010). Knowing its importance, this review mainly focuses on the diversity, distribution, medicinal and food values, phytochemicals constituent, and pharmacological activities of the *Actinidia* species, focusing on *A. chinensis*, *A. deliciosa* and *A. macrosperma,* to provide the reference for its further research and relevant knowledge towards drug development (Figure 17.1).

Figure 17.1a–e. This showing flowering to fruiting stages of kiwifruit.

Taxonomic Description

The taxonomic description of Kiwifruit (*Actinidia*) is based on Flora of China 2007 (Morgan 1988, Soejarto et al. 2007).

Actinidia Lindley, Nat. Syst. Bot., ed. 2. 439. 1836

Kalomikta Regel; *Trochostigma* Siebold & Zuccarini.

The genus *Actidinia* includes climbing shrubs growing to about 6 m tall and vigorous, strong-growing vines, growing up to 30 m in tree canopies; glabrous or hairy, indumentum of stellate or simple hairs; pith solid or lamellate. Branches usually with linear, lengthwise lenticels; winter buds small, enclosed in swollen base of the petiole or exposed. Leaves often long petiolate; stipules minute, obsolete, or absent; leaf blade membranous, papery, or leathery, venation penniveined, veinlets reticulate, usually in cross-bars, margin serrate or dentate, rarely entire. Inflorescences cymose, axillary, often pseudo-umbellate, few- or many-flowered, or flowers solitary; bracts present, minute. Flowers white, pink, red, yellow, or green, bisexual, plants polygamous or functionally dioecious. Sepals (2–)5(or 6), distinct or connate at base, imbricate, rarely valvate, persistent or not. Petals (4 or)5(or more than 5), imbricate. Stamens numerous, in functionally female flowers often with shorter filaments and smaller sterile anthers; filaments slender; anthers yellow, brown, purple, or black, versatile, attached at middle, 2-celled, dehiscing lengthwise, usually divaricate at base. Disk absent. Ovary ovoid, cylindrical, or bottle-shaped, glabrous or hairy, many located; ovules numerous per locule; styles as many as carpels (15–30), usually reflexed, persistent, radiating, in functionally female flower elongating after anthesis; rudimentary ovary in functionally male flower very small, with minute styles. Fruit a berry, globose, ovoid, or oblong, spotted with lenticels or not, glabrous or hairy. Seeds numerous, oblong, small, immersed in pulp; embryo comparatively large, cylindrical; cotyledons short (Morgan 1988, Soejarto et al. 2007).

The description of *Actinidia chinensis, A. deliciosa* and *A. macrosperma* is also based on Flora of China 2007 (Morgan 1988, Soejarto et al. 2007).

Actinidia chinensis Planchon, London J. Bot. 6: 303. 1847

Climbing vine, large, deciduous, cultivated. Branchlets reddish, with paler oblong lenticels, young branchlets white pubescent to brownish long strigose or densely ferruginous hispid-setose; pith whitish to brown, large, lamellate. Petiole 3–6(–10) cm, white pubescent or brownish long strigose to densely ferruginous hispid-setose; leaf blade abaxially pale green, adaxially dark green, broadly ovate to broadly obovate or suborbicular, 6–17 × 7–15 cm, papery, abaxially whitish or brownish stellate tomentose, adaxially usually glabrous, occasionally ± puberulent, especially more densely so on midvein and lateral veins, or densely scabrid-hispid throughout, midvein and lateral veins conspicuous abaxially, subconspicuous adaxially, lateral veins 5–8 pairs, furcate above middle, veinlets in parallel cross-bars, conspicuous abaxially, base rounded to truncate to cordatulate, margin setose-serrulate with teeth terminating tips of veinlets, apex truncate to emarginate to abruptly cuspidate or shortly acuminate. Inflorescences cymose, 1–3-flowered, white silky-tomentose or yellowish-brown velutinous; peduncles 0.7–1.5 cm; pedicels 0.9–1.5 cm; bracts linear, ca. 1 mm. Flowers orange-yellow. Sepals (3–)5(–7), broadly ovate to oblong-ovate, 6–10 mm, both surfaces densely yellowish tomentose. Petals (3–)5(–8), rarely with the second whorl of 2 smaller petals, broadly obovate, 1–2 cm, shortly clawed at the base, rounded at apex. Filaments 5–10 mm; anthers yellow, oblong, 1.5–2 mm, sagittate at the base or not, apex acute to rounded. Ovary globose, ca. 5 mm in diam., densely golden villous. Fruit subglobose to cylindric to obovoid or ellipsoidal, 4–6 cm, densely tomentose when young, glabrous when mature to densely hispid, with many brownish lenticels; persistent sepals reflexed. Fl. Apr–May, fr. Sep (Morgan 1988, Soejarto et al. 2007).

Actinidia deliciosa (A.Chev.) C.F.Liang & A.R.Ferguson

Actinidia deliciosa is a cultivated vigorous, woody, twining vine reaching 9 m. Young branchlets and petioles brownish strigose, hair not easily lost. Leaves usually glabrous adaxially, occasionally ± puberulent, especially more densely so on midvein and lateral veins. The flowers are fragrant, dioecious or unisexual, borne singly or in threes in the leaf axils, are five- to six-petalled, white at first, changing to buff-yellow, 2.5–5 cm broad, and both sexes have central tufts of many stamens, though those of the female flowers with no viable pollen. Fruit subglobose to cylindric or ovoid, 5–6 cm, densely hispid, when mature (Morgan 1988, Soejarto et al. 2007, www.wikepedia.org).

Actinidia macrosperma C. F. Liang

It is a medium-sized wild species of kiwifruit, which is climbing, deciduous vine, grows below 800 m above sea level (Liang et al. 1984). Branchlets glabrous; pith white, solid. Petiole 1–2.2 cm, glabrous to setose; leaf blade abaxially pale green, adaxially green, ovate to elliptic, 3–8 × 1.7–5 cm, membranous when young, leathery when mature, abaxially glabrous to strigillose on midvein, barbate in vein axils or not, adaxially glabrous, midvein and lateral veins subconspicuous abaxially, inconspicuous adaxially, lateral veins 4 or 5 pairs, base broadly cuneate to rounded, equal or oblique, margin serrate or crenate, subentire when mature, apex acute to acuminate, occasionally rounded. Inflorescences often 1-flowered; peduncles 6–7 mm; pedicels 0.9–1.5 cm; bracts linear to lanceolate, minute. Flowers white. Sepals 2 or 3, ovate to narrowly ovate, 0.6–1.2 cm, glabrous, rostrate at the apex. Petals 5–12, orbicular-obovate, 1–1.5 cm. Filaments 3–7 mm; anthers yellow, ovoid, 1.5–2.5 mm, sagittate at the base. Ovary bottle-shaped, 6–8 mm, glabrous; styles ca. 5 mm. Fruit orange when mature, globose to ovoid, 3–3.5 cm, glabrous, mammilliform-rostrate at apex, lenticels absent; sepals persistent or caducous. Seeds 4–5 mm, ca. 3 mm in diameter (Morgan 1988, Soejarto et al. 2007, www.wikepedia.org).

Classification

Kingdom: Plantae; Sub-kingdom: Tracheobionta; Super-division: Spermatophyta; Phylum: Magnoliophyta; Class: Magnoliopsida; Order: Ericales; Family: Actinidiaceae; Genus: *Actinidia.*

History, Origin, and Production of Kiwifruit

The 12th-century report documented that Kiwifruit is native to north-central and eastern China (Morton 2011, Ward and Courtney 2013), rarely cultivated and regularly collected from the wild for medicinal usage (Huang and Ferguson 2003). The first cultivation spread from China to New Zealand in the early 20th century (Morton 2011), and was later first exported to Great Britain and then to California in the 1960s (Emily 2002, Morton 2011). In New Zealand during the 1940s and 1950s, the fruit became an agricultural product through the development of commercially viable cultivars (Knowles 2017). Much of the breeding to improve the green kiwifruit and to develop the gold Zespri was undertaken by the Plant and Food Research Institute during 1970–1999 (Mike 2017). In 1990, the New Zealand Kiwifruit Marketing Board opened an office for Europe in Antwerp, Belgium, which became the headquarters for European marketing of Zespri gold kiwifruit in 2010 (Knowles 2017, Zespri 2018, www.en.wikipedia.org).

It was cultivated on a small scale at least 300 years ago, but today most of the 1,000 tones crop is derived from wild vines. Specimens of the plant were collected by the agent for the Royal Horticultural Society, London, in 1847 and described the plant from herbarium specimens. In 1900, seeds gathered in Hupeh were sent to England; the resulting plants flourished and bloomed in 1909. Seeds from China were introduced into New Zealand in 1906 and some vines bore fruits in 1910. Several growers raised numerous seedlings and selected the best fruiting types. By 1940, there were many plantings especially on the eastern coast of the North Island (Knowles 2017, Zespri 2018, www.en.wikipedia.org).

The kiwifruit was already being grown in Cambodia, Vietnam and southern Laos, France, Spain, Belgium, Italy where plantings were first made in the late 1960s and commercial growing started in the late 1970s. Italy advanced to third place in world production by 1983. By 1971, there were small plantings scattered around south-western and south-eastern areas of the country–valleys of the Garonne, Dordogne, Rhone, and Loire rivers. Greece is now producing kiwifruits for export to other European countries. The vine was introduced into the Philippines at Baguio in 1923. Large plantings are being made in Chile (Knowles 2017, Zespri 2018, www.en.wikipedia.org). The current worldwide production appears limited to ten main countries, which sustain over 95% of total production (Lippi and Mattiuzzi 2019). Although China remains the largest producer (2.4 million metric tons in 2016), New Zealand is the largest exporter (0.58 million metric tons in 2016), followed by Italy, Belgium and Chile (www.mordorintelligence.com). The top ten kiwifruit producer countries in the world in 2012 and 2013 are given in Table 17.1. Traditionally, Italy, New Zealand, and Chile have been the largest exporters and have also dominated world kiwifruit production. However, China has joined these three countries as a major producer (Knowles 2017, Zespri 2018, www.en.wikipedia.org).

Table 17.1. Top kiwifruit producer countries in the world (2012 and 2013).

SN	Country	Production (in 000 Tones)		Productivity based on 2013 (T/ha)
		2012	2013	
1	China	1452767	1765847	12.71
2	Italy	384844	447560	17.98
3	New Zealand	376400	382337	33.0
4	Chile	240000	255758	23.0
5	Greece	161400	162800	17.50
6	France	65253	55999	14.75
7	Turkey	36781	41635	-
8	Iran	32000	31603	13.5
9	Japan	28000	29225	-
10	United States	26583	27300	10.27

Source: www.wikipedia.org, FAO 2014, Gotame et al. 2016.

Diversity and Distribution

About 55 species were reported from East and South Asia. Fifty-two species are found in China and among these 44 species are endemic in China (Soejarto et al. 2007). But verifying all the recent literature and website information, the total number of recorded species (including one variety) of kiwifruit is 68. The names of 68 species and their distribution are given in detail in Table 17.2 (www.plantlist.org, www.worldflora.online.org, www.gbif.org, www.efloras.org, www.floraofchina.org).

Table 17.2. Distribution of kiwifruit.

SN	Name of species	Distribution
1	*Actinidia acuminata* Budisch. ex Trautv.	Native range is NE China
2	*Actinidia arguta* (Siebold & Zucc.) Planch. ex Miq.	Mountain forests, thickets, streamsides, moist places, widely cultivated; 700–3600 m; Anhui, Chongqing, Fujian, Gansu, Guangxi, Guizhou, Hebei, Heilongjiang, Henan, Hubei, Hunan, Jiangxi, Jilin, Liaoning, Shaanxi, Shandong, Shanxi, Sichuan, Taiwan, Yunnan, Zhejiang [Japan, Korea]
3	*Actinidia arisanensis* Hayata	Taiwan, in thickets and forests, 1160–2260 m; E. Asia–S. China
4	*Actinidia asymmetrica* F.Chun	Low mountain forests, thickets, slopes and valleys, 1000–1400 m; E. Asia–S. China
5	*Actinidia callosa* Lindl.	Forests, forest margins, thickets, slopes, valleys, streamsides, moist places, widely cultivated; 400–2600 m; Anhui, Chongqing, Fujian, Gansu, Guangdong, Guangxi, Guizhou, Henan, Hubei, Hunan, Jiangxi, Shaanxi, Sichuan, Taiwan, Xizang, Yunnan, Zhejiang [Bhutan, India, Nepal]
6	*Actinidia carnosifolia* C.Y.Wu	China (Guangdong, Guangxi, Guizhou, Hunan, Yunnan)
7	*Actinidia chengkouensis* C.Y.Chang	Forests; 1000–2000 m; Chongqing (Chengkou, Wushan, Wuxi), Hubei (Badong)
8	*Actinidia chinensis* Planch.	Mountain forests, sparse secondary forests, tall grassy thickets on low mountains, thickets; 200–2600 m; Anhui, Chongqing, Fujian, Gansu, Guangdong, Guangxi, Guizhou, Henan, Hubei, Hunan, Jiangsu, Jiangxi, Shaanxi, Sichuan, Taiwan, Yunnan, Zhejiang
9	*Actinidia chrysantha* C.F.Liang	Sparse forests, thickets, open sunny places in forests; 900–1300 m; Guangdong, Guangxi, Hunan, Jiangxi

Table 17.2 Contd. ...

...Table 17.2 Contd.

SN	Name of species	Distribution
10	*Actinidia cylindrica* C.F.Liang	Low mountain forests, thickets; 400–800 m; Guangxi
11	*Actinidia deliciosa* (A.Chev.) C.F.Liang & A.R.Ferguson	600–2000 m; E. Asia–China, Native to southern China
12	*Actinidia eriantha* Benth.	Forests, tall grassy thickets on low mountains; 200–1000 m; Fujian, Guangdong, Guangxi, Guizhou, Hunan, Jiangxi, Zhejiang
13	*Actinidia fanjingshanensis* S.D.Shi & Q.B.Wang	Mountain forests, forest margins, thickets, valleys; 400–2000 m; Anhui, Fujian, Guangdong, Guangxi, Guizhou, Hunan, Jiangxi, Sichuan, Taiwan, Yunnan, Zhejiang
14	*Actinidia farinosa* C.F.Liang	Sunny places, roadsides; 1000–1200 m; NW Guangxi (Tianlin)
15	*Actinidia fasciculoides* C.F.Liang	Sparse mountain forests; 400–1500 m; Guangxi, Yunnan
16	*Actinidia fortunatii* Finet & Gagnep.	Low mountain forests, thickets, slopes, valleys; ca. 1000 m; Guangdong, Guangxi, Guizhou, Hunan
17	*Actinidia fulvicoma* Hance	Mountain forests, sparse forests, thickets, mountain slopes; 100–1800 m; Fujian, Guangdong, Guangxi, Guizhou, Hunan, S Jiangxi, Yunnan
18	*Actinidia glabra* H.L.Li	Southern China, Kwangsi, thickets
19	*Actinidia glaucocallosa* C.Y.Wu	Broad-leaved evergreen forests, valleys; 2300–2800 m; Yunnan (Jingdong, Longling, Tengchong)
20	*Actinidia gracilis* C.F.Liang	Low mountain forests, thickets, slopes, valleys; ca. 1000 m; Guangdong, Guangxi, Guizhou, Hunan
21	*Actinidia grandiflora* C.F.Liang	Mountain forests; ca. 1800 m; Sichuan (Tianquan)
22	*Actinidia hemsleyana* Dunn	Low mountain forests; 500–900 m; N Fujian, E Jiangxi, S Zhejiang
23	*Actinidia henryi* Dunn	Mountain forests, thickets; 1400–2500 m; Guangdong, Guangxi, Guizhou, Hunan, Yunnan
24	*Actinidia holotricha* Finet & Gagnep.	1400–2000 m; Southwestern China (Yunnan and southwestern Szechuan)
25	*Actinidia hubeiensis* H.M.Sun & R.H.Huang	Hubei (Yichang)
26	*Actinidia indochinensis* Merr.	Dense mountain forests; 600–1300 m; Fujian, Guangdong, Guangxi, Yunnan [N Vietnam]
27	*Actinidia jiankouensis* S.D.Shi & Z.S.Zhang	Sparse mountain forests, thickets; 800–2000 m; Guizhou (Jiangkou, Yingjiang), Hubei (Xuan'en)
28	*Actinidia jijiangensis* C.F.Liang & Y.X.Lu	NE Guangxi
29	*Actinidia kolomikta* (Rupr. & Maxim.) Maxim.	Mountain mixed forests in open places, widely cultivated; 1600–2900 m; Chongqing, Gansu, Hebei, Heilongjiang, Hubei, Jiangsu, Jilin, Liaoning, Shaanxi, Shanxi, Sichuan, Yunnan [Japan, Korea, Russia]
30	*Actinidia kungshanensis* C.Y.Wu & S.K.Chen	Mountain forests; ca. 2000 m; Yunnan (Gongshan).
31	*Actinidia laevissima* C.F.Liang	Sparse mountain forests, thickets; 800–2000 m; Guizhou (Jiangkou, Yingjiang), Hubei (Xuan'en)
32	*Actinidia lanceolata* Dunn	Sparse forests, forest margins, tall grassy thickets; 200–800 m; Anhui, Fujian, Guangdong, Hunan, Jiangsu, Jiangxi, Zhejiang
33	*Actinidia latifolia* (Gardner & Champ.) Merr.	Forests, thickets, mountain slopes, valleys; 400–1700 m; Anhui, Fujian, Guangdong, Guangxi, Guizhou, Hainan, Hunan, Jiangxi, Sichuan, Taiwan, Yunnan, Zhejiang [Cambodia, Laos, Malaysia, Thailand, Vietnam]

Table 17.2 Contd. ...

...Table 17.2 Contd.

SN	Name of species	Distribution
34	*Actinidia leptophylla* C.Y.Wu	Mountain mixed forests in open places, widely cultivated; 1600–2900 m; Chongqing, Gansu, Hebei, Heilongjiang, Hubei, Jiangsu, Jilin, Liaoning, Shaanxi, Shanxi, Sichuan, Yunnan [Japan, Korea, Russia]
35	*Actinidia liangguangensis* C.F.Liang	Forests in sunny places, thickets, mountain valleys; 200–1000 m; Guangdong, Guangxi, Hunan (Jianghua)
36	*Actinidia lijiangensis* C.F.Liang & Y.X.Lu	NE Guangxi
37	*Actinidia linguiensis* R.G.Li & X.G.Wang	Guangxi (Lingui)
38	*Actinidia longicarpa* R.G.Li & M.Y.Liang	Guangxi (Longsheng or Ziyuan)
39	*Actinidia macrosperma* C.F.Liang	Forest margins on low mountains, thickets, low mountain slopes; Anhui, Guangdong, Hubei, Jiangsu, Jiangxi, Zhejiang
40	*Actinidia maloides* H.L.Li	2000 m; Western China (Sikang), on mountain slopes
41	*Actinidia melanandra* Franch.	Broad-leaved forests, mountain forests, moist places, also widely cultivated; 1000–1600 m. Anhui, Chongqing, Fujian, Gansu, Guizhou, Henan, Hubei, Hunan, Jiangxi, Shaanxi, Sichuan, Yunnan, Zhejiang
42	*Actinidia melanandra* var. *melanandra* Synonym: *Actinidia changii*; *A. globosa*; *A. henanensis*; *A. hypoglauca*; *A. viridiflava*	Broad-leaved forests, moist places, also widely cultivated; 1000–1600 m; Chongqing, Fujian, Gansu, Guizhou, Henan, Hubei, Hunan, Jiangxi, Shaanxi, Sichuan, Yunnan, Zhejiang
43	*Actinidia melliana* Hand.-Mazz.	Mountain forests, thickets; 200–1300 m; Guangdong, Guangxi, Hainan, Hunan, Jiangxi
44	*Actinidia obovata* Chun ex C.F.Liang	1200–1600 m; Guizhou (Jiangkou, Qingzhen), Yunnan (Shizong)
45	*Actinidia persicina* R.G.Li & L.Mo	Guangxi (Rongshui)
45a	*Actinidia petelotii* Diels	1800 m; Indo-China, known from Chapa, Tonkin
46	*Actinidia pilosula* (Finet & Gagnep.) Stapf ex Hand.Mazz.	Mountain forests; 2000–3000 m; Yunnan (Gongshan) and Northern Burma
47	*Actinidia polygama* (Siebold & Zucc.) Maxim.	Mountain forests, also widely cultivated; 500–1900 m; Anhui, Chongqing, Gansu, Guizhou, Hebei, Heilongjiang, Henan, Hubei, Hunan, Jilin, Liaoning, Shaanxi, Shandong, Sichuan, Yunnan [Japan, Korea, Russia]
48	*Actinidia rongshuiensis* R.G.Li & X.G.Wang	Guangxi (Rongshui)
49	*Actinidia rubricaulis* Dunn	Broad-leaved forests; 300–2900 m; Chongqing, NW Guangxi, Guizhou, Hubei, W Hunan, Sichuan, Yunnan [Thailand]
50	*Actinidia rubus* H.Lév.	Mountain forests, valleys; 2000–2100 m; Sichuan, Yunnan (Zhaotong)
51	*Actinidia rudis* Dunn	Sparse forests, thickets, stream sides, moist places, roadsides; 1200–2300 m; Yunnan
52	*Actinidia rufa* (Siebold & Zucc.) Planch. ex Miq.	Mountain forests; 1000–2000 m; Taiwan [Japan, Korea]
53	*Actinidia rufotricha* C.Y.Wu	Broad-leaved evergreen forests, valleys, roadsides; 900–1500 m; NW and SW Guangxi, Guizhou, Yunnan
54	*Actinidia scabiifolia* Dunn	Sparse forests on mountain slopes and summits; above 1000 m; Anhui, Fujian, Hunan, Jiangxi
55	*Actinidia sorbifolia* C.F.Liang	Dense forests, mountain thickets, along rivers; 800–1600 m; Guizhou, Hunan (Chengbu), Sichuan (Qianwei)

Table 17.2 Contd. ...

...Table 17.2 Contd.

SN	Name of species	Distribution
56	*Actinidia stellatopilosa* C.Y.Chang	Mountain thickets; ca. 1200 m; Chongqing (Chengkou)
57	*Actinidia strigosa* Hook.f. & Thomson	2100–3300 m; India (Sikkim)
58	*Actinidia styracifolia* C.F.Liang	600–900 m; Fujian, Guizhou, Hunan, Jiangxi
59	*Actinidia suberifolia* C.Y.Wu	Dry thickets; 900–1000 m; Yunnan (Mengzi, Pingbian)
60	*Actinidia tetramera* Maxim.	Mountain forests, thickets, moist places; 1100–2700 m; Chongqing, Gansu, Henan, Hubei, Shaanxi, Sichuan, Yunnan
61	*Actinidia trichogyna* Franch.	Mountain forests; 1000–1800 m; Chongqing (Chengkou, Wushan, Wuxi), Guizhou (Yanhe), Hubei (Hefeng, Lichuan), Hunan, Jiangxi (Jingdezhen, Lichuan), Sichuan (Wanyuan)
62	*Actinidia truncatifolia* C.Y.Chang & P.S.Liu	Dense forests, mountain thickets, along rivers; 800–1600 m; Guizhou, Hunan (Chengbu), Sichuan (Qianwei).
63	*Actinidia ulmifolia* C.F.Liang	Mountain forests; ca. 900 m; Sichuan (Pingshan)
64	*Actinidia umbelloides* C.F.Liang	Mixed forests; 1800–2000 m; Yunnan
65	*Actinidia valvata* Dunn	Sparse forests, thickets, low mountain valleys; ca. 1000 m; Anhui, Fujian, Guangdong, Hubei, Hunan, Jiangsu, Jiangxi, Zhejiang
66	*Actinidia venosa* Rehder	Mountain forests, in thickets, 1000–3650 m; Western China (Sikang, Szechuan, and northern Yunnan) and north-eastern India (Khasia)
67	*Actinidia vitifolia* C.Y.Wu	Broad-leaved forests, forest margins, limestone mountains; 1600–1900 m; Sichuan (Ebian, Leibo, Mabian), Yunnan
68	*Actinidia zhejiangensis* C.F.Liang	Fujian, Zhejiang

Source: www.plantlist.org, www.worldflora.online.org, www.gbif.org, www.efloras.org.

Actinidia is a dioecious plant native to temperate eastern Asia, occurring throughout most of China, Taiwan, Korea, and Japan, and extending north to southern areas of Russian Far East and south into Indochina. This interesting species is native to the provinces of Hupeh, Szechuan, Kiangsi, and Fukien in the Yangtze Valley of northern China and Zhejiang Province on the coast of eastern China. Plants and seeds have been distributed from New Zealand to the United States of America and to Australia, South Africa, Germany, Netherlands, and Denmark. The fruits were very popular, going mainly to Japan, North America and Europe, with small quantities to Australia, the UK and Scandinavia. Today, West Germany is New Zealand's biggest customer for kiwifruit (Tables 17.1 and 17.2).

Health Benefits of Kiwifruit

(i) Food and nutritional benefit

Kiwifruits are hand-picked, served as appetizers, used in bread, beverages, salads, fish, fowl, meat dishes, pies, puddings, arranged as cake-filling, and ice cream may be topped with kiwifruit sauce or slices (Morgan 1988). Overripe fruits are utilized for flavouring ice cream and juice production. The fruits so used are not peeled but put through a processing machine that removes the hairs, skin and seeds (Morgan 1988). Slightly under-ripe fruits, which are high in pectin, must be chosen for making jelly, jam and chutney. Freeze-dried kiwifruit slices are shipped to health food outlets in Sweden and Japan and sometimes covered with chocolate. The peeled fruits may be pickled with vinegar, brown sugar and spices. Cull fruits can be made into wine (Morgan 1988). Kiwifruit is a nutrient-dense fruit. The updated information on the nutritional composition of kiwifruit is presented in Table 17.3 (USDA 2016, Richardson et al. 2018).

The total ascorbic acid content in the Hayward green cultivar is typically between 80 and 120 mg per 100 g fresh weight (Beever and Hopkirk 1990), the most nutritional characteristic of kiwifruit (Boland

Table 17.3. Nutritional composition of kiwifruit.

Nutrient	Units/100 g	Green kiwifruit	Gold kiwifruit
Proximates			
Water	g	83.1	82.4
Energy	kcal	61	63
Energy	kj	255	262
Protein	g	1.14	1.02
Total lipid (fat)	g	0.52	0.28
Ash	g	0.61	0.47
Carbohydrate, by difference	g	14.7	15.8
Fibre, total dietary	g	3	1.4
Sugars, total	g	9.0	12.3
Minerals			
Calcium, Ca	mg	34	17.0
Iron, Fe	mg	0.31	0.21
Magnesium, Mg	mg	17	12.0
Phosphorus, P	mg	34	25
Potassium, K	mg	312	315
Sodium, Na	mg	3	3
Zinc, Zn	mg	0.14	0.08
Copper, Cu	mg	0.13	0.103
Manganese, Mn	mg	0.098	0.05
Selenium, Se	µg	0.2	0.44
Vitamins			
Vitamin C, total ascorbic acid	mg	92.7	161.3
Vitamin B1-Thiamin	mg	0.027	< 0.01
Vitamin B2-Riboflavin	mg	0.025	0.074
Vitamin B3-Niacin	mg	0.341	0.231
Vitamin B5-Pantothenic acid	mg	0.`183	0.12
Vitamin B6-Pyridoxin	mg	0.063	0.079
Vitamin B9-Folate	µg, DFE	25	31.0
Choline	mg	7.8	1.9
Vitamin B-12	µg	0	0.08
Vitamin A, RAE	µg_RAE	4	1
Vitamin A	1U	87	23
Vitamin E (α-tocopherol)	mg	1.46	1.51
Vitamin K	µg	40.3	6.1
Others			
Carotene, beta	µg	52	14
Lutein + zeaxanthin	µg	122	24
Scientific name		*Actinidia deliciosa*	*Actinidia chinensis*
Cultivar		Hayward	SunGold
USAD NDB No		09148	09520

Source: USDA National Nutrient Database for Standard Reference Release 28, green and gold raw kiwifruit per 100 g (USDA 2016).

2013). High levels of vitamin C in kiwifruit can improve iron bioavailability (Beck et al. 2010). The variation of vitamin C content in fruit is due to various factors (Lee and Kader 2000). Vitamin C (ascorbic acid) is an essential dietary nutrient for humans (Nishikimi et al. 1994). The vitamin C content of green and gold kiwifruit is 92.7 and 161.3 mg per 100 g, respectively (USDA 2016). In the European Union, the reference intake for vitamin C for labelling purposes is 80 mg (European Commission 2011). Kiwifruit contains relatively high levels of vitamin E (Ferguson and Ferguson 2003, Boland 2013). SunGold and green kiwifruit contain 1.40 and 1.46 mg per 100 g (USDA 2016), respectively. Fiorentino et al. (2009b) identified that α-tocopherol found in the flesh of kiwifruit is a new form of vitamin E (δ-tocomonoenol) (Chang and Liu 2009, Hunter et al. 2011a,b). Kiwifruit is often referred to as being a good source of dietary folate. The folate content of 31 µg per 100 g in gold kiwifruit meets the criteria of EU Regulation (Richardson 2015, Richardson et al. 2018). The folate requirements are 600 µg/day during pregnancy (Richardson 2015). As folate is extremely labile and its presence in green leafy vegetables is easily destroyed by cooking, fresh kiwifruit can make a useful contribution to the total diet, especially during pregnancy (Richardson 2015).

Green and gold kiwifruit are good sources of potassium, containing typically around 301–315 mg per 100 g. The sodium to potassium (Na^+/K^+) ratio of kiwifruit is consistent with recommendations to increase potassium intake through increased consumption of fruit and vegetables (Hsieh et al. 2016). Potassium-rich diets or interventions with potassium can lower blood pressure, especially in individuals with hypertension (Drewnowski et al. 2012, Rust and Ekmekcioglu 2016). Kiwifruit contains about 2–3% of fresh weight non-starch polysaccharides (Ferguson and Ferguson 2003), providing a valuable contribution of both soluble and insoluble fibre. Analysis of dietary fibre of kiwifruit has shown that they comprise about one-third soluble and two-thirds insoluble fibres (Mishra and Monro 2012). The high swelling and water retention of kiwifruit fibre in comparison with other forms of dietary fibre such as wheat bran, commercial preparations of sugar beet fibre and apple fibre, accentuate the value of consuming kiwifruit as a natural whole product (Sims and Monro 2013).

The predominant sugars present in *Actinidia* are glucose and fructose with a small amount of sucrose (Pérez et al. 1997, Nishiyama et al. 2008). As kiwifruit begins to ripen, starch concentration decreases and total sugars increases and the concentration of soluble solids also increases before the fruit is ripe for eating. In fruit that is ready for consumption, the sugars provide the appealing sweet flavour of kiwifruit (Pérez et al. 1997, Nishiyama et al. 2008). From a physiological perspective, the sugar content of kiwifruit, like all fruits, may potentially influence the management of blood sugar levels following their consumption; however, current research suggests that the glycaemic response effects of kiwifruit as a whole food are potentially different from that which could be expected of individual components (Mishra et al. 2017). Interestingly, the glycaemic index (GI) of kiwifruit is relatively low (green kiwifruit, 39.3 ± 4.8 and gold kiwifruit, 48.5 ± 3.1 (Rush and Drummond 2009). The low GI value of kiwifruit is observed in both healthy human subjects and those with Type 2 diabetes (Chen et al. 2011). Kiwifruit contains several unique proteins, and the cysteine protease actinidin is the most abundant protein (Lewis and Luh 1988, Arcus 1959), having potential role in human health (Grozdanovic et al. 2014, Bayer et al. 2017) to influence protein digestion, and intestinal permeability (Cavic et al. 2014). Green and gold kiwifruit have been known to cause allergic reactions (Lucas et al. 2004). Kiwellin is another protein in kiwifruit that as a function of the ripening stage and postharvest treatment of the fruit is susceptible to actinidin activity producing the peptide kissper, and KiTH (Tuppo et al. 2008, Hamiaux et al. 2014, Ciacci et al. 2014).

(ii) Medicinal benefits

The health benefits of consuming fruits are well documented (Boeing 2012). Health benefits of kiwifruit are attributed to a high content of specific minerals, vitamins and phytonutrients like carotenoids and essential fatty acids. Studies have shown that increased consumption of plant foods like kiwifruit decreases the risk of obesity, overall mortality, reduced risk of heart disease, diabetes, cancer, and other conditions (www.medicalnewstoday.com). There are many medicinal uses of Kiwifruit (Morgan 1988, Lal et al. 2010, Chawla et al. 2016).

A. chinensis has been used as ethnomedicine in China (He et al. 2017, Wei et al. 2018, Fang et al. 2019). The ripe kiwifruit tastes sweet and sour, acts on the spleen, stomach, and kidney meridians, and has improving properties on dyspepsia, loss of appetite, and vomiting. The branches and leaves have been used to treat arthronalgia, bleeding, empyrosis, and ulcer. The vine has appetizing, heat-clearing, and wind-dampness dispelling effects and is used to treat indigestion, jaundice, and urolithiasis. The root and bark of *A. chinensis* taste bitter and astringent, and they have various medical effects such as dispelling wind and heat, improving blood circulation, and detumescence properties, and are used for the treatment of rheumatoid arthritis, bruises, furuncle, swelling, filariasis, hepatitis, and dysentery (Xie 1975, He et al. 2019).

It is reported that the roots and stems of *A. macrosperma* (Wild Kiwifruit) have been extensively employed to treat various ailments, such as leprosy, abscess, rheumatism, arthritis inflammation, jaundice, and abnormal leucorrhoea (Lu et al. 2011, Hettihewa et al. 2018, Lai and Zhang 2002) and were also useful for the treatment of cancers, especially those of lung, liver, and digestive system including oesophageal cancer (Jiangsu New Medicine College 1984, Yao and Wang 1989). It has been enrolled in traditional Chinese medicine standards of Zhejiang Province (Health Bureau of Zhejiang Province 1994).

Cancer

Kiwifruit contains an anti-mutagenic component, helping to prevent mutations of genes that may initiate the cancer process. Lutein, an important phytochemical found in kiwifruit, has been linked to the prevention of prostate and lung cancer. The FDA, an excellent source of dietary fibre in kiwifruit, is important for colorectal cancer prevention. The benefit of this laxative action is to decrease the build-up of cancer-promoting metabolites (Lal et al. 2010). It has been found beneficial in the treatment of gastric and breast cancer, as well as cancers of oesophagus and liver (Chawla et al. 2016).

Cardiovascular disease

Kiwifruit is also particularly high in two amino acids: arginine and glutamate. Arginine may help improving blood flow important for heart health. Fruits and seeds of kiwifruit, due to the presence of vitamin E and omega-3 fatty acids, have the property of blood thinner, reduce clotting and lower fats in the blood (triglycerides) (Lal et al. 2010, Chawla et al. 2016). Kiwifruit contains a wealth of carotenoids, phenolic compounds, and antioxidants, including vitamin C and E. The antioxidants may help to prevent the oxidation of good cholesterol (Lal et al. 2010, Chawla et al. 2016). Cardiologists believe the sodium to potassium ratio is critical for heart health. The ratio is extremely favourable in kiwifruit. Kiwifruit contains magnesium at 6% DV. Poor magnesium status is associated with heart disease, myocardial infarction and hypertension (Lal et al. 2010, Chawla et al. 2016). The fibre and potassium in kiwis support heart health. An increase in potassium intake along with a decrease in sodium can help reduce their risk of cardiovascular disease. Due to their high potassium content, kiwifruit can help reduce high blood pressure. It is well known that high sodium intake is a risk factor for high blood pressure (Chawla et al. 2016).

Depression and diabetes

Inositol is also found in kiwifruit and because of its function as a precursor of an intracellular second messenger system, kiwifruit is beneficial for the treatment of depression (Lal et al. 2010). Inositol, sugar alcohol naturally occurring in kiwifruit, plays a positive role in regulating diabetes and also improves nerve conducting velocity in diabetic neuropathy (Lal et al. 2010).

Eye health/macular degeneration

Kiwifruit is rich in xanthophylls and leutin. These important compounds are important to prevent macular degeneration, the leading cause of irreversible blindness (Lal et al. 2010).

Nutrient-rich

Kiwifruit is the most nutrient-dense fruit among all fruits which is beneficial in many ways to human beings. According to a recent report from New Zealand, the kiwifruit is rich in folic acid, potassium, chromium and Vitamin E (Morgan 1988, Chawla et al. 2016).

Immunity and impotence

Kiwifruit is an immune promoter likely due to its extremely high Vitamin C content and its complement of antioxidant compounds (Lal et al. 2010). Kiwifruit contains the amino acids arginine, a well-known vasodilator that has been used to treat impotence in men (Lal et al. 2010).

Physical fitness and weight control

Kiwifruit contains a wide range of minerals essential for replenishing those lost during exercise, especially in hot environments. It is also a naturally significant source of electrolytes for a pre-workout regimen (Lal et al. 2010). Kiwifruit is one of the most nutrient-rich fruits. It has the best balance of nutrients per calorie (Lal et al. 2010).

Stress reduction and respiratory health

Kiwifruit contains a relatively high level of serotonin. Serotonin causes claiming effects in most individuals (Lal et al. 2010). The consumption of vitamin C rich foods such as kiwifruit may reduce respiration related health problems in childhood (Lal et al. 2010). It is specifically useful in treating asthma in children (Lal et al. 2010).

Constipation

The high fibre content also has a mild laxative effect (Lal et al. 2010). Ripe kiwifruits contain the proteolytic enzyme actinidin, which is said to aid digestion. Regular consumption of kiwifruit was shown to promote bulkier, softer and more frequent stool production (Lal et al. 2010, Chawla et al. 2016, www.medicalnewstoday.com).

Anaemia and cell development

It facilitates the absorption of iron, thus preventing anaemia. Folate in the kiwi plays an important role in cell development in pregnant women (Lal et al. 2010).

Metabolism and nerve function

Consumption of kiwifruit strengthens metabolism and improves nerve function (Lal et al. 2010).

Fluid balance and reduce aging

It also helps to maintain proper fluid balance in the body (Lal et al. 2010). Kiwifruits are rich in antioxidants that reduce signs of aging considerably. Ascorbic acid present in this fruit provides great results in tightening the skin and pores on it (Lal et al. 2010).

Treatment of stone and better sleep

In China, the fruit and the juice of the stalk are esteemed for expelling gravel (Morgan 1988). Different parts of kiwi plants such as fruit, stems and roots have been used in the treatment of stones in the urinary tract. One study found that eating kiwis improved sleep according to self-rated measures (Chawla et al. 2016).

Treating mange

Mange is a type of skin disease caused by parasitic mites. The branches and leaves are boiled in water and the liquid is used for treating mange in dogs (Morgan 1988).

Additional medicinal usage

Kiwifruit is mainly used as a tonic for growing children and post-partum mothers and is useful in hepatic injury and gingival inflammation. Roots of the kiwi plant are anti-hepatotoxic and anti-pyorrheal. Different parts of kiwi plants such as fruit, stems and roots have been used in the treatment of hepatitis, oedema, and rheumatoid arthritis. High potassium intakes are also associated with protection against loss of muscle mass and preservation of bone mineral density (Lal et al. 2010, Chawla et al. 2016).

Other usages

The scraped stems of the vine are used as a rope in China, and paper has been made from the leaves and bark. If the bark at the base of the vine, close to the roots, is removed in one piece and placed in hot ashes, it will roll into a firm tube which can be used as a pencil. It can be extracted and purified as a powder for tenderizing meat (Morgan 1988).

Phytochemical Constituents of *Actinidia chinensis* and *Actinidia deliciosa*

The genus *Actinidia* (mostly *A. chinensis* and *A. deliciosa*) contains a range of phytochemicals varying in structures, including triterpenoids, phenolic compounds (flavonoids, polyphenols, anthraquinones, and coumarins), glucosides, volatile compounds, and essential oils, etc. in their fruits, roots, leaves, etc.

Triterpenoids

The triterpenoids have been the major research focus of *Actinidia*. It is noteworthy that some of these triterpenoids have significant antitumour activity and deserve further research and development (He et al. 2019). The following triterpenoides were isolated.

(2α, 3β, 4α)-2, 3, 23-Trihydroxyursa-12, 20(30)-dien-28-oic acid (Actinidic acid); Maslinic acid; Ursolic acid acetate; 23-Hydroxyursolic acid; Ergosta-4, 6, 8(14), 22-tetraen-3-one; 2α, 3β, 24-Trihydroxyurs-12-en-28-oic acid; 2α, 3α, 24-Trihydroxyurs-12, 20(30)-dien-28-oic acid; Pygenic acid A (3-*epi*-corosolic acid); 2α,3β-Dihydroxyurs-12-en-28,30-olide; 2α,3β,24-Trihydroxyurs-12-en-28,30-olide; 3β-Hydroxyurs-12,18-dien-28-oic acid; 2α,3α,23-Trihydroxyursa-12, 20(30)-dien-28-oic acid; 2α,3α,19α,23, 24-Pentahydroxyurs-12-en-28-oic acid; Ursolic acid; Pseudotaraxasterol; 2α,3α,23-Trihydroxyurs-12-en-28-oic acid; 2α,3β,24-Trihydroxyurs-12-en-28-oic acid; 2α,3β,19α, 23-Tetrahydroxyurs-12-en-28-oic acid; 2α,3α,19α, 24-Tetrahydroxyurs-12-en-28-oic acid; Oleanolic acid acetate; Corosolic acid; Arjunic acid; Euscaphic acid; Oleanolic acid; 2α, 3α, 24-Trihydroxyolean-12-en-28-oic acid; 2α, 3α, 19α, 24-Tetrahydroxyurs-12-en-28-oic acid; Jacoumaric acid; 3β-Hydroxystigmast-5-en-7-one; (2α, 3α)-2, 3, 23, 24-Tetrahydroxyurs-12-en-28-oic acid; 2α, 3α, 23, 24-Tetrahydroxyursan-12-en-28-acid; Oleanan-28-oic acid, 12-chloro-2, 3, 13, 23-tetrahydroxy-, γ-lactone, (2α, 3β, 4α, 12α)-; Urs-13(18)-en-28-oic acid, 2, 3, 23-trihydroxy-, (2α, 3β, 4α)-; Urs-13(18)-en-28-oic acid, 2, 3, 19, 23-tetrahydroxy-, β-D-glucopyranosyl ester, (2α, 3β, 4α)-; Pygenic acid B (2α,3α,24-trihydroxyurs-12-en-28-oic acid); 2α,3α,23,24-Tetrahydroxyursa-12, 20(30)-dien-28-oic acid; 2α,3β,23,24-Tetrahydroxyurs-12-en-28-oic acid; 2α,3β,23-Trihydroxyurs-12-en-28-oic acid; 3β-Hydroxyurs-12-en-28-oic acid; 3β-Hydroxyolean-12-en-28-oic acid; 2β,3α,23-Trihydroxyurs-12-en-28-oic acid; 2β,3β,23-Trihydroxyurs-12-en-28-oic acid; Spathodic acid 28-O-β-glucopyranoside; Fupenzic acid; 2α-hydroxyoleanolic acid; 2α-hydroxyursolic acid; 3β-O-acetylursolic acid; ergosta-4,6,8,(14),22-tetraen-3-one; 3-epi-corosolic acid; 23-hydroxyursolic acid; 2α,3α,19,24-tetrahydroxyurs-12-en-28-oic acid; 2α,3α,23-trihydroxyursa-12,20,(30)-dien-28-oic acid; 12α-chloro-2α,3β,13β, 23-tetrahydroxyolean-28-oic acid-13-lactone; 2α,3α,19α,23,24-pentahydroxyurs-12-en-28-oic acid; 2α,3β-dihydroxyolean-12-en-28-oic acid; 2α,3β-dihydroxyurs-12-en-28-oic acid; 2α,3α,24-trihydroxyolean-12-en-28-oic acid; 2α,3α,19α,24-tetrahydroxyurs-12-en-28-oic acid; 2α,3β,19α,23-tetrahydroxyurs-12-en-28-oic acid; 2α,3α,24-trihydroxyurs-12,20(30)-dien-28-oic acid; γ-quinide; 2α,3α,19-trihydroxyurs-12-en-28-oic acid; 2α,3α,23-trihydroxyurs-12-en-28-oic acid; Asiatic acid; 2α,3β,19-trihydroxyolean-12-en-28-oic acid; 2α,3α,24-trihydroxyolean-12-en-28-oic acid; 3β-O-acetyloleanolic acid; 2α,23-dihydroxylmicromeric acid; 2β,3β,23-trihydroxyurs-12-ene-28-olic acid; (2α, 3β)-2,3,23-trihydroxyurs-13(18)-en-28-oic acid; 28-O-β-Dglucopyranoside, 2α,3β,23-trihydroxyursa-12,20(30)-dien-28-oic acid; 2α,3α,23,24-tetrahydroxyurs-12-en-28-oic acid, spathodic acid-28-O-β-D-glucopyranoside; 12-diene-30-oic acid; 3-β-(2-carboxybenzoyloxy) oleanolic acid (Ji and Liang 1985, Lahlou et al. 2001, Chang and Case 2005,

Cui et al. 2007, Zhou et al. 2009, Xu et al. 2010, 2016, Chen et al. 2011, Peng et al. 2013, Zhu et al. 2013, He et al. 2014, 2015a, 2019, Cheng et al. 2015, Cui 2016, Wei et al. 2018, Zhang et al. 2018).

Phenolic compounds

Phenolic compounds are abundantly present in different botanical parts of *A. chinensis*, and they have been drawing increased attention. These compounds include polyphenols, flavonoids, anthraquinones and coumarins, and are characterized by antitumour, antioxidant and free radicals scavenging properties (He et al. 2019) which is supported by other literatures (Michaud and Ane-Margail 1977, Ji and Liang 1985, Motohashi et al. 2001, Chang and Case 2005, Wurms and Cooney 2006, Parveen et al. 2008, Lee et al. 2010, Zhou et al. 2010, Chen et al. 2011, Pinelli et al. 2012, Parameswaran and Murthi 2014, He et al. 2014, 2015a, b, 2016, Xu et al. 2016, Henriques et al. 2018, Kim et al. 2018, Tingting et al. 2018, Zhang et al. 2018).

Planchol A; Planchol B; Planchol C; Planchol D; Benzeneacetic acid; 2-[(3, 4-dihydroxybenzoyl) oxy]-4, 6-dihydroxy-, methyl ester; Vanillic acid; 1-O-(β-D-glucosyl)-2-[2-methoxy-4-(ω-hydroxypropyl)-phenoxy]-propan-3-ol; Protocatechualdehyde; rel-(1R, 2R)-1, 2-Bis(4-hydroxy-3-methoxyphenyl)-1, 3-propanediol; rel-(1R, 2S)-1, 2-Bis(4-hydroxy-3-methoxyphenyl)-1, 3-propanediol; p-Hydroxyl benzoic acid; Chlorogenic acid; Caffeic acid; Cryptochlorogenic acid; Neochlorogenic acid; 5-O-Coumaroylquinic acid; Dihydroxy-dihydrochalcone-2'-O-β-D-glucopyranoside; β-vanillic acid glucoside; Danshensu (Salvianic acid A).

Hydroxy-cinnamic acid

Ferulic acid glucoside; Ferulic acid derivative; ferulic acid; Caffeoylthreonic acid dimer; Caffeoylthreonic acid; caffeic acid glucosyl derivatives; Caffeic acid glucoside; Dimethyl caffeic acid; Caffeic acid derivative; hydroxyl cinnamic acid.

Flavonoids

Rutin, Quercetin 3-O-β-D-rhamnoside; Procyanidin trimer, Procyanidin tetramer, Catechin, Epicatechin, Quercetin, protocatechuic acid; 3,5-Dihydroxy-2-(Methoxycarbonylmethyl) phenyl 3,4-Dihydroxybenzoate; phlorizin, L-epicatechin, (−)-epigallocatechin gallate, (−)-epicatechingallate, p-coumaric acid, gallic acid, gentisic acid, salicylic acid, stilbene (trans-resveratrol), Epicatechin, epi-Afzelechin,, 2-(3, 4-Dihydroxyphenyl)-3, 4-dihydro-4-[(phenylmethyl) thio]-2H-1-benzopyran-3, 5, 7-triol, 2, 2'-Bis(3, 4-dihydroxyphenyl)-3, 3', 4, 4'-tetrahydro-4'-[(phenylmethyl) thio] [4, 8'-bi-2H-1-benzopyran]-3, 3', 5, 5', 7, 7'-hexol, Afzelechin, Procyanidin B3, Procyanidol B2, Afzelechin-(4α→8)–afzelchin, (2R, 2'R, 3R, 3'R, 4R)-3, 3', 4, 4'-Tetrahydro-2, 2'-bis(4-hydroxyphenyl) [4, 8'-bi-2H-1-benzopyran]-3, 3', 5, 5', 7, 7'-hexol, Quercetin, (+)-Catechin, (−)-Epicatechin-5-O-β-D-glucopyranoside; Kaempferol, (−)-epicatechin; procyanidol B2; procyanidol C1; (−)-epi-afzelechin; benzylthio-(−)-epi-catechin; 4'-benzylthioprocyanidol B2; (+)-afzelechin; (−)-epi-afzelechin(4β→8)(−)-epi-afzelechin; (−)-epicatechin(4β→8)(−)-catechin; (+)-afzelechin(4α→8)(+)-afzelechin; (+)-catechin(4α→8)(+)-catechin, formononetin; 7,4'-dihydroxyflavone; (+)-catechins-7-phytol.

Coumarin

5-methoxy-coumarin-7-β-D-glycosidase; Esculin; Isofraxoside; 5-Hydroxy-6-methoxy-7-O-β-D-glucosyl coumarin; 6-hydroxy-7-(β-D-glucopyranosyloxy) coumarin; Coumarin glucosydes; Isoscopoletin, Fraxetin, Aesculetin, Umbelliferone.

Anthraquinone

Emodic acid, Hydroxyemodin, Emodin, Emodin 3-methyl ether, Questin, emodin-6-methyl ether, emodin 8-β-D-glucoside, emodin 8-methyl ether, ω-hydroxyemodin.

Carotenoids and chlorophyll

9′-cis-neoxanthin, violaxanthin, antheraxanthin, lutein, zeaxanthin, β-cryptoxanthin, β-carotene, provitamin A, anthocyanins-carotenoids, Chlorophyll a, chlorophyll b (Sommerburg et al. 1998, McGhie and Ainge 2002, Montefiori et al. 2005).

Glucosides

Methoxyhydroquinone-3-O-β-D-glucopyranoside, methoxyhydroquinone-1-O-β-D-glucopyranoside, Fraxin, 1-O-(β-D-glucosyl)-2-[2-methoxy-4-(ω-hydroxypropyl)-phenoxy]-propan-3-ol, n-butyl-β-D-fructopyranoside, sucrose, stearyl-β-D-glucopyranoside, 4,4′-dihydroxyl-dihydrochalcone-2'-O-β-D-glucopyranoside, 4-*O*-*β*-D-glucopyranoside, acetylated monosaccharides, polysaccharide liposomes (Xu et al. 2010, 2016, Zhou et al. 2010, Chen et al. 2011, Liu et al. 2013, Ding et al. 2014, Wei et al. 2018).

Other compounds

β-Sitosterol, Butyl β-D-fructopyranoside, Lignoceric acid, (–)-Quinic acid, γ-lactone, Daucosterol, Indole-3-carboxylic acid, Stigmastane-3,6-diol, Sitoindoside I, 3β-hydroxystigmast-5-en-7-one, uracil, adenine, myoinositol, (Z)-9,10,11-trihydroxy-12-Octadecenoic acid, Syringaldehyde, Dotriacontanic acid, n-stearic acid, 5,7-dihydroxychromone,8-dimethyl-2-(4, 8, 12-trimethyltridec-11-enyl) chroman-6-ol, alpha- and delta-tocopherol, stigmasterol, Sitosterol, Germacrene D, (E)-β-ocimene, Histidine, Arginine, Tyrosine, Valine, Phenylalanine, Starch, Cellulose, Pectin, Sugars (fructose, glucose), Dietary fibre, Vitamin B_1 (Thiamine), Vitamin B_2 (Riboflavin), Vitamin B_3 (Niacin), Vitamin B_6, Vitamin B_9 (Folate), Vitamin C, Vitamin E, and Vitamin K, Magnesium (Mg), Phosphorous (P), Manganese (Mn), Potassium (K), Sodium (Na), Zinc (Zn), Citric acid, Quinic acid, Maleic acid, 3-penten-2-ol, 3-hydroxy-2-butanone, 3-methyl-2-butenal, 2-hexanol, 3-methyl-1-butanol, 2-methyl-1-butanol, 3-methyl-2-butanone, 3-methyl 3-buten-2-one (Ji and Liang 1985, Takeoka et al. 1986, Bolzoni et al. 1988, Cossa et al. 1988, Young and Paterson 1995, Motohashi et al. 2001, Mariäa et al. 2002, Cui et al. 2007, 2016, Wen-Wei et al. 2010, Zhou et al. 2010, Chen et al. 2011, Shastri et al. 2012, Parameswaran and Murthi 2014, He et al. 2014, 2016, Cui 2016, Xu et al. 2016).

Steroids

Campesterol (Wen-Wei et al. 2010).

Pharmacological Activities

Actinidia contains a variety of bioactive compounds accounting for natural pharmacological properties including antitumour, antioxidant, anti-inflammatory, immunoregulatory, hypolipemic, antidiabetic, cardiovascular protective, etc., and most of these biological activities validate its traditional ethnomedicinal use (He et al. 2019).

Antitumour activity

The roots of *A. chinensis* are usually used as a crude drug in China for the treatment of all kinds of cancer (Zuo et al. 2012). The polysaccharide isolated from *A. chinensis* revealed anti-tumour effect (Lin 1998), and polysaccharide derived from *A. chinensis* root inhibited the growth of human gastric cancer (Song et al. 2014). Zhang and Han (2012) reported that the polysaccharide of *A. chinensis* root can reduce the H22 tumour cell proliferation. The cytotoxicity of phenolic constituents of *A. chinensis* root showed significant cytotoxic effect against leukaemia P-388 and adenocarcinomic human alveolar basal epithelial (A549) cell lines (Yu et al. 2009, Peng et al. 2013, Olatunde et al. 2019). The round extraction of *A. chinensis* root showed potential inhibition of the proliferation of colon cancer and gastric cancer, leukaemia and hepatic cancer (Huo et al. 2013, Yang et al. 2016).

Cheng et al. (2015) evaluated 2β, 3β, 23-trihydroxy-urs-12-ene-28-olic acid against the lung cancer cell lines (Olatunde et al. 2019), which showed that this compound can be regarded for the treatment of lung cancer. Xu et al. (2016) studied the rude extraction and six main isolated compounds from *A. chinensis* root against P450 enzyme where five compounds exhibited distinct inhibition of cytochrome P450 enzyme and one compound showed a significant inhibitory effect on the CYP3A4 (Olatunde et al. 2019). The rude EtOH extraction of the root of *A. chinensis* showed strong inhibition against SW480 cells and could trigger apoptosis of tumour cell *in vitro* (Yang et al. 2016). Wang et al. (2017) obtained the total triterpenoids from the *A. chinensis* root and evaluated its inhibition of the proliferation rates of human colon cancer cells and LOVO cells, human hepatoma cells and human breast cancer (Olatunde et al. 2019) and the results showed that the total triterpenoids inhibited cancer cells proliferation at Gap and Synthesis phase, while the rude extraction from the root of *A. chinensis* possessed anti-tumour efficacy (He et al. 2017, Olatunde et al. 2019). Wei et al. (2018) investigated the cytotoxicity of pure compounds isolated from the root of *A. chinensis* to HepG2, A549, MCF-7, SK-OV-3, and Hela cell lines which showed that 2α, 3α, 23,24-tetrahydroxyursa-12,20(30)-dien-28-oic acid is the most active compound. Hou et al. (2018) tested the rude extraction from the roots of *A. chinensis* against Hepatocellular Carcinoma, which revealed that the extraction inhibited HepG2 proliferation. Fang et al. (2019) studied the extraction of *A. chinensis* root's anticancer effect and the results showed that the extraction could significantly inhibit the proliferation, migration, clonality, invasion and epithelial mesenchymal transition of hepatocellular carcinoma. Lv et al. (2018) validated that the extraction of *A. chinensis* root could significantly inhibit lung cancer (Olatunde et al. 2019).

Hepatoprotective activity

Certain bioactive compounds such as Asiatic acid and Ursolic in *A. chinensis* root possess good hepatoprotective activity (Liu et al. 1994, Gao et al. 2006). *A. chinensis* root contains these compounds in abundance. Zhou et al. (2009, 2010) reported that 2α, 3β-dihydroxyurs-12-en-28,30-olide and 2α,3β,24-trihydroxyurs-12-en-28,30-olide isolated from *A. chinensis* showed strong hepatoprotective activity. Oleanolic acid found in *A. chinensis* root has been used for the treatment of chronic liver and human acute diseases (Liu 1995, Olatunde et al. 2019).

Antimicrobial activity

Zhang et al. (2018) tested certain compounds isolated from *A. chinensis* root for their antiviral activity and the results revealed that spathodic acid-28-O-β-D-glucopyranoside and 5-methoxy-coumarin-7-β-D-glycosidase exhibited higher anti-phytoviral activity than other isolated compounds, with inactivation effect ranging from 56.76% to 42.56% (Olatunde et al. 2019). The 24-kd protein (a potential allergen of kiwi) has a partial N-terminal sequence, which revealed that it is a thaumatin-like protein (TLP) with concanavalin A-binding ability. The protein, TLP, showed antifungal activity toward Saccharomyces carlsbergensis and Candida albicans (Gavrovic-Jankulovic et al. 2002b).

The *in vitro* antimicrobial activity of essential oil of *A. macrosperma* leaves showed that the oil possessed mild antibacterial activity against the two Gram-positive bacteria and a noteworthy activity against *Escherichia coli*. However, it was ineffective on *Pseudomonas aeruginosa* (Lu et al. 2007b). In the antifungal assay, the oil displayed a higher *in vitro* fungicidal activity against the tested fungi. The results indicated that the most profuse constituent of *A. macrosperma* leaf volatile oils is linalools, which possess high levels of antimicrobial activities. According to Knobloch et al. (1988), linalool exhibits a strong antimicrobial activity, especially on whole cells. Therefore, high linalool content seems to be responsible for the bacteriostatic and fungistatic properties and, hence, *A. macrosperma* is a potential source of antimicrobial ingredients for the food and pharmaceutical industry (Lu et al. 2007b).

Antioxidant activity

Potential benefits of kiwi in asthma might be due to antioxidant properties of vitamin C or other fruit constituents (Hatch 1995, Forastier et al. 2000). The consumption of fruits or fruit juices like kiwi may reduce damage from oxidative stress, and that this effect may be a consequence of the antioxidant activity of fruits in scavenging the reactive oxygen species generated in human plasma (Ko et al. 2005). Kiwi fruit was taken as an example of a food with putative antioxidant properties, and its effectiveness at decreasing oxidative DNA damage was assessed in *ex vivo* as well as *in vitro* tests (Collins et al. 2009). The comet assay (single-cell gel electrophoresis) was used to measure DNA damage in lymphocytes collected during a human supplementation trial with a single 500 mL drink of kiwi fruit juice (with water as a control). The comet assay was also modified to assess the antioxidant effect of kiwi fruit *in vitro* by measuring the ability of an extract to interfere with oxidative damage to DNA induced by H_2O_2. The consumption of kiwi fruit led to increased resistance of DNA to oxidative damage induced by H_2O_2 in isolated lymphocytes, in comparison with lymphocytes collected after a control drink of water. No effect was seen on endogenous DNA damage (Collins et al. 2009). A simple extract of kiwi fruit, buffered to pH 7, was more effective than a solution of vitamin C (of equivalent concentration) at protecting DNA from damage, whereas at the highest concentrations tested, neither kiwi extract nor vitamin C had a protective effect. The authors concluded that significant antioxidant activity of kiwi fruit, *ex vivo* and *in vitro*, is not attributable entirely to the vitamin C content of the fruit (Collins et al. 2009). Yang and He (2012) revealed that the extract from the root of *A. chinensis* could remove the DPPH, superoxide anion radicals effectively and strongly inhibit the hydroxyl radical's production.

Antiangiogenic Activities

Zhu et al. (2013) assessed the bioactive compounds isolated from the root of *A. chinensis* to the human umbilical vein endothelial cells proliferation and tube formation, which revealed that 2α,3β-dihydroxyurs-12-en-28-oicacid, 2α,3α,23-trihydroxyurs-12-en-28-oic acid, asiatic acid and 2α,3α,24-trihydroxyolean-12-en-28-oicacid demonstrated strong antiangiogenic effect (Olatunde et al. 2019).

Cytotoxic activity

Kiwi gold fruits were extracted successively with hexane, acetone, methanol and 70% methanol, and were further fractionated for biological activities assays (Motohashi et al. 2002). Five fractions—H1, H2 (hexane extract), Al, A2 (acetone extract) and M2 (methanol extract)—showed selective cytotoxic activity against human oral tumour cell lines, which was more sensitive than human gingival fibroblasts. More hydrophilic fractions of 70% methanol extract displayed higher anti-HIV activity, radical generation and O_2-scavenging activity. The antibacterial activity of 70% methanol extracts was generally lower than that of more lipophilic fractions (hexane, acetone, methanol extracts) (Motohashi et al. 2002).

The hexane, EtOAc and CHC13 extracts of *A. macrosperma* showed cytotoxicity against human hepatoma SMMC-7721 tumour cell lines ($IC_{50} < 100\ \mu g/mL$) with a dose-dependent manner. Methanol and aqueous extracts, however, showed weak activities ($IC_{50} > 180\ \mu g/mL$) in this test. Thereinto, a very significant cytotoxic activity, not significantly different from the positive control of quercetin ($P > 0.05$), was observed in CHC13 extract ($IC_{50} = 50.97\ \mu g/mL$). Since the plant was extracted successively with hexane, EtOAc, and MeOH, a slight cytotoxic activity of the ethyl acetate extract indicated that the cytotoxic compound in the hexane extract differed from that in the methanol extract (Lu et al. 2012).

Dermatological activity

Potential dermatological activity from *A. chinensis* was characterized and tested under *in vitro* conditions for regulating activities on cell physiology of human keratinocytes, fibroblasts, and skin equivalents (Deters et al. 2005). Ten micrograms per milliliter of raw polysaccharide, neutral type-II-arabinogalactans, and acidic arabinorhamnogalacturonans of kiwifruits stimulated cell proliferation of human keratinocytes (NHK, HaCaT) up to 30% significantly stimulated, while mitochondrial activity was stimulated for nearly 25% in regard to control cells. Fibroblasts were not as sensitive as keratinocytes but > 130 mcg/mL kiwifruit polysaccharides increased proliferation and ATP-synthesis significantly. In three-dimensional skin equivalents, it was shown that the polysaccharides led to a double collagen synthesis of fibroblasts compared to the normally strongly reduced biosynthetic activity.

Platelet effects

Consuming two or three kiwi fruit per day for 28 days reduced platelet aggregation response to collagen and ADP by 18% compared with the controls. In addition, consumption of kiwi fruit lowered blood triglycerides levels by 15% compared with control ($p < 0.05$), whereas no such effects were detected in the case of cholesterol levels (Duttaroy and Jorgensen 2004).

Allergy constituents

The major allergen of kiwifruit is actinidin (Pastorello et al. 1998), which may be responsible for adverse effects. Actinidin, a proteolytic enzyme belonging to the class of thiolproteases appears to sensitize digestive tract via the GI tract (Mills et al. 2004). Three proteins of the kiwi extract were found considered to contain a carbohydrate moiety (Rudeschko et al. 1998). Profilin seems to be relevant in cross-reactivity of kiwi allergens. Twelve IgE-binding components with molecular weights ranging from 12 to 64 kd were identified in the kiwi extract, but only a 30-kd component acted as a major allergen (Pastorello et al. 1996) and other allergens have been proposed, such as 24-kd TLP (Gavrovic-Jankulovic et al. 2002a, b). It is known that within 16 hours, kiwi allergens were largely degraded by simulated intestinal fluid (Yagami et al. 2000).

Ethnopharmacology of Wild Kiwifruit (*Actinidia macrosperma*)

Kiwifruit are nutritious and medicinally important (Ferguson 2007, Nishiyama 2007, Lu et al. 2007a,b, 2012, Fiorentino et al. 2009a,b, Hunter et al. 2011a,b, Stonehouse et al. 2012), little varieties are grown commercially, and many varieties are being evaluated for commercialization. *Actinidia macrosperma*, a naturally wild kiwi, endemic to eastern China, has been used as a stimulant and to heal wounds (Jiangsu New Medicine College 1984). It grows in thickets, mixed forest, forest margins on low mountains, and moist places in Zhejiang, Jiangxi, Hubei and Jiangsu Provinces (Liang 1984). *A. macrosperma* fresh leaves and twigs are used by cats to excite themselves and cure wounds and so is called 'Ginseng of cat'. *A. macrosperma* is a non-commercial kiwifruit, which is orange and small-sized with large seeds, and has relatively thick, highly-coloured, hairless skin (Hettihewa et al. 2018). The wild resource of this species has decreased rapidly because of its huge demand and thus its protection in its natural habitat has been suggested. To address this issue, in recent years, some research projects focusing on the chemistry and tissue culture (Jiang and Li 2003, Feng et al. 2004, Lu et al. 2004, Zhao et al. 2006, Yun-peng et al. 2006) of this species have been conducted.

A. macrosperma significantly improved appetite, alleviated weakness, increased weight and stabilized white blood cell counts (Zhou 2000). It could improve the quality of life by enhancing physical function and healthy transition without any adverse side effects in the patients with cancers of gastric, oesophageal, liver and lung (Lai and Zhang 2002). The injection isolated from this plant had *in vitro* antitumour efficacy to three liver cancer cell lines (Wan et al. 2004). However, information regarding its active components and biological activities remains unclear.

A total of 28 compounds were identified from the leaves of *A. macrosperma* (Yun-peng et al. 2006). The main constituents of the essential oil were straight-chain fatty acid derivatives, among which α-Linolenic acid was the major one (63.97%), followed by triterpenes. Short-chain enol derivatives may act as messengers in chemical communication (Du 2001), and these compounds were common in *A. macrosperma*. It was speculated that artificial habitat and immature growth stage might lead to the absence of these enol derivatives but the differences of behaviour in chemical communication between wild and regenerated plants need further investigation (Yun-peng et al. 2006).

The essential oil found in *A. macrosperma* are: 2-hexenal, 3-hexenol, 2-hexenol, 3-hexen-1-ol, acetate, (Z)-, 3-methyl-3-heptanol, Benzyl alcohol, β-Linalool, 3-Decyn-2-ol, β-phenylethyl alcohol, Dodecane, *cis*-α-terpineol, (*R*)-citronellol, (*E*)-geraniol, 2-methyl bicyclo[3,2,1]octane, 1,2-dimethylindoline, Dodeca-1,6-dien-12-ol-6,10-dimethyl, 3-hexadecyne, Tetradecane, Limonene-1,2-epoxide, Dihydronepetalactone, Iridomyrmecin, Dihydroactinidiolide, Hexadecane, 3,7,11,15-tetramethyl-2-hexadecenol, 2-pentadecanone-6,10,14-trimethyl, (*Z*)-11-hexadecenoic acid, *n*-hexadecanoic acid, Linolenic acid, methylester, Phytol, α-Linolenic acid, Linolenic acid ethylester, Squalene, Linolenic acid, phenylmethyl ester (Yun-peng et al. 2006).

Plants phenolic compounds constitute one of the major groups of compounds acting as primary antioxidants and free radical terminators (Dziezak 1986). The phenolic compounds extracted from the fruits of *A. macrosperma* have gained increased attention. The subclass flavonol was the most abundant group of flavonoids detected. Quercetin-3-O-galactoside, quercetin-3-O-glucoside, quercetin-3-O-rhamnoside, quercetin-3-O-rutinoside, quercetin-3-O-arabinoglucoside, catechin, epigallocatechin gallate, epigallocatechin, chlorogenic, ferulic, isoferulic, and caffeic acid were prominent phenolics found in *A. macrosperma* (Hettihewa et al. 2018). The antioxidant potential of *A. macrosperma* can increase their therapeutic applications towards the prevention of degenerative and neoplastic diseases (Lu et al. 2012).

Growing concern in flavonoids in kiwifruit is due to its health-promoting properties. Inhibition of the Angiotensin-Converting Enzyme (ACE) is one of the main therapeutic targets in controlling hypertension, which was reported in one study. Kiwifruits, namely *A. deliciosa*, *A. chinensis* and *A. macrosperma*, have been evaluated for their pharmacological applications towards the management and treatment of human diseases, including antimicrobial activity (Lu et al. 2012), antioxidant activity (Ferguson 2007, Nishiyama 2007, Fiorentino et al. 2009a, Latocha et al. 2010), immune modulatory activity (Lu et al. 2007a), and anticancer activity (Hunter et al. 2011a), etc. There are no reports on ACE inhibitory activity found in kiwifruits (Hettihewa et al. 2018).

All tested extracts inhibited ACE activity in a dose-dependent manner with different IC_{50} ranging from 0.49 to 69.54 mg/mL of flavonoid-rich extracts. The potential anti-hypertensive activities of kiwifruit cultivars showed that *A. macrosperma* possesses the lowest IC_{50} compared to the other two commercially grown cultivars. This, in theory, indicates that the flavonoid-rich extract obtained from *A. macrosperma* possesses strong anti-hypertensive agents. These observations were supported by the significantly higher total phenolic and total flavonoid contents determined by LC-MS/MS (Hettihewa et al. 2018).

It was investigated that the ACE inhibitory activity of flavonoid-rich extracts acquired from different kiwifruit genotypes of *Actinidia macrosperma*, *Actinidia deliciosa* and *Actinidia chinensis* (Hettihewa et al. 2018) were prepared by steeping the lyophilized fruit samples in 70% aqueous acetone, followed by partitioning the crude extracts with hexane. The composition of each extract was analysed using ultrahigh-performance liquid chromatography-mass spectrometry and finally, ACE inhibitory activity of the fruit extracts was performed using a fluorescence-based biochemical assay (Hettihewa et al. 2018).

Generally, the flavonoid-rich extract from *A. macrosperma* showed a significantly ($p < 0.05$) high percentage of inhibition ($IC_{50} = 0.49$ mg/mL), and enzyme kinetic studies recommended that it inhibits ACE activity *in vitro*. The tested kiwifruit extracts were found to be moderately effective as ACE inhibitors *in vitro* when compared to other plant extracts (Hettihewa et al. 2018). Several biological activities have already been described for kiwifruit cultivars, such as antioxidant, anticancer, anti-inflammatory, and antimicrobial activities. However, this is the principal report supporting the potential anti-hypertensive activities of kiwi fruits. The outcomes of this research show that the flavonoid-rich extract from *A. macrosperma* shows potential as a food or nutraceutical source of anti-hypertensive agents. Overall, the flavonoid-rich extract from *A. macrosperma* showed a significantly ($p < 0.05$) high percentage of inhibition ($IC_{50} = 0.49$ mg/mL) *in vitro*. Kinetic determinations advised that the flavonoids-rich extract obtained from *A. macrosperma* possibly inhibits enzyme activity (Hettihewa et al. 2018).

Conclusion and Future Perspectives

Kiwifruits are gradually becoming popular throughout the globe due to its commercial and nutritional values and health benefits. Kiwifruit is one of the important fruits in the international market. It is an important source of phenolic compounds, vitamin C, carbohydrates, sugars, amino acids, minerals, etc. *Actinidia deliciosa*, *A. chinensis*, and *A. macrosperma* are usually available and primarily used for the treatment of different diseases including cardiovascular diseases, depression, diabetes, eye health/ macular degeneration, hepatitis, oedema, rheumatoid arthritis, constipation, stress reduction, respiratory health, physical fitness, weight control, immunity, impotence, anaemia, cell development, metabolism, nerve function, cancer, etc.

Kiwifruit is a nutrient-dense fruit, commonly eaten fresh, but can also be used in beverages, desserts, and as a flavouring. Kiwifruit is a good source of carotenoids, vitamin C, vitamins A and E, potassium, etc. *Actinidia deliciosa* is a rich source of vitamin C. High levels of vitamin C in kiwifruit can improve iron bioavailability. It also contains vitamin E and a small amount of vitamin A. Kiwifruit should be considered as a part of a natural and effective dietary strategy to tackle some of the major health problems.

Different parts of kiwifruit (mainly from *A. deliciosa*, *A. chinensis*, and *A. macrosperma*) may have added-value and be used with benefit for the isolation of important and useful bioactive compounds. The research findings and consciousness about the emerging values of kiwifruit increase interests for their regular consumption as a part of healthy food and diet. This review highlighted the emerging values of kiwifruit and suggested further scientific research, which will be useful for drug development. It is well-known fact that kiwifruit has remarkable attractiveness, holding unexpected potential for the future and thus clinical experiments need to be carried out to exploit the therapeutic utility in fighting several diseases. In summary, the scientific research results for the health benefits of kiwifruit need to be extended through well-designed research.

Acknowledgments

The first (SB) and second (MD) authors are thankful to Nepal Agriculture Research Council, Khumaltar, Lalitpur, Nepal and Nepal Academy of Science and Technology, Khumaltar, Lalitpur, Nepal, respectively, for various supports.

References

Arcus, A.C. 1959. Proteolytic enzyme of *Actinidia chinensis*. Biochim. Biophys. Acta, 33(1): 242–244.

Bayer, S.B., Gearry, R.B., Drummond, L.N. 2017. Putative mechanisms of kiwifruit on maintenance of normal gastrointestinal function. Crit. Rev. Food Sci. Nutr., 58(14): 2432–2452.

Beck, K., Conlon, C.A., Kruger, R., Coad, J., Stonehouse, W. 2010. Gold kiwifruit consumed with an iron-fortified breakfast cereal meal improves iron status in women with low iron stores: a 16-week randomised controlled trial. Br. J. Nutr., 105(1): 101–109.

Beever, D.J., Hopkirk, G. 1990. Fruit development and fruit physiology. pp. 97–126. *In*: Warrington, I.J., Weston, G.C. (eds.). Kiwifruit: Science and Management. The New Zealand Society for Horticultural Science and Ray Richards Publisher, Auckland.

Benzie, I.F., Choi, S.W. 2014. Antioxidants in food: content, measurement, significance, action, cautions, caveats, and research needs. Adv. Food Nutr. Res., 71: 1–53.

Boland, M.J. 2013. Kiwifruit proteins and enzymes: actinidin and other significant proteins. pp. 59–80. *In*: Boland, M., Moughan, P.J. (eds.). Advances in Food and Nutrition Research: Nutritional Benefits of Kiwifruit. Vol. 68, Academic Press.

Bolzoni, L., Careri, M., Gherardi, S. 1988. Study of the aroma components of kiwifruit. Ind. Conserve., 63(4): 347–353.

Canright, J.E. 1955. The comparative morphology and relationships of the Magnoliaceae. IV. Wood and nodal anatomy. J. Arnold Arbor, 36: 119–140.

Cavic, M., Grozdanovic, M.M., Bajic, A., Jankovic, R., Andjus, P.R., Gavrovic-Jankulovic, M. 2014. The effect of kiwifruit (*Actinidia deliciosa*) cysteine protease actinidin on the occludin tight junction network in T84 intestinal epithelial cells. Food Chem. Toxicol. Int. J. Publ. Br. Ind. Biol. Res. Assoc., 72c: 61–68.

Chang, J., Case, R. 2005. Cytotoxic phenolic constituents from the root of *Actinidia chinensis*. Planta Med., 71: 955–959.

Chang, W.-H., Liu, J.-F. 2009. Effects of kiwifruit consumption on serum lipid profiles and antioxidative status in hyperlipidemic subjects. Int. J. Food Sci. Nutr., 60(8): 709–716.

Chawla, H., Parle, M., Yadav, M. 2016. Medicinal potential and phytopharmacology of *Actinidia deliosa*. Int. J. Expt. Pharm., 6(1): 20–25.

Chen, X., Yang, S., Bai, S. 2011. Chemical constituents in root of *Actinidia chinensis*. Zhongcaoyao., 42: 841–843.

Cheng, Q.L., Li, H.L., Huang, Z.Q., Chen, Y.J., Liu, T.S. 2015. 2β, 3β, 23-trihydroxy-urs-12-ene-28-olic acid (TUA) isolated from *Actinidia chinensis* radix inhibits NCI-H460 cell proliferation by decreasing NF-κB expression. Chem. Biol. Interact., 240: 1–11.

Ciacci, C., Russo, I., Bucci, C., Iovino, P., Pellegrini, L., Giangrieco, I., Tamburrini, M., Ciardiello, M.A. 2014. The kiwi fruit peptide kissper displays anti-inflammatory and anti-oxidant effects in *in-vitro* and *ex-vivo* human intestinal models. Clin. Exp. Immunol., 175(3): 476–484.

Collins, B.H., Horska, A., Hotten, P.M., Riddoch, C., Collins, A.R. 2009. Kiwifruit protects against oxidative DNA damage in human cells and *in-vitro*. Nutr. Cancer, 39(1): 148–153.

Cossa, G., Trova, C., Gandolfo, G. 1988. Extraction and identification of kiwifruit volatile constituents. Ind. Aliment., 27(261): 531–534.

Cui, Y., Zhang, X., Chen, J., Zhang, Y., Lin, X., Zhou, L. 2007. Chemical constituents from root of *Actinidia chinensis*. Zhongguo Zhongyao Zazhi., 32(16): 1663–1665.

Cui, Y. 2016. Study on chemical constituents in ethyl acetate extracts from the root of *Actinidia chinensis*. Hubei Agri. Sci., 55: 5359–5360.

Deters, A.M., Schroder, K.R., Hensel, A. 2005. Kiwi fruit (*Actinidia chinensis* L.) polysaccharides exert stimulating effects on cell proliferation via enhanced growth factor receptors, energy production, and collagen synthesis of human keratinocytes, fibroblasts, and skin equivalents. J. Cell Physiol., 202(3): 717–722.

Ding, M.H., Zhu, Y.Q., Wu, J.J., Li, C.Y., Ge, W.H., Shi, S.L., Song, S.C. 2014. Process research of formulation and preparation on polysaccharide liposomes in Chinese *Actinidia* root. Chin. J. Trad. Chin. Med. Pharm. (in Chinese), 29: 3942–3945.

Drewnowski, A., Maillot, M., Rehm, C. 2012. Reducing the sodium-potassium ratio in the US diet: a challenge for public health. Am. J. Clin. Nutr., 96(2): 439–444.

Duttaroy, A.K., Jorgensen, A. 2004. Effects of kiwi fruit consumption on platelet aggregation and plasma lipids in healthy human volunteers. Platel., 15(5): 287–292.

Dziezak, J.D. 1986. Antioxidants. Int. J. Food Sci. Techno., 40(7): 94–102.

eFloras.org. www.efloras.org. Retrieved on October 2019.

Emily, G. 2002. Kiwi Act II. https://www.latimes.com/archives/la-xpm. Retrieved on November 2018.

Esau, K. 1977. Anatomy of Seed Plants. John Wiley and Sons, New York.

European Commission. 2011. Regulation (EU) no. 1169/2011, Provision of Food Information to Consumers. Official Journal of the European Union L 304/18.

Fang, T.T., Fang, Y., Xu, X.J., He, M.Y., Zhao, Z.Y., Huang, P.X., Yuan, F., Guo, M., Yang, B., Xia, J. 2019. *Actinidia chinensis* Planch., root extract attenuates proliferation and metastasis of hepatocellular carcinoma by inhibiting epithelial-mesenchymal transition. J. Ethnopharm., 231: 474–485.

FAO. 2014. http://www.faostat.org/docrep/004/ab985e/ab985e09.htm. Retrieved on November 2019.

Feng, Y.T., Chen, S.Y., Jiang, W.M., Fu, C.X. 2004. Study on the effective components of *Actinidia macrosperma* (II): analysis on the amino acids. J. Zhej. Univ. (Agriculture & Life Science Edition), 30(2): 189–190 (in Chinese).

Ferguson, A.R., Ferguson, L.R. 2003. Are kiwifruit really good for you? Acta Hort., 610: 131–138.

Ferguson, A.R. 2007. The need for characterization and evaluation of germplasm: Kiwifruit as an example. Euphyt., 154: 371–382.

Fiorentino, A., D'Abrosca, B., Pacifico, S., Mastellone, C., Scognamiglio, M., Monaco, P. 2009a. Identification and assessment of antioxidant capacity of phytochemicals from kiwi fruits. J. Agric. Food Chem., 57: 4148–4155.

Fiorentino, A., Mastellone, C., D'Abrosca, B., Pacifico, S., Scognamiglio, M., Cefarelli, G., Caputo, R., Monaco, P. 2009b. δ-Tocomonoenol: a new vitamin E from kiwi (*Actinidia chinensis*) fruits. Food Chem., 115(1): 187–192.

Flora of China. www.floraofchina.org. Retrieved on October 2019.

Forastiere, F., Pistelli, R., Sestini, P., Fortes, C., Renzoni, E., Rusconi, F., Dell'Orco, V., Ciccone, G., Bisanti, L. 2000. Consumption of fresh fruit rich in vitamin C and wheezing symptoms in children. SIDRIA Collaborative Group, Italy (Italian Studies on Respiratory Disorders in Children and the Environment). Thorax, 55(4): 283–288.

Gao, J., Chen, J., Tang, X., Pan, L., Fang, F., Xu, L., Zhao, X., Xu, Q. 2006. Mechanism underlying mitochondrial protection of asiatic acid against hepatotoxicity in mice. J. Pharm. Pharmacol., 58: 227–233.

Gavrovic-Jankulovic, M., Cirkovic, T., Burazer, L., Vuckovic, O., Jankov, R.M. 2002a. IgE cross-reactivity between meadow fescue pollen and kiwi fruit in patients' sera with sensitivity to both extracts. J. Investig. Allergol. Clin. Immunol., 12(4): 279–286.

Gavrovic-Jankulovic, M., Cirkovic, T., Vuckovic, O., Atanaskovic-Markovic, M., Petersen, A., Gojgic, G., Burazer, L., Jankov, R.M. 2002b. Isolation and biochemical characterization of a thaumatin-like kiwi allergen. J. Allergy Clin. Immunol., 110(5): 805–810.

Global Biodiversity Information Facility, free and open access to biodiversity data. www.gbif.org. Retrieved on October 2019.

Gotame, T.P., Gautam, I.P., Subedi, G.D., Dhakal, M., Sapkota, S. 2016. Kiwiphel Kheti Prabidhi (in Nepali). Nepal Agriculture Research Council, Horticulture Research Division, Khumaltar, Lalitpur.

Grozdanovic, M.M., Ostojic, S., Aleksic, I., Andjelkovic, U., Petersen, A., Gavrovic-Jankulovic, M. 2014. Active actinidin retains function upon gastro-intestinal digestion and is more thermo sgota than the E-64-inhibited counterpart. J. Sci. Food Agric., https://doi.org/10.1002/jsfa.6656.

Gupta, P.K. 2007. Genetics Classical to Modern. Rastogi Publication, pp. 2–134.

Hamiaux, C., Maddumage, R., Middleditch, M.J., Prakash, R., Brummell, D.A., Baker, E.N., Atkinson, R.G. 2014. Crystal structure of kiwellin, a major cell-wall protein from kiwifruit. J. Struct. Biol., 187(3): 276–281.

Hatch, G.E. 1995. Asthma, inhaled oxidants, and dietary antioxidants. Am. J. Clin. Nutr., 61(3 Suppl): 625S–630S.

He, G.N., Hu, X.M., Wang, H., Fan, L., Wang, B.C. 2015a. Chemical constituents in root of *Actinidia chinensis* Planch. Zhonghua Zhongyiyao Zazhi., 30: 498–500.

He, J., Ma, B.Z., Zhao, T., Wang, W., Wei, F.L., Lu, J., Zhang, X.L. 2014. Chemical constituents from the roots of *Actinidia chinensis*. Zhongguo Yaoxue Zazhi., 49(3): 184–186.

He, J., Ma, B.Z., Wang, X.X., Liu, F., Qin, W.J., Zhang, X.L., Zhao, T. 2015b. Chemical constituents from the root of *Actinidia chinensis* (II). Chinese Journal of Pharmacy (in Chinese), 50(22): 1960–1963.

He, M., Zeng, J., Zhai, L., Liu, Y., Wu, H., Zhang, R., Li, Z., Xia, E. 2017. Effect of *in-vitro* simulated gastrointestinal digestion on polyphenol and polysaccharide content and their biological activities among 22 fruit juices. Food Res. Int. (Ottawa Ont), 102: 156–162.

He, X., Fang, J., Chen, X., Zhao, Z., Li, Y., Meng, Y., Huang, L. 2019. *Actinidia chinensis* Planch.: A review of chemistry and pharmacology. Front. Pharm., 10: 1–18.

Health Bureau of Zhejiang Province. 1994. TCM Making Standards of Zhejiang Province. Zhejiang Science and Technology Press, Hangzhou, p.89 (in Chinese).

Henriques, J., Falé, P.L., Pacheco, R., Florêncio, M.H., Serralheiro, M.L. 2018. Phenolic compounds from *Actinidia deliciosa* leaves: Caco-2 permeability, enzyme inhibitory activity and cell protein profile studies. J. King Saud. Univ.–Sci., 30: 513–518.

Hettihewa, S.K., Hemar, Y., Rupasinghe, H.P.V. 2018. Flavonoid-rich extract of *Actinidia macrosperma* (A wild kiwifruit) inhibits angiotensin-converting enzyme *in-vitro*. Foods, 7(146): 1–8.

Hou, J.J., Wang, L.Y., Wu, D.J. 2018. The root of *Actinidia chinensis* inhibits hepatocellular carcinomas cells through LAMB3. Cell Biol. Toxicol., 34: 321–332.

Hsieh, C.L., Huang, S.M., Chen, L.I., Yu, C.M., Wong, C.H., Peng, R.Y. 2014. Novel approach of using nutraceutic-directed caloric antioxidant density and ion-ratio for evaluating fruit's health quality. J. Food Sci., 81(8): H2059–2068.

Huang, H., Ferguson, A.R. 2003. Kiwifruit (*Actinidia chinesis* and A. deliciosa) plantings and production in China, 2002. New Zealand J. Crop and Hort. Sci., 31(3): 197–202.

Hunter, D.C., Greenwood, J., Zhang, J., Skinner, M.A. 2011a. Antioxidant and natural protective properties of kiwifruit. Curr. Top. Med. Chem., 11: 1811–1820.

Hunter, D.C., Skinner, M.A., Wolber, F.M., Booth, C.L., Loh, J.M.S., Wohlers, M., Stevenson, L.M., Kruger, M.C. 2011b. Consumption of gold kiwifruit reduces severity and duration of selected upper respiratory tract infection symptoms and increases plasma vitamin C concentration in healthy older adults. Br. J. Nutr., 108(7): 1235–1245.

Huo, J., Qin, F., Cai, X., Ju, J., Hu, C., Wang, Z., Lu, W., Wang, X., Cao, P. 2013. Chinese medicine formula "Weikang Keli" induces autophagic cell death on human gastric cancer cell line SGC-7901. Phytomed., 20(2): 159–165.

Ji, Z., Liang, X. 1985. Chemical constituents of the root of *Actinidia chinensis* Planch. Yaoxue Xuebao, 20: 778–781.

Jiang, W.M., Li, F.Y. 2003. Establishment of plantlet regeneration system of *Actinidia macrosperma*. J. Zhej. Univ. (Agriculture & Life Science Edition), 29(3): 295–299 (in Chinese).

Jiangsu New Medicine College. 1984. Dictionary of Chinese Traditional Medicines. Shanghai Science and Technology Press, Shanghai, China.

Kim, G.-D., Lee, J.Y., Auh, J.-H. 2019. Metabolomic screening of anti-inflammatory compounds from the leaves of *Actinidia arguta* (Hardy Kiwi). Foods, 8: 47.

Knobloch, K., Pauli, A., Iberl, B., Weis, N., Weigand, H. 1988. Mode of action of essential oil components on whole cells of bacteria and fungi in plate tests. pp. 287–299. *In*: P. Schreier (Ed.), Bioflavour'87, Walter de Gruyter, Berlin.

Knowles, M. 25 April 2017. NZ Kiwifruit: How a Major Brand Emerged. Eurofruit, Market Intelligence Ltd.—Fruitnet.com. Retrieved 10 September 2018.

Ko, S.H., Choi, S.W., Ye, S.K., Cho, B.L., Kim, H.S., Chung, M.H. 2005. Comparison of the antioxidant activities of nine different fruits in human plasma. J. Med. Food., 8(1): 41–46.

Lahlou, E.H., Hirai, N., Kamo, T., Tsuda, M., Ohigashi, H. 2001. Actinidic acid, a new triterpene phytoalexin from unripe kiwi fruit. Biosci. Biotech. Bioch., 65: 480–483.

Lai, P.F., Zhang, H.Y. 2002. The research progress of TCM Cat Ginseng in Zhejiang province location. J. Zhej. Col. TCM, 26(1): 77–78.

Lal, S., Ahmed, N., Singh, S.R., Singh, D.B., Mir, J.I., Kumar, R. July 2010. Kiwifruit: Miracle berry. Sci. Repor., pp. 52–54.

Latocha, P., Krupa, T., Wołosiak, R., Worobiej, E., Wilczak, J. 2010. Antioxidant activity and chemical difference in fruit of different *Actinidia* sp. Int. J. Food Sci. Nutr., 61: 381–394.

Lee, D.E., Shin, B.J., Hur, H.J., Kim, J.H., Kim, J., Kang, N.J., Kim, D.O., Lee, C.Y., Lee, K.W., Lee, H.J. 2010. Quercetin, the active phenolic component in kiwifruit, prevents hydrogen peroxide-induced inhibition of gap-junction intercellular communication. Brit. J. Nutr., 104: 164–170.

Lee, S.K., Kader, A.A. 2000. Preharvest and postharvest factors influencing vitamin C content of horticultural crops. Posthar. Biol. Technol., 20: 207–220.

Lewis, D.A., Luh, B.S. 1988. Development and distribution of Actinidin in kiwifruit (*Actinidia chinensis*) and its partial characterization. J. Food Biochem., 12(2): 109–116.

Lewis, R.A. 2002. CRC Dictionary of Agricultural Sciences. CRC Press.

Li, J., Li, X., Soejarto, D.D. 2007. *Actinidia* Lindley. *In*: Flora of China, 12: 334–355.

Liang, C., Chen, Y.C., Wang, Y.S. 1984. Flora of China. Science Press, Beijing, China.

Liang, C.F. 1984. *Actinidia*. pp. 196–268. *In*: Feng, K.M. (ed.). Flora Reipublicae Popularis Sinicae. Science Press, Beijing (in Chinese).

Lin, P.F. 1988. Antitumor effect of *Actinidia chinensis* polysaccharide on murine tumor. Zhonghua zhong liu za zhi, 10: 441–444 (in Chinese).

Lippi, G., Mattiuzzi, C. 2019. Kiwifruit and cancer: an overview of biological evidence. Nut. Can. https://doi.org/10.1080/01635581.2019.1650190.

Liu, J., Liu, Y.P., Mao, Q., Klaassen, C.D. 1994. The effects of 10 triterpenoid compounds on experimental liver-injury in mice. Fundam. Appl. Toxicol., 22: 34–40.

Liu, J. 1995. Pharmacology of oleanolic acid and ursolic acid. J. Ethnopharmacol., 49: 57–68.

Liu, Z.-x., Yin, H., Xiao, L.-l., Tang, M., Li, J.-x., Xiao, W.-j., Huang, C. 2013. Antifatigue and Antioxidant Activities and monosaccharide composition of polysaccharide from roots of *Kiwifruit* (*Actinidia deliciosa*). Food Sci., 34(13): 239–242.

Lu, Y., Chen, S.Y., Wu, P., Chan, G., Fu, C.X. 2004. Study on the effective components of *Actinidia macrosperma* (I): analysis on the inorganic elements. J. Zhej. Univ. (Agriculture & Life Science Edition), 30(2): 185–188 (in Chinese).

Lu, Y., Fan, J., Zhao, Y., Chen, S., Zheng, X., Yin, Y., Fu, C. 2007a. Immunomodulatory activity of aqueous extract of *Actinidia macrosperma*. Asia Pac. J. Clin. Nutr., 16(Suppl. 1): 261–265.

Lu, Y., Zhaoy, Y.P., Wang, Z.C., Chen, S.Y., Fu, C.X. 2007b. Composition and antimicrobial activity of the essential oil of *Actinidia macrosperma* from China. Nat. Prod. Res., 21(3): 227–233.

Lu, Y., Zhao, Y., Fu, C. 2011. Biological activities of extracts from a naturally wild kiwifruit, *Actinidia macrosperma*. Afr. J. Agric. Res., 6: 2231–2234.

Lu, Y., Du, X., Lai, L., Jin, H. 2012. *Actinidia macrosperma* C. F. Liang (a Wild Kiwi): Preliminary study of its antioxidant and cytotoxic activities. Evid.-Bas. Complem. Altern. Med., 1–7. doi:10.1155/2012/180262.

Lucas, J.S.A., Grimshaw, K.E.C., Collins, K., Warner, J.O., Hourihane, J.O.B. 2004. Kiwi fruit is a significant allergen and is associated with differing patterns of reactivity in children and adults. Clin. Exp. Allergy, 34(7): 1115–1121.

Lv, J.P., Wang, L.Y., Shen, H., Wang, X.D. 2018. Regulatory roles of OASL in lung cancer cell sensitivity to *Actinidia chinensis* Planch. root extract (acRoots). Cell Biol. Toxicol., 34: 207–218.

Mariäa, J.J.N., Carlos, A.M., Philip, E.S., Kevin, L.G. 2002. Aroma active components in aqueous kiwi fruit essence and kiwi fruit puree by GC-MS and multidimensional GC/GC-O. J. Agric. Food Chem., 5386–5390.

Mauseth, J.D. 2003. Botany: An Introduction to Plant Biology. Jones and Bartlett, pp. 271–72.

McGhie, T.K., Ainge, G.D. 2002. Color in ruit of the genus *Actinidia*: carotenoid and chlorophyll compositions. J. Agri. Food Chem., 50: 117–121.

Medical News Today: source for health news since 2003. www.medicalnewstoday.com. Retrieved on November 2019.

Melville, R. 1962. On the application of biometrical methods in plant taxonomy. Proc. Linn. Soc. Lond., 162: 153–159.

Michaud, J., Ane-Margail, M. 1977. Analytical study of catechuic tannins I Flavanolic oligomers of *Actinidia chinensis* Planch. Babin. Bull. Soc. Parma. Bord., 116: 52–64.

Mills, E.N., Jenkins, J.A., Alcocer, M.J., Shewry, P.R. 2004. Structural, biological, and evolutionary relationships of plant food allergens sensitizing via the gastrointestinal tract. Crit. Rev. Food Sci. Nutr., 44(5): 379–407.

Mishra, S., Monro, J.A. 2012. Kiwifruit remnants from digestion *in vitro* have functional attributes of potential importance to health. Food Chem., 135(4): 2188–2194.

Mishra, S., Edwards, H., Hedderley, D., Podd, J., Monro, J. 2017. Kiwifruit non-sugar components reduce glycaemic response to co-ingested cereal in humans. Nutrients, https ://doi.org/10.3390/nu9111195.

Montefiori, M., McGhie, T.K., Costa, G., Ferguson, A.R. 2005. Pigments in the fruit of red-fleshed kiwifruit (*Actinidia chinensis* and *Actinidia deliciosa*). J. Agric. Food Chem., 53(24): 9526–9530.

Mordor Intelligence. www.mordorintelligence.com. Retrieved on November 2019.

Morgan, E.D. 1988. The Architecture and Radiation Regime of A Kiwifruit Stand. Unpublished MSc. Thesis, Massey University, New Zealand.

Morton, J. 2011. Kiwifruit: *Actinidia deliciosa*. *In*: Fruits of Warm Climates, 1987. Center for New Crops and Plant Products at Purdue University. Retrieved 2 November 2019.

Motohashi, N., Shirataki, Y., Kawase, M., Tani, S., Sakagmi, H., Satoh, K., kurihara, T., Nakashima, H., Wolfard, K., Miskolci, C., Molnar, J. 2001. Biological activity of Kiwifruit peels extract. Phyto. Res., 15(4): 337–343.

Motohashi, N., Shirataki, Y., Kawase, M., Tani, S., Sakagami, H., Satoh, K., Kurihara, T., Nakashima, H., Mucsi, I., Varga, A., Molnar, J. 2002. Cancer prevention and therapy with kiwifruit in Chinese folklore medicine: a study of kiwifruit extracts. J. Ethnopharmacol., 81(3): 357–364.

Naik, V.N. 1984. Taxonomy of Angiosperms. Tata McGraw-Hill Education, pp. 304.

Nishikimi, M., Fukuyama, R., Minoshima, S., Shimizu, N., Yagi, K. 1994. Cloning and chromosomal mapping of the human nonfunctional gene for l-gulono–gamma–lactone oxidase, the enzyme for l-ascorbic acid biosynthesis missing in man. J. Biol. Chem., 269(18): 13685–13688.

Nishiyama, I. 2007. Fruits of the *Actinidia* Genus. Adv. Food Nutr. Res., 52: 293–324.

Nishiyama, I.I., Fukuda, T., Shimohashi, A., Oota, T. 2008. Sugar and organic acid composition in the fruit juice of different actinidia varieties. Food Sci. Technol. Res., 14(1): 67–73.

Olatunde, O.Z., Yang, Y., Yong, J., Lu, C. 2019. The progress of chemical constituents isolated from the root of *Actinidia chinensis* Planch and their biological activities. J. Biomed. Res. Rev., 2(2): 12–21.

Ozenda, P. 1949. Recherches sur lex dictyledones apocarpiques contribution a Petude des angiosperms dites primitives. Ecole Normale Super Publ. Lab. Biol., Paris.

Parameswaran, I., Murthi, V.I. 2014. Comparative study on physico & phytoChemical analysis of *Persea americana* & *Actinidia deliciosa*. Int. J. Sci. Res. Pub., 4(5): 1–5.

Parveen, I., Winters, A., Threadgill, M.D., Hauck, B., Morris, P. 2008. Extraction, structural characterisation and evaluation of hydroxycinnamate esters of orchard grass (*Dactylis glomerata*) as substrates for polyphenol oxidase. Phytochem., 69: 2799–2806.

Pastorello, E.A., Pravettoni, V., Ispano, M., Farioli, L., Ansaloni, R., Rotondo, F., Incorvaia, C., Asman, I., Bengtsson, A., Ortolani, C. 1996. Identification of the allergenic components of kiwi fruit and evaluation of their cross-reactivity with timothy and birch pollens. J. Allergy Clin. Immunol., 98(3): 601–610.

Pastorello, E.A., Conti, A., Pravettoni, V., Farioli, L., Rivolta, F., Ansaloni, R., Ispano, M., Incorvaia, C., Giuffrida, M.G., Ortolani, C. 1998. Identification of actinidin as the major allergen of kiwi fruit. J. Allergy Clin. Immunol., 101(4 Pt 1): 531–537.

Peng, S., Cheng, J., Zhang, M., Chen, M., Zhou, H., Cheng, M. 2013. Analysis of two anti-tumor active ingredients in radix *Actinidiae chinensis* by dispersive liquid–liquid micro extraction coupled to high performance liquid chromatography–mass spectrometry. Anal. Metho., 5: 5227–5233.

Pérez, A.G., Olías, R., Espada, J., Olías, J.M., Sanz, C. 1997. Rapid determination of sugars, nonvolatile acids, and ascorbic acid in strawberry and other fruits. J. Agric. Food Chem., 45(9): 3545–3549.

Pinelli, P., Romani, A., Fierini, E., Remorinib, D., Agati, G. 2013. Characterisation of the polyphenol content in the kiwifruit (*Actinidia deliciosa*) exocarp for the calibration of a fruit-sorting optical sensor. Phytochem. Anal., 24: 460–466.

Rajiv Gandhi University of Knowledge Technologies-AP Catering to the Educational Needs of Gifted Rural Youth of Andhra Pradesh. www.rkv.rgukt.in. Retrieved on October 2019.

Richardson, D.P. 2015. Developing the right public health strategies for folic acid and reduction of risk of neural tube defects (NTDs) in the United Kingdom. Eur. J. Nutr. Food Saf., 5(4): 242–249.

Richardson, D.P., Ansell, J., Drummond, L.N. 2018. The nutritional and health attributes of kiwifruit: a review. Europ. J. Nut., 57: 2659–2676.

Rudeschko, O., Fahlbusch, B., Steurich, F., Schlenvoigt, G., Jager, L. 1998. Kiwi allergens and their cross-reactivity with birch, rye, timothy, and mugwort pollen. J. Investig. Allergol. Clin. Immunol., 8(2): 78–84.

Rush, E., Drummond, L.N. 2009. The glycaemic index of kiwifruit. N Z Kiwifruit J., 192(May/June): 29–33.

Rust, P., Ekmekcioglu, C. 2016. Impact of salt intake on the pathogenesis and treatment of hypertension. Adv. Exp. Med. Biol., https://doi.org/10.1007/5584_2016_147.

Schlegel, R.H.J. 2003. Encyclopedic Dictionary of Plant Breeding and Related Subjects. Haworth Press, pp. 177.

Scognamiglio, M., Cefarelli, G., Caputo, R., Monaco, P. 2009. δ-Tocomonoenol: a new vitamin E from kiwi (*Actinidia chinensis*) sensitized to Kiwi with different clinical symptoms. Clin. Exp. Aller., 38(7): 1220–1228.

Shastri, K.V., Bhatia, V., Parikh, P.R., Chaphekar, V.N. 2012. *Actinidia deliosa*: a review. IJPSR, 3(10): 3543–3549.

Sims, I.M., Monro, J.A. 2013. Fiber: Composition, structures, and functional properties. pp. 81–99. *In*: Boland, M., Moughan, P.J. (eds.). Advances in Food and Nutrition Research: Nutritional Benefits of Kiwifruit. vol 68. Academic Press.

Singh, G. 2004. Plants Systematics: An Integrated Approach. Science Publishers, pp. 83.

Singh, S.P. 1995. Commercial Fruits. Kalyani Publishers, Ludhiana, New Delhi.

Skipworth, J.P., Philipson, W.R. 1966. The cortical vascular system and the interpretation of Magnolia flower. Phytomor., 16: 463–469.

Sommerburg, O., Keunen, J.E., Bird, A.C., van Kuijk, F.J. 1998. Fruits and vegetables that are sources for lutein and zeaxanthin: the macular pigment in human eyes. Br. J. Ophthalmol., 82(8): 907–910.

Song, W.Y., Xu, G.H., Zhang, G.J. 2014. Effect of *Actinidia chinensis* planch polysaccharide on the growth and apoptosis, and p-p38 expression in human gastric cancer SGC-7901 cells. Zhongguo Zhong xi yi jie he za zhi, 34: 329–333 (In Chinese).

Stonehouse, W., Gammon, C.S., Beck, K.L., Conlon, C.A., von Hurst, P.R., Kruger, R. 2012. Kiwifruit: Our daily prescription for health. Can. J. Physiol. Pharmacol., 91: 442–447.

Takeoka, G.R., Guntert, M., Flath, R.A., Wurtz, R.E., Jennings, W. 1986. Volatile constituents of kiwifruit (*Actinidia deliciosa*). J. Sci. Food Agric., 34: 576–578.

The Plant List, a working list of all plant species. www.plantlist.org. Retrieved on October 2019.

Tingting, M., Tian, L., Tonghui, G., Yanlun, J., Guo, C., Zhiluo, Q., Guitian, G., Yulin, F., Xiangyu, S. 2019. Nutritional properties and biological activities of kiwifruit (*Actinidia*) and kiwifruit products under simulated gastrointestinal *in-vitro* digestion. Food & Nut. Res., 63: 1674.

Tuppo, L., Giangrieco, I., Palazzo, P., Bernardi, M.L., Scala, E., Carratore, V., Tamburrini, M., Mari, A., Ciardiello, M.A. 2008. Kiwellin, a modular protein from green and gold kiwi fruits: evidence of *in-vivo* and *in-vitro* processing and IgE binding. J. Agric. Food Chem., 56(10): 3812–3817.

US Department of Agriculture. 2016. Green Kiwifruit. 2016. USDA National Nutrient Database for Standard Reference, Release 28 (slightly revised). Version: May 2016. US Department of Agriculture (USDA), Agricultural Research Service (ARS), Nutrient Data Laboratory, Beltsville (MD).

Wan, X.Y., Zhang, Y.N., Zhang, C. 2004. Experimental study on antihepatocarcinoma effect of Maorenshen injection *in vitro*. J. Zhej. Coll. TCM, 28(2): 45–47.

Wang, M.Y., MacRae, E., Wohlers, M., Marsh, K. 2011. Changes in volatile production and sensory quality of kiwifruit during fruit maturation in *Actinidia deliciosa* 'Hayward' and *A. chinensis* 'Hort16A'. Postharvest. Biol. Technol., 59: 16–24.

Wang, Q., Lei, X.X., Yang, X.D., Xin, T., Li, Y.Q. 2017. Purification of total triterpenes from the root of *Actinidia chinensis* Planch. and its antitumor activity. Guangdong Med. J., 38: 514–518.

Ward, C., Courtney, D. 2013. Kiwifruit: taking its place in the global fruit bowl. pp. 1–15. *In*: Boland, M., Moughan, P.J. (eds.). Advances in Food and Nutrition Research: Nutritional Benefits of Kiwifruit. Vol. 68. Academic Press.

Wei, L.B., Ma, S.Y., Liu, H.X., Huang, C.S., Liao, N. 2018. Cytotoxic triterpenoids from roots of *Actinidia chinensis*. Chem. Biodivers., 15: e1700454.

Wen-Wei, Fu., Tan, C.-H., Lu, L.-L., Meng, X.-X., Luo, H.-F., Zhu, D.-Y. 2010. Chemical constituents from the root of *Actinidia deliciosa*. Chinese J. Nat. Med., 247–249.

Wikipedia, kiwifruit. https://en.wikipedia.org/wiki/Kiwifruit. Retrieved on November 2019.

Wikipedia, the free encyclopedia. www.en.wikipedia.org. Retrieved on October 2019.

World Flora Online. www.worldflora.online.org. Retrieved on October 2019.

Wurms, K.V., Cooney, J.M. 2006. Isolation of a new phenolic compound, 3, 5-dihydroxy-2-(methoxycarbonylmethyl) phenyl 3,4-dihydroxybenzoate, from leaves of *Actinidia chinensis* (kiwifruit). Asian J. Biochem., 1: 325–332.

Xie, Z.W. 1975. National Compilation of Chinese Herbal Medicine Vol. I. Beijing: People's Health Publishing House, 820–821.

Xu, Y.X., Xiang, Z.B., Jin, Y.S., Shen, Y., Chen, H.S. 2010. Two new triterpenoids from the roots of *Actinidia chinensis*. Fitote., 81: 920–924.

Xu, Y.X., Xiang, Z.B., Jin, Y.S., Xu, W., Sun, L.N., Chen, W.S., Chen, H.S. 2016. Constituents from the roots of *Actinidia chinensis* and their cytochrome P450 enzyme inhibitory activities. Chem. Biodivers., 13(11): 1454–1459.

Yagami, T., Haishima, Y., Nakamura, A., Osuna, H., Ikezawa, Z. 2000. Digestibility of allergens extracted from natural rubber latex and vegetable foods. J. Allergy Clin. Immunol., 106(4): 752–762.

Yang, Q., Wang, F., Rao, J. 2016. Effect of putrescine treatment on chilling injury, fatty acid composition and antioxidant system in kiwifruit. PLoS One, 11: e0162159.

Yang, Y.J., He, H.S. 2012. Kiwi Root Extract *in-vitro a*ntioxidant effect research. Food Res. Dev., 33: 21–24.

Yao, G., Wang, T.S. 1989. Medicinal plants of *Actinidia* genus in East China. J. Chin. Med. Mat., 12(2): 15–16.

Young, H., Paterson, V.J. 1995. Characterisation of bound flavour components in kiwifruit. J. Sci. Food Agric., https://doi.org/10.1002/jsfa.274068021.

Yu, A.N., Tian, D.T., Qu, W.Y., Tan, Z.W., Song, X.J. 2009. Chemical composition of the essential oil of the roots of *Actinidia chinensis* from China. Chem. Natu. Comp., 45: 108–109.

Yun-peng, Z., Wang, X.-y., Wang, Z.-c., Lu, Y., Fu, C.-x., Chen, S.-y. 2006. Essential oil of *Actinidia macrosperma*, a catnip response kiwi endemic to China.

Zespri (Official Site). Zespri Group Ltd. 2018. Retrieved 10 September 2018.

Zhang, G.J., Han, J.Y. 2012. Mechanism study of kiwifruit root polysaccharide anti-tumor injection on tumor *in-vivo*. CJTCMP, 27: 2177–2179.

Zhang, X.Y., Zhou, Y., Wei, Z.P., Shen, J., Wang, L.K., Ma, Z.Q., Zang, X. 2018. Antiphytoviral toxins of *Actinidia chinensis* root bark (ACRB) extract: laboratory and semi-field trials. Pest Manag. Sci., 74(7): 1630–1636.

Zhao, Y.P., Lu, Y., Wang, Z.C., Chen, S.Y., Fu, C.X. 2006. Pharmacognostic discrimination on stems of nine *Actinidia* species. Chin. Pharma. J., 41(14): 17–21 (in Chinese).

Zhou, X.F., Zhang, P., Pi, H.F., Zhang, Y.H., Ruan, H.L., Wang, H., Wu, J.Z. 2009. Triterpenoids from the roots of *Actinidia chinensis*. Chem. Biodivers., 6: 1202–1207.

Zhou, X.F., Liu, Y.H., Tang, L., Zhang, P., Wu, J.Z. 2010. Chemical constituents from the roots of *Actinidia chinensis*. Chem. Nat. Compd., 46: 308–309.

Zhou, X.Z. 2000. Tang Fu'an talk about lung cancer treatment variation. J. Zhej. Coll. TCM, 24(2): 45–46.

Zhu, W.-J., Yu, D.-H., Zhao, M., Lin, M.-G., Lu, Q., Wang, Q.-W., Guan, Y.-Y., Li, G.-X., Luan, X., Yang, Y.-F., Qin, X.-M., Fang, C., Yang, G.-H., Chen, H.-Z. 2013. Antiangiogenic triterpenes isolated from Chinese herbal medicine *A. chinensis* Planch. Anticancer Ag. Med. Chem., 13: 195–198.

Zuo, L.L., Wang, Z.Y., Fan, Z.L., Tian, S.Q., Liu, J.R. 2012. Evaluation of antioxidant and antiproliferative properties of three *Actinidia* (*A. kolomikta*, *A. arguta*, *A. chinensis*) extracts *in-vitro*. Inter. J. Mol. Sci., 13: 5506–5518.

Chapter 18
Rhubarb
The King of Herbs with Diverse Bioactivities

*Gan B. Bajracharya** and *Richa K. Gupta*

Introduction

The plants included in the genus *Rheum*, belonging to the family Polygonaceae, are collectively called as rhubarb (Agarwal et al. 2001, Wang and Ren 2009, Tayade et al. 2012, Zheng et al. 2013, Akram and Azhar 2016, Malik et al. 2016, Ghorbani et al. 2019) (Figure 18.1). Rhubarb is one of the oldest and most frequently used medicinal herbs that is mainly distributed in the mountainous regions of the Qinghai-Tibetan Plateau area and Asian interior (Losina-Losinskaya 1936). It has been in use for more than 2000 years in the traditional medical systems and is known as the 'Lord or King of herbs' (Wang et al. 2005). They are perennial herbs and bear large leaves with thick succulent petioles. Wild rhubarb is used in herbal medicines. Some rhubarb is grown as a vegetable and used for culinary arts, while few others are grown as ornamental plants. Rhubarb is a member of vegetables, which is enriched with vitamin C and dietary fiber, but is often thought to be a fruit (Lloyd 2008). Particularly, *R. x hybridum* is an ancient garden rhubarb of unknown origin (possibly Siberia) that was grown as a vegetable in England and Scandinavia in the early 18th century (Bratsch and Mainville 2009, Lee et al. 2017). Stem, flower and leaf stalk of *R. officinale* and *R. rhaponticum* are also used as food (Agarwal 1986). The red stalks, but not the roots or leaves, are widely used for puddings, preservation and wine. The leaves of rhubarb are non-edible because of their oxalic acid (soluble oxalates) contents, which are poisonous. The flavor present in rhubarb is also used in the preparation of jam, jelly and sauce (Prior 2009).

Rheum acuminatum *Rheum australe* *Rheum nobile* *Rheum spiciforme*

Figure 18.1. Rhubarbs in their natural habitat (*Photo sources*: Polunin and Stainton 1997, Ghimire et al. 2008 and the authors).

Faculty of Science, Nepal Academy of Science and Technology, Khumaltar, Lalitpur, Nepal.
* Corresponding author: ganbajracharya@yahoo.com, gan.bajracharya@nast.gov.np

Almost all rhubarbs contain anthraquinones. Many species of *Rumex* dock roots also constitute anthraquinone derivatives and hence they have been used as a substitute for rhubarb (Wallis 1985). *Rumex nepalenis* is often used for adulteration of the official rhubarbs. *R. hotaoense* and *R. rhabarbarum* have been also used as commercial substitutes of official rhubarb (Zheng et al. 2013).

In ancient times, people believed that the dwelling of evil spirits cause illness. They used plants with unpleasant odor and taste, accompanied by enchantments and ceremonial dances, to get rid of the demons and the person would then recover from his illness. They ignored the etiology of diseases and the mechanism by which the medicinal agents eradicated the diseases. However, with the advancement of science, studies have shaped the rationale of the selection of plants and proved their medicinal efficacies. Therefore, the basis of modern medicine actually relies on ancient knowledge on the treasure of medicinal herbs. In particular, the plants grown in forests and at high altitudes are biologically and chemically diverse. They biosynthesized a wide range of secondary metabolites that exhibit promising biological activities due to their existence in the harsh climatic conditions and hence they are prone to important sources of therapeutic agents.

Today, new diseases are appearing and drug-resistance problems are emerging. This exerts tremendous pressure to search for more efficacious and safe agents from nature. Rhubarb possesses a very high potential for natural healings. Although a large body of literature proved the medicinal importance of rhubarb and determined its bioactive constituent, it is more essential to conduct further experimental and clinical studies on rhubarb for the betterment of humanity. Therefore, this chapter is designed to bridge the research gap on rhubarb by providing a comprehensive review of history, economic importance, worldwide distribution, traditional formulation, pharmacognosy, pharmacology, phytochemistry and pharmacokinetic studies on rhubarb.

History and Nomenclature

Apparently, the Chinese were the first to realize the medicinal values of rhubarb, then called 'Pen-king' (Azhar and Anjum 2019). Pen-king is one of the most ancient plants used in Chinese herbal medicine that is attributed to the Chinese emperor Shen-Nung, the father of Chinese agriculture and medicine, who reigned approximately 2695 BC. It is said that he had tasted hundreds of herbs to test their medicinal value and taught his people to cultivate. He authored 神农本草经 (*Shen-nung pen tśao ching*) as the earliest extant Chinese pharmacopeia and died from a toxic herb overdose as he was unable to find an appropriate antidote in time. Rhubarb was named 'Huang-hang' there, which means excellent yellow, and 'Ta-huang', which means the great yellow. During the Han dynasty, the wide uses of rhubarb were first documented in the *Divine Husbandman's Classic of Materia Medica*, around 200 AD (Clementi and Misiti 2010).

Pedanius Dioscorides (40–90 AD) was a Greek physician in the Roman army during the period of the Roman Emperors Caligula, Claudis and Nero (Yildirim 2013). He studied medicinal plants while travelling around Anatolia, Arabia, Caucasia, Egypt, Gallia, North Africa and Persia. Wonderfully, he created a five-volume Greek encyclopedia Περί ύλης ιάτρικης (*Peri hyles iatrikes*) describing hundreds of herbal medicines as well as animals and minerals. Later, his monumental work received Latin title *De Materia Medica*, which became the basis for all modern pharmacopeias. The books were translated in many languages by several translators, such as Arabic (by Stephanos Basilos, Mihran Mansur Mihran, Ali Serif el-Huseyni and Osman Abdurrahman), Syriac (by Gabriel Bocht-Ichô, Huneyn Ibrahim El-Hasan Hurshid et-Tabari en-Natili, Priest Nicolas and Abu Salim), Persian (by Ali Sherif al-Huseyni), Turkish (by Osman Abdurrahman) and English (by John Goodyer in 1655). The Arabic translation *Kitabul-Hasaish* had described rhubarb.

The word rhubarb possibly derives from Latin word *rha,* which means a river (the ancient name of Volga river in European Russia) and *barb,* which means barbarian land referring to Dioscorides (Alam and Khan 2015). It was brought from beyond Bosporus and called Rha-ponticum (Pontus Euxinus was the ancient name of Black sea). It was transported via Barbarike (possibly Turkey) and then known as Reu-barbarum, which later changed to the name rhubarb in English. Due to the purgative properties of the root, rhubarb also received the name from the Greek *rheo* meaning to flow (Rehman et al. 2014).

Rhubarb was named 'Ribas or Ravind' in the Persian language and the root without corky layer was called 'Rawand'. *R. ribes* is still called as Ribas. It is the only species growing in Turkey and is locally known as Iskin, Usgun or Ucgun (Cullen 1966). It is one of the most important sources of crude drugs in the Middle East (Kashiwada et al. 1988). In old Ayurveda, rhubarb was called as 'Amlavetes' and was later described as 'Revan', which was actually taken from the Unani literature.

In the classical Unani literature, rhubarbs are classified according to their origin and morphological features of the plants (Azhar and Anjum 2019). Some varieties are reported as Revand-chini, Revand-turki, Revand-khurasanee, Revand-shammi, Revand-hindi, Revand-rusi, Revand-franceece, Revand-japanee, Revand-farsi, Revand-junji, Ribas-e-pahan (*R. palmatum*), Ribas-e-tibbi (*R. officinale*), Ribas-e-barbari (*R. rhaponticum*), Ribas-e-sulab (*R. compactum*), Ribas-e-mawaaj (*R. undulatum*), etc. In 11th century Ibn-e-jazla differentiated two kinds of rewand, viz., Chinese and Khorasan rhubarbs. The Chinese rhubarb is used by human beings for different therapeutic effects, while the Khorasan rhubarb (also named as 'Rewand-aldawaab') is used in veterinary practices (Akram and Azhar 2016).

Rhubarb reached to the West via Turkey and Russia, and was first planted in England around 1770s and named as 'Hayward' (Foust 1992). Cultivation of rhubarb started in Europe in 1860s and it was then named as English, German, Bucharest, Dutch and French rhubarbs. *R. undulatum* is commonly known as Korean rhubarb that is mainly distributed and cultivated in South Korea (Rashid et al. 2014). Today, rhubarb is known in various languages in various regions, some of them are Chukri, Rawash (Afghanistan); Rebas, Rewand (Arabic); Amlaparni, Amlavetasa (Ayurvedic); Thuza (Bhutanese); Huang-hang, Da-huang (Chinese); Rhubarb (English); Rhubarbe (French); Rhabarber (German); Reon (Greek); Rubaba, Hind-revand-chini, Revand-hindi (Hindi); Daiou (Japanese); Chutial, Pumb chalen (Kashmir); Lachu (Ladakh); Padamchal (Nepali); Chontal, Chotial (Pakistan); Tursak, Ravind, Rebas (Persian); Gandhini, Pitamulika, Amla-vetasa (Sanskrit); Rivalchinikattai, Nattirevaichini (Siddha); Churtsa, Gyasa (Tibetan); Bekh-e-ashqoon (Turkish); Rewan-chini (Unani); Rewand-chini (Urdu); etc. (Rokaya et al. 2012, Azhar and Anjum 2019).

Trading and Markets of Rhubarb

R. palmatum, *R. officinale* and *R. tanguticum* are considered as Chinese rhubarbs. *R. palmatum* is the main source of rhubarb in China and marketed by the name Chinghai. As early as 114 BC, rhubarb was brought from Shensi in North China to Bokhara in Russian Turkestan and then entered into Europe via the Black sea (Singh 2001). The roots and rhizomes of rhubarb became one of the major exporting materials of Asia (particularly from China) in the 10th century. From 12th to 18th century, it was carried from China to Persia by caravan and then to Aleppo, Alexandria, Tripoli, and Smyrna. From these ports, rhubarb was shipped to Europe and became known as Turkey rhubarb (Evans 2007). Some rhubarb was also shipped from China to Europe by way of India and Singapore, and became East Indian rhubarb, whereas some were directly sent from China to Russia from Canton port and were called Canton rhubarb. Eventually, rhubarb appeared in Western Europe in 1608, probably coming from Italy or Spain (Lee et al. 2017, Azhar and Anjum 2019). Garden rhubarb was first cultivated at the Botanical Garden of Padua by Prosper Alpinus (1553–1617) and was widely grown in the 18th century. In Europe, the price of rhubarb parts was more than that of opium. In France, rhubarb was sold ten times expensive than cinnamon and four times than saffron (Prior 2009).

China established its monopoly on the supply of rhubarb until 1800. From 1750 to 1850, Russia was making fortune by rhubarb trading and sent rhubarb via the Baltic. At that period, rhubarb seeds were more valuable than gold. Dr. James Mounsey (1710–1773), the First Royal Physician of Empress Elizabeth, smuggled several pounds of rhubarb seeds from Russia and then introduced them into Great Britain (Lee et al. 2017). He presented a pound of rhubarb seed to Sir Alexander Dick (1703–1785), then President of the Royal College of Physicians of Edinburg. He planted rhubarb seeds in the garden of Prestonfield. They both received the gold medal of the Royal Society of Arts in London in 1770 for their contributions. Prof. John Hope (1725–1786) also received rhubarb seeds from Mounsey in 1763. He planted the seeds in the botanic garden at Leith Walk and began to sell rhubarb. Sometime around 1800, a gardener at Chelsea Physic Garden accidentally noticed that the rhubarb massively shot up in dark (etiolation), producing a vibrant reddish-yellow coloration with much sweetening. Thereafter, the era of the commercial growth of rhubarb as a foodstuff in the UK began and a number of cultivars produced hybrid rhubarbs.

Rhubarb entered in America in the 1820s via Maine and Massachusetts (Castleman 1991). In 1842, Russian trade rapidly diminished with the opening of the North Chinese ports and then the bulk of rhubarb began to be supplied directly from China to Europe (Alam and Khan 2015). The Silk Road was used for transportation to the Levantine ports and then on to Venice and so to Europe. Now the price of rhubarb gradually began to fall and the monopoly of Russia in rhubarb trading withered away (Lee et al. 2017).

The Indian or Himalayan rhubarb gained importance during World War II as the supply of Chinese rhubarb became scarce. In the Pharmacopoeia of India, the Indian rhubarb includes the dried rhizomes of *R. emodi*, *R. webbianum* and *R. moorcroftianum* (CSIR 1972). *R. emodi* being the chief source, about 11,000 kg of rhizomes and roots are annually collected from Kulu and Kangra of India. According to a report from the Ministry of Forests and Soil Conservation, Government of Nepal, a total of 75,290.5 kg of *R. emodi* was traded in 2016, with received revenue amounting of 568,080 Nepali rupees (NRs) (MoFSC 2016). Rhubarb is listed in the prioritized medicinal and aromatic plants of Nepal and is highly recommended for scientific research and cultivation (DPR 2006). Like most medicinal plants, the wild rhubarb is generally used for medicinal purpose and the cultivated variety is usually an inferior kind (Gupta et al. 2015, Azhar and Anjum 2019). Because of the over-harvesting of Himalayan rhubarb for commercial exploitation, *R. emodi* is classified as vulnerable endangered species by the Governments of Nepal and India (Bhattarai et al. 2002, Nautiyal et al. 2003, Singh et al. 2017).

The rhizome of *R. rhaponticum*, probably indigenous to Southern Siberia and China, is considered as Rhapontic rhubarb (CSIR 1972). It is not official in the Pharmacopoeias and is extensively cultivated for its edible petioles all over Europe and America (Bratsch and Mainville 2009).

Taxonomy

Kingdom: Plantae (plants); Sub-kingdom: Tracheobionta (vascular plants); Super-division: Spermatophyta (seed plants); Division: Magnoliophyta (flowering plants); Class: Magnoliopsida (dicotyledons); Sub-class: Caryophyllidae; Order: Caryophyllales; Family: Polygonaceae; Genus: *Rheum*.

General Morphology

Rhubarb is a perennial herb that grows from thick, short rhizomes. The height varies (up to 3 meters). Flowering occurs in July–August and fruiting in September–October. The rhizomes are collected in October–November.

The leaves are large, soft, orbicular or broadly ovate, toothed or lobed, obtuse apex, cordate base, entire margin, 5–7 nerved, papillose beneath subscaberulous. Leaves are in basal clumps with scary stipules and have a long fleshy petiole. The stem is stout, erect, glabrous, sulcate, branch and panicle leafy. There is a green and brown color streak. Flowers are small (~ 3.2 mm diameter), dark purple or pale red or greenish-white in color, grouped, pedicellate and axillary panicle inflorescence. Flowers are bisexual or polygamo-monoecious having 5 sepals and 6–9 stamens. Ovary comprises 2–4 angled rhomboid-obovoid ovaries. Stigma is dilated capitates, oblate, muricate or horse-shoe shaped. The nut is 2–4 winged, very much larger than the usually unchanged sepals. Fruits are ovoid-oblong, about 13 mm long, purple in color, cordate base, notched apex and wings narrower than the disk. Seeds are flat and grey in color when ripened. The root is thick, woody, stout, long, and darker in color with a peculiar pungent aroma. The rhizome is long (~ 6–12 inch long) and dull orange to yellowish-brown in color with a cambium line. The outer surface is irregularly longitudinally wrinkled, covered with the brownish or yellowish-brown cortex. The transverse section shows white and red parallel lines and continuous rings of star-spots with starch and calcium oxalate in parenchymatous cells. The morphology description above is collected from the literature (CSIR 1972, Polunin and Stainton 1997, Li et al. 2003, Ghimire et al. 2008).

The matK gene sequence could be valuable in identifying the rhubarb species, predicting their production areas and evaluating the quality (Komatsu et al. 2005).

Distribution

Rhubarb is distributed in the temperate and sub-tropical regions of the world, chiefly in mountains and desert regions of Asian and European countries. Naturally, it is widely distributed in Northern and Central Asia (China, Nepal, Bhutan and India). It is distributed in the Himalayas from North-Eastern Afghanistan, Northern Pakistan, Kashmir, Nepal, Tibet, Sikkim, Bhutan to Myanmar at 2750–5000 meter (CSIR 1972, Hooker 1855, Ghimire et al. 2008). Rhubarb grows well in rocky drained, porous and humus-rich soil of alpine and sub-alpine areas of the Himalaya (Nautiyal et al. 2003). Qinghai, Sichuan and Gansu provinces at altitudes varying from 2300 to 4200 meter are the major producing areas of Chinese rhubarb (Li et al. 2003, Zheng et al. 2013, Web 1).

Rhubarb comprises about 60 species (Wang et al. 2005, Zheng et al. 2013). *The Plant List* includes 114 scientific names of species for the genus *Rheum*, among which 41 names were accepted (Table 18.1) (Web 2).

Table 18.1. Global distribution of rhubarb.

S.N.	Name of ***Rheum*** species (synonym)	**Distribution** (Web 2, Web 3)
1	*R. acuminatum* Hooker & Thomson (*R. orientalixizangense* Y.K. Yang, J.K. Wu & Gasang)	Slopes, forests, 2800–4000 m. South Gansu, Sichuan, Xizang, Yunnan (China); Kashmir; Sikkim; Nepal; Bhutan; Myanmar.
2	*R. alexandrae* Batalin	Slopes, 3000–4600 m. West Sichuan, East Xizang, North-West Yunnan (China).
3	*R. altaicum* Losinskaja (*R. rhaponticum* Herder)	Forests, valleys, 1900–2400 m. North Xinjiang (China), Kazakhstan, Mongolia, Russia.
4	*R. australe* D. Don (*R. emodi* Wallich ex Meisner)	Grassy slopes, 2750–5200 m. Xizang (China), India, Myanmar, Nepal, Pakistan, Sikkim (India).
5	*R. compactum* Linnaeus (*R. nutans* Pallas; *R. orientale* Losinskaja)	Slopes, 2000 m. North Xinjiang (China), Kazakhstan, Mongolia, Far East Siberia.
6	*R. delavayi* Franchet (*R. strictum* Franchet)	Slopes, 3000–4800 m. West Sichuan, North Yunnan (China); Bhutan; Nepal.
7	*R. forrestii* Diels	Slopes, 3000 m. Xizang, Yunnan (China).
8	*R. glabricaule* Samuelsson	Among rocks, 3000–3500 m. Gansu (China).
9	*R. globulosum* Gage	Slopes, 4500–5000 m. Central and South Xizang (China), Sikkim.
10	*R. hotaoense* C.Y. Cheng & T.C. Kao	Slopes, gullies, 1000–1800 m. Gansu, Shaanxi, Shanxi (China).
11	*R.* × *hybridum* Murray (*R. x dentatum* Mart; *R. x collinianum* Baill.)	Low land. UK, Ireland.
12	*R. inopinatum* Prain	Slopes, 4000–4200 m. Cental and South Xizang (China).
13	*R. kialense* Franchet (*R. micranthum* Samuelsson; *Rumex cacaliifolius* H. Léveillé)	Slopes, forests, 2800–3900 m. Gansu, Sichuan, Yunnan (China).
14	*R. laciniatum* Prain	Hills, 3000 m. North Sichuan (China).
15	*R. lhasaense* A.J. Li & P.G. Xiao	Grassy slopes, 4200–4600 m. Lhasa, Xizang (China).
16	*R. likiangense* Samuelsson (*R. ovatum* C.Y. Cheng & T.C. Kao)	Forests, shrubby meadows, 2500–4000 m. South-West Sichuan, East Xizang, North-West Yunnan (China).
17	*R. lucidum* Losinskaja (*R. korshinskyi* Titov ex Losinskaja)	Tadzhikistan.

Table 18.1 Contd. ...

...Table 18.1 Contd.

S.N.	Name of *Rheum* species (synonym)	Distribution (Web 2, Web 3)
18	*R. macrocarpum* Losinskaja (*R. ferganense* Titov; *R. lobatum* Litv. ex Losinskaja; *R. nuratavicum* Titov; *R. plicatum* Losinskaja; *R. vvedenskyi* Sumner; *R. zergericum* Titov)	Kirgizstan, Tadzhikistan, Uzbekistan.
19	*R. maculatum* C.Y. Cheng & T.C. Kao	Sichuan (China).
20	*R. nanum* Sievers ex Pallas (*R. cruentum* Sievers ex Pallas; *R. leucorrhizum* Pallas)	Slopes, valleys, 700–2000 m. Gansu, Central and West Nei Mongol, North-East Xinjiang (China); Kazakhsta; Mongolia; West Siberia.
21	*R. nobile* Hooker & Thomson	Slopes, 4000–4800 m. South Xizang (China), Afghanistan, Bhutan, Sikkim, Myanmar, Nepal, Pakistan.
22	*R. officinale* Baillon	Hills, forests, 1200–4000 m. Guizhou, South-West Henan, West Hubei, Shaanxi, Sichuan, Yunnan (China).
23	*R. palmatum* Linnaeus (*R. potaninii* Losinskaja; *R. qinlingense* Y.K. Yang, J.K. Wu & D.K. Zhang)	Slopes, valleys, naked mountains, 1500–4400 m. Gansu, Hubei, Nei Mongol, Qinghai, Shaanxi, Sichuan, Xizang, Yunnan (China); cultivated in Russia.
24	*R. przewalskyi* Losinskaja	Slopes, forests, 1500–5000 m. Gansu, Qinghai, North-West Sichuan (China).
25	*R. pumilum* Maximowicz	Slopes, 2800–4500 m. Gansu, Qinghai, Sichuan, Xizang (China).
26	*R. racemiferum* Maximowicz	Slopes, grasslands, 1300–2000 m. Gansu, Nei Mongol, Ningxia (China).
27	*R. rhabarbarum* Linnaeus (*R. franzenbachii* Münter; *R. franzenbachii* var. *mongolicum* Münter; *R. undulatum* Linnaeus; *R. undulatum* var. *longifolium* C.Y. Cheng & T.C. Kao)	Slopes, 1000–1600 m. Hebei, Heilongjiang, Hubei, Jilin, Nei Mongol, Shanxi (China); Mongolia; East Siberia; South Korea; cultivated in Europe.
28	*R. rhaponticum* Linnaeus (*R. rotundatum* Stokes)	2560–2560 m. Bulgeria, Colombia, Czechoslovakia, Norway, Russia.
29	*R. rhizostachyum* Schrenk (*R. aplostachyum* Karelin & Kirlov)	Slopes, 2600–4200 m. Xinjiang (China), Kazakhstan.
30	*R. rhomboideum* Losinskaja	Slopes, grasslands, 4700–5400 m. Central and East Xizang (China).
31	*R. spiciforme* Royle (*R. moorcroftianum* Royle; *R. reticulatum* Losinskaja; *R. scaberrimum* Lingelsheim ex Limpricht)	Slopes, 4000–5000 m. West Xizang (China), Afghanistan, Bhutan, Kashmir, Pakistan, Sikkim, Nepal.
32	*R. subacaule* Samuelsson	Hilltops, slopes, 3500–4300 m. West Sichuan (China).
33	*R. sublanceolatum* C.Y. Cheng & T.C. Kao	Slopes, 2400–3000 m. Gansu, Qinghai, Xinjiang (China).
34	*R. tanguticum* Maximowicz ex Balfour	Valleys, 1600–3000 m. Gansu, Qinghai, Shaanxi, Xizang (China).
35	*R. tataricum* Linnaeus (*R. caspicum* Pallas; *R. songaricum* Schrenk)	Grasslands, deserts, plains, 500–1000 m. West Xinjiang (China), Afghanistan, Kazakhstan, Russia (European part).
36	*R. tibeticum* Maximowicz ex Hooker	Slopes, 4000–4600 m. Xizang (China), Afghanistan, Kashmir, Pakistan.
37	*R. turkestanicum* (*R. megalophyllum* Sumner; *R. renifolium* Sumner; *R. rupestre* Litv. ex Losinskaja; *R. turanicum* Litv.) Janischew	Kazakhstan, Tadzhikistan, Turkmenistan, Uzbekistan.

Table 18.1 Contd. ...

...*Table 18.1 Contd.*

S.N.	Name of *Rheum* species (synonym)	Distribution (Web 2, Web 3)
38	*R. uninerve* Maximowicz	Slopes, roadsides, 1100–2300 m. Gansu, Nei Mongol, East Qinghai (China); Mongolia.
39	*R. webbianum* Royle	Slopes, 2400–3600 m. South-West Xizang (China), North-West India, Kashmir, Pakistan, Nepal.
40	*R. wittrockii* C. E. Lundström	Grassy slopes, forests, gullies, 1200–2600 m. Xinjiang (China), Kazakhstan, Kyrgyzstan.
41	*R. yunnanense* Samuelsson	Slopes, ca. 4000 m. North-West Yunnan (China), Myanmar.

Rhubarb as Foodstuff

Wild rhubarbs have found their importance as medicinal herbs, whereas garden rhubarbs are grown for their fleshy petioles and thick stalks for culinary arts. Because of greenhouse production, today rhubarbs are grown in many areas of the US and England for both the fresh-shipping and processing markets (Bratsch and Mainville 2009). *R. rhabarbarum*, *R. rhaponticum*, *R. x hybridum* and *R. x cultorum* are used as edible vegetables, and are therefore also named as 'false rhubarb'. The number of hybrid rhubarbs has now roughly reached to 120 (Lee et al. 2017). Stalks of the rhubarbs can be eaten raw or cooked. The color may vary from green to dark red. The pleasantly acidic taste of stalks is due to the presence of malic, oxalic and citric acids along with the saturated and unsaturated C_6 aldehydes (Clementi and Misiti 2010). Some people consume the stalks after dipping in sugar to remove the tartness. The addition of sugar is practiced in many preparations of rhubarbs, including puddings, jams, jellies, and sauces, rendering them more palatable. However, at the present time, tartness present in rhubarb has become popular. Rhubarb is also used to make wine and as an ingredient in baked foods and non-alcoholic beverages (Uphof 1968, Duke 1985). These common uses provide the name of 'pie plant' to rhubarbs. The nutritional factors in 100 g of fresh rhubarb stalks are: carbohydrate (1.1 g), fat (0.2 g), protein (0.9 g), Vitamin A (5 μg of retinol activity equivalents), Vitamin B1 (0.02 mg), Vitamin B2 (0.03 mg), Vitamin B3 (0.3 mg), Vitamin B5 (0.085 mg), Vitamin B6 (0.024 mg), Vitamin B9 (7 μg), Vitamin C (8 mg), Vitamin E (0.38 mg), Vitamin K (41 μg), β-carotene (61 μg), Ca (86 mg), Cu (0.021 mg), Fe (0.22 mg), K (288 mg), Mg (12 mg), Mn (0.196 mg), Na (4 mg), P (14 mg), Se (1.1 μg) and Zn (0.1 mg) (Clementi and Misiti 2010). Rhubarb contains anthraquinones, which possess natural laxative and cathartic properties. It is enriched with dietary fiber, minerals, vitamins and antioxidants. Furthermore, 100 g of rhubarb produces about 21 calories. Therefore, synergistic effects of its ingredients and low-calorie diet ensure improvement in the body's metabolism rate and elimination of unwanted fat, making it a preferred vegetable for those who are on a strict low fat and low-calorie diet (Xinga et al. 2011).

The roots, rhizomes, and stems of rhubarb are bitter, astringent and aromatic with a pungent smell. They are consumed especially for medicinal purposes. The young stems are eaten fresh or pickled (Ghimire et al. 2008). Dried rhizomes are used to prepare hot beverages and are drunk for internal medicinal effects. The medicinal values of the rhizomes vary with the altitude where the plant materials are grown. With a higher altitude, the quality of rhubarb rhizomes improves. The physicochemical parameters in 100 g of dried rhizome are ash (6.81–8.30 g), water-soluble ash (4.83–6.32 g), water-insoluble ash (1.71–1.98 g), acid insoluble ash (1.08–1.55 g), carbohydrate (12.3–25.8 g), fat (3.6–4.2 g), Fe (77–91 mg), K (2.4–3.6 g), Mn (5–19 mg) and Na (72–232 mg) (Gupta et al. 2016).

The leaves of rhubarbs are toxic as they contain potassium oxalate, calcium oxalate, and oxalic acid, and therefore should not be eaten. These soluble constituents can cause human and animal poisoning leading to death. Consumption of about 1 g of pure oxalic acid per kg body weight can cause death. Rhubarb leaves contain about 0.5% of oxalic acid. Hence, leaves must be trimmed from the edible petiole prior to consumption (Clementi and Misiti 2010).

Formulations and Herbal Remedies

Formulation of more than 800 herbal preparations has been estimated in the *Chinese Materia Medica* (Zheng et al. 2013). *R. palmatum*, *R. tanguticum*, *R. officinale*, *R. coreanum*, *R. hotaoense*, *R. franzenbachil* and Imo-daio (Japanese rhubarb) are considered to be the main commercial and/or official rhubarbs. The rhizomes of species *R. palmatum*, *R. tanguticum* and *R. officinale* are official in *Chinese Pharmacopoeia*, *Indian Pharmacopoeia* recognizes *R. emodi* and *R. webbianum*, and *British Pharmacopoeia* includes the drugs obtained from *R. palmatum* and *R. emodi* (Kurian and Sankar 2007). According to the *Chinese Pharmacopoeia*, the rhubarbs are used in constipation, hematemesis, swelling, abdominal pain, injuries, dysentery, jaundice, hematuria, burn, etc. (Chinese Pharmacopoeia Commission 2010). The wine-fried rhizome is used as a tonic and charred rhizome is used to resolve bleeding symptoms.

Unani medicines use rhubarb to prepare Habb-e-usara (as laxative and diuretic), Itrifal-mulayyin (as laxative), Habb-e-shifa (as antipyretic, antispasmodic, anti-fatigue, bronchodilator), Habb-e-zeequn-nafas (as expectorant), Sufoof-e-mudirr-e-haiz (as diuretic, emmenagogue), Hab-shabyar (as pain reliever), Haba-shafa (as decongestant), Qur-didan (as antihelmentic), Roughan-aqrab (as astringent), etc. (Ministry of Ayush 2016, Azhar and Anjum 2019). In Japanese traditional herbal constitutional medicine, rhubarb is used to prepare Choi-joki-to (neuroprotective), Daio-kanzo-to (neuroprotective), Dai-saiko-to (for liver disease), San'o-shashin-to (for hypertension), etc. (Iizuka et al. 2004).

Traditionally, rhubarbs are used for several therapeutic actions to defeat abscess, acidity, amenorrhea, anodyne, anorexia, anuria, ascites, asthma, atonic dyspepsia, bilious fever, boils, bronchitis, bruises, calculi, cancers, catarrh of biliary duct, cold, conjunctivitis, constipation, coryza, cough, cuts, diarrhea, dysmenorrhea, duodenal catarrh, dysentery, earache, eczema, emenagogue, fever, flatulence, fracture, frost bite, gastritis, goiter, hemoptysis, headache, hematochezia, hemorrhage, hepatitis, hernia, hiccup, hypochondriasis, indigestion, inflammations, insect bites, jaundice, kidney problems, lumbago, lymphocytic leukemia, malaria, mammary carcinoma, mania, menorrhagia, migraine, mumps, ophthalmic diseases, pains, palpitation, paralysis, pigmentation, piles, ring worm, scabies, sciatica, scorpion bite, skin diseases, sores, tetanus, thread worm, tinnitus, tonsillitis, tuberculosis, ulcer, uticaria, vertigo, wounds, etc. (Rokaya et al. 2012, Rehman et al. 2014, Alam and Khan 2015, Azhar and Anjum 2019). A mixture of rhubarb root, magnesia, mentha and ginger root, known as Gregory's mixture, is used as a mild laxative, antacid, cathartic and carminative. It has been mentioned that drinking of rhubarb juice is beneficial in delaying the onset of Alzheimer's disease, reducing hepatitis B, preventing sagging skin and wrinkles, relieving pain, reducing edema, treating diarrhea and constipation, antibrowning, etc. (Son et al. 2000, Sanghi and Tiwle 2016). It is also mentioned that drinking rhubarb juice helps in fast recovery after surgery (Tian et al. 1991). The juice also had positive effects on the patients suffering from osteoporosis and abdominal cancers. Using rhubarb is prohibited for the patients suffering from gout, rheumatism, epilepsy and uric acid troubles (CSIR 1972). People with a history of calculi should avoid rhubarb as it contains oxalic acid, which when consumed, can combine with Ca^{2+}, Fe^{2+} and Mg^{2+} in the blood-forming corresponding insoluble oxalates and may deposit in the kidneys and gall bladder (Nadkarni 2009). Although rhubarb is used to treat various ailments, official Chinese and Himalayan rhubarbs should be prescribed only by a trained herbalist. It should not be given to children below 12 years of age and pregnant women. There is insufficient information for pharmacological effects in the breast-fed infants; therefore, it should be avoided during the lactation period (Fok 2001). It is not suitable for hemorrhoids and old patients as well (Duke 1985). Despite this, some authors have mentioned in their review articles that rhubarb (*R. emodi*) is a safe herb without any adverse effect or toxicity and can be given to children as well as old people (Agarwal et al. 2001, Rokaya et al. 2012). It is better to use rhubarb for less than two weeks at a time. Prolonged use can imbalance body fluids and minerals, which can trigger bone deterioration, irregular heart rhythms, kidney problems, and produce a laxative dependence (Clementi and Misiti 2010). Chronic use may lead to reversible pigment deposition in the intestinal mucosa.

Petioles of the rhubarbs are directly chewed or eaten fresh or dried (CSIR 1972). Raw petioles are used as an appetizer with common salt and also used to make fresh pickles. They are cooked with common salt and pepper. Roots and rhizomes of the rhubarbs are powdered and sprinkled over the wounds for quick healing. It is also used for cleaning teeth. The paste is prepared with turmeric powder and then applied to the affected parts of ulcers. The paste is also prepared with a little flour. It is used as a poultice to fix dislocated bones and to treat pain and rheumatism. Powders are consumed with eggs, warm milk or yogurt. Watery extracts are used for general applications and also taken orally. Rhizomes are roasted with butter, ghee or oil. To make pills, they are mixed with other herbs and drunk with water. Stems are also used for decoction and preparation of hot beverages (Dutta 2007, Ghimire et al. 2008, Rokaya et al. 2012). According to the clinical conditions and severity, the oral dose of rhubarb may vary. For constipation, an average dose is 0.5–1.5 g of dried roots and rhizomes. Some authors have mentioned that rhubarb produces constipation in a low dose and purgation in a high dose (Clark 1995, Qin et al. 2011, Azhar and Anjum 2019). Perhaps, constipation resulted due to the presence of rheotannic acid and the purgative activity of rhubarb is due to chrysorobin. A tablespoon of the powder in a cup of boiling water can be used daily. A tincture (1–2 mL) can be taken thrice a day. The therapeutic dose of rhubarb 2–7 g is mentioned by many authorities, and the dose of rhubarb extract is 125–500 mg (Azhar and Anjum 2019). Noteworthy is the fact that this dose information should be taken as a range used in common herbal practices. Bowel irritation with spasm as a side effect of rhubarb is also mentioned. When given internally, the roots impart a deep tinge to urine (Nadkarni 2009). It has been mentioned that rhubarb may produce dryness (Azhar and Anjum 2019). Besides the medicinal uses, the roots and rhizomes are also used for coloration of fabrics, textiles, hair and wooden materials (Das 2008, Rokaya et al. 2012). They are soaked with water and used for the herbal bath to remove heat from the body (Rokaya et al. 2010).

Rhubarb leaves are poisonous as they are constituted of oxalic acid, tannic acid, etc. (Lewis and Elvin-Lewis 1977, Duke 1985). Consumption of rhubarb leaves causes nausea, vomiting, abdominal pain, dyspnoea, internal bleeding and occasionally death.

Pharmacological Activities of Rhubarb

Oxidative stress results due to the imbalance between the production of reactive free radicals and antioxidant defense in the body (Betteridge 2000). Oxidative stress is one of the primary factors for the occurrence and progression of many diseases and disorders such as atherosclerosis, cancer, diabetes, hepatotoxicity, hyperlipidemia, neural degeneration, renal failure, etc. (Bilal et al. 2013). Antioxidants present in the plant sources are useful in their prevention and treatment. Rhubarbs act as antioxidants (Yuan and Gao 1997, Iizuka et al. 2004, Kogure et al. 2004, Oktay et al. 2007, Öztüerk et al. 2007, Singh et al. 2013, Gupta et al. 2014). The anthraquinone derivatives of rhubarb are considered to be responsible for the purgative and cathartic effects. Different pharmacological studies proved that anthraquinones are responsible for several therapeutic actions (de Witte 1993). It was described in several studies that rhubarb could suppress lipid peroxides production (Chen et al. 1994.), reduce malondialdehyde (Yin et al. 2003, Zhang et al. 2007b), inhibit myeloperoxidase and phospholipase A2 activities, reduce production of platelet-activating factor (Yang et al. 1998), increase superoxide dismutase activity (Ma et al. 2012), suppress gene expression, decrease secretion of ILs, regulate catalase and glutathione peroxidase activities (Su 2014), decrease endotoxin level (Zhao et al. 1992), etc. (Lai et al. 2015). These diverse therapeutic actions of rhubarb prove that it is really the king of herbs as mentioned in ancient times.

Analgesic and anti-inflammatory activities

The anti-inflammatory activity of rhubarb has been demonstrated (Moon et al. 2006, Liu et al. 2008). Evaluation of analgesic and anti-inflammatory activities of the rhubarb extracts (400–500 mg/kg) on animal models revealed that the inhibitory activity was comparable to the standard drugs, viz., paracetamol (75 mg/kg), aspirin (100 mg/kg) and ibuprofen (50 mg/kg) (Chauhan et al. 1992, Gupta et al. 2017). Possibly, both peripheral and central mechanisms of pain inhibition are acting for the analgesic effect. Strongest inhibition of edema within 30 minutes of oral administration of *R. nobile* extract may suggest that it effectively suppresses the exudative phase of acute inflammation.

Antimicrobial activity

The methanolic extract of *R. emodi* exhibited antifungal activity against *Aspergillus fumigatus*, *Candida albicans* and *Trichophyton mentagrophytes* (Agarwal et al. 2000). The rhubarb extract was effectively used to control *Fusarium solani* in brinjal (Joseph et al. 2008). Aqil and Ahmad (2003) reported antifungal activity against *Alternaria alternate*, *Aspergillus niger*, *Fusarium chlamydosporum*, *Rhizoctonia bataticola* and *Trichoderma viride*. They further reported that the ethanolic extract inhibited the growth of *Shigella dysenteriae* (microorganism responsible for dysentery), but not *Escherichia coli*, *Pseudomonas aeruginosa*, *Salmonella paratyphi*, *Salmonella typhi*, *Salmonella typhimurium*, and *Staphylococcus aureus*. Others reported inhibition of *E. coli*, *S. aureus*, *P. aeruginosa*, *S. typhimurium* as well as *Bacillus subtilis*, *Bacteroides fragilis*, *Enterbacter cloacae*, *Klebsiella pneumoniae*, *Saccharomyces cerevisiae*, *Salmonella enteritidis*, *Serratia marcescens* and *Shiegella flexineri* by the rhubarb extracts (Cyong et al. 1987, Han et al. 2013, Gupta et al. 2014, 2017, Hou et al. 2015). The extracts of *R. undulatum* showed antibacterial activity against *Streptococcus mutans* and *Streptococcus sorbrinus*; therefore, it can be used for the treatment of dental plaque and cavities (Song et al. 2006).

The extract of rhubarb displayed a promising activity against *Helicobacter pylori,* indicating its effectiveness in gastric ulcer (Bae et al. 1998, Ibrahim et al. 2006). The antiulcer effect of the rhizome extract on pyloric ligation-induced ulcers in rats by reducing ulcer index and increasing the pH of gastric fluid was reported (Kaur et al. 2012). In a study, 312 cases of gastric and duodenal ulcer bleeding were treated with alcoholic extract rhubarb (*R. officinale*, *R. palmatum* and *R. tanguticum*) tablets, which effectively cured the upper gastrointestinal hemorrhage (Zhou and Jiao 1990). Antiviral activity of rhubarb against *Herpes simplex virusrubella* and *Varicella zoster* with a minimal inhibitive dose of 100–5000 mg/L is also mentioned (Wang et al. 1996).

Antioxidant and anticancer potentiality

Rhubarb possess antioxidants and anticancer potential (Huang et al. 2007, Öztürk et al. 2007, Rajkumar et al. 2011a, Singh et al. 2013, Gupta et al. 2014, 2015). They effectively scavenged 2,2-diphenyl-1-picrylhydrazyl-, hydroxyl- and superoxide-radicals, reduced Fe^{3+} and Cu^{2+}, and inhibited lipid peroxidation in a dose-dependent manner. The extracts of *R. emodi* are also tested in the human breast, liver, and prostate carcinoma cell lines (Rajkumar et al. 2011a, b). It is demonstrated that rhubarb acts against tumors by reducing the neoplastic growth and malignancy that are often caused by oxidative stress (Zargar et al. 2011). Antiangiogenic action of rhubarb has also been shown (Huang et al. 2007, Srinivas et al. 2007, Cai et al. 2008).

Antidiabetic effect

Administration of rhubarb extracts reduces the sugar level in normoglycemic, alloxan-induced and streptozotocin-induced diabetic rats (Li et al. 1997, Babu et al. 2004, Choi et al. 2006, Radhika et al. 2010, 2012, Gupta et al. 2017). The antidiabetic effect observed is due to the utilization of peripheral glucose, decreasing carbohydrate digestion, restoring enzymes to control level and correcting impaired liver and kidney functions by limiting gluconeogenesis similar to insulin.

Nephroprotective activity

Rhubarb extracts have shown significant nephroprotective effect (Nagasawa et al. 1980, Yokozawa et al. 1984, 1985, 1986, 1991, Gu et al. 2005). The extracts reduce proteinuria and the severity of glomerulosclerosis. Rhubarb can improve nitrogen metabolism that may delay hemodialysis (Sadana 1996). Investigation of the *R. emodi* extract on cadmium chloride, mercuric chloride, potassium dichromate and gentamycin-induced nephrotoxicity in rats by monitoring the levels of urea nitrogen and creatinine in serum revealed that the extract exhibited a nephroprotective effect on all the proximal tubule segments (S1, S2, and S3) (Alam et al. 2005). The nephroprotective effect of rhubarb appeared possibly through antioxidant action. *R. officinale* was recommended in the treatment of diabetic nephropathy as its administration in streptozotocin-induced diabetic rats helped to normalize renal hypertrophy and glomerular filtration rate (Yang and Li 1995).

Hepatoprotective effect

Rhubarb is used alone or in combination with other crude drugs for the treatment of cholestatic hepatitis (Hu 1986). Rhubarb could prevent and treat hyperlipidemia and fatty liver by decreasing total cholesterol (decreases low-density lipoprotein cholesterol and increases high-density lipoprotein cholesterol) (Xu et al. 2007). Modern pharmacological studies have confirmed that rhubarb had significant effects in reducing liver cell injury (Chen et al. 2002, Dong et al. 2005). Rhubarb possesses hepatoprotective principles and shows a significant liver protective effect on paracetamol- or CCl_4-injured liver in rats (Akhtar et al. 2009, Xinga et al. 2011). CCl_4 administration caused a significant increase of aspartate transaminase, alanine transaminase, and alkaline phosphatase activities, and increased the total bilirubin. Oral administration of *R. emodi* at 3 g/kg dose significantly decreased the activities and restored the normal liver function (Ibrahim et al. 2008). It has also been noted that the hepatoprotective effect would be weakened or even toxic with the increase of the dose. The canonical correlation between the toxicity-attenuating effects was studied (Wang et al. 2009).

Immuno-enhancing activity

R. emodi rhizome extract has displayed immuno-enhancing activity in murine macrophage cell lines via Th-1 and Th-2 cytokine regulation (Kounsar et al. 2011). A dose-dependent increase in the release of nitric oxide, cytokine TNF-α, and IL-12, and a decrease in IL-10 by RAW 264.7 were considered for the effect.

Cathartic effect

Rhubarb is well known for its cathartic and diuretic effects, which are closely correlated with water adjustment of colon and kidney by the theory of traditional Chinese medicine (Li et al. 2008). Researches have indicated that the anthraquinone glycosides can bring about fairly obvious effects of 'watery diarrhea'. Rhubarb regulates aquaporins in the colon that lead to less absorption of water and more secretion of intestinal juice. It is likely that aquaporin 2 is regulated through protein kinase A signal pathway (Zhang et al. 2008).

Other activities

The beneficial effects of rhubarb on Alzheimer's and Parkinson's diseases have been mentioned elsewhere (Clementi and Misiti 2010, Sanghi and Tiwle 2016). Rhubarb extract has been used as a dental desensitizer (Sauro et al. 2006). Studies showed that rhubarb extracts possess estrogenic activity (Usui et al. 2002, Kang et al. 2008). Rhubarb reduced fat and weight in animal experiments (Lee et al. 2012). *R. emodi* rhizome extract showed antidyslipidemic activity in Triton-WR-1339 and high-fat diet-induced dyslipidemic rats (Mishra et al. 2014). Rhubarb extracts were effective on the inhibition of degranulation in antigen-induced RBL-2H3 cells (Wang et al. 2007). *R. undulatum* rhizome extract produced vascular relaxant effect (Yoo et al. 2007). *R. nobile* rhizome extract produced a significant central nervous system depressant effect at dose 250 mg/kg on mice (Gupta et al. 2017). The extract possibly acted through direct activation of GABA receptor. It produced a profound effect in comparison to the standard drug clonazepam (1 mg/kg). A dose-dependent antifertility effect of the extract of *R. nobile* rhizome is also disclosed. Molluscicidal activity against the snails of the aqueous extract of *R. palmatum* and *Rumex dentatus* roots is reported (Liu et al. 1997). Rhubarb could be used to prevent bacterial translocation, maintain intestinal environment and protect the gut barrier (Chen et al. 2009a, Chen and Wang 2009). Rhubarb has also been used in clinical trials and reported that it can improve senile person's memory ability, abate menstrual pain, treat diabetic nephropathy and prevent progression of renal failure (Zhang et al. 1990, Tian et al. 1997, Huang et al. 2007, Srinivas et al. 2007, Rehman et al. 2015).

Phytochemical Constituents

About 270 compounds have been isolated from different species of rhubarb and are listed in Table 18.2 (Schmidt et al. 1995, Agarwal et al. 2001, Shikishima et al. 2001, Iwashina et al. 2004, Komakine et al. 2005, Medyńska and Smolarz 2005, Jin et al. 2006, Ashnagar et al. 2007, Yoo et al. 2007, El-Lakany et al. 2008, Ngoc et al. 2008, Gao et al. 2012, Rokaya et al. 2012, Wang et al. 2012, Nazir et al. 2013, Singh et al. 2013, Zheng et al. 2013). Chemical structures of some important constituents isolated from rhubarb are shown in Figures 18.2–18.9.

Anthraquinones

Anthraquinones are the most important phytochemicals present nearly in all rhubarbs. Among them, chrysophanol (**1**), physcion (**2**), emodin (**3**), aloe-emodin (**4**) and rhein (**5**) are the main free anthraquinones. These anthraquinones are known for various biological activities including antioxidant (Yen et al. 2000, Cai et al. 2004), cathartic (Li et al. 2008), estrogenic relative potency (Kang et al. 2008), antidiabetic (Arvindekar et al. 2015), anti-cancer (Huang et al. 2007), etc.

Chrysophanol (**1**), physcion (**2**), aloe-emodin (**4**) and rhein (**5**) exhibited antifungal activity against *C. albicans*, *T. mentagrophytes* and *A. fumigatus* (Agarwal et al. 2000). Emodin (**3**), aloe-emodin (**4**) and rhein (**5**) showed noticeable antibacterial activity against methicillin-resistant *S. aureus* and *Bifidobacterium adolescentis* (Hatano et al. 1999, Wang et al. 2010b). Rhein (**5**) was also effective against *Escherichia coli* as well as *Bacillus megaterium* (Tegos et al. 2002), *Bacteroides fragilis* (Cyong et al. 1987) and *S. typhimurium* (Hou et al. 2015). Emodin (**3**), aloe-emodin (**4**) and rhein (**5**) displayed potent anti-angiogenic activity by preventing blood vessel formation in zebrafish embryos (He et al. 2009).

Chrysophanol (**1**) possesses antibacterial and spasmolytic activities (Khetwal and Pathak 1988, Coopoosamy and Magwa 2006). It protects brain tissue damage during cerebral ischemia/reperfusion by inhibiting NALP3 inflammasome and could protect cerebral ischemic stroke (Zhang et al. 2014).

Emodin (**3**) was found effective against herpes simplex virus (Xiong et al. 2011). It had an inhibitory effect on the pathogenicity of *Trichomonas vaginalis* in mice (Wang 1993). It blocks both the binding of severe acute respiratory syndrome-CoV spike (S) protein to angiotensin-converting enzyme 2 and the infectivity of S protein-pseudotyped retrovirus to Vero E6 cells (Ho et al. 2007). It inhibits monoamine oxidase B and hence can be used for the treatment of Parkinson's disease (Kong et al. 2004). Emodin (**3**) inhibited the proliferation of ovarian, breast, lung, liver and prostate cancer cells. It showed a dual role in the inhibition of drug-resistant ovarian tumor growth by increasing paclitaxel cellular concentration and re-sensitizing the resistant cells to paclitaxel (Li et al. 2009). Emodin (**3**) also has the capacity to inhibit the proliferation in prostate cancer cell lines by androgen receptor and p53-p21 pathways (Yu et al. 2008). It inhibited the proliferation of human gastric cancer cell line MKH45 by arresting the cell cycle at G2/M phase (Chihara et al. 2015). It induces apoptosis through the up-regulation of caspase-3 and -9. Ameliorative effect of emodin (**3**) was studied on cisplatin-induced nephrotoxicity in rats (Ali et al. 2013). The treatment significantly increased urine volume and *N*-acetyl-β-D-glucosaminidase activity and decreased osmolarity and protein concentration. Emodin (**3**) acted as a potent free radical scavenger and provided nephroprotection against oxidative stress (Waly et al. 2013). It augmented the cisplatin-induced inhibition of antioxidant enzymes (catalase, glutathione peroxidase, glutathione S-transferase, glutathione reductase and superoxide dismutase). Emodin (**3**) selectively inhibited casein kinase II and a serine-threonine kinase as a competitive inhibitor against adenosine triphosphate (ATP) (Yim et al. 1999). It is a major active drug present in rhubarb that is responsible for the hepatoprotective effect. It treats cholestatic hepatitis via an anti-inflammatory pathway (Ding et al. 2008). Emodin (**3**) showed significant lipid-lowering activity in the high diet-fed model (Mishra et al. 2014). It enhanced cutaneous wound healing in rats (Tang et al. 2007).

Aloe-emodin (**4**) showed antiviral activity against herpes simplex virus type I and type II, varicella-zoster virus, pseudorabies virus, influenza virus, adenovirus and rhinovirus (Sydiskis et al. 1991). It inhibited cell proliferation and induced apoptosis by enhancing expression of Bax, hence possessing anticancer effect (Kuo et al. 2002). It was able to inhibit the growth of neuroectodermal tumors and cervical cancer cells representing a new antitumor drug (Pecere et al. 2000, Guo et al. 2007).

Table 18.2. List of compounds isolated from rhubarb.

Anthraquinones**:** Chrysophanol (**1**); physcion (**2**); emodin (**3**); aloe-emodin (**4**); rhein (**5**); chrysaron (**6**); citreorosein (**7**); istizin (**8**); laccaic acid D (**9**); revandchinone-3 (**10**); rheinal (**11**); ribesin (**12**); 6-methyl-aloe-emodin (**13**); 6-methyl-aloe-emodin-triacetate (**14**); 6-methyl-rhein (**15**); 6-methyl-rhein-diacetate (**16**); chrysophanol-8-methyl ether (**17**); 1-methyl-8-hydroxyl-9,10-anthraquinone-3-*O*-β-D-(6′-*O*-cinnamoyl)glucopyranoside (**18**); aloe-emodin-1-*O*-β-D-glucopyranoside (**19**); aloe-emodin-3-(hydroxymethyl)-*O*-β-D-glucopyranoside (**20**); aloe-emodin-8-*O*-β-D-glucopyranoside (**21**); chrysophanol-1-*O*-β-D-glucopyranoside (**22**); chrysophanol-8-*O*-β-D-(6′-*O*-acetyl) glucopyranoside (**23**); chrysophanol-8-*O*-β-D-(6′-*O*-galloyl)glucopyranoside (**24**); chrysophanol-8-*O*-β-D-(6′-*O*-malonyl) glucopyranoside (**25**); chrysophanol-8-*O*-β-D-glucopyranoside (**26**); emodin-1-*O*-β-D-glucopyranoside (**27**); emodin-6-*O*-β-D-glucopyranoside (**28**); emodin-8-*O*-(2′,3′,4′,6′-tetracetyl)glucopyranoside (**29**); emodin-8-*O*-β-D-glucopyranoside (**30**); emodin-8-*O*-β-D-glucopyranosyl-6-*O*-sulfate (**31**); emodin-gemtiobioside (**32**); physcion-1-*O*-β-D-glucopyranoside (**33**); physcion-8-*O*-β-D-gentiobioside (**34**); physcion-8-*O*-β-D-glucopyranoside (**35**); rhein-11-*O*-β-D-glucoside (**36**); rhein-5-*O*-β-D-glucopyranoside (**37**); rhein-8-*O*-β-D-[6′-*O*-(3″-methoxylmalonyl)]glucopyranoside (**38**); rhein-8-*O*-β-D-glucopyranoside (**39**).

Anthrones**:** Anthralin (**40**); cascaroside C (**41**); cascaroside D (**42**); 10-hydroxycascaroside C (**43**); 10-hydroxycascaroside D (**44**); 10*R*-chrysaloin-1-*O*-β-D-glucopyranoside (**45**); cassialoin (**46**); revandchinone-1 (**47**); revandchinone-2 (**48**); revandchinone-4 (**49**); rheinoside A (**50**); rheinoside B (**51**); rheinoside C (**52**); rheinoside D (**53**); palmidin A (**54**); palmidin B (**55**); palmidin C (**56**); rendin A (**57**); rendin B (**58**); rendin C (**59**); sennidin C (**60**); sennoside A (**61**); sennoside B (**62**); sennoside C (**63**); sennoside D (**64**); sennoside E (**65**); sennoside F (**66**).

Stilbenes**:** Resveratrol (**67**); rhapontigenin (**68**); desoxyrhapontigenin (**69**); cis-3,3′,5-trihydroxy-4′-methoxystilbene (**70**); piceatannol (**71**); 3,5,4′-trimethyl piceatannol (**72**); rhaponticin (**73**); isorhaponticin (**74**); desoxyrhaponticin (**75**); resveratrol-3-*O*-β-D-glucopyranoside (**76**); 3,4′,5-trihydroxystilbene-4′-*O*-β-D-(2″-*O*-galloyl)glucopyranoside (**77**); 3,4′,5-trihydroxystilbene-4′-*O*-β-D-(6″-*O*-galloyl)glucopyranoside (**78**); 3,4′,5-trihydroxystilbene-4′-*O*-β-D-glucopyranoside (**79**); cis-3,3′,5-trihydroxy-4′-methoxystilbene-3-*O*-β-D-(2″-*O*-galloyl)glucopyranoside (**80**); cis-3,3′,5-trihydroxy-4′-methoxystilbene-3-*O*-β-D-glucopyranoside (**81**); cis-3,3′,5-trihydroxy-4′-methoxystilbene-3-*O*-β-D-(6″-*O*-galloyl)glucopyranoside (**82**); cis-3,5-dihydroxy-4′-methoxystilbene-3-*O*-β-D-glucopyranoside (**83**); piceid (**84**); 4′-*O*-methylpiceid (**85**); desoxyrhaponticin-6″-*O*-gallate (**86**); piceatannol-3′-*O*-β-D-glucopyranoside (**87**); piceatannol-3′-*O*-β-D-xylopyranoside- (**88**); piceatannol-3-*O*-β-D-(6″-*O*-galloyl)glucopyranoside (**89**); piceatannol-3-*O*-β-D-glucopyranoside (**90**); piceatannol-4′-*O*-(6″-*O*-galloyl)-β-D-glucopyranoside (**91**); piceatannol-4′-*O*-(6″-p-coumaroyl)-β-D-glucopyranoside (**92**); piceatannol-4′-*O*-β-D-glucopyranoside (**93**); piceatannol-3,4′-*O*-β-D-diglucopyranoside (**94**); resveratrol-4′-*O*-β-D-(6″-*O*-galloyl)-glucopyranoside (**95**); resveratrol-4′-*O*-β-D-glucopyranoside (**96**); rhaponticin-2″-*O*-gallate (**97**); rhaponticin-2″-p-coumarate (**98**); rhaponticin-6″-*O*-gallate (**99**); rhapontigenin-3′-3-*O*-β-D-glucopyranoside (**100**); gentin C (**101**); maximol A (**102**); maximol B (**103**); ε-viniferin (**104**); ampelopsin B (**105**).

Flavonoids: Cyanidin-3-*O*-glucoside, chrysanthemin (**106**); cyanidin-3-*O*-rutinoside (**107**); cyanin (**108**); (–)-epicatechin (**109**); (–)-epicatechin-3-*O*-gallate (**110**); (–)-epigallocatechin-3-*O*-gallate (**111**); (+)-catechin (**112**); (+)-catechin-3′,4′-di-*O*-β-D-glucopyranoside (**113**); (+)-catechin-3′,5-di-*O*-β-D-glucopyranoside (**114**); (+)-catechin-3′,7-di-*O*-β-D-glucopyranoside (**115**); (+)-catechin-3′-*O*-β-D-glucopyranoside (**116**); (+)-catechin-4′,5-di-*O*-β-D-glucopyranoside (**117**); (+)-catechin-4′-*O*-β-D-glucopyranoside (**118**); (+)-catechin-5-*O*-β-D-glucopyranoside (**119**); (+)-catechin-6-*C*-β-D-glucopyranoside (**120**); (+)-catechin-7-*O*-β-D-glucopyranoside (**121**); (+)-catechin-8-*C*-β-D-glucopyranoside (**122**); 3-O-galloylepicatechin-(4β→6)-3-*O*-galloylepicatechin-(4β→6)-catechin (**123**); 3-*O*-galloylepicatechin-(4β→6)-3-*O*-galloylepicatechin-(4β→8)-catechin (**124**); 3-*O*-galloylepicatechin-(4β→6)-3-*O*-galloylepicatechin-(4β→8)-catechin (**125**); 3-*O*-galloylepicatechin-(4β→8)-3-*O*-galloylepicatechin-(4β→6)-catechin (**126**); 3-*O*-galloylepicatechin-(4β→8)-3-*O*-galloylepicatechin-(4β→8)-catechin (**127**); epicatechin-(4β→6)-epicatechin-(4β→6)-catechin (**128**); epicatechin-(4β→6)-epicatechin-(4β→8)-catechin (**129**); epicatechin-(4β→8)-epicatechin-(4β→6)-catechin (**130**); epicatechin-(4β→8)-epicatechin-(4β→8)-catechin (**131**); procyanidin B-1 (**132**); procyanidin B-1-3-*O*-gallate (**133**); procyanidin B-1-6-*C*-β-D-glucopyranoside (**134**); procyanidin B-1-8-*C*-β-D-glucopyranoside (**135**); procyanidin B-2 (**136**); procyanidin B-2-3,3′-di-*O*-gallate (**137**); procyanidin B-2-3′-*O*-gallate (**138**); procyanidin B-2-6-*C*-β-D-glucopyranoside (**139**); procyanidin B-2-8-*C*-β-D-glucopyranoside (**140**); procyanidin B-3 (**141**); procyanidin B-3-7-*O*-β-D-glucopyranoside (**142**); procyanidin B-4 (**143**); procyanidin B-4-3′-*O*-gallate (**144**); procyanidin B-5 (**145**); procyanidin B-5-3,3′-di-*O*-gallate (**146**); procyanidin B-7 (**147**); procyanidin B-7-3-*O*-gallate (**148**); procyanidin C-1 (**149**); procyanidin C-1-3,3′,3″-tri-*O*-gallate (**150**); procyanidin C-1-3,3′-di-*O*-gallate (**151**); rhatannin (**152**); RG-tannin (**153**); (+)-taxifolin (**154**); carpusin (**155**); mesopsin (**156**); daidzein (**157**); puerarin (**158**); kaempferol (**159**); kaempferol-3-*O*-(2′,6′-di-*O*-rhamnopyranosyl)-β-D-galactopyranoside (**160**); kaempferol-3-*O*-rhamnoside (**161**); kaempferol-7-*O*-glucoside (**162**); meratin (**163**); myricetin (**164**); quercetin (**165**); quercetin-3-*O*-[6″-(3-hydroxy-3-methylglutaroyl)]-glucoside (**166**); quercetin-3-*O*-arabinoside, guaijaverin (**167**); quercetin-3-*O*-galactoside, hyperin (**168**); quercetin-3-*O*-glucoside, isoquercitrin (**169**); quercetin-3-*O*-rutinoside, rutin (**170**); quercetin-7-*O*-glucoside (**171**); phloretin-3′-*C*-β-D-glucoside (**172**).

Table 18.2 Contd. ...

...Table 18.2 Contd.

Pyranones: 2-(2′-Hydroxypropyl)-5-methyl-7-hydroxychromone (**173**); 2-(2′-hydroxypropyl)-5-methyl-7-hydroxychromone-7-*O*-β-D-glucopyranoside (**174**); 2,5-dimethyl-7-hydroxychromone (**175**); 2-methyl-5-(2′-oxo-4′-hydroxypentyl)-7-hydroxychromone-7-*O*-β-D-glucopyranoside (**176**); 2-methyl-5-acetonyl-7-hydroxychromone (**177**); 2-methyl-5-carboxymethyl-7-hydroxychromone (**178**); 3-(3-(3,5-dihydroxyphenyl)acryloyl)-5-hydroxy-3,4-dihydro-2-pyrone (**179**); 5,7-dihydroxy-2-methylchromone, noreugenin (**180**); aloesone-7-*O*-β-D-glucopyranoside (**181**).
Phenols: (–)-Epicatechol (**182**); (–)-epicatechol-gallate (**183**); (–)-epigallocatechol (**184**); (–)-epigallocatechol-gallocatechol (**185**); (+)-catechol (**186**); (+)-gallocatechol (**187**); (±)-gallocatechol-gallate (**188**); (*S*)-4-(4-hydroxyphenyl)-2-butanol-2-*O*-(6-*O*-acetyl)-β-D-glucopyranoside (**189**); (*S*)-4-(4-hydroxyphenyl)-2-butanol-2-*O*-(6-*O*-galloyl)-β-D-glucopyranoside (**190**); (*S*)-4-(4-hydroxyphenyl)-2-butanol-2-*O*-[6-*O*-(3,5-dimethyloxy-4-*O*-α-L-rhamnopyranosylgalloyl)-β-D-glucopyranoside] (**191**); (*S*)-4-(4-hydroxyphenyl)-2-butanol-2-*O*-sulfate (**192**); epirhondodendrin (**193**); 4-(4′-hydroxyphenyl)-2-butanone (**194**); 4-hydroxybenzoic acid (**195**); carvacrol (**196**); chlorogenic acid (**197**); coumaric acid (**198**); eugenol (**199**); ferulic acid (**200**); 1-*O*-feruloyl-β-D-glucopyranoside (**201**); gallic acid (**202**); 4-methoxygallic acid (**203**); glucogallin (**204**); methyl 3,4-dihydroxybenzoate (**205**); pyrogallol (**206**); syringic acid (**207**); vanillic acid (**208**); β-asarone (**209**); β-resorcilic acid (**210**).
Acylglucosides: 1,2,6-Tri-*O*-galloylglucose (**211**); 1,2-di-*O*-galloyl-6-*O*-cinnamoyl-β-D-glucose (**212**); 1,2-di-*O*-galloyl-6-*O*-p-coumaroyl-β-D-glucose (**213**); 1,2-di-*O*-galloyl-β-D-glucose (**214**); 1,6-di-*O*-galloyl-2-*O*-p-coumaroyl-β-D-glucose (**215**); 1,6-di-*O*-galloylycerol-β-D-glucopyranoside (**216**); 1′-*O*-galloylsucrose (**217**); 1-*O*-galloyfructose (**218**); 1-*O*-galloyl-6-*O*-cinnamoyl-β-D-glucose (**219**); 1-*O*-galloylycerol (**220**); 1-*O*-galloyl-β-D-glucose (**221**); 1-ophloroglucinyl-2-*O*-galloyl-6-*O*-cinnamoyl-β-D-glucoside (**222**); 2,6-di-*O*-galloylglucose (**223**); 2-*O*-cinnamoyl-1,6-di-*O*-galloy-β-D-glucose (**224**); 2-*O*-cinnamoyl-1-*O*-galloy-β-D-glucose (**225**); 2-*O*-cinnamoyl-β-D-glucose (**226**); 2-*O*-galloylsucrose (**227**); 2-*O*-galloyl-β-D-glucose (**228**); 2-*O*-p-coumaroyl-1-*O*-galloy-β-D-glucose (**229**); 3,5-dihydroxyphenol 1-*O*-β-D-(6-O-galloyl)glucopyranoside (**230**); 4-(4′-hydtoxyphenyl)-2-butanone-4′-*O*-β-D-(2″-*O*-galloyl-6″-*O*-cinnamoyl)glucopyranoside (**231**); 4-(4′-hydtoxyphenyl)-2-butanone-(2″-*O*-galloyl-6″-*O*-p-coumaroyl) glucopyranoside (**232**); 4-(4′-hydtoxyphenyl)-2-butanone-4′-*O*-β-D-glucopyranoside (**233**); 4′-*O*-galloylsucrose (**234**); 4′-*O*-β-D-(2″,6″-di-*O*-galloyl)glucopyranoside (**235**); 6′-*O*-galloylsucrose (**236**); 6-hydroxymusizin glucoside (**237**); 6-*O*-galloylglucose (**238**); 6-*O*-galloylsucrose (**239**); coumaroyl-*O*-galloylglucose (**240**); gallic acid-3-*O*-β-D-(6′-*O*-galloyl)glucopyranoside (**241**); gallic acid-3-*O*-β-D-glucopyranoside (**242**); gallic acid-4-*O*-β-D-(6′-*O*-galloyl)-glucopyranoside (**243**); gallic acid-4-*O*-β-D-glucopyranoside (**244**); glucopyranosyl-galloyl-glucose (**245**); isolindleyin (**246**); lindleyin (**247**); p-coumaric acid glucoside (**248**); p-hydroxybenzoic acid glucose ester (**249**); rheumone A (**250**); sucrose (**251**); torachrysone-8-*O*-β-D-glucopyranoside (**252**).
Sterols and terpenes: Aldosterone (**253**); daucosterol (**254**); lup-20(29)-en-2α,3β,28-triol (**255**); β-sitosterol (**256**); β-sitosterol-D-glucoside (**257**).
Alkaloids: (+)-3-Cyanomethyl-3-hydroxyindole (**258**); 3-cyanomethyl-3-hydroxyoxindole-*N*-β-D-glucoside (**259**).
Pyrrolidinones: 5-Oxo-pyrrolidine-2-carboxylic acid methyl ester (**260**); pyrrolidin-2-one (**261**).
Organic acids: Acetic acid (**262**); citric acid (**263**); formic acid (**264**); fumaric acid (**265**); lactic acid (**266**); malic acid (**267**); oxalic acid (**268**); palmitic acid (**269**); succinic acid (**270**); succinic acid methyl ester (**271**); 2-hydroxy-succinic acid-1-methyl ester (**272**); trans-cinnamic acid (**273**).

Rhein (**5**) displayed antitumor effects (Lin et al. 2007, He et al. 2011). Studies suggested that it induces the apoptosis in squamous carcinoma cell 4 via caspase, reactive oxygen species and mitochondrial death pathways (Lin et al. 2003). It might have an inhibitory activity on apoptosis of L-02 cells induced by TNF-α and triglyceride (Wan et al. 2006). Rhein (**5**) reduced the Ca^{2+} by inhibiting the TNF-α production in lipopolysaccharide-stimulated peritoneal macrophages of rats and acted as an analgesic and anti-inflammatory agent (Yang et al. 2003). It showed anti-hyperuricemic and nephroprotective effects on urate nephropathy induced by adenine and ethambutolin in mice (Meng et al. 2015). It was effective in ameliorating diabetic renal pathological changes and attenuating hyperlipidemia (Gao et al. 2010). Rhein (**5**) might be a major poisonous ingredient in rhubarb (Fang et al. 2011).

Chrysophanol-1-*O*-β-D-glucopyranoside (**22**), chrysophanol-8-*O*-β-D-glucopyranoside (**26**) and physcion-8-*O*-β-D-glucopyranoside (**35**) exhibited moderate cytotoxic activity against HeLa epithelioid carcinoma cells (Kubo et al. 1992). Chrysophanol-8-*O*-β-D-glucopyranoside (**26**) exhibited a potent inhibitory effect on collagen thrombin-induced platelet aggregation and mammalian α-glucosidase (Babu et al. 2004, Seo et al. 2012).

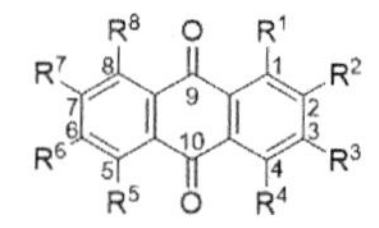

Compound	R^1	R^2	R^3	R^4	R^5	R^6	R^7	R^8
1	OH	H	Me	H	H	H	H	OH
2	OH	H	OMe	H	H	Me	H	OH
3	OH	H	Me	H	H	OH	H	OH
4	OH	H	CH_2OH	H	H	H	H	OH
5	OH	H	CO_2H	H	H	H	H	OH
6	H	H	OH	OH	H	OH	Me	H
7	OH	H	OH	H	H	CH_2OH	H	OH
8	OH	H	H	H	H	H	H	OH
9	Me	CO_2H	OH	H	H	OH	H	OH
10	OH	H	Me	H	H	OH	H	OH
11	OH	H	CHO	H	H	H	H	OH
12	OH	OMe	Me	H	H	OH	H	OH
22	O-Glc	H	Me	H	H	H	H	OH
26	OH	H	Me	H	H	H	H	OGlu
31	OH	H	Me	H	H	OSO_3H	H	OGlu
32	OH	H	Me	H	H	OH	H	O-Glc^{6}-^{1}Glc
35	OH	H	OMe	H	H	Me	H	O-Glc

Figure 18.2. Some anthraquinones isolated from rhubarb.

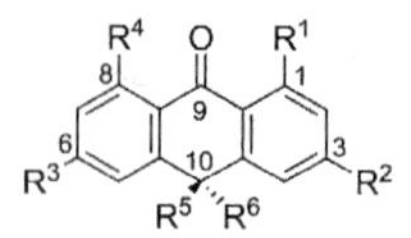

Compound	R^1	R^2	R^3	R^4	R^5	R^6
40	OH	Me	H	OH	H	H
41	OH	Me	H	O-Glc	Glc	H
42	OH	Me	H	O-Glc	H	Glc
45	O-Glc	Me	H	OH	H	Glc
46	OH	Me	H	OH	OH	Glc
47	OH	Me	OMe	OH	H	$OCO(CH_2)_7CHCH(CH_2)_7CH_3$
48	OH	Me	H	OH	H	$OCO(CH_2)_{26}CH_3$
49	OH	CH_2OH	OH	OH	OH	$O(CH_2)_{17}CH_3$
50	OH	CO_2H	H	O-Glc	Glc	OH
51	OH	CO_2H	H	O-Glc	OH	Glc
52	OH	CO_2H	H	O-Glc	Glc	H
53	OH	CO_2H	H	O-Glc	H	Glc

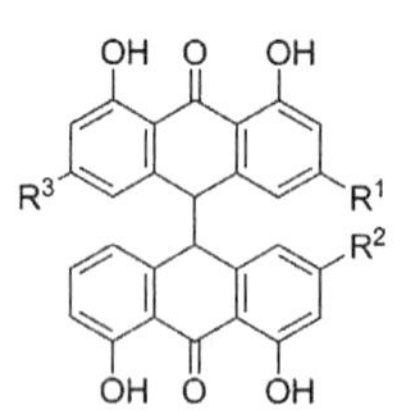

Compound	R^1	R^2	R^3
54	Me	CH_2OH	OH
55	Me	CH_2OH	H
56	Me	Me	OH
57	Me	CO_2H	OH
58	Me	CO_2H	H
59	Me	CO_2H	OMe
60	CO_2H	CH_2OH	H

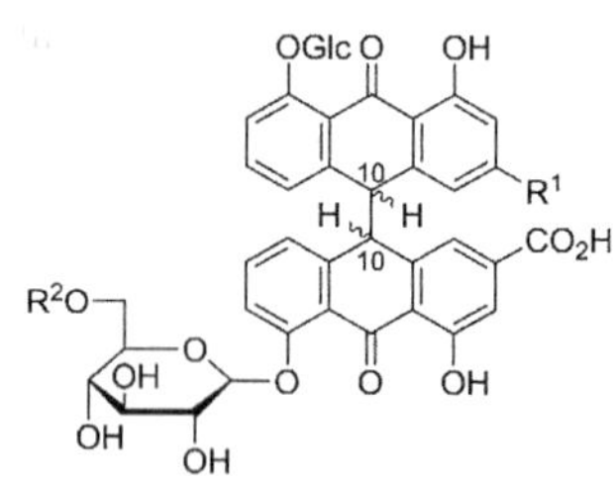

Compound	R^1	R^2	
61	CO_2H	H	10-10' *trans*
62	CO_2H	H	10-10' *meso*
63	CH_2OH	H	10-10' *trans*
64	CH_2OH	H	10-10' *meso*
65	CO_2H	oxalyl	10-10' *trans*
66	CO_2H	oxalyl	10-10' *meso*

Figure 18.3. Some anthrones isolated from rhubarb.

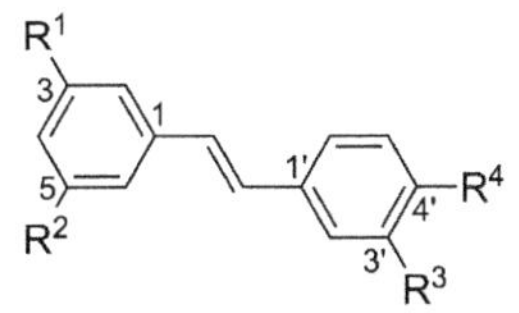

Compound	R^1	R^2	R^3	R^4
67	OH	OH	H	OH
68	OH	OH	OH	OMe
69	OH	OH	H	OMe
71	OH	OH	OH	OH
72	OMe	OMe	OH	OMe
73	OH	O-Glc	OH	OMe
74	OH	O-Glc	OMe	OH
75	OH	O-Glc	H	OMe
84	OH	O-Glc	H	OH
85	OH	O-Glc	H	OMe

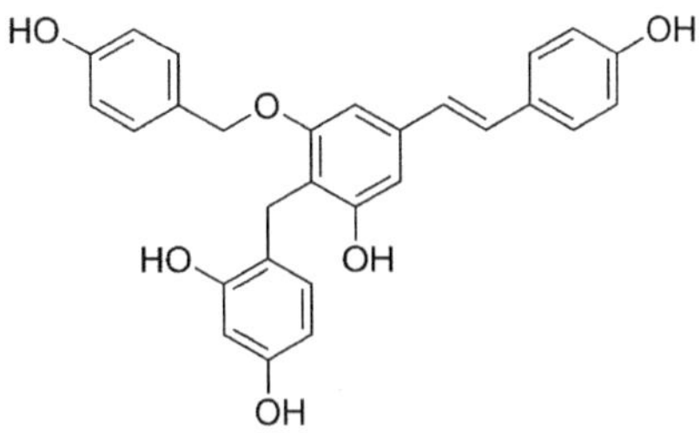

101

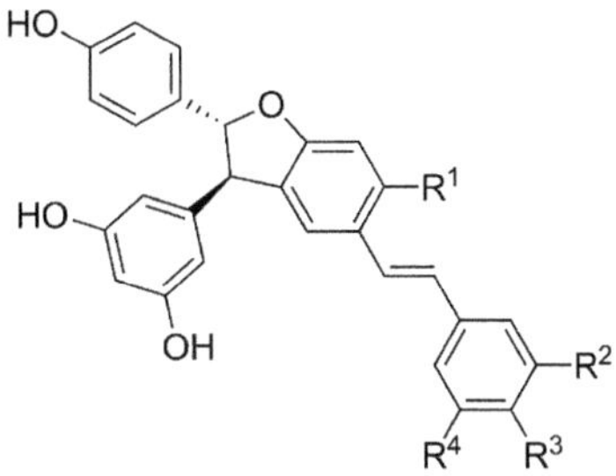

102 $R^1 = R^3 = H$, $R^2 = R^4 = OH$
104 $R^1 = R^3 = OH$, $R^2 = R^4 = H$

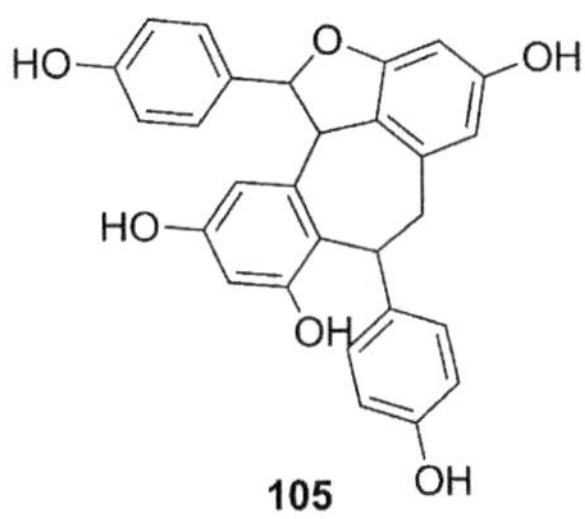

105

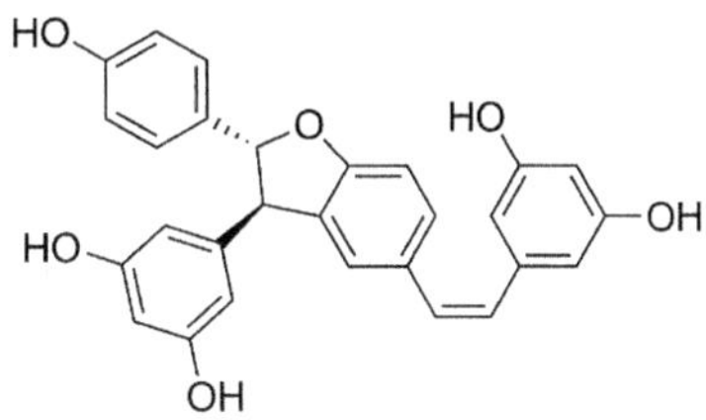

103

Figure 18.4. Some stilbenes isolated from rhubarb.

Anthrones

Twenty-seven anthrones have already been isolated from different species of rhubarb. Revandchinone-4 (**49**) was found to be a more effective antibacterial agent than revandchinone-1 (**47**) and revandchinone-3 (**10**) (Babu et al. 2003). Rheidin A (**57**), rheidin B (**58**), rheidin C (**59**), sennidin C (**60**), etc. are dianthrones composed of two molecules of different anthraquinones. Dimeric anthrones sennosides A–F (**61**–**66**) are major active principles for purgative action of rhubarb (Nishioka 1991).

Stilbenes

Resveratrol (**67**) showed anticancer, anti-inflammatory, anti-diabetic and other beneficial cardiovascular effects (Kundu et al. 2008, Markus and Morris 2008, Raval et al. 2008). It may enhance the prognosis of neurological disorders such as Alzheimer's, Huntington's and Parkinson's diseases, and stroke. It reduced liver damage induced by chronic ethanol administration (Raal et al. 2009).

Resveratrol (**67**), rhapontigenin (**68**) and piceatannol (**71**) were studied as inhibitors of nitric oxide production in lipopolysaccharide-activated-macrophages (Kageura et al. 2001). Rhapontigenin (**68**) and desoxyrhapontigenin (**69**) have shown strong inhibition of aggregation induced by arachidonic acid and collagen (Ko et al. 1999). Antioxidative rhapontigenin (**68**) could protect V79-4 cells against oxidative damage by modulating the cellular signal pathways (Zhang et al. 2007a). It was found effective in inhibiting mammalian α-glucosidase (Babu et al. 2004). Piceatannol (**71**) inhibited the migration and invasion of prostate cancer cells (Kwon et al. 2012). It was also responsible for the vasorelaxant property

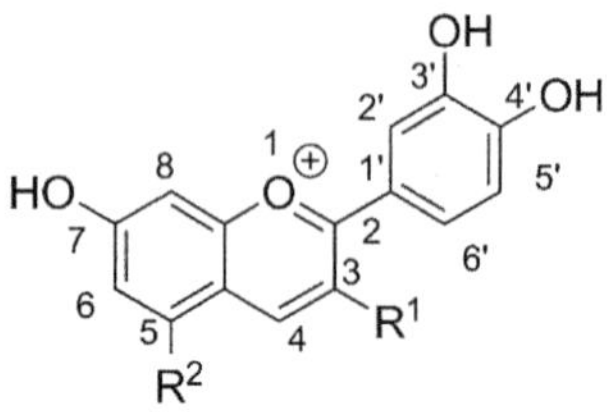

106 R^1 = O-Glc, R^2 = OH
107 R^1 = O-Rut, R^2 = OH
108 R^1 = O-Glc, R^2 = O-Glu

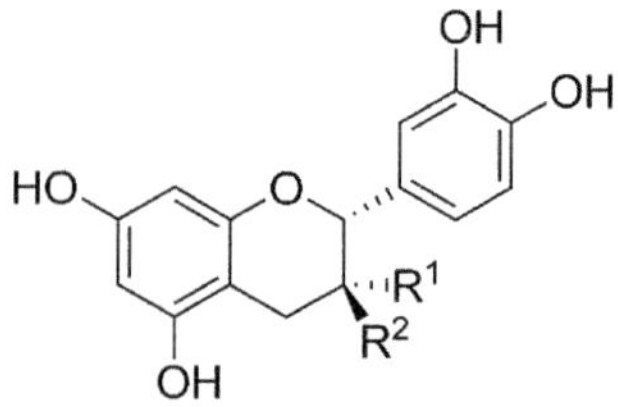

109 R^1 = OH, R^2 = H
110 R^1 = OG, R^2 = H
112 R^1 = H, R^2 = OH

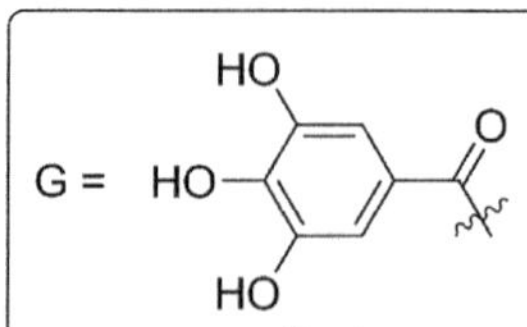

galloyl group

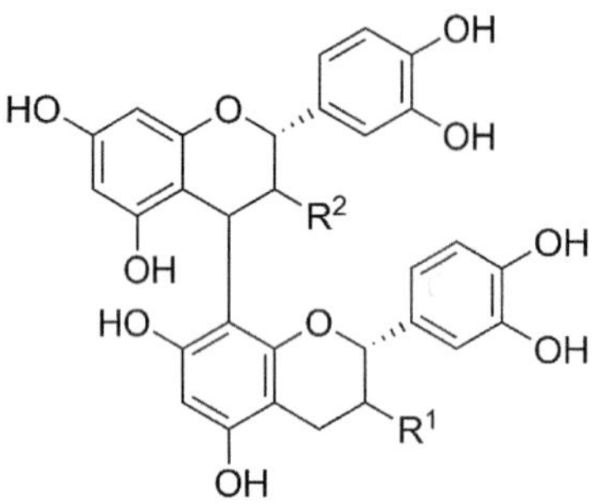

132 R^1 = β-OH, R^2 = α-OH
136 R^1 = α-OH, R^2 = α-OH
137 R^1 = α-OG, R^2 = α-OG

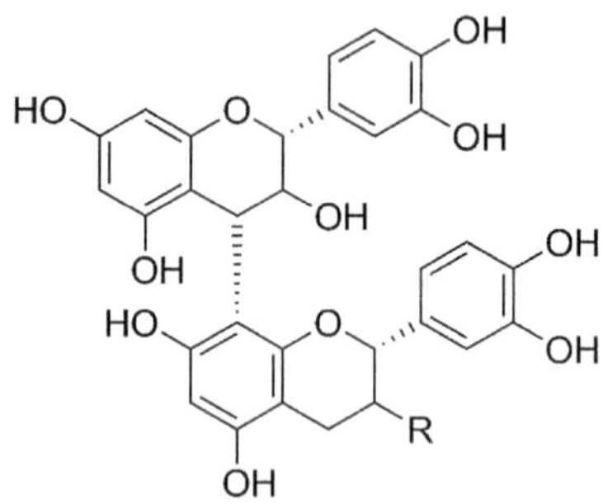

141 R = β-OH
143 R = α-OH

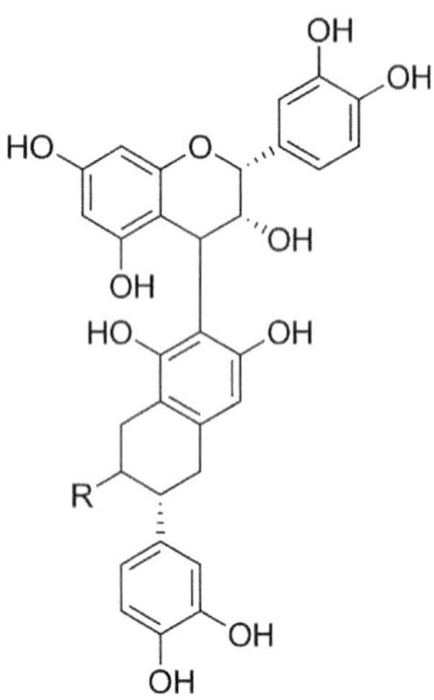

145 R = α-OH
147 R = β-OH

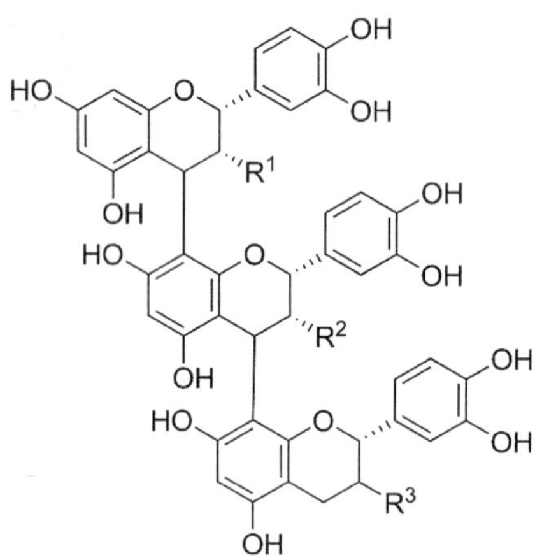

149 R^1 = OH, R^2 = OH, R^3 = α-OH
150 R^1 = OG, R^2 = OG, R^3 = α-OG

Figure 18.5 Contd. ...

...Figure 18.5 Contd.

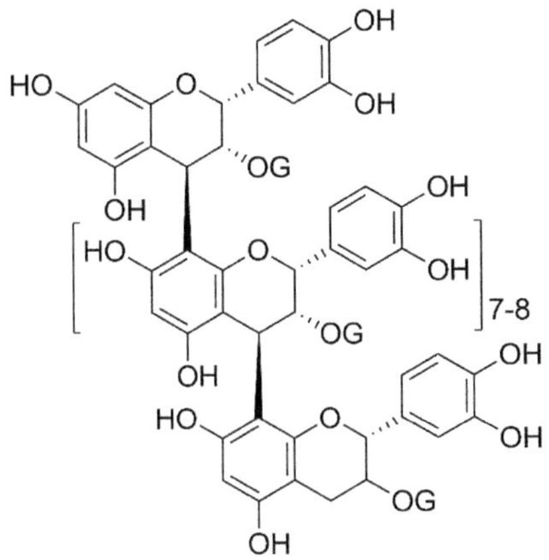

152

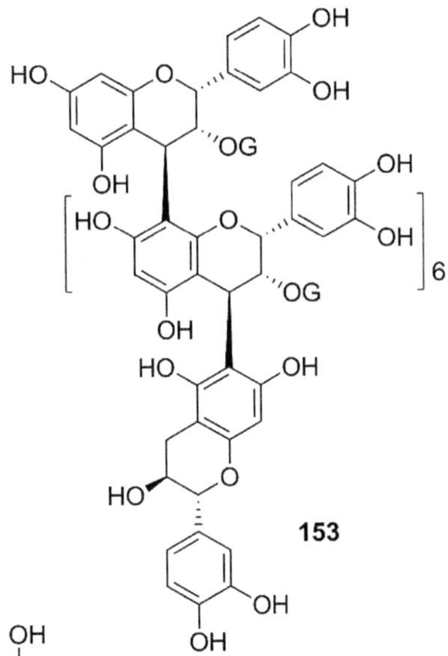

153

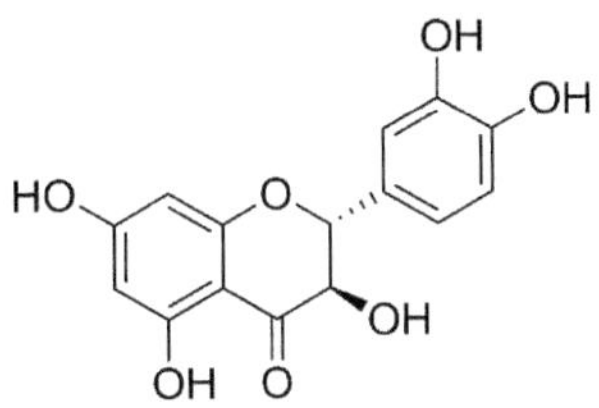

154

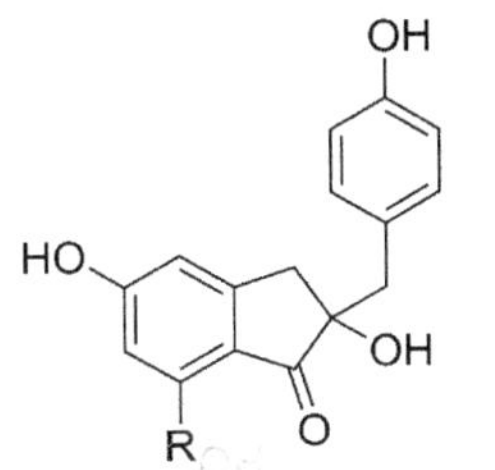

155 R = OMe
156 R = OH

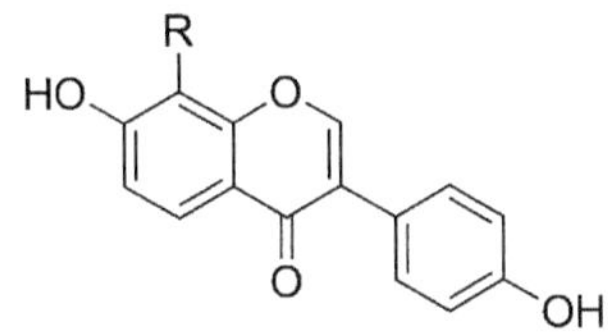

157 R = H
158 R = Glc

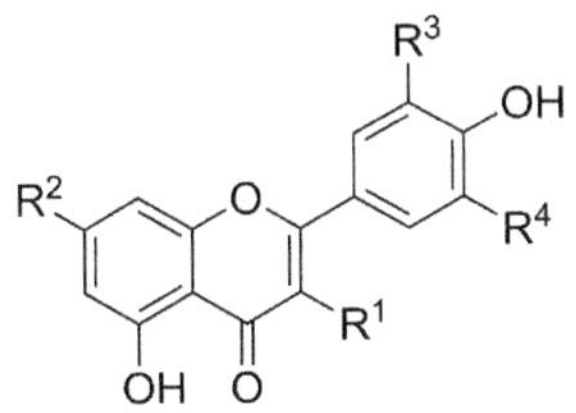

Compound	R^1	R^2	R^3	R^4
159	OH	OH	H	H
161	O-Rha	OH	H	H
162	OH	O-Glc	H	H
163	O-Glc[4]-[1]Glc	OH	OH	H
164	OH	OH	OH	OH
165	OH	OH	OH	H
167	O-Ara	OH	OH	H
168	O-Gal	OH	OH	H
169	O-Glc	OH	OH	H
170	O-Rut	OH	OH	H
171	OH	O-Glc	OH	H

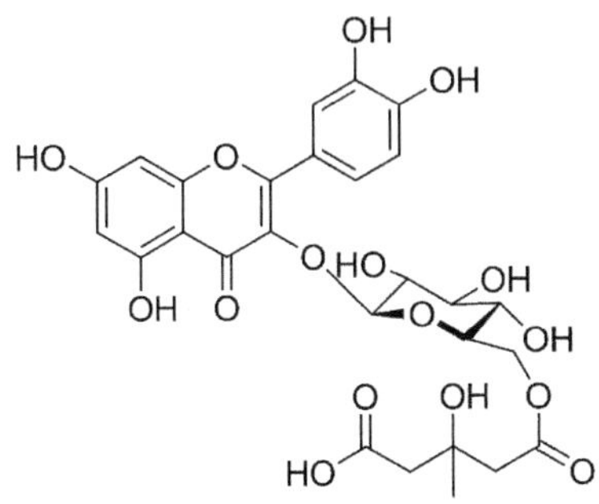

166

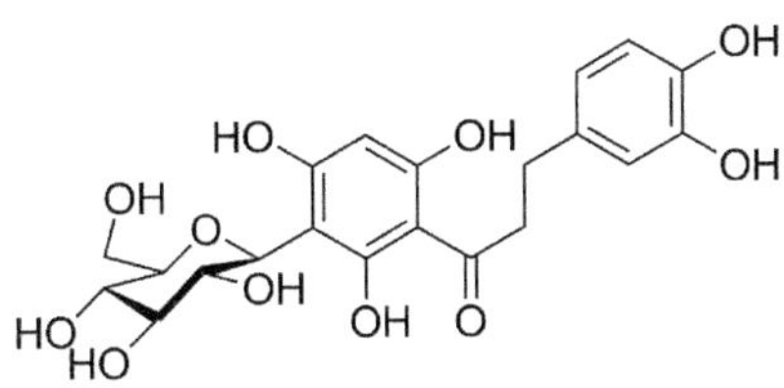

172

Figure 18.5. Some flavonoids isolated from rhubarb.

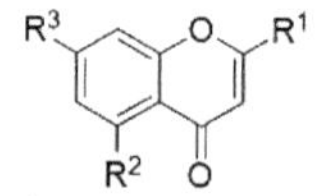

Compound	R^1	R^2	R^3
173	CH_2CHCH_3OH	Me	OH
174	CH_2CHCH_3OH	Me	O-Glc
175	Me	Me	OH
176	Me	$CH_2COCH_2CHCH_3OH$	O-Glc
177	Me	CH_2COCH_3	OH
178	Me	CH_2CO_2H	OH
180	Me	OH	OH
181	CH_2COCH_3	Me	O-Glc

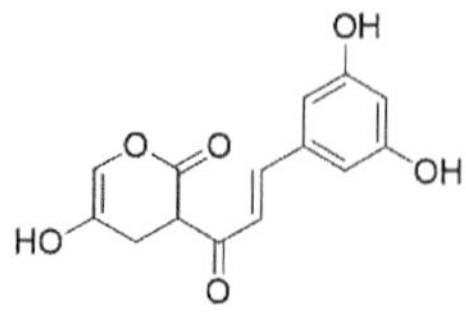

179

Figure 18.6. Some pyranones isolated from rhubarb.

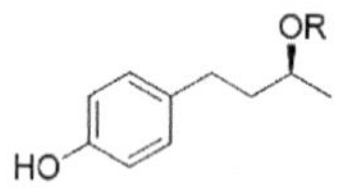

189 R = 6-O-acetyl-β-D-Glc
190 R = 6-O-galloyl-β-D-Glc
191 R = α-L-Rha
192 R = SO_3H

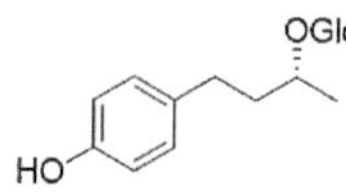

193

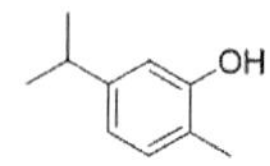

196

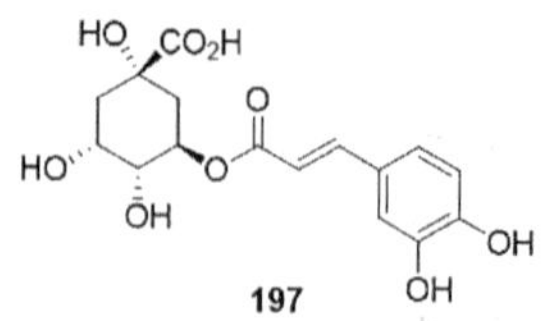

197

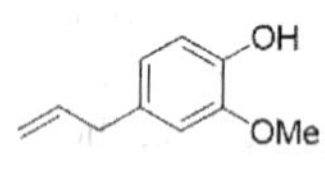

199

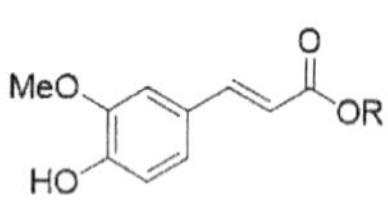

200 R = H
201 R = Glc

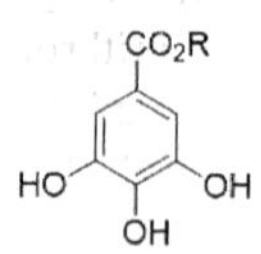

202 R = H
204 R = Glc

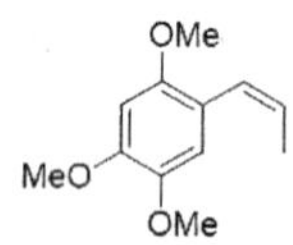

209

Figure 18.7. Some phenols isolated from rhubarb.

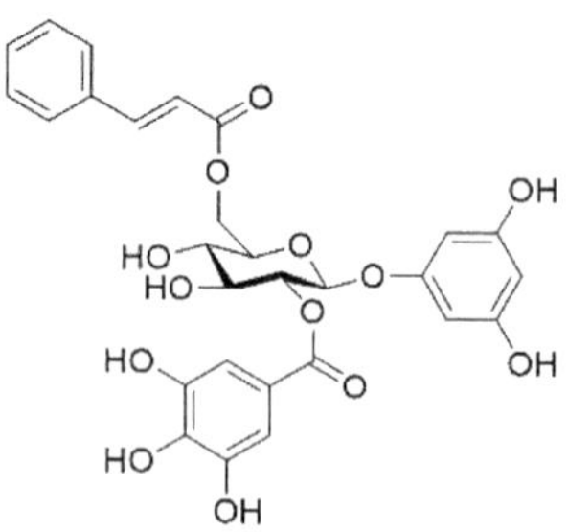

222

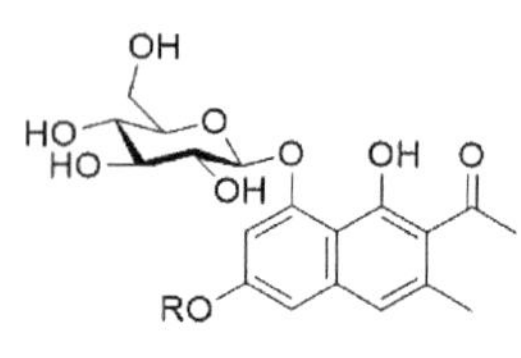

237 R = H
252 R = Me

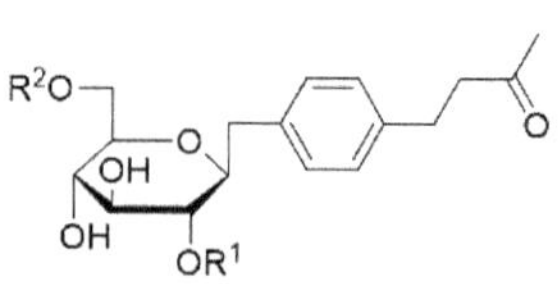

246 R^1 = galloyl, R^2 = H
247 R^1 = H, R^2 = galloyl

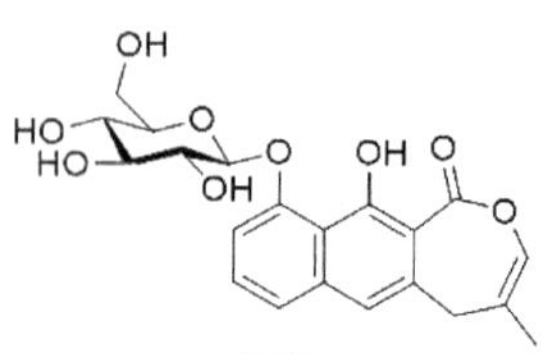

250

Figure 18.8. Some acyl glucoside isolated from rhubarb.

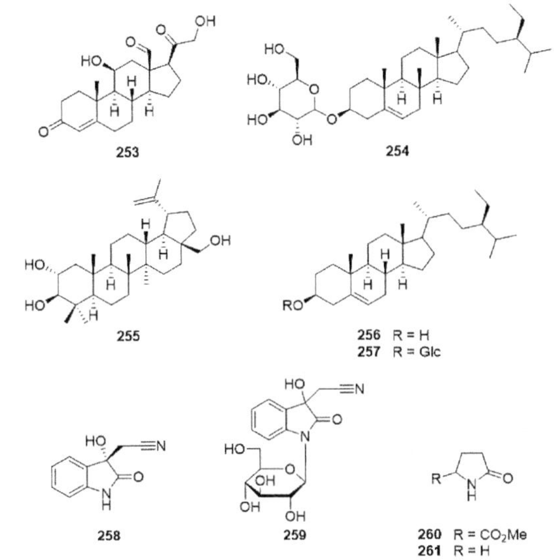

Figure 18.9. Some sterols, terpenes, alkaloids, and pyrrolidinones isolated from rhubarb.

of the rhubarb (Yoo et al. 2007). It possessed anticancer, antiatherogenic, anti-inflammatory, antimicrobial and estrogenic activities (Piotrowska et al. 2012).

Rhapontigenin (**68**), 3,5,4′-trimethylpiceatannol (**72**) and rhaponticin (**73**) produced potent inhibitory activity against histamines; therefore, they are considered to be a prodrug for anti-allergic actions (Kim et al. 2000, Matsuda et al. 2004).

The quality of rhubarb can also be determined by the presence of stilbene glucoside, rhaponticin (**73**) (Kashiwada et al. 1984a). It has antiallergic, antithrombotic and antidiabetic properties, inhibits copper-induced lipid peroxidation, produces neuroprotective effect against β-amyloid induced neurotoxicity, and acts as an oral hemostatic agent (Matsuda et al. 2001, Park et al. 2002, Misiti et al. 2006, Yoo et al. 2007, Chen et al. 2009b). Rhaponticin (**73**) showed a protective effect on amyloid-beta (1-42) peptide-induced apoptosis in IMR-32 cells, suggesting that it could be developed as an agent for the management of Alzheimer's disease (Misiti et al. 2006). It induced apoptosis in the stomach cancer KATO III cells (Hibasami et al. 2007). Rhaponticin (**73**) reported restoring the oestrus cycle in castrated female rats (Khare 2007).

Stilbenes ε-viniferin (**104**) and ampelopsin B (**105**) showed a remarkable inhibitory effect on lipoxygenase and exhibited strong free radical scavenging activity against $ABTS^{\bullet+}$ radical cation (Ngoc et al. 2008).

Flavonoids

Flavonoids are especially deposited in the flowers, fruits, buds and bracts of rhubarb (Agarwal et al. 2001, Iwashina et al. 2004). Flavonoids are well known for their antioxidative properties. Anthocyanins such as cyanidin-3-*O*-glucoside (**106**), cyanidin-3-*O*-rutinoside (**107**) and cyanin (**108**) were isolated from *R. rhaponticum* and *R. tataricum* (Wrolstad et al. 1968, 1971). (–)-Epicatechin-3-*O*-gallate (**110**) increased glomerular filtration rate, renal plasma flow, and renal blood flow in rats with adenine-induced renal failure (Yokozawa et al. 1993). Other tannins, procyanidin B-2-3,3′-di-*O*-gallate (**137**) increased renal functional parameters and procyanidin C-1-3,3′,3″-tri-*O*-gallate (**150**) caused aggravation of renal function. Strong antioxidative activity of (–)-epicatechin-3-*O*-gallate (**110**), equal to antiatherosclerotic agent probucol, was reported (Iizuka et al. 2004). Carpusin (**155**) and mesopsin (**156**) are reported as antioxidants (Krenn et al. 2004).

Meratin (**163**), quercetin (**165**), isoquercitrin (**169**) and rutin (**170**) were isolated from *R. tataricum* and *R. wittrockii* (Chumbalov and Nurgalieva 1967, Alyukina et al. 1970). Rutin (**170**) has also been isolated from *R. emodi*, *R. officinale*, *R. nobile*, *R. rhaponticum*, *R. undulatum*, etc. (Hörhammer and Müller 1954, Iwashina et al. 2004). Several kaempferol- and quercetin-glycosides (**162**, **166**–**171**) including quercetin (**165**) have been isolated from the translucent bracts of *R. nobile* (Iwashina et al. 2004).

Pyranones

Nine pyranones (**173**–**181**) have been isolated from *R. hotaoense*, *R. glabricaule* and/or commercial rhubarbs (Kashiwada et al. 1984b, 1990, Li et al. 1998, Wei et al. 2005).

Phenols

Gallic acid (**202**), present as glucogallin (**204**), acts as an antioxidant. It is considered to be an astringent principle of rhubarb (Ahmad et al. 1998). Total phenolic content in rhubarb has been estimated by several researchers (Jahan et al. 2013, Wasim et al. 2013, Gupta et al. 2015, 2016). Phenolic acids present in *R. palmatum*, *R. rhaponticum* and *R. undulatum* were comparatively studied (Medyńska and Smolarz 2005).

Acylglucosides

Among acylglucosides isolated from rhubarbs, lindleyin (**163**) was considered to be a new phytoestrogen that showed estrogenic activity (Usui et al. 2002).

Sterols and terpenes

Rhubarb is a considerably poor source of steroids, which were reported from *R. hotaoense* and *R. emodi* (Li et al. 1998, Liu et al. 2007, Wang et al. 2010a). Daucosterol (**254**) showed the neuroprotective activity by reducing neuron loss and apoptotic rate (Jiang et al. 2015). β-sitosterol (**256**) bore antidiabetic and antioxidant potentialities (Gupta et al. 2011).

Alkaloids

Indole alkaloids, (+)-3-cyanomethyl-3-hydroxyindole (**258**) and its glycoside 3-cyanomethyl-3-hydroxyoxindole-*N*-β-D-glucoside (**259**), were isolated from *R. maximowicizii* seeds (Komakine et al. 2005). They inhibited the production of IL-10 induced by lipopolysaccharide.

Pyrrolidinones

5-Oxo-pyrrolidine-2-carboxylic acid methyl ester (**260**) and pyrrolidin-2-one (**261**) were also isolated from *R. maximowicizii* seeds (Komakine et al. 2005).

Organic acids

Organic acids play an important role in the plant's growth, such as structure of cell wall, phosphates store as well as building blocks for the formation of lignin. The content of organic acid in rhubarb determines the color, appearance and smell of the food prepared. Rhubarb species comprises several organic acids (Maeda 1964). Malic acid (**267**) is predominant in all the varieties of rhubarb, while citric acid (**263**) and oxalic acid (**268**) are present in smaller quantities (Salunkhe et al. 1991). Large quantities of oxalic acid (**268**) are present in the plants of Polygonaceae family making them inedible. The content of oxalic acid (**268**) present in rhubarb is approximately 40–50 mg/100 g (Bennet-clark 1937). Oxalic acid (**268**) content in Indian rhubarb was found to be 1.34% (Chopra et al. 1978).

Vitamins

Vitamins A, B, C, D, E, and K have been reported from rhubarb (Allison 1966, Chopra et al. 1978, Munzuroğlu et al. 2000).

Essential oil

About 0.05% of essential oil, with a characteristic odor, was obtained from the rhizomes of *R. emodi* (Chopra et al. 1978). The presence of eugenol (**199**) and methyl heptyl ketone in the oil is responsible for the odor. An essential oil from the root of *R. palmatum* constituted 108 volatile compounds, including terpenoids (27.3%), palmitic acid (22.5%), paeonol (16.2%), α-copaene (9.8%), methyl stearate (9.3%), δ-cadinene (5.4%), methyl-eugenol (5.4%), etc. (Miyazawa et al. 1996). Fifty-nine volatile components were identified from the stalks of *R. rhabarbarum* (Dregus and Engel 2003).

Inorganic elements

Atomic absorption spectroscopy studies showed that rhubarb contains Al, Br, Ca, C, Cl, Co, Cu, Fe, Li, Mg, Mn, O, P, K, Si, Na, S, Se, Th, Ti, Zn, etc. (Hameed et al. 2008, Singh et al. 2010, 2012, Haq and Ullah 2011, Gupta et al. 2015, 2016).

Pharmacokinetics

Rhubarb extract was administered to rats after they were modeled by CCl_4. Their blood samples were used to determine chrysophanol (**1**), physcion (**2**) emodin (**3**), aloe-emodin (**4**) and rhein (**5**), and the rate of removing superoxide anion (Lu et al. 2012). A good correlation between integrated pharmacokinetics rhubarb anthraquinones and the pharmacodynamics of antioxidant effects was found. In the rat single-pass intestinal perfusion experiment, anthraquinones, viz., chrysophanol (**1**), physcion (**2**) emodin (**3**), aloe-emodin (**4**) and rhein (**5**), were found absorbed in the different segments of intestine (Wang et al. 2011b). In earlier experiments, rhein (**5**) was considered to be the best anthraquinone that is absorbed by the body (Lee et al. 2003). HPLC method for the determination of anthraquinones in rat plasma has been documented (Yan and Ma 2007, Peng and Dai 2010, Gong et al. 2011). HPLC-mass spectrometry methods were also developed for pharmacokinetics studies (Zhao et al. 2012).

The toxic potentiality of the rhubarb extract was evaluated. Aged and immature rats responded reversely and the aged rats were found susceptible to the toxicity at a high dose (Wang et al. 2011a). LD_{50} value of 500 mg/kg of the *R. nobile* extract in mice is reported (Gupta et al. 2017).

Intravenous administration of chrysophanol (**1**) in animals showed that it rapidly distributes and shows the effect, the retention time is long and the elimination is the main process (Tan et al. 2006). Thermal characterization and decomposition kinetics of free anthraquinones were investigated by thermogravimetry (Zhang et al. 2006).

Conclusions

Historically, the use of rhubarb as traditional herbal medicine began around 5000 BC and was first documented in the Chinese pharmacopeia. Specifically, rhizomes and roots of Chinese and Indian or Himalayan rhubarbs, viz., *R. emodi*, *R. officinale*, *R. palmatum*, and *R. webbianum,* have been found to have immense medicinal importance. Some species of rhubarbs such as *R. rhaponticum* and *R. x hybridum* are grown as vegetables in Europe for their edible leaf petioles; however, their leaves are poisonous due to oxalic acid content. Rhubarb has been used in several ailments, particularly for constipation, dysentery, inflammation and blood disorders. Anthraquinones and stilbenes are the major chemical constituents in rhubarb and are considered to be responsible for the therapeutic effects. Although several studies on rhubarb (despite many rhubarb species not yet scientifically explored!) have proved its medicinal efficacies, about 270 compounds have been isolated, pharmacokinetics and quality analysis methods were developed, and clinical trials were performed, but still a lot of information is missing—why did rhubarb gain a high medicinal attention as the king of herbs during the ancient time and which continues to the present world?

Abbreviations

ABTS: 2,2'-azinobis-3-ethyl benzothiazoline-6-sulfonic acid; GABA: γ-aminobutyric acid; HPLC: high-performance liquid chromatography; IL: interleukin; LD_{50}: lethal dose for 50% mortality; TNF: tumor necrosis factor.

References

Agarwal, S.K., Singh, S.S., Verma, S., Kumar, S. 2000. Antifungal activity of anthraquinone derivatives from *Rheum emodi*. J. Ethnopharmacol., 72: 43–46.

Agarwal, S.K., Singh, S.S., Lakshmi, V., Verma, S., Kumar, S. 2001. Chemistry and pharmacology of rhubarb (Rheum species)—a review. J. Sci. Ind. Res., 60: 1–9.

Agarwal, V.S. 1986. Economic Plants of India. Kailash Prakashan, Calcutta, India.

Ahmad, Z., Jafri, M.A., Javed, K. 1998. Quantitative evaluation of free and combined states of rhein and chrysophanol in Indian rhubarb by TLC: colourimetric method. Indian Drugs, 35(2): 75–78.

Akhtar, M.S., Amin, M., Ahmad, M., Alamgeer. 2009. Hepatoprotective effect of *Rheum emodi* roots (revandchini) and Akseer-e-jigar against paracetamol-induced hepatotoxicity in rats. Ethnobot. Leaflets, 13: 310–315.

Akram, M., Azhar, M. 2016. Revend (rhubarb): an important unani drug for prevention of nephrotoxicity. Int. J. Pharma Professional's Res., 7(3): 1333–1340.

Alam, M.M.A., Javed, K., Jafri, M.A. 2005. Effect of *Rheum emodi* (revand hindi) on renal functions in rats. J. Ethnopharmacol., 96: 121–125.

Alam, S., Khan, N.A. 2015. Rhubarb (rewand): a review. Hamdard Medicus, 58: 84–95.

Ali, B.H., Al-Salam, S., Husseini, I.S.A., Al-Lawati, I., Waly, M., Yasin, J., Fahim, M., Nemmar, A. 2013. Abrogation of cisplatin-induced nephrotoxicity by emodin in rats. Fundamental Clin. Pharmacol., 27: 192–200.

Allison, R.M. 1966. Soluble oxalates, ascorbic and other constituents of rhubarb varieties. J. Sci. Food Agric., 17: 554–557.

Alyukina, L.S., Baigozina, V.P., Ramazanova, Z. 1970. *Rheum wittrockii* flavonoids. Trudy Instituta Botaniki Akademiya Nauk Kazakhskoi SSR, 28: 170–178.

Aqil, F., Ahmad, I. 2003. Broad-spectrum antibacterial and antifungal properties of certain traditionally used Indian medicinal plants. World J. Microbiol. Biotechnol., 19: 653–657.

Arvindekar, A., More, T., Payghan, P.V., Laddha, K., Ghoshal, N., Arvindekar, A. 2015. Evaluation of anti-diabetic and alpha glucosidase inhibitory action of anthraquinones from *Rheum emodi*. Food Funct., 6: 2693–2700.

Ashnagar, A., Naseri, N.G., Nasab, H.H. 2007. Isolation and identification of anthralin from the roots of rhubarb plant (*Rheum palmatum*). E-J. Chem., 4: 546–549.

Azhar, M., Anjum, N. 2019. Revad chini (Chinese rhubarb): a review on historical and unani classical prospect. Int. J. Unani Integr. Med., 3: 11–18.

Babu, K.S., Srinivas, P.V., Praveen, B., Kishore, K.H., Murty, U.S., Rao, J.M. 2003. Antimicrobial constituents from the rhizomes of *Rheum emodi*. Phytochem., 62: 203–207.

Babu, K.S., Tiwari, A.K., Srinivas, P.V., Ali, A.Z., Raju, B.C., Rao, J.M. 2004. Yeast and mammalian α-glucosidase inhibitory constituents from Himalayan rhubarb *Rheum emodi* Wall.ex Meisson. Bioorg. Med. Chem. Lett., 14: 3841–3845.

Bae, E.-A., Han, M.J., Kim, N.-J., Kim, D.-H. 1998. Anti-*Helicobacter pylori* activity of herbal medicines. Biol. Pharm. Bull., 21: 990–992.

Bennet-Clark, T.A. 1937. Organic acids of plants. Annual Rev. Biochem., 6: 579–594.

Betteridge, D.J. 2000. What is oxidative stress? Metabolism, 49(Suppl. 1): 3–8.

Bhattarai, N., Tandon, V., Ved, D.K. 2002. Highlights and outcomes of the conservation assessment and management planning (CAMP) workshop, Pokhara, Nepal. pp. 46–53. *In*: Bhattarai, N., Karki, M. (eds.). Sharing Local and National Experience in Conservation of Medicinal and Aromatic Plants in South Asia. International Development Research Centre (IDRC), Nepal. URL: idl-bnc-idrc.dspacedirect.org/bitstream/handle/10625/29240/120935.pdf?sequence=1.

Bilal, S., Mir, M.R., Parrah, J.D., Tiwari, B.K., Tripathi, V., Singh, P., Mehjabeen, Abidi, A.B. 2013. Rhubarb: the wondrous drug. A review. Int. J. Pharm. Bio. Sci., 3: 228–233.

Bratsch, A., Mainville, D. 2009. Specialty crop profile: rhubarb. Virginia Cooperative Extension Publication 438-110. URL: https://www.pubs.ext.vt.edu/438/438-110/438-110.html.

Cai, J., Razzak, A., Hering, J., Saed, A., Babcock, T.A., Helton, S., Espat, N.J. 2008. Feasibility evaluation of emodin (rhubarb extract) as an inhibitor of pancreatic cancer cell proliferation *in vitro*. J. Parenteral Enteral. Nutr., 32: 190–196.

Cai, Y., Sun, M., Xing, J., Corke, H. 2004. Antioxidant phenolic constituents in roots of *Rheum officinale* and *Rubia cordifolia*: structure-radical scavenging activity relationships. J. Agric. Food Chem., 52: 7884–90.

Castleman, M. 1991. The Healing Herbs: The Ultimate Guide to the Curative Powers of Nature's Medicine. Rodale Press, Emmaus, PA, pp. 305–307.

Chauhan, N.S, Kaith, B.S., Mann, S.K. 1992. Anti-inflammatory activity of *Rheum australe* roots. Int. J. Pharmacog., 30: 93–96.

Chen, D.C., Jing, B.W., Chen, J.D., Zhao, L., Yan, M. 1994. Experimental study on the effect of rhubarb on enterogenous infection caused by endotoxin. J. Emerg. Tradit. Chin. Med., 3: 84–86.

Chen, D.C., Li, H.J., Qiao, L., Yang, X.Y., Jing, B.W. 2002. The effects of rhubarb on the function of mitochondria of hepatic cell in posttrauma sepsis. Chin. J. Integr. Tradit. Western Med. Intensive Crit. Care, 9: 9–11.

Chen, D.C., Wang, L. 2009. Mechanisms of therapeutic effects of rhubarb on gut origin sepsis. Chin. J. Traumatol., 12: 365–369.

Chen, D.C., Ma, L.-Q., Liu, S.-Z. 2009a. Effect of rhubarb on intestinal flora and bacterial translocation in rats with sepsis. Chin. Crit. Care Med., 21: 17–20.

Chen, J., Ma, M., Lu, Y., Wang, L., Wu, C., Duan, H. 2009b. Rhaponticin from rhubarb rhizomes alleviates liver steatosis and improves blood glucose and lipid profiles in KK/Ay diabetic mice. Planta Med., 75: 472–477.

Chihara, T., Shimpo, K., Beppu, H., Yamamoto, N., Kaneko, T., Wakamatsu, K., Sonoda, S. 2015. Effects of aloe-emodin and emodin on proliferation of the MKN45 human gastric cancer cell line. Asian Pac. J. Cancer Prev., 16: 3887–3891.

Chinese Pharmacopoeia Commission. 2010. Pharmacopoeia of the People's Republic of China. Vol 1. China Medical Science Press, Beijing, China, pp. 375–376.

Choi, S.B., Ko, B.S., Park, S.K., Jang, J.S., Park, S. 2006. Insulin sensitizing and α-glucoamylase inhibitory action of sennosides, rheins and rhaponticin in *Rheum* rhizoma. Life Sci., 78: 934–942.

Chopra, R.N., Nayar, S.L., Chopra., I.C. 1978. Glossary of Indian Medicinal Plants. Council of Scientific and Industrial Research, New Delhi, p. 212.

Chumbalov, T.K., Nurgalieva, G.M. 1967. Flavonoids of *Rheum tataricum*. V. Khimiya Prirodnykh Soedinenii, 3: 345–346. Chem. Abstr., 1968. 68: 27516p.

Clark, J.H. 1995. A dictionary of practical materia medica. Vol. 3. B. Jain Publisher Ltd., New Delhi, pp. 973–978.

Clementi, E.M., Misiti, F. 2010. Potential health benefits of rhubarb. pp. 407–423. *In*: Watson, R.R., Preedy, V.R. (eds.). Bioactive Foods in Promoting Health. Academic Press Inc., San Diego.

Coopoosamy, R.M., Magwa, M.L. 2006. Antibacterial activity of chrysophanol isolated from *Aloe excelsa* (Berger). African J. Biotechnol., 5: 1508–1510.

CSIR. 1972. The wealth of India. A dictionary of Indian raw materials and industrial products. Raw materials. Vol. IX: Rh–So. Publications and Information Directorate, Council of Scientific and Industrial Research (CSIR), New Delhi, pp. 3–6.

Cullen, J. 1966. *Rheum* L. pp. 268-269. *In*: Davis, P.H. (ed.). Flora of Turkey and the East Aegean Islands. Vol. 2. Edinburgh University Press, Edinburgh.

Cyong, J.-C., Matsumoto, T., Arakawa, K., Kiyohara, H., Yamada, H., Otsuka, Y. 1987. Anti-*Bacteroides fralilis* substance from rhubarb. J. Ethnopharmacol., 19: 279–283.

Das, D. 2008. Colouration of wool and silk with *Rheum emodi*. Indian J. Fibre Textile Res., 33: 163–170.

de Witte, P. 1993. Metabolism and pharmacokinetics of anthranoids. Pharmacol., 47: 86–97.

Ding, Y., Zhao, L., Mei, H., Zhang, S.-L., Huang, Z.-H., Duan, Y.-Y., Ye, P. 2008. Exploration of emodin to treat alpha-naphthylisothiocyanate-induced cholestatic hepatitis via anti-inflammatory pathway. Eur. J. Pharmacol., 590: 377–386.

Dong, H., Lu, F.-E., Gao, Z.-Q., Xu, L.-J., Wang, K.F., Zou, X. 2005. Effects of emodin on treating murine nonalcoholic fatty liver induced by high caloric laboratory chaw. World J. Gastroenterol., 11: 1339–1344.

DPR. 2006. Nepal ko arthik bikaska lagi prathamikta prapta jadibutiharu. Department of Plant Resources (DPR), Ministry of Forest and Soil Conservation, Kathmandu, Nepal. URL: https://dpr.gov.np/category/publication/book/page/3/. (In Nepali).

Dregus, M., Engel, K.-H. 2003. Volatile constituents of uncooked rhubarb (*Rheum rhabarbarum* L.) stalks. J. Agric. Food Chem., 51: 6530–6536.

Duke, J.A. 1985. Hand Book of Medicinal Herbs. CRC Press, Florida, p. 404.

Dutta, I.C. 2007. Non timber forest products of Nepal. Identification, classification, ethnic uses and cultivation. Hill Side Press Pvt. Ltd., Kathmandu, Nepal, pp. 206–207.

El-Lakany, A.M., Abdul-Ghani, M.M., El-Boukhary, R.M. 2008. Anthraquinones from *Rheum ribes* growing in Lebanon. Alex. J. Pharm. Sci., 22: 117–119.

Evans, W.C. 2007. Trease and Evans Pharmacognosy. Edn. 15. WB Saunders Company Ltd., London, UK. pp. 230–231, 238–240.

Fang, F., Wang, J.-B., Zhao, Y.-L., Jin, C., Kong, W.-J., Zhao, H.-P., Wang, H.-J., Xiao, X.-H. 2011. Tissue distribution of free anthraquinones in SD rats after orally administered extracts from raw and prepared rhubarbs. Acta Pharm. Sin., 46: 350–354.

Fok, T.F. 2001. Neonatal jaundice—traditional Chinese medicine approach. J. Perinatol., 21(Suppl. 1): S98–S100.

Foust, C.M. 1992. Rhubarbs: The Wondrous Drug. Princeton University Press, Princeton, New Jersey.

Gao, L., Xu, X., Nan, H., Yang, J., Sun, G., Wu, H., Zhong, M. 2012. Isolation of cinnamic acid derivatives from the root of *Rheum tanguticum* Maxim.ex Balf. and its significance. J. Med. Plants Res., 6: 929–931.

Gao, Q., Qin, W.-S., Jia, J.-H., Zheng, J.-M., Zeng, C.-H., Li, L.-S., Liu, Z.-H. 2010. Rhein improves renal lesion and ameliorates dyslipidemia in db/db mice with diabetic nephropathy. Planta Med., 76: 27–33.

Ghimire, S.K., Sapkota, I.B., Oli, B.R., Parajuli, R.R. 2008. Non-timber forest products of Nepal Himalaya. WWF Nepal, Kathmandu, Nepal, pp. 115–118.

Ghorbani, A., M.S. Amiri, R.R., Hosseini, A. 2019. Pharmacological properties of *Rheum turkestanicum* Janisch. Heliyon, 5: e01986.

Gong, H.L., Tang, W.F., Wang, H., Xia, Q., Huang, X. 2011. Effects of food and gender on the pharmacokinetics of rhein and emodin in rats after oral dosing with Da-cheng-qi decoction. Phytother. Res., 25: 74–80.

Gu, J.-W., Hasuo, H., Takeya, M., Akasu, T. 2005. Effects of emodin on synaptic transmission in rat hippocampal CA1 pyramidal neurons *in vitro*. Neuropharmacol., 49: 103–111.

Guo, J.-M., Xiao, B.-X., Liu, Q., Zhang, S., Liu, D.-H., Gong, Z.-H. 2007. Anticancer effect of aloe-emodin on cervical cancer cells involves G_2/M arrest and induction of differentiation. Acta Pharm. Sin., 28: 1991–1995.

Gupta, R., Sharma, A.K., Dobhal, M.P., Sharma, M.C., Gupta, R.S. 2011. Antidiabetic and antioxidant potential of β-sitosterol in streptozotocin-induced experimental hyperglycemia. J. Diabetes, 3: 29–37.

Gupta, R.K., Bajracharya, G.B., Jha, R.N. 2014. Antibacterial activity, cytotoxicity, antioxidant capacity and phytochemicals of *Rheum australe* rhizomes of Nepal. J. Pharmacog. Phytochem., 2(6): 125–128.

Gupta, R.K., Gupta, P.K., Jha, R.N., Bajracharya, G.B. 2015. Pharmacognostical comparison and standardization of wild and cultivated rhizomes of *Rheum australe* (Padamchal) of Nepal. J. Pharmacog. Phytochem., 3(6): 112–116.

Gupta, R.K., Bajracharya, G.B., Jha, R.N. 2016. Quantitative and qualitative chemical analyses of the rhizome of *Rheum nobile*. J. Nepal Chem. Soc., 34: 51–57.

Gupta, R.K., Bharati, L., Shakya, K.S., Regmi, B.M., Bajracharya, G.B., Jha, R.N. 2017. *In vitro* and *in vivo* pharmacological activities of the extracts of *Rheum nobile* Hook. f. & Thomson rhizomes. Indian J. Nat. Prod. Res., 8: 230–239.

Hameed, I., Dastagir, G., Hussain, F. 2008. Nutritional and elemental analyses of some selected medicinal plants of the family Polygonaceae. Pak. J. Bot., 40: 2493–2502.

Han, C., Wang, J., Li, Y., Cui, Y. 2013. *In vitro* antimicrobial activity and effect on *E. coli* integrity of cinnamon essential oil and rhubarb ethanol extract. Food Sci. Technol. Res., 19: 1155–1163.

Haq, F., Ullah, R. 2011. Comparative determination of trace elements from *Allium sativum*, *Rheum australe* and *Terminalia chebula* by atomic absorption spectroscopy. Int. J. Biosci., 1(5): 77–82.

Hatano, T., Uebayashi, H., Ito, H., Shiota, S., Tsuchiya, T., Yoshida, T. 1999. Phenolic constituents of Cassia seeds and antibacterial effect of some naphthalenes and anthraquinones on methicillin-resistant *Staphylococcus aureus*. Chem. Pharm. Bull., 47: 1121–1127.

He, Z.-H., He, M.-F., Ma, S.-C., But, P.P.-H. 2009. Anti-angiogenic effect of rhubarbs and its anthraquinone derivatives. J. Ethnopharmacol., 121: 313–317.

He, Z.-H., Zhou, R., He, M.-F., Lau, C.B.-S., Yue, G.G.-L., Ge, W., But, P. P.-H. 2011. Anti-angiogenic effect and mechanism of rhein from *Rhizoma rhei*. Phytomed., 18: 470–478.

Hibasami, H., Takagi, K., Ishii, T., Tsujikawa, M., Imai, N., Honda, I. 2007. Induction of apoptosis by rhapontin having stilbene moiety, a component of rhubarb (*Rheum officinale* Baillon) in human stomach cancer KATO III cells. Oncol. Rep., 18: 347–351.

Ho, T.-Y., Wu, S.-L., Chen, J.-C., Li, C.-C., Hsiang, C.Y. 2007. Emodin blocks the SARS coronavirus spike protein and angiotensin-converting enzyme 2 interaction. Antiviral Res., 74: 92–101.

Hooker, J.D. 1855. Illustration of Himalayan Plants. Lovell Reeve, London.

Hörhammer, L., Müller, K. 1954. Occurrence of rutin in *Rheum* species. Arch. Pharm. Berl., 287: 126–128. Chem. Abstr., 1955. 49: 570f.

Hou, Y., Zhao, Y., Wu, W., Wu, X., Xu, J. 2015. Antibacterial activities of rhubarb extract and the bioactive compounds against *Salmonella*. Int. J. Nutr. Sci. Food Tech., 1(1): 1–9.

Hu, L.H. 1986. Experimental study of *Rheum officinale* Baill. in treating severe hepatitis and hepatic coma. Chin. J. Modern Develop. Tradit. Med., 6: 41–42.

Huang, Q., Lu, G., Shen, H.-M., Chung, M.C.M., Ong, C.N. 2007. Anti-cancer properties of anthraquinones from rhubarb. Med. Res. Rev., 27: 609–630.

Ibrahim, M., Khan, A.A., Tiwari, S.K., Habeeb, M.A., Khaja, M.N., Habibullah, C.M. 2006. Antimicrobial activity of *Sapindus mukorossi* and *Rheum emodi* extracts against *H. pylori*: *in vitro* and *in vivo* studies. World J. Gastroenterol., 12: 7136–7142.

Ibrahim, M., Khaja, M.N., Aara, A., Khan, A.A., Habeeb, M.A., Devi, Y.P., Narasu, M.L., Habibullah, C.M. 2008. Hepatoprotective activity of *Sapindus mukorossi* and *Rheum emodi* extracts: *in vitro* and *in vivo* studies. World J. Gastroenterol., 14: 2566–2571.

Iizuka, A., Iijima, O.T., Kondo, K., Itakura, H., Yoshie, F., Miyamoto, H., Kubo, M., Higuchi, M., Takeda, H., Matsumiya, T. 2004. Evaluation of rhubarb using antioxidative activity as an index of pharmacological usefulness. J. Ethnopharmacol., 91: 89–94.

Iwashina, T., Omori, Y., Kitajima, J., Akiyama, S., Suzuki, T., Ohba, H. 2004. Flavonoids in translucent bracts of the Himalayan *Rheum nobile* (Polygonaceae) as ultraviolet shields. J. Plant Res., 117: 101–107.

Jahan, N., Khalil-ur-Rahman, Ali, S., Asi, M.R. 2013. Phenolic acid and flavonol contents of gemmo-modified and native extracts of some indigenous medicinal plants. Pak. J. Bot., 45: 1515–1519.

Jiang, L.-H., Yuan, X.-L., Yang, N.-Y., Ren, L., Zhao, F.-M., Luo, B.-X., Bian, Y.-Y., Xu, J.-Y., Lu, D.-X., Zheng, Y.-Y., Zhang, C.-J., Diao, Y.-M., Xia, B.-M., Chen, G. 2015. Daucosterol protects neurons against oxygen-glucose deprivation/reperfusion-mediated injury by activating IGF1 signaling pathway. J. Steroid Biochem. Mol. Biol., 152: 45–52.

Jin, W., Ge, R.-L., Huang, Z.-Q., Wei, Q.-J., Bao, T.-Y., Tu, P.-F. 2006. Chemical constituents of *Rheum tanguticum* Maxim. ex Balf. J. Chin. Pharm. Sci., 15: 206–210.

Joseph, B., Dar, M.A., Kumar, V. 2008. Bioefficacy of plant extracts to control *Fusarium solani* F. sp. *melongenae* incitant of brinjal wilt. Global J. Biotech. Biochem., 3: 56–59.

Kageura, T., Matsuda, H., Morikawa, T., Toguchida, I., Harima, S., Oda, M., Yoshikawa, M. 2001. Inhibitors from rhubarb on lipopolysaccharide-induced nitric oxide production in macrophages: structural requirements of stilbenes for the activity. Bioorg. Med. Chem., 9: 1887–1893.

Kang, S.C., Lee, C.M., Choung, E.S., Bak, J.P., Bae, J.J., Yoo, H.S., Kwak, J.H., Zee, O.P. 2008. Anti-proliferative effects of estrogen receptor-modulating compounds isolated from *Rheum palmatum*. Arch. Pharm. Res., 31: 722–726.

Kashiwada, Y., Nonaka, G.-I., Nishioka, I. 1984a. Studies on rhubarb (Rhei rhizoma). VI. Isolation and characterization of stilbenes. Chem. Pharm. Bull., 32: 3501–3517.

Kashiwada, Y., Nonaka, G.-I., Nishioka, I. 1984b. Studies on rhubarb (Rhei rhizoma). V. Isolation and characterization of chromone and chromanone derivatives. Chem. Pharm. Bull., 32: 3493–3500.

Kashiwada, Y., Nonaka, G.-I., Nishioka, I., Yamagishi, T. 1988. Galloyl and hydroxycinnamoylglucoses from rhubarb. Phytochem., 27: 1473–1477.

Kashiwada, Y., Nonaka, G.-I., Nishioka, I. 1990. Chromone glucosides from rhubarb. Phytochem., 29: 1007–1009.

Kaur, A., Kumar, S., Sharma, R. 2012. Assessment of anti-ulcer activity of *Rheum emodi* rhizomes extracts. Indo Global J. Pharm. Sci., 2: 333–341.

Khare, C.P. 2007. Indian Medicinal Plants. An Illustrated Dictionary. Springer-Verlag Berlin, Heidelberg.

Khetwal, K.S., Pathak, R.P. 1988. Active anthraquinoids from *Rheum webbianum*. J. Indian Chem. Soc., 65: 227.

Kim, D.-H., Park, E.-K., Bae, E.-A., Han, M. 2000. Metabolism of rhaponticin and chrysophanol 8-O-β-D-glucopyranoside from the rhizome of *Rheum undulatum* by human intestinal bacteria and their anti-allergic actions. Biol. Pharm. Bull., 23: 830–833.

Ko, S.K., Lee, S.M., Whang, W.K. 1999. Anti-platelet aggregation activity of stilbene derivatives from *Rheum undulatum*. Arch. Pharm. Res., 22: 401–403.

Kogure, K., Yamauchi, I., Tokumura, A., Kondou, K., Tanaka, N., Takaishi, Y., Fukuzawa, K. 2004. Novel antioxidants isolated from plants of the genera *Ferula, Inula, Prangos* and *Rheum* collected in Uzbekistan. Phytomed., 11: 645–651.

Komakine, N., Takaishi, Y., Honda, G., Ito, M., Takeda, Y., Kodzhimatov, O.K., Ashurmetov, O. 2005. Indole alkaloids from *Rheum maximowiczii*. Nat. Med., 59: 45–48.

Komatsu, K., Yang, D.-Y., Fushimi, H., Cai, S.-Q. 2005. Authentication of Rhei rhizoma. J. Tradit. Med., 22(Suppl. 1): 70–85.

Kong, L.D., Cheng, C.H.K., Tan, R.X. 2004. Inhibition of MAO A and B by some plant-derived alkaloids, phenols and anthraquinones. J. Ethnopharmacol., 91: 351–355.

Kounsar, F., Rather, M.A., Ganai, B.A., Zargar, M.A. 2011. Immuno-enhancing effects of the herbal extract from Himalayan rhubarb *Rheum emodi* Wall ex. Meissn. Food. Chem., 126: 967–971.

Krenn, L., Pradhan, R., Presser, A., Reznicek, G., Kopp, B. 2004. Anthrone C-glucosides from *Rheum emodi*. Chem. Pharm. Bull., 52: 391–393.

Kubo, I., Murai, Y., Soediro, I., Soetarno, S., Sastrodihardjo, S. 1992. Cytotoxic anthraquinones from *Rheum pulmatum*. Phytochem., 31: 1063–1065.

Kundu, J.K., Surh, Y.-J. 2008. Cancer chemopreventive and therapeutic potential of resveratrol: mechanistic perspectives. Cancer Lett., 269: 243–261.

Kuo, P.-L., T.-C. Lin, Y.-J., Lin, C.-C. 2002. The antiproliferative activity of aloe-emodin is through p53-dependent and p21-dependent apoptotic pathway in human hepatoma cell lines. Life Sci., 71: 1879–1892.

Kurian, A., Sankar, M.A. 2007. Medicinal plants. pp. 301–302. *In*: Peter, K.V. (ed.). Horticulture Science Series. Vol. 2. New India Publishing Agency, Pitam Pura, New Delhi.

Kwon, G.T., Jung, J.I., Song, H.R., Woo, E.Y., Jun, J.-G., Kim, J.-K., Her, S., Park, J.H.Y. 2012. Piceatannol inhibits migration and invasion of prostate cancer cells: possible mediation by decreased interleukin-6 Signaling. J. Nutr. Biochem., 23: 228–238.

Lai, F., Zhang, Y., Xie, D.-P., Mai, S.-T., Weng, Y.-N., Du, J.-D., Wu, G.-P., Zheng, J.-X., Han, Y. 2015. A systematic review of rhubarb (a traditional Chinese medicine) used for the treatment of experimental sepsis. Evid. Based Complement. Alternat. Med. Article ID 131283. URL: http://dx.doi.org/10.1155/2015/131283.

Lee, J.-H., Kim, J.M., Kim, C. 2003. Pharmacokinetic analysis of rhein in *Rheum undulatum* L. J. Ethnopharmacol., 84: 5–9.

Lee, M.R., Hutcheon, J., Dukan, E., Milne, I. 2017. Rhubarbs (*Rheum* species): the role of Edinburg in its cultivation and development. J. R. Coll. Physicians Edinb., 47: 102–109.

Lee, W., Yoon, G., Hwang, Y.R., Kim, Y.K., Kim, S.-N. 2012. Anti-obesity and hypolipidemic effects of *Rheum undulatum* in high-fat diet-fed C57BL/6 mice through protein tyrosine phosphatase 1B inhibition. BMB Rep., 45: 141–146.

Lewis, W.H., Elvin-Lewis, P.F. 1977. Medical Botany: Plants Affecting Man's Health. John Wiley and Sons, New York.

Li, A., Bao, B., Grabovskaya-Borodina, A.E., Hong, S.-P., McNeill, J., Mosyakin, S.L., Ohba, H., Park, C.-W. 2003. Polygonaceae. pp. 277–350. *In*: Wu, Z.-Y., Raven, P.H., Hong, D.-Y. (eds.). Flora of China. Vol. 5 (Ulmaceae through Basellaceae). Science Press, Beijing, and Missouri Botanical Garden Press, St. Louis.

Li, F., Wang, S.-C., Wang, X., Ren, Q.-Y., Wang, W., Shang, G.-W., Zhang, L., Zhang, S.-H. 2008. Novel exploration of cathartic pharmacology induced by rhubarb. China J. Chin. Mater. Med., 33: 481–484.

Li, J., Wang, Z., Li, J. 1997. Studies on hypoglycemic action of *Rheum emodi*. J. Chin. Med. Materials, 20: 249–250.

Li, J., Wang, A., Li, J., He, W., Kong, M. 1998. Studies on the non-anthraquiones of *Rheum hotaoense*. Chin. Tradit. Herbal Drugs, 29: 721–723. Chem. Abstr., 1999. 130: 234656y.

Li, J., Liu, P., Mao, H., Wanga, A., Zhang, X. 2009. Emodin sensitizes paclitaxel-resistant human ovarian cancer cells to paclitaxel-induced apoptosis *in vitro*. Oncol. Rep., 21: 1605–1610.

Lin, M.-L., Chen, S.-S., Lu, Y.-C., Liang, R.-Y., Ho, Y.-T., Yang, C.-Y., Chung, J.-G. 2007. Rhein induces apoptosis through induction of endoplasmic reticulum stress and Ca^{2+}-dependent mitochondrial death pathway in human nasopharyngeal carcinoma cells. Anticancer Res., 27: 3313–3322.

Lin, S., Fujii, M., Hou, D.-X. 2003. Rhein induces apoptosis in HL-60 cells via reaction oxygen species-independent mitochondrial death pathway. Arch. Biochem. Biophys., 418: 99–107.

Liu, B., Yang, J., Wang, S. 2007. The chemical constituents in rhubarb rhizomes and roots derived from *Rheum emodi* Wall. West China J. Pharm. Sci., 22: 33–34.

Liu, S.Y., Sporer, F.W., Wink, M., Jourdane, J., Henning, R., Li, Y.-L., Ruppel, A. 1997. Anthraquinones in *Rheum palmatum* and *Rumex dentatus* (Polygonaceae), and phorbol esters in *Jatropha curcas* (Euphorbiaceae) with molluscicidal activity against the schistosome vector snails *Oncomelania*, *Biomphalaria* and *Bulinus*. Tropical Med. Int. Health, 2: 179–188.

Liu, Y., Yan, F., Liu, Y., Zhang, C., Yu, H., Zhang, Y., Zhao, Y. 2008. Aqueous extract of rhubarb stabilizes vulnerable atherosclerotic plaques due to depression of inflammation and lipid accumulation. Phytother. Res., 22: 935–942.

Lloyd, J.U. 2008. Origin and history of all the pharmacopeial vegetable drugs, chemicals and preparations with bibliography. American Drug Manufacturers Association.

Losina-Losinskaya, A.S. 1936. The genus *Rheum* and its species. Acta Instituti Botanici Academiae Scientiarum Unionis Rerum Publicarum Soveticarum Socialisticarum, Ser., 1: 5–141.

Lu, J.-L., Yuan, H.-L., Jin, S.-Y., Fu, S.-S., Xiao, X. 2012. Research on the daily administration times of rhubarb. Global J. Tradit. Chin. Med., 5: 161–164.

Ma, C.Y., Peng, C.R., Geng, Y., Yang, C., Tu, X.Q. 2012. Experimental study on protective effects of rhein treating endotoxemia in rats. Chin. Arch. Tradit. Chin. Med., 30: 288–289.

Maeda, Y. 1964. Quantitative studies of organic acid in rhubarb. J. Japan Soc. Food Sci. Technol., 11: 444–446. Chem. Abstr., 1966. 64: 1269a.

Malik, M.A., Bhat, S.A., Fatima, B., Ahmad, S.B., Sidiqui, S., Shrivastava, P. 2016. *Rheum emodi* as valuable medicinal plant. Int. J. General Med. Pharm., 5(4): 35–44.

Markus, M.A., Morris, B.J. 2008. Resveratrol in prevention and treatment of common clinical conditions of aging. Clin. Interv. Aging, 3: 331–339.

Matsuda, H., Tomohiro, N., Hiraba, K., Harima, S., Ko, S., Matsuo, K., Yoshikawa, M., Kubo, M. 2001. Study on anti-oketsu activity of rhubarb. II. Anti-allergic effects of stilbene components from Rhei undulati rhizoma (dried rhizome of *Rheum undulatum* cultivated in Korea). Biol. Pharm. Bull., 24: 264–267.

Matsuda, H., Tewtrakul, S., Morikawa, T., Yoshikawa, M. 2004. Anti-allergic activity of stilbenes from Korean rhubarb (*Rheum undulatum* L.): structure requirements for inhibition of antigen-induced degranulation and their effects on the release of TNF-α and IL-4 in RBL-2H3 cells. Bioorg. Med. Chem., 12: 4871–4876.

Medyńska, E., Smolarz, H.D. 2005. Comparative study of phenolic acids from underground parts of *Rheum palmatum* L., *R. rhaponticum* L. and *R. undulatum* L. Acta Societatis Botanicorum Poloniae, 74: 275–279.

Meng, Z., Yan, Y., Tang, Z., Guo, C., Li, N., Huang, W., Ding, G., Wang, Z., Xiao, W., Yang, Z. 2015. Anti-hyperuricemic and nephroprotective effects of rhein in hyperuricemic mice. Planta Med., 81: 279–285.

Ministry of Ayush. 2016. The unani pharmacopoeia of India. Part II, Vol. III (Formulations). Ministry of Ayush, Government of India, Pharmacopoeia Commission for Indian Medicine and Homoeopathy, Ghaziabad. URL: https://pcimh.gov.in/show_content.php?lang=1&level=1&ls_id=58&lid=56.

Mishra, S.K., Tiwari, S., Shrivastava, A., Srivastava, S., Boudh, G.K., Chourasia, S.K., Chaturvedi, U., Mir, S.S., Saxena, A.K., Bhatia, G., Lakshmi, V. 2014. Antidyslipidemic effect and antioxidant activity of anthraquinone derivatives from *Rheum emodi* rhizomes in dyslipidemic rats. J. Nat. Med., 68: 363–371.

Misiti, F., Sampaolese, B., Mezzogori, D., Orsini, F., Pezzotti, M., Giardina, B., Clementi, M.E. 2006. Protective effect of rhubarb derivatives on amyloid beta (1-42) peptide-induced apoptosis in IMR-32 cells: a case of nutrigenomic. Brain Res. Bull., 71: 29–36.

Miyazawa, M., Minamoto, Y., Kameoka, H. 1996. Volatile components of the rhizomes of *Rheum palmatum* L. Flavour Fragrance J., 11: 57–60.

MoFSC. 2016. Hamro Ban (FY 2072/73). Department of Forest, Ministry of Forests and Soil Conservation (MoFSC), Government of Nepal, Nepal. pp. 202–203. URL: http://dof.gov.np/image/data/publication/All_Yearly_Publications/Hamro%20ban_2073.pdf (In Nepali).

Moon, M.K., Kang, D.G., Lee, J.K., Kim, J.S., Lee, H.S. 2006. Vasodilatory and anti-inflammatory effects of the aqueous extract of rhubarb via a NO-cGMP pathway. Life Sci., 78: 1550–1557.

Munzuroğlu, Ö., Karataş, F., Gür, N. 2000. A study of the levels of vitamins A, E and C and selenium in rhubarb (*Rheum ribes* L.). Turk. J. Biol., 24: 397–404.

Nadkarni, K.M. 2009. Indian plants and drugs. Srishti Book Distributors, New Delhi, pp. 1056–1058.

Nagasawa, T., Shibutani, S., Oura, H., Shoyama, Y., Nishioko, I. 1980. Effect of extract of Rhei rhizome on urea nitrogen concentration in rat serum. Chem. Pharm. Bull., 28: 1736–1739.

Nautiyal, B.P., Prakash, V., Maithani, U.C., Chauhan, R.S., Purohit, H., Nautiyal, M.C. 2003. Germinability, productivity and economic viability of *R. emodi* Wall. ex Meissn. cultivated at lower altitude. Curr. Sci., 84: 143–148.

Nazir, S., Sharma, M., Saxena, M., Abrar, M., Ajaz, M. 2013. *Rheum emodi*: phytochemistry, bioactive compounds and their biological activity. Int. J. Phytopharmacol., 4(4): 272–276.

Ngoc, T.M., Minh, P.T.H., Hung, T.M., Thuong, P.T., Lee, I., Min, B.-S., Bae, K. 2008. Lipoxygenase inhibitory constituents from rhubarb. Arch. Pharm. Res., 31: 598–605.

Nishioka, I. 1991. Biological activities and the active components of rhubarb. Int. J. Oriental Med., 16: 193–212.

Oktay, M., Yildirim, A., Bilaloğlu, V., Gülcin, I. 2007. Antioxidant activity of different parts of isgin (*Rheum ribes* L.). Asian J. Chem., 19: 3047–3055.

Öztürk, M., Aydoğmuş-Öztürk, F., Duru, M.E., Topçu, G. 2007. Antioxidant activity of stem and root extracts of rhubarb (*Rheum ribes*): an edible medicinal plant. Food Chem., 103: 623–630.

Park, E.-K., Choo, M.-K., Yoon, H.-K., Hyun, K.D. 2002. Antithrombotic and antiallergic activities of rhaponticin from Rhei rhizoma are activated by human intestinal bacteria. Arch. Pharm. Res., 25: 528–533.

Pecere, T., Gazzola, M.V., Mucignat, C., Parolin, C., Vecchia, F.D., Cavaggioni, A., Basso, G., Diaspro, A., Salvato, B., Carli, M., Palu, G. 2000. Aloe-emodin is a new type of anticancer agent with selective activity against neuroectodermal tumors. Cancer Res., 60: 2800–2804.

Peng, W.-W., Dai, Z.-Y. 2010. Simultaneous determination of concentration of rhein and emodin in plasma by RP-HPLC. China J. Modern Med., 20: 231–233.

Piotrowska, H., Kucinska, M., Murias, M. 2012. Biological activity of piceatannol: leaving the shadow of resveratrol. Mutation Res., 750: 60–82.

Polunin, O., Stainton, A. 1997. Flowers of the Himalaya. Oxford University Press, Delhi.

Prior, M. 2009. Rhubarbaria: Recipes for Rhubarb. Prospect Books, UK.

Qin, Y., Wang, J.-B., Kong, W.-J., Zhao, Y.-L., Yang, H.-Y., Dai, C.-M., Fang, F., Zhang, L., Li, B.-C., Jin, C., Xiao, X.-H. 2011. The diarrhoeogenic and antidiarrhoeal bidirectional effects of rhubarb and its potential mechanism. J. Ethnopharmacol., 133: 1096–1102.

Raal, A., Pokk, P., Arend, A., Aunapuu, M., Jõgi, J., Õkva, K., Püssa, T. 2009. Trans-resveratrol alone and hydroxystilbenes of rhubarb (*Rheum rhaponticum* L.) root reduce liver damage induced by chronic ethanol administration: a comparative study in mice. Phytother. Res., 23: 525–532.

Radhika, R., Krishnakumari, Sudarsanam, D. 2010. Antidiabetic activity of *Rheum emodi* in alloxan induced diabetic rats. Int. J. Pharma Sci. Res., 1: 296–300.

Radhika, R., Ragavan, B., Pawar, S.D., Sudarsanam, D. 2012. Action of marker enzymes of *Rheum emodi* in alloxan induced diabetic rats. Asian J. Exp. Biol. Sci., 3: 420–423.

Rajkumar, V., Guha, G., Kumar, R.A. 2011a. Antioxidant and anti-cancer potentials of *Rheum emodi* rhizome extracts. Evid. Based Complement. Alternat. Med. Article ID 697986. URL: http://dx.doi.org/10.1093/ecam/neq048

Rajkumar, V., Guha, G., Kumar, R.A. 2011b. Apoptosis induction in MDA-MB-435S, Hep3B and PC-3 cell lines by *Rheum emodi* rhizome extracts. Asian Pacific J. Cancer Prev., 12: 1197–1200.

Rashid, S., Kaloo, Z.A., Singh, S., Bashir, I. 2014. Medicinal importance of genus *Rheum*—a review. Int. J. Adv. Res., 2(7): 261–267.

Raval, A.P., Lin, H.W., Dave, K.R., Defazio, R.A., Morte, D.D., Kim, E.J., Pérez-Pinzón, M.A. 2008. Resveratrol and ischemic preconditioning in the brain. Curr. Med. Chem., 15: 1545–1551.

Rehman, H., Begum, W., Anjum, F., Tabasum, H. 2014. *Rheum emodi* (rhubarb): a fascinating herb. J. Phrmacog. Phytochem., 3(2): 89–94.

Rehman, H., Begum, W., Anjum, F., Tabasum, H., Zahid, S. 2015. Effect of rhubarb (*Rheum emodi*) in primary dysmenorrhoea: a single-blind randomized controlled trial. J. Complement. Integr. Med., 12: 61–69.

Rokaya, M.B., Münzbergová, Z., Timsina, B. 2010. Ethnobotanical study of medicinal plants from the Humla district of western Nepal. J. Ethnopharmacol., 130: 485–504.

Rokaya, M.B., Münzbergová, Z., Timsina, B., Bhattarai, K.R. 2012. *Rheum australe* D. Don: a review of its botany, ethnobotany, phytochemistry and pharmacology. J. Ethnopharmacol., 141: 761–774.

Sadana, H. 1996. Study on the clinical effect of rhubarb on nitrogen-metabolism abnormality due to chronic renal failure and its mechanism. Japanese J. Nephrol., 38: 379–387.

Salunkhe, K.D., Bolin, H.R., Reddy, N.R. 1991. Storage, Processing and Nutritional Quality of Fruits and Vegetables. CRC Press, Boston, p. 162.

Sanghi, D.K., Tiwle, R. 2016. Importance of village plant rhubarb: review. Intl. J. Pharma Res. Health Sci., 4: 1438–1143.

Sauro, S., Gandolfi, M.G., Prati, C., Mongiorgi, R. 2006. Oxalate-containing phytocomplexes as dentine desensitisers: an *in vitro* study. Arch. Oral Biol., 51: 655–664.

Schmidt, J., Himmelreich, U., Adam, G. 1995. Brassinosteroids, sterols and lup-20(29)-en-2α,3β,28-triol from *Rheum rhabarbarum*. Phytochem., 40: 527–531.

Seo, E.J., Ngoc, T.M., Lee, S.-M., Kim, Y.S., Jung, Y.-S. 2012. Chrysophanol-8-O-glucoside, an anthraquinone derivative in rhubarb, has antiplatelet and anticoagulant activities. J. Pharmacol. Sci., 118: 245–254.

Shikishima, Y., Takaishi, Y., Honda, G., Ito, M., Takeda, Y., Kodzhimatov, O.K., Ashurmetov, O. 2001. Phenylbutanoids and stilbene derivatives of *Rheum maximowiczii*. Phytochem., 56: 377–381.

Singh, P., Negi, J.S., Rawat, M.S.M., nee Pant, G.J. 2010. Quantification of mineral elements of *Rheum emodi* Wall. (Polygonaceae). Biol. Trace Element Res., 138: 293–299.

Singh, P., Negi, J.S., Rawat, M.S.M., nee Pant, G.J. 2012. Determination of trace elements in *Rheum moorcroftianum* by AAS. Nat. Prod. Res., 26: 161–166.

Singh, P.P., Ambika, Chauhan, S.M.S. 2013. Activity-guided isolation of antioxidants from the roots of *Rheum emodi*. Nat. Prod. Res., 27: 946–949.

Singh, R. 2001. Encyclopaedic dictionary of biomedicine. Vol 2 (H-Z). Sarup and Sons, New Delhi, pp. 425–431.

Singh, R., Tanuja, T., Chaturvedi, P. 2017. *Rheum emodi* Wall ex. Meissn (Indian rhubarb): highly endangered medicinal herb. J. Med. Plants Studies, 5(4): 13–16.

Son, S.M., Moon, K.D., Lee, C.Y. 2000. Rhubarb juice as a natural antibrowning agent. J. Food Sci., 65: 1288–1289.

Song, J.-H., Yang, T.-C., Chang, K.-W., Han, S.-K., Yi, H.-K., Jeon, J.-G. 2006. *In vitro* anti-cariogenic activity of dichloromethane fraction from *Rheum undulatum* L. root. Arch. Pharm. Res., 29: 490–496.

Srinivas, G., Babykutty, S., Sathiadevan, P.P., Srinivas, P. 2007. Molecular mechanism of emodin action: transition from laxative ingredient to an antitumor agent. Med. Res. Rev., 27: 591–608.

Su, H. 2014. Protective effect of emodin on rats with sepsis and acute lung injury and its molecular mechanism. Pract. J. Cardiac Cerebral Pneumal Vascular Dis., 22(7): 39–41.

Sydiskis, R.J., Owen, D.G., Lohr, J.L., Rosler, K.H., Blomster, R.N. 1991. Inactivation of enveloped viruses by anthraquinones extracted from plants. Antimicrob. Agents Chemother., 35: 2463–2466.

Tan, X.-H., Zhang, D.-S., Zhang, L., An, F., Jiang, H., Wang, S.-H. 2006. Pharmacokinetics of chrysophanol in rabbits. Chin. J. Hospital Pharm., 26: 947–949.

Tang, T., Yin, L., Yang, J., Shan, G. 2007. Emodin, an anthraquinone derivative from *Rheum officinale* Baill, enhances cutaneous wound healing in rats. Eur. J. Pharmacol., 567: 177–185.

Tayade, A., Dhar, P., Ballabh, B., Kumar, R., Chaurasia, O.P., Bhatt, R.P., Srivastava, R.B. 2012. *Rheum webbianum* Royal: a potential medicinal plant from trans-Himalayan cold deserts of Ladakh, India. Plant Arch., 12: 603–606.

Tegos, G., Stermitz, F.R., Lomovskaya, O., Lewis, K. 2002. Multidrug pump inhibitors uncover remarkable activity of plant antimicrobials. Antimicrob. Agents Chemother., 46: 3133–3141.

Tian, J., Du, H., Yang, H., Liu, X., Li, Z. 1997. A clinical study on compound Da huang (radix et rhizoma rhei) preparations for improvement of senile persons' memory ability. J. Tradit. Chin. Med., 17(3): 168–173.

Tian, J.-C., Chen, X.-M., Li, L.-S. 1991. Effects of *Cordyceps sinensis*, rhubarb and serum renotropin on tubular epithelial cell growth. Chin. J. Integr. Tradit. Western Med., 11: 547–549.

Uphof, J.C.T. 1968. Dictionary of economic plants. Edn. 2. J. Cramer, Lehre; Stechert-Hafner, New York, pp. 448–450.

Usui, T., Ikeda, Y., Tagami, T., Matsuda, K., Moriyama, K., Yamada, K., Kuzuya, H., Kohno, S., Shimatsu, A. 2002. The phytochemical lindleyin, isolated from Rhie rhizome, mediates hormonal effects through estrogen receptors. J. Endocrinol., 175: 289–296.

Wallis, T.E. 1985. Text Book of Pharmacognosy. Edn. 5. CBS Publishers and Distributors, Delhi, pp. 365–372.

Waly, M.I., Ali, B.H., Al-Lawati, I., Nemmar, A. 2013. Protective effects of emodin against cisplatin-induced oxidative stress in cultured human kidney (HEK 293) cells. J. Appl. Toxicol., 33: 626–630.

Wan, X.-Q., Li, X.-S., Zeng, Q.-G., Liu, L.Y. 2006. Effect of rhein on hepatocyte line L-02 apoptosis induced by TNF-α and TG. Chin. J. Gastroenterol. Hepatol., 15: 383–386.

Wang, A., Yang, M., Liu, J. 2005. Molecular phylogeny, recent radiation and evolution of gross morphology of the rhubarb genus *Rheum* (Polygonaceae) inferred from chloroplast DNA trnL-F sequences. Annals Bot., 96: 489–498.

Wang, A.-Q., Li, J.-L., Li, J.-S. 2010a. Chemical constituents of *Rheum emodi*. Chin. Tradit. Herbal Drugs, 41: 343–347.

Wang, H.-H. 1993. Antitrichomonal action of emodin in mice. J. Ethnopharmacol., 40: 111–116.

Wang, J., Zhao, H., Kong, W., Jin, C., Zhao, Y., Qu, Y., Xiao, X. 2010b. Microcalorimetric assay on the antimicrobial property of five hydroxyanthraquinone derivatives in rhubarb (*Rheum palmatum* L.) to bifidobacterium adolescentis. Phytomed., 17: 684–689.

Wang, J.-B., Ma, Y.-G., Zhang, P., Jin, C., Sun, Y.-Q., Xiao, X.-H., Zhao, Y.-L., Zhou, C.-P. 2009. Effect of processing on the chemical contents and hepatic and renal toxicity of rhubarb studied by canonical correlation analysis. Acta Pharm. Sin., 44: 885–890.

Wang, J.-B., Kong, W.-J., Wang, H.-J., Zhao, H.-P., Xiao, H.-Y., Dai, C.-M., Xiao, X.-H., Zhao, Y.-L., Jin, C., Zhang, L., Fang, F., Li, R.-S. 2011a. Toxic effects caused by rhubarb (*Rheum palmatum* L.) are reversed on immature and aged rats. J. Ethnopharmacol., 134: 216–220.

Wang, P., Xian, Y., Xu, S., Wang, G., Sony, Y., Xu, H., Yao, P., Wang, Z., Wang, Y. 1996. Study on antiviral action of *Rheum officinale* Baill. J. Shandong Univ. Health Sci., 34(2): 166–169. Chem. Abstr., 1996. 125: 316327c.

Wang, P., Meng, X.-L., Wang, J.-R., Liu, H., Yang, Y.-M., Liu, R. 2011b. Intestinal absorption kinetics of rhubarb mixture free anthraquinones in rats. Lishizhen Med. Mater. Med. Res., 22: 790–792.

Wang, Q., Yuan, D., Matsuda, H., Yoshikawa, M. 2007. Inhibitory effects of Rheum officinale and anthraquinone aglycones on antigen-induced degranulation in RBL-2H3 Cells. Asian J. Tradit. Med., 2: 189–197.

Wang, X.M., Ren, Y. 2009. *Rheum tanguticum*, an endangered medicinal plant endemic to China. J. Med. Plants Res., 3: 1195–1203.

Wang, Z.-W., Wang, J.-S., Luo, J., Wei, D.-D., Kong, L.-Y. 2012. Three new phenolic glucosides from the roots of *Rheum palmatum*. Chem. Pharm. Bull., 60: 241–245.

Wasim, A., Mohd, M., Sayeed, A., Zaidi, S.M.A. 2013. Current strategy for research on quality identification of *Rheum emodi* Wall. rhizome. Int. J. Drug Dev. Res., 5: 297–304.

Web 1. http://www.efloras.org/florataxon.aspx?flora_id=2&taxon_id=10717. (25 September, 2019)

Web 2. http://www.theplantlist.org/browse/A/Polygonaceae/Rheum/. (25 September, 2019)

Web 3. http://www.plantsoftheworldonline.org/. (25 September, 2019)

Wei, Y.-H., Wu, X.-A., Zhang, C.-Z., Li, C. 2005. Studies on chemical constituents of *Rheum glabricaule* Sam. II. J. Chin. Med. Mater., 28: 658–660.

Wrolstad, R.E., Heatherbell, D.A. 1968. Anthocyanin pigments of rhubarb, Canada red. J. Food Sci., 33: 592–594.

Wrolstad, R.E., Struthers, B.J. 1971. Polyvinylpyrrolidone column chromatography of strawberry, rhubarb, and raspberry anthocyanins. J. Chromatogr. A., 55: 405–408.

Xinga, X.-Y., Zhao, Y.-L., Konga, W.-J., Wang, J.-B., Jia, L., Zhang, P., Yan, D., Zhong, Y.-W., Li, R.-S., Xiao, X.-H. 2011. Investigation of the dose-time-response relationships of rhubarb on carbontetrachloride-induced liver injury in rats. J. Ethnopharmacol., 135: 575–581.

Xiong, H.-R., Luo, J., Hou, W., Xiao, H., Yang, Z.-Q. 2011. The effect of emodin, an anthraquinone derivative extracted from the roots of *Rheum tanguticum*, against herpes simplex virus *in vitro* and *in vivo*. J. Ethnopharmacol., 133: 718–723.

Xu, Z.-P., Lu, Z.-J., Chen, J.-H., Deng, X.-Y., Mao, Y.-Z., Huo, X. 2007. The effect of rhubarb ethanol-extract on hyperlipidemia and liver fatty in rabbits. Chin. J. Appl. Physiol., 23: 375–379.

Yan, D., Ma, Y. 2007. Simultaneous quantification of five anthraquinones in rat plasma by high-performance liquid chromatography with fluorescence detection. Biomed. Chromatogr., 21: 502–507.

Yang, J.-D., Chen, D.-C., Jing, B.-W. 1998. The influence of rhubarb on phospholipase A2 and platelet activation factor release in rats after endotoxic shock. Chin. Crit. Care Med., 10: 470–473.

Yang, J.-W., Li, L.-S. 1995. Effects of *Rheum officinale* on renal hypertrophy and hyperfiltration in the streptozotocin-induced diabetic rats. Chin. J. Integr. Tradit. Western Med., 1: 125–130.

Yang, W.-X., Li, X.-D., Liu, B.-H., Chen, L.-J., Wang, H., Zhao, Y.-L., Qi, Q.-H. 2003. The inhibition of rhein on increase in $[Ca^{2+}]_i$ and TNF-α release in lipopolysaccharide-stimulated macrophages. Acta. Sci. Nat. Univ. Nankaiensis Nat. Sci. Edit., 36: 111–115.

Yen, G.-C., Duh, P.-D., Chuang, D.-Y. 2000. Antioxidant activity of anthraquinones and anthrone. Food Chem., 70: 437–441.

Yildirim, R.V. 2013. Studies on de materia medica of dioscorides in the islamic era. Asclepio-Revista de Historia de la Medicina y de la Ciencia, 65: 80–87. URL: http://dx.doi.org/10.3989/asclepio.2013.07.

Yim, H., Lee, Y.H., Lee, C.H., Lee, S.K. 1999. Emodin, an anthraquinone derivative isolated from the rhizomes of *Rheum palmatum*, selectively inhibits the activity of casein kinase II as a competitive inhibitor. Planta Med., 65: 9–13.

Yin, G., Wang, Z.-Q., Huang, M.-R. 2003. Effects of rhubarb extract and cistanche extract on lung injury in septic shock rats. Shandong J. Tradit. Chin. Med., 22: 691–693.

Yokozawa, T., Suzuki, N., Zheng, P.D., Oura, H., Nishioka, I. 1984. Effect of orally administered rhubarb extract in rats with chronic renal failure. Chem. Pharm. Bull., 32: 4506–4513.

Yokozawa, T., Suzuki, N., Okuda, I., Oura, H., Nishioka, I. 1985. Uremia-preventing effect of rhubarb extract in rats. J. Med. Pharm. Soc. Wakan-yaku, 2: 344–350.

Yokozawa, T., Fujioka, K., Oura, H., Nonaka, G.-I., Nishioka, I. 1986. Effects of extract obtained from rhubarb in rats with chronic renal failure. Chem. Pharm. Bull., 34: 4718–4723.

Yokozawa, T., Fujioka, K., Oura, H., Nonaka, G.-I., Nishioka, I. 1991. Effects of rhubarb tannins on uremic toxins. Nephron, 58: 155–160.

Yokozawa, T., Fujioka, K., Oura, H., Nonaka, G.-I., Nishioka, I. 1993. Effects of rhubarb tannins on renal function in rats with renal failure. Japanese J. Nephrol., 35: 13–18.

Yoo, M.Y., Oh, K.-S., Lee, J.W., Seo, H.W., Yon, G.H., Kwon, D.Y., Kim, Y.S., Ryu, S.Y., Lee, B.H. 2007. Vasorelaxant effect of stilbenes from rhizome extract of rhubarb (*Rheum undulatum*) on the contractility of rat aorta. Phytother. Res., 21: 186–189.

Yu, C.-X., Zhang, X.-Q., Kang, L.-D., Zhang, P.-J., Chen, W.-W., Liu, W.-W., Liu, Q.-W., Zhang, J.-Y. 2008. Emodin induces apoptosis in human prostate cancer cell LNCaP. Asian J. Androl., 10: 625–634.

Yuan, Z., Gao, R. 1997. Antioxidant actions of anthraquinolines contained in *Rheum*. Pharm. Pharmacol. Lett., 7: 9–12.

Zargar, B.A., Masoodi, M.H., Ahmed, B., Ganie, S.K. 2011. Phytoconstituents and therapeutic uses of *Rheum emodi* wall. ex Meissn. Food Chem., 128: 585–589.

Zhang, J.-H., Li, L.-S., Zhang, M.-J. 1990. Clinical effects of rheum and captopril on preventing progression of chronic renal failure. Chin. Med. J., 103: 788–793.

Zhang, L.-M., Li, X., Dai, Y.-J. 2006. Thermal characterization and decomposition kinetics of free anthraquinones from rhubarb. Chin. Chem. Lett., 17: 699–702.

Zhang, N., Zhang, X. Liu, X., Wang, H., Xue, J., Yu, J., Kang, N., Wang, X. 2014. Chrysophanol inhibits NALP3 inflammasome activation and ameliorates cerebral ischemia/reperfusion in mice. Mediators Inflamm., Article ID 370530. URL: http://dx.doi.org/10.1155/2014/370530.

Zhang, R., Kang, K.A., Piao, M.J., Lee, K.H., Jang, H.S., Park, M.J., Kim, B.J., Kim, J.S., Kim, Y.S., Ryu, S.Y., Hyun, J.W. 2007a. Rhapontigenin from *Rheum undulatum* protects against oxidative-stress-induced cell damage through antioxidant activity. J. Toxicol. Environ. Health A, 70: 1155–1166.

Zhang, W.-S., Li, F., Bao, J.-Q. 2008. Regulatory effect of anthraquinone derivatives from rhubarb on aquaporin 4 expression in colon of rats and in LoVo cell line. Chin. J. Integr. Tradit. Western Med., 28: 818–823.

Zhang, Y.-Q., Liu, J.-H., Li, D.-G., Sun, B. 2007b. Study on protective effect and mechanism of rhubarb on intestinal barrier of piglet with sepsis. China J. Tradit. Chin. Med. Pharm., 22: 843–847.

Zhao, S.-Y., Zhang, S.-W., Liu, F.-K., Wang, B.-E. 1992. Effects of rhubarb on acute infectious endotoxemia. Beijing J. Tradit. Chin. Med., 11: 46–47.

Zhao, Y.-Y., Su, Q., Cheng, X.-L., Tan, X.-J., Bai, X., Lin, R.-C. 2012. Pharmacokinetics, bioavailability and metabolism of rhaponticin in rat plasma by UHPLC-Q-TOF/MS and UHPLC-DAD-MSn. Bioanalysis, 4: 713–723.

Zheng, Q.-X., Wu, H.-F., Guo, J., Nan, H.-J., Chen, S.-L., Yang, J.-S., Xu, X.-D. 2013. Review of rhubarbs: chemistry and pharmacology. Chin. Herbal Med., 5: 9–32.

Zhou, H., Jiao, D. 1990. 312 cases of gastric and duodenal ulcer bleeding treated with 3 kinds of alcoholic extract rhubarb tablets. Chin. J. Modern Develop. Tradit. Med., 10: 150–151, 131–132.

Index

R

S

T

W

About the Editors

Dr. Mahendra Rai is a Senior Professor and UGC-Basic Science Research Faculty at the Department of Biotechnology, Sant Gadge Baba Amravati University, Maharashtra, India. He was a visiting scientist at the University of Geneva, Debrecen University, Hungary; University of Campinas, Brazil; Nicolaus Copernicus University, Poland; VSB Technical University of Ostrava, Czech Republic, and National University of Rosario, Argentina. He has published more than 400 research papers in national and international journals. In addition, he has edited/authored more than 55 books and 6 patents.

Dr. Shandesh Bhattarai is a Senior Scientific Officer at Nepal Academy of Science and Technology, Khumaltar, Lalitpur, Nepal. He worked in the collaborative research projects funded by NUFU-Norway, VW Foundation-Germany, Darwin Initiative-UK, DFID-UK, IUCN-Bangkok, UGC-Nepal, etc. He is a coordinator of the Flora of Nepal Project, an international initiative of the institutions of Nepal, UK, and Japan. He has published more than 35 research papers in national and international journals and has edited/authored three books.

Dr. Chistiane M. Feitosa is a Professor at the Department of Chemistry, Federal University of Piaui, Teresina-Piaui, Brazil. Her research area includes natural products with the potential for the treatment of neurodegenerative diseases. She has published more than 50 research papers in national and international journals. In addition, she has edited/authored more than 10 books/chapters and 5 patents.